PLANT ANATOMY

INCLUDING EMBRYOLOGY OF ANGIOSPERMS, MORPHOGENESIS OF ANGIOSPERMS AND DIVERSITY AND MORPHOLOGY OF FLOWERING PLANTS

For Undergraduate and Postgraduate Students

Dr. B.P. PANDEY

MSc, PhD, FPSI

Former Reader and Head
Department of Botany
J.V. College, Baraut, Uttar Pradesh

S. CHAND PUBLISHING

S. Chand And Company Limited
(ISO 9001 Certified Company)

S Chand And Company Limited

(ISO 9001 Certified Company)

Head Office: D-92, Sector–2, Noida – 201301, U.P. (India), Ph. 91-120-4682700
Registered Office: A-27, 2nd Floor, Mohan Co-operative Industrial Estate, New Delhi – 110 044, Phone: 011-49731800
www.schandpublishing.com; e-mail: info@schandpublishing.com

Branches

Chennai	:	Ph: 23632120; chennai@schandpublishing.com
Guwahati	:	Ph: 2738811, 2735640; guwahati@schandpublishing.com
Hyderabad	:	Ph: 40186018; hyderabad@schandpublishing.com
Jalandhar	:	Ph: 4645630; jalandhar@schandpublishing.com
Kolkata	:	Ph: 23357458, 23353914; kolkata@schandpublishing.com
Lucknow	:	Ph: 4003633; lucknow@schandpublishing.com
Mumbai	:	Ph: 25000297; mumbai@schandpublishing.com
Patna	:	Ph: 2260011; patna@schandpublishing.com

© S. Chand And Company Limited, 1978

All rights reserved. No part of this publication may be reproduced or copied in any material form (including photocopying or storing it in any medium in form of graphics, electronic or mechanical means and whether or not transient or incidental to some other use of this publication) without written permission of the copyright owner. Any breach of this will entail legal action and prosecution without further notice.

Jurisdiction: *All disputes with respect to this publication shall be subject to the jurisdiction of the Courts, Tribunals and Forums of New Delhi, India only.*

First Edition 1978
Subsequent Editions and Reprints 1978, 82, 84, 86, 87, 89, 91, 93, 94, 96, 97, 98, 2001 (Twice),2002, 2004, 2005, 2007, 2009, 2010
Revised Edition 2012; Reprints 2020, 2021 (Twice), 2022 (Twice)

Reprint 2024

ISBN: 978-81-219-0145-1　　　　　　　　　**Product Code:** H6PAT68BOTN10ENAG0XO

PRINTED IN INDIA

By Vikas Publishing House Private Limited, Plot 20/4, Site-IV, Industrial Area Sahibabad, Ghaziabad – 201 010 and Published by S. Chand And Company Limited, A-27, 2nd Floor, Mohan Co-operative Industrial Estate, New Delhi – 110 044.

WITH ALL DEVOTION
TO LORD KRISHNA OF SHRIMADBHAGWADGITA

The vast sky
And so the vast land
The vast ocean
And the great mountains
The great people
And the majestic animals
The beautiful birds
And the colourful butterflies
All the insects
And all the moths
The Creater has created them
Why?
Because they are need of creation
They are components of creation.

B.P.P.

PREFACE TO THE SEVENTH EDITION

The text of this edition of **Plant Anatomy** has been thoroughly revised. Most of the diagrams have been redrawn and relabelled. A new chapter, entitled **Diversity and Morphology of Flowering Plants** has been added in this edition, and therefore, the utility of the book has been augmented to a great extent for teacher and taught alike.

My sincere thanks are due to my colleagues, friends, botanists and educationists who have accorded their full co-operation and well wishes during preparation of the text of this edition.

'Neel Kanth Kuteer'
Subhash Nagar   **Dr. B.P. Pandey**
BARAUT-250611.

Disclaimer : While the author of this book have made every effort to avoid any mistake or omission and have used their skill, expertise and knowledge to the best of their capacity to provide accurate and updated information. The author and S. Chand does not give any representation or warranty with respect to the accuracy or completeness of the contents of this publication and are selling this publication on the condition and understanding that they shall not be made liable in any manner whatsoever. S.Chand and the author expressly disclaim all and any liability/responsibility to any person, whether a purchaser or reader of this publication or not, in respect of anything and everything forming part of the contents of this publication. S. Chand shall not be responsible for any errors, omissions or damages arising out of the use of the information contained in this publication.
Further, the appearance of the personal name, location, place and incidence, if any, in the illustrations used herein is purely coincidental and work of imagination. Thus the same should in no manner be termed as defamatory to any individual.

PREFACE TO THE FIRST EDITION

The study of gross internal structure of plant organs by the technique of section-cutting is called plant anatomy (*ana* = asunder; *temnein* = to cut).

This text is a compilation work and embodies a fairly comprehensive treatment of the fundamental facts and aspects of plant anatomy. This book will serve as an introduction to plant anatomy to the beginners in this field. Actually the book is intended to fulfil the long felt need of students of graduate, honours and post-graduate level of all Indian universities. The syllabi of all the universities have been kept in view during the preparation of the manuscript of this text. This book may also serve as laboratory manual. The present text provides a background of facts, terminology and internal structure of common plants around us and may safely be used as laboratory guide. It is not always necessary that the same plants which have been described in this text be used in the laboratory. Any available materials may be used and their internal structure may be compared with the illustrations given in the text, to make the study of plant anatomy more effective. The present text deals with the anatomy of vascular plants only. Much emphasis has been given on the anatomical study of angiosperms. Ecological anatomy has been dealt with in sufficient details. There is comprehensive anatomical study of some controversial topics such as—anomalous structure of stem and the anatomy of floral parts. Study of cell structure and its components in nutshell is also dealt with. A few general topics such as—anatomy in relation to taxonomy, wood anatomy in relation to phylogeny and the mechanical tissue have been added to make the text more interesting and attractive too for the post-graduate students.

The text has been written in simple way and profusely illustrated with self-explanatory diagrams and micrographs. The figures are provided with descriptive legends. Many diagrams have been drawn by author himself from prepared slides and most of the figures have been quoted from the authentic works of various authors.

As already mentioned the book is primarily an elementary text for degree and post-graduate students and not a text for researchers, and therefore, the bibliography has been kept at minimum. Only in the end of the text a list of important books and articles is being given. A detailed index is also worked out and given in the last which makes the consultation of the subject easier.

In conclusion, I wish to express my deep sence of gratitude and indebtedness to those who helped me directly or indirectly during the preparation of the manuscript of this text. Especially I am indebted to Dr. N.P. Saxena, D.N. College, Meerut who allowed the diversion from research to this work. I also wish to express my appreciation to Professor Sudhir Kumar and Dr. Shankar Nath Shukla who gave the time to review the manuscript. My sincerest thanks are also due to my friend Professor Sheobir Singh who always encouraged and inspired me for the task. I am really grateful to Sri T.N. Goel, my publisher, who managed to bring out the book in this shape. In the last but not the least I take the opportunity to thank Dr. H.S. Harindra, Principal, J.V. College, Baraut for the encouragement and appreciation of such work. My son Sanjeeva also deserves appreciation who always assisted me in several ways.

The suggestions and criticisms for further improvement of the book will be most welcomed and thankfully acknowledged.

Department of Botany
J.V. College, Baraut—250611
January 15, 1978

B.P. Pandey

CONTENTS

Pages

1. **Plant Anatomy—Introduction**   1—10
   Historical sketch, 1; the plant body and its development, 2; fundamental parts of the plant body; 3; development, of the plant body, 3; primary and secondary growth, 4; internal organization of vascular plant, 6; cell types and tissues, 6.

2. **The Cell-Structure and its Components**   11—59
   Historical account, 11; the cell theory, 11; the protoplasm, 13; composition and chemical nature of protoplasm, 13; physical nature of protoplasm, 14; properties of protoplasm, 15; the plasma membrane 16; cell structure, 18; cytoplasm, 20; the nucleus, 21; the mitochondria, 31; the plastids, 34; endoplasmic reticulum, 40; ribosomes, 41; lysosomes, 42; sphaerosomes, 42; lomasomes, 42; peroxisomes, 42; glyoxysomes, 42; microtubules, 43; Golgi complex, 43; non-protoplasmic components, 45; vacuoles, 45; ergastic substances, 46.

3. **The Cell Wall**   60—78
   The wall layers, 60; plasmodesmata, 65; thickening of the cell wall, 65; pits, 67; chemical nature of the cell wall, 71; intercellular spaces, 72; microscopic and sub microscopic structure of cell walls, 73; formation of walls, 75; growth of walls, 77.

4. **The Cell Division**   79—88
   Mitosis, 79; Meiosis, 83; comparison of mitosis and meiosis, 87.

5. **The Tissue**   89—116
   Meristems or meristematic tissue, 89; meristems and growth of plant body, 89; classification of meristems, 90; meristem and meristematic, 93; meristems and permanent tissues, 93, simple tissues, 94; complex tissues, 101; secretory tissue, 113.

6. **Apical Meristems**   117—133
   Initials and derivatives, 117; vegetative shoot apex, 118; other theories of shoot apex organization, 121; root apex, 122; theories of structural development and differentiation, 125; origin of leaves, 130; origin of branches, 130; origin of reproductive shoot apex, 132.

7. **The Tissue System**   134—169
   The dermal or epidermal tissue system, 135; the fundamental or ground tissue system, 148; the vascular tissue system, 155; stelar system, 160; nodal anatomy, 163; branch traces and branch gaps, 168; closing of leaf gaps, 169.

8. **The Cambium**   170—174
   Origin of cambium, 170; fascicular and interfascicular cambium, 171; duration of cambium, 171; function of cambium, 172; structure of cambium, 172; cellular structure of cambium, 172; cell division in cambium, 173; cambium growth about wounds, 173; cambium in budding and grafting, 174; cambium in monocotyledons, 174.

9. **The Root—Primary and Secondary Structure**   175—209

General characteristics of the roots, 175; anatomical characteristics of the root, 175; anatomy of dicotyledonous roots, 181; anatomy of monocotyledonous roots, 185; formation of lateral roots, 192; mycorrhiza, 192; formation of adventitious roots, 195; anatomy of epiphytic roots, 196; anatomy of storage roots, 199; heteroarchy in roots, 204; differences between dicotyledonous and monocotyledonous roots, 204; secondary growth in dicotyledonous root, 204; periderm, 209.

10. **The Stem—Primary and Secondary Structure**   210—261

Origin of the stem, 210; root-stem transition, 211; anatomy of dicotyledonous stems, 213; variations in stem structure, 218; anatomy of monocotyledonous stems, 226; anatomy of the phylloclade, 234; secondary growth in dicotyledonous stems, 236; periderm, 250; bark, 252; rhytidome, 253; lenticels, 255; secondary xylem and secondary phloem in conifers, 256; secondary growth in the monocotyledons, 259; stem of *Pinus*—primary and secondary structure, 260; differences between dicotyledonous and monocotyledonous stems, 261.

11. **The Stem—Anomalous Structure**   262—294

Anomalous secondary growth, 262; anomalous position of cambium, 262; abnormal behaviour of normal cambium, 264; accessory cambium formation and its activity, 268; extrastelar cambium, 273; interxylary or included phloem, 276; absence of vessels in the xylem, 282; scattered vascular bundles in dicots, 282; presence of exclusive phloem and xylem bundles, 284; presence of medullary bundles, 285; presence of cortical bundles, 286; intraxylary phloem, 286; vascular bundles arranged in a ring in monocots, 291; secondary growth in monocots, 291.

12. **Anatomy of the Leaf and the Petiole**   295—315

Anatomy of dicotyledonous leaf, 296; anatomy of monocotyledonous leaf, 304; anatomy of gymnosperm leaf, 307; anatomy of the petiole, 309; anatomy of the phyllode, 312; abscission of leaves, 313.

13. **Ecological Anatomy**   316—348

Hydrophytes, 316; mesophytes, 323; xerophytes, 323; epiphytes, 338; saprophytes, 339; parasites, 340; halophytes, 341.

14. **Anatomy of the Floral Parts**   349—356

Structure—parts of the flower and their arrangement, 349; vascular anatomy, 351; inferior ovary, 355.

15. **Anatomy of the Embryo and Young Seedling**   357—360

Anatomy of the cotyledons, 357; anatomy of the mesocotyl, 358; anatomy of the sheath of cotyledon, 359; anatomy of the hypocotyl, 359; anatomy of the seedling root, 360.

16. **The Mechanical Tissue**   361—374

Collenchyma, 361; sclerenchyma, 364; xylem, 368; phloem, 370; principles governing the construction of mechanical system, 370; arrangement or distribution of mechanical tissues in different plant organs, 372.

17. **Anatomy in Relation to Taxonomy.**   375—378

Hairs, 375; stomata, 375; epidermal cells and hypoderm, 376; veins, 376; petiole, 376; microchemistry, 376; cork, 377; endodermis, 377; sclerenchyma of pericycle, 377; width of medullary rays, 377; bicollateral bundles, 377; cortical and medullary bundles and anomalous secondary thickening, 377; wood, 377.

18. **Wood Anatomy in Relation to Phylogeny**   379—382
    Vessels, 380; parenchyma, 381; rays, 381; fibres, 381; intercellular canals, 381.

19. **Development of Plant Anatomy in India**   383—387
    Angiosperms, 383; gymnosperms, 386; pteriodophytes, 387.

20. **The Fruit, The Fruit Wall, And the Seed Coat**   388—397
    The fruit, 388; the fruit wall, 388; dry fruit wall, 388; fleshy fruit wall, 391; the seed coat, 393.

21. **Embryology of Angiosperms**   398—438
    Introduction, 398; alternation of generations, 398; the flower and its parts, 401; stamen or microsporophyll, 402; carpel or megasporophyll, 411; female gametophyte or megagametophyte, 416; male gametophyte or microgametophyte, 422; fertilization, 425; the endosperm, 427; embryo and its development (embryogenesis), 432; seed and fruit, 436; apomixis, 436; polyembryony, 437.

22. **Morphogenesis: Tissue and Organ Culture**   439—458
    Introduction, 439; organ culture, 440; tissue culture, 446; culture technique, 453; plastochron, 458.

23. **Comparative Account**   459—464
    Comparison of simple pits and bordered pits, 459; comparison of latex cells and latex vessels, 459; comparison of sap wood and heart wood, 460; differences between sieve cells and sieve tubes, 460; differences between tracheids and vessels, 461; differences between protoxylem and metaxylem, 461; differences between protophloem and metaphloem, 462; differences between primary and secondary phloem, 462; differences between primary and secondary xylem, 463; comparative anatomy of *Helianthus* and *Cucurbita*, 463.

24. **Microscopy and Micrometry**   465—469
    Resolving power of microscope, 465; the light microscope, 465; phase contrast microscope, 466; electron microscope, 467; micrometry, 467.

25. **Models of Plasma Membrane**   470—474
    Ultrastructure, 470; molecular structure, 471; membrane models, 472.

26. **Diversity and Morphology of Flowering Plants**   475—547
    Diversity and Morphology in the Plant Body, 475; Diversity and Morphology in the Root, 476; Diversity and Morphology in the Stem, 486; Diversity and Morphology in the Leaf, 496; Diversity and Morphology in the Inflorescence, 507; Diversity and Morphology in the Flower, 513; Diversity and Morphology in the Fruit, 519; Diversity and Morphology in Seed, 528; Dispersal of Fruits and Seeds, 530; Defence Mechanisms in Flowering Plants, 536; Modular Growth in Plants, 537; Convergence of Evolution of Tree Habit, 541; Trees : Largest and Longest Lived Organisms, 545.

Glossary	548—563
Further Reading	564
References	565—573
Question Bank	574—582
Index	583—591

1

Plant Anatomy – Introduction

"The study of gross internal structure of plant organs by the technique of section cutting is called **plant anatomy** (*ana* = asunder ; *temnein* = to cut)".

Plant Anatomy deals with the study of internal structure of the various organs of the plant. It includes the structure of cell, which makes the basic unit of all living organisms including plants. The cells constitute tissues and their detailed study is called **histology**. To study plant anatomy one should be quite familiar with the study of tissues, *i.e.*, histology. In modern days, with the discovery of transmission electron microscope and the scanning electron microscope many interesting features have been discovered.

HISTORICAL SKETCH

The study of the plant anatomy began in the seventeenth century in the hands of two men who worked quite independently of each other; Marcello Malpighi (1628–94) and Nehemiah (Grew 1641–1712). It is also remarkable that preliminary treatises by the two men were presented to the Royal Society of London, of which Grew was Secretary, on the same day, 7th December 1671. Sherwin Carlquist writes in his article plant Anatomy's 300th Anniversary. "As the fore page of Grew's *The Anatomy of Vegetables Begun* indicates, Grew's paper was read on November 9, 1671, before the Royal Society of London. Sachs (1890) tells us that Grew presented his manuscript in May 1671, and that Malpighi's work was received by the Royal society on December 7, 1671. However, Malpighi's (1675) *Anatome Plantarum* gives the date and place of writing as November 1, 1671 and Bologna. Grew's *The Anatomy of Vegetables Begun* was published on December 7, 1671, according to Sachs, although the title page bears the date 1672. We know that both Malpighi and Grew were studying aspects of plant Anatomy several years before 1671. The various 1671 dates cited above are all close, however, and we may, along with Sachs, cite 1671 as the year in which the '*Anatomy of vegetables*' was indeed '*begun*', although I am citing his book here as Grew (1672)".

Grew never reached a correct idea of cell structure and believed that the plant was composed of lace-work of interwoven threads (hence the term "*tissue*"), which were no doubt the profiles of the cell walls as seen in sections.

Sherwin Carlquist quotes further in his article, "*Hooke's Micrographia of 1667 cannot be called the first volume to be published in the field of plant anatomy, despite Hooke's figuring of cork cells, and giving them that term.*"

In eighteenth century Kaspar Friedrich Wolff (1733—94), studied meristems and tried to formulate a theory of apical development, and Sir John Hill (1716–75), published a book on the structure of timber in 1770.

In the beginning of nineteenth century C.F. Brisseau-Mirbel (1776–1854), published a theory of plant organization in 1802. J.J. Bernhardi (1774–1850), published his work on angiospermic vessels. The other workers of nineteenth century are Kurt Sprengel (1766–1833); K.A. Rudolphi (1771–1832), H.F. Link (1767–1851) and L.C. Treviranus (1779–1864). Hugo von Mohl (1805–72) and Carl von Naegeli (1817–91) were held responsible for modern outlook founded on a clear preception of cellular structure, based upon the cell theory, as it was elaborated by Schleiden and Schwann.

THE PLANT BODY AND ITS DEVELOPMENT

The plant body consists of a number of organs, *i.e.*, root, stem, leaf and flower. The flower consists of sepals, petals, stamens, carpels and sometimes also sterile members. Each organ, is made up of a number of tissues. Each tissue consists of many cells of one kind.

The complex multicellular body of the seed plant is a result of evolutionary specialization of long duration. This specialization has given rise to the establishment of morphological and

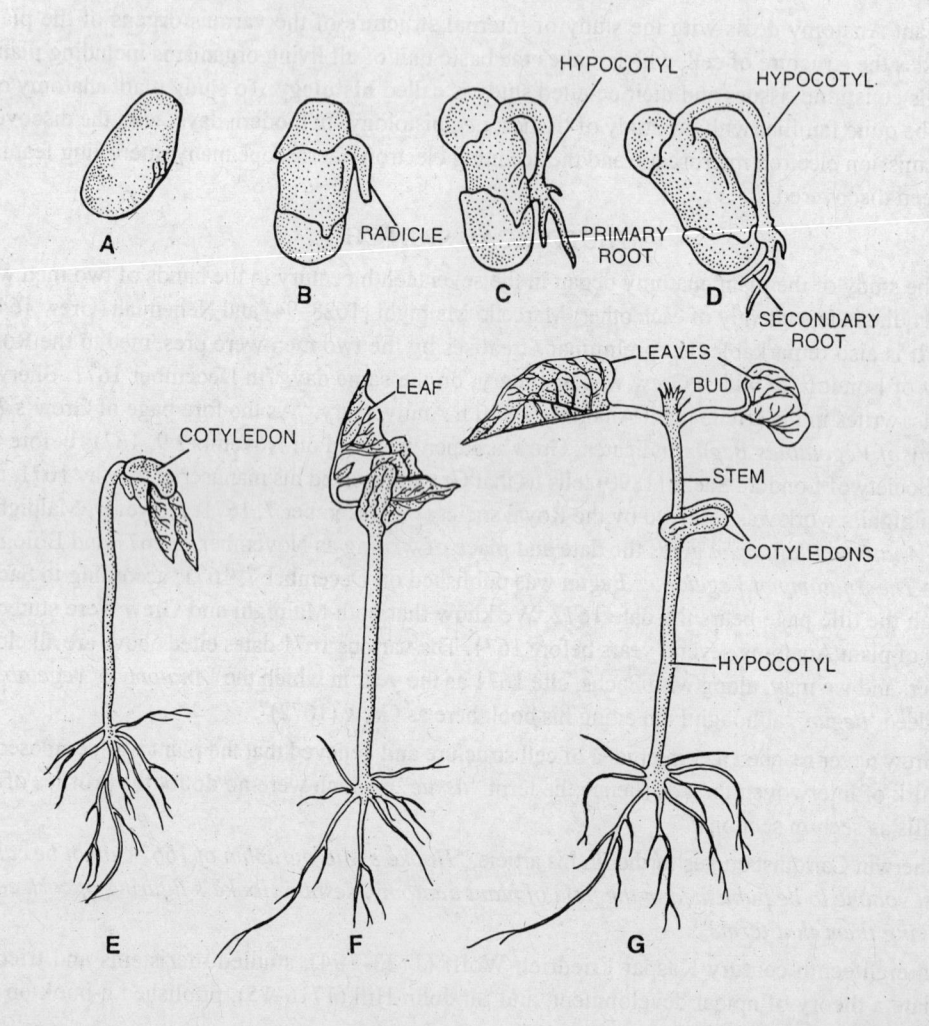

Fig. 1.1. Different stages in the germination of the seed leading to the formation of young seedling (A–G); G, seedling with well developed roots, two young leaves, hypocotyl, cotyledons, stem and young apical bud.

Introduction

physiological differences between the various parts of the plant body and also caused the development of the concept of *plant organs*.

The organization of the plant body of the oldest known land plants, the Psilophytales, suggests that the differentiation of the vegetative plant into leaf, stem and root is a result of evolutionary development from an originally simple axial structure (Arnold, 1947; Eames, 1936).

As regards the morphologic nature of the flower it is thought that the flower is homologous with a shoot and the floral parts with leaves.

Fundamental parts of the plant body. The axis, consists of two parts—that portion which is normally aerial is know as the *stem*, and the portion which is subterranean is called the *root*. There are three types of appendages arising from the axis. 1. *Leaves*—The strands of vascular tissue pass through the leaves. The leaves are characteristic of the stem and do not occur on the root. The leaves are found to be arranged on the stem in a definite manner, and bear an intimate structural relation to the skeleton of the axis. The leaf is looked upon as the lateral expansion of the stem, continuous with it. All fundamental parts of the stem are concerned with the formation of the leaf. 2. *Emergences*—In the appendages of the second rank only the outermost layers of stem, the cortex and the epidermis, are usually present which are known as emergences. The prickles of the rose make a good example of it. 3. *Hairs*—The appendages of the third rank are hairs. These are projections of the outermost layer of the cells. The emergences and hairs occur on both axis and leaves, usually without definite arrangement.

DEVELOPMENT OF THE PLANT BODY

A vascular plant begins its existence as a morphologically simple unicellular zygote (2n). The zygote develops into the embryo and thereafter into the mature sporophyte. The development of the sporophyte involves division and differentiation of cells, and an organization of cells into the tissues and tissue systems. The embryo of the seed plant possesses relatively a simple structure as compared with the mature sporophyte. The embryo bears a limited number of parts—generally only an axis bearing one or more cotyledons. The cells and tissues of this structure are less differentiated. However the embryo grows further, because of the presence of

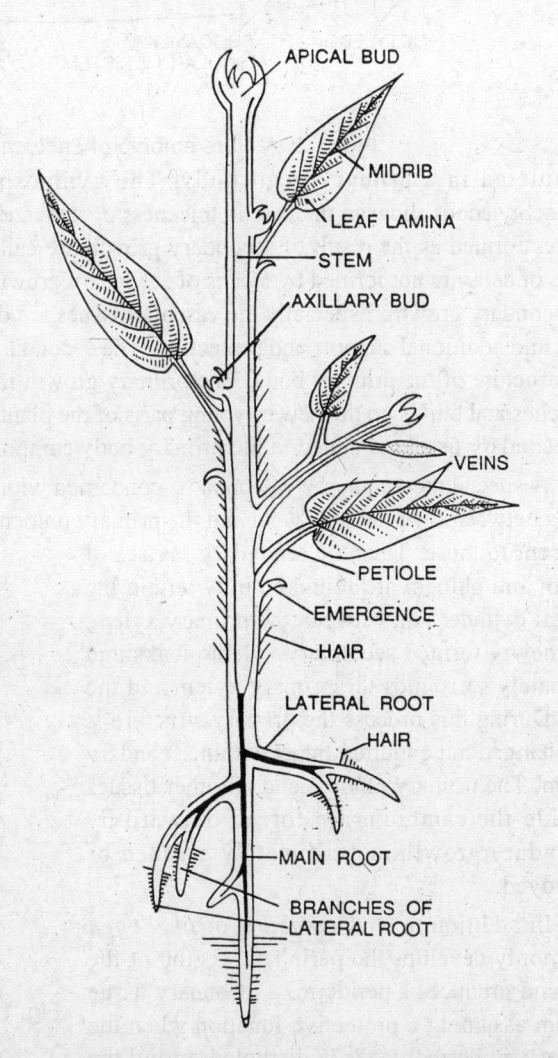

Fig. 1.2. The plant body, showing fundamental parts.

meristems, at two opposite ends of the axis, of future shoot and root. After the germination of the seed, during the development of shoot and root, the new apical meristems appear which cause a repetitive branching of these organs. After a certain period of vegetative growth, the reproductive stage of the plant is attained.

Primary and secondary growth. As mentioned above, this first-formed plant body is known as the *primary plant body*, since it is built up by means of first or *primary growth*. The tissues of this first-formed body are known as *primary tissues* ; for example the first-formed xylem is called *primary xylem*. In most vascular cryptogams and monocotyledons, the entire life cycle of the sporophyte is

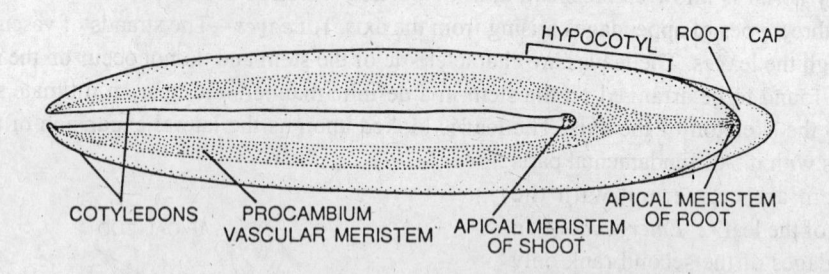

Fig. 1.3. Mature embryo of *Lactuca sativa* (After Esau).

completed in a primary plant body. The gymnosperms, most dicotyledons, and some monocotyledons show an increase in thickness of stem and root by means of *secondary growth*. The tissues formed as the result of secondary growth are called *secondary tissues*. Generally the new types of cells are not formed by means of secondary growth. The bulk of the plant increases because of secondary growth. Especially the vascular tissues are developed which provide new conducting cells and additional support and protection. The secondary growth does not fundamentally change the structure of the primary body. The primary growth increases the length of the axis, forms the branches and builds up the new or young parts of the plant body. Thus, a secondary body composed of secondary tissues is added to the primary body composed of primary tissues.

A special meristem, the *cambium,* is concerned with the secondary thickening. The cambium arises between the primary xylem and the primary phloem, and lays down new xylem and phloem adjacent to these. Thus the secondary masses of xylem and phloem are found entirely within the central cylinder and between the primary xylem. The newly formed secondary xylem cloaks and ultimately surrounds the primary xylem and the pith. During this process the primary structure is not changed but engulfed intact within secondary xylem. The primary phloem and all other tissues outside the cambium are forced outward by secondary growth and ultimately crushed or destroyed.

In addition, a *cork cambium* or *phellogen* commonly develops the peripheral region of the axis and produces a periderm, a secondary tissue system assuming a protective function when the primary epidermal layer is disrupted during the secondary growth in thickness.

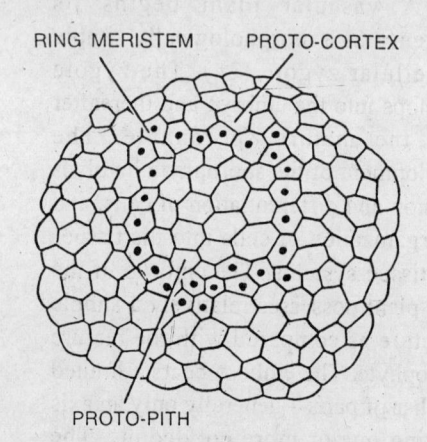

Fig. 1.4. T.S. through the shoot apex of *Ranunculus acris*.

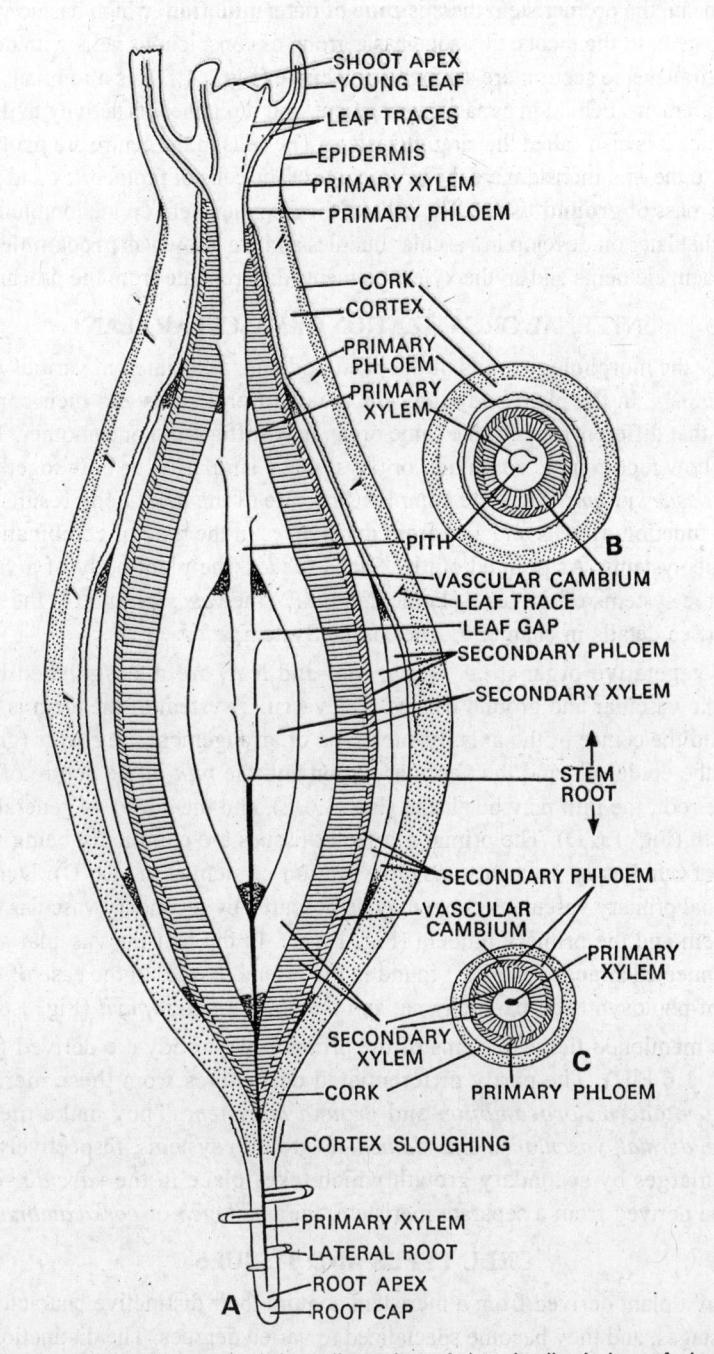

Fig. 1.5. Primary and secondary growth in a dicot plant A, longitudinal view of plant; B, transverse section of stem; C, transverse section of root.

The primary growth of an axis is completed in a relatively short period, whereas the secondary growth persists for a longer period and in a perennial axis the secondary growth continues indefinitely.

The stem apex like the root apex consists of a meristematic zone of cells that remain in a continuous and rapid state of division. This is called **promeristem** having cells with very thin walls.

Immediately beneath the promeristem there is **zone of determination** which has no visible boundary with the promeristem. In the dicots, this zone has a group of conspicuous cells with dense cytoplasm. These cells in a transverse section are arranged in a circle (fig. 1.4.). It is a remnant of a primordial meristem, which remains behind in a maturing segment, and it retained its activity to divide. Due to its circular appearance it is also called the **ring meristem**. The cells in the centre are **protopith** and those that are external to the ring meristem are the **protocortex**. The cells of protocortex and protopith divide and build up the mass of **ground tissue**. The cells of the ring meristem divide longitudinally and form elongated cells that later on develop in vascular bundles and are known as **procambical strands**. The first formed phloem elements and so the xylem elements differentiate from the procambial strands.

INTERNAL ORGANIZATION OF VASCULAR PLANT

The *cells* or the morphologic units of the plant body are associated in various ways with each other forming *tissues*. In the plant body the cells are of several kinds and their combinations into tissues are such that different parts of the same organ may differ from one another. The larger units of tissues may show topographic continuity or physiologic similarity, or both together. Such tissue units are called *tissue systems*. Thus the complex structure of the plant body results from variation in the form and function of cells and also from differences in the type of combination of cells into tissues and tissue systems. As pointed out by Sachs (1875), the plant body of a vascular plant is composed of three systems of tissues—(1) the *dermal* (2) the *vascular* and (3) the *fundamental* or *ground* system (See details in chapter 7, The Tissue System).

The three vegetative organs, *i.e.*, stem, root and leaf, are distinguished by the relative distribution of the vascular and ground tissues. The vascular system of the stem is found between the epidermis and the centre of the axis. In such type of arrangement the *cortex* (ground tissue) is found between the epidermis and the vascular region and the *pith* in the centre of the stem (Fig. 1.6 B,C.). In the root, the pith may be absent (Fig. 1.6 E), and the cortex is generally shed during secondary growth (Fig. 1.6 D). The primary vascular tissues are commonly being arranged in the form of a ring of bundles as seen in transverse section of stem (Fig. 1.6 B). During secondary growth the original primary vascular system may be obscured by secondary vascular tissues between the primary xylem and the primary phloem (Fig. 1.6 C). In the leaf the vascular system consists of many interconnected strands (bundles) found in the ground tissue. In the case of leaf the ground tissue consists of photosynthetic parenchyma, and is known as *mesophyll* (Fig. 1.6 G).

The above mentioned tissue systems of the primary plant body are derived from the apical meristems (Fig. 1.6 F,H). The partly differentiated derivatives from these meristems may be classified as—*protoderm, procambium* and *ground meristem*. They make the meristematic precursors of the *dermal, vascular* and *fundamental* (ground) systems, respectively. The vascular tissue system enlarges by secondary growth which takes place in the *vascular cambium*. The periderm may be derived from a separate meristem, the *phellogen* or *cork cambium* (Fig. 1.6 D).

CELL TYPES AND TISSUES

The cells of a plant derived from a meristem acquire their distinctive characteristics through developmental stages, and they become specialized to varied degrees. The distinctions among cells and tissues are summarized here:

Epidermis. The epidermal cells make a continuous layer on the surface of the plant body in the primary state. Very often they are tabular in shape. Other specific epidermal cells are guard cells, various trichomes and root hairs. The cuticle is present on the outer wall of the epidermal cells of the aerial parts of the plant. The epidermis is commonly supplanted by the periderm after secondary growth of stems and roots.

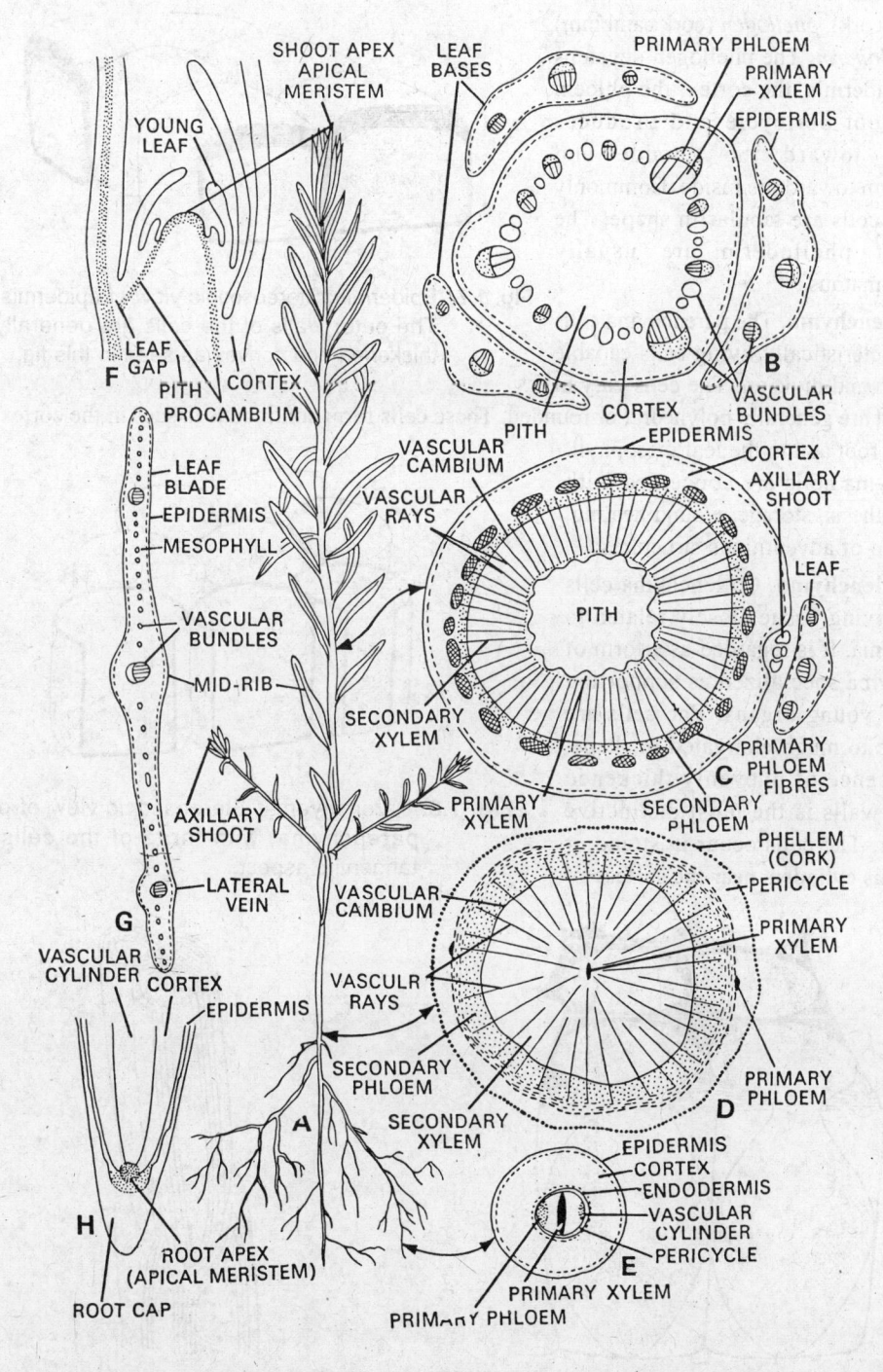

Fig. 1.6. Organization of a vascular plant. A, habit sketch of linseed plant (*Linum usitatissimum*) in vegetative state; B-C, transverse sections of stem; D-E, transverse sections of root; F, L.S. of shoot apex with apical meristem and developing leaves; G, transection of leaf lamina; H, L.S. of root apex with apical meristem and other root regions.

Periderm. The periderm consists of *phellem* (cork), *phellogen* (cork cambium) and *phelloderm*. The phellogen develops in the epidermis, the cortex, the phloem or the root pericycle and produces phellem toward the outside and phelloderm toward the inside. Commonly the cork cells are tabular in shape. The cells of phelloderm are usually parenchymatous.

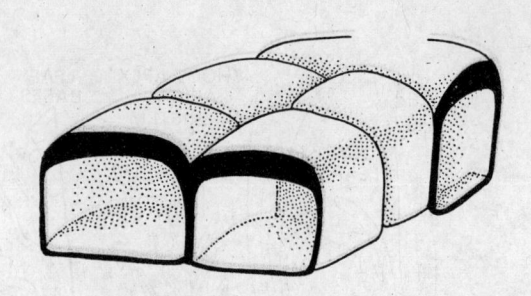

Fig. 1.7. Epidermis. Stereoscopic view of epidermis. The outer walls of the cells are generally thickened and convex as seen in this fig.

Parenchyma. The parenchyma cells are characteristically living cells capable of growth and division. The cells vary in shape and are generally polyhedral or rounded. These cells form continuous tissues in the cortex of stem and root and in the leaf mesophyll. Parenchyma cells are concerned with photosynthesis, storage, wound healing, and origin of adventitious structures.

Collenchyma. Collenchyma cells make a living tissue closely related to parenchyma. It is regarded as a form of parenchyma specialized as supporting tissue of young organs. The cells are prismatic to much elongated in shape. The presence of unevenly thickened primary walls is the most distinctive character. The cells occur in strand or continuous cylinders near the surface of

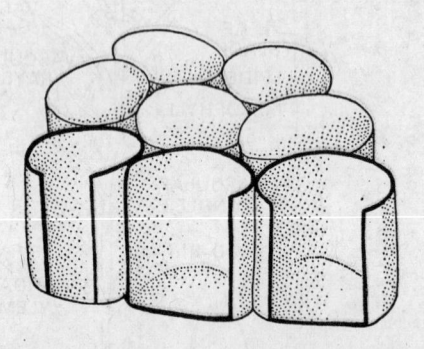

Fig. 1.8. Parenchyama. Stereoscopic view of pith parenchyma, with three of the cells in tangential aspect.

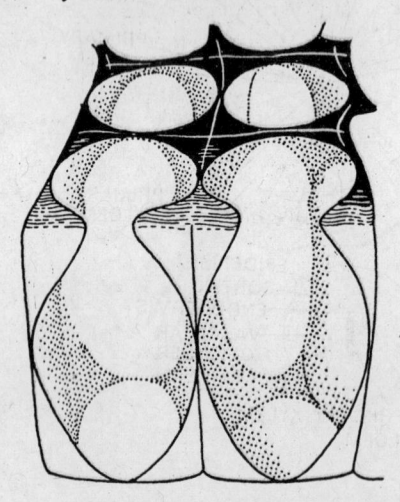

Fig.1.9. Collenchyma. Stereoscopic view of collenchyma, with two cells in tangential aspect.

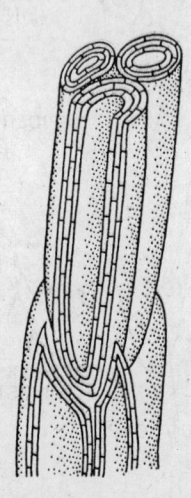

Fig. 1.10. Sclerenchyma. Stereoscopic view of sclerenchymatous fibres, with one cell tangentially sectioned.

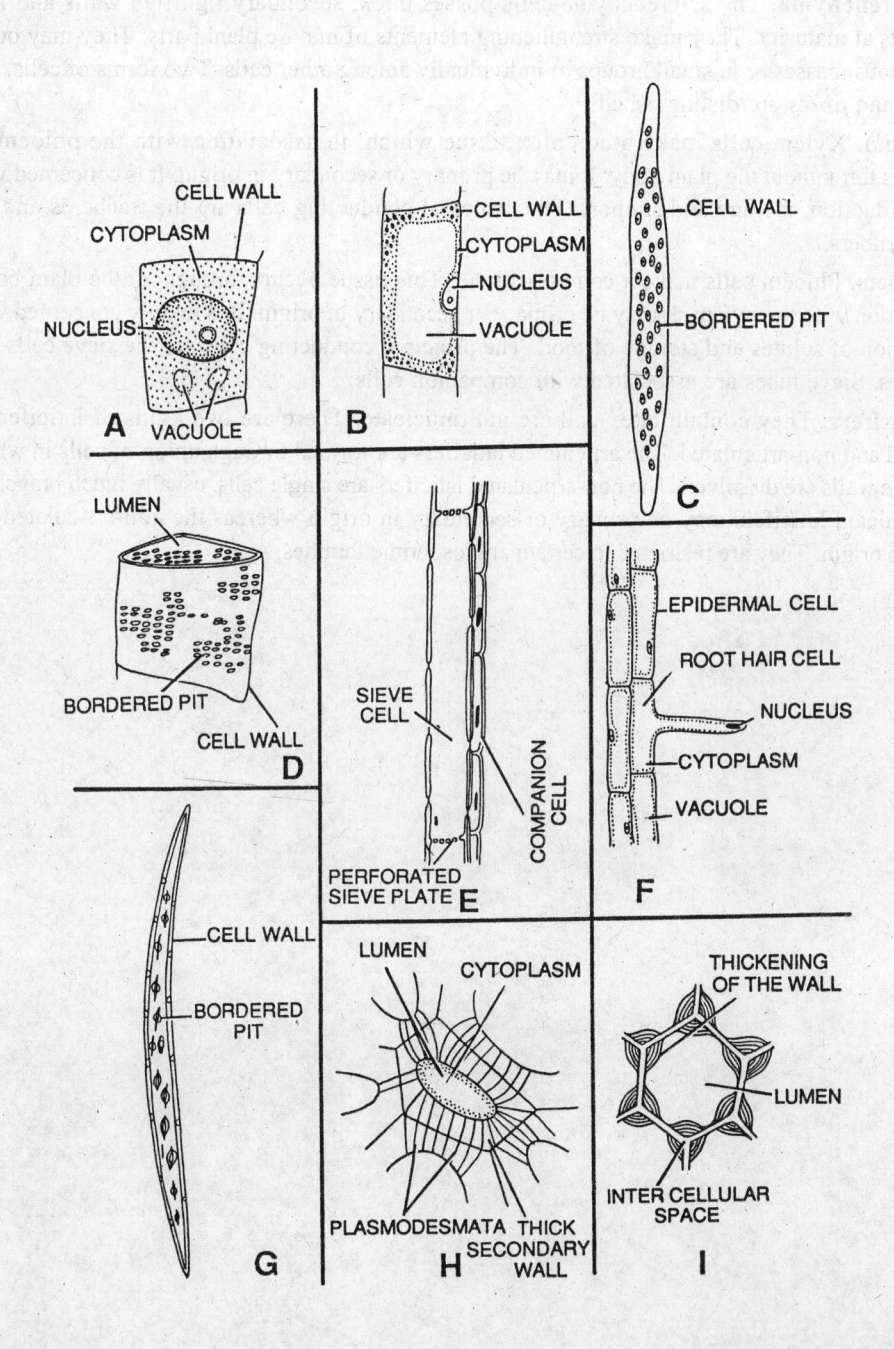

Fig. 1.11. The cell structure. Various types of cells in plants. A, meristematic cell; B, parenchymatous cell; C, tracheid; D, vessel cell; E, sieve cell and companion cell; F, epidermal cell and root hair cell; G, fibre; H, stone cell; I, collenchyma cell.

the cortex in stems and petioles and along the veins of leaves.

Sclerenchyma. The sclerenchyma cells posses thick, secondary lignified walls and lack protoplasts at maturity. They make strengthening elements of mature plant parts. They may occur in continuous masses or in small groups or individually among other cells. Two forms of cells, *i.e., sclereids* and *fibres* are distinguished.

Xylem. Xylem cells make a complex tissue which, in association with the phloem, is continuous throughout the plant body. It may be primary or secondary in origin. It is concerned with water conduction, storage and support. The principal conducting cells are the tracheids and the vessel members.

Phloem. Phloem cells make a complex tissue. This tissue occurs throughout the plant body, in association with the xylem. It may be primary or secondary in origin. It is mainly concerned with translocation of solutes and storage of food. The principal conducting cells are the sieve cells and sieve tubes. Sieve tubes are associated with companion cells.

Laticifers. They contain latex and are multinucleate. There are two kinds of laticifers—articulated and non-articulated. The articulated laticifers are formed through union of cells in which parts of the walls are dissolved. The non-articulated laticifers are single cells, usually much branched. The articulated laticifers may be primary or secondary in origin whereas the non-articulated are primary in origin. They are restricted to certain angiospermic families.

2

The Cell – Structure and its Components

The plant body consists of a number of small, microscopic, box-like compartments called *cells*. In other words, *the cells are the universal elementary units of organic structure*. A plant call may be defined as microcosm having a definite boundary or the cell-wall within which complicated chemical reactions are going on. A cell devoid of this chemical reaction is inert and is considered to be a dead cell.

HISTORICAL ACCOUNT

Robert Hooke (1635–1703), an English microscopist for the first time in seventeenth century (1665), studied the internal structure of a thin slice of bottle cork with the help of crude microscope designed by himself, and discovered a honeycomb like structure in it, and to each compartment he recognized a cell (latin, *cellula* = a small apartment). Grew (1641–1712) published his first hand description of the plant tissues in 1671 and thereafter in 1682. Malpighi (1628–1694), an Italian botanist studied various plant tissues and published his original work in 1675. The cell theory, initiated by Schleiden (a botanist) and Schwann, (a zoologist) in the years 1838–39, that all animals and plants are made up of the cells and their products, and that growth and reproduction are fundamentally due to division of cells. Robert Brown (1773–1838), an English botanist could recognize for the first time in 1831, the general occurrence of a spherical body in each cell and named it, '*the nucleus*'. In the year 1846 Hugo Von Mohl made a clearcut demarcation in between the protoplasm and the cell sap. Kolliker (1861), gave the name '*cytoplasm*' to the substance found around the nucleus. Rudolf Virchow (1858), stated that the cells come from pre-existing cells. Hanstein in 1880 gave the term protoplast to designate one unit of the protoplasm found within a single cell and also suggested that this term may be used instead of the term cell. Strasburger (1881) gave a clear account of the structure of nucleus and nuclear division. In 1888, he also revealed the occurrence of reduction division in angiosperms. Farmer and Moore (1905), used the term '*meiosis*' for such division.

THE CELL THEORY

Schleiden (1838) a German botanist and Schwann (1839) a German zoologist sponsored a theory that all animal and plant organisms are composed of cells. Prior to them several other workers also put forward the similar views. Mirbel (1808) suggested that, "*plants are formed by a membranous cellular tissue*". Lamark (1819) also suggested that, "*no body can have life, if its constituent parts are not cellular tissue or are not formed by cellular tissue*". Turpin (1826), Meyen (1830) and Von Mohl (1831) also put forward the similar views. In the beginning of nineteenth century Dutrochet, however, started the work of systematic comparison of plant and animal tissues. He was very close to a theory and he stated, "*all the tissues of animals united only by cohesion*". This theory advocates that all the tissues and all the organs of plants and animals are in real sense a cellular tissue diversely modified.

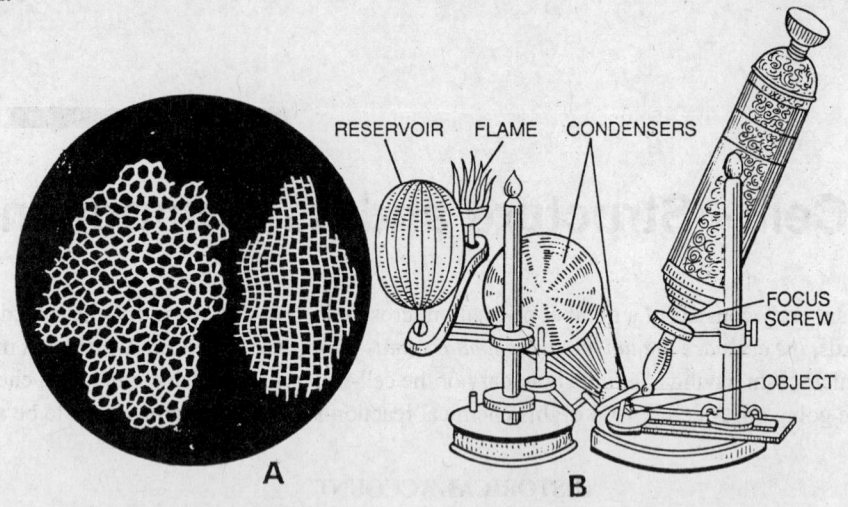

Fig. 2.1. The cell., A, Robert Hooke's drawing of the microscopic structure of cork; B, the microscope with which he observed it.

Schleiden (1838) and Schwann (1839) for the first time used the term *Cell Theory* and stated,"*The cells are organisms, and animals as well as plants are aggregates of these organisms, arranged in accordance with definite law*". Schwann established the cell theory in a difinite form. He had clear ideas both of morphological and physiological significance of the cell. According to him the cellular phenomena consists of two groups—(1) the *plastic phenomena* and (2) *the metabolic phenomena*. The former one corresponds to morphology of cell and also includes in it "*the combination of molecules which form the cell*". On the other hand the metabolic phenomena (physiologic phenomena) includes "*the chemical changes, whether in the particles constituting the cell itself, or in the surrounding "cytoblastema"*". Thus, about a century back Schwann expressed the present view point of cytology and, therefore, he is generally considered as 'father of modern cytology'. R. Virchow (1858), the great German physician, applied the cell theory to pathology and demonstrated that pathologic processes take place in the cells and tissues. He also made another important generalization—that cells come only from pre-existing cells. When biologists further recognized that sperm and ova are also cells that unite with each other in the process of fertilization, it gradually became clear that life from one generation to another is an uninterrupted succession of cells. Growth, development, inheritance, evolution, disease, aging, and death are, therefore, but varied, aspects of cellular behaviour.

Andre Lwoff, the French microbiologist, has expressed the gist of cell theory as follows:

When the living world is considered at the cellular level, one discovers unity. *Unity of plan*: each cell possesses a nucleus embedded in protoplasm. *Unity of function*: the metabolism is essentially the same in each cell. *Unity of composition*: the main macromolecules of all living beings are composed of the same small molecules. For, in order to build the immense diversity of the living systems, nature has made use of a strictly limited number of building blocks. The problem of diversity of structures and functions, the problem of heredity and the problem of diversification of species have been solved by the elegant use of a small number of building blocks organized into specific macromolecules. Each macromolecule is endowed with specific function.

The cell theory postulates, in brief that the cells from the structural units of living organisms and the latter show a characteristic cellular construction. In addition, they are also functional units in that all the vital functions of the living organism are traceable ultimately to the living substance, the protoplasm contained in them.

THE PROTOPLASM

According to Huxley *the protoplasm is the physical basis of life*. Inside the cell wall of living cell the living substance is known as *protoplasm*. The protoplasm is a thick fluid or jelly-like substance. These two forms are not much differentiable from each other and may easily be transformed from one form to another. When found fluid state it seems to be more active. Generally it is greyish or somewhat yellowish in colour. It is always transparent. Many small granules found in it are food granules. It is a very complex substance and found to be dispersed in the medium of water. It consists of ninety percent of water. The protoplasm found in the cells of the seeds contains less percentage of water and, therefore, it is of thick consistency. In such cases the protoplasm is somewhat inactive and becomes active only when sufficient amount of water is absorbed. In addition to water other substances are also found in the protoplasm. Among these substances the most important ones are proteins which are found in large quantity. The proteins remain dispersed in the water. The proteins are highly complex chemical compounds. The protein molecules are very large and of very high molecular weight. These complex compounds always contain carbon, hydrogen, oxygen and nitrogen. In addition to these elements sulphur and phosphorus are generally found. The proteins of protoplasm always contain sulphur and phosphorus. The protoplasm contains the proteins of various types. The protoplasm also contains some other organic compounds, such as fats and carbohydrates. In addition to these, some inorganic salts also found in it.

COMPOSITION AND CHEMICAL NATURE OF PROTOPLASM

Composition. Generally the protoplasm consists of oxygen carbon, hydrogen and nitrogen. Approximately the oxygen is 62%, carbon 20%, hydrogen 10% and nitrogen 3%. The remainder of 5% part contains about thirty elements, of which calcium (Ca), iron (Fe), magnesium (Mg), chlorine (Cl), phosphorus (P), potassium (K), sulphur (S), etc., are important ones. In addition to these, boron (Bo), copper (Cu), fluorine (Fl), manganese (Mn) and silicon (Si) are found in small traces. In certain special cells alcohol, cobalt (Co) and zinc (Zn) are also found. All these elements are found in ionic state or essentially found in adenosine triphosphate (ATP). All chemical reactions going on in the protoplasm obtain energy for their performance from ATP. The protoplasm contains 67-75% of water. Moreover, certain gases such as carbon-dioxide and oxygen remain dissolved in it.

The protoplasm of each cell contains several organic substances of which *carbohydrates, fats, proteins,* and *nucleoproteins* are important ones. These organic substances make protoplasm by molecular combination.

Carbohydrates. About thirteen percent part of protoplasm consists of carbohydrates. The carbohydrates contain carbon, hydrogen and oxygen. The important carbohydrates are—glucose, sucrose, starch, cellulose, glycogen, etc. The granules of carbohydrates either remain suspended or dissolved in the cytoplasm. They are mainly responsible for the production of energy.

Fats. The fats or lipids consist of carbon, hydrogen and oxygen elements. They are formed by the combination of glycerol and fatty acids. They contain lesser amount of oxygen than carbohydrates. When they are chemically decomposed the energy is being liberated. They contain much more energy in comparison to carbohydrates. The cell membrane consists of fat.

Proteins. About fifteen per cent of the protoplasm consists of proteins. In addition to carbon, hydrogen and oxygen elements, the proteins essentially contain nitrogen. Usually they contain sulphur and sometimes phosphorus too. The proteins are formed by the combination of the molecules of amino-acids. About twenty amino-acids are found in the nature, that give rise to different kinds of proteins by their molecular combination. A protein molecule is made of hundreds or thousands of amino-acid molecules joined together by peptide links into one or more chains, which are variously folded. There are 20 different kinds of amino-acids commonly found in proteins, and most of these usually occur in any one protein molecule; they are arranged in the chain in a sequence which is exactly the same in all molecules of a given kind of protein. The possible different arrangement of the amino-acids are evidently practically infinite, and the diversity is fully exploited by living things, every species having kinds of protein molecule peculiar to itself. A protein molecule is very large (molecular weight from about 20,000 up to several millions), and dissolved proteins form therefore colloidal solutions. Proteins are not soluble in fat solvents. Many are soluble in water or dilute salt solutions (*e.g.*, globulins); others, with elongated (fibrous) molecules, are insoluble in these solvents (*e.g.*, scleroproteins, myosin). Proteins are synthesized from amino-acids by all living things; the precise sequence of the amino-acids being determined by the sequence of nucleotides in nucleic acids. The proteins are destroyed by proteolytic enzymes. They are frequently combined with other substances, especially nucleic acids (nucleo-proteins), carbohydrates (glycoproteins), fats (lipoproteins).

Nucleoproteins. They are most complex substances ever found. They are compounds of nucleic acid and protein. The protoplasm contains two types of nucleic acids—ribonucleic acid (RNA) and deoxyribonucleic acid (DNA). The RNA is found in the complete cell, whereas DNA remains confirmed in the nucleus. Ribonucleic acid (RNA), a molecule consisting of a large number of nucleotides attached together to form a long strand one nucleotide thick. Deoxyribonucleic acid (DNA) is mainly found in the genes of chromosomes. The DNA, a compound consisting of large number of nucleotides attached together in a single file to form a long strand. Usually two such strands are linked together parallel to each other by base pairing, and coiled into a helix. In each cell the DNA and RNA are mainly concerned with the metabolic activities. Numerous RNA and DNA are formed within the living cells. The nucleic acids of two different living beings are never identical. The RNA and DNA control all the metabolic activities going on within the living cells. They play an important role in the origin of life. DNA is the material of inheritance of almost all living beings.

Other chemical substances. Besides above mentioned organic substances several other inorganic substances are also found in the living cells in small quantities. These substances are specially concerned with the cell metabolism. Some of these substances are found in all cells whereas some are confined to certain special cells. Various pigments, latex, alkaloids, vitamins, hormones, antibiotics and some other substances are found within the plant cells.

Thus, the carbohydrates, fats, proteins, nucleoproteins and several other chemical substances make the protoplasm by their molecular combination. The molecules of all these substances are well organized and form protoplasm. In addition to these the molecules of living cells possess special characteristics by which various chemical reactions take place and with the result the energy liberates. This energy is utilized in the performance of the metabolic activities of the cells. Thus, this becomes an established fact that all constituents of the cell form a very powerful organization by their combination, which is living and quite active.

PHYSICAL NATURE OF PROTOPLASM

Colloidal state. The proteins of protoplasm are found in water in colloidal state. The proteins are formed in the form of very minute granules. Each granule consists of several molecules. Generally the protoplasm is an amulsoid, in which the proteins remain dispersed in the water. Many colloidal particles possess the power of holding and absorption of water. This is called *hydration* and the colloids which absorbed the water are said to be *hydrated*. The colloids of protoplasm remain hydrated, and the amount of hydration depends upon various conditions.

Distribution. Generally the protoplasm remains divided into two parts—*the nucleus* and the *cytoplasm*. Usually the young cells remain filled up with the protoplasm. In mature cells, the cytoplasm forms the inner layer of the cell wall and a layer around the nucleus. The nucleus remains embedded in this layer of cytoplasm. All the protoplasm of a cell excluding the nucleus is known as *cytoplasm*. The cytoplasm is usually a transparent slightly viscous fluid with inclusions of various sizes.

Permeability. Permeability of membrane, exhibits to which molecules of a given kind can pass through it. Substances differ greatly in the case with which they pass through a given membrane, *e.g.*, water or fatty substances pass easily, many ions and proteins with great difficulty, through plasma membrane. Permeability on any biological membrane is itself variable within wide limits. The very numerous membranes of animals and plants, *e.g.*, plasma membranes, produce many osmotic effects and are of great importance in determining local differences in composition of the organism. According to Virgin (1953), the permeability is the quality of a living membrane through which different ions and molecules may pass, and during this process the energy obtained from metabolic activities is not utilized. This is the permeability of the protoplasm which is dependable upon the physiochemical properties of the protoplasmic membrane of the cell. Virgin has also propounded that the permeability of the protoplasm is the control of active and passive forces on the particles passing through the living protoplasmic membrane. The cellulose cell wall is permeable for water and other solutes, whereas the protoplasmic membrane is permeable only to water and not to other solutes.

PROPERTIES OF PROTOPLASM

The protoplasm is a living matter and makes the physical basis of life. On the basis of the properties given below it may be differentiated from non-living matter. The properties are as follows:

Absorption and excretion. The protoplasm takes essential substances for its growth by absorption and it gets rid of from non-required substances by excretion. The protoplasm of green plants absorbs water and minerals from the soil and carbon dioxide from the atmosphere. Thus, the substances quite different from the protoplasm are being absorbed by it.

Metabolism. In general the chemical processes occurring within an organism, or within part of one. They involve breaking down of organic compounds from complex to simple (catabolism) with liberation of energy available for the organism's many activities; and building up of organic compounds from simple to complex (anabolism) using energy liberated by catabolism and in the case of autotropic organisms, energy from external non-organic sources (particularly from sunlight). Catabolism is by no means confined to the breaking down of the 'food material' necessary for the organism's energy requirements. It involves, continuously, very many constituents of the organism, though they are broken down at various rates, and on the other hand anabolism involves, continuously, synthesis of many constituents. These processes go on even in an apparently static tissue. This metabolic whirlpool is a system of reactions predominantly controlled by enzymes. In most of the aspects of metabolism which have been sufficiently investigated, such as catabolism of carbohydrates, there is a fundamental similarity throughout a wide variety of organisms including plants, animals and bacteria.

Growth and reproduction. Growth means increase in size. It includes differentiation of cells, increase in plant size, even though such differentiation contributes nothing to size change. The growth of protoplasm is the result of the separation and growth of such a small part which has been derived from one or parental living beings. In other words, the reproduction is another form of growth.

Movement. The protoplasm has the power of movement and usually it remains in dynamic condition. The dynamic state of the protoplasm of the cell may be seen under the microscope. This phenomenon is known as *Brownian movement*. The protoplasm does not only possess the property of movement, but it also helps the living being or its organs to remain in dynamic state. Such movements are more clearly seen in animals than in plants, but in certain plants such as *Mimosa pudica*, it is quite obvious. Besides, the young parts of several plants bend towards the light; several unicellular and other types of water plants move like the animals from one place to another such as *Chlamydomonas, Volvox*, etc.

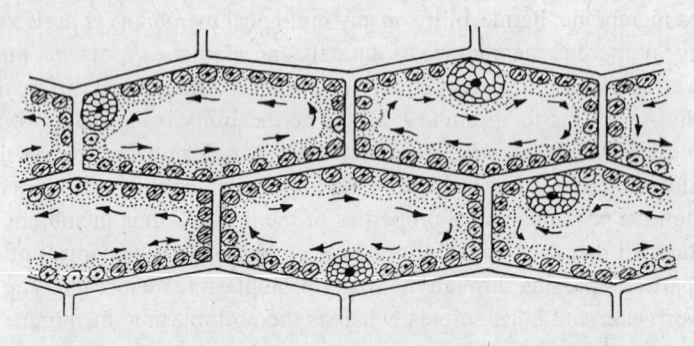

Fig. 2.2. Movement of protoplasm. Rotation in the leaf of *Vallisneria*.

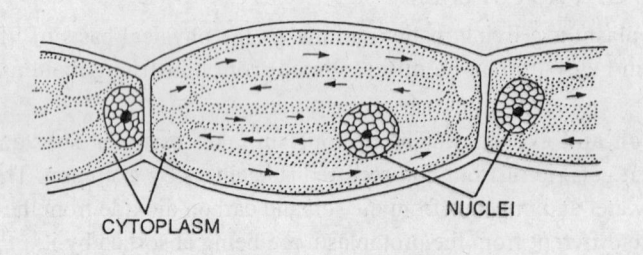

Fig. 2.3. Movement of protoplasm. Circulation in the staminal hair of *Commelina*.

Irritability. This is a universal property of living things (protoplasm). The reponsiveness of protoplasm to outer stimulus is known as irritability. Temperature and several chemicals have definite effect upon the movement of the protoplasm. At ordinary temperature the rate of movement increases whereas at low temperature it decreases. The effect of certain chemicals increases the rate of movement while the effect of certain other chemicals decreases it. With the result of the stimulus upon the protoplasm, a complete organ of the plant may be affected, *e.g., Mimosa pudica*.

THE PLASMA MEMBRANE
(Cell Membrane)

All cells remain covered by a *plasma membrane*. The plasma membrane is quite separate and should not be confused with the cell wall which lies outside the cytoplasm. Generally a cell wall

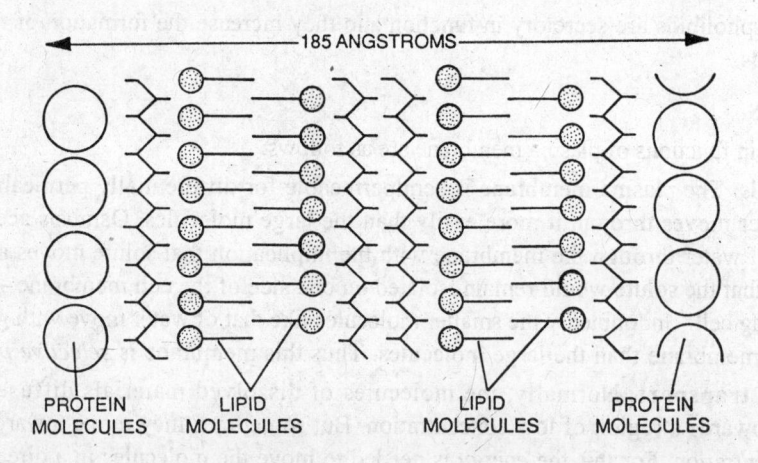

Fig. 2.4. The cell. The plasma membrane in molecular terms showing protein and lipid molecules.

is regarded as protoplasmic product and is not believed to be truly living. However, in certain cases the cell wall remains capable of growth.

The plasma membrane is generally considered to be the outer living limit of the cell, but not necessarily its outer boundary. Such outer boundaries are most commonly found in plant cells, many of which possess heavy walls of cellulose, but animal cells and unicellular organisms also exhibit a number of somewhat comparable external substances.

The cell membrane is double in nature and about 10 Å (Å = *Angstrom* unit) thick. This membrane remains connected with the nuclear membrane through the endoplasmic reticulum.

The double nature and molecular make-up of the plasma membrane are depicted in figure. The membrane has a lipoprotein (fat + protein) composition. The protein constituent of the membrane gives the cell wettability and flexibility. Since protein molecules, which are long and complex, can fold or unfold, the membrane can expand or contract, thus providing through molecular spacing a possible means of control over which molecules can enter the cell from the outside environment, or pass to the environment from inside the cell. Such a membrane is known as *selectively permeable*.

CHEMICAL NATURE OF PLASMA MEMBRANE

The plasma membrane consists of a protein called *stromatin*, which is fibrous in nature and possesses a high molecular weight. These are also glycoproteins and mucoproteins. It also contains large amount of arginine and lysine. In addition to it, histidine, tyrosine, trypotophan, methionine and many other aminoacids are also present. The protein is acidic in nature and forms greater percentage of plasma membrane as compared to lipid components. The lipids are mainly phospholipids and cholesterol. The phospholipids mainly consist of lecithin and cephalin. On an average the membrane possesses 75 to 90 molcoules of lipid for every molecule of protein. Five phospholipids have been identified from the cell membrane. One is *phosphatidic acid* and the remaining four are more complex being comprising either cholein, inositol ethanolamine or sercine attached to the phosphate groups in addition. The one phospholipid containing choline is called *lecithin*. The phospholipid molecules that are made up of glycerol and phosphate are water soluble, whereas the fatty acid molecules that constitute the tail of the whole structure are water insoluble. According to Parpart and Ballantine (1952) the structure of the plasma membrane is different in different cells. Salts are also present in the cell membrane. Lowell and Hokin (1965) demonstrated that the phospholipids are secretory in function and they increase the formation of the membrane phospholipids.

FUNCTIONS

The main functions of plasma membrane are as follows:

Osmosis. The plasma membrane is semipermeable for differentially permeable membrane because water moves through it more easily than the large molecules. Osmosis accounts for the movement of water through the membrane with the implication that solute moves at slower rate. This means that the solute would remain isolated on one side of the cell membrane—this does not occur in living cell. Undoubtedly the smaller molecules like that of water move with greater facility through the membrane than the large molecules. Thus this membrane is *selective permeable*.

Active transport. Normally, the molecules of dissolved materials diffuse through the membrane towards a region of low concentration. But sometimes they move towards a region of higher concentration. For this the energy is needed to move the molecules in a direction opposite to that in which they would move by diffusion. This type of movement is called *active transport*. In brief, here the substance does not move by itself but is transported through the cell membrane. It can be defined as, *"the movement of molecule of a substance against a concentration gradient and (or) against an electrochemical."*

CELL STRUCTURE

The protoplast of each cell contains the protoplasmic and non-protoplasmic components. The protoplasmic or living components are—nucleus, mitochondria, plastids, endoplasmic reticulum, ribosomes, lysosomes, sphaerosomes, microtubules and Golgi bodies.

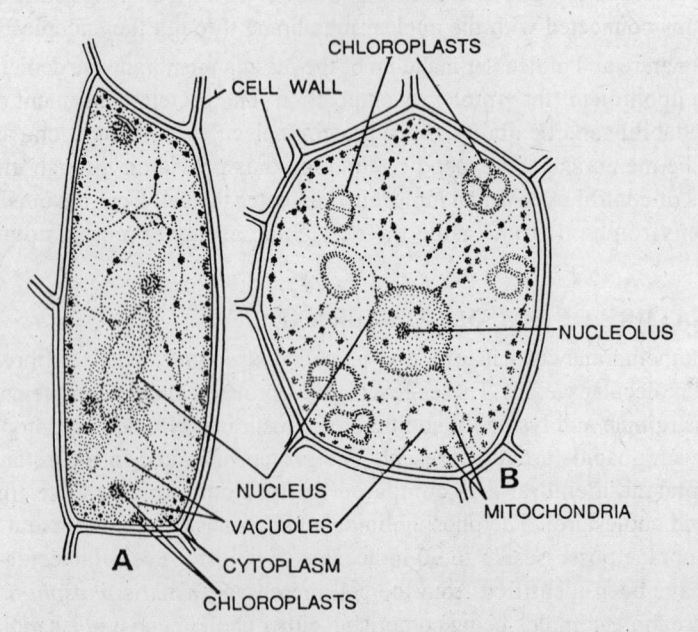

Fig. 2.5. Plant cells. A, cell from sugar-beet leaf with vacuolated cytoplasm, a nucleus and granular chloroplasts; B, a cell from starch-sheath of tobacco stem (after K. Esau).

The chief non-protoplasmic or non-living components are—vacuoles, food products, secretory products and waste products. The protoplasmic and non-protoplasmic components of the cell are summarised on next page.

The Cell—Structure and its Components

```
                                    CELL
                         _____|_____
                        |                       |
             Protoplasmic Components    Non-Protoplasmic Components
             |                           _____|_____
             1. Nucleus                 |                       |
             2. Mitochondria         Vacuoles            Ergastic Substances
             3. Plastids                         _____|_____
                (i)   Leucoplasts                |           |           |
                (ii)  Elaioplasts             Food products  Secretory   Waste products
                (iii) Chromoplasts                           products
                (iv)  Chloroplasts     _____|_____   |           |
             4. Endoplasmic          |        |        |    (i)  Enzymes (i)  Tannins
                reticulum      Non-nitrogenous Nitrogenous Fats and (ii) Pigments (ii) Mineral crystals
             5. Ribosomes    (Carbohydrates)            fatty oils (iii) Nectar  (iii) Latex
             6. Lysosmes       (i)  Starch   (i)  Proteins                       (iv) Essential oils
             7. Sphaerosomes   (ii) Inulin   (ii) Amino
             8. Microtubules   (iii) Hemicellulose  compounds
             9. Golgi bodies   (iv) Cellulose
                               (v)  Sugars
```

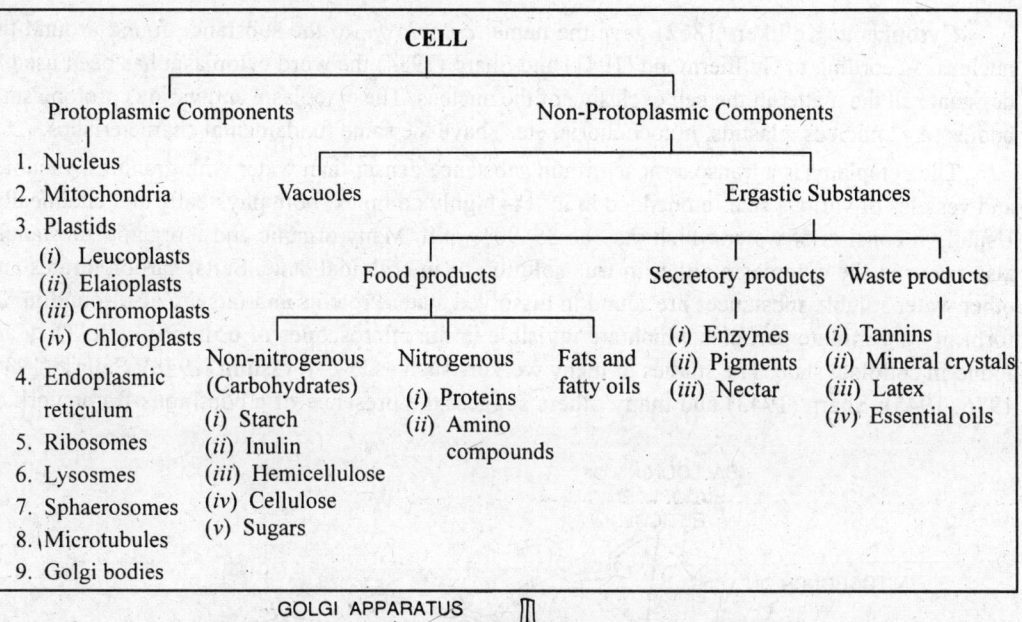

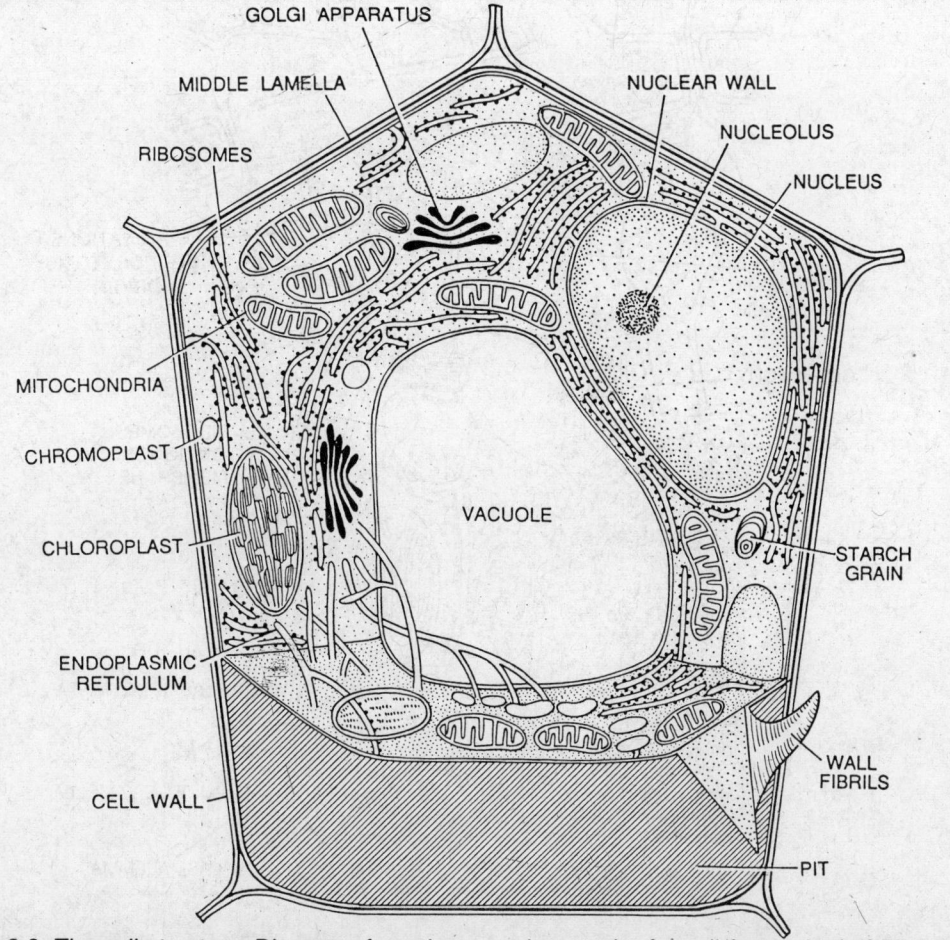

Fig. 2.6. The cell structure. Diagram of an electron micrograph of a cell from a leaf of Phaseolus vulgaris.

Cytoplasm. Kolliker (1862) gave the name '*cytoplasm*, to the substance found around the nucleus. According to Guilliermond (1941) and Sharp (1934) the word cytoplasm has been used to designate all the matter in the cell exclusive of the nucleus. The cytoplasm and various protoplasmic bodies (*e.g.*, nucleus, plastids, mitochondria, etc.,) have the same fundamental characteristics.

The cytoplams is a transparent semifluid substance denser than water with granules, vacuoles and vesicles of various sizes embedded in it. It is highly complex, both physically and chemically. Usually it consists of water which may be 85–90% in it. Many organic and inorganic substances also occur in the cytoplasm either in true solution or in colloidal state. Salts, carbohydrates and other water soluble substances are found in dissolved state. Proteins and fats are also found in the form of very minute particles which are invisible in the microscopes of ordinary light. They are found in colloidal state. The studies of many workers such as Fry–Wyssling (1948); Seifriz (1935, 1936, 1945); Sharp (1943) and many others suggest the presence of a continous framework of

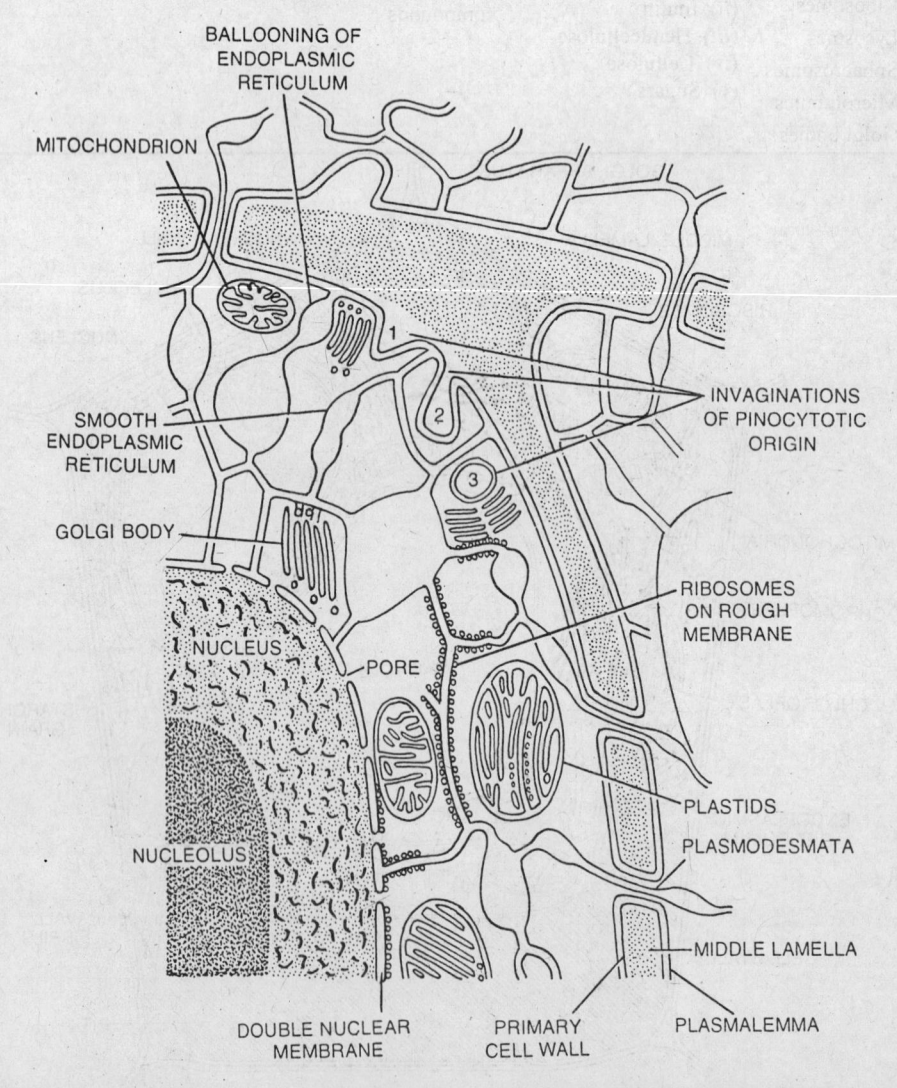

Fig. 2.7. Portion of a plant cell as seen under electron microscope. The invaginations of plasma membrane and ballooning of endoplasmic membrane are clearly seen.

proteins. It remains always in a dynamic state due to constant phase inversions. It is bounded by a monomolecular layer or a multimolecular layer on the outside called *ectoplast* (plasma membrane). Similar second layer on inside called *tonoplast* (vacuolar membrane). The protoplasmic layers are semipermeable in nature which involve in differential absorption. In the cytoplasm, fatty substances such as lipids and certain proteins take part in the formation of ectoplast (plasma membrane). The tonoplast (vacuolar membrane) also develops from the same substance and possesses similar structure to that of plasma membrane.

The electron microscope reveals membranous differentiations in the interior of the cytoplasm notably, the **endoplasmic reticulum** and the **dictyosomes**. The cytoplam also includes granules of various sizes. Granules of 0.25 to 1 micron in diameter, containing lipids and proteins, constitute the **spherosomes**. These granules are highly mobile in living cells. At the submicroscopic level, a granule about 150 Å in diameter, the **ribosome** a globular macromolecule of ribonucleoprotein that takes part in protein synthesis. Ribosomes occur free in the cytoplasm or they remain associated with the endoplasmic reticulum.

Some of the resolvable entities in the protoplast, such as the **nucleus, plastids** and **mitochondria**, are referred to as **organelles**, whereas the endoplasmic recticulum and the dictyosomes are called sometimes membrane systems and sometimes organelles.

Cytoplasmic membranes. Among the cytoplasmic membranes, ectoplast and tonoplast, have long been associated with the important physiological characteristics of the protoplast. These membranes are difficult to recognize with the light microscope, but the electron microscopy has confirmed their morphologic identity (Mercer, 1960). The tonoplast is thinner than the ectoplast.

The endoplasmic reticulum is a system of membrane-bound cavities or **cisternae**. The cisternae are commonly much flattened and therefore, their sections appear as double lines. The endoplasmic reticulum is responsible for providing the cell with a large internal membrane surface along which enzymes are orderly distributed.

Dictyosomes are stacks of flattened sacks or cisternae, circular in out line, each surrounded by vesicles. The vesicles originate from the margins of the cisternae and pass into the cytoplasm. Dictyosomes are concerned with secretory activities and wall formation.

THE NUCLEUS
(The Controlling Centre)

The controlling centre of a cell is the *nucleus*. The *chromosomes* and *genes* are found within it which determine the character, activities and destiny of each individual cell.

Each nucleus is surrounded on the outside by a *nuclear membrane*. This membrane quite resembles the cell membrane. It is double-layered and consists of proteins and lipids. The main part of the nucleus consists of chromatin, which during cell division becomes more distinctly visible as a definite number of individual chromosomes. Certain coarser granules are also visible in the chromatin. They are known as *chromocentres* which stain darker than the rest of the net work of the chromatin. Each nucleus consists of *nuclear membrane, nuclear sap (nucleoplasm), nucleolus or nucleoli* and *chromatin*.

Structure of the nucleus. Each nucleus remains surrounded by a limiting membrane as the *nuclear membrane*. As electron microscopy reveals the nuclear membrane consists of two membranes each being 90Å thick and the space in between the two, the *perinuclear space* being 100-115Å wide. The nuclear membrane possesses pores here and there varying in diameter from 400-600Å. Through these pores the nucleoplasm communicates with cytoplasm. At the margins of the pores the two membranes are continuous. The number of pores varies from species to species

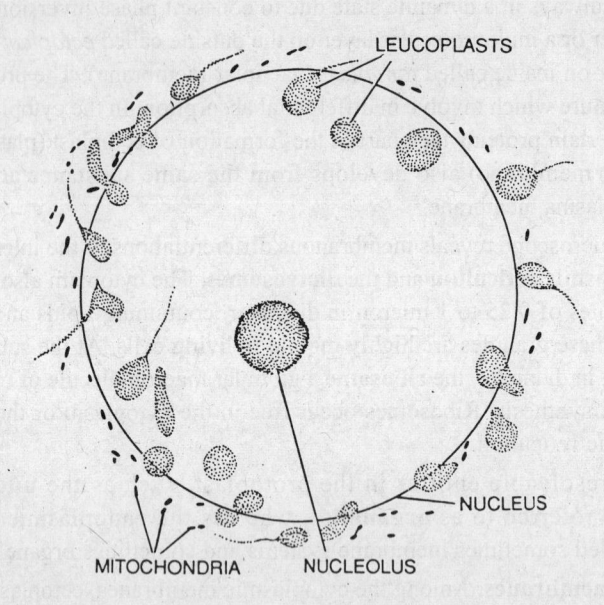

Fig. 2.8. The nucleus. Nucleus, amoeboid leucoplasts, and mitochondria from the cell of sugar-beet hypocotyl.

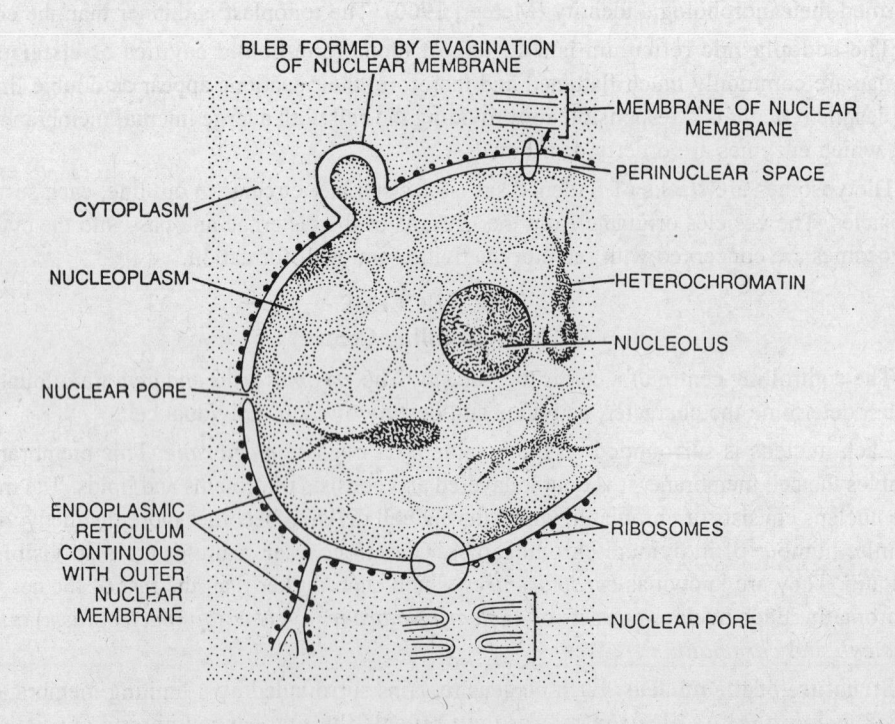

Fig. 2.9. The structure of the interphase nucleus (diagrammatic).

and even from cell to cell. According to some workers, each pore remains encircled by a circular structure, the *annulus*. According to Watson (1959) these pores and annuli constitute the *pore complex*. The outer surface of the nucleus membrane which remains in contact of cytoplasm has RNA containing granules attached to it, the ribosomes. The outer membrane at places gives out

The Cell—Structure and its Components

tubular structures which get branched and form the endoplasmic reticulum. The inner membrane exposed to nucleo-plasm lacks ribosomes but may have aggregated chromatin. The nuclear membrane consists of lipoproteins.

Inside the nuclear membrane or nuclear envelope their lies the *nuclear sap* or *karyolymph* or *nucleoplasm* in which are found nuclear ribosomes and chromatin network that develops into well defined chromosomes during cell-division. Nucleoplasm is some-what a granular and homogeneous semifluid which contains nucleic acids, nucleoprotein, etc.

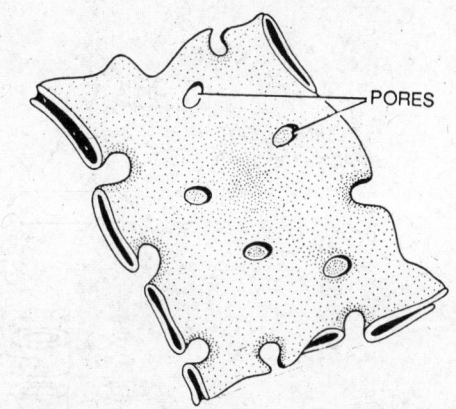

Fig. 2.10. Nucleus. Structure of a part of nuclear membrane. The membrane is two layered and the pores are clearly visible.

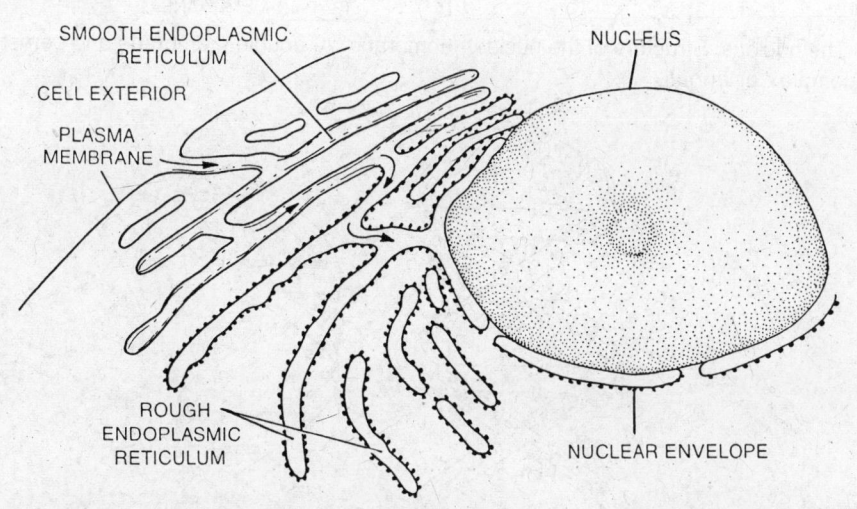

Fig. 2.11. The diagram depicts a eukaryotic cell that shows how the endoplasmic reticulum (ER) connects with both the plasma membrane and the double layered nuclear envelope. Connections of this type may form channels through which materials may pass to and from the cell's surroundings without penetrating the plasma membrane. However, such channels are formed only in cells with a highly developed system of endoplasmic reticulum (ER).

Certain distinct one or more rounded bodies known as *nucleoli* (singular-*nucleolus*) are also found within the nucleus. The number of nucleoli per nucleus of a particular species is definite. For example, nucleus of onion cell possesses four nucleoli, which sometimes fuse together.

The *nucleus* is a relatively large, generally spherical ball-like body found inside the nucleus. The nucleoli are made up of small granules containing proteins and RNA which has got some part to play in the activity of the nucleus. It is thought that nucleoli help in the synthesis of nucleoprotein. Bonner (1965) is of opinion that ribosomes have a nuclear origin and nucleolus plays a role for the same. There is no membrane around the nucleolus. It is now an established fact that nucleus is related to the biogenesis of the cytoplasmic ribosomes. During mitotic division the nucleoli undergo cyclic changes. The nucleoli of the interphase nucleus seem to disappear at the

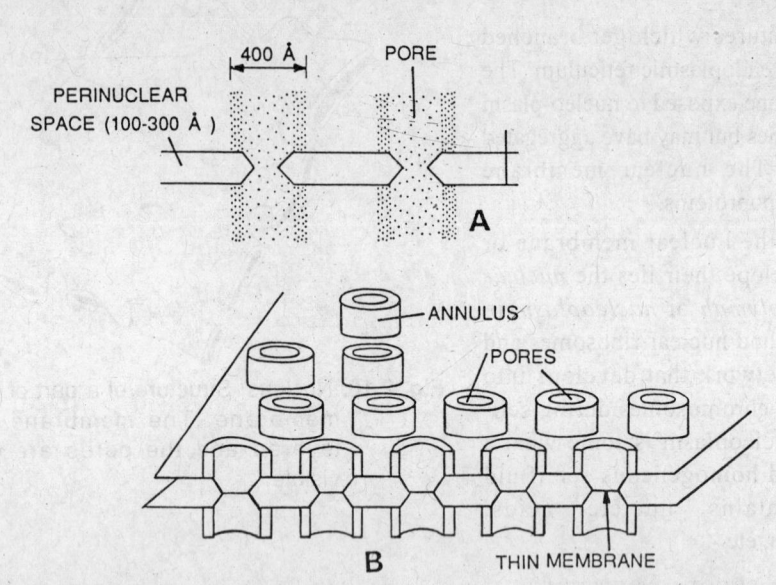

Fig. 2.12. The nucleus. Structure of the nuclear membrane. A, double membrane and pores; B, pore complex or annuli.

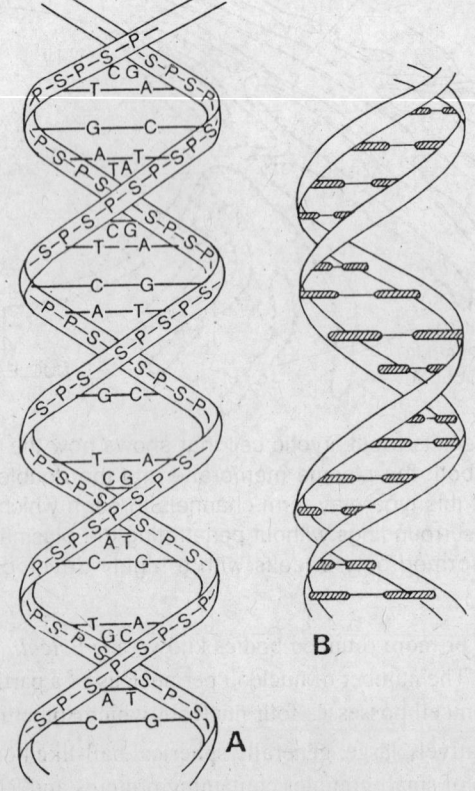

Fig. 2.13. DNA structure. A, schematic diagram of the double helix in a portion of a DNA molecule. Alternating sugar deoxyribose (S) and phosphate (P) groups make up the back bones of the strands. Attached to each sugar unit are one of the purine or pyrimidine bases, adenine (A), thymine (T), guanine (G), cytosine (C), the strands of the helix are together through hydrogen bonds between the base pairs, adenine to thymine and guanine to cytosine; B, diagram of the parallel strands in the helix.

The Cell–Structure and its Components 25

beginning of cell division, *i.e.*, prophase at the same time as chromosomes increase their staining property. At the end of mitosis, *i.e.* telophase the nucleoli reappear. Each nucleolus has contact with a chromosome and possesses at the point of union a special region to which the name nucleolar organizing region has been given.

The *chromatin* is the hereditary material of the cell. The question arises, what determines the hereditary powers of the chromatin. The science of heredity known as *Genetics*, informs us that the genes found on the chromosomes are responsible for the determination of the characters of the species.

An experiment has been done, to analyse the chromatin chemically and on the basis of this experiment : the idea of the molecular basis of heredity has been gained. According to this experiment the chromatin has been resolved into four major molecules. They are—(*i*) *histone*, a low molecular weight protein (2000), (*ii*) a more complex protein than histone; (*iii*) deoxyribonucleic acid (DNA) and (*iv*) ribonucleic acid (RNA). These four molecules constitute the chromatin. But now, with the result of extensive studies this has been demonstrated that DNA is the molecule whose structure determines hereditary uniqueness on the cell and the individual. In other words, this

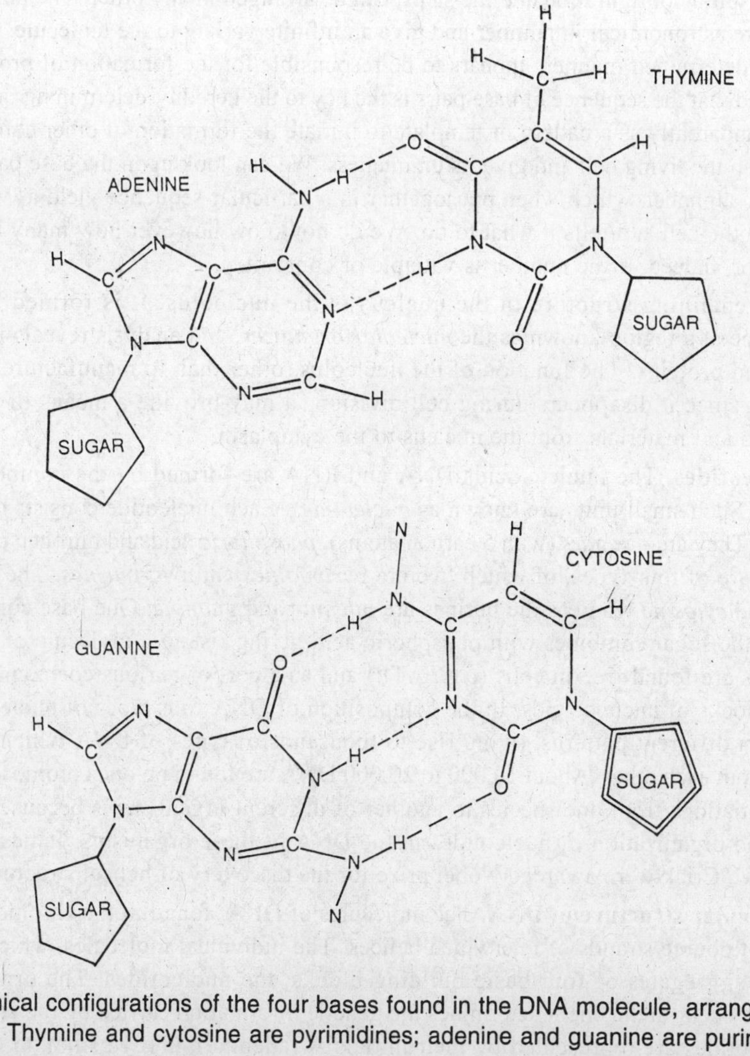

Fig. 2.14. Chemical configurations of the four bases found in the DNA molecule, arranged as base pairs. Thymine and cytosine are pyrimidines; adenine and guanine are purines.

can be said that DNA is the hereditary material of the cell. The proteins and RNA are necessary for the functioning of the nucleus but not so important from the point of view of the heredity. In certain plant viruses only RNA is found in their structure and no DNA, in such cases the RNA acts as hereditary molecule.

DNA. The *deoxyribonucleic acid* (DNA) is the compound of high molecular wt. (over 1,000,000), *i.e.*, made up of a number of molecules linked together. These molecules include a sugar, *deoxyribose, phosphoric acid* and four bases of which two are *pyrimidines* (*thymine* and *cytosine*) and two are *purines* (*adenine* and *guanine*).

Chemical structure of DNA. The alternating sugar and phosphate arrangement forms the outside boundaries of DNA while base pairs link the two sides together. The base pairs, however, are not at random, for adenine and thymine are always paired, as are guanine and cytosine. Hydrogen bonds tie the bases to each other. X-ray analysis of the molecular arrangements reveals that DNA is not a flat structure, as one might suspect, but a double helix, *i.e.*, a sort of sipral' stair case with alternating sugar-phosphate 'bannisters' and 'steps' of base pairs. This is the Watson-Crick model of DNA, so named after its discoverers. The DNA molecule may have thousand of turns in its spiral configuration and the steps, can be arranged in any order. The possible variations therefore are astronomical in number and give an infinite variety to the molecule. Since the DNA, in some undetermined manner, appears to be responsible for the formation of proteins and RNA, it is believed that the sequence of base pairs is the key to the heredity-determining qualities of DNA. These act apparently as a pattern or template to initiate the formation of other complex molecules that make up the living cell and give its uniqueness. We can look upon the base pairs as the letters in a genetic alphabet, which when put together in a particular sequence yield a 'word' which has meaning to the cell and tells it what to do. We do not know however how many base pairs make up a gene or, indeed, if the number is variable or constant.

The remaining structure in the nucleus is the nucleolus. It is formed by a particular chromosome at a region known as the *nucleolar organizer*, and analysis reveals it to be made up of RNA and proteins. The function of the nucleolus, other than to manufacture proteins, is not known but since it disappears during cell division, it may provide a means of passing genetic information and materials from the nucleus to the cytoplasm.

Nucleotides. The nucleic acids DNA and RNA are formed by the composition of small molecules. Such small units are known as *nucleotides*. Each nucleotide consists of three types of molecules. They are—*sugars* (with 5 carbon atoms), *phosphoric* acid and nitrogen containing bases. The bases are of four types, of which two are *pyrimidines* and two *purines*. The pyrimidines are *thymine* and *cytosine* whereas the purines are *adenine* and *guanine*. One base combines with one sugar and the sugar combines with phosphoric acid giving rise to a molecule of nucleotide. The nucleotides are found free in cells (*e.g.*, ATP) and as a part of various coenzymes, and as three building blocks of nucleic acids. In the composition of DNA four types of nucleotides are being arranged in different patterns giving rise to thousands of types of DNA which are chemically different from each other. About 10,000 to 20,000 DNA are found on one chromosome. The reason of differentiations from one species to another of different organisms is because of the different number and organization of nucleotides in the DNA of these organisms. James D. Watson and Francis H.C. Crick were awarded Nobel prize for the discovery of helical structure of DNA.

Molecular structure of DNA. The molecules of DNA, found in the chromosomes, occur in the form of double strands of interwined helices. The individual molecules, which are very long, are linear aggregates of four basic building blocks, the nucleotides. The order in which the nucleotides occur in the chain determines the genetic information carried by the chain. Thus, if the four nucleotides are represented by their initials A (adenine), C (cytosine), G (guanine) and T

The Cell–Structure and its Components

(thymine), a linear aggregate in the order ACGT will constitute different genetic information from—AGCT—or—ATCG—or ACCA–GTGT—chains. The inheritance or every organism consists in its essence of some repeated pattern of nucleotide units in the DNA molecules of the chromosomes contained in nuclei.

DNA REPLICATION.

It is an established fact that the chemical nature and total amount of DNA remains constant in similar kind of cells from generation to generation. It means that quality and quantity of DNA remain same in similar cells, derived from the same parent cell. While explaining the DNA replication one has to keep in the view these both requirements. The weak hydrogen bond that holds together the two strands of the double helix works some-what like a zipper. If one starts at one end of the molecule, one can 'unzip' each purine from it pyrimidine partner one by one. This will leave unpaired purine and pyrimidine ends on both strands. It is well known that the cell is a storehouse of precursors of the

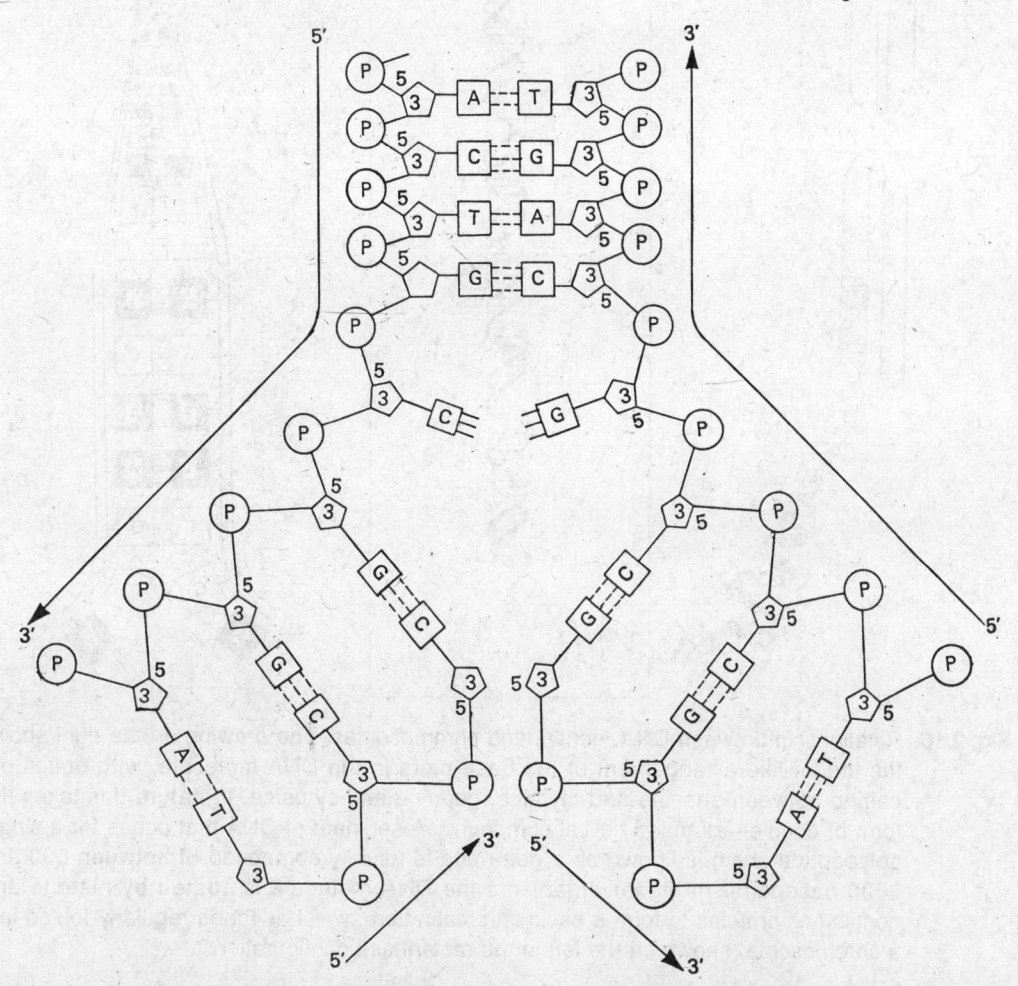

Fig. 2.15. DNA structure. The three components of DNA are the bases adenine, guanine, thymine and cytosine; a sugar molecule, deoxyribose; and phosphate groups. The phosphate groups are attached to the 3' and 5' carbons of the deoxyribose and serve to connect successive pairs of nucleotides. The replication process must one DNA strand by adding bases in the 5' → 3' direction and the other DNA strand, apparently by adding bases in the opposite direction.

four sorts of building blocks needed to make a new chain. These precursors are nucleotides. As such when the DNA helix 'unzips' new nucleotides from the cell of the proper kind fall into place, the improper ones are rejected as they will not fit. An adenine group will become bounded only with a new thymine group whereas the thymine group of the other strand will become bounded to a new adenine nucleotide thus completing the double chain in the two original strands. In the words of Crick the two strands of DNA 'unzip' as follows—'it is the growth of the two new chain that unwinds the original pair. This is likely in the terms of energy because, for every hydrogen bond that has to be broken, two new ones will be forming".

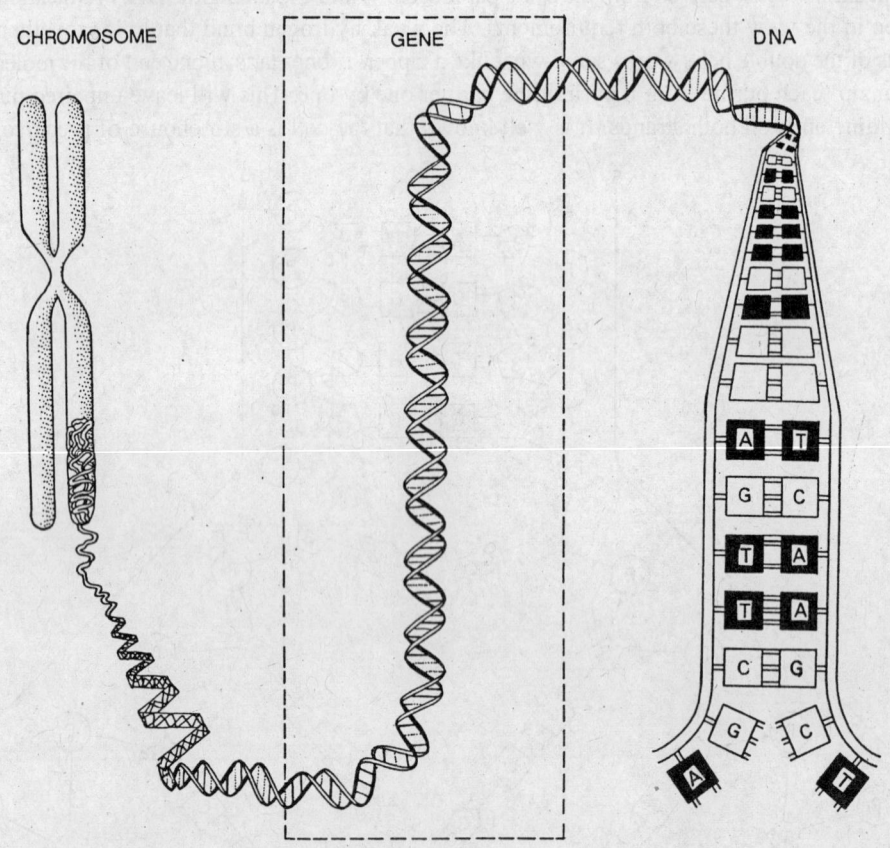

Fig. 2.16. Relationship between DNA, genes, and chromosomes. The drawing on the right shows the ladder-like arrangement of the base pairs in the DNA molecule, with deligatory pairing between adenine and thymine, guanine and cytosine. In nature this takes the form of double—stranded helical compound. A segment of DNA that codes for a single polypeptide chain is known as a gene and is usually composed of between 500 and 2000 basepairs. In higher organisms the DNA molecule is coated by histone and nonhistone proteins to form a basic chromatin fibre, which is then irregularly folded into a chromosome, shown on the left in the metaphase configuration.

The unzipping of DNA molecule at the weak forces (H-bonds) leaves the two strands apart, each strand with a purine or pyrimidine base attached to the sugar molecule. Each strand has not the capacity to replicate and exactly identical strand like the partner from which it has separated.

CLASSIFICATION OF DNA

The DNA in the eucaryotic cells may be classified as follows:

The Cell–Structure and its Components

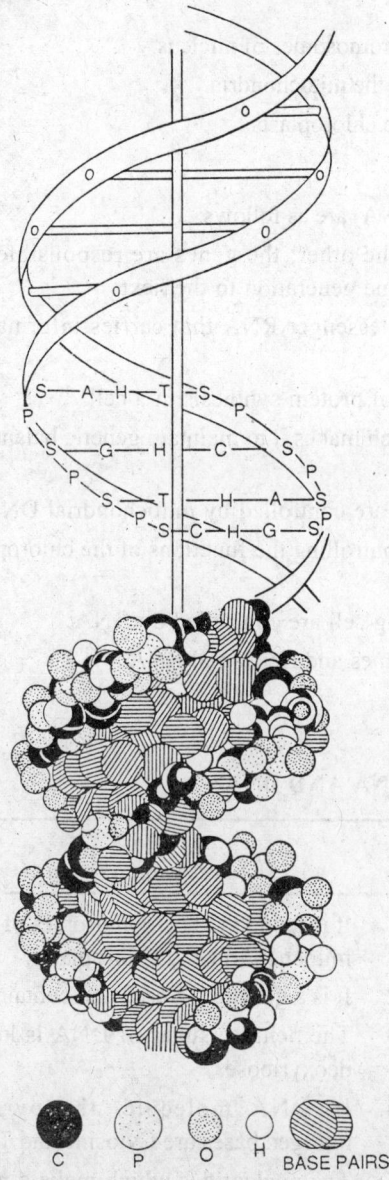

Fig. 2.17. The helix of DNA, with three different ways of representing the molecular arrangement. Top, general diagram of the double helix, with the phosphate sugar combinations making up the outside spirals and the base pairs the cross bars. Middle, a somewhat more detailed representation—phosphate (P), sugar deoxyribose (S), adenine (A), thymine (T), guanine (G), cytosine (C) and hydrogen (H). Bottom, detailed structure showing how the space is filled with atoms—carbon (C), oxygen (O), hydrogen (H), phosphorus (P) and base pairs.

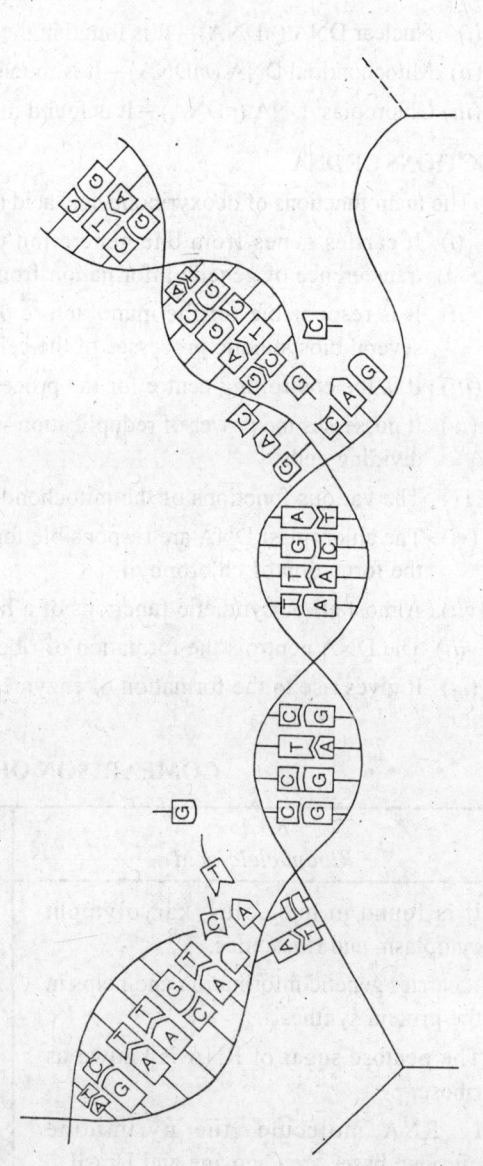

Fig. 2.18. Replication of DNA molecule. Diagrammatic representation.

(i) Nuclear DNA (nDNA)—It is found in the chromosomes of nucleus.
(ii) Mitochondrial DNA (mDNA)—It is found in the mitochondria.
(iii) Chloroplast DNA (cDNA)—It is found in the chloroplasts.

FUNCTIONS OF DNA

The main functions of deoxyribonucleic acid (DNA) are as follows:

(i) It carries genes from one generation to the other; the genes are responsible for the transference of genetic information from one generation to the next.
(ii) It is responsible for the manufacture of messenger RNA that carries information for several biosynthetic processes of the cell.
(iii) It is the controlling centre for the process of protein synthesis in a cell.
(iv) It possesses the power of reduplication which makes it to maintain genetic balance in the dividing cells.
(v) The various functions of the mitochondria are controlled by mitochondrial DNA.
(vi) The chloroplast DNA are responsible for controlling the functions of the chloroplast and the formation of chlorophyll.
(vii) Almost all biosynthetic functions of a living cell are controlled by DNA.
(viii) The DNA controls the formation of ribosomes and ribosomal RNA.
(ix) It gives rise to the formation of enzymes.

COMPARISON OF RNA AND DNA

RNA Ribonucleic acid	DNA Deoxyribonucleic acid
1. It is found in nucleolus, karyolymph, cytoplasm and ribosomes.	1. It is found in the chromosomes of nucleus, mitochondria and chloroplasts.
2. It carries genetic information and helps in the protein synthesis.	2. It is a genetic material and contains genes.
3. The pentose sugar of RNA is known as ribose.	3. The pentose sugar of DNA is known as deoxyribose.
4. In RNA molecule, the pyrimidine nitrogen bases are Cytosine and Urasil.	4. In DNA molecule, the pyrimidine nitrogen bases are Cytosine and Thymine.
5. The nucleotides which make a molecule of RNA are – (i) adenosine monophosphate, (ii) guanosine monophosphate, (iii) cytidine monophosphate and (iv) Uridine monophosphate.	5. The nucleotides which make a molecule of DNA are – (i) deoxyadenosine monophosphate, (ii) deoxyguanosine monophosphate, (iii) deoxycytidine monophosphate and (iv) deoxythymidine monophosphate.
6. The nucleotides of a RNA molecule are arranged in a single chain.	6. The paired nucleotides of a DNA molecule are arranged in a double stranded helix.
7. Three main types of RNA are messenger (mRNA), transfer (tRNA) and ribosomal (rRNA).	7. Three main types of DNA are — nuclear (nDNA), mitochondrial (mDNA) and chloroplast (cDNA).

The Cell–Structure and its Components

RNA.

The molecule of RNA (ribonucleic acid) consists of a single strand, but it may be wound back upon itself in several places, producing helices in these places as found in DNA. In the region of helix formation, the bases are hydrogen bonded. The RNA differs chemically from the DNA in that it contains a ribose sugar instead of deoxyribose sugar. The ribose sugar possesses an additional oxygen molecule which is not found in deoxyribose sugar. The base pairs in RNA are same, i.e., three of the four bases—adenine, guanine and cytosine are the same as found in DNA however, the fourth base, thymine is replaced in RNA by *uracil* which possesses one methyl group less. There are three kinds of RNA—ribosomal RNA (rRNA), soluble or transfer RNA (tRNA) and messenger RNA (mRNA). The RNA contained in the ribosomes are known as rRNA. Electron microscopy reveals that ribosomes are composed of two sub-units—the larger and the smaller. The small subunits are attached to mRNA and the appropriate tRNA are specially bound to the large subunits.

CLASSIFICATION OF RNA

The RNA in the eucaryotic cells may be classified as follows:
(*i*) The messenger RNA (mRNA)— This carries message from the DNA to the actual site of protein synthesis.
(*ii*) The transfer RNA (tRNA)— This carries a specific amino acid to the site of protein synthesis.
(*iii*) The ribosomal RNA (rRNA)— It is found in the ribosomes. The function of this is not very clear.

THE MITOCHONDRIA
(Power house of cell)

They are microscopic bodies occurring in cytoplasm of every cell in varying numbers (upto 2500 per cell in rat liver) except in bacteria, blue-green algae and human r.b.c. Most are granular, rod or thread shaped. They are generally known as 'powerhouses of cell'. They are the site of the chemical events that supply the energy to the cell. They are also held responsible for fat synthesis. They range in their size from 0.2μ to 3.0μ. They also vary in their shape from spheres to rods. They are always found in dynamic state.

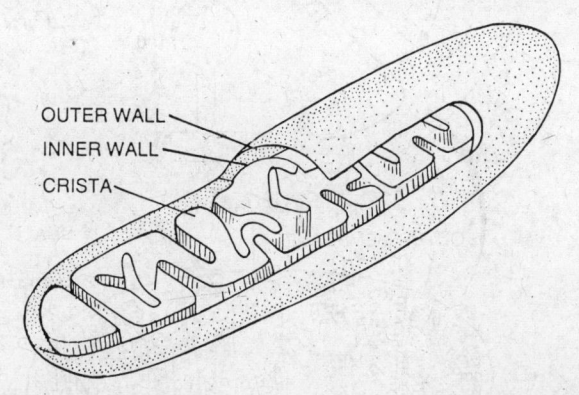

Fig. 2.19. Mitochondrion. Schematic diagram of a typical mitochondrion.

Each mitochondrion has a double-layered envelope with an outer and inner membranes. Each membrane is a typical unit membrane, being about 50–70 Å thick. The two membranes are generally 60–100 Å apart from each other. In between them remain filled the fluid rich in coenzymes. Extending from the inner membrane into the interior of the cavity, there are present a series of folds, called the *cristae*. The cristae are shorter in mitochondria of plant origin that in those

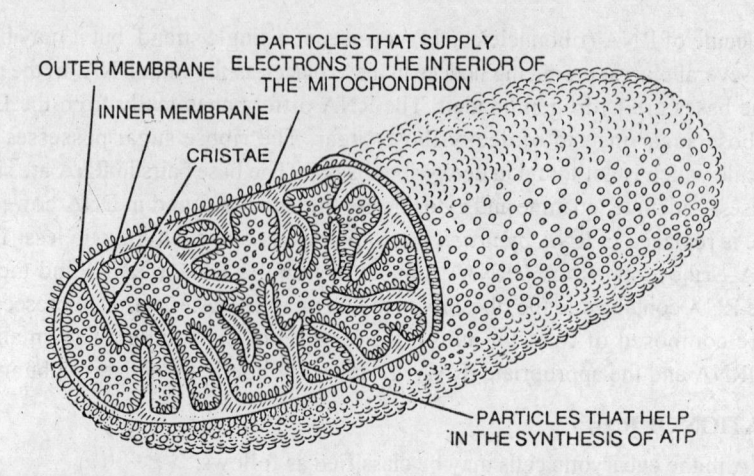

Fig. 2.20. Mitochondrion. Structure as viewed in electron microscope. The particles that supply electrons to the interior of the mitochondrion are clearly seen.

of animal cells. In all the cases the purpose of the presence of cristae is to increase the surface area of the interior of the mitochondrion. The number of folds per unit of volume of a mitochondrion is variable. The cristae may be branched, tubular rather than lamellate, and may be arranged parallel with long axis of mitochondrion, not, as is usual, at right angles. Thus each mitochondrion contains two cavities. The outer cavity found in between the two membranes is called '*outer compartment*' while the cavity limited by the folds of inner membrane is known as '*inner compartment* or *mitochondrial matrix*'. The matrix is normally homogeneous but it contains granules of 300–500 Å diameter. The outer covering membrane of the mitochondrion is quite elastic and may be stretched by swelling sometimes 200 times its normal dimensions. These facts advocate that the protein in molecules constituting the membrane can be greatly folded or stretched.

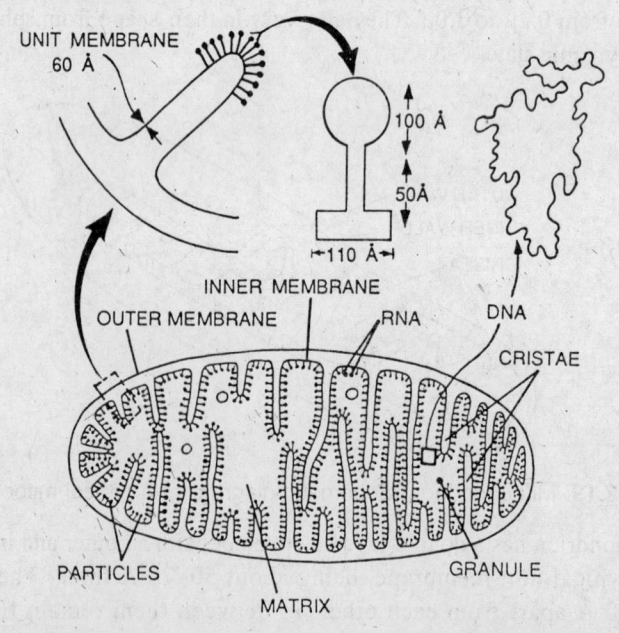

Fig. 2.21. Structure of mitochondrion as seen in electron microscope. Above, the various parts of mitochondrion magnified.

The Cell–Structure and its Components

Electron microscopy has also revealed the presence of very small particles adhered to the outside of the outer membrane and the inside of the inner membrane. These particles were first described by Humberto Fernandez-Moran (1962). The particles of the two membranes differ considerably. Depending upon the size and type of the mitochondrion there are from 10,000 to 100,000 particles per mitochondrion. Each particle consists of a base, a stalk and a head. The base which remains attached to the inner membrane has about 80 Å diameter. The stalk is 50 Å long and 33 Å wide while the head has the same diameter as does the base. The length of the complete particle is about 160 Å.

The mitochondria contain DNA and many oxidative enzyme systems.

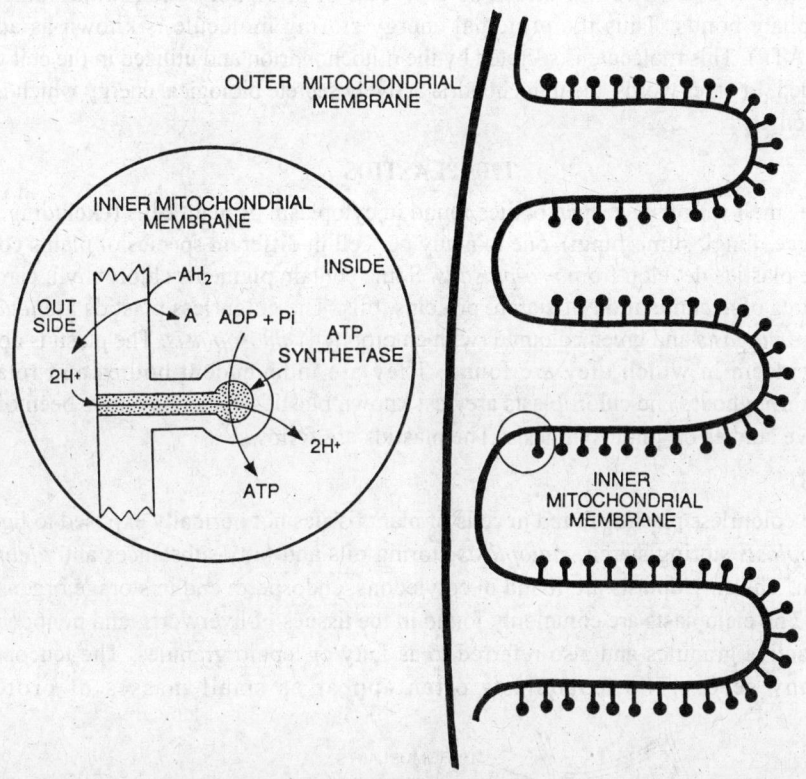

Fig. 2.22. The mitochondria. The figure shows the principle of the chemiasmotic hypothesis of energy conservation. The insert represents a magnified view of the encircled portion of the inner mitochondrial membrane.

Development. The mitochondria are essential structures where combustion of organic substances by oxygen takes place. Since this form of energy supply is essential for the life of most higher organisms the presence of functional mitochondria is essential for life. There are certain organisms, such as yeast, which can grow in the absence of oxygen. When such anaerobic cells are examined critically it is found that they contain small organelles having double membranes but the inner membrane does not have any of the invaginations (*i.e.*, cristae) typical of normal mitochondria. The characteristic enzymes and coenzymes which take part in aerobic respiration are also absent. Upon exposure to oxygen these structures soon develop to become normal respiring organelles. This observation shows that mitochondria can exist in an *embryonic* form, called a *promitochondrion*, and that upon stimulation by environmental factor (*i.e.*, oxygen) they undergo a process of development to become functional structures.

FUNCTIONS

As regards the functions of the mitochondria they perform two important tasks. They are as follows.

(*i*). The carbohydrates, proteins and fats are broken down gradually into smaller and smaller molecules. During this process there is transference of energy. This process takes place through a series of steps, and each step is controlled by an enzyme. All these energy giving reactions are known as *oxidations*.

(*ii*). The energy is not yielded out in the form of heat but is passed on to other molecules that contain phosphate (PO_4) where it is stored by a process of phosphorylation in the form of high-energy phosphate bonds. Thus the principal energy storing molecule is known as adenosine triphosphate (ATP). This molecule is secreted by the mitochondrion and utilized in the cell wherever energy is needed. In other words, the mitochondria actively secrete biological energy which is utilized in metabolic activity.

THE PLASTIDS

They are small variously shaped bodies found in cytoplasm of plant cells (excluding bacteria, blue-green algae, fungi, slime-fungi). one to many per cell in different species of plants containing pigments. The plastids develop from *pro-plastids*. Some contain pigments chlorophyll, carotenoids, some are centres of accumulation of starch, proteins, oils. The colourless plastids are *leucoplasts*, pigmented *chromoplasts* and green coloured (with chlorophyll) *chloroplasts*. The plastids do not mix with the cytoplasm in which they are found. They are independent bodies and retain their individuality throughout. The chloroplasts are best known plastids since they have been of interest in the extensive studies on photosynthesis. The plastids are follows:

LEUCOPLASTS

They are colourless plastids found in cells of plant tissues not normally exposed to light. They include *amyloplasts* storing starch, *elaioplasts* storing oils and fatty substances and *aleuroplasts*, storing protein. The amyloplasts are found in cotyledons, endosperm and in storage organs such as potato tubers. The elaioplasts are commonly found in the tissues of liverworts and nonocotyledons. They are refractive granules and also referred to as fatty or lepoid granules. The leucoplasts are found in many seeds. The leucoplasts often appear as small masses of protoplams,

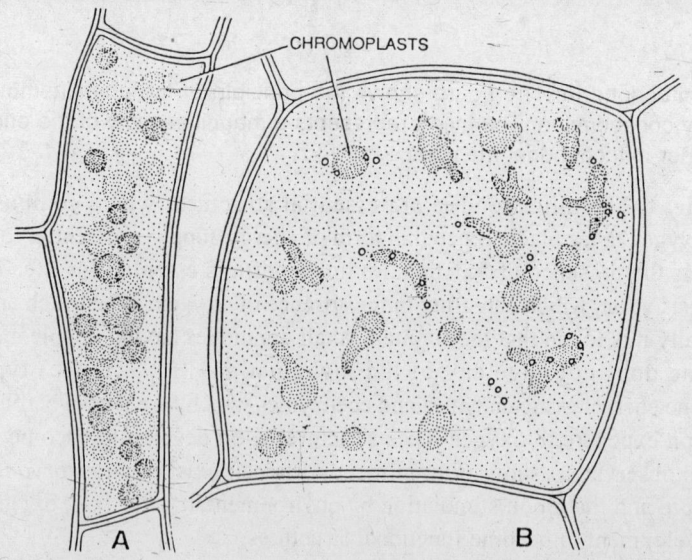

Fig. 2.23. Chromoplasts. A, from *Calendula* petal; B from *Pyracantha* fruit.

variable and unstable in form. They commonly aggregate near the nucleus. The plastids of the epidermis frequently appear nonpigmented and are then classified as leucoplasts.

CHROMOPLASTS

They are pigmented plastids of plant cells. They my be red, orange or yellow, *e.g.*, tomato fruits, carrot roots containing carotenoid pigments. They may appear rather granular with the pigments enclosed in it in a dispersed condition.

They show great variety in shape but are chiefly irregular; granular, angular, acicular and forked types occur. The irregular and sharp-pointed shapes are believed to be caused in part by the presence of the coloured substances, especially carotin and carotenoids, in crystalline form, as in the root of *Daucus*. They are associated with colour in

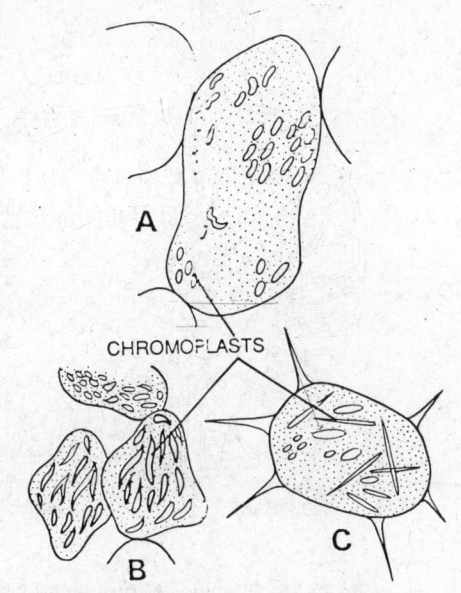

Fig. 2.24. Plastids. the chromoplasts. A, chromopolasts in pulp cells of tomato; B, chromoplasts in pulp cells of fruit of *Arisaema*; C, chromoplasts in the cells of *Daucus* root.

flowers, fruits and roots. Usually they represent transformed chloroplasts, but may form directly from small leucoplasts. The development of chromoplasts with globular and fibrous inclusions from chloroplasts involves the destruction of the original grana system.

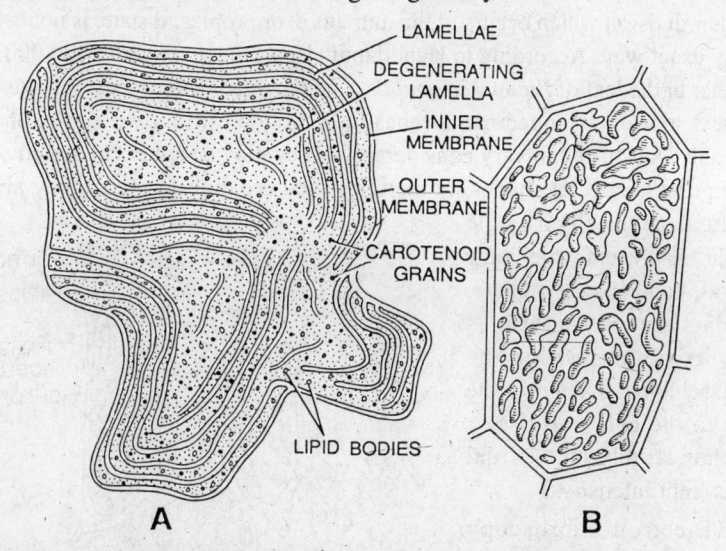

Fig. 2.25. Plastids. A, a chromoplast as seen in electron mircoscope; B, many chromoplasts as seen under light microscope.

CHLOROPLASTS

The sequence of the trapping of light energy, its conversion into chemical energy, and its storage in molecules derived from CO_2 and water, is known as *photosynthesis*. The photosynthesis is initiated by the capture of light energy through absorption in the green pigment *chlorophyll*. The *chloroplast* is the cytoplasmic particle in which this takes place.

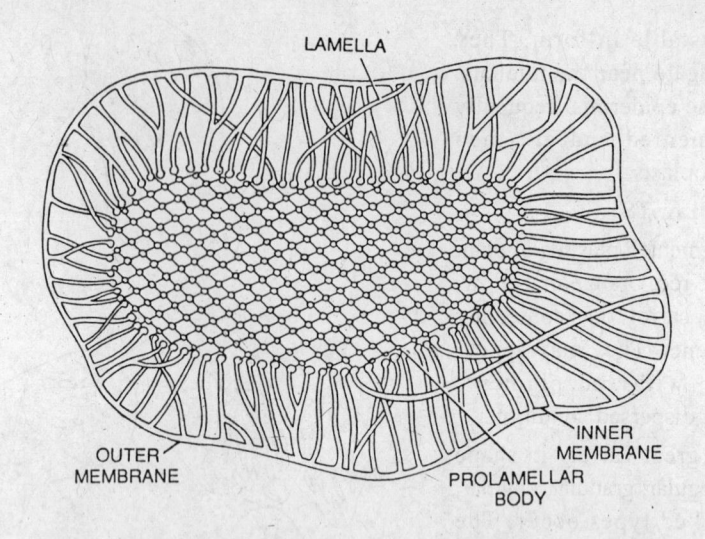

Fig. 2.26. Plastids. A, proplastid which gives rise to a plastid.

Morphology. The chloroplast is one of the largest cytoplasmic structures which can very well be seen under the low power of compound microscope. The size, shape and distribution of chloroplasts vary in different cells and species, but they remain relatively constant within the same tissue. The average size varies from 4 to 6 μ in diameter and 1 to 3 μ in thickness. The chloroplasts may assume many forms, and vary widely in number per cell in different plants. In some algae, such as the filamentous *Spirogyra*, only a single chloroplast is present in each cell; when the cell divides, it divides at the same time. On the other hand, a cell in the spongy tissue of a grass leaf may have 30 to 50 chloroplasts; their division which occurs in the immature, or proplastid state, is not correlated with cell division in any exact way. According to Haberlandt (1914) there are about 400,000 chloroplasts per square millimeter in the leaf of *Ricinus communis*. The blue-green algae lack definite chloroplasts; instead they possess loosely arranged membranes in the cytoplasm on which the photosynthetic pigments are layered. The shape too vary considerably. They may be spherical, ovoid or discoid. In certain cells they possess special shapes. Sometimes they are club-shaped. They are of various shapes in algae. In algae usually a single large size chloroplast is present which may be reticulate, spiral, band like or stellate. The chloroplast are sometimes homogeneously distributed within the cytoplasm, but are seldom packed near the nucleus or close to the cell wall. Their distribution depends largely on external conditions such as light intensity.

Structure. Electron microscopy reveals that the chloroplast is a structure of considerable complexity. A mature chloroplast remains surrounded by a semipermeable membrane. The membrane comprises of two separate layers, each being 40 to 60 Å thick and the space

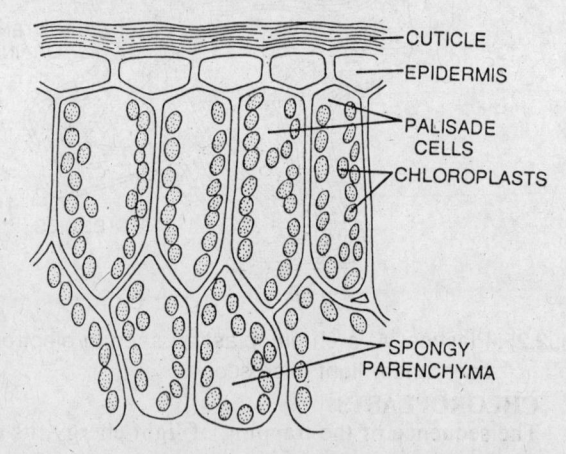

Fig. 2.27. Chloroplasts. Discoid chloroplasts in palisade cells of *Andromeda*.

The Cell–Structure and its Components

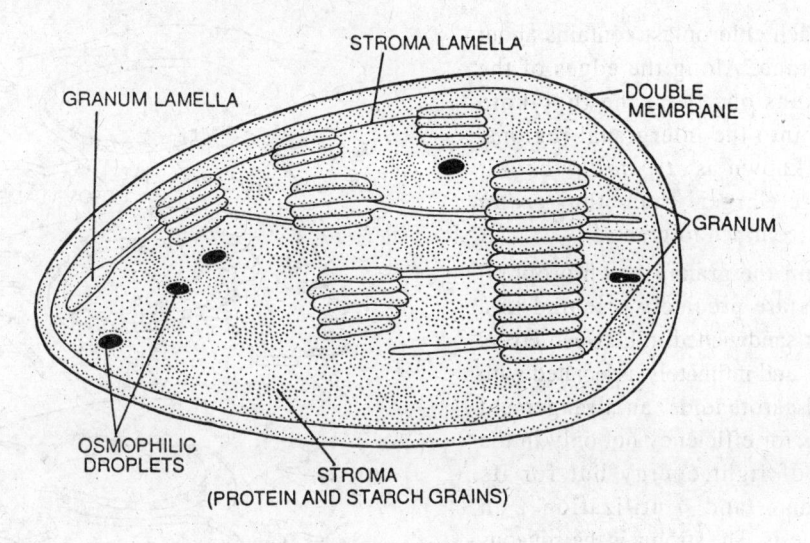

Fig. 2.28. A mature chloroplast (ultrastructure).

between them vary from 25 to 75 Å. It is organized internally into series of lamellar areas (*grana*) and nonlamellar areas (*stroma*). Numerous small platelets, the *grana* remain embedded in stroma. The grana can be visualized as pieces of many-layered plywood lying in a less well-organized stroma. The number of grana is variable in different chloroplasts. The mesophyll cell of spinach has 40 to 60 grana per chloroplast whereas one granum per chloroplast is found in *Euglena*. Each granum consists of double membrane *discs* or *lamellae* which vary in thickness and are of two types, *i.e., granum lamellae* and *stroma lamellae*.

Various particles and molecules may be found in the stroma; 175 Å diameter chloroplast *ribosomes*; the proteinaceous *stroma center*, starch grains *pyrenoids* in lower plants; osmophilic globule; and in some cases *phytoferritin* as well as fine fibrils of DNA—different from the nuclear DNA. The osmophilic globuli are generally referred to as *plastoglobuli*. They contain various lipid materials but no chlorophyll or caroteniod pigments.

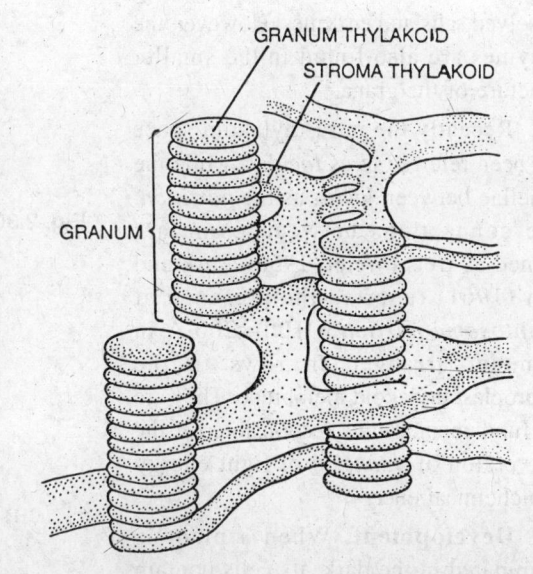

Fig. 2.29. Chloroplast. Diagram showing the three dimensional arrangement and inner relationships of grana (the scheme given in this diagram does not necessarily apply to all types of chloroplast). (After Weier, 1963).

In the stroma is also found suspended a chlorophyll containing lipoprotein membrane system. They are the site of light reactions as well as of the electron transport system operating during photosynthesis. It is usually found in the form of flattened sacs called *lamellae* or *thylakoids*. In many algae they lie in parallel array and run the length of the plastid. In higher plants, the structure varies and comprise of *grana*, connected by membranes. Each granum consists of *thylakoids* resembling a pile of coins, and they remain connected to each other by membranes running in the stroma. Each chloroplast contains about 20–100 grana. Along the edges of the grana extensions from the thylakoid penetrate into the intergranal regions. They are known as *stroma thylakoids* which are larger than the *grana thylakoids* confined to grana.

Within the grana, the chlorophyll molecules are precisely oriented in a monolayer sandwiched in between layers of proteins and intimately associated with lipids and carotenoids, an arrangement that makes for efficiency not only in the trapping of light energy but for its conduction and utilization in photosynthesis. The stroma is the aqueous part of the chloroplast, containing dissolved salts and enzymes. However, the enzymes are also found in the smaller structure of the grana.

Recently the intra-thylakoid space has been referred to as *loculus*, while the lamellae between loculi as the *partition*. The connections between grana are termed as frets (Weier, 1966). Park and Pon (1961) reported the presence of *quantosome particles*, 100—200 Å in diameter arranged in rows in the chloroplast thylakoid membrane. They are the fundamental units responsible for the conversion of *quantum* of light energy into chemical energy.

Development. When a plant is germinated in the dark, its cells contain small double membraned structures. These structures appear colourless but they can be shown to contain a very low concentration of substances which are precursors of chlorophyll. Upon exposure to light these precursor substances are immediately converted to chlorophylls. At the same time a process of growth and development starts which results in the transformation of the small *prochloroplast* into a typical photosynthetic chloroplast. The entire process takes place in less than

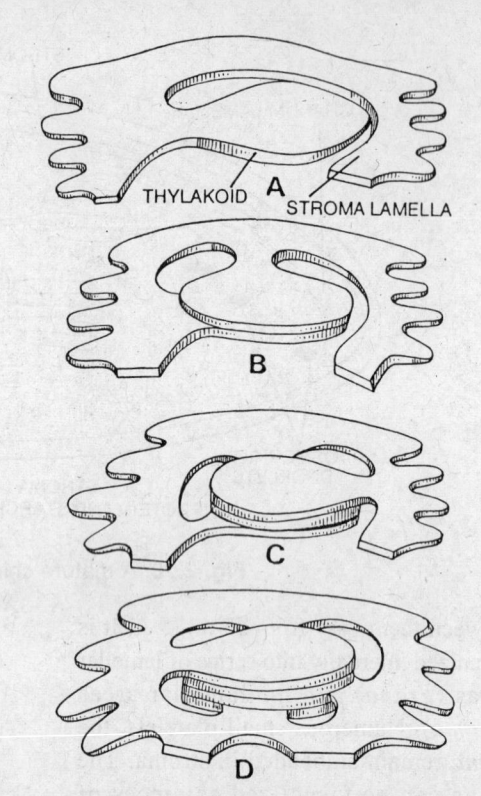

Fig. 2.30. Chloroplasts. A—D, stages showing the formation of thylakoids as outgrowths of stroma lamellae. The thylakoids are arranged one above the other forming the granum. (After Heslop Harrison.)

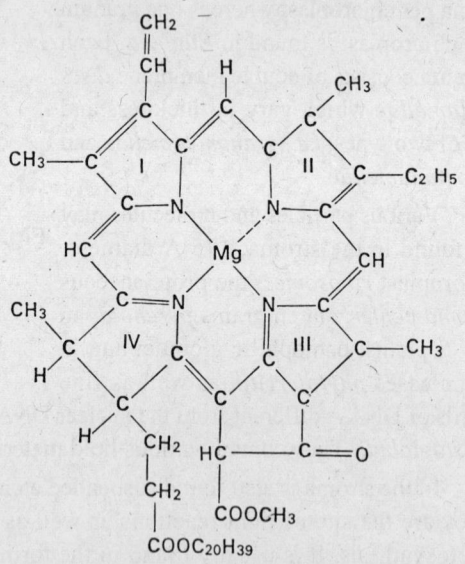

Fig. 2.31. Chlorophyll. Chemical structure of the chlorophyll— a molecule. The four pyrrole rings (1—4) attached to it.

The Cell–Structure and its Components 39

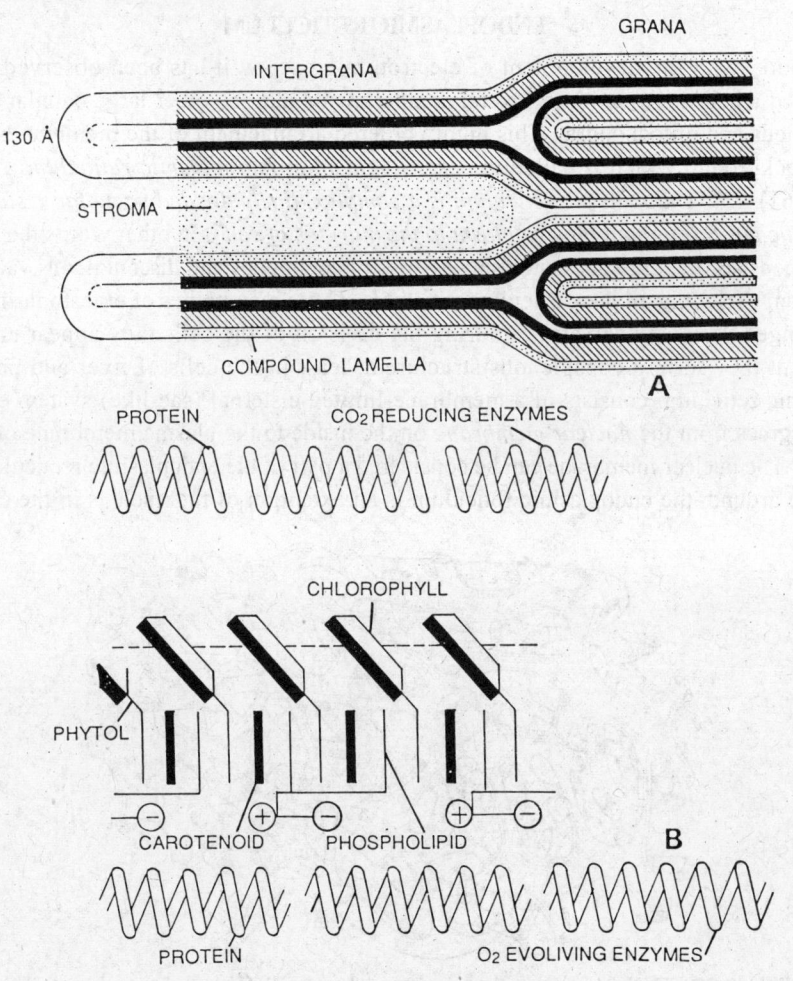

Fig. 2.32. Chloroplast. A, schematic representation of the layered arrangement of lamella, in the grana; B, possible arrangement of molecules in lamellae. The enzymes involved in photosynthesis are part of the protein layers, and the caroteniod and phospholipid molecules assist in the transfer of energy captured by the chlorophyll. (After A.J. Hodge and M. Calvin).

twenty four hours. It seems quite obvious that light induces the activation of synthetic processes within the cell that result in the construction of the complex photosynthetic apparatus (*i.e.*, chloroplast). During the induced development process many new enzymes which were not present in the prochloroplasts are synthesized and organized. This phenomenon shows that control systems operate within the cells.

FUNCTIONS

The main function of chloroplast is to take active part in photosynthesis. The photosynthesis is initiated by the capture of light energy through absorption in the green pigment chlorophyll inside the chloroplast.

The light energy trapped by chlorophyll can also be funneled through a series of enzymatically controlled reactions, into an energy compound called *adenosine triphosphate* (ATP). The chloroplast is, therefore, a dual energy converter, since the energy of sugars and ATP can be utilized by the cell in a variety of ways.

ENDOPLASMIC RETICULUM

Recently with the development of electron microscopy it has been observed that ground substance of a cell is divided into several compartments like the other large cellular bodies, such as the nucleus and mitochondria. This highly ordered arrangement of the membranes constitute a cellular background which is called the *ergastoplasm* or *endoplasmic reticulum*. According to Porter (1961), *"the endoplasmic reticulum is a complex, finely divided vacuolar system extending from the nucleus throughout the cytoplasm to the margins of cell"*. In other words the endoplasmic reticulum consists of membranes enclosing a series of continuous and discontinuous vacuoles, found in cytoplasm excluding all the particulate structures. These membranes of ergastoplasm may found to be arranged variously sometimes during the division of the cells they appear and disappear rapidly. But they form a conspicuous structure at least in the cells of liver and pancreas. The endoplasmic reticulum consists of a membrane-limited cisternal (sac-like) system extending, in various degrees from the *nuclear membrane* on the inside to the plasma membrane on the outside of the cell. The nuclear membrane can be considered a part of the endoplasmic recticulum or to turn the picture around, the endoplasmic reticulum is an extention of the nucleus in the cytoplasm.

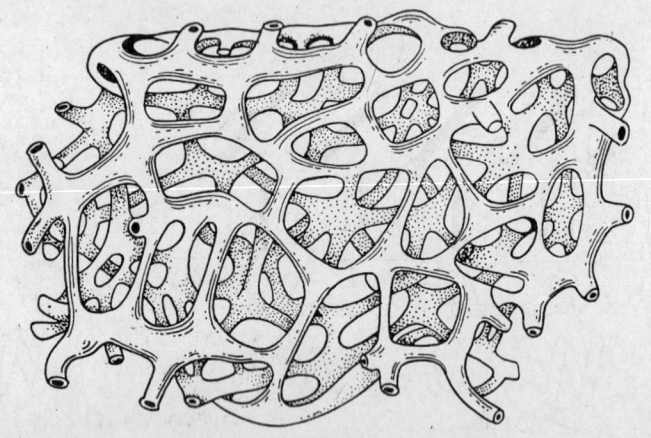

Fig.2.33. Endoplasmic reticulum (ergastoplasm). Three dimensional view.

Morphology. The endoplasmic reticulum is of a variable morphology, and each kind of cell has a characteristic ER. The ER can be loosely organized or its membranes can be tightly packed in the cytoplasm. Furthermore, the membranes can be either *rough* or *smooth*.

The rough, or *granular*, ER is found in great abundance in those cells engaged in protein synthesis. This chemical capacity resides in electron-dense particles (microsomes) that are rich in ribose nucleic acid (RNA). Since ribosomes live free in the cytoplasm, their association with ER is not necessary. On the other hand, the association of ribosomes with membranes, does not provide the intact cell with a means of compartmentalizing specific chemical reactions. Further the membranes of ER also provide a tremendous increase in surface area within the cell. If this is accepted that enzymes are part of membrane systems, then the cell can possess local patterns of synthesis.

The smooth, or *agranular*, ER, lacks the ribosomes. There is probably no sharp morphological discontinuity between smooth and rough kinds of ER, but the former is particularly prevalent in those cells engaged in the synthesis of fatty substances, *i.e.*, lipids in the cells of sebaceous glands, or steroid hormones in certain endocrine glands. The enzymes necessary for these syntheses appear to be a part of the membrane itself since they cannot be separated physically from the membrane fragments.

The Cell–Structure and its Components

FUNCTION

Thus ER is a kind of cytoskeleton providing surfaces for chemical reactions, pathways for the transport of materials, and collection depots for synthesized materials. We also find the smooth ER in particular, in these cells that must maintain a particular shape; in these the ER is a structural component as well. The RNA-rich particles, called *microsomes*, are particularly active in the synthesis of proteins, while the membrane itself appears to be involved in the synthesis of *steroids*, a group of compounds which include certain of the hormones. We therefore look upon the endoplasmic reticulum as the principal manufacturing portion of our factory (cell), although we are far from a complete comprehension of its operation as whole.

RIBOSOMES

For the first time in 1955 Palade discovered the ribosomes in the cytoplasm. They are the intracellular sites of protein synthesis and are present in all the cells. They are very small granules (smaller than 1μ) and are visible only under the electron microscope. They are found frequently associated with membranes of endoplasmic reticulum. They are found attached to the outer surface of the double fold endoplasmic reticulum membrane. Sometimes this association is very tight. Their association with the endoplasmic reticulum has given rise to a number of different names, such as '*rough endoplasmic reticulum*' or '*granular endoplasmic reticulum*' as opposed to the '*smooth endoplasmic reticulum* ' or *agranular endoplasmic reticulum.*' In the smooth or agranular endoplasmic reticulum the membranes remain devoid of ribosomes. In tangential view the ribosomes seem to be arranged in definite patterns such as rosettes, spirals and circles. Though these two components (ER and ribosomes) are closely associated yet they are quite different in their functions and properties. They contain RNA (ribonucleic acid), nucleoproteins and enzymes needed in many synthetic reactions. The RNA is of very high molecular weight. The ribosome globules may be extracted out by centrifuging the cells. It is believed that they are mainly concerned with the synthesis of cellular proteins. These proteins control the chemical reactions going on within the cells. They are enzymic in nature and act as catalysts. The proteins of ribosomes appear to be basic in nature.

As electron microscopy reveals they appear as spherical to ellipsoidal bodies. The size varies from 150 Å in bacteria and chloroplast to 200 Å in the cytoplasm of plant and animal cells. Each ribosome consists of two distinctive sub-units, which differ in sizes. The two subunits are held together by bonds which appear to need magnesium ions for their continued stability.

Sub-units of ribosomes. The ribosome sizes are known in terms of speed with which they sediment in the centrifugal field. *Svedberg unit* (S) is the unit in measuring this speed. In this context, each plant cytoplasmic ribosome is typically a 80 S which possesses typically 60 S and 36

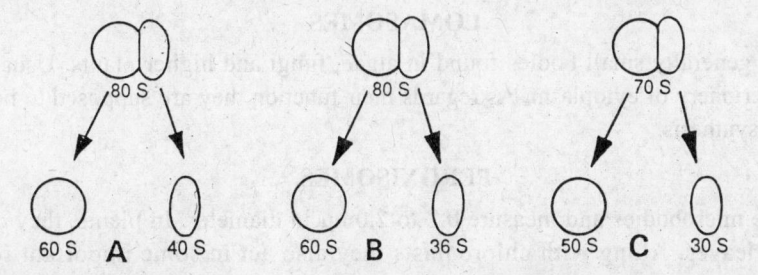

Fig. 2.34. Ribosomes. On the basis of sedimentation constant three classes of ribosomes can be distinguished—A, animal cytoplasmic ribosomes; B, plant cytoplasmic ribosomes; C, organelle bacterial ribosomes. The differences can be noted in the above mentioned diagrams. (S—Svedberg unit).

S particles or sub-units which in the natural conditions are probably held together by bonds which appear to need magnesium ions for their continued stability. The 60 S sub-unit (larger one) contains an RNA molecule whose molecular weight is about 1.5 million, *i.e.*, twice than those of 36 S sub-unit (smaller one) RNA. Further, the larger sub-unit contains about 30 proteins—all different from each other while the smaller sub-unit contains only 20.

The smaller particles are not active in protein synthesis. It is only the larger particles that are active. As regards protein synthesis, mRNA is specifically held by the smaller sub-units while tRNA seems to be specifically held by the larger sub-units. As a matter of fact, a portion of tRNA molecule is bound to the larger sub-unit and portion of it is bound to mRNA on the smaller sub-unit.

LYSOSOMES

They represent a new discovery among the cytoplasmic particles. They are found in most of the animal cells and in some meristematic cells of the plants. C. De Duve (1955) for the first time observed a fluffy layer of particles in the cytoplasm in which number of hydrolytic enzymes were present. The electron microscopy revealed that fluffy layer consists of membrane bounded structures whose appearance is markedly affected by the osmotic conditions present during preparation. De Duve called these structures, the lysosomes because of their contents of hydrolytic enzymes. As regards their origin it has been thought that they have been originated from the golgi complex and the ribosomes. The lysosomes vary in size from 0.4μ to 0.8μ in diameter, but sometimes they are as large as 5μ in diameter, *e.g.*, in a mammalian kidney. They may be spherical, rod like or irregular in shape, being covered by a single layered membrane composed of lipoprotein. They may be uniformly solid or some of them possess a very dense outer zone and airless dense core. The vacuoles or cavities are found within the granular material. They are distinguished from the mitochondria by their lack of internal cristae and by their characteristic enzyme content. The breakdown products diffuse out of the lysosomes and into the mitochondria, where they are further broken down in the process of respiration. They are also concerned with the digestion of proteins.

SPHAEROSOMES

The sphaerosomes are usually spherical bodies. They remain surrounded by membranes and range in diameter from 0.5 to 1.0μ. They are associated with the synthesis of fats and other such substances. They contain lipids and various types of lysosome-like enzymes. Sorokin (1966) reported the presence of a limiting membrane around each sphaerosome, however, James and Hilton (1972) do not agree with this view; they are in agreement with Yatsu that partial bounding membrane surrounds the sphaerosomes. Besides lipid storage and their transport, they are also thought to be intermediaries in the biosynthesis of waxes, cutin and other allied compounds, which result in cuticularization and suberization of cell walls.

LOMASOMES

They are generally small bodies found in algae, fungi and higher plants. Usually they are found in the periphery of cytoplasm. As regards their function they are supposed to be concerned with cell wall synthesis.

PEROXISOMES

These are microbodies and measure 0.2 to $2.0m\mu$ in diameter. In plants, they occur in the mesophyll of leaves. Along with chloroplasts they also act in some important reactions of photosynthesis.

GLYOXYSOMES

They are small bodies and occur only in plant cells. They are supposed to contain enzymes which convert fatty acids into sugars.

MICROTUBULES

For the first time Ledbetter and Porter (1963) demonstrated the presence of fine elongated tubular structures in the peripheral cytoplasm of some plant cells. These tubules are usually found to be attached to the cell wall. They are thought to be responsible for the orientation of cellulose synthesis. These structures are very clearly visible during the process of mitosis as cytoplasmic fibrils. They are several microns in length and their diameter ranges from 150 to 250 Å with a wall of about 60 Å in thickness. According to some workers they are thought to be fibres of spindle apparatus.

GOLGI COMPLEX

Camillio Golgi (1898), an Italian neurologist, first discovered these organelles in animal cells. The *Golgi complex*, so named after its discoverer is characteristic system of cytoplasmic membranes. The term, *dictyosome*, is also used to describe this structure. It is comparable to the smooth endoplasmic reticulum, but discontinuous, smaller and more compact in nature. It is still debatable that there is a direct morphological relation between the endoplasmic reticulum and the Golgi complex. However, it was early recognized by its affinity for osmium of silver-containing stains. In modern days its existence has been clearly demonstrated with the electron microscope. They are found in living cells of both animals and plants, immersed in their cytoplasm.

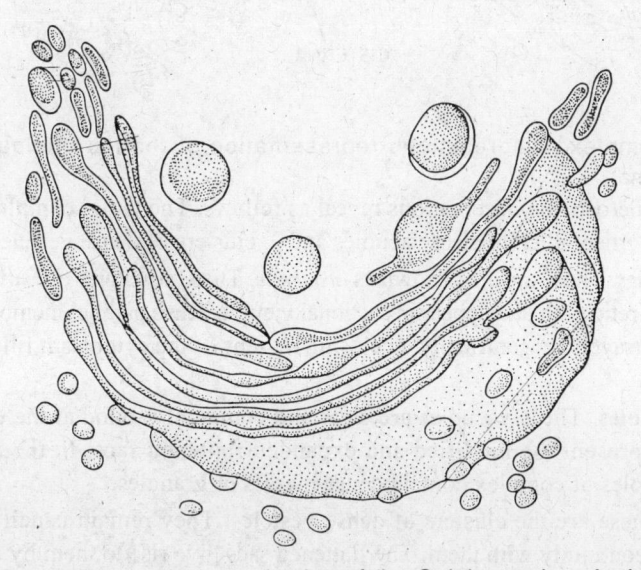

Fig. 2.35. The diagram indicates the organisation of the Golgi complex. At the convex "forming face" small vesicles containing the products of the synthetic activity of the cell are shown coalescing with the outermost saccule or cisterna. At the concave or "maturing face" of the complex, saccules containing-concentrated secretory product are rounding up to form secretory droplets or granules.

Various workers have given various names to this complex, such as *dictyosome*, idiosome, lipochondria, Golgi body, Golgi substance. Golgi apparatus and Golgi complex. Generally the name Golgi complex is used for the material in vertebrates and dictyosome in invertebrates and plants. However, the Golgi complex is not found in bacteria and blue green algae.

STRUCTURE

The shape of Golgi complex is quite variable in somatic cells of plants and animals. Even in the same cell there are variations with functional stage. However, the shape is characteristic for each cell type. Their appearance depends partly on the position of cells within the plant or animal body and partly on the way in which cells are prepared for microscopic study. These bodies look like the complexes of the droplets and therefore, termed as Golgi complex. They appear thin plate like layers of irregular arrangement. Each complex consists of stacks of parallel lamellae. In some cases, it occurs as a dense reticulum of anastamosing trabeculae while in others as an irregular plaque, a ring, hollow spheres united together. In nerve cells it occurs as a recticulum of wide meshes around the nucleus. The size of Golgi complex is also variable. It is small in muscle cells but quite large in the nerve and gland cells. Generally the complex is well developed in an active cell while in old cells the complex diminishes in size and ultimately disappears.

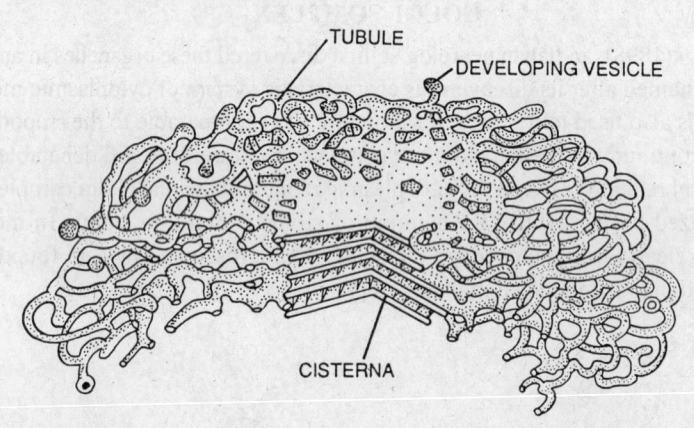

Fig. 2.36. Golgi complex. Diagrammatic representation of the morphology of Golgi body (dictyosome),

The electron microscopic observations reveal as follows. The Golgi complex comprises of 1. flattened sacs or cisternae, 2. large clear vacuoles and 3. clusters of dense vesicles.

1. Flattened sacs. These are also known as *cisternae*. These structures are similar to the smooth surface endoplasmic reticulum and appear in sectional view as dense parallel membranes. The packed cisternae are often arrayed concentrically, enclosing regions of the cytoplasm filled with numerous large vesicles.

2. Large vacuoles. These vacuoles are clear and commonly found at the edge of Complex. These structures represent the modified and expanded flattened sacs. In the cells of liver and pancreas these vacuoles of complex contain dense masses or granules.

3. Vesicles. These are the clusters of dense vesicles. They remain usually associated with cisternae and show continuity with them. The flattened sacs give rise to them by means of budding or pinching off. These structures remain dispersed in the surrounding hyaloplasm.

FUNCTIONS

The Golgi bodies are believed to function in the manufacture of cellular secretion products. Since its membranes have little RNA it is not connected with protein synthesis. The general consensus of modern days is that it is somehow involved in the storage and possible modification of lipid substances, a point of view supported by observations that the appearance of the Golgi complex in animal cells can be modified greatly by changes in fat diet.

During the maturation of sperm, the Golgi complex plays a role in the formation of acrosome (Burgos and Fawcett, 1955).

The complex also activates mitochondria to produce the ATP which is utilized in respiratory cycle, nervous transmission, and nucleic acid and protein synthesis; the latter occurs in the vicinity of endoplasmic reticulum.

NON PROTOPLASMIC COMPONENTS: VACUOLES

The vacuoles (latin, *vacuus* = empty) are non-living inclusions of cytoplasm. Some of these are found in the form of minute droplets in the cytoplasm of meristematic cells. Each droplet is a small vacuole filled up with cell-sap. These small vacuoles enlarge and finally coalesce to give rise to a large vacuole. The enlargement of the vacuoles is accompanied by the entry of water into the cell. It becomes quite evident that with the entry of water into the cell, the size of the cell will increases, and, in fact, the cell attains a volume many times greater than the volume of the original meristematic cell. During this process of cell enlargement by vacuolation the nucleus and cytoplasm do not greatly increase in size, and in the fully vucuolated cell the cytoplasm forms a parietal layer which lines the cellwall internally, and a number of strands which extend from this parietal layer across the central vacuole. The nucleus may be suspended in the centre of the vacuole by these cytoplasmic strands or it may be embedded in the parietal layer of cytoplasm.

The cell-sap which fills the vacuole consists of an aqueous solution of various inorganic and organic compounds. Inorganic ions are always present, especially those resulting from the dissociation of the *nitrates, sulphates* and *phosphates* of certain alkaline and alkaline-earth metals, and *carbondiozide* in solution is also present.

A number of sugars have been isolated from the cell-sap of various plants. *Glucose* and *fructose*, are specially common in the cell-sap. Sometimes *maltose* is also present in the sap of leaf cells. *Sucrose*, or cane sugar is widely distributed in plants, being abundant in the root of the sugar-beet and the stem of the sugar-cane. A complex soluble polysaccharide, *inulin*, occurs in the cell-sap of the tubers of dahlia. Other carbohydrates of less common occurrence in the cell-sap include *dextrin, mannitol* and various *pentosans* while soluble *pectin*, with other substances, sometimes occurs in appreciable amount. *Glycosides* are also of common occurrence in the cell-sap. *Mucilaginous compounds* also occur in cell-sap and they give the slimy character to the cytoplasm *Mucilages* are found abundantly in the cell-sap of many bulbs, *e.g.,* onion, and in the leaves of many succulent plants. They are polysaccharides in nature. *Tannins* may also be present in cell-sap. They are complex substances and soluble in water and alcohol. *Anthocyanin pigments* are frequently present in cell-sap. These pigments give blue or red colours to the cell-sap. *Malic, citric* and *tartaric acids* are all of common occurrence, and the acidity of many fruits is due to the presence of these acids in the cell-sap. *Oxalic acid* is also found in the cell-sap, in the form of its insoluble calcium-salt, and as such is usually regarded as a waste product of cell metabolism. *Calcium oxalate* occurs in a crystalline form and the crystals may be of various types. *Calcium carbonate* occurs only infrequently in plants in the form of *cystoliths*. Soluble *nitrogenous compounds* may also be present in the cell-sap. *Ammonium compounds* may be present too.

Besides the proteins which form a part of living protoplasm, *non-living proteins* may occur either dissolved in the cell-sap or in the form of crystal-like bodies, *crystalloids*. In many seeds, especially those containing oil, the vacuoles contain large amounts of *dissolved proteins*, and as the seed dries out and ripens these proteins may be transformed into *aleurone grains*. Fats and oils are frequently present in plant cells, especially in certain seeds. They occur as droplets or globules, and may be present either in the cell-sap or protoplasm or both. Other substances sometimes present include basic substances such as the *purine bases* and the *alkaloids*. *Resins* as found also, but are of limited distribution, being confined to special cells or ducts in certain plants and essential oils are produced by some plants. These essential oils are usually mixtures of complex organic compounds such as *alcohols, esters, ketones, aldehydes* and *hydrocarbons. Resin* also is not itself a simple chemical substance, but a mixture of substances insoluble in water and containing *resin acids*, resin esters and *resenes*, all of which are complex substances of high molecular weight.

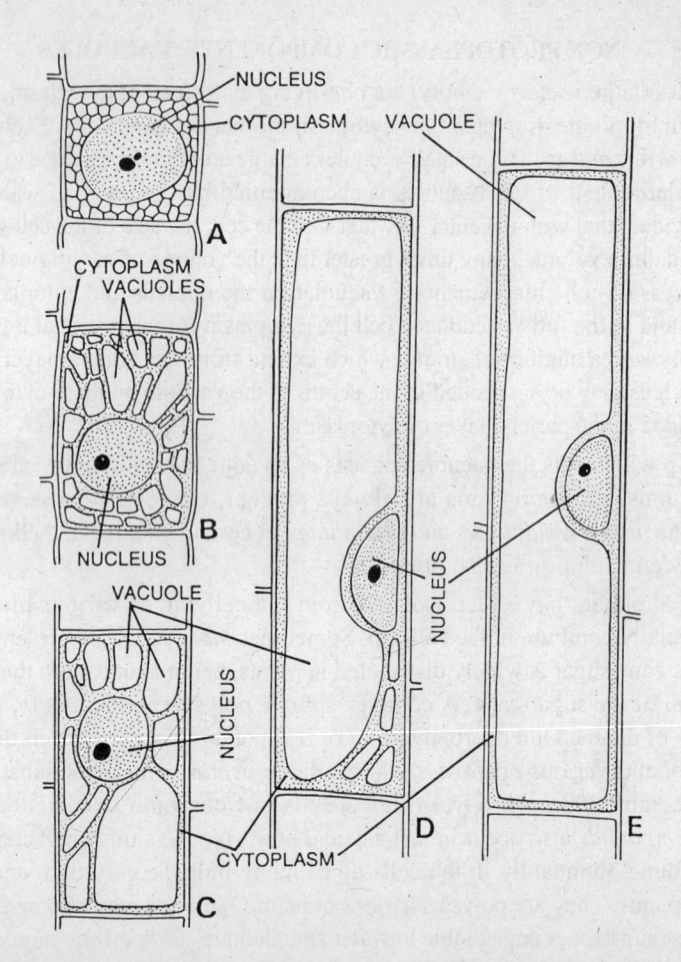

Fig. 2.37. The cell. A, the young cell without vacuoles; B—E. various stages in the formation and enlargement of the vacuole.

FUNCTIONS

The vacuoles have many functions to their credit. They maintain the rigidity of the plant cell. They are responsible for rapid exchange of material by pushing the cytoplasm towards the periphery of the cell. They contain various substances in their cell sap such as enzymes, acids, salts, etc., which act in several ways. They store waste products. The vacuoles contain ions and molecules, and therefore, they maintain osmotic relationships of the cell. They help in the enlargement of the cell. Several metabolic activities take place in the cell sap of vacuoles.

ERGASTIC SUBSTANCES

Several chemical compounds occur as non-living inclusions in the cytoplasm. These inclusions remain dispersed either in cell-sap or in cytoplasm. Principally they are classified as follows:

A. FOOD PRODUCTS

These are the substances manufactured by the protoplasm of the cells from the simple inorganic substances like carbondioxide and water and stored in the cells as food materials. The food thus manufactured is partly used up to make new protoplasm and partly it is broken down to provide necessary energy, and remainder of it is stored as reserve food material in protoplasm of cells. The reserve food materials may be classified as follows:

Ergastic Substances

```
                              Ergastic Substances
              ┌──────────────────────┼──────────────────────┐
      A. Food Products          B. Secretory Products    C. Waste Products
    ┌─────────┼─────────┐              │              ┌─────────┴─────────┐
Non-Nitrogenous Nitrogenous Fats and  Enzymes      Non-Nitrogenous   Nitrogenous
Starch         Proteins    Fatty     Pigments      Tannins           Alkaloids
Inulin         Amino-      acids     Nectar        Mineral Crystals
Hemicellulose  Compounds.                          Latex
Cellulose                                          Essential Oils
Sugars.                                            Gums
```

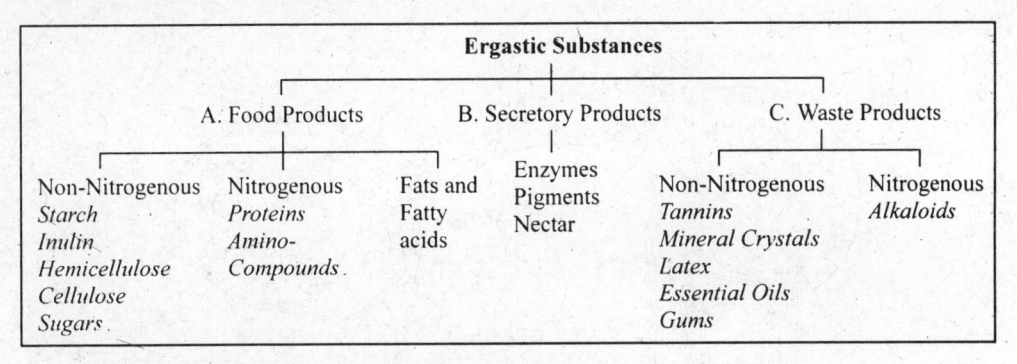

1. Carbohydrates (non-nitrogenous products).
2. Nitrogenous products
3. Fats and fatty oils.

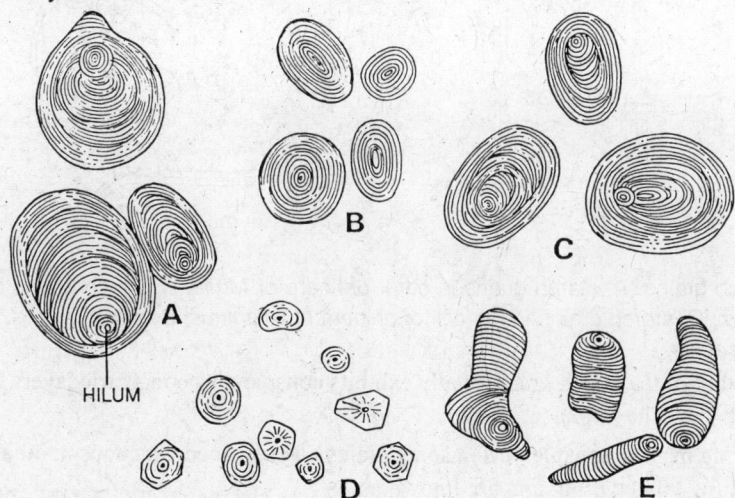

Fig. 2.38. Starch grains of different shapes. A, from root of *Marantia*; B, from seed of *Phaseolus*; C, from potato tuber; D, from grain of *Zea mays*; E, from fruit of *Musa*.

1. CARBOHYDRATES

They are non-nitrogenous food products. These are compounds of carbon (C), hydrogen (H), and oxygen (O). They are derived more or less directly from carbon dioxide (CO_2) and water (H_2O) during photosynthesis. Some carbohydrates are insoluble while some are soluble in water. The most important insoluble product is starch and the soluble products are inulin, sugar, etc.

(*i*) **Starch.** It is an insoluble carbohydrate of polysaccharide type formed by the condensation of simple sugars like glucose. The starch is usually found in the form of starch grains of various shapes. The starch grains are abundantly found in the storage organs of plants, *e.g.*, tuberous roots, underground stems, cortex of stems, endodermis, grains of cereals, banana fruits, etc. The starch grains vary in their shape and may be used for the identification of plants. The starch grains are not found in fungi and certain algal groups. The starch grains have different shapes that are characteristic of the plant types, *e.g.*, they are oval-shaped in potato; flat in wheat; polygonal in maize; spherical in pulses and dumbell or rod-shaped in the latex cells of some *Euphorbias*. The starch grains of rice are smallest and those of *Canna* largest. The starch grains vary from 5–100μ in size. The starch is always derived either from chloroplasts of green cells or from leucoplasts (amyloplasts) of storage tissue.

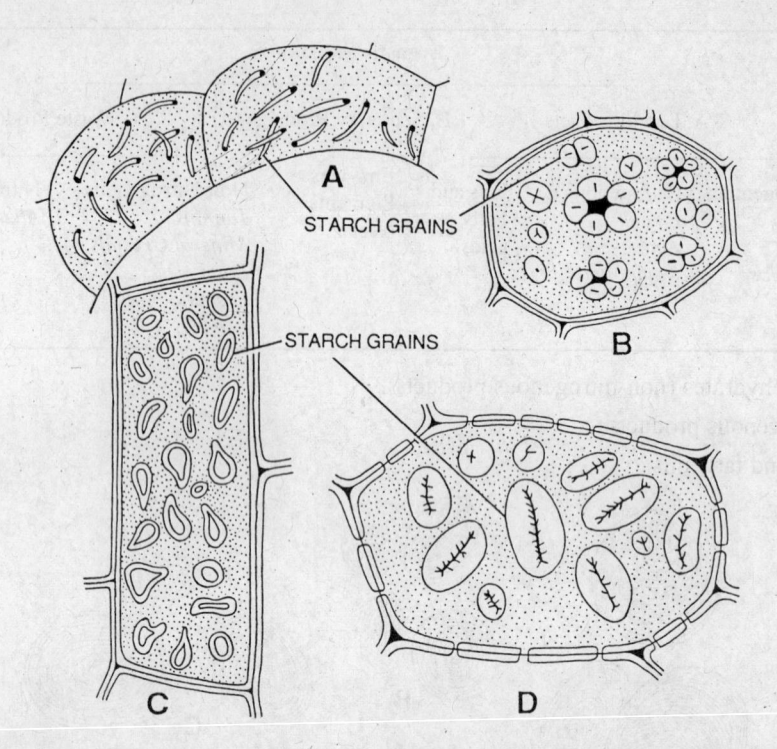

Fig. 2.39. Starch grains. A, starch grains in outer pericarp of *Musa*; B, starch grains in cotyledon of *Pisum*; C, starch grains in ray cell of phloem of *Ailanthus*; D, starch grains in cotyledon of *Phaseolus*.

The structure of the starch grain usually exhibits conspicuous concentric layers formed around a dark roundish spot, the *hilum*.

The layering may be conspicuous in some grains whereas inconspicuous in others. Most of the starch grains show this layering and are known as stratified starch grains. If the concentric layers of the starch are formed on one side of the hilum of starch grain, the grain is said to be *eccentric* (*e.g.* potato) and when the layers are deposited concentrically around the hilum (*e.g.*, wheat) the grain is known as *concentric*. Concentric types of starch grains are quite common in most of the plants. If the starch grain possesses a single hilum, it is known as *simple*. Sometimes two, three or many grains, arranged in a group with as many hila as the starch grains, they are known as *compound* grains. Compound grains are commonly found in potato, sweet potato, rice and oats. The starch is turned blue or black in aqueous solution of iodine.

(ii) Inulin. This is also a carbohydrate of polysaccharide type. It is soluble carbohydrate usually found in the cell-sap. Inulin has been reported from the roots of many Compositae. It is

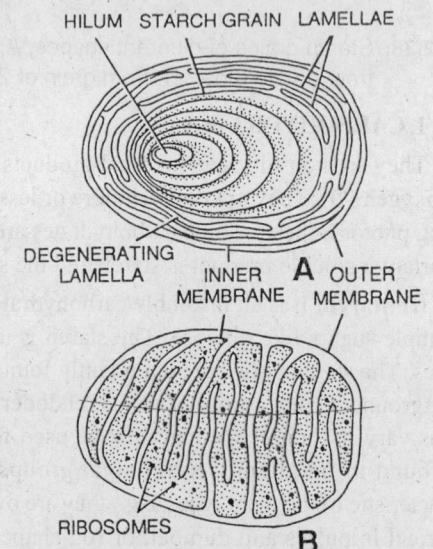

Fig. 2.40. Plastids. A, amyloplast with a starch grain; B, a leucoplast.

commonly found in the tuberous roots of *Dahlia* and *Helianthus tuberosus*. It can easily be precipitated by keeping the *Dahlia* roots for 6–7 days, in alcohol, in the form of spherical, star-shaped or wheel-shaped crystals.

(*iii*) **Hemicellulose.** In some seeds food is stored in thickened cell walls in the form of hemicellulose. Food is stored in this form, however, much more rarely than as sugar or starch. Hemicellulose (reserve cellulose) is found in some palm seeds and also in the seeds of some other plants.

(*iv*) **Cellulose.** This is a carbohydrate with a general formula similar to that of starch, *i.e.*, $(C_6H_{10}O_5)n$. However, the atoms are arranged differently in the molecule, and starch and cellulose have very different properties. In the plant, cellulose is made from sugars. It serves as building material in the formation of the cell wall.

(*v*) **Sugars.** The sugars that are found most abundantly in plants are *glucose* (grape sugar), *fructose* (fruit sugar) and *sucrose* (cane sugar). Glucose and fructose have the formula $C_6H_{12}O_6$. They are thus composed of the same kinds of atoms in the same proportion, but the arrangement in the molecule is different. The simplest natural sugars are monosaccharides. Sucrose ($C_{12}H_{22}O_{11}$) has twice as many carbon atoms and is a disaccharide. Sucrose is known as the ordinary sugar, which is obtained from sugarcane or sugar beets. The sugars are soluble and simplest carbohydrates. As glucose and fructose are in solution and have relatively simple molecules, they are good material for the building up

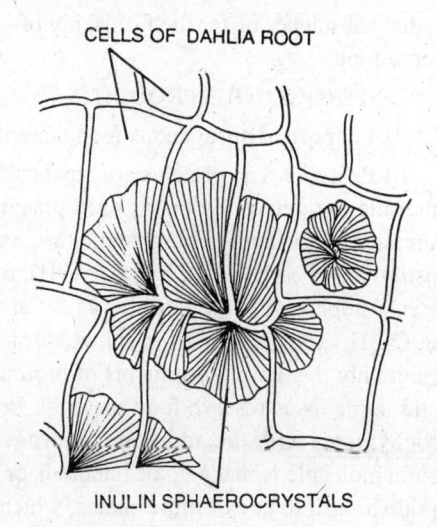

Fig. 2.41. Inulin. Sphaerocrystals of inulin in cells of dahlia root.

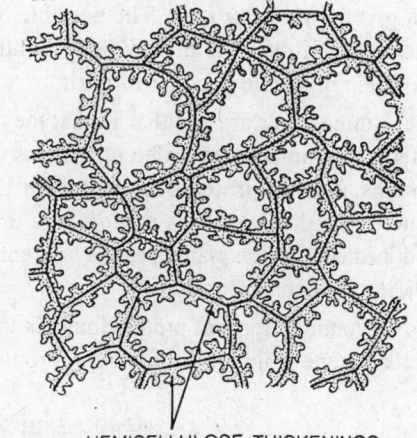

Fig. 2.42. Hemicellulose. Thickened walls of hemicellulose from *Areca* seed.

Fig. 2.43. Cellulose. Reserve cellulose in the form of thickened cell walls in the seeds of *Diospyros*.

of other substances or for the furnishing of energy. In sugar cane and sugar beet sucrose is found as reserve food.

2. NITROGENOUS PRODUCTS

The important nitrogenous food materials are proteins and amino compounds.

(i) Proteins. The proteins are the most important group of compounds found in plants, as they constitute the active matter of protoplasm, and the chemical phenomena of life processes are associated with them. The proteins are exceedingly complex organic nitrogenous substances constituting of carbon (C), hydrogen (H), oxygen (O) and nitrogen (N). In certain more complex proteins sulphur (S) and phosphorus (P) are also present. For example, representative proteins as zein $C_{736}H_{1161}N_{184}O_{208}S_3$ from Indian corn and gliadin $C_{685}H_{1068}N_{196}O_{211}S_5$ from wheat. Proteins are not only the main constituents of protoplasm but, in the form of solid granules, are frequently found in plants as reserve food material. Proteins are formed by a rearrangement of the atoms of carbohydrates with the addition of nitrogen, commonly sulphur and sometimes phosphorus. A protein molecule is made up of hundreds or thousands of amino acid molecules joined together by peptide links into one or more chains, which are variously folded. There are twenty different kinds of amino-acids commonly found in proteins, and most of these usually occur in any one protein molecule; they are arranged in the chain in a sequence which is exactly the same in all molecules of a given kind of protein. The possible different arrangements of the aminoacids are evidently practically infinite, and the diversity is fully exploited by living things every species having kinds of protein molecule peculiar to itself.

Amino-acids are peculiar in that they have both basic and acid properties. The proportion of different amino-acids in different proteins varies; and some proteins lack amino-acid found in other proteins. Animal proteins are better for human food than plant protein because the amino-acid content of animal proteins is more like that of human proteins than is the amino acid content of plant proteins. Some plant proteins lack entirely some of the amino acids which are essential for the building of human proteins.

A common form of protein found in the endosperm of castor seed is known as *aleurone grains*. Each aleurone grain is a solid, ovate or rounded body which usually encloses a crystal like body in it,

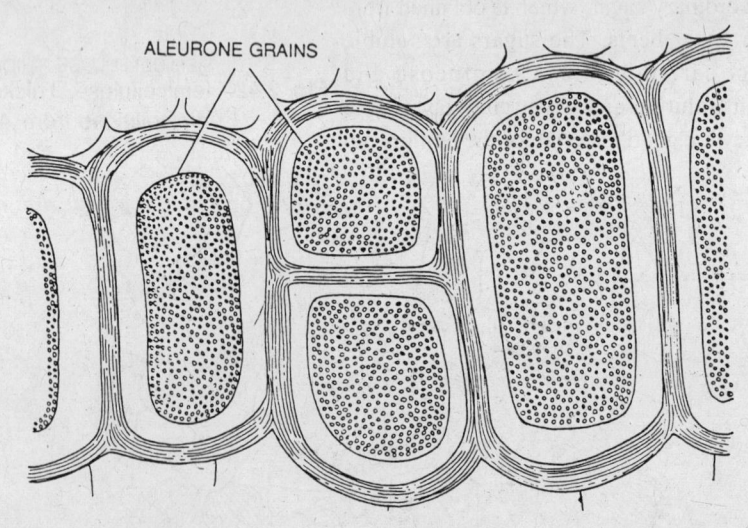

Fig. 2.44. Nitrogenous inclusions. The figure indicates the presence of aleurone grains in endosperm of *Zea mays*.

The Cell–Structure and its Components 51

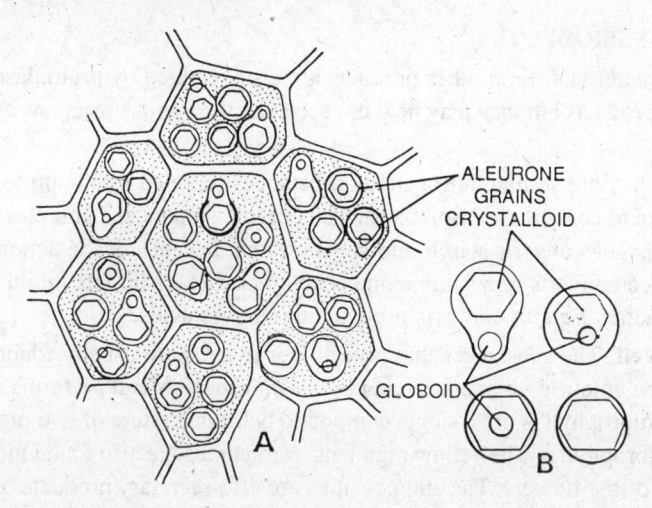

Fig. 2.45. Proteins. A, aleurone grains in the endosperm of castor seed; B, few grains magnified containing crystalloids and globoids in them.

known as *crystalloid* and a globule like body known as *globoid*. The crystalloid is proteinaceous in nature and occupies the major portion of aleurone grain, whereas the globoid is double phosphate of calcium (Ca) and magnesium (Mg) and occupies the narrower part of the grain. The aleurone grains vary in their shape and size in different plants. When they are found in oily seed of castor they are larger in size, while when found with starch they are very minute in size.

(ii) **Amino compounds.** They are simple nitrogenous food materials. They are found in the form of amino-acids and amines which occur in the cell sap. These are abundantly found in the growing apices of the plants, while less frequently occur in storage tissues. There are about twenty known amino acids. Amino acids are peculiar in that they have both basic and acid properties. They are constituted of carbon, hydrogen, nitrogen, oxygen, with formula R—CH (NH$_2$)—COOH, where R is a variable grouping of atoms, an amino group always being attached to the carbon atom next to the carboxyl group.

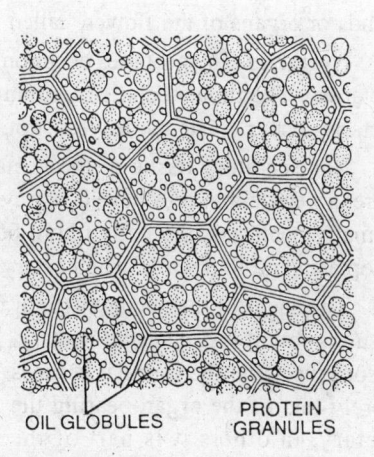

Fig. 2.46. Cells of endosperm of coconut. The large globules are oil and the small granules are protein.

3. FATS AND FATTY OILS

The fats and fatty oils in plants are composed of glycerine and organic acids. They occur in the form of minute globules in the protoplasm. Special types of fats and fatty oils are found in the seed and fruits of flowering plants. Fats and fatty oils are composed of carbon, hydrogen and oxygen and are characterized by a small percentage of oxygen as may be seen from the formula of such common fats as stearin $C_{57}H_{110}O_6$, and palmitin $C_{51}H_{98}O_6$, olein $C_{57}H_{104}O_6$, and linolein $C_{57}H_{98}O_6$. Owing to the very small percentage of oxygen contained fats, in the oxidation of fats produces large amounts of energy. They are insoluble in water but soluble in ether, chloroform and petroleum. Fats, may be solid or liquid (fatty oils), according to the temperature.

B. SECRETORY PRODUCTS

Besides food material several other products are also secreted by protoplasm which are not useful as nutritional products but they may help or accelerate the various reactions in the cells. These are as follows:

***(i)* Enzymes.** They are soluble nitrogenous substances secreted by the protoplasm. They are digestive in function and convert the insoluble substances into soluble and complex compounds into simple ones, *e.g. diastase* converts starch into sugar, so that because of the action of this enzyme an insoluble substance converts into soluble one. The *lipase* breaks up fats into their components, glycerine and fatty acids. *Papain* converts proteins into amino acids.

***(ii)* Colour in cell.** The substances that give colour to cells are usually found in the plastids. Chlorophyll is a green colouring matter secreted by the chloroplasts and performs the phenomenon of photosynthesis. Chlorophyll is not a single compound but it is mixture of two pigments known as chlorophyll *a* and chlorophyll *b*. The yellow pigments, carotenoids are also found in cell sap and give colours to the petals of the flowers. The anthocyanins are also secretary products of the protoplasm and are stored in the cell sap; they also impart colour to the petals of the flowers.

***(iii)* Nectar.** The nector is another useful secretion of protoplasm. It is secreted by special glands or organs of the flower, called nectaries.

Nectaries. The nectaries occur on flowers (floral nectaries) and on vegetative parts (extrafloral nectaries). The floral nectaries are found in various positions on the flower, whereas the extrafloral nectaries occur on stems, leaves, stipules and pedicels of flowers.

The secretory tissue of the nectary is generally found in the epidermal layer. Usually the secretory epidermal cells have dense cytoplasm and may be papillate or elongated like palisade cells. In many nectaries, the cells found beneath the epidermis are also secretory. They have dense cytoplasm and thin walls. The nectary is covered by a cuticle.

The sugar of the nectaries (both floral and extrafloral), is derived from the phloem. Vascular tissue is found just near the nectary. In some nectaries the vascular tissue is merely that of the organ bearing the nectary, in others it is part of the nectary.

The nectar is excreted either through the cell wall and the ruptured cuticle or sometimes through stomata.

***(iv)* Osmophors.** The fragrance of flowers, is generally produced by volatile substances distributed throughout the epidermal region of perianth parts. However, in some plants the fragrance originates in special glands known as *osmophors* (Vogel, 1962). Examples of such special glands are commonly found in Asclepiadaceae, Aristolochiaceae, Araceae, Burmanniaceae and Orchidaceae. Various floral parts are differentiated as osmophors and they

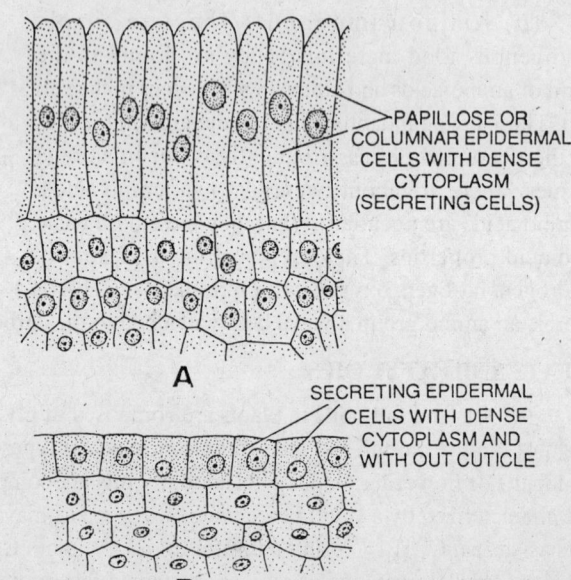

Fig. 2.47. Nectary. A, section of surface of nectary of *Euphorbia pulcherrima*; B, section of floral nectary of *Malus pumila*.

The Cell–Structure and its Components

assume the form of flaps, cilia or brushes. The prolongation of the spadix of Araceae and the insect-attracting tissue in the flowers of Orchidaceae are osmophors. The osmophors have a secretory tissue usually several layers in depth.

C. WASTE PRODUCTS

They are excretions of the plant cells. Usually these products are stored in the dead cells. These products are formed as a result of metabolic activities of protoplasm, and therefore, they are known as metabolic wastes of the plants. Usually there are two types of metabolic wastes. They are 1. Non-nitrogenous waste products and 2. Nitrogenous waste products.

1. Non-nitrogenous waste products.

They are tannins, mineral crystals, latex, essential oils, gums, resins, and organic acids. They are either found in the cytoplasm or in the cell-sap.

(a) **Tannins.** They are non-nitrogenous complex compounds, commonly found dissolved in the cell-sap. They are the derivatives of phenol and usually related to glucosides. They are found in the cell walls, in the dead cells, in the heart wood and in the bark. They are also found in abundance in leaves and unripe fruits. On the ripening of the fruits the tannins are being converted into glucose and other substances. The tannins possess a bitter taste and their presence in tea leaves make the tea decoction bitter. They are used in

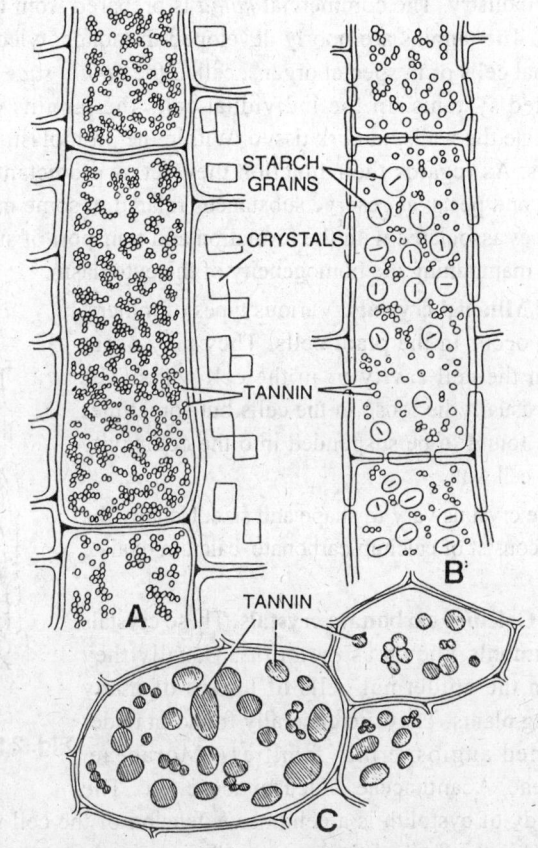

Fig. 2.48. Tannin, starch grains and crystals, A, Tannin and crystals in phloem parenchyma of *Pinus*; B, tannin and starch grains in ray cells of wood of *Malus pumila*; C, tannin in pith cells of *Fragaria*.

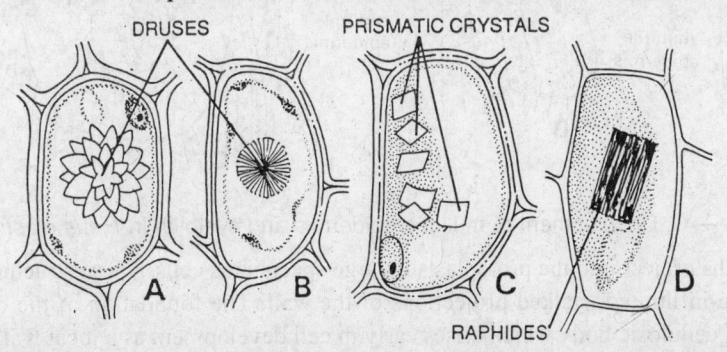

Fig. 2.49. Cells with different types of crystals. A, and B, druses from the cells of *Gnetum gnemon*; C, prismatic crystals from cell of *Gnetum indicum*; D, a bundle of raphides in leaf cell of *Vitis vinifera*.

tanning industry. The commercial *katha* is prepared from the tannins of the heart wood of *Acacia catechu*. The tannins are poorly developed in monocotyledonous plants. They are either found in individual cells or in special organs, called the *tannin sacs*. The tannin-containing cells often form connected systems. In the individual cells the tannins occur in the protoplast and may also impregnate the walls, in cork tissue. Within the protoplast, tannins are a common ingredient of the vacuoles. As regards their function they act as protectants of protoplasm against injury, decay termites and pests; as reserve substances related in some manner to the starch and metabolism; as substances associated with the formation and transport of sugars; as antioxidants; and as protective colloids maintaining the homogeneity of the cytoplasm.

(*b*) **Mineral crystals.** Various types of mineral crystals occur in the plant cells. They may occur either in the cell cavity or in the cell walls. The crystals usually lie loose in the cells but sometimes they are found to be suspended into the cell cavity from the cell wall.

The crystals vary in shape and size. Usually the crystals consist of calcium carbonate, calcium oxalate or silica.

(*i*) **Calcium carbonate crystals.** These crystals are commonly known as *cystoliths*. Usually they occur in the epidermal cells of leaves of many flowering plants. They are generally found in under mentioned angiospermic families—Moraceae, Urticaceae, Acanthaceae, Cucurbitaceae, etc. The main body of cystolith is a cellulose extension of the cell wall in which the calcium carbonate is deposited in the form of fine granules. In certain plants double cytoliths are also found.

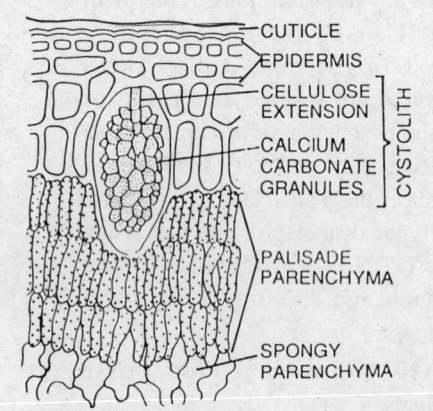

Fig. 2.50. Cystolith (calcium carbonate) in leaf of *Ficus elastica* (India rubber).

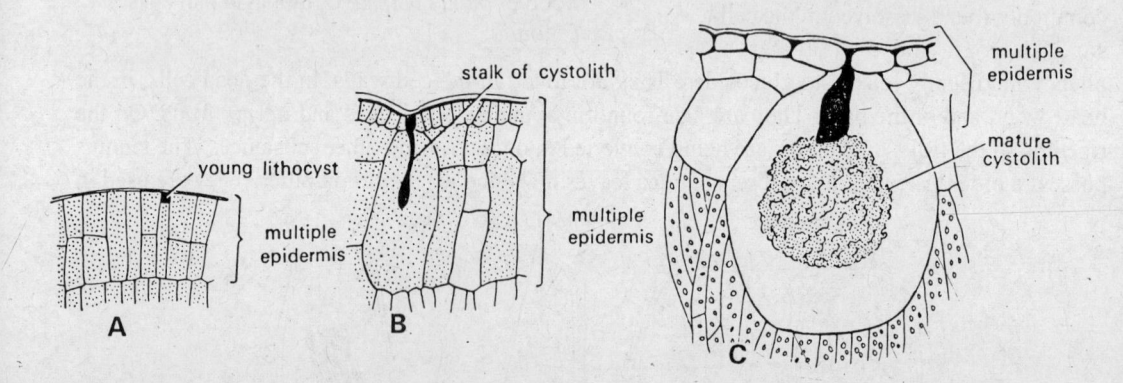

Fig. 2.51. A — C, Development of multiple epidermis and cystolith in *Ficus elastica* leaf.

The cystoliths project into the protoplasts in large specialized cells. These structures are lime-encrusted or lime-infiltrated, stalked projections of the wall. The foundation of the cystolith is a stalked stratified, cellulosic body which arises early in cell development as a local wall thickening. With the addition of large amounts of calcium carbonate this becomes an irregular body which may nearly fill the cell. In shape, cystoliths vary greatly in different genera and families.

The Cell–Structure and its Components 55

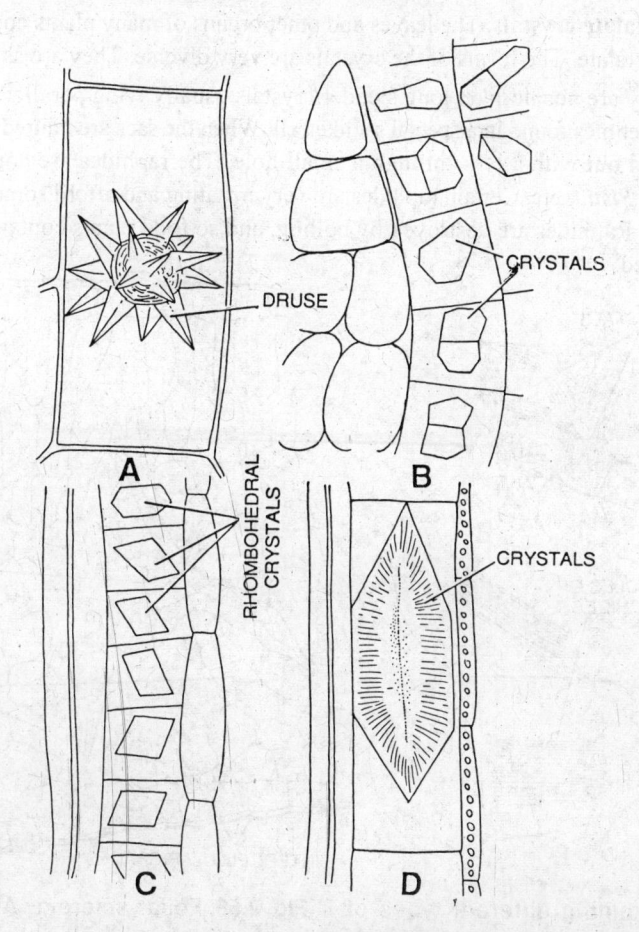

Fig. 2.52. Crystals. A, druse with organic centre, in phloem parenchyma cell of *Juglans nigra*; B, various forms of crystals in phloem parenchyma of *Malus pumila*; C, rhombohedral crystals in phloem parenchyma of *Salix nigra*; D, L.S. of crystal in phloem parenchyma of *Tilia americana*.

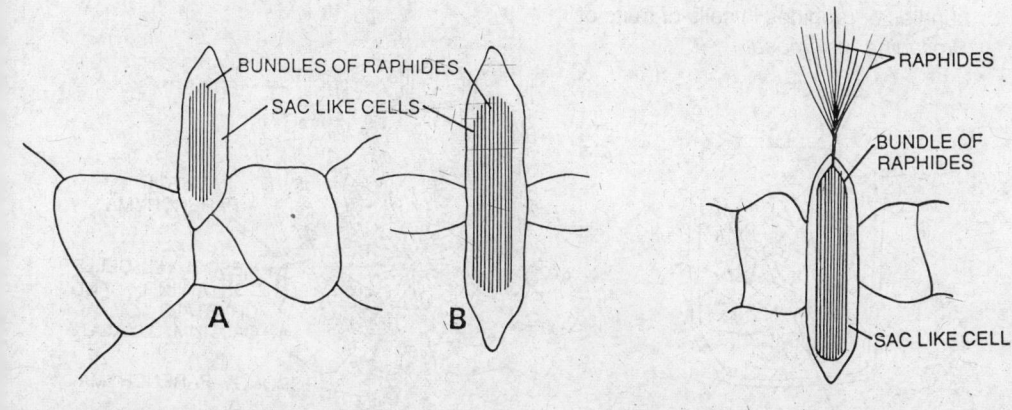

Fig. 2.53. Raphides (calcium oxalate needle like crystals). A, from leaf of *Colocasia*; B, from leaf of *Pistia*.

Fig. 2.54. Raphides. Ejection of raphides from saclike cell of leaf of *Colocasia*.

(ii) Calcium oxalate crystals. The leaves and other organs of many plants contain conspicuous crystals of calcium oxalate. The forms of the crystals are very diverse. They are as follows:

Raphides. They are needle-like, long slender crystals usually lying parallel to each other in a bundle, which is sometimes found in a special saclike cell. When the sacs are injured mechanically, the raphides are released out with a jerk through a small hole. The raphides are commonly found in *Alocasia, Colocasia, Pistia,* etc. Certain raphides are very irritating and afford some protection to the plants from animals. Raphides are destroyed by boiling, and so food plants containing them are not irritating when cooked.

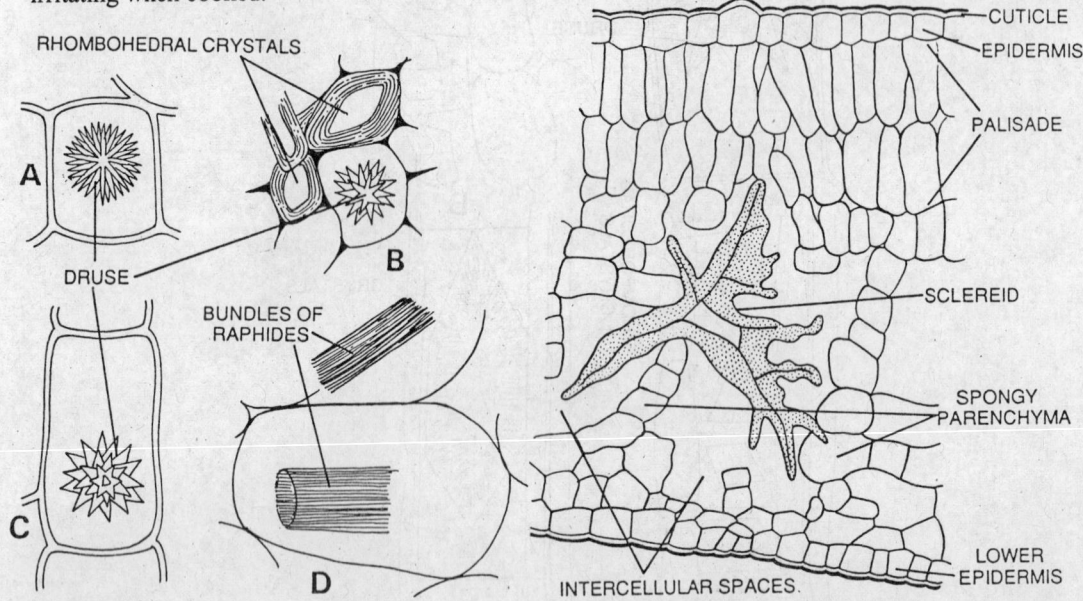

Fig. 2.55. Cell containing different types of crystals, A, druse in cortex cell of stem of *Viburnum lentago*; B, druse and rhombohedral crystals in stone cell of *Carya glabra*; C, druce in cortical cell of *Carica papaya*; D, bundles or raphides in cells of fruits of *Smilacina racemosa*.

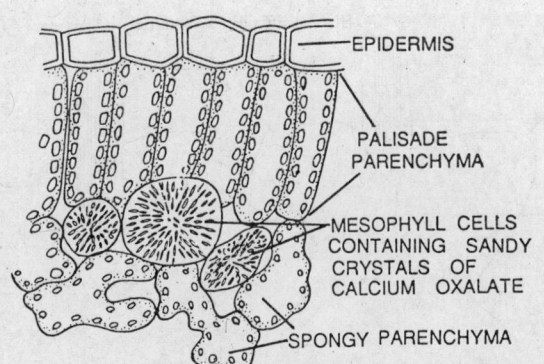

Fig. 2.56. Foliar sclereid. A branched foliar sclereid from the leaf blade of *Trochodendron*.

Fig. 2.57. Rosette crystal (calcium oxalate) from cell

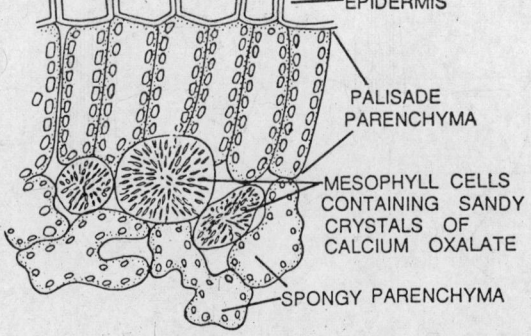

Fig. 2.58. Calcium oxalate crystals. Sandy crystals in leaf of *Atropa belladona*.

Idioblasts. They are stellate calcium oxalate crystals usually found in the aerenchyma of aquatic plants to give support to the tissue. They are quite common in *Limnanthemum, Nymphaea, Trapa,* etc.

Druses, rosette, crystals or cluster crystals. This is one of the commonest compound crystal having the appearance of a rosette and known as a *rosette crystal*. Such crystals are quite common in *Eucalyptus, Nerium, Ixora,* etc.

Prismatic crystals. They are single calcium oxalate crystals found in various plants. They may be square, prismatic, rhomboidal or pyramid-like in shape.

Sand crystals. They are commonly found in masses of micro sphenoidal crystals packed into a cell. Such crystals are generally found in the leaves and roots of *Atropa belladona* of Solanaceae.

Silica is deposited mostly in cell walls, but sometimes it forms bodies in the lumen of the cell. The Gramineae is the best-known example of a plant group having silica in both the walls and the cell lumina.

(c) **Latex.** It is the milky or watery juice found in long branching tubes known as latex tubes. In many cases neighbouring tubes become connected, thus forming a network. When these tubes are cut or injured, the latex exudes. Rubber, opium, chewing gum and other valuable substances are derived from coagulated latex. The latex contains starch grains, proteins, oils, tannins, gums, resins, alkaloids, salts, enzymes and some poisonous substances.

The dispersed particles commonly belong to the terpenes (hydrocarbon) which include such substances as essential oils, balsams, resins, camphors, carotenoids and rubber. Among these substances the resins and particularly rubber $(C_5H_8)n$ are characteristic components of latex in many plants. The latex may contain a large amount of protein (*Ficus callosa*), Sugar (Compositae), or tannins (*Musa,* Araceae). The latex of some Papaveraceae is well known for alkaloids (*Papaver somniferum*) and that of *Carica papaya* for proteolytic enzyme papain. The latex of *Euphorbia* species is rich in vitamin B_1. Crystals of oxalates and malates are abundantly found in latex. Certain plants contain starch grains in their latex.

The best-known latex is of various rubber-yielding plants. In *Hevea brasiliensis*, rubber constitutes 40 to 50 per cent of the latex. The rubber particles suspended in latex vary in size and shape. When the latex is released from the plant, the particles clump together; that is, the latex coagulates. This property is utilized in the commercial separation of rubber from latex.

The latex of various plants may be transparent or clear (*e.g., Morus, Nerium,* etc.) or milky (*e.g.,* in

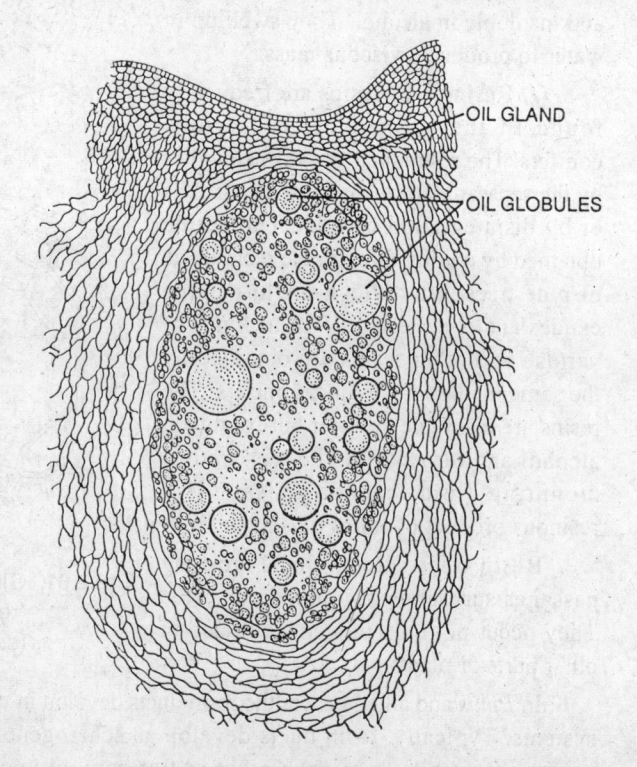

Fig. 2.59. L.S. of oil gland in orange peel.

Asclepias, Calotropis, Euphorbia, Ficus, etc.). It is yellow-brown in *Cannabis* and yellow or orange in several members of Papaveraceae.

(*d*) **Essential or volatile oils.** The essential or volatile oils frequently occur in oil glands. These oils are volatile and are usually very odoriferous. Well-known examples are eucalyptus oil and the oil from orange peel. The orange peel contains large oval glands. These glands originate in the splitting apart of certain cells, but are formed largely by the breaking down of the cells containing the oil. The disintegration of the cells brings the oil into the large cavity of the gland. The oil glands are also found in the petals of rose, jasmines and many other flowers.

(*e*) **Gums.** The gums are exuded out from the stems of many plants. The species of *Acacia* produce best gum. These spacies are *Acacia senegel, A modesta* and *A. arabica*. The gums are soluble in water and insoluble in alcohol. They swell up in water to produce a viscous mass.

(*f*) **Resins.** The resins are frequently found in the resin ducts of various conifers. The resin ducts are formed either by the separation of the neighbouring cells or by disintegration of cells. Turpentine obtained by cutting through the resin ducts of pine trees, after which the turpentine exudes and is collected. Various kinds of varnish and other resins are obtained by the same method from other conifers. The resins are insoluble in water but soluble in alcohol and turpentine. The well known mounting fluid, canada balsam is a resinous product of *Abies balsamea*.

Resin ducts. Resin ducts are long passages surrounded by glandular cells. They occur not only in stems but also in other parts of plants.

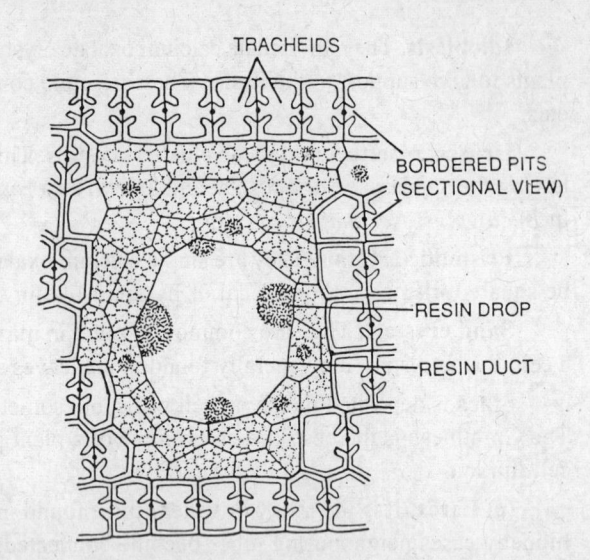

Fig. 2.60. Schizogenous cavities. Resin-duct of pine stem with resin.

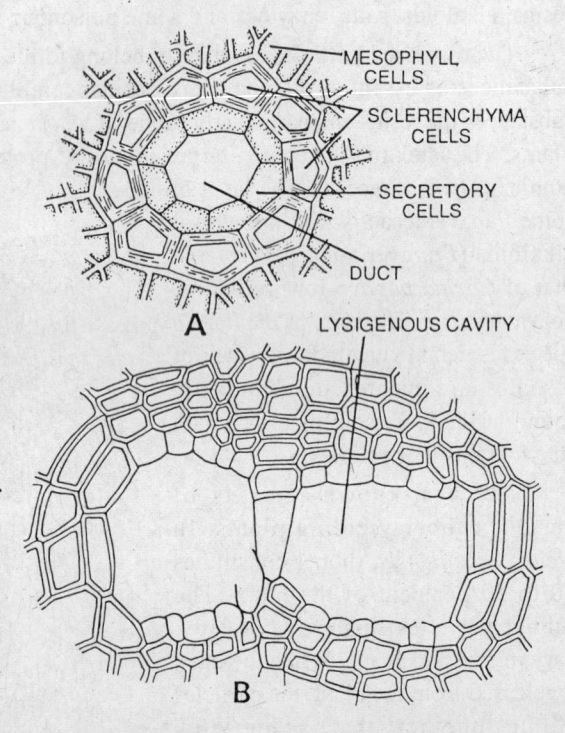

Fig. 2.61. Glands. A, internal schizogenous gland (resin duct of *Pinus*); B, lysigenous gland (secondary xylem of *Copaifera*).

In *Pinus* and allied genera the resin ducts develop in the axial system or in both the axial and ray systems. Typically, resin ducts develop as schizogenous intercellular spaces by separation of parenchyma cells from each other. After some divisions these cells form the lining, or the *epithelium*, of the resin ducts and excrete resin. In *Pinus* the epithelial cells are thin-walled, remain active for several years, and produce abundant resin. In *Abies* and *Tsuga* the epithelial cells have thick lignified walls and most of them die during the year of origin. These genera produce little resin. Eventually a resin duct may become closed by enlarging epithelial cells. These tylosis like intrusions are called *tylosoids* (Record, 1934). They differ from tyloses in that they do not grow through pits. According to some workers there is a distinction between resin ducts that are *normal* and those that are *traumatic* (Greek *trauma*; a wound), that is, arise in response to injury. Normal ducts are elongated and occur singly; traumatic ducts are cyst-like and occur in tangential series.

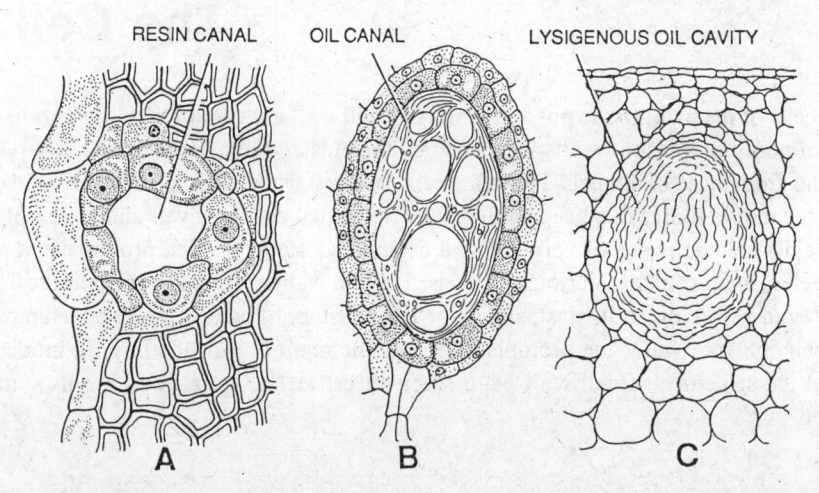

Fig. 2.62. Secretory tissue. A, C.S. of resin canal of *Pinus*; B, C.S. of oil canal of young fruit of *Angelica*; C, sectional view of lysigenous oil cavity of rind of *Citrus*.

(g) Organic acids. They are found in several vegetable and fruit juices and are often combined with peculiar bases and alkaloids. Several organic acids are found in various plants, *e.g.*, citric acid in citrus fruits; tartaric acid in grapes and tamarinds; malic acid in the fruits of apple; gallic acid in mango seeds; oxalic acid in *Rumex, Oxalis* and *Nepenthes*.

2. Nitrogenous waste products. *Alkaloids.* They make a group of nitrogen-containing, basic organic compounds present in plants of a few families of Dicotyledons, *e.g.*, Solanaceae, Papaveraceae. They are thought to be end products of nitrogen metabolism. They are of great importance because of their poisonous and medicinal properties, *e.g.*, atropine (found in *Atropa belladona* of Solanaceae.), cocaine, morphine, nicotine, quinine, strychnine, etc.

3

The Cell Wall

The cells of the plant tissue possess protective cell wall in near about all flowering plants. The presence of non-protoplasmic walls is considered the outstanding characteristic differentiating the cell of plants from those of animals. Few plant cells lack cell walls. Examples of cells without walls in plants are the motile spores in algae and fungi and the sexual cells in lower and higher plants. As we know the cells are composed of very soft and delicate substance, called protoplasm and thus they need protection against other injurious agencies. The cell walls protect the protoplasts. The cell wall was discovered before the protoplast, and in the early history of botany has received more attention than the cell contents. Later the protoplast became the main object of study. In modern days the origin and the structure of cell wall have received critical attention and are now much better understood.

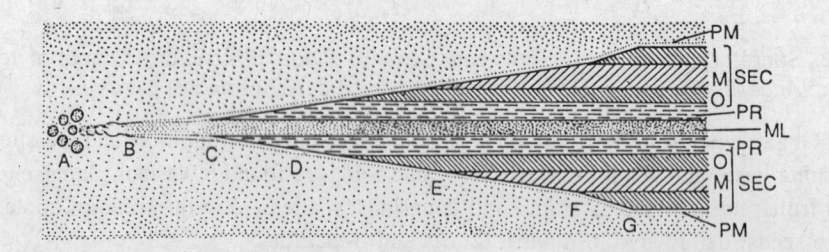

Fig. 3.1. The cell wall. Diagrammatic representation of successive stages (from lef to right) in formation of cell wall with secondary layers. Walls of both daughter cells are shown in the figure. A, origin of cell plate; B, beginning of transformation of cell plate into middle lamella; C, beginning of deposition of primary wall D, E and F, beginning of deposition of outer, middle and inner layers (O, M.I.); G, completion of wall thickening.

THE WALL LAYERS

Each cell within a tissue has its own cell wall. The cell walls of plants vary much thickness in relation to age and type of cells. Generally, young cells have thinner cell walls than the fully developed ones, but sometimes the wall does not thicken much after the cell ceases to grow. In mature thick-walled cells, a concentric layering is usually distinct in the cell wall. The layers differ from one another in physical and chemical nature. The cell wall is complex in its structure and usually consists of three layers, the *primary wall, the intercellular substance* or *middle lamella,* and the *secondary wall*. The intercellular substance cements together the primary walls of two contiguous cells, and the secondary wall is laid over the primary, that is, next to the lumen of the cell. The primary wall usually consists of a single layer, whereas the secondary wall is made up of one to many layers, most frequently of three.

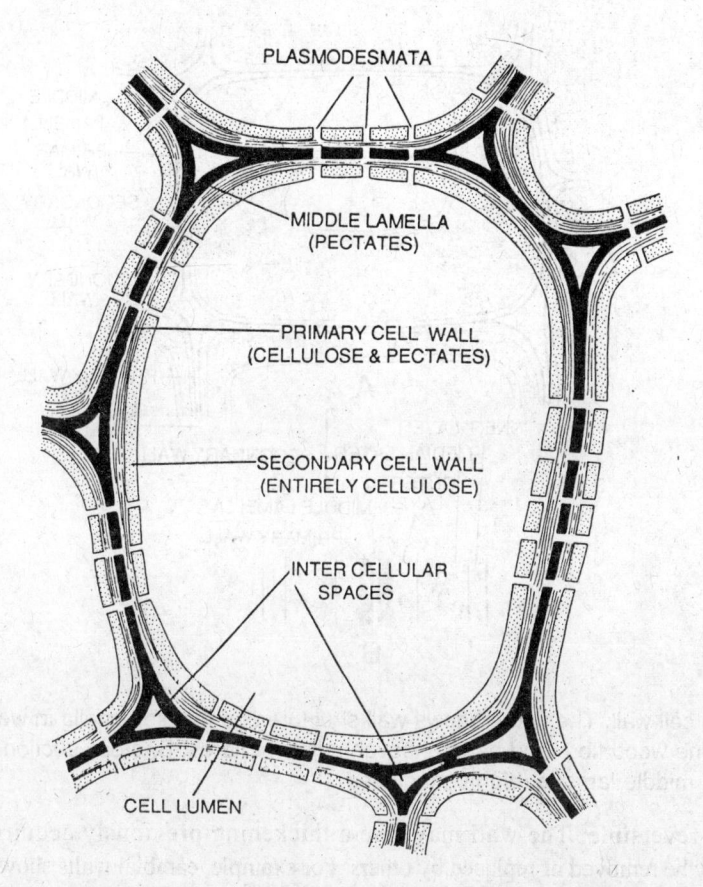

Fig. 3.2. The cell wall. Schematic representation of structure and composition of plant cell walls.

THE PRIMARY WALL

The membrane developed on the surface of the cell plate denotes the first stage of the primary wall. Generally the primary wall is formed in the developing cells, and in many types of cells it is the only wall. The primary wall consists of cellulose, pectic compounds, noncellulosic polysaccharides and hemicelluloses (Bonner, 1950). Sometimes it becomes lignified. The primary walls are usually associated with living protoplasm. The walls of dividing and growing meristematic cells are primary. The primary wall may change its shape and volume according to the growth of young protoplast. This wall is not uniformly thickened and may maintain its thickness and unevenness, or be alternately thickened and thinned.

As already mentioned, the walls of dividing and growing meristematic cells are primary, and so are those of most of the cells which retain living protoplast. The changes that occur in primary walls

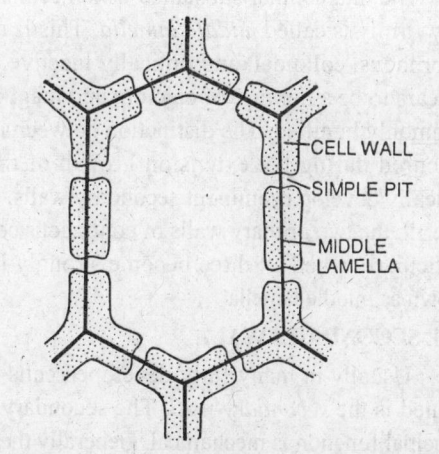

Fig. 3.3. The cell wall. Simple pits. A cell in sectional view showing simple pits in its wall.

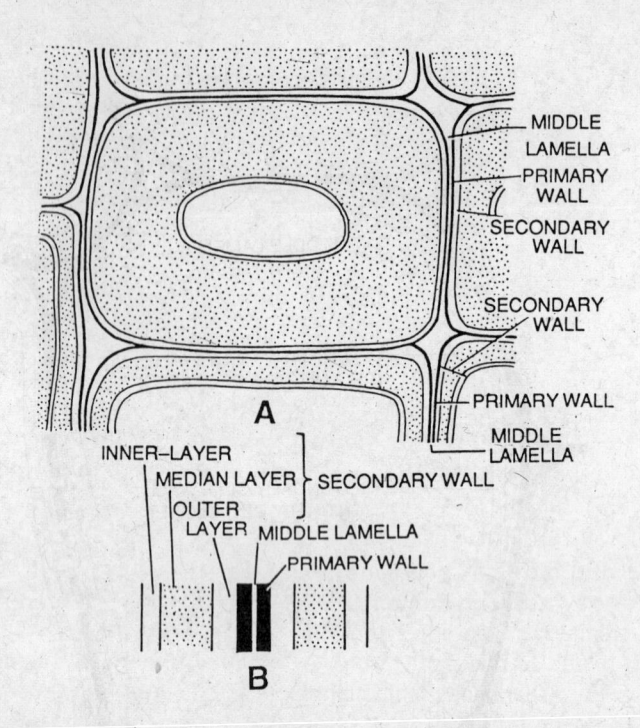

Fig. 3.4. The cell wall. The figure shows wall structure and middle lamella in wood fibres. A, T.S. of one wood fibre and parts of seven adjacent fibres; B, cross section of adjacent walls and middle lamella (highly magnified).

are therefore reversible. The wall may lose a thickening previously acquired and chemical substances may be removed or replaced by others. For example, cambial walls show seasonal changes in thickness and in colloidal properties (Kerr and Bailey, 1934). In other cases the thick primary walls of the endosperm in many seeds are digested during germination.

THE MIDDLE LAMELLA

The intercellular substance which cements together the primary walls of two contiguous cells very firmly is called *middle lamella.* This is a complex layer in its structure and morphology. It is amorphous, colloidal, and optically inactive. It is composed mainly of a pectic compound which appears to be a mixture of calcium and magnesium pectate. In woody tissues the middle lamella is commonly lignified. The distinction between intercellular lamella and the primary wall is frequently obscured during the extension growth of the cell. In such cells as tracheids and fibres, which typically develop prominent secondary walls, the intercellular layer becomes extremely tenuous. As a result, the two primary walls of contiguous cells and the intervening middle lamella appear as a unit, particularly when all three become strongly impregnated with lignin. This triple structure is often known as middle lamella.

THE SECONDARY WALL

Usually in many fully developed cells further thickening of cell wall occurs. The wall then formed is the *secondary wall.* The secondary wall maybe considered a supplementary wall whose principal function is mechanical. Generally the cells with secondary walls are devoid of protoplasts at maturity. These walls are most characteristic of cells that are highly specialized and undergo irreversible changes in their development. However, cells with active, living protoplasts, such as the xylem ray and xylem parenchyma cells, also may have secondary walls. The secondary wall is laid

The Cell Wall

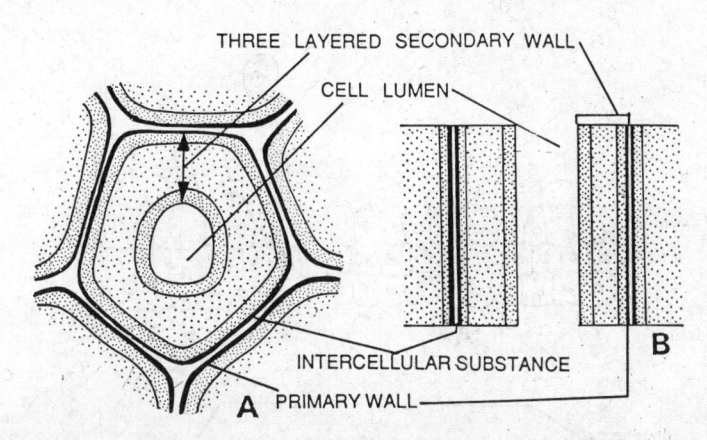

Fig. 3.5. A cell wall. A, cell with secondary wall layers in T.S.; B, cell with secondary wall layers in L.S.

down over the primary wall except over the pit membranes. In the tracheids and vessels of protoxylem the secondary wall covers much less of the primary wall; it forms only as rings, spiral bands. and bars over the delicate primary wall. The secondary wall is more massive than the primary, and in most thick-walled cells it constitutes the major part of the wall. Usually the secondary wall constitutes of three layers—*inner, middle* and *outer*. This wall may consist of cellulose or mixtures of cellulose, non-cellulosic polysaccharides and hemicelluloses. Sometimes the number of layers of secondary wall may exceed three, and the innermost sometimes only of a helical band and such bands are called *tertiary spirals* or *spiral thickenings* (Eames and Mac Daniels, 1947).

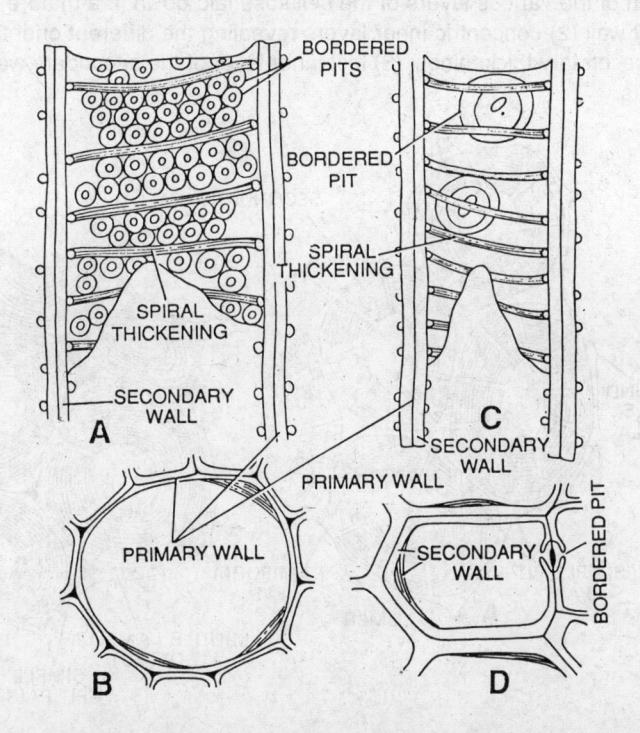

Fig. 3.6 The cell wall. This figure shows the primary wall and secondary wall with tertiary spirals. A and B, L.S. and T.S. of vessel of *Tilia* C and D, L.S. and T.S. of tracheid of *Taxus* (a gymnosperm).

Plant Anatomy

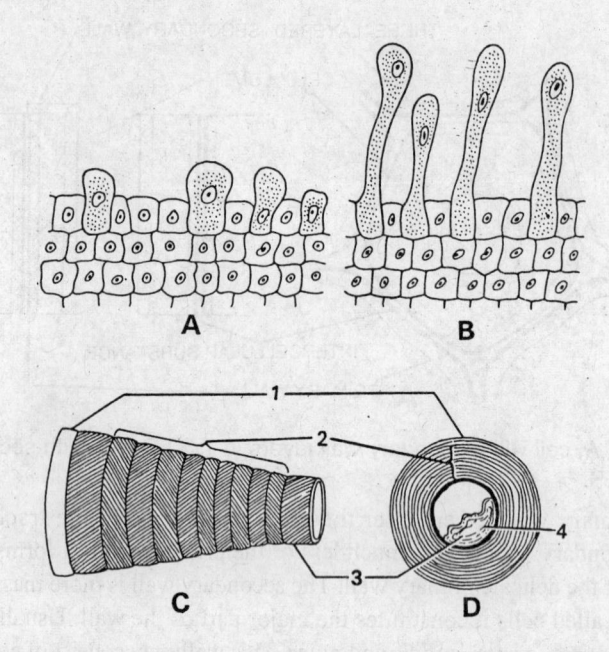

Fig. 3.7. Structure and growth of cotton fibre. A, outer layer of cells of young cotton seed showing the beginning enlargement of the fibre on the day of flowering; B, same after 24 hours, C, diagram of the various layers of the cellulose laid down in a mature cotton fibre (1) outer primary wall (2) concentric inner layers revealing the different orientation of the cellulose in the secondary thickenings, (3) last inner layer of the secondary wall, (4) remains of cell contents.

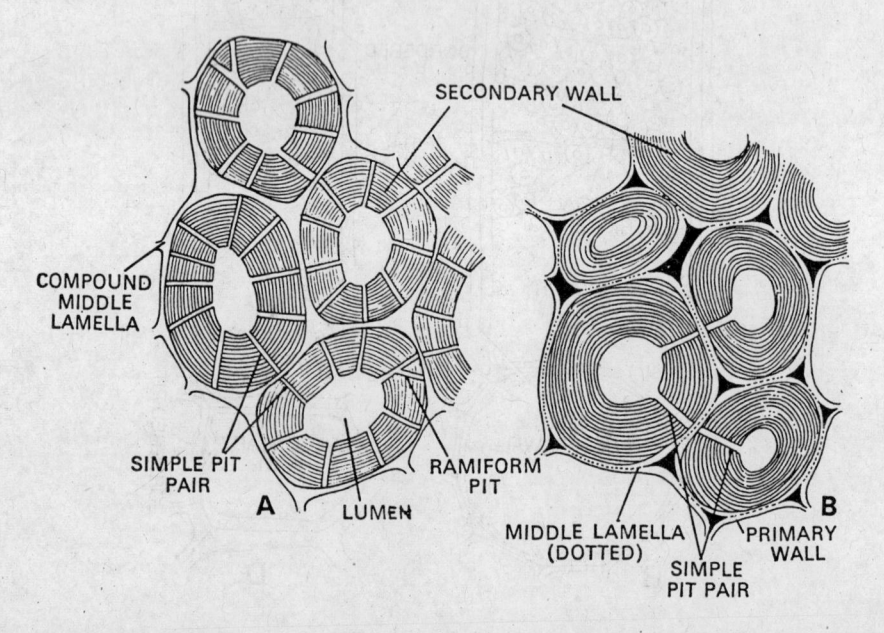

Fig. 3.8. The cell wall. Secondary cell wall sclereids from a T.S. of *Cydonia* fruit; B, phloem fibres from a transection of *Nicotiana* stem.

PLASMODESMATA

The relation between cell wall and protoplast is quite close. The protoplast of one cell remains connected with that of the adjacent or neighbouring cell by delicate threads of cytoplasm. These threads known as *plasmodesmata*, fill minute passages which constitute the only breaks in the primary wall of two cells. The presence of plasmodesmata is the characteristic of living cells and they are found in the cell walls of all living cells to maintain the continuity of the protoplasm.

The name plasmodesmata (singular plasmodesma—from the Greek *desmos*, strand), dates back to Strasburger (1901). Livingston and Bailey (1946) reported that plasmodesmata have been recognized in the living cambium and ray cells of certain gymnosperms. This gives the best support to their interpretation as cytoplasmic structures.

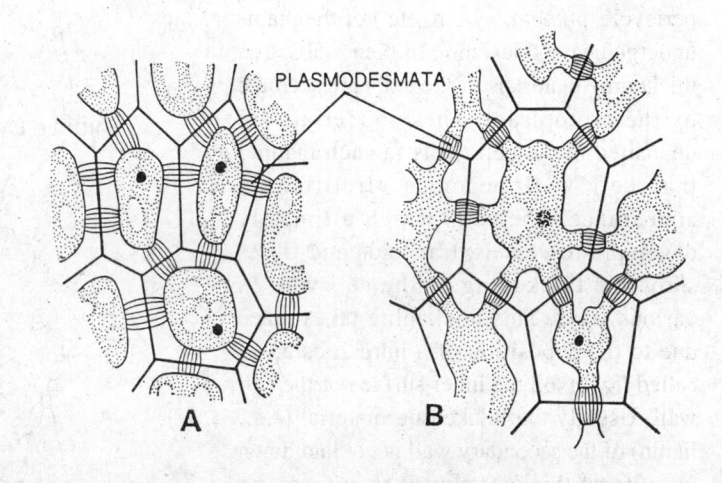

Fig. 3.9. Plasmodesmata. A, in edosperm of *Diospyros;* B, in endosperm of *Phoenix.*

The plasmodesmata have been seen in red algae, liverworts, mosses, vascular cryptogams, gymnosperms and angiosperms. They are found throughout all living tissues of a plant including the meristematic tissues. Plasmodesmata either occur in groups or are distributed throughout a wall. The plasmodesmata found in groups are frequently localized in the primary pit fields. It is commonly known that all primary pit fields of living cells are traversed by plasmodesmata. The relation of the plasmodesmata to primary pit fields is characteristic in two contiguous cells, cytoplasmic strands extend into the cavities of a pair of pit fields, and the thin wall of the pit field is traversed by very fine threads connecting the two small masses of cytoplasm filling the depression of the pit field.

As regards their origin, it has been suggested that young, growing walls are permeated with cytoplasm which later withdraws except in localized areas where it then makes the plasmodesmata. It is also thought that plasmodesmata are formed *de novo* in older walls.

As regards their function, they are thought to be concerned with material transport and conduction of stimuli. The plasmodesmata are also regarded as channels permitting the movement of viruses from cell to cell. In the case of the angiospermic parasites such as *Cuscuta, Viscum* and *Orobanche*, the presence of plasmodesmata between the haustoria-like structures and the cells of their host plants is thought to be related to food and virus movement (Esau, 1948).

Plasmodesmata are most readily seen in the endosperm of some seeds (*e.g., Phoenix, Aesculus, Diospyros*) where food storage has greatly increased wall thickness.

THICKENING OF THE CELL WALL

The cells cease to increase in size and become mature. At this stage, the cells become modified according to the function they have to perform. During the process of their maturation, the cells undergo additional or secondary thickening deposition of new materials such as cellulose, lignin, suberin, cutin, etc. Some of the cells (*e.g.,* parenchymatous cells) remain unthickened. The cells of certain parts (*e.g.,* pericycle, phloem, xylem, etc.) of the plants undergo heavy thickening of their walls. The thickening materials of the cells are secreted by the protoplasm. These materials are deposited in the cell walls in such manner that cell wall becomes stratified in appearance. The cells which ultimately develop into vessels, tracheids and fibres show the thickening of the cell wall in various ways. This thickening takes place due to the deposition of a hard substance, called *lignin,* on the inner surface of the cell wall. Usually the thickening material (*e.g.,* lignin) of the secondary wall is not laid down in uniform thickness but it may form the special patterns such as *annular, spiral, scalariform, reticulate* and *pitted.* In such patterns the whole of the wall is not thickened. Only those portions of the cell walls are thickened where the thickening material is laid down in the form of above mentioned patterns and the remaining portion of the wall remains thin. The details of these thickenings are as follows:

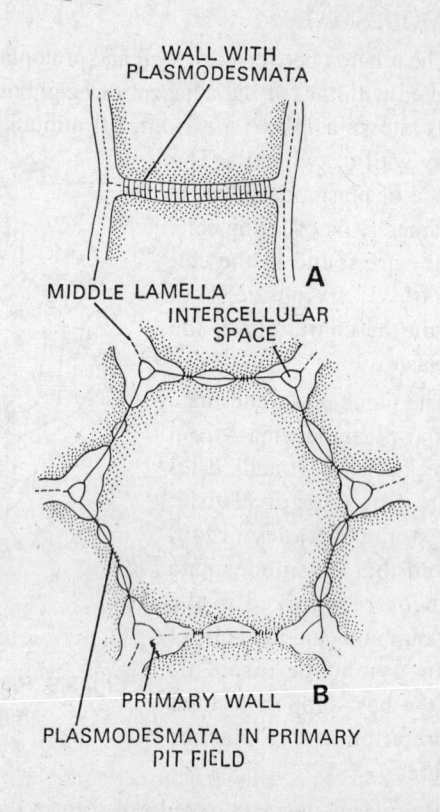

Fig. 3.10. Plasmodesmata. A and B, parenchyma cells without secondary walls from *Nicotiana* stem. Plasmodesmata dispersed throughout the wall in A and restricted to primary pit fields in B.

Annular or ring-like thickenings:
The deposition of lignin takes place in the form of rings on the inner surface of the cell wall. These lignified rings are placed one above the other like coins leaving sufficient space in between each other. The gaps of the walls remain unthickened. Such thickenings are commonly found in the vessels and tracheids.

Spiral thickenings. In such cases the deposition of thickening material (lignin) takes place in the form of complete spiral bands. The number of such bands may be one or more than one. This type of thickening is commonly found in the vessels or tracheae of angiosperms.

Scalariform or ladder-like thickenings. In such thickenings of the cell wall the lignin is being deposited in the form of the transverse rods of the ladder, and thus known as scalariform or ladder-like. The unthickened areas between the successive thickening layers appear as elongated transverse pits. This type of thickenings are common in xylem vessels and tracheids of protoxylem.

Reticulate or net-like thickenings. In such thickenings of the cell wall the thickening matter or lignin is being deposited in the form of a net or reticulum and thus known as reticulate or net like thickening of the cell wall. In such cases the unthickened areas of the cell wall are irregular in shape. These thickenings are commonly found in the vessels of the stems, roots and leaves of angiosperms and in the tracheids of protoxylem.

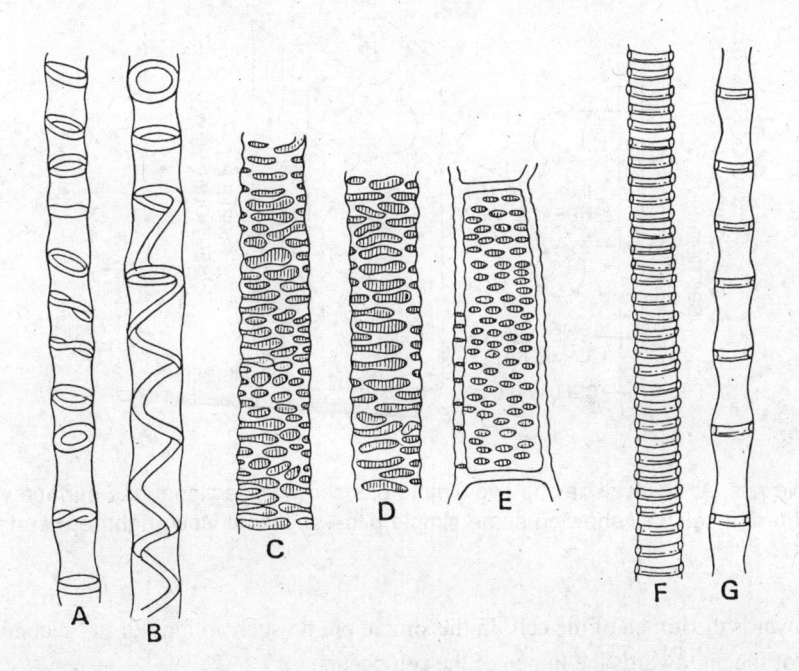

Fig. 3.11. Xylem-vessels. Secondary wall structure in primary tracheary (vessel) elements, the thickenings are visible. A, annular; B, partly annular, partly helical; C, reticulate; D, scalariform reticulate; E, pitted; F, annular in an unstretched vessel element; G, annular in a stretched vessel element.

Pitted thickenings. In such thickenings of the cell wall, the whole inner wall is more or less uniformly thickened, leaving here and there some small unthickened areas, the *pits*.

PITS

The pits are formed in pairs lying against each other on the opposite sides of the wall, and morphologically more correct they are called '*pit pairs*'. A pit pair is structural and directional unit constituted by two pits lying opposite to each other of contiguous cells.

Structure. The space found inside the pit is called the *pit cavity* or *pit chamber*. The separating membrane which separates the two chambers or cavities of a pit pair, is called the *pit membrane*, or *pit aperture*. The pit cavity opens internally in the lumen of the cell and is closed by the closing or pit membrane along the line of junction of two contiguous cells. A pit has two pit cavities, two pit apertures, and one pit or closing membrane. The pit membrane is common to both pits of a pit pair and consists of two primary walls and a middle lamella or intercellular substance. Usually two types of pits are met with in the cells of various plants, *viz.*, *simple pits* and *bordered pits*. Two bordered pits make up a bordered pit pair, two simple pits form a simple pit pair. A bordered pit and a simple pit lying opposite to each other in contiguous cells, constitute a *half bordered pit pair*. A pit occurs opposite an intercellular space has no complementary pit and is known as *blind pit*. Fundamentally the bordered pit differentiates from a simple pit in having a secondary wall arching over the pit cavity, which constitutes the actual border and becomes narrow

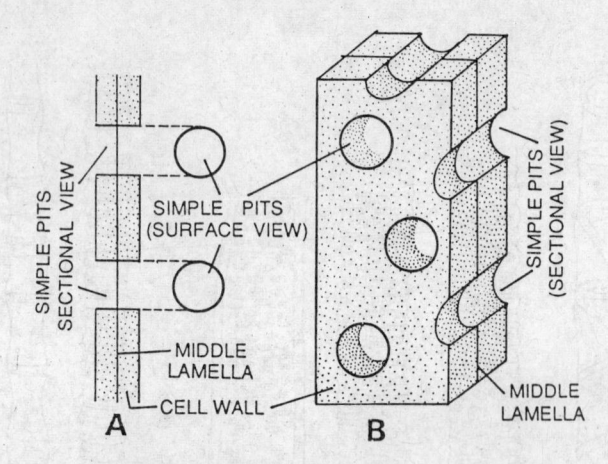

Fig. 3.12. Simple pits. A, cell-wall having two simple pits showing sectional and surface views; B, portion of a cell wall showing some simple pits—sectional view (right top) surface view (front).

like a funnel towards the lumen of the cell. In the simple pit, no such arching of the secondary wall and narrowing of the pit towards the lumen of the cell occurs.

Simple pits. Simple pit pairs occur in parenchyma cells, in medullary rays, in phloem fibres, companion cells, and in tracheids of several flowering plants. In the simple pits, the pit cavity remains of the same diameter and the pit or closing membrane also remains simple and uniform in its structure. The simple pit may be circular, oval, polygonal, elongated or somewhat irregular in its facial view. The simple pits occurring in the thin walls are shallow, whereas in thick wall the pit cavity may have the form of a canal passing from the lumen of the cell towards the closing or common pit membrane. The diffusion of protoplasm takes place through these pits.

Bordered pits. They are abundantly found in the vessels of many angiosperms and in the tracheids of many conifers. They are more complex and variable in their structure than simple pits. The overarching secondary wall which encloses a part of the pit cavity is called, the *pit border*, which opens outside by a small rounded mouth known as *pit aperture*. The overarching rim forms a border around the aperture and thus named 'bordered pits'. The pit aperture may be of various shapes in the facial view. It may be circular, lenticular, linear or oval. In the case of relatively thick secondary walls, the border divides the cavity into two parts. The space between the closing membrane and the pit aperture may be called the pit chamber and the canal leading from pit chamber to the lumen of the cell may be termed as *pit canal*. The pit canal opens in the pit chamber by an outer aperture and at the same time it opens in the lumen of the cell by an inner aperture. The closing membrane of a bordered pit pair which consists of the parts of two primary walls and the intercellular substance or middle lamella, is somewhat thickened in its central part. This thickening is called, *torus* which remains surrounded by a delicate margin. In many bordered pits, the closing membrane may change its position within pit cavities. The torus may remain in central position or it may shift to the lateral position. As the torus is shifted to the lateral position the pit aperture closes, and the passage of the protoplasm may take place only by diffusion through torus.

The Cell Wall

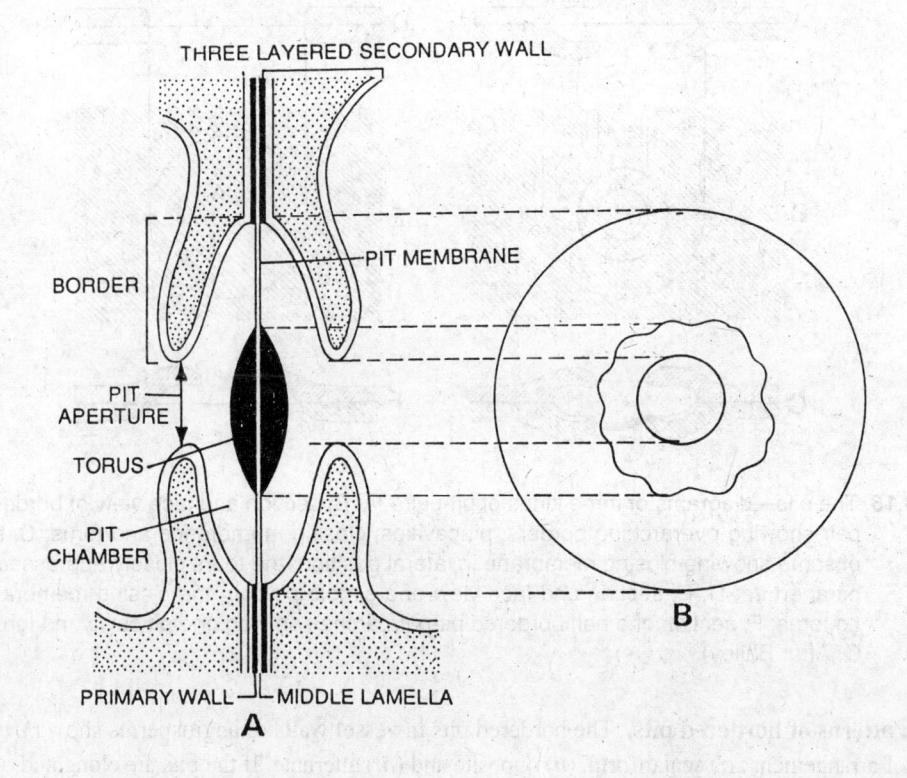

Fig. 3.13. Pits. Diagrams of bordered pit pairs. A, sectional view; B, front view. Pit membrane is made up of two primary walls and the intercellular lamella. The tours is formed by thickenings of primary wall. The front view shows that outline of torus is uneven.

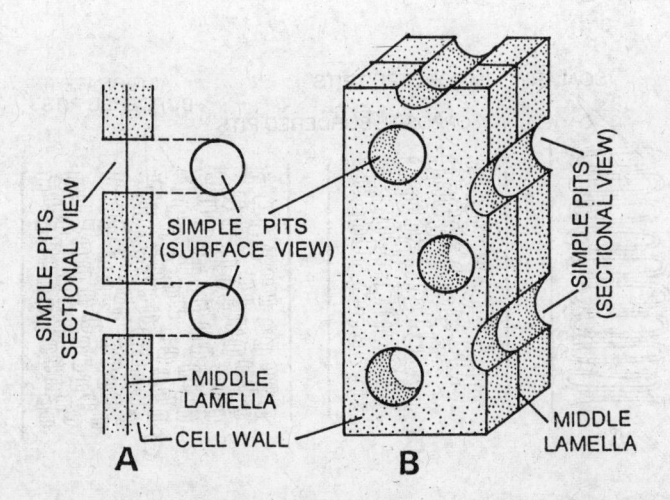

Fig. 3.14. Bordered pits. A, cell-wall with two bordered pits—sectional and surface views; B, a portion of the cell-wall with two bordered pits—sectional view (bottom) and surface view (top).

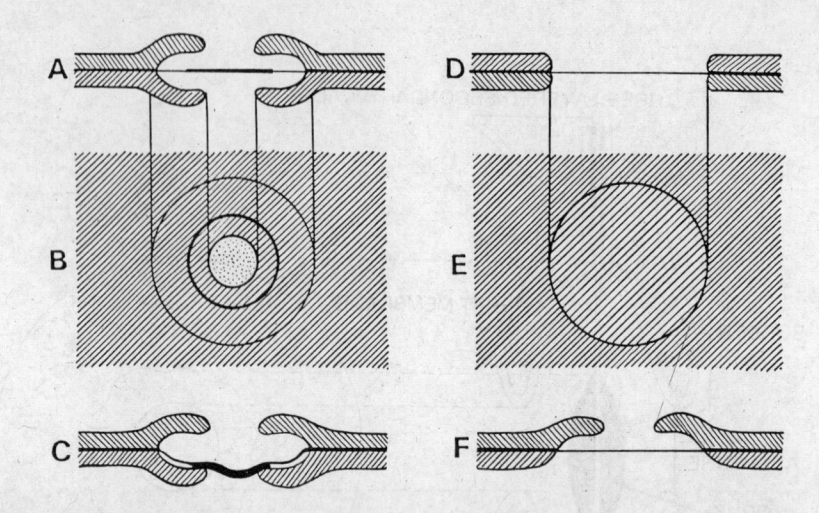

Fig. 3.15. The pits—diagrams of three kinds of pit-pairs. A, B, section and face view of bordered pit-pair showing overarching borders, pit cavities, closing membrance and torus; C, section of same showing closing membrane in lateral position, the tours closely appressed to the pit aperture; D, E, section and face view of a simple pit-pair, the closing membrane has no torus; F, section of a half bordered pit-pair, arches only on one pit cavity, no torus (A—C, After Bailey).

Patterns of bordered pits. The bordered pits in vessel walls of angiosperms show *three* main types of arrangement : (*i*) scalariform, (*ii*) opposite and (*iii*) alternate. If the pits are elongated or linear and make ladder-like arrangement, it is called *scalariform pitting* (Fig. 3.16 A). When the pits are arranged in horizontal pairs, it is called *opposite* pitting (Fig. 3.16 B). If such pits are crowded, their borders assume rectangular outlines in face view. When the pits occur in diagonal rows, the arrangement is called *alternate pitting* (Fig. 3.16C). Generally small simple pits are arranged in clusters, and such arrangement is called *sieve pitting*.

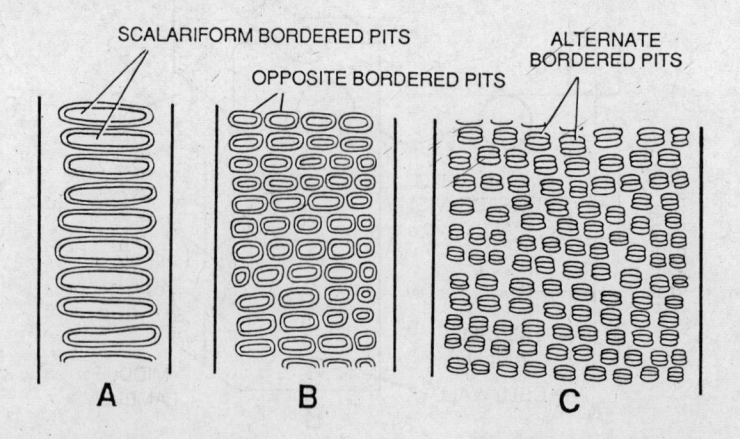

Fig. 3.16. Pits. Arrangement of bordered pits in vessel walls of angiosperms as seen in face view. A, scalariform in *Magnolia*; B, opposite in *Liriodendron*; C, alternate in *Salix*.

CHEMICAL NATURE OF THE CELL WALL

The most-common compound in plant cell wall is the carbohydrate cellulose. It is associated with various substances, most often with other compound carbohydrates. The common carbohydrate constituents of the cell walls other than cellulose are noncellulosic polysaccharides, hemicellulose, and pectic compounds. The fatty compounds, cutin, suberin and waxes, occur in varying amounts in the walls of many types of cells. Various other organic compounds and mineral substances are also present. Water is a common constituent of cell walls.

Cellulose. It is a hydrophilic crystalline compound with the general empirical formula $(C_6H_{10}O_5)_n$. It is a polysaccharide and closely related to starch. Its molecules are chain-or ribbon-like structures with 100 or more of the glucose residues held together by oxygen bridges.

Non-cellulosic polysaccharides. The mannans, the galactans, the xylans and the arabans which are closely allied to cellulose are commonly found in the cell walls.

Pectic substances. They are derivative of polygalacturonic acid and occur in three general types, protopectin, pectin and pectic acid. Pectic compounds are amorphous colloidal substances plastic and highly hydrophilic. The pectic substances constitute the intercellular substance that binds together the wall of individual cells. It also occurs associated with the cellulose in other wall layers, mainly the primary.

Gums and mucilages. They are also the compound carbohydrates of the cell walls. They are related to the pectic compounds, and both possess the property of swelling in water (Bonner, 1950). Gums exude as a result of physiological or pathological disturbances that induce a break down of walls and cell contents, whereas the mucilages occur in some gelatinous or mucilaginous type of cell walls.

Lignin. It is an organic compound of high carbon content, distinct from the carbohydrates. Lignin is found in all three wall layers—the primary wall, the middle lamella, and the secondary wall. The impregnation with lignin usually starts in the intercellular lamella and then spreads centripetally through the primary and secondary walls. The process of conversion of cellulose into lignin is called *lignification*.

Cutin, suberin and waxes. The cutin and suberin are fatty substances which are not meltable and exhibit considerable insolubility in fat solvents. Cutin and suberin are closely related, highly polymerized compounds consisting of fatty acids. Suberin occurs in association with the cellulose in cork cells of the periderm. Suberin also occurs in the endodermis and exodermis of many plants. The phenomenon of impregnation of walls with suberin is referred to as *suberization*. Usually the cutin forms a continuous layer, the cuticle, on the surface of the epiderms of various plants. Cutin also occurs with the cellulose in the outer walls of the epidermis. The process of impregnation of walls with cutin is known as *cutinization*, and the formation of cuticle as *cuticularization*.

The waxes are also fatty substances which are meltable and are easily extracted by fat solvents. Waxes are associated with suberin and cutin and are found to be deposited on the surface of the cuticle in various forms. This type of deposition of wax is responsible for the glaucous condition of many fruits, leaves and stems.

The deposition of these fatty substances (cutin, suberin and waxes) checks the transpiration and protects the leaves from leaching effects of rain. The hard varnish-like cuticle, may protect against penetration of living tissues by potential parasites against mechanical injuries.

Mineral substances. The mineral substances like silica, calcium carbonate, calcium oxalate and several organic compounds like tannins, resins, fatty substances, volatile oils, etc., may impregnate cell walls. Silica or sand particles are fairly common in the walls of grasses and horsetails (*Equisetum* sp.). Calcium carbonate is being deposited in the form of cystoliths in many specis of *Ficus* and other plants. Organic compounds are frequently deposited in the xylem walls when this tissue changes from sapwood into heartwood.

INTERCELLULAR SPACES

Usually the meristematic or young cells do not possess any intercellular spaces among them but as they become older their walls split at certain places giving rise to small spaces called intercellular spaces, usually these are filled up with air or water. Generally there are two types of intercellular spaces—1. Schizogenous intercellular spaces and 2. Lysigenous intercellular spaces.

Schizogenous intercellular spaces. The most common intercellular spaces result from separation of cell walls from each other along more or less extended areas of their contact. In such cases, the intercellular substance dissolves partly and an intercellular space develops. Ultimately this becomes quite big in size and is known as schizogenous cavity. The ordinary intercellular spaces and schizogenous cavities form an intercommunicating system of long intercellular canals which facilitate diffusion of gases and liquids from one part of the plant body to the other. The resin ducts in the Coniferales, and the secretory ducts in the Compositae and Umbelliferae are the typical examples. The cells lining the cavity are secretory in nature and release their product in the intercellular cannal.

Lysigenous intercellular spaces. This type of intercellular space arises through dissolution of entire cells, which are therefore called lysigenous intercellular spaces (*lysis*, loosening, Greek). These cavities of intercellular spaces store up water, gases and essential oils in them. The examples are commonly found in water plants and many monocotyledonous plants. The secretory cavities in *Eucalyptus*, *Citrus* and *Gossypium* are good examples.

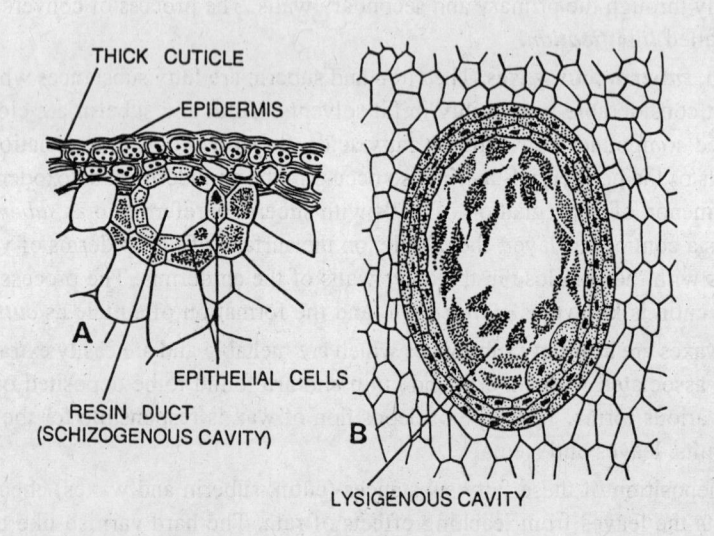

Fig. 3.17. Intercellular spaces. A, a schizogenous intercullular space in *Bryophyllum* stem; B, a schizogenous intercellular space in leaf of *Sequoia sempervirens*. After K. Esau.

MICROSCOPIC AND SUBMICROSCOPIC STRUCTURE OF CELL WALLS

The chemical substances of cell walls remain combined physically and chemically with each other. Several physical and chemical methods have been employed for such investigations. At first the secondary wall was the main object of study, but with the refinement of methods the primary wall has been also investigated successfully. The significance of the study of primary walls is that it yields information on the methods of growth of cell walls in surface area. Investigators combine observations on differential staining; differential solubilities; coarse and fine structure variations; reaction to polarized and fluorescent light, to X-rays, and to dark-field illumination; refractive indices; and composition of ash. In modern days, the electron microscope is the main tool for the study of cell walls.

MICELLAR AND INTERMICELLAR SYSTEMS

The organization of the structure of the cell walls is based on cellulose. The fundamental units of the system are the chain-like cellulose molecules of different lengths. These chains are not dispersed at random but occur in aggregates and generally known as *micelles*. The chain molecules possess a parallel arrangement in a micelle and the glucose residues within a chain are spaced at uniform distances from each other. Thus, a bundle of cellulose molecules, the micelle, may be compared to a crystal in that its units are arranged symmetrically. This way, the bundles of cellulose molecules are interconnected by means of the lower chain molecules and from a porous coherent system, the *micellar system*, interpenetrated by an equally coherent *intermicellar system* in which various wall substances other than the cellulose are present.

Frey Wyssling (1959) graphically described these structural elements and their interrelations on the basis of the secondary wall of the fibre of *Boehmeria*. Here one cellulose melecule is 8 Å wide. Cellulose molecules are combined into an *elementary microfibril* which has a widest diameter of 100 Å and is discernible with the electron microscope. It contains 100 cellulose molecules in a transverse section. Both the cellulose molecules and the elementary fibrils are ribbon like structures. Elementary fibrils form a bundle known as *microfibril* which is 250 Å wide and contains 2,000 cellulose molecules in a transverse section. Electron microscope studies on cell walls are connected mainly with this unit. Microfibrils are combined into macrofibrils 0.4 micron (μ) wide and containing 500,000 cellulose molecules in transverse section. Finally, 2,000,000,000 cellulose molecules make up a transverse section of the secondary wall of the fibre.

MICROFIBRILLAR AND MICROCAPILLARY STRUCTURE

As mentioned earlier the cellulose of plant cell walls is being interpreted as a combination of two interpenetrating systems, the micellar and the intermicellar. They are submicroscopic. The walls contain a porous matrix of cellulose consisting of very fine coalesced fibrils, the *microfibrils* and an interfibrillar system of *microcapillaries* containing various noncellulosic wall constituents. However, within the microfibrils the micelles and consequently also the chain molecules occur approximately parallel to the long axis of the fibrils. The microcapillaries within the cellulose framework may contain liquids, waxes, lignin, cutin, hemicelluloses, suberin, pectic substances, other less common organic compounds, and even crystals and silica.

MICROFIBRILLAR ORIENTATION IN THE CELL WALL

In the three layered walls of certain vessels, tracheids and wood fibres the fibrillar orientations of the inner and outer layers vary between transverse and helical, the helices being of comparatively low pitch, and those of the central layer fluctuate between longitudinal and relatively steeply pitched helical. Characteristic patterns occur about the large bordered pits of the early wood tracheids.

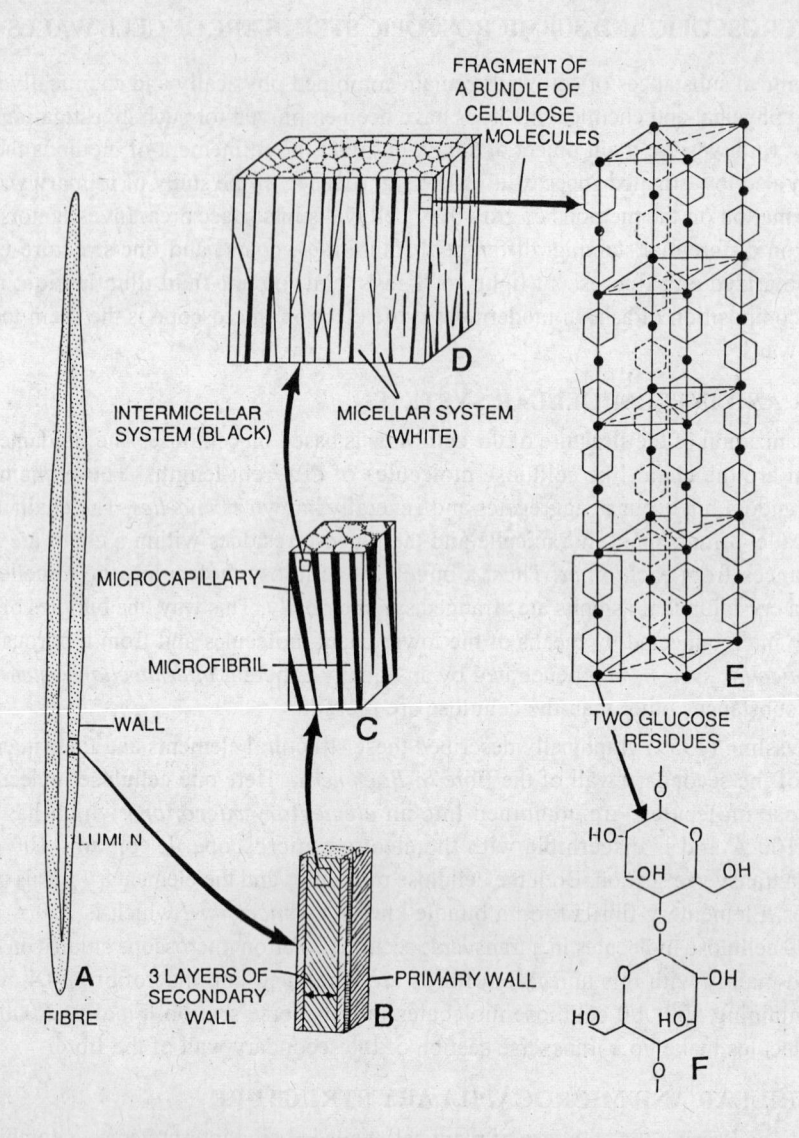

Fig. 3.18. Diagrams illustrating the common concept of microscopic and submicroscopic structure of secondary cellulose wall. Fibre (A) has a three layered secondary wall (B), In a fragment of the central layer of this wall (C), the microfibrils of cellulose (white), the other of micro capillaries (black) containing various noncellulosic wall materials. Microfibrils consist of inter micellar (black) and micellar (white) systems (D). The intermiceller system contains noncellulosic wall constituents, whereas the micellar system is composed of bundles of chain like cellulose molecules (E). The bundles of cellulose chains, the micelles, show an orderly arrangement of the glucose residues (F) that make the basic units of the cellulose molecules.

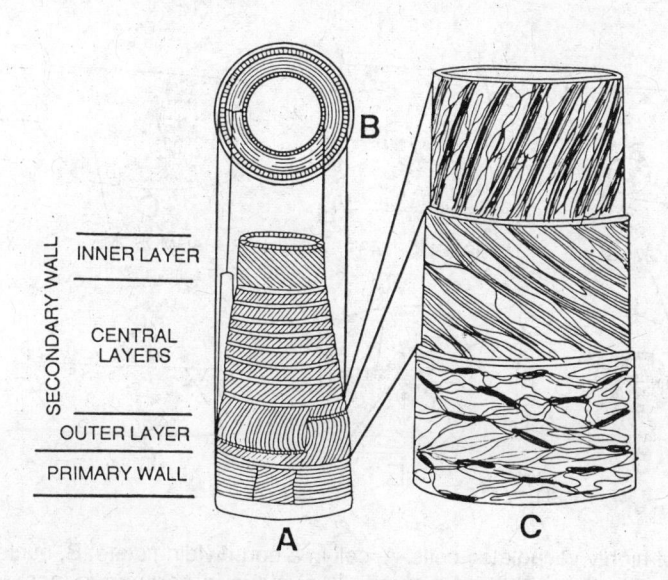

Fig. 3.19. Cotton fibre's wall structure. A, telescoped segment; B, T.S. of fibre showing spatial relation of the various layers and orientation of microfibrils; C, showing the primary wall with its reticulate microfibrillar structure the outer layer of the secondary wall combining a reticulate a parallel orientation of microfibrils, and the first central layer of secondary wall with predominantly parallel microfibrillar structure.

In the cotton fibre the bulk of the secondary wall consists of microfibrils oriented at an angle of 45 degrees and less with respect to the longitudinal axis of the fibre.

In the consecutive lamella of the flax fibre the helices are wound in opposite directions. In tracheary cells with annular and scalariform secondary thickenings the crystalline regions of these thickenings have a horizontal, ring-like orientation.

The pitch of the helices of microfibrils varies in the secondary walls of different cells. However, among the layers of the same wall, within a given layer the microfibrils are usually parallel to one another and always parallel to the surface of the cell. The secondary walls possess a parallel texture.

FORMATION OF WALLS

INITIATION OF WALLS IN CELL DIVISION

The process of somatic division of a protoplast may be separated into two parts—(*i*) The division of the nucleus (*i.e., mitosis* or *karyokinesis*) and (*ii*) The division of the part of the protoplast other than nucleus (*i.e., cytokinesis*)., In the cells which have cell walls the new wall develops during cytokinesis. The partition between the newly formed protoplasts, when first appears, is referred to as the *cell plate*. If cytokinesis takes place immediately after the nuclear division, the cell plate develops in the equatorial plane of a fibrous spindle, the *phragmoplast*, extending between the two groups of chromosomes that move apart during the anaphase of mitosis. As these two groups develop into the telophase nuclei, the phragmoplast becomes wider in the

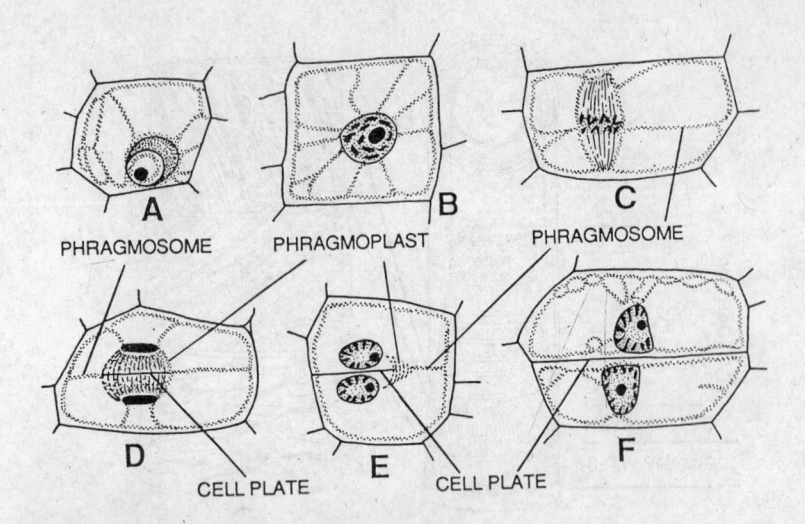

Fig. 3.20. Division of highly vacuolated cells. A, cell in a non-dividing state; B, nucleus in prophase and located in the middle of the cell; C, nucleus in early anaphase; laterally mitotic spindle is connected to parietal cytoplasm by a cytoplasmic layer the, phragmosome; D, daughter nuclei in telophase, the barrel-shaped spindle between the nuclei is the phragmoplast; cell plate appears in its equatorial plane; E, cell plate intersects one of the walls of mother cell; F, cell division is completed and the cell plate occupies the former position of phragmosome.

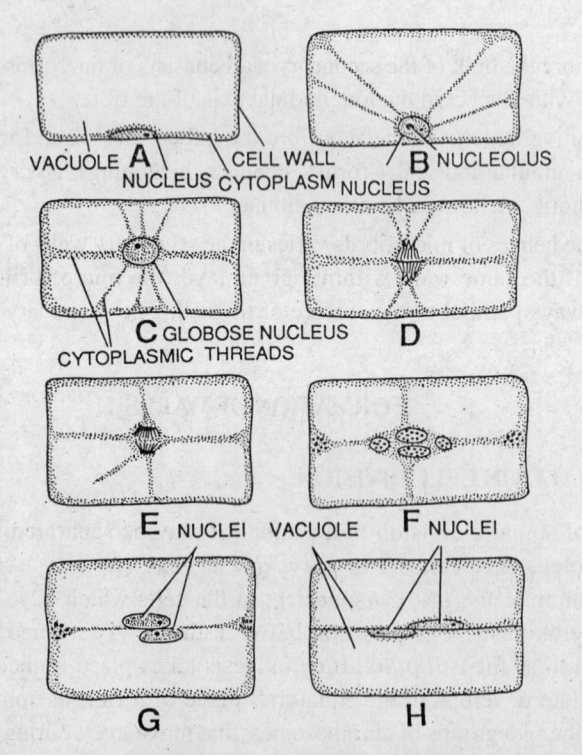

Fig. 3.21. Origin of the cell wall. Division in cell with large central vacuole. The cytoplasm increases in amount forming layer in position of new wall; nucleus becomes rounded and shifts to central position; normal nuclear division takes place with extending phragmoplast.

The Cell Wall

equatorial plane and becomes barrel-shaped. When the cell plate appears in the median part of the equatorial plane of the phragmoplast, the fibres of the phragmoplast disappear in this position but remain evident at the margin's, until the cell plate appears here too.

Cytokinesis is not limited to meristematic cells with dense protoplasts. Some vacuolated meristematic cells and other vacuolated cells also divide actively during the growth of roots, shoots, leaves and fruits of higher plants. The process of the development of the phragmoplast through vacuolated cells is complicated by the fact that the new cell plate eventually occurs in the region formerly occupied by the vacuole. In some cases, long before the beginning of cytokinesis, a cytoplasmic plate, the phragmosome is formed across the cell in the plane of cell division. The phragmsome is derived from the strands of parietal cytoplasm and thus forms a living medium in which the phragmoplast and the cell plate develop.

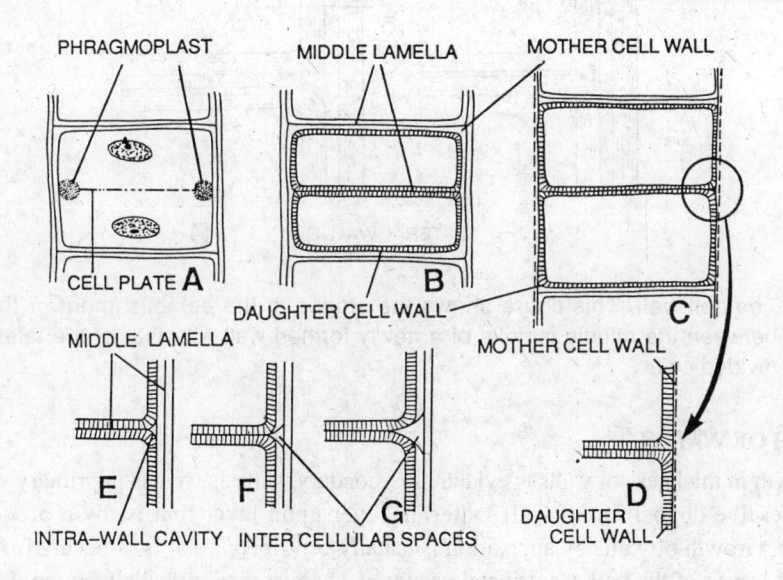

Fig. 3.22. After cell division, concepts regarding the adjustments between new and old cell walls. A, cell plate has been formed; B, two primary walls cemented by intercellular substance occupy the position of cell plate; primary daughter cell walls have been laid down on the inside of primary mother-cell wall; C, D, daughter cells have expanded vertically and the mother cell wall has been stretched and ruptured opposite the new wall which separates the daughter protoplasts; E—G, establishment of continuity between old and new middle lamellae through formation of intercellular space.

Division in cells with a large central vacuole involves predivision changes in form and position of nucleus and in structure of the vacuole. The cytoplasm increases in amount extending in strands through the vacuole; the nucleus becomes more rounded and moves away from the wall along one of the cytoplasmic strands; the strands gradually become concentrated in the *'phragmosome'* in the plane where the new wall is to be formed. Nuclear division and wall formation then take place as in other meristematic cells or cells with enough cytoplasm.

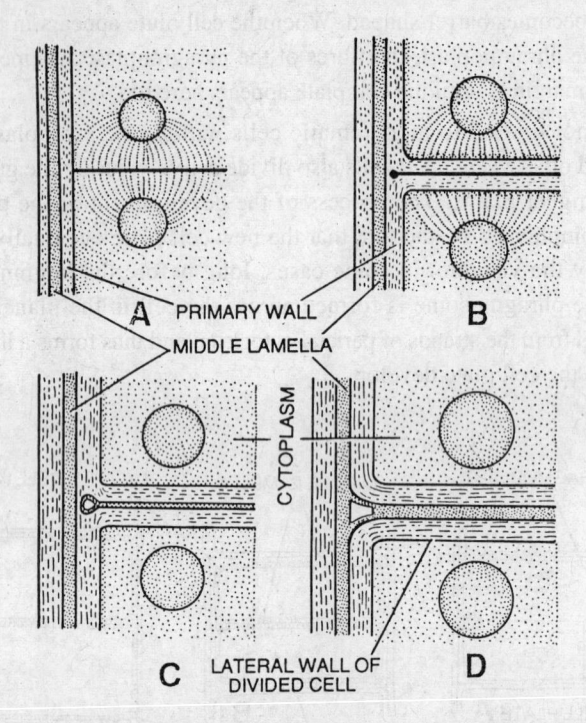

Fig. 3.23. The cell wall. This figure shows the stages in the establishment of the connection between the middle lamella of a newly formed wall with that of the lateral wall of the divided cell.

GROWTH OF WALLS

Growth in thickness of walls is evident in secondary walls as well as in primary walls. It occurs by a successive deposition of wall material, layer upon layer that is, by a process known as *apposition*. Growth of walls by apposition is usually centripetal, that is, it occurs from outside and toward the lumen of the cell. Centripetal growth is characteristic of cells forming tissues. Another important aspect of wall growth establishment of continuity between the new intercellular lamella and that located outside the primary wall of the mother cell. There is an extension and break down of the parent wall opposite the new middle lamella. The formation of intercellular spaces is thought to be associated with this phase of wall growth.

4

The Cell Division

The details of mitotic (mitosis) and meiotic (meiosis) types of cell division have been discussed here.

MITOSIS

The division of the cell is initiated by the division of the nucleus. In the ordinary method of division a nucleus passes through many stages, and the whole complicated process is known as *mitosis*. The details of mitosis were worked out in the later part of the nineteenth century by W. Flaming and others. Commonly this type of cell division is found in the vegetative parts of the plant body. In the process of mitosis prior to the cell division, the number of chromosomes is always duplicated. For example, if a corn plant possesses 20 chromosomes in the somatic cell, then prior to each cell division the 20 chromosomes are duplicated, then the division takes place resulting in the formation of two daughter cells each containing 20 chromosomes again. The different stages of mitosis may easily be recognised in the root apices of onion. From the study point of view, the process of mitosis may be differentiated into two main phases — The *karyokinesis* and the *cytokinesis*. The actual division of the nucleus is known as karyokinesis whereas the division of the cytoplasm of the cell is called cytokinesis. Thus, during the process of mitosis the nucleus undergoes several changes which may easily be studied in the onion root apex by special cytological techniques. The chief function of mitosis seems to be to divide all parts of the chromatin equally between the two daughter nuclei. The important phases of mitosis are — *prophase*, *metaphase*, *anaphase* and *telophase*.

Prophase. In the resting nucleus the chromatin is spread out as a reticulum. It is actually composed of a number of separate units, the *chromosomes*. The number of chromosomes in the nuclei is definite in different species. Gradually the chromosomes become thick and condensed and each of them splits lengthwise forming two *chromatids*. The chromatids remain coiled around each other throughout their length. Gradually, they become much more thick and smooth. The chromatids coil around each other spirally and each chromosome itself remains surrounded by a membrane. In well fixed chromosomes some unstained gaps or constrictions are seen; there are the attachment regions, called *centromeres*. The nucleoli lose their staining power and disappear completely. The nucleus then rapidly passes into the next stage, the metaphase, through a complicated series of changes.

Metaphase. The nuclear membrane disappears and simultaneously a new structure, the *spindle*, appears in the cytoplasm, which chemically, consists of long chain protein molecules oriented longitudinally between two poles. The *fibres* of the spindle, however, are really fine tubules, not just protein threads. Chemical analysis of the cells has indicated that approximately 15 per cent of the

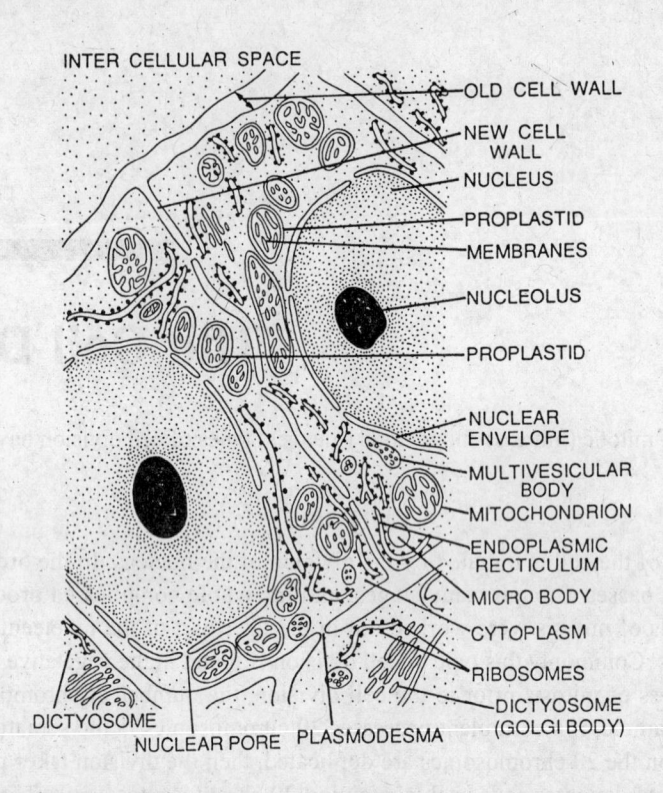

Fig. 4.1. The cell division. Diagram of an electron micrograph of metristematic cells in shoot apex of *Chenopodium album*.

cytoplasmic proteins go into its make-up. Once the spindle is formed, the chromosomes move through the cytoplasm to it, and become fastened by their *centromeres* to a region midway between the poles called the *equator* of the spindle, a position of apparent equilibrium. The centromere of each chromosome always contacts the spindle at the equator; the arms of the chromosomes, not being so restricted, are randomly oriented.

The centromere is the organ of movement. Without it, a chromosome cannot orient on the spindle, and the chromatids cannot separate from each other later. The position of centromere is visible in a chromosome during metaphase by a constriction, and since the position of the constriction is characteristic for each chromosome, the centromere divides the chromosome into two arms of varying lengths. Very few chromosomes have strictly terminal centromeres.

Anaphase. Anaphase follows metaphase. At the end of the metaphase the centromeres of each pair of chromatids appear to repel each other. The centromeres now divide so that each chromatid has its own centromere; they then move apart from each other to initiate a slow movement that will take sister chromatids to opposite poles. Termination of anaphase movement occurs when the chromosomes form a densely packed group at the two poles.

Telophase. As soon as the chromosomes reach the poles, they collect into a more or less solid-appearing mass. This marks the beginning of telophase. The mass of chromosomes gradually

The Cell Division

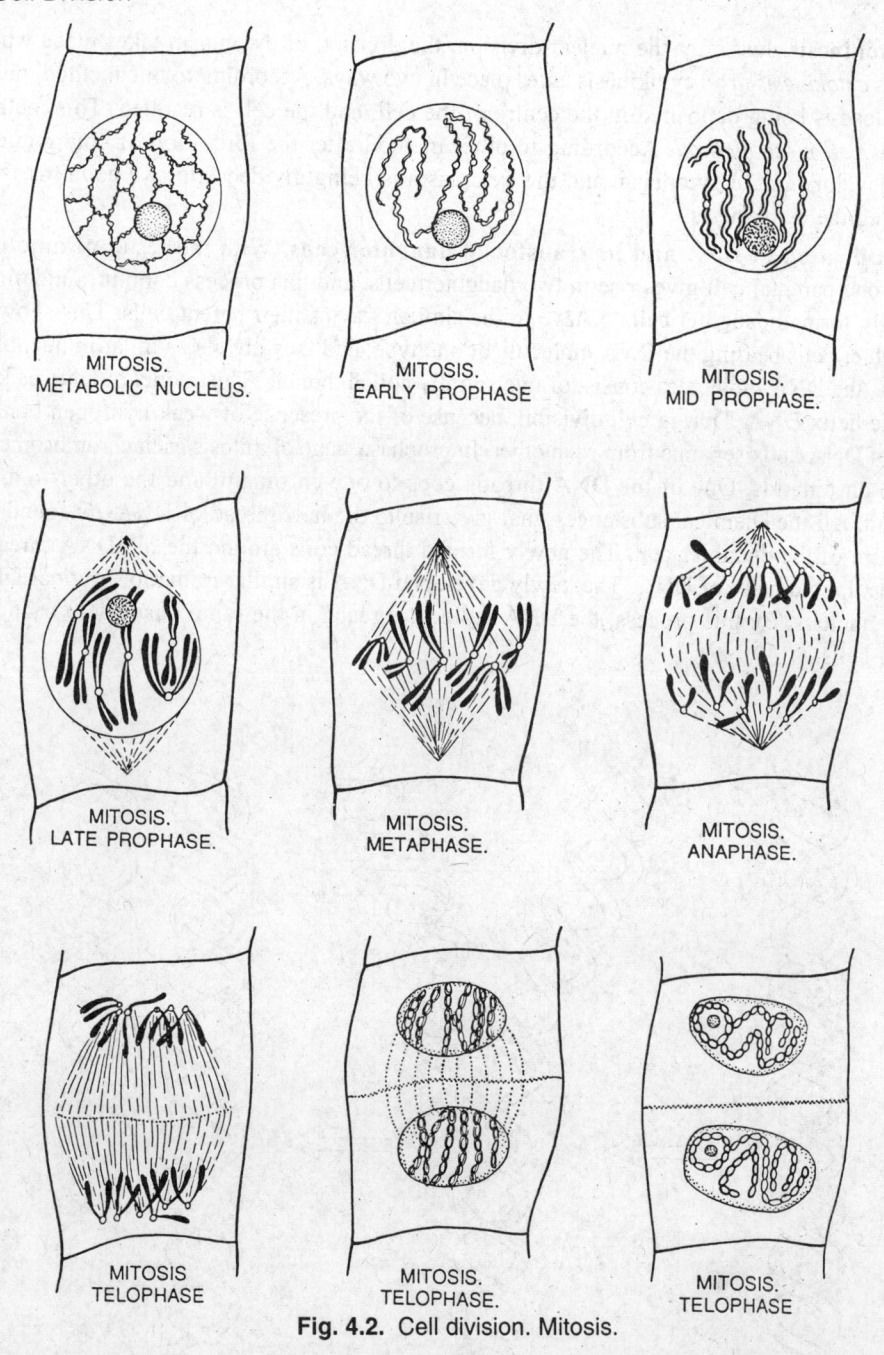

Fig. 4.2. Cell division. Mitosis.

converts into a nucleus. A new nuclear membrane forms. Spindle gradually disappears. The formation and enlargement of the spaces containing nucleoplasm continue until the chromosomes again become scattered in the form of a network typical of the resting stage. As the mass of chromosomes becomes more and more spread on by the formation of nucleoplasm a new nucleolus makes its appearance. The newly formed nucleus contains the same number of chromosomes, as this was in parent nucleus.

Cytokinesis. Just after the nuclear division, the division of cytoplasm takes place which is known as *cytokinesis*. The cytokinesis takes place in two ways. According to one method, much of the cellulose is being deposited in the centre of the cell, and the cell is resulted. This method is known as *cell plate method*. According to other method after the formation of young nuclei, a furrow develops in the cytoplasm and the cytoplasm is being divided into two equal parts, thus completing the cytokinesis.

Duplication of DNA and its transfer to daughter cells. With the result of mitotic cell division one parental cell gives rise to two daughter cells, and this process continues indefinitely. The newly formed daughter cells behave in the similar way as their parent cells. This shows that the daughter cells bearing the DNA molecule of one type, and they are also similar in quantity. As we know, the DNA molecules consist of two spirally coiled threads. This model of DNA is known as double helix DNA. During cell division, because of the presence of weak hydrogen bonds the threads of DNA helix separate from each other. In prophase stage of mitosis each chromosome splits into two chromatids. One of the DNA threads goes to one chromatid and the other to another chromatid. All the chemical substances, that give rise to the new thread of DNA are found in the protoplasm of the daughter cell. The newly formed thread coils around the old DNA thread and forms the double helix of DNA. The newly developed DNA is similar to that of original DNA of parental nucleus. By this process, the DNA molecules reach in the same quantity to each of the daughter cells.

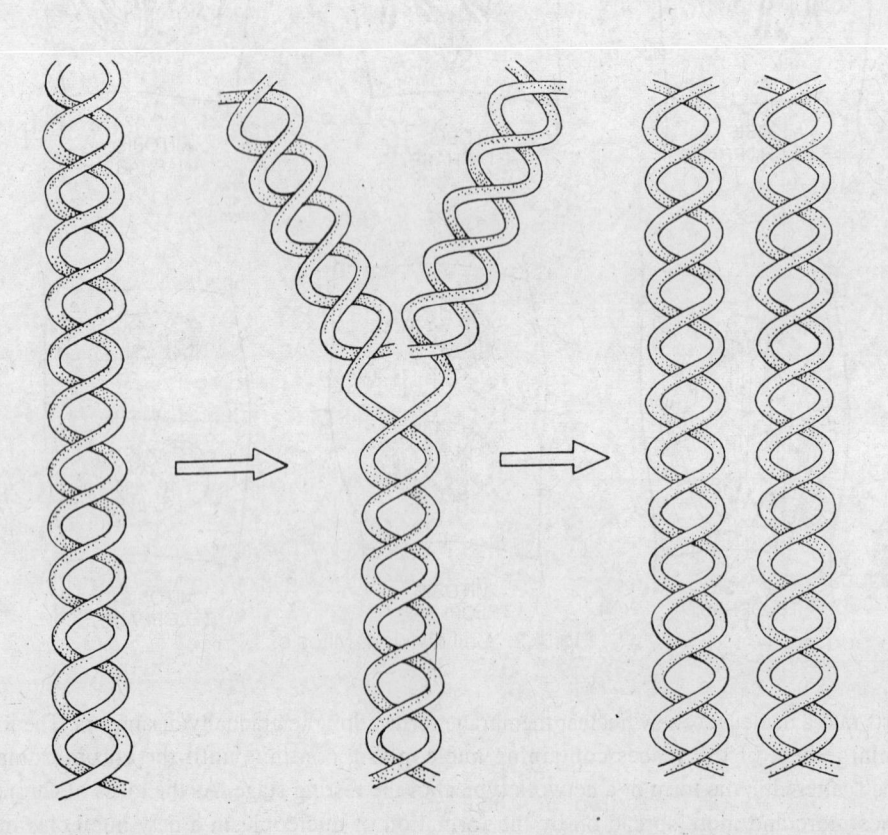

Fig. 4.3. Replication of the DNA molecule. The old helix unwinds (middle) and the two new helices are being formed.

The Cell Division

SIGNIFICANCE OF MITOSIS

With the result of mitosis, the chromosmes split lengthwise into two chromatids. Each chromatid bears all those characteristics which were present in mother chromosome. In other words, with the result of mitosis, two identical cells have the same genetic constitution, qualitatively and quantitatively, as the parental cell from which they arose. Thus, the maintenance of the genetic integrity of the cell population and ultimately of the organism and its descendent depends upon the mechanism of mitosis. This process has been proved to be beneficial to vegetative reproduction. In the similar way, the characters of the plants grown by vegetative reproduction may be preserved for long time.

MEIOSIS

The division during which the number of chromosomes are reduced are known as the reducing or meiotic divisions, and the process of reduction division as *meiosis*. For the first time in 1888,

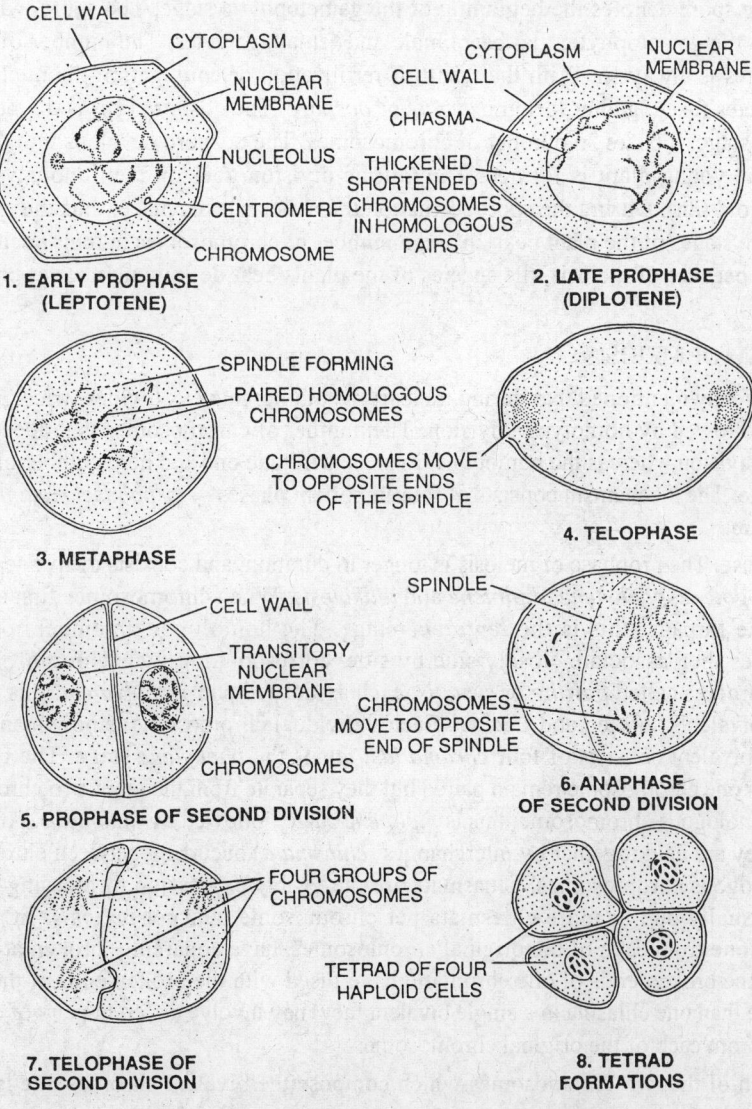

Fig. 4.4. The Cell Division. Stages in Meiosis.

Strasburger discovered this phenomenon. There are two phases in the life-cycle of lower and higher plants — *sporophyte* and *gametophyte*. The sporophyte contains the 2n number of chromosomes, whereas the gametophyte possesses n number of chromosomes. If the nucleus of sporophyte contains 6 chromosomes, then its nucleus of gametophyte contains only 3 chromosomes, *i.e.*, half the number of the chromosomes of the nucleus of the sporophyte. With the result of meiotic division (meiosis), from one nucleus of the sporophyte four nuclei of the gametophyte are formed, and the number of chromosomes for each nucleus becomes half of the number of chromosomes of the nucleus of the sporophyte. In all the plants which are sexually reproduced, the meiosis takes place. In the sporophyte, the spore mother cells represent the last phase of the sporophytic status. Thus, each spore mother cell bears 2n number of chromosomes. Each spore mother cell eventually divides reductionally producing four haploid (n) spores. In such division, the first division is known as *heterotypic division*. The number of chromosomes is reduced to half during this division. The division is immediately followed by another ordinary division, which is known as *homotypic division*. The spore denotes the beginning of the gametophytic stage. The spore, when germinates gives rise to the gametophyte, that bears male and female gametes. The number of chromosomes in the gametes is always n. With the result of fertilization or conjugation the nuclei of male and female gametes fuse together forming *zygote* or *oospore*. The oospore (2n) is the beginning of the sporophytic stage. It bears 2n number of chromosomes. The oospore develops into the embryo and thereafter the young plant is formed. The mitotic division goes on continuously, and the plant remains diploid (2n). During the spore formation the meiosis takes place, thus the sporophytic and gametophytic stages of the plant bear 2n and n number of chromosomes in their nuclei respectively. This is only because of meiosis, the species of the plants bear definite number of chromosomes in their nuclei.

MECHANISM OF MEIOSIS

It is already mentioned that during meiosis the mother nucleus divides twice, *i.e.*, heterotypic division is followed by homotypic division. The number of chromosomes is reduced to half in the heterotypic division whereas the homotypic division is mitotic one and here number of chromosomes remains same. The mechanism consists of four important phases — *prophase, metaphase, anaphase* and *telophase*.

Prophase. The prophase of meiosis is longer in duration, and consists of five separate states — *leptotene, zygotene, pachytene, diplotene* and *diakinesis*. When chromosomes first appear they are separate fine threads; this is the *leptotene* stage. The homologous members of each pair of chromosomes then associate closely side by side, corresponding loci (chromosomes) adhering together, a process called *pairing* or *synapis*; each associated pair is a *bivalent*; this is the *zygotene* stage. The bivalents shorten and thicken. Each individual chromosome is now manifestly double; hence each bivalent consists of four *chromatids*; this is the *pachytene* stage. The two chromatids derived from one chromosome remain paired but they separate from the other two chromatids derived from the homologous chromosome; this is *diplotene* stage, followed by *diakinesis*. At certain places, however, they are held together by interchanges (*chiasmata*) occurring between chromatids derived from homologous chromosomes. Chiasmata are the visible expression of crossing-over of genes. There are usually one or more chiasmata per chromosome pair per meiosis. In a chiasma two chromatids, one from each of the original chromosomes, have apparently broken at corresponding places; and the broken ends of one chromatid have fused with the broken ends of the other. Where there is more than one chiasma in a single bivalent they may involve different pairs of chromatids, but always one from each of the original chromosomes.

As each of the two chromosomes which compose the bivalent chromosome is split into two chromatids, the bivalent chromosome is really a quadruple structure and is called a tetrad chromosome. After the formation of tetrad chromosomes, the nucleus has only half the number of

The Cell Division

MEIOSIS–I LEPTOTENE.	MEIOSIS–I LEPTOTENE.	MEIOSIS–I LEPTOTENE.
MEIOSIS–I ZYGOTENE.	MEIOSIS–I PACHYTENE.	MEIOSIS–I DIPLOTENE.
METOSIS–I DIAKINESIS	MEIOSIS–I METAPHASE–I	MEIOSIS–I ANAPHASE–I	MEIOSIS–I TELOPHASE–I

Fig. 4.5. Cell division. Meiosis I.

| MEIOSIS–II | MEIOSIS–II | MEIOSIS–II | MEIOSIS–II |
| PROPHASE–II | METAPHASE–II | ANAPHASE–II | TELOPHASE–II |

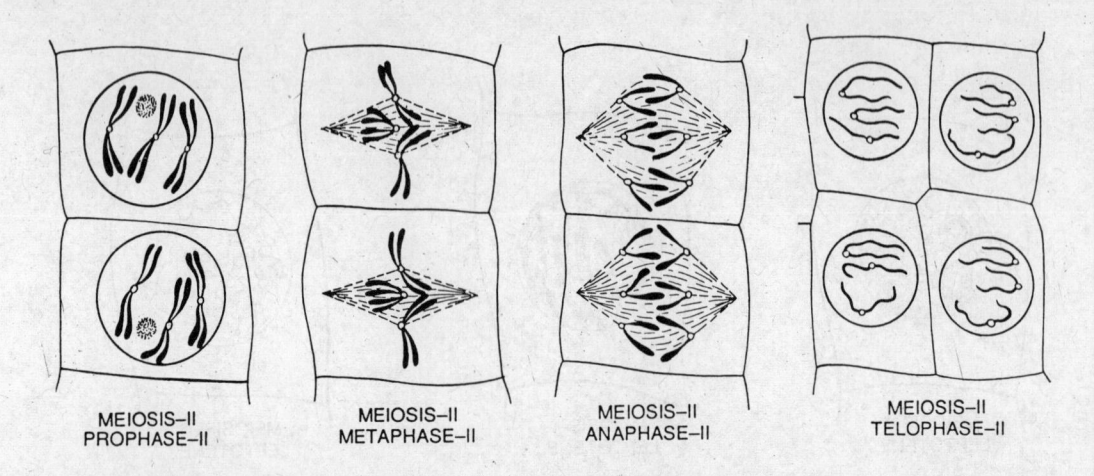

Fig. 4.6. Cell division. Meiosis II.

chromosomes that are found in ordinary vegetative cells. The subsequent events in this and the next division distribute one of the chromatids of the tetrad to each of the four cells.

Metaphase. At metaphase each of tetrad chromosomes is separated longitudinally into two daughter chromosomes which pass to opposite poles. Each of these chromosomes is, however, already split into two chromatids. At metaphase of the second of the reduction divisions these chromatids are separated from each other. We can now see why there are two reduction divisions. This is connected with the fact that the chromosomes which are formed in the prophase of the first meiotic division are quadruple and that it takes two divisions to divide them up into their constituent chromatids. At the end of the second reduction division not only has each chromosome been reduced to a single chromatid but the total number of chromosomes is also haploid (n).

Anaphase. At anaphase two of the four chromatids from each bivalent go to one pole of the spindle, and the other two to the other pole. The chromatids go in pairs because the spindle attachment of each original chromosome has not yet duplicated, so that two chromatids are united at that point. Owing to the effects of chiasmata, however, the united chromatids are not usually derived throughout their length from the same chromosome but are mixtures of one or more pieces from each of the original chromosomes. It is a matter of chance, and quite uninfluenced by behaviour of other bivalents, which spindle attachment with its chromatids, goes to which pole of the spindle.

Telophase. After anaphase there may be a short telophase and resting stage, or the second meiotic division may follow immediately, usually in both the daughter-cells formed by the first division. It is like a mitosis, except that it starts prophase with only half the normal number of chromosomes, each already divided into two chromatids before the previous prophase. The two chromatids separate at anaphase, one going to each daughter cell. Again, the distribution of the chromatids is a matter of chance, except that the two from each chromosome must go to opposite poles. On this chance distribution of the chromatids at both anaphase depends the law of independent assortment. Each daughter-cell, therefore, eventually receives only the haploid number of chromosomes. The reduction from the diploid number of the original cell to the haploid number of the four products is one of the bases of genetic segregation.

SIGNIFICANCE OF MEIOSIS

During the process of fertilization the male nucleus and the egg nucleus come together and the contents of the two nuclei come to be enclosed by a single nuclear membrane, a part of which comes from the male nucleus and a part from the egg nucleus. In this way, the chromosomes which were in

The Cell Division

the male nucleus are added to those which were in the egg nucleus. The fertilized egg thus contains the chromosomes of both the egg nucleus and the male nucleus.

If the same number of chromosomes as are found in the vegetative nuclei went into the egg and sperm nuclei, the fertilized egg of each generation would contain twice as many chromosomes as the nuclei of the preceding generation. This result is prevented by the presence of two successive divisions in the life cycle in which the number of chromosome is reduced to half (meiosis) that found in the ordinary vegetative nuclei.

The characters of the species are determined by means of genes present on the chromosomes, and therefore, to establish the characters of species, it becomes essential that the number of chromosomes may remain fixed, and this is possible only by means of meiosis.

COMPARISON OF MITOSIS AND MEIOSIS

Mitosis	*Meiosis*
1. It takes place in all the somatic cells, except those of spore mother cells, or germinal cells.	1. It always takes place either in the spore mother cells or in germinal cells.
2. The whole process is being completed in one sequence; two daughter cells or nuclei are produced from the mother cell.	2. The whole process is being completed in two sequences; with the result four daughter cells or nuclei are produced from one spore mother cell or germinal cell.
3. The daughter cells are identical to the parent cell as each cell possesses the similar number of chromosomes, *i.e.*, diploid (2n).	3. The daughter cells are dissimilar to that of parent cells as crossing over takes place and the number of chromosomes is reduced to half in each of daughter cells or nuclei, *i.e.*, haploid (n).
4. The prophase stage is of short duration and it does not bear any sub-stages. From the very beginning the chromosomes are double structures each formed of two chromatids. The split homologous chromosomes are not attracted towards each other. There is no chiasma formation and the crossing over does not take place. Synapsis does not occur.	4. The prophase is of long duration and it consists of several sub-stages. The chromosomes are granular and single structures in the beginning. In zygotene and later sub-stages, the homologous chromosomes attract towards each other forming bivalents. Each chromosome splits into two but the chromatids are not separated. The chromatids of one chromosome exchange with the chromatids of the second chromosome during the process of diplotene. The process of chiasma formation and crossing over takes place during diplotene. Synapsis takes place in the homologous chromosomes.
5. During metaphase the centromere of each chromosome divides, and the chromatids travel towards the opposite poles. No	5. During metaphase the centromeres of chromosome do not divide, and the homologous chromosomes travel towards

change takes place in the number of chromosomes and the resultant daughter cells remain similar to parent cells. In the last stage of metaphase, the chromatids are separated from each other in the divided parts of centromeres. This separation takes place because of repulsion.	the opposite poles themselves. The number of chromosomes travelling towards one of the poles remain half to the normal number. The resultant daughter cells are dissimilar to parent cell. The maternal and paternal chromosomes are separated because of repulsion of homologous chromosomes.
6. In anaphase the chromosomes travel upto poles.	6. As in mitosis.
7. As the result of telophase the two nuclei bear diploid (2n) number of chromosomes.	7. Two haploid (n) nuclei are formed during first telophase stages.
8. There is no second division in mitosis.	8. The second division occurs in meiosis. The reduction division takes place during the first telophase, whereas during second telophase ordinary mitotic division takes place.
9. As the result of mitosis, the number of chromosomes, their characters and constitution do not change.	9. As the result of meiosis, four daughter cells are formed. Each daughter cell or nucleus is haploid (n), *i.e.*, it contains half number of the chromosomes, thus the characters and the constitution of the chromosomes also change.
10. Cytokinesis always takes place after each division.	10. Cytokinesis does not necessarily occur in telophase one, normally it occurs in second telophase.

5

The Tissue

Groups or masses of the cells that are alike in origin, structure, and function form tissue. The plant body consists of *vegetative tissue* and *reproductive tissue*. In the higher plants the plant body is somewhat more complex in its cellular structure. The cells differ very much in their kind, form and origin in higher plants. Morphologically, a tissue is a group or a mass of cells or vessels, alike in form, origin and function. From the study point of view, the tissues may be grouped into three principal groups — 1. *Meristems* or *meristematic tissue*; 2. *Permanent tissue*, and 3. *Secretory tissue*.

1. MERISTEMS OR MERISTEMATIC TISSUE

A meristematic tissue consists of a group of cells which remain in a continuous state of division or they retain their power of division. The characteristic features of meristematic tissue are as follows:

1. They are composed of immature cells which are in a state of division and growth.
2. Usually the intercellular spaces are not found among these cells.
3. The cells may be rounded, oval or polygonal in shape; they are always living and thin-walled.
4. Each cell of meristematic tissue possesses abundant cytoplasm and one or more nuclei in it.
5. The vacuoles in the cells may be quite small or altogether absent.

MERISTEMS AND GROWTH OF PLANT BODY

Beginning with the division of the oospore, the vascular plant generally produces new cells and forms new organs until it dies. In the beginning of the development of the plant embryo cell division occurs throughout the young organism. But as soon as the embryo develops and converts into an independent plant the addition of new cells is gradually restricted to certain parts of the plant body, while the other parts of the remain concerned with activities other than growth. This shows that the portions of embryonic tissue persists in the plant throughout its life, and the mature plant is a composite of adult and juvenile tissues. These juvenile tissues are known as the *meristems*. The presence of meristems remarkably differentiates the plant, from the animal. In the growth resulting from meristematic activity is possible throughout the life of the organism, whereas in animal body the multiplication of the cells mostly ceases when the organism attains adult size and the number of organs is fixed.

The term meristem (Greek *meristos*, meaning divisible) emphasizes the cell-division activity characteristic of the tissue which bears this name. It is obvious that the synthesis of new living substance is a fundamental part of the process of the formation of new cells by division. The living tissues other than the meristems may also produce new cells, but the meristems carry on such activity indefinitely, because they not only add cells to the plant body, but also perpetuate themselves, that is, some of the products of division in the meristems do not develop into adult cells but remain meristematic.

The meristems usually occur at the apices of all main and lateral shoots and roots and thus their number in a single plant becomes quite large. In addition, plants bearing secondary increase in thickness possess extensive meristems, the vascular and cork cambia, responsible for the secondary growth. The combined activities of all these meristems give rise to a complex and large plant body. The primary growth, initiated in the apical meristems expands the plant body and produces the reproductive parts. On the other hand, the cambia, aid in maintenance of the expanding body by increasing the volume of the conducting system and forming supporting and protecting cells.

CLASSIFICATION OF MERISTEMS

Various systems of classifying meristems have been proposed by many eminent workers which are based on the characteristics such as stage of development, position in plant body, origin, function and topography. No system is exclusive and rigid. A few important types have been discussed here:

1. Meristems Based on Stage of Development

Promeristem or primordial meristem. Promeristem is the region of new growth in a plant body where the foundation of new organs or parts of organs is initiated. Sometimes, it is also called *primordial meristem*, *urmeristem* and *embryonic meristem*. From the view point of its structure, this region consists of the initials and their immediate derivatives. The cells of this region are isodiametric, thinwalled, vacuolate, with active cytoplasm and early stages of pits. Prominent nuclei and inconspicuous intercellular spaces may be seen. As soon as the cells of this region begin to change in size, shape, and character of wall and cytoplasm, setting off the beginning of tissue differentiation, they are no longer a part of typical meristem; they have passed beyond that earliest stage.

2. Meristems Based on Origin of Initiating Cells

Primary and secondary meristems. The meristems are classified as *primary* and *secondary*, on the basis of type of tissue in which origin occurs.

The primary meristems are those that build up the primary part of the plant and consist in part of promeristem. In primary meristems, promeristem is always the earliest stage. The possession of promeristem continuously from a early embryonic origin is characteristic of primary meristems. The main primary stems are the apices of roots, stems, leaves and similar appendages.

The secondary meristem appears later at a stage of development of an organ of a plant body. Secondary meristems always arise in permanent tissues and they are always found lying lateral along the side of the stem and root. Sometimes, some of the primary permanent tissues acquire the power of division and become meristematic. These tissues build up the secondary meristem. Secondary meristems are so called because they arise as new meristems in tissue which is not meristematic. The most striking example of secondary meristem is phellogen or cork cambium. It is formed from mature cells — cortical, epidermal or phloem cells.

The primary meristems build up the early and structurally and functionally complete plant body. The secondary meristems later add to that body forming supplementary tissues that functionally replace the early formed tissues or serve in protection and repair of wounded regions.

The Tissue

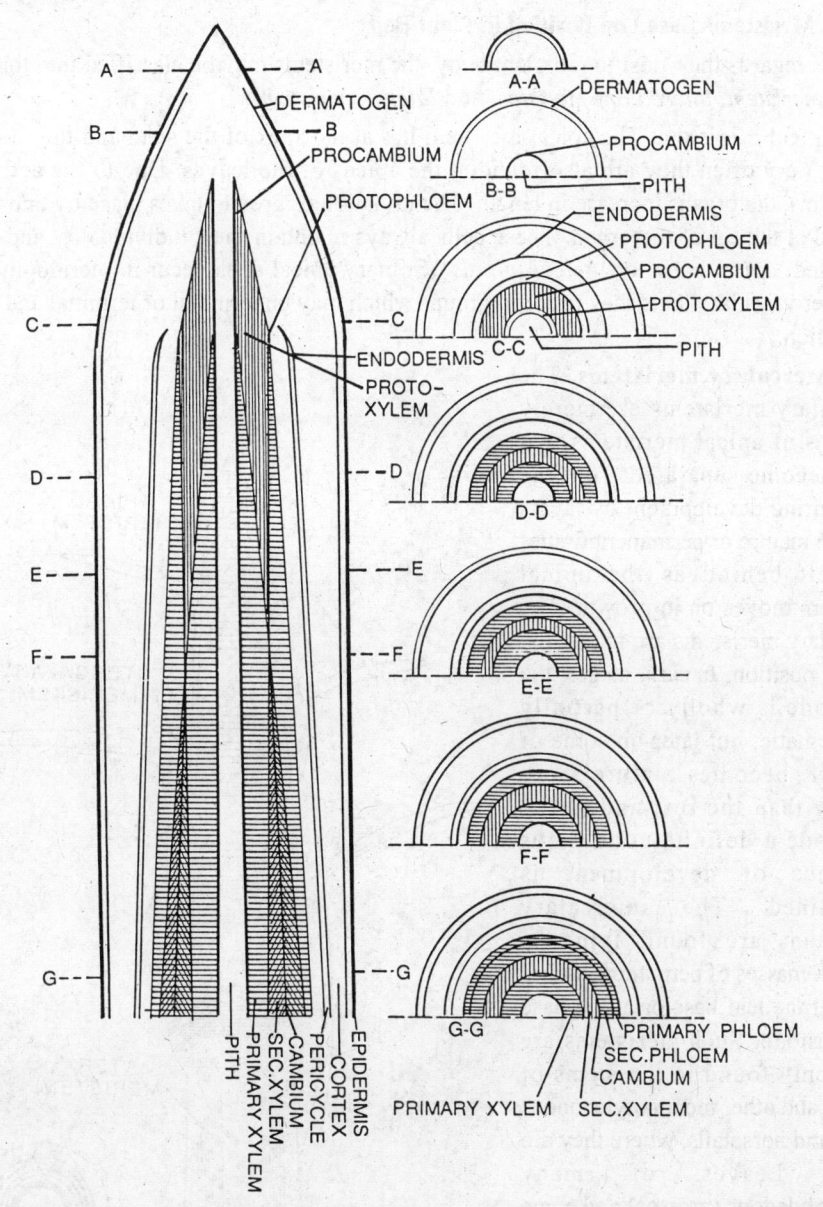

Fig. 5.1. Development of the stem. Longitudinal and cross-section of tip of elongating axis. At A – A, entire axis promeristem; B – B, dermatogen, procambium and pith in early stages of development; C – C, procambium formed a complete cylinder, outermost and innermost procambium has become protophloem and protoxylem; D – D, large portions of procambium adjacent to the protophloem and protoxylem become phloem and xylem; E – E, remaining layer of procambium becomes the cambium and forms the first secondary phloem and secondary xylem cells; F – F, secondary tissues increased in amount; G – G, further secondary growth occurred. (After Eames and MacDaniels).

The cambium does not fall definitely in either group (primary and secondary). It arises from apical meristem of which it is late and specialized stage. However, the accessory cambia are secondary. The tissues formed by the cambium are secondary, whereas the primary meristems form only primary tissues.

3. Meristems Based on Position in Plant Body

As regards their position in plant body, the meristems may be classified into three groups — *apical meristem*, *intercalary meristem* and *lateral meristem*.

Apical meristem. The apical meristem lies at the apex of the stem and the root of vascular plants. Very often they are also found at the apices of the leaves. Due to the activity of these meristems, the organs increase in length. The initiation of growth takes place by one or more cells situated at the tip of the organ. These cells always maintain their individuality and position and are called '*apical cells*' or '*apical initials*'. Solitary apical cells occur in pteridophytes, whereas in higher vascular plants they occur in groups which may be terminal or terminal and sub-terminal in position.

Intercalary meristems. The intercalary meristems are merely portions of apical meristems that have become separated from the apex during development by layers of more mature or permanent tissues and left behind as the apical meristem moves on in growth. The intercalary meristems are internodal in their position. In early stages, the internode is wholly or partially meristematic, but later on some of its part, becomes mature more rapidly than the rest and in the internode a definite continuous sequence of development is maintained. The intercalary meristems are found lying in between masses of permanent tissues either at the leaf base or at the base of internode. Such meristems are commonly found in the stems of grasses and other monocotyledonous plants and horsetails, where they are basal. Leaves of many monocotyledons (grasses) and some other plants, such as *Pinus*, have basal meristematic regions. These meristematic regions are short living and ultimately disappear, ultimately, they become permanent tissues.

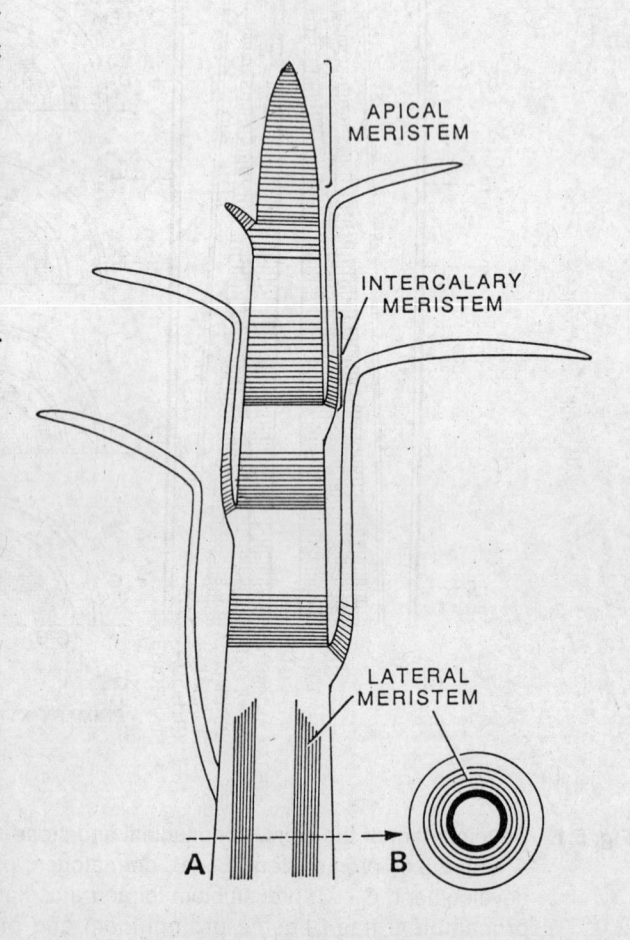

Fig. 5.2. Position of meristems. A, longitudinal view; B, cross-section.

Lateral meristems. The lateral meristems are composed of such initials which divide mainly in one plane (periclinally) and increase the diameter of an organ. They add to the bulk of existing tissues or give rise to new tissues. These tissues are responsible for growth in thickness of plant body. The cambium and the cork cambium are the examples of this type.

4. Meristems Based on Function

As regards their function a system of classification of meristems was proposed by Haberlandt in the end of nineteenth century. He suggested that the primary meristem at the apex of the stem and root is distinguished into three tissues — *protoderm*, *procambium* and *ground* or *fundamental meristem*. The protoderm is the outermost tissue which develops into epidermis. The procambium develops into primary vascular tissue. It forms isolated strands of elongated cells very near to the central region; in cross-section each procambium appears as a small group of cells in the ground or fundamental meristem, but in longitudinal section the cells appear to be long and pointed. The ground or fundamental meristem develops into ground tissue and pith; the cells of this region are large, thin walled, living and isodiametric. In later stages, they become differentiated into hypodermis, cortex, endodermis, pericycle, pith rays and pith.

Meristem and meristematic. The terms meristem and meristematic, as applied to developing cells and tissues, are somewhat loose. According to Eames and MacDaniels, the term '*meristem*' is applied to regions of more or less continuous cell and tissue initiation; the adjective '*meristematic*' is used to indicate resemblance in an important way to a meristem, but not necessarily as consisting of or constituting meristem, *i.e.*, it is applied to those cells, tissues and regions that have characteristics of developing structures — especially cell division — but do not themselves strictly constitute meristems. For example, the apices of the stems and the cambium are regions of tissue initiation, developing xylem and phloem are meristematic tissues, because they form some new cells and are immature but they are not permanent or semipermanent initiating regions (meristems). On the other hand, cells in mature tissue, such as the primary cortex of stems, may divide. Such cells are meristematic, but neither they nor the tissues of which they are a part constitute a meristem.

MERISTEMS AND PERMANENT TISSUES

In active meristems there occurs a continuous separation between cells that remain meristematic — the *initiating cells* — and those that develop into the various tissue elements — the *derivatives* of the initiating cells. During this process of developing the derivatives gradually change, physiologically and morphologically, and assume more or less specialized characteristics. In other words, the derivatives differentiate into the specific elements of the various tissue-systems. As the cells of vascular plants vary in their physiologic and morphologic characteristics, they also vary in details of differentiation. Different types of cells attain different degrees of differentiation as compared with their common meristematic precursors. For example, various parenchyma cells diverge relatively little from their meristematic precursors and retain the power of division to a high degree. On the other hand, such as sieve elements, fibres, tracheary elements are more thoroughly modified and lose most, or all of their former meristematic potentialities. These variously differentiated cells are known as *mature* or *permanent* in the sense that they have reached the degree of specialization and physiological stability that normally characterizes them as components of certain tissues of an adult plant part. During the differentiation of tissues from meristems the derivatives of meristematic cells synthesize protoplasm, enlarge and divide. It is difficult to delimit the meristem proper from its recent derivatives. The development of meristematic derivatives into mature cells also is gradual. In other words, the differentiation is a continuous process.

2. PERMANENT TISSUES

The permanent tissues are those in which growth has stopped either completely or for the time being. Sometimes, they again become meristematic partially or wholly. The cells of these tissues may be living or dead and thin-walled or thick-walled. The thin-walled permanent tissues are generally living whereas the thick-walled tissues may be living or dead. The permanent tissues may be *simple* or *complex*. A simple tissue is made up of one type of cells forming a uniform of homogeneous system of cells. The common simple tissues are — *parenchyma, collenchyma* and *sclerenchyma*. A complex tissue is made up of more than one type of cells working together as a unit. The complex tissues consist of parenchymatous and sclerenchymatous cells; collenchymatous cells are not present in such tissues. The common examples are — the *xylem* and the *phloem*.

SIMPLE TISSUES

1. Parenchyma. *The parenchyma tissue is composed of living cells which are variable in their morphology and physiology, but generally having thin walls and a polyhedral shape, and concerned with vegetative activities of the plant.* The individual cells are known as *parenchyma cells*. The word parenchyma is derived from the Greek *para*, beside and *enchein*, to pour. This combination of words expresses the ancient concept of parenchyma as a semi-liquid substance *poured beside* other tissues which are formed earlier and are more solid. Phylogenetically the parenchyma is a primitive tissue since the lower plants have given rise to the higher plants through specialization and since the single type or the few types of cells found in the lower plants have become by specialization the many and elaborate types of the higher plants. The unspecialized meristematic tissue is parenchyma and is often called parenchyma thus it can be said that, ontogenetically parenchyma is a primitive tissue.

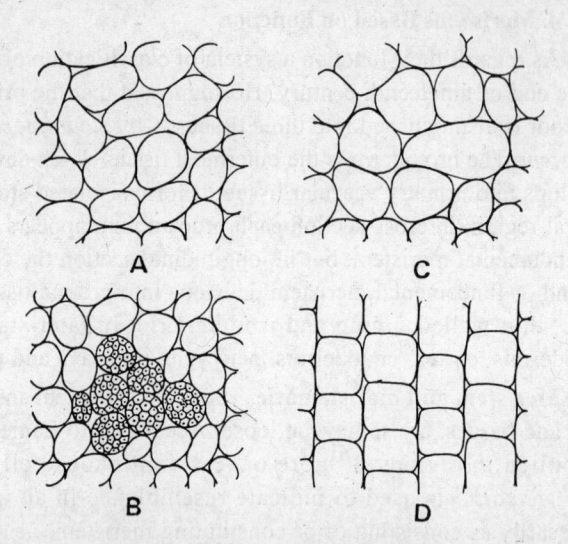

Fig. 5.3. Parenchyma. A, from pith of rhizome of *Polypodium*; B, from cortex of root of *Asclepias*, the cells containing starch grains; C and D, from pith of *Zea* in transverse and longitudinal sections.

The parenchyma consists of isodiametric, thin-walled and equally expanded cells. The parenchyma cells are oval, rounded or polygonal in shape having well developed spaces among them. The cells are not greatly elongated in any direction. The cells of this tissue are living and contain sufficient amount of cytoplasm in them. Usually each cell possesses one or more nuclei.

Parenchyma makes up large parts of various organs in many plants. Pith, mesophyll of leaves, the pulp of fruits, endosperm of seeds, cortex of stems and roots, and other organs of plants consist mainly of parenchyma. The parenchyma cells also occur in xylem and phloem.

In the aquatic plants, the parenchyma cells in the cortex possess well developed air spaces (intercellular spaces) and such tissue is known as *aerenchyma*. Parenchyma may be specialized as water storage tissue in many succulent and xerophytic plants. In *Aloe, Agave, Mesembryanthemum, Hakea* and many other plants chlorophyll-free, thin-walled and water-turgid cells are found which represent water storage tissue. When the parenchyma cells are exposed to light they develop chloroplasts in them, and such tissue is known as *chlorenchyma*. The chlorenchyma possesses well developed aerating system. Intercellular spaces are abundant in the photosynthetic parenchyma (chlorenchyma) of stems too.

The Tissue

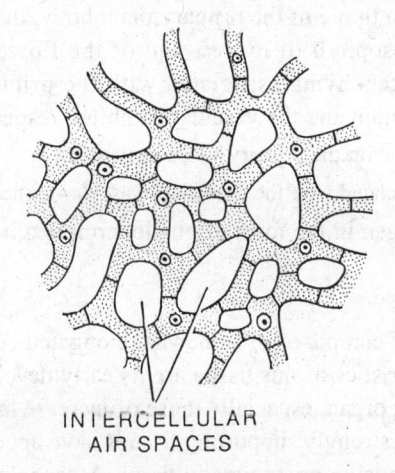

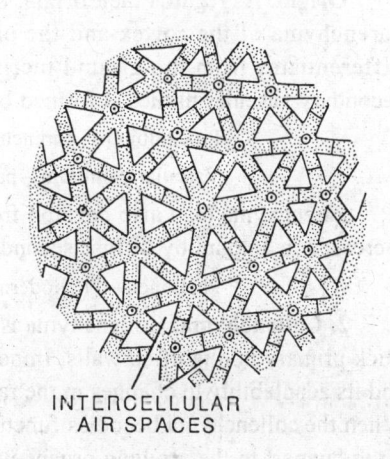

Fig. 5.4. Aerenchyma—in the petiole of *Canna*. **Fig. 5.5.** Aerenchyma—in the petiole of *banana*.

Commonly the parenchyma cells have thin primary walls. Some such cells may have also thick primary walls. Some storage parenchyma develop remarkably thick walls and the carbohydrates deposited in these walls, the hemicellulose, are regarded by some workers as reserve materials. Thick walls occur, in the endosperm of *Phoenix dactylifera*, *Diospyros*, *Asparagus* and *Coffea arabica*. The walls of such endosperm become thinner during germination.

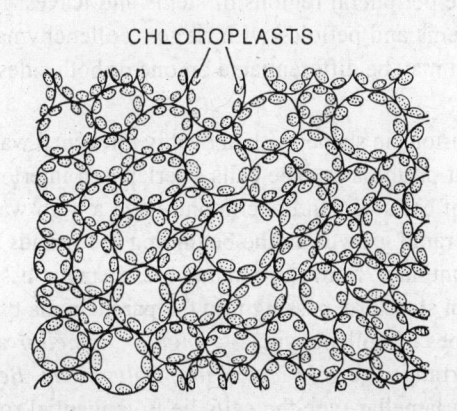

Fig. 5.6. Palisade parenchyma. Section cut parallel with the epidermis. Each cell contains many chloroplasts.

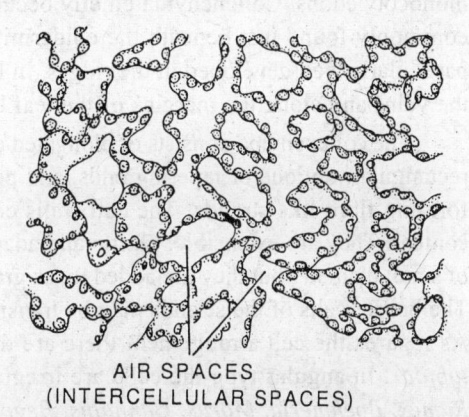

Fig. 5.7. Spongy chlorenchyma. Section cut parallel with the epidermis showing irregular cells with chloroplasts and well developed intercellular spaces.

The turgid parenchyma cells help in giving rigidity to the plant body. Partial conduction of water is also maintained through parenchymatous cells. The parenchyma acts as special storage tissue to store food material in the form of starch grains, proteins, fats and oils. The parenchyma cells that contain chloroplasts in them make chlorenchyma which are responsible for photosynthesis in green plants. In water plants the aerenchyma keep up the buoyancy of the plants. Such air spaces also facilitate exchange of gases. In many succulent and xerophytic plants such tissues store water and known as water storage tissue. Vegetative propagation by cuttings takes place because of meristematic potentialities of the parenchyma cells which divide and develop into buds and adventitious roots.

Origin. As regards their origin, the parenchyma tissue of the primary plant body, that is, the parenchyma of the cortex and the pith, of the mesophyll of leaves, and of the flower parts, differentiates from the ground meristem. The parenchyma associated with the primary and secondary vascular tissues is formed by the procambium and the vascular cambium respectively.

Procambium—parenchyma associated with the primary vascular tissues.

Vascular cambium—parenchyma associated with the secondary vascular tissues.

Parenchyma may also develop from the phellogen in the form of phelloderm, and it may be increased in amount by diffuse secondary growth.

Phellogen—Phelloderm (parenchyma).

2. Collenchyma. Collenchyma is a living tissue composed of somewhat elongated cells with thick primary nonlignified walls. Important characteristics of this tissue are its early development and its adaptability to changes in the rapidly growing organ, especially those of increase in length. When the collenchyma becomes functional, no other strongly supporting tissues have appeared. It gives support to the growing organs which do not develop much woody tissue. Morphologically, collenchyma is a simple tissue, for it consists of one type of cells.

Collenchyma is a typical supporting tissue of growing organs and of those mature herbaceous organs which are only slightly modified by secondary growth or lack such growth completely. It is the first supporting tissue in stems, leaves and floral parts. It is the main supporting tissue in many dicotyledonous leaves and some green stems. Collenchyma may occur in the root cortex, particularly, if the root is exposed to light. It is not found in the leaves and stems of monocotyledons. Collenchyma chiefly occurs in the peripheral regions of stems and leaves. It is commonly found just beneath the epidermis. In stems and petioles with ridges, collenchyma is particularly well developed in the ridges. In leaves it may be differentiated on one or both sides of the veins and along the margins of the leaf blade.

The collenchyma consists of elongated cells, various in shape, with unevenly thickened walls, rectangular, oblique or tapering ends, and persistent protoplasts. The cells overlap and interlock, forming fibre-like strands. The cell walls consist of cellulose and pectin and have a high water content. They are extensible, plastic and adapted to rapid growth. In the beginning the strands are of small diameter but they are added to, as growth continues, from surrounding meristematic tissue. The border cells of the strands may be transitional in structure, passing into the parenchyma type. As regards the cell arrangement there are three types of collenchyma — *angular, lamellar* and *tubular*. In angular type the cells are irregularly arranged (*e.g., Ficus, Vitis, Polygonum, Beta, Rumex, Boehmeria, Morus, Cannabis, Begonia*); in lamellar type the cells lie in tangential rows (*e.g., Sambucus, Rheum, Eupatorium*) and in tubular type the intercellular spaces are present (*e.g.,* Compositae, *Salvia, Malva, Althaea*). The common typical condition, is that with thickenings at the corners. The three forms of collenchyma have been named by Muller (1890) angular (Eckencollenchym), lamellar (Plattencolenchym), and tubular or lacunate (Luchencollenchym), respectively. The word lamellar has reference to the plate like arrangement of the thickenings; the lacunate (tubular) to the presence of intercellular spaces.

The walls of collenchyma are chiefly composed of cellulose and pectic compounds and contain much water (Majumdar and Preston, 1941). In some species collenchyma walls possess an alternation of layers rich in cellulose and poor in pectic compounds with layers that are rich in pectic compounds and poor in cellulose. In many plants collenchyma is a compact tissue lacking intercellular spaces. Instead, the potential spaces are filled with intercellular material (Majumdar, 1941).

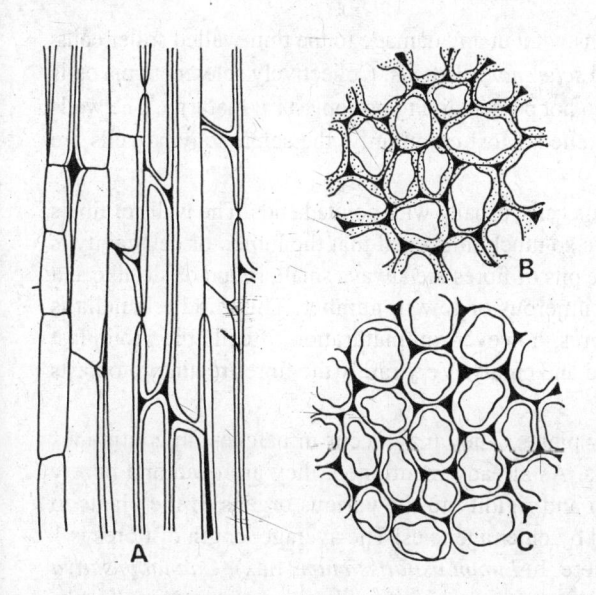

Fig. 5.8. Collenchyma. A and B, longitudinal and transverse sections from stem of *Solanum tuberosum*; C, transverse section from stem of *Abutilon*.

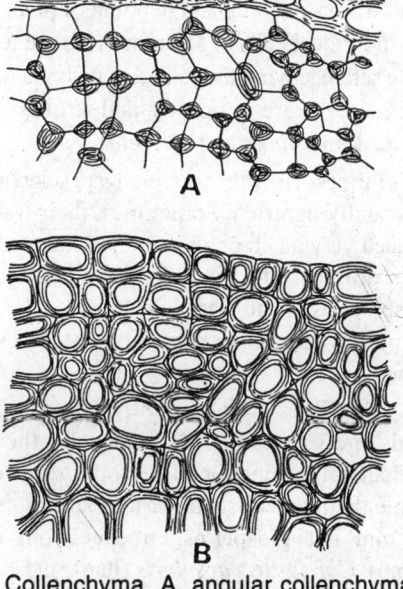

Fig. 5.9. Collenchyma. A, angular collenchyma of *Cucurbita* stem (thickenings in the angles); B, lacunar collenchyma of *Lactuca* stem where intercellular spaces are present and the thickenings are located next to these spaces.

The mature collenchyma cells are living and contain protoplasts. Chloroplasts also occur in variable numbers. They are found abundantly in collenchyma which approaches parenchyma in form. Collenchyma consisting of long narrow cells contains only a few small chloroplasts or none. Tannins may be present in collenchyma cells.

Ontogenetically collenchyma develop from elongate, procambium like cells that appear very early in the differentiating meristem. In the beginning, small intercellular spaces are present among these cells, but they disappear in angular and lamellar types as the cells enlarge, either by the enlarging cells or filled by intercellular substance.

The chief primary function of the tissue is to give support to the plant body. Its supporting value is increased by its peripheral position in the parts of stems, petioles and leaf mid-ribs. When the chloroplasts are present in the tissue, they carry on photosynthesis.

3. Sclerenchyma. The sclerenchyma (Greek, *sclerous*, hard; *enchyma*, an infusion) consists of thick walled cells, often lignified, whose main function is mechanical. This is a supporting tissue that withstands various strains which

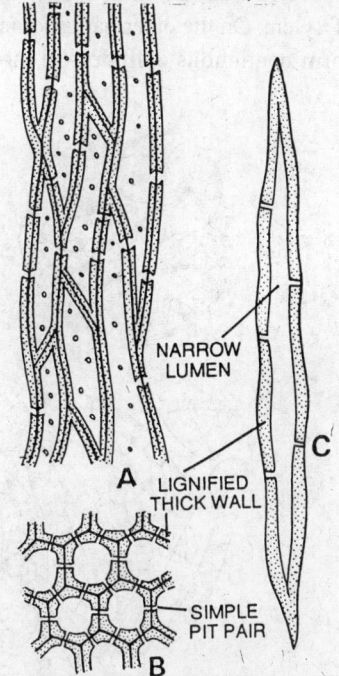

Fig. 5.10. Sclerenchyma. A, L.S. of fibres; B, T.S. of fibres; C, a single fibre as seen in longitudinal section.

result from stretching and bending of plant organs without any damage to the thin-walled softer cells. The individual cells of sclerenchyma are termed *sclerenchyma cells*. Collectively sclerenchyma cells make sclerenchyma tissue. Sclerenchyma cells do not possess living protoplasts at maturity. The walls of these cells are uniformly and strongly thickened. Most commonly, the sclerenchyma cells are grouped into *fibres* and *sclereids*.

Fibres. The fibres are elongate sclerenchyma cells, usually with pointed ends. The walls of fibres are usually lignified. Sometimes, their walls are so much thickened that the lumen or cell cavity is reduced very much or altogether obliterated. The pits of fibres are always small, round or slit-like and often oblique. The pits on the walls may be numerous or few in number. The middle lamella is conspicuous in the fibres. In most kinds of fibres, however, on maturation of cells the protoplast disappears and the permanent cell becomes dead and empty. Very rarely the fibres retain protoplasts in them.

The fibres are abundantly found in many plants. They may occur in patches, in continuous bands and sometimes singly among other cells. As already mentioned, they are dead and purely mechanical in function. They provide strength and rigidity to the various organs of the plants to enable them to withstand various strains caused by outer agencies. The average length of fibres is 1 to 3 mm. in angiosperms, but exceptions are there. In *Linum usitatissimum* (flax), *Cannabis sativa* (hemp), *Corchorus capsularis* (jute), and *Boehmeria nivea* (ramie), the fibres are of excessive lengths ranging from 20 mm. to 550 mm. Such long, thick-walled and rigid cells constitute exceptionally good fibres of commercial importance. In addition to these plants common long fibre yielding plants are — *Hibiscus cannabinus* (Madras hemp), *Agave sisalana* (sisal hemp), *Sansevieria* and many others.

The fibres are divided into two large groups — *xylem fibres* and *extraxylary fibres*. The xylem fibres develop from the same meristematic tissues as the other xylem cells and constitute an integral part of xylem. On the other hand, some of the extraxylary fibres are related to the phloem. The fibres that form continuous cylinders in monocotyledonous stems arise in the ground tissue under the

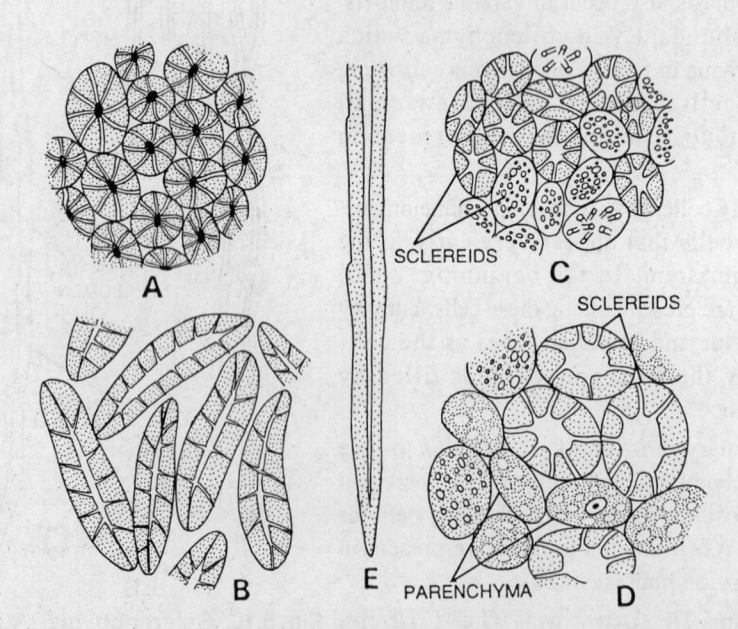

Fig. 5.11. Sclerenchyma. A and B, transverse and longitudinal sections of the sclereids from endocarp of *Cocos nucifera*; C, sclereids from the pericarp of *Pyrus communis*; D, sclereids and parenchyma cells from stem cortex of *Dracaena fragrans*; E.L.S. of fibres (After Eames and MacDaniels).

epidermis at variable distances. They are known as *cortical fibres*. The fibres forming sheaths around the vascular bundles in the monocotyledonous stems arise partly from the same procambium as the vascular cells, partly from the ground tissue. The fibres present in the peripheral region of the vascular cylinder, often close to the phloem are known as *pericyclic fibres*. The extraxylary fibres are sometimes combined into a group termed *bast fibres*. Generally the term extraxylary fibres is used for bast fibres, which are classified as follows — *phloem fibres*, fibres originating in primary or secondary phloem; *cortical fibres*, fibres originating in the cortex; *perivascular fibres* (Van Fleet, 1948), fibres found in the peripheral region of the vascular cylinder inside the innermost cortical layer but not originating in the phloem. The extraxylary fibres may vary in length, and their ends are sometimes blunt, rather than tapering, and may be branched. The longest fibres (primary phloem fibres) measured *Boehmeria nivea* (ramie). The cell walls of extraxylary fibres are very thick. The pits are simple or slightly bordered. Some possess lignified walls, others non-lignified. The fibres of *Linum usitatissimum* are non-lignified, and their secondary walls consist of pure cellulose. On the other hand, the extraxylary fibres of the monocotyledons, are strongly lignified. Concentric lamellations are found in extraxylary fibres. In the fibres of *Linum usitatissimum* the individual lamellae vary in thickness from 0.1μ to 0.2μ.

Xylem fibres typically possess lignified secondary walls. They vary in size, shape, thickness of wall, and structure and abundance of pits.

Sclereids. The sclereids are widely distributed in the plant body. They are usually not much longer than they are broad, occurring singly or in groups. Usually these cells are isodiametric but some are elongated too. They are commonly found in the cortex and pith of gymnosperms and dicotyledons, arranged singly or in groups. In many species of plants, the sclereids occur in the leaves. The leaf sclereids may be few to abundant. In some leaves the mesophyll is completely permeated by sclereids. Sclereids are also common in fruits and seeds. In fruits they are disposed in the pulp singly or in groups (*e.g., Pyrus*). The hardness and strength of the seed coat is due to the presence of abundant sclereids.

The secondary walls of the sclereids are typically lignified and vary in thickness. In many sclereids the lumina are almost filled with massive wall deposits and the secondary wall shows prominent pits. Commonly the pits are simple and rarely bordered pits may also occur.

The sclereids are grouped into four categories (Foster, 1949). They are as follows — *brachysclereids, macrosclereids, osteosclereids* and *astrosclereids*.

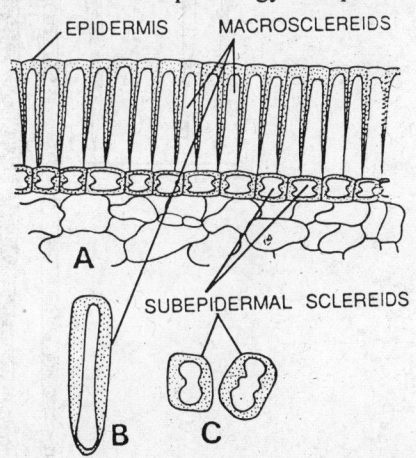

Fig. 5.12. Sclereids of leguminous seed coats. A, the epidermis of a solid layer of macrosclereids; B, single epidermal macrosclereid; C, sub-epidermal sclereids. (After K. Esau).

Brachysclereids. These stone cells or sclereids are short and more or less isodiametric. They are commonly distributed in cortex, phloem and pith of stem and in the pulp of fruits.

Macrosclereids. They are more or less rod-like cells forming palisade-like epidermal layer of many seeds (of Leguminosae) and fruits and frequently found in xerophytic leaves and stem cortices.

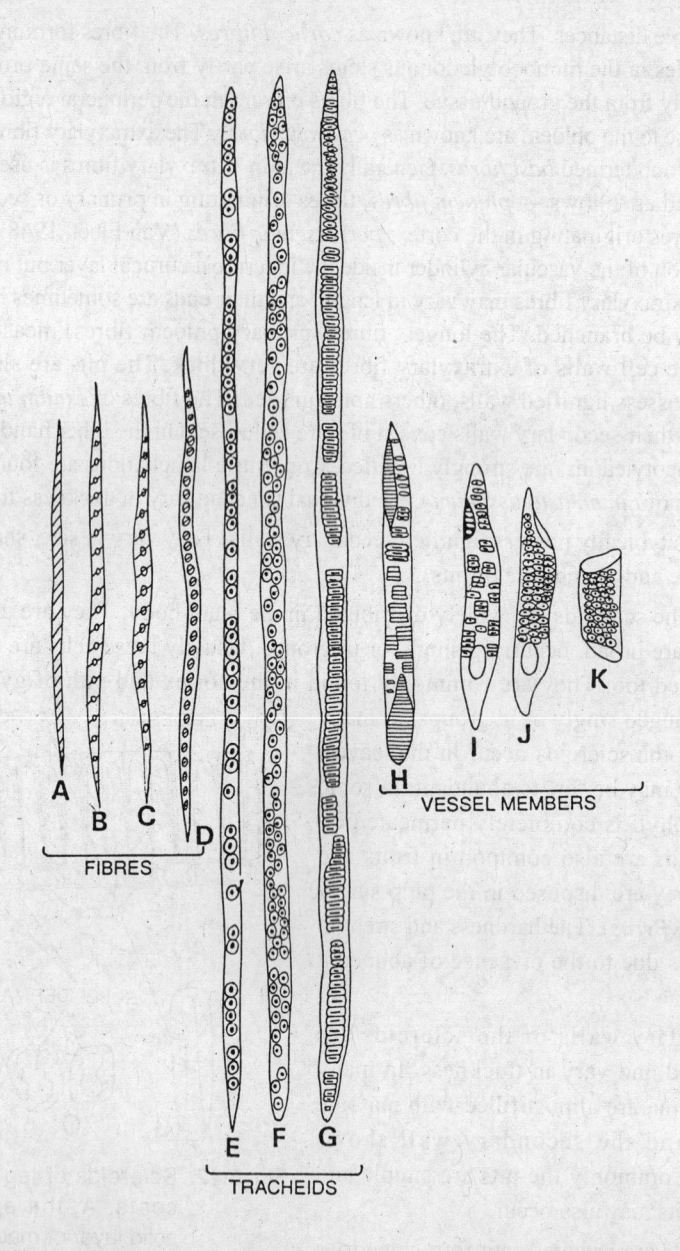

Fig. 5.13. Xylem. A – D, fibres; E – G, tracheids and H – K, vessel members. (After Bailey and Tupper).

Osteosclereids. They are bone-shaped sclereids, *i.e.*, columnar cells are enlarged at their ends. Such sclereids are commonly found in the hypodermal layers of many seeds and fruits. They are also found in xerophytic leaves.

Astrosclereids. They are star-shaped sclereids; such sclereids with lobes projecting, like hairs are commonly found in the intercellular spaces of the leaves and stems of hydrophytes.

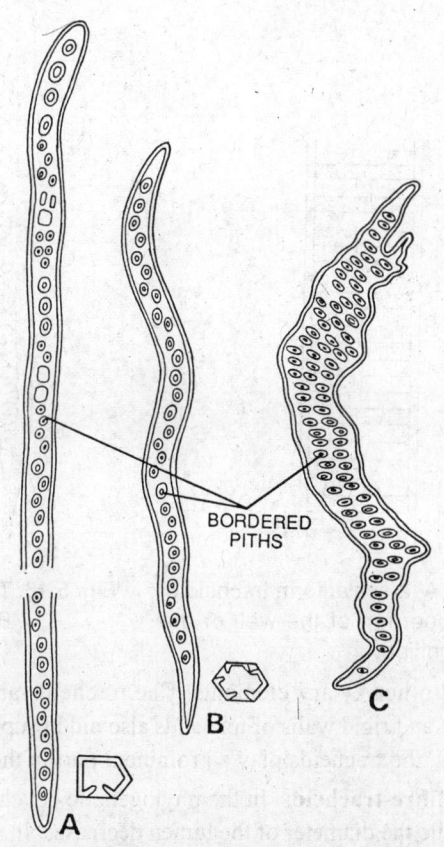

Fig. 5.14. Tracheids. A, L.S. and T.S. of tracheid from *Pinus strobus*; B, L.S. and T.S. of normal tracheid from *Quercus alba*; C, flattened and distorted tracheid from the spring wood of *Quercus alba*. (After E & M).

COMPLEX TISSUES

Here the vascular tissues have been treated as *complex tissues*. The most important complex tissues are — *xylem* and *phloem*.

XYLEM

Xylem is a conducting tissue, which conducts water and mineral nutrients upward from the root to the leaves. The xylem is composed of different kinds of elements. They are: (*a*) tracheids, (*b*) fibres and fibre-tracheids, (*c*) vessels or tracheae, (*d*) wood fibres and (*e*) wood parenchyma. The xylem is also meant for mechanical support to the plant body.

(*a*) **Tracheids.** The tracheid is a fundamental cell type in xylem. It is an elongate tube like cell having tapering, rounded or oval ends and hard and lignified walls. The walls are not much thickened. It is without protoplast and non-living on maturity. In transverse section the tracheid is typically angular, though more or less rounded forms occur. The tracheids of secondary xylem have fewer sides and are more sharply angular than the tracheids of primary xylem. The end of a tracheid of secondary xylem is somewhat chisel-like. They are dead empty cells. Their walls are provided with abundant, bordered pits arranged in rows or in other patterns. The cell cavity or lumen of a tracheid is large and without any contents. The tracheids possess various kinds of thickenings in them and they may be distinguished as annular, spiral, scalariform, reticulate or pitted tracheids. Tracheids alone make the xylem of ferns and gymnosperms, while in the xylem of angiosperms they occur associated

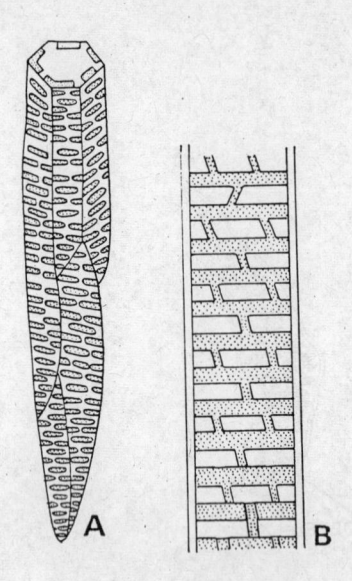

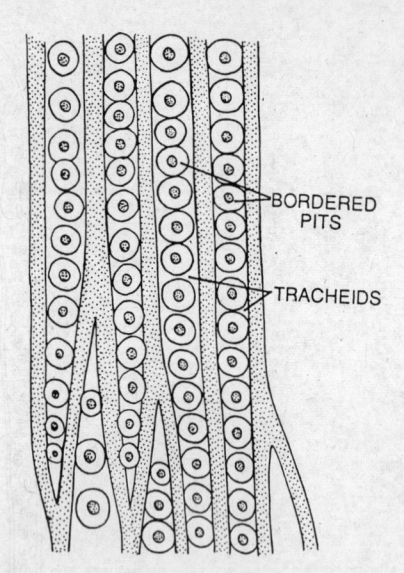

Fig. 5.15. Tracheids. A, a scalariform tracheid of fern; B, a portion of the wall of the same magnified.

Fig. 5.16. Tracheids. Tracheids of pine stem in R.L.S. with bordered pits.

with the vessels and other xylary elements. The tracheids are specially adapted to function of conduction. The thick and rigid walls of tracheids also aid in support and where there are no fibres or other supporting cells, the tracheids play a prominent part in the support of an organ.

(*b*) **Fibres and fibre-tracheids.** In the phylogenetic development of the fibre, the thickness of the wall increases while the diameter of the lumen decreases. In most types the length of the cell also

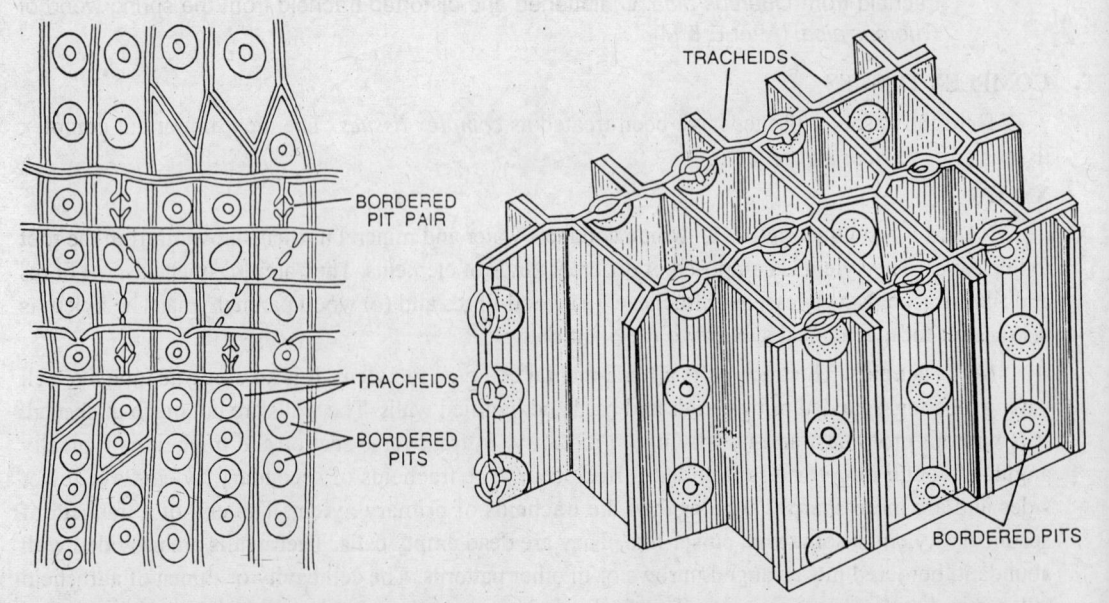

Fig. 5.17. Tracheids with bordered pits of pine stem in R.L.S.

Fig. 5.18. Tracheids with bordered pits of pine stem (diagrammatic).

decreases and the number and size of the pits found on the walls also decrease. Sometimes the lumen of the cell becomes too much narrow or altogether obliterated and simultaneously pits become quite small in size. At this stage it is assumed that either there is very little conduction of water or no conduction through such type of cells, typical *fibres* are formed. Between such cells (*i.e.*, fibres) and normal tracheids there are many transitional forms which are neither typical fibres nor typical tracheids. These transitional types are designated as *fibre-tracheids*. The pits of fibre-tracheids are smaller than those of vessels and typical tracheids. However, a line of demarcation cannot be drawn in between tracheids and fibre-tracheids and between fibre-tracheids and fibres. When the fibres possess very thick walls and reduced simple pits, they are known as libriform wood fibres because of their similarity to phloem fibres (liber = phloem fibres). The libriform wood fibres chiefly occur in woody dicotyledons (*e.g.*, in Leguminosae). The walls of fibre tracheids and fibres of many genera of different families possess gelatinous layers. The cells possessing such layers are known as *gelatinous tracheids*, *fibre-tracheids* and *fibres*. In certain fibre-tracheids the protoplast persists after the secondary wall is mature and may divide to produce two or more protoplasts. These protoplasts are separated by thin transverse partition walls and remain enclosed within the original wall. Such fibre tracheids are called *septate fibre-tracheids*. In fact, they are not individual cells but rows of cells. Here, the transverse partitions are true walls, and each chamber has a protoplast with nucleus.

(c) Vessels. In the phylogenetic development of the tracheid the diameter of the cell has increased and the wall has become perforated by large openings. Due to these adaptations and specializations water can move from cell to cell without any resistance. In the more primitive types of vessels, the general form of the tracheid is retained

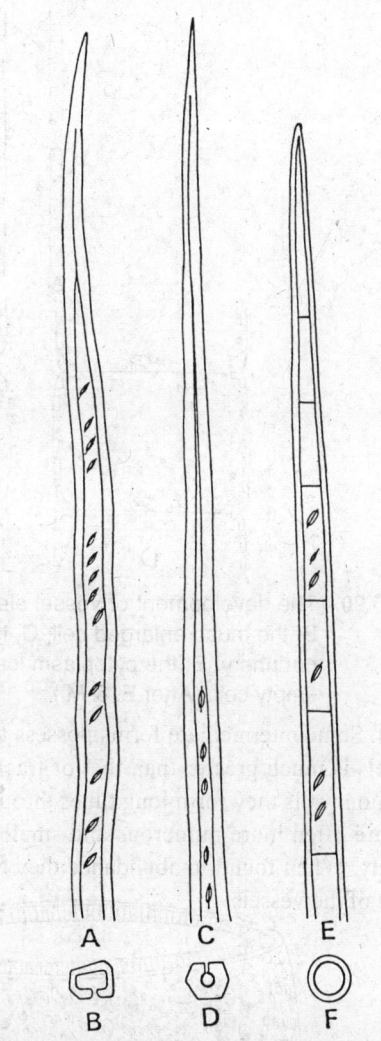

Fig. 5.19. Wood fibres and fibre-tracheids. A and B, L.S. and T.S. of fibre-tracheid from *Malus pumila*; C and D, L.S. and T.S. of libriform fibre from *Quercus alba*; E and F, L.S. and T.S. of septate fibre-tracheid from *Swietenia mahagoni*. (After E. and M.)

and increase in diameter is not much. In the most advanced types, increase in diameter is much and the cell becomes drum-shaped (*e.g., Quercus alba*). The tracheid is sufficiently longer than the cambium cell from which it is derived. The primitive vessel is slightly longer than the cambium cell. The most advanced type of vessel retains the length of cambium cell or is somewhat shorter, with a diameter greater than its length (drum-shaped vessel). The ends of the cells change in shape in the series from least to highest specialization. The angle formed by the tapering end wall becomes greater and greater until the end wall is at right angles to the side walls (as in drum-shaped vessel in *Quercus*

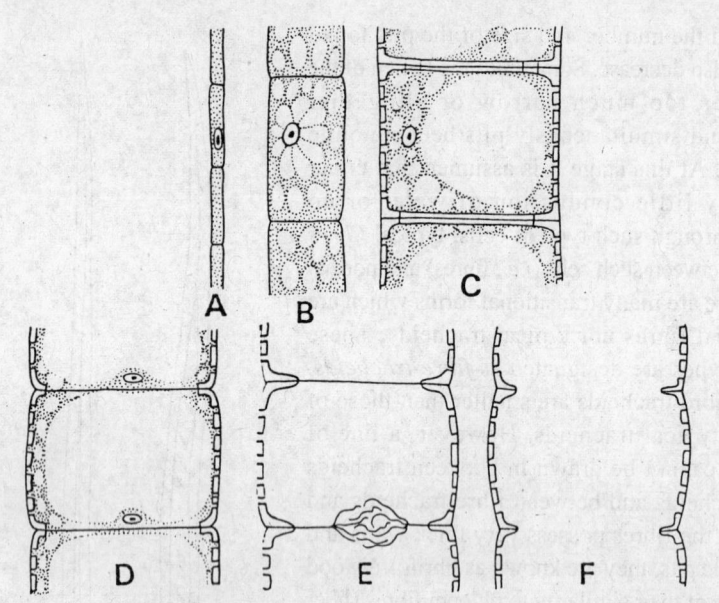

Fig. 5.20. The development of vessel element. (Ontogeny of the vessel) — A, the cambium initial; B, the much enlarged cell; C, the cell further enlarged; D, the cytoplasm restricted to the periphery; E, the cytoplasm lost, thin end walls disintegrating; F, the mature, perforated, empty cell. (After E. & M.)

alba). Some intermediate forms possess tail-like lips beyond the end wall. Usually the diameter of vessels is much greater than that of tracheids and because of the presence of perforations in the partition walls they form long tubes through which water is being conducted from root to leaf. The pits are often more numerous and smaller in size than are those of tracheids and cover the wall closely. When found in abundance they are either scattered or arranged in definite patterns on the walls of the vessels.

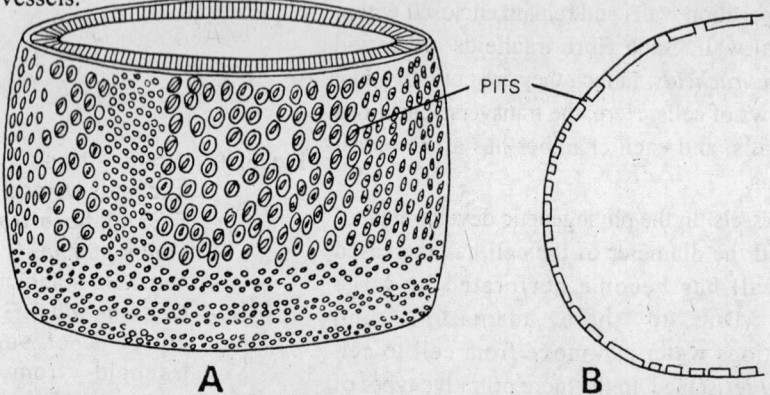

Fig. 5.21. Vessels. A and B, L.S. and T.S. of vessels from *Quercus alba*.

The openings in vessel-element walls are known as *perforations*. These openings are restricted to the end walls except in certain slender, tapering types. The area in which the perforations occur is known as *perforation plate*. Commonly this is an end wall. The stripes of cell wall between scalariform perforations are the *perforation bars*. The perforation plate when bears single opening is described as having *simple perforation*. If there are two or more openings, they are known as *multiple perforations*.

The Tissue

The secondary walls of vessel-elements develop in a wide variety of patterns. Generally, in the first-formed part of the primary xylem a more limited area of the primary wall is covered by secondary wall layers than in the later-formed primary xylem and in the secondary xylem. The secondary thickenings are deposited in the vessels as rings, continuous spirals or helices, with the individual coils of a helix here and there interconnected with each other, giving the wall a ladder-like appearance. Such secondary thickenings are called — *annular*, *spiral* or *helical* and *scalariform* respectively. In a still later ontogenetic type of vessel elements, the *reticulate* vessel elements, the secondary wall appears like a reticulum. When the meshes of the reticulum are transversely elongated, the thickening is called *scalariform-reticulate*. The *pitted* elements are characteristic of the latest primary xylem and of the secondary xylem.

Vessels are characteristic of the angiosperms. However, certain angiospermic families lack the vessels — the Winteraceae, Trochodendraceae and Tetracentraceae. In many monocotyledons (*e.g.*, *Yucca*, *Dracaena*) they are absent from the stems and leaves. They are found in some species of *Selaginella*, in two

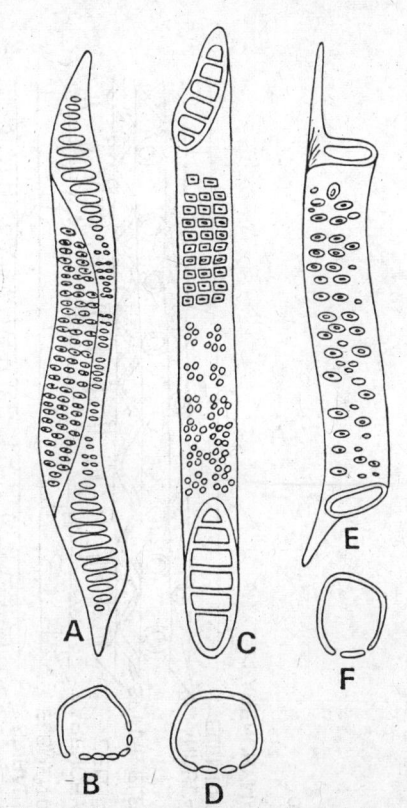

Fig. 5.22. Vessel elements. A and B, L.S. and T.S. of vessel from *Betula*; C and D, L.S. and T.S. of vessel from *Liriodendron*; E and F, L.S. and T.S. of vessel from *Malus*.

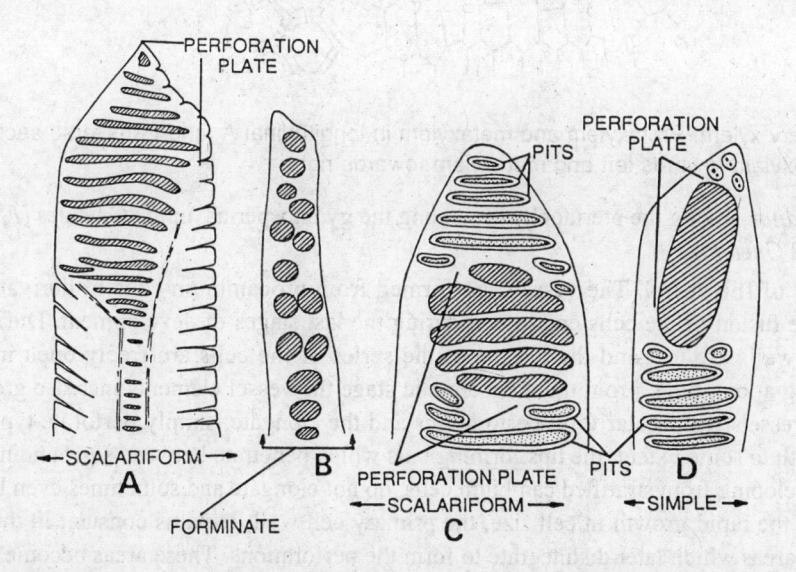

Fig. 5.23. End walls of vessels with perforation plates. A scalariform vessel of *Pteridium*; B, foraminate type of *Ephedra*; C, scalariform vessel of *Vitis* and D, simple type of *Vitis*.

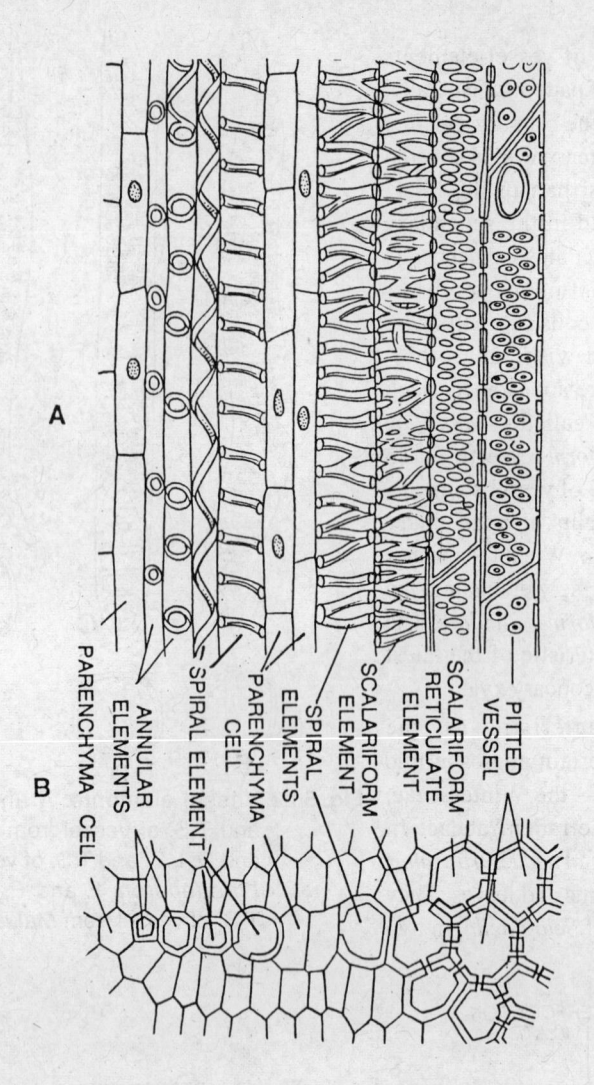

Fig. 5.24. Primary xylem. Protoxylem and metaxylem in longitudinal A, and transverse sections B, Protoxylem towards left and metaxylem towards right.

species of *Pteridium* among the pteridophytes; among the gymnosperms, in the Gnetales (*Ephedra*, *Welwitschia* and *Gnetum*).

Ontogeny of the vessel. The vessels are formed from procambium cells or derivatives of cambium by the fusion of the cells end to end during the last stages of development. During this fusion the end walls are lost and the lumina of the series of the cells are freely open into one another, forming a long tube. From the meristematic stage the vessel elements increase greatly in diameter. The vessels with scalariform perforations and the elongate, simply perforate types may increase in length to some extent, the tips forming tails which penetrate between surrounding cells. The vessels developing from stratified cambium cells, do not elongate and sometimes even become shorter. During the rapid growth in cell size, the primary cell wall, remains constant in thickness except in those areas which later disintegrate to form the perforations. These areas become thicker and limited in their margins. In the sectional view they are lens-shaped or plate like and can be seen to be three layered, composing of the primary walls of the two adjacent cells and the middle

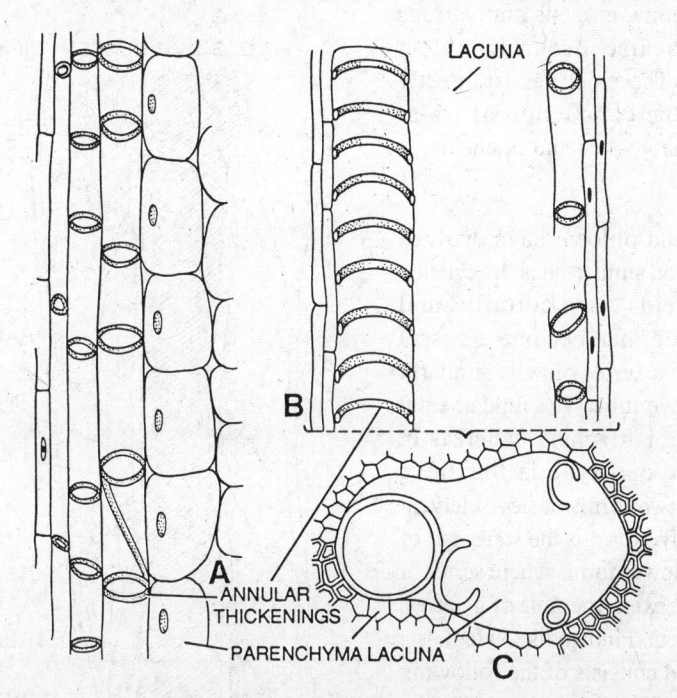

Fig. 5.25. Protoxylem. A, L.S. of protoxylem from *Arisaema* fruit, the elements are all annular; B – C, L.S. and T.S. of protoxylem from *Zea*, the surrounding parenchyma pulled away from the protoxylem elements, leaving a large lacuna.

lamella. When the cell reaches its maturity, the cytoplasm of the cell begins to disintegrate. In certain woody plants the nucleus becomes quite small and flat and lies in scant cytoplasm against the wall where perforation is about to occur. As soon as the primary wall becomes mature, the perforation of the end wall and loss of the protoplast begin. The wall in the perforation area becomes thinner and thinner and ultimately disintegrates. The maturation in all members of a vessel series takes place from one end to the other and not simultaneously (see Fig. 5.22).

(*d*) **Wood parenchyma.** The parenchyma cells which frequently occur in the xylem of most plants. In secondary xylem such cells occur vertically more or less elongated and placed end to end, known as *wood* or *xylem parenchyma*. The radial transverse series of the cells form the wood rays and are known as *wood* or *xylem ray parenchyma*. The xylem parenchyma cells may be as long as the fusiform initials, or they may be several times shorter, if a fusiform derivative divides transversely before differentiation into parenchyma (wood parenchyma). The shorter type of xylem parenchyma cells is the more common. The ray and the xylem parenchyma cells of the secondary xylem may or may not have secondary walls. If a secondary wall is present, the pit pairs between the parenchyma cells and the tracheary elements may be simple, half-bordered or bordered. In between, parenchyma cells only simple pit pairs occur. The xylem parenchyma cells are noted for storage of food in the form of

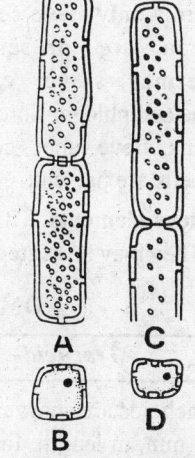

Fig. 5.26. Wood parenchyma. A and B, L.S. and T.S. of wood parenchyma from *Quercus alba*; C and D, L.S. and T.S. of wood parenchyma from *Carya ovata*.

starch or fat. Tannins, crystals and various other substances also occur in xylem parenchyma cells. These cells assist directly or indirectly in the conduction of water upward through the vessels and tracheids.

2. PHLOEM

The xylem and phloem have evolved along more or less on similar lines. In xylem a series of tracheids, structurally and functionally united, has become a vessel whereas in phloem a series of cells similarly united, forms a sieve tube. The fundamental cell type of xylem is tracheid, whereas in phloem the basic cell type is the sieve element. There are two forms of sieve element — the more primitive form is the *sieve cell* of gymnosperms and lower forms where series of united cells do not exist, the unit of a series, the sieve tube element. Phloem like xylem, is a complex tissue, and consists of the following elements — (*a*) sieve elements, (*b*) companion cells, (*c*) phloem fibres and (*d*) phloem parenchyma. In the pteridophytes and gymnosperms only sieve cells and phloem parenchyma are present. In some gymnosperms, sieve cells, phloem parenchyma and phloem fibres are present. In angiosperms, sieve tubes, companion cells, phloem parenchyma, phloem fibres, sclereids and secretory cells are present.

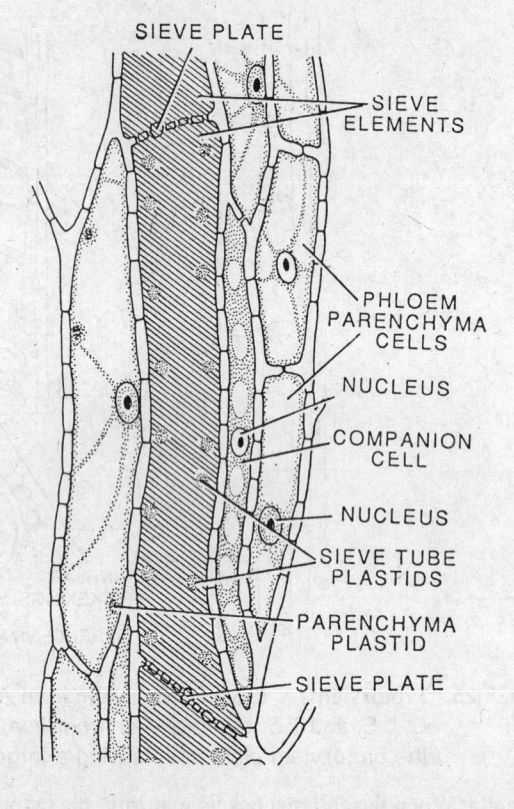

Fig. 5.27. Phloem. Phloem tissue from the stem of *Nicotiana*.

(*a*) **Sieve elements.** The conducting elements of the phloem are collectively known as *sieve elements*. They may be segregated into the less specialized *sieve cells* and the more specialized *sieve*

COMPARISON OF TRACHEID AND VESSEL

Tracheid	*Vessel*
1. The tracheids are short and are generally upto 1 mm. in length. In rare cases their length becomes upto 1.2 cm. or so.	1. They are comparatively longer and may reach upto 10 cm. in length. In rare case they attain the length upto 2–6 metres (*e.g.*, in *Eucalytus*, *Quercus*, etc.)
2. It consists of a single elongated cell which possesses tapering end walls.	2. The vessel consists of a row of cells placed one above the other. Their intervening walls are absent.
3. The tracheids are not tubular. The tracheids found one above the other are separated by cross walls which bear bordered pits. They are not perforated	3. The vessels are tubular and have no cross walls. They are well adapted for the conduction of water. They may be perforated by small or large pores.

tubes or *sieve tube elements*. The morphologic specialization of sieve elements is expressed in the development of sieve areas on their walls and in the peculiar modifications of their protoplasts. The *sieve areas* are depressed wall areas with clusters of perforations, through which the protoplasts of the adjacent sieve elements are interconnected by *connecting strands*. In a sieve area each connecting strand remains encased in a cylinder of substance called *callose*. The wall in a sieve area is a double structure consisting of two layers of primary wall, one belonging to one cell and the other to another, cemented together by intercellular substance. Like the pits in the tracheary elements, the sieve areas occur in various numbers and are variously distributed in sieve elements of different plants. The wall parts bearing the highly specialized sieve areas are called *sieve plates* (Esau, 1950). If a sieve plate consists of a single sieve area, it is a *simple sieve plate*. Many sieve areas, arranged in scalariform, reticulate, or any other manner, constitute a *compound sieve plate*. However,

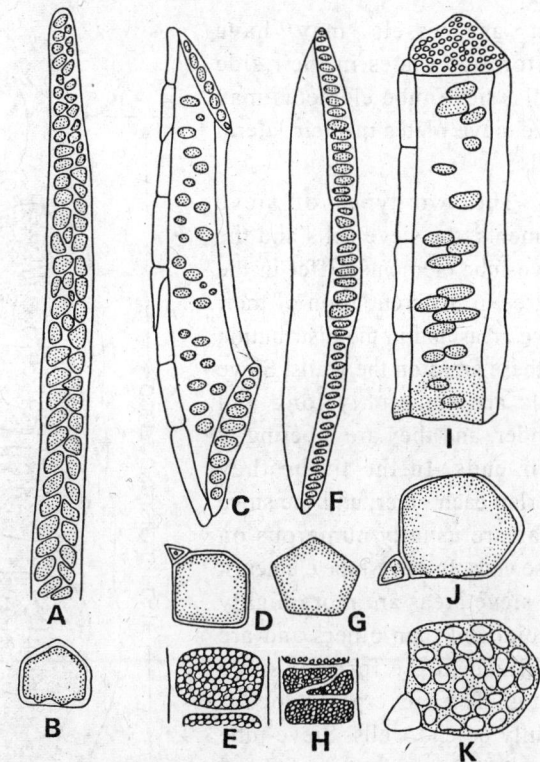

Fig. 5.28. The Phloem. Sieve cells and sieve-tube elements in side view and cross-section, with detailed structure of sieve plates. A and B, L.S. and T.S. of sieve cell of *Pteridium*; C and D, L.S. and T.S. of sieve tube with companion cells attached in *Liriodendron*; E, detail of sieve plate in the same; F and G, sieve tube from *Malus pumila* in L.S. and T.S.; H, detail of sieve plate in same; I and J sieve tube from *Robinia* in L.S. and T.S. with companion cells attached; K, detail of sieve plate in same.

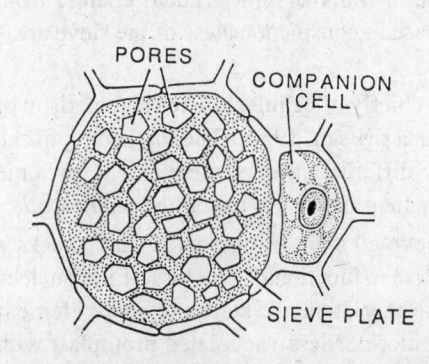

Fig. 5.29. Phloem. T.S. through a sieve plate and companion cell.

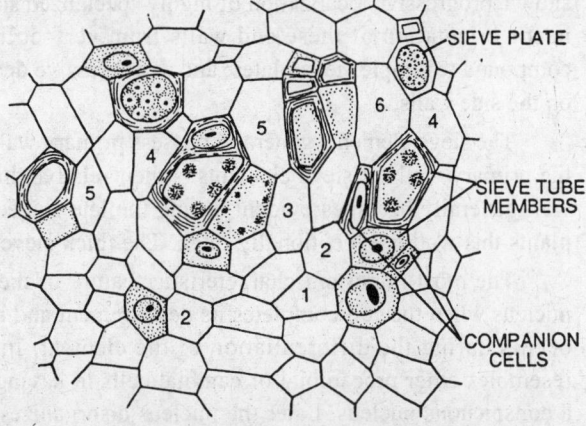

Fig. 5.30. Phloem. Differentiation of sieve tube members in *Cucurbita*.

just as vessels may have perforation plates in their side walls, sieve tube elements may have sieve plates in their lateral walls.

The two types of sieve elements, the sieve cells and the sieve-tube elements differ in the degree of differentiation of their sieve areas and in the distribution of these areas on the walls. Sieve cells are commonly long and slender, and they are tapering at their ends. In the tissue they overlap each other, and the sieve areas are usually numerous on these ends. In sieve-tube elements, the sieve areas are more highly specialized than others and are localized in the form of sieve plates. The sieve plates occur mainly on end walls. Sieve-tube elements are usually disposed end to end in long series, the common wall parts bearing the sieve plates. These series of sieve-tube elements are *sieve-tubes*.

The lower vascular plants and the gymnosperms generally have sieve cells, whereas most angiosperms have sieve-tube elements. The sieve-tube elements

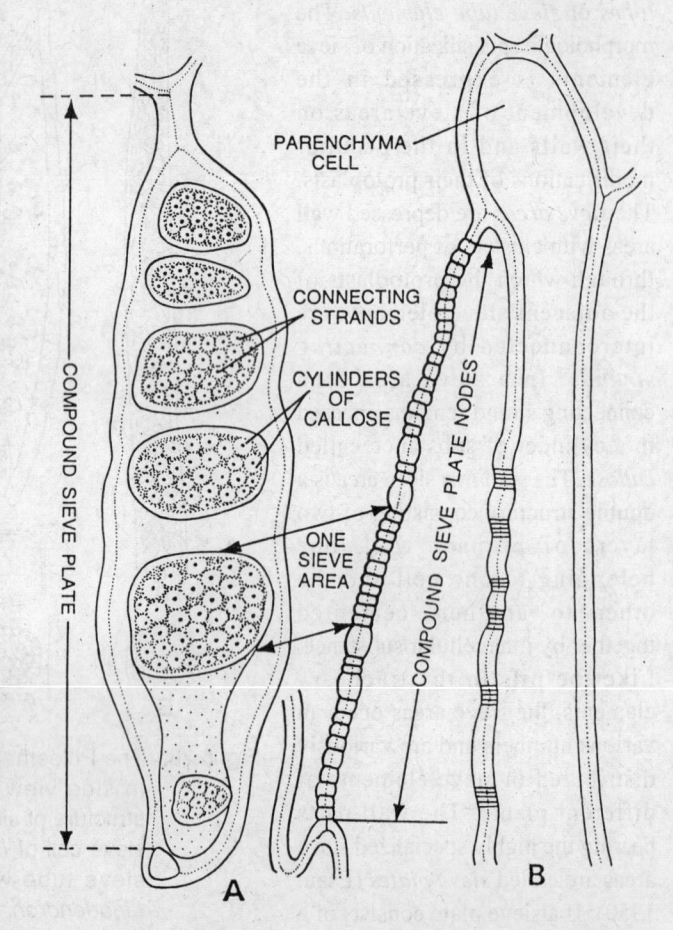

Fig. 5.31. Phloem. Structure of sieve areas. A, compound sieve plate of *Nicotiana* in surface view; B, parts of sieve tube members and a phloem parenchyma cell.

show a progressive localization of highly specialized sieve areas on the end walls; a gradual change in the orientation of these end walls from very oblique to transverse; a gradual change from compound to simple sieve plates; and a progressive decrease in conspicuousness of the sieve areas on the side walls.

The sieve elements generally possess primary walls, chiefly of cellulose. The characteristic of the primary walls of sieve elements is their relative thickness (Esau, 1950). The thickening of the wall generally becomes evident during the late stages of differentiation of the element. In some plants this wall is exceptionally thick. The thick sieve element is usually called the *nacre wall*.

The most important characteristic feature of the sieve-element protoplast is that it lacks a nucleus when the cell completes its development and becomes functional. The loss of the nucleus occurs during the differentiation of the element. In the meristematic state the sieve element resembles other procambial or cambial cells in having a more or less vacuolated protoplast with a conspicuous nucleus. Later the nucleus disorganizes and disappears.

The important property of the sieve-element protoplast of dicotyledons is the presence of variable amounts of a relatively viscous substance, the *slime*. The slime is proteinaceous in nature.

The slime appears to be located mainly in the cell-sap together with various organic and inorganic ingredients. The slime originates in the cytoplasm in the form of discrete bodies, the *slime-bodies*. They may be spherical, or spindle shaped, or variously twisted and coiled. They occur singly or in multiples in one element.

(b) Companion cells. The *companion cell* is a specialized type of parenchyma cell which is closely associated in origin, position and function with sieve-tube elements. When seen in transverse section the companion cell is usually a small, triangular, rounded or rectangular cell beside a sieve-tube element. These cells are living having abundant granular cytoplasm and a prominent elongated nucleus which is retained throughout the life of the cell. Usually the nuclei of the companion cells serve for the nuclei of sieve tubes as they lack them. The companion cells do not contain starch. They live only so long as the sieve-tube element with which they are associated and they are crushed with those cells. The companion cells are formed by longitudinal division of the mother cell of the sieve tube element before specialization of this cell begins. One daughter cell becomes a companion cell and other a sieve tube element. The companion cell initial may divide transversely several times producing a row of companion cells so that one to several companion cells may accompany each sieve-tube element. A companion cell or a row of companion cells formed by the transverse division of a single companion-cell initial may extend the full length of the sieve-tube element. The number of companion cells accompanying a sieve-tube element is fairly constant for a particular species. The solitary and long companion cells occur in primary phloem of herbaceous plants whereas numerous companion cells occur in the secondary phloem of woody plants.

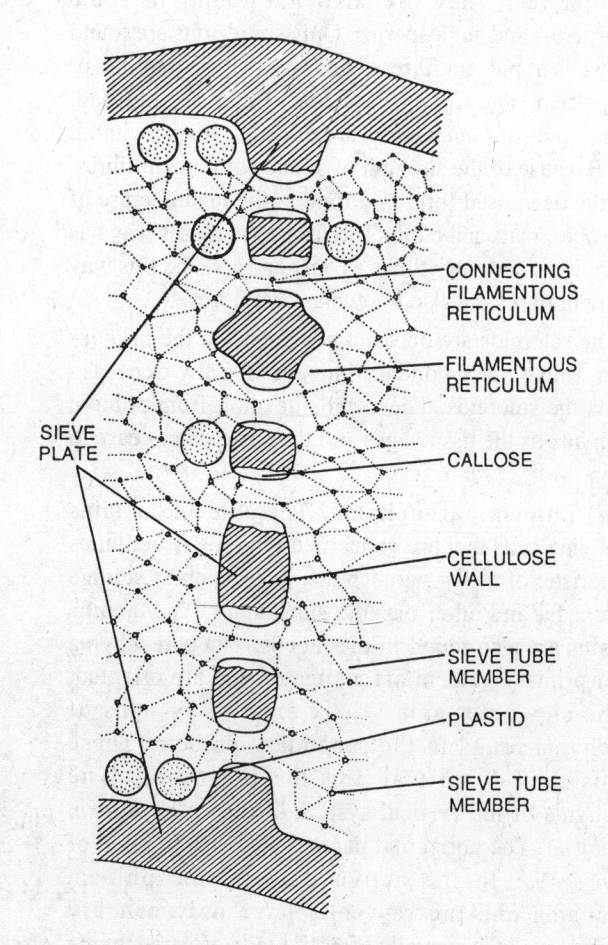

Fig. 5.32. Phloem. Diagrammatic representation of the ultra-structure of a sieve plate.

The companion cells occur only in angiosperms where they accompany most sieve-tube elements. In the phloem of many monocotyledons, they are abundant, together with sieve tubes making up the entire tissue. The sieve cells of the gymnosperms and vascular cryptogams have no companion cells.

(c) Phloem fibres. In many flowering plants, fibres form a prominent part of both primary and secondary phloem. The phloem fibres are rarely found or absent in phloem of living pteridophytes. They are also not found in some gymnosperms and angiosperms. Only simple pits are found on the walls of phloem fibres. The walls may be lignified or non-lignified. The *Cannabis* (hemp) fibres are lignifed, whereas fibres of *Linum* (flax) are of cellulose and without lignin. Because of the strength of strands of phloem fibres, they have been used for along time in the manufacture of cords, ropes, mats and cloth. The fibre used in this way has been known since early times as *bast* or *bass*, and this way the phloem fibres are also known as bast fibres.

The sclereids are occasionally found in the primary phloem. The older secondary phloem of many trees also contains the sclereids. These cells develop from phloem parenchyma as the tissue ages and the sieve tubes cease to function.

(d) Phloem parenchyma. The phloem contains parenchyma cells that are concerned with many activities characteristic of living parenchyma cells, such as storage of starch, fat and other organic substances. The tannins and resins are also found in these cells. The parenchyma cells of primary phloem are elongated and are oriented, like the sieve elements. There are two systems of parenchyma found in the secondary phloem. These systems are — vertical and the horizontal. The parenchyma of the vertical system is known as *phloem parenchyma*. The horizontal parenchyma is composed of *phloem rays*. In the active phloem, the phloem parenchyma and the ray cells have only primary unlignified walls. The walls of both kinds of parenchyma

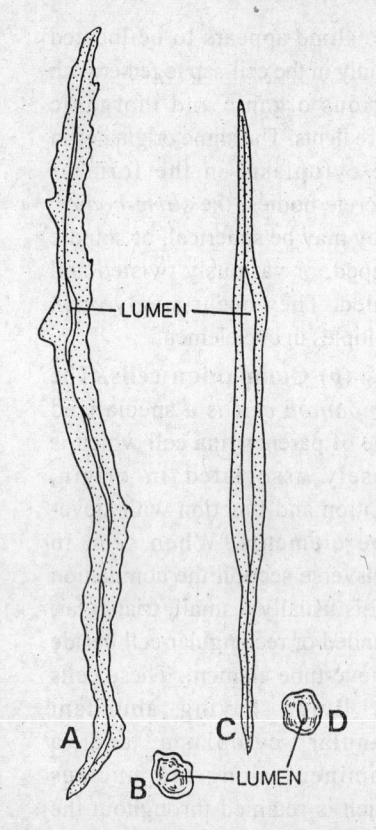

Fig. 5.33. Phloem fibres. A and B, L.S. and T.S. of phloem fibres from *Malus pumila*; C and D, L.S. and T.S. of phloem fibre from *Robinia*. (After E. & M.).

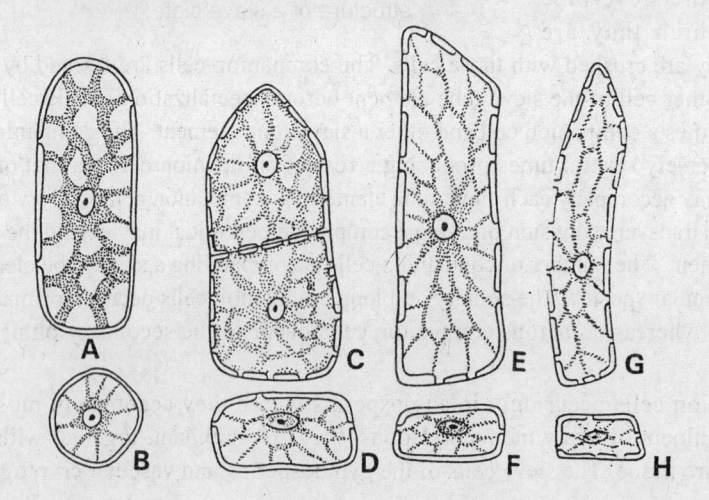

Fig. 5.34. Phloem parenchyma in L.S. and T.S. A – B, from *Salix*; C – D, from *Robinia*; E – F, from *Liriodendron*; G – H, from *Malus*.

cells have numerous primary pit fields. The phloem parenchyma is not found in many or most of monocotyledons.

3. SECRETORY TISSUE

The tissues that are concerned with the secretion of gums, resins, volatile oils, nectar, latex and other substances are called *secretory tissues*. These are further subdivided into two groups — (a) laticiferous tissue and (b) glandular tissue.

(a) **Laticiferous tissue.** Usually latex is present in the families of many flowering plants. This substance may be white, yellow or pinkish in colour. This is a viscous fluid and established to be colloidal in nature. Many substances like sugars, proteins, gums, alkaloids, enzymes, rubber, etc., remain suspended in a matrix of watery fluid. Starch grains may be abundantly present. The latex of some plants is of great importance, especially as a source of rubber (*e.g., Haevea, Ficus,* etc.), chicle (*Achras*), and papain (*Carica*). The laticiferous ducts, in which latex is found may be of two types — *latex cells* or *non-articulate latex ducts* and *latex vessels* or *articulate latex ducts*. The functions and the contents of the two are same but they differ in their nature and morphology. They contain numerous nuclei in the thin layer of cytoplasm along the cell wall. The function of these tissues is not yet clearly understood. They may act as food storage organs or as reservoirs of waste products.

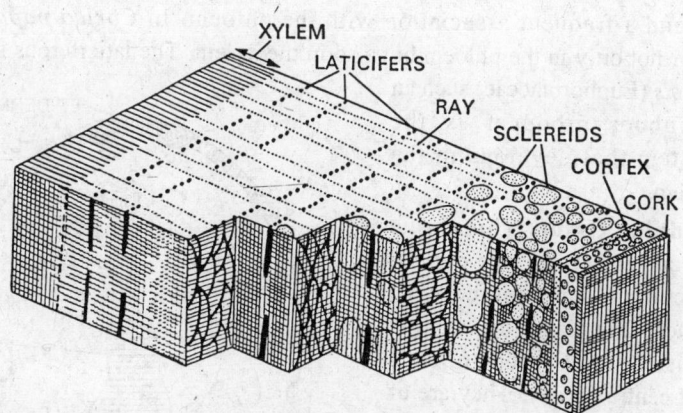

Fig. 5.35. Laticiferous tissue. Three dimensional diagram of bark of *Haevea brasiliensis* showing arrangement of articulated laticifer in secondary phloem. In tangential sections, laticifers form a reticulum.

Nonarticulate latex ducts or latex cells. These ducts are independent units which extend as branched structures for long distances in the plant body. They originate as minute structures, elongate quickly and ramify in all directions of the plant body by repeated branching, but they do not fuse together, thus no netted structures are formed as they are formed in articulate ducts. The walls of the ducts are soft and very often thick. Such ducts are commonly found in *Calotropis*, *Euphorbia*, *Nerium*, *Vinca,* etc.

Branched nonarticulated laticifers commonly occur in leaves. Here they travel through the vascular bundles, ramify in the mesophyll, and often reach the epidermis.

Fig. 5.36. Laticiferous tissue. Latex cells.

The unbranched nonarticulated laticifers of *Vinca* and *Cannabis* occur in the primary phloem but are apparently absent in the secondary tissues.

Articulate latex ducts or latex vessels. These ducts or vessels are the result of anastamosing of many cells together. They originate in the meristems from rows of cells by the abortion, of the separating walls early in the ontogeny of the cells. They grow more or less as parallel ducts which by means of branching and frequent anastamoses form a complex network. A duct of this type resembles with the xylem vessel only in the respect that it is made up of a series of cells united to form a tube otherwise the latex tube is living and coenocytic. Such latex vessels are commonly found in angiospermic families — Papaveraceae, Compositae, Euphorbiaceae, Moraceae, etc.

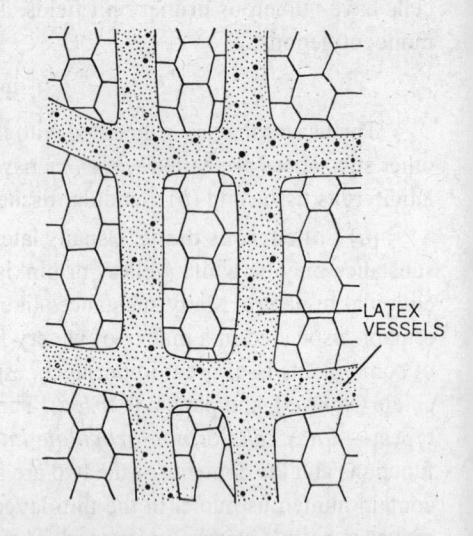

Fig. 5.37. Laticiferous tissue. Latex vessels.

Articulated laticifers show various arrangements and a frequent association with the phloem. In *Carica papaya* the laticifers apparently occur not only in the phloem but also in the xylem. The laticiferous system that makes *Hevea brasiliensis* (Euphorbiaceae) such an outstanding rubber producer is the secondary system that develops in the secondary phloem.

(b) Glandular tissue. This tissue consists of special structures, the *glands*. These glands contain some secretory or excretory products. The glands may consist of isolated cells or small groups of cells with or without central cavity. They are of various kinds, and more common types are those which secrete digestive enzymes, called *digestive glands* and those which secrete nectar, known as *nectaries*. They may be *internal* or *external*. The common internal glands which are usually lying embedded in the interior tissues of plant body are *oil glands, mucilage secreting, glands secreting gums, resin* and *tannin*, etc., *digestive glands* secreting enzymes, *water secreting glands* also known as *hydathodes*. The common external glands which occur on the epidermis may be *glandular epidermal hairs, nectaries,* etc.

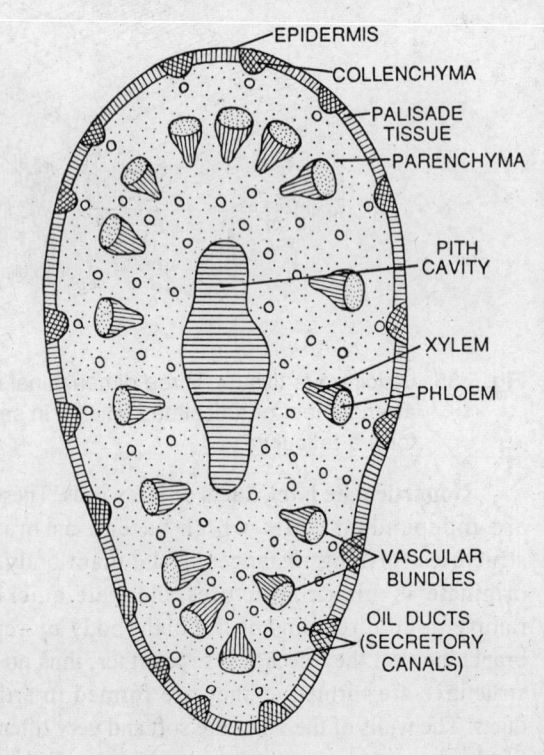

Fig. 5.38. Oil glands. T.S. of the petiole of *Foeniculum vulgare* (dicot-Umbelliferae) — diagrammatic. Note the oil ducts or secretory canals.

Oil glands. These are internal glands which frequently contain essential oils in

The Tissue

them. These oils are volatile and odoriferous. These glands originate due to split of certain cells, but they are formed in abundance by the breaking down of cells containing the volatile oil. On the disintegration of the cells the oil stores up in the large cavities of glands. These cavities are lysigenous in nature. Such oil glands are commonly found in *Citrus*, *Eucalyptus* and other plants.

Glands secreting resins, gums, etc. In the gymnosperms and in many angiospermic families, resins, gums, oils and many other substances are secreted and conducted in ducts. In many gymnosperms (*e.g.*, in *Pinus*) these ducts or canals form extensive systems extending both vertically and horizontally. In other plants (*e.g.*, in *Umbelliferae*) the ducts are local in occurrence and limited in extent. In *Pinus* and many other gymnosperms, the resin ducts are schizogenous in nature, and when mature, have the structure of a tube with an epithelial lining. These glands are internal.

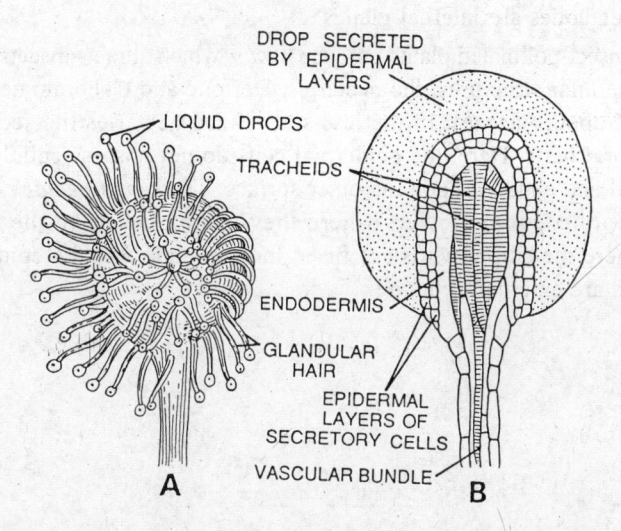

Fig. 5.39. Glands. A, leaf of *Drosera* plant covered with numerous glandular hairs; B, a digestive gland of sundew (*Drosera*).

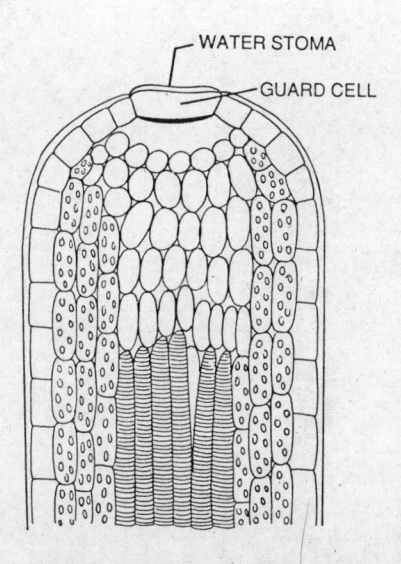

Fig. 5.40. Hydathode of *Pistia*.

Digestive glands. In certain insectivorous plants there are special glands which secrete protein-digesting enzymes. These enzymes act upon insects so that the products of digestion can be absorbed by the plant. In *Drosera*, the secretory tissue is at the tips of the leaf tentacles. Here in addition to the digestive enzymes, there are secreted viscid substances which hold the insects. These are internal glands.

Hydathodes (water secreting glands). Many plants possess special structures which exudate water under conditions of low transpiration and abundant soil moisture. These modified special structures are known as *hydathodes* or *water stomata*. The water stoma resembles an ordinary stoma, and morphologically it is considered to be enlarged stoma which has lost the power of movement and serves for water secretion. Commonly, the hydathodes occur at the tips of the leaves of *Pistia* and other aroids, water hyacinth, grasses, garden nasturtium and many other plants of humid climate. The hydathodes are internal glands.

Nectaries. Many insect-pollinated plants produce *nectar* which attracts insects. This substance is secreted by special cellular structures, the nectaries. Definite and elaborate nectaries occur in certain families (*e.g.*, Euphorbiaceae). In the less specialized nectaries the secreting cells are superficial upon the floral parts. Here, the epidermal cells do not possess cuticle. The nectar is exuded through the wall and exposed upon the outer surface. The *septal nectaries* are commonly found in many monocotyledonous flowers where they make pockets in the septal walls of syncarpous ovaries where the carpel walls are fused incompletely and the epidermal cells are glandular. These glands are external in nature.

6

Apical Meristems

The apical meristem includes the meristematic initials and their immediate derivatives at the apex of a shoot or root. The apical meristem, thus delimited corresponds approximately to the promeristem, and to contrast with the partly developed derivatives of the promeristem, *i.e.*, the protoderm, the ground meristem, and the procambium. This seems quite impracticable, to think of the apical meristem as consisting of the initiating cells only because cells may be poorly differentiated from their most recent derivatives.

The terms *shoot apex* and *root apex* are more convenient to use instead of apical meristem of the shoot and apical meristem of the root, respectively. In the similar way, the terms shoot apex and root apex are more conveniently used as the substitutes of growing points. Growth in the sense of cell division, which is characteristic of the meristematic state, is not restricted to the so-called growing point but occurs abundantly — and may be even more intense — at some distance from the apical meristem (Wardlaw, 1945; Goodwin and Stepka, 1945). On the other hand, growth in the sense of increase in size of cells, tissues, and organs is most pronounced, not in the apical meristem, but in its derivatives.

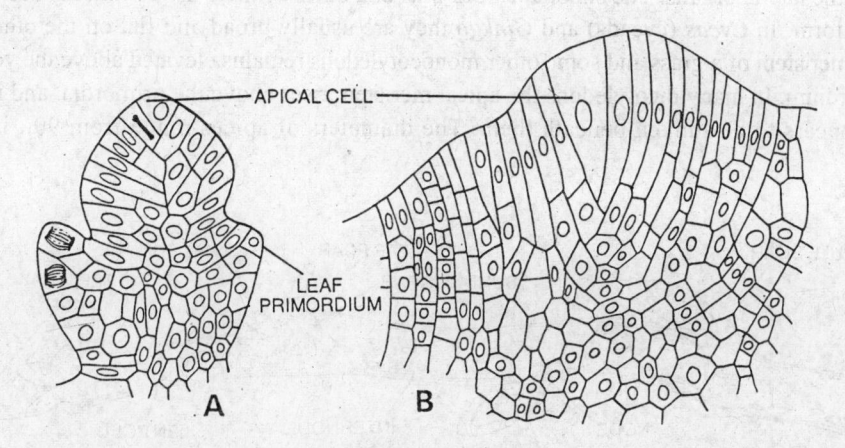

Fig. 6.1. Apical meristems. A, shoot apex with apical cell of *Equisetum* shoot; B, shoot apex of *Pteridium*.

INITIALS AND DERIVATIVES

An *initial* or initiating cell, is a cell that remains within the meristem indefinitely with the addition of cells to the plant body by combining self-perpetuation. The concept regarding meristematic initials, implies that a cell is an initial, not because of its inherent characteristics, but simply because of its particular position in the meristem, a position that cannot be treated as permanent.

The number of initials in root and shoot apices is variable. In most of pteridophytes a single initial cell occurs at the apex. In the lower vascular plants, as well as in the higher, several initials are present. The single initial in its morphology is quite distinct from its derivatives and is commonly known as the *apical cell*. If the initials are numerous, they are called apical initials.

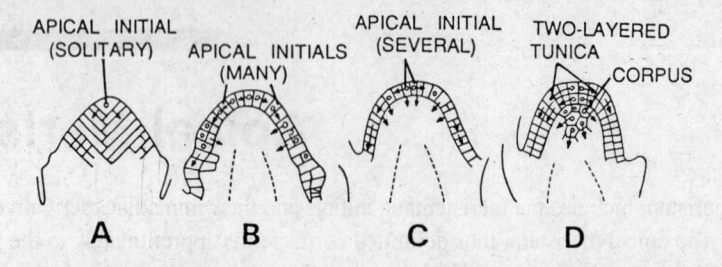

Fig. 6.2. Position and planes of division of stem-apex initials. A, initial solitary, with oblique and anticlinal divisions; B, initials many, superficial, divisions both anticlinal and periclinal; C, initials several, superficial, divisions both anticlinal and periclinal; D, initials in three tiers, the two outer, with anticlinal divisions, forming a two-layered tunica and the innermost, with divisions in all planes, forming a corpus. (After Eames and MacDaniels).

Usually the apical initials occur in one or more tiers. If there is only one tier, all cells of a plant body are ultimately derived from it. On the other hand, different parts of a plant body are derived from different groups of initials.

VEGETATIVE SHOOT APEX

The vegetative shoot apices vary in shape, size and cytohistologic structure, and in their relation to the lateral organs. The shoot apex of *Pinus* and other conifers are commonly narrow and conical in form. In *Cycas* (cycads) and *Ginkgo* they are usually broad and flat on the other hand the apical meristem of a grass and some other monocotyledons remains elevated above the youngest leaf primordium. In many dicotyledons the apical meristem rises above the primordia, and in other cases it appears to be sunken beneath them. The diameters of apices range from 90µ in some

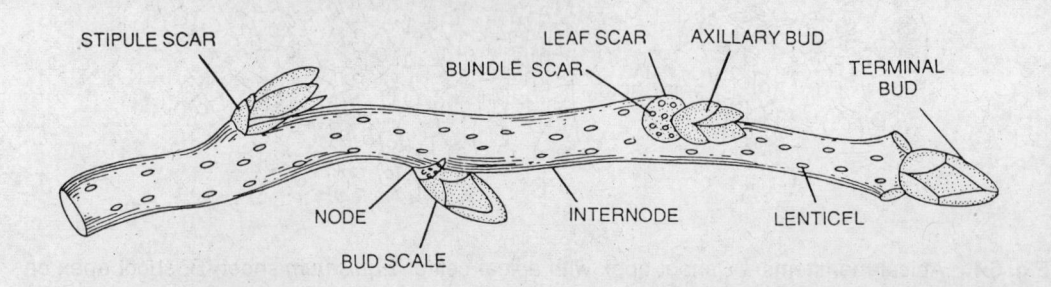

Fig. 6.3. Apical Meristem. Stem of *Aesculus hippocastanum* in winter condition showing *terminal, axillary* and *adventitious* buds. Axillary buds may be classified according to the number at a node into (1) alternate with a single bud at a none, (2) opposite, with two buds at a node on opposite sides of the stem, and (3) whorled with more than two buds at each node. Buds which arise anywhere on the plant except at the tips of stems or in the axils of leaves are called *adventitious buds*. In this figure, terminal bud is located at the tip. The leaf scars are left by the breaking away of the leaf stalks. Leaf trace scars are the broken ends of leaf traces. When the scales of terminal bud are shed a *ring* or *girdle* is left.

Apical Meristems 119

angiosperms to 3.5 mm in *Cycas revoluta* (Foster, 1949). The size and shape of the apex marked by change during the plant development.

Pteridophytes (vascular cryptogams). In the vascular cryptogams (lower tracheophyta) growth at the apex proceeds either from one or few initial cells, which are usually distinctive in their morphology (Bower, 1889; Hartel, 1938; Wardlaw, 1945). Most commonly the single apical cell is tetrahedral (pyramid like) in shape (*e.g.*, in Psilotales, Equisetaceae, and ferns). The base of this pyramid is turned towards the free surface, and the new cells are formed at the other three sides. The *Salvinia* and *Azolla* (water ferns) have three sided apical cells with two sides from which new cells are cut off. In *Selaginella* apical growth occurs from a single three or four-sided apical cell or from a group of initials, and the two situations may be found in the same plant. The eusporangiate ferns have two or four initials, the leptosporangiate ferns have one, but there is no sharp line of division between the two groups with reference to this character. As regards their

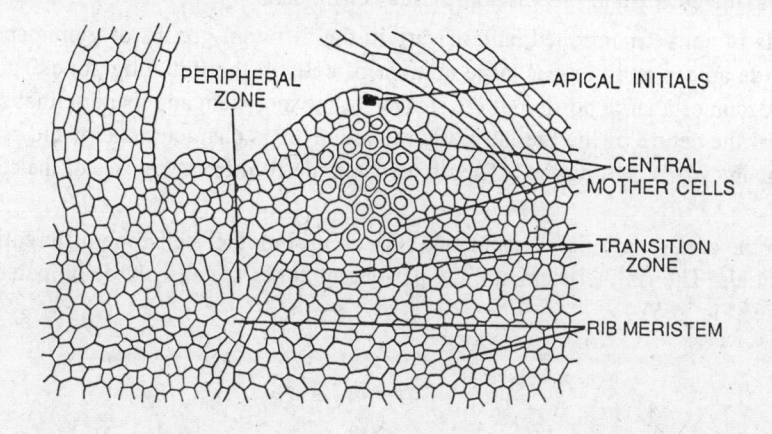

Fig. 6.4. The shoot apex. L.S. of shoot apex of *Pinus strobus*. Apical initials divide anticlinally contributing the cells to the surface layer, also divide periclinally adding the cells to central mother cell zone. The central mother cells divide actively and contribute to the transition zone. The products of these divisions form the rib meristem. (After K. Esau).

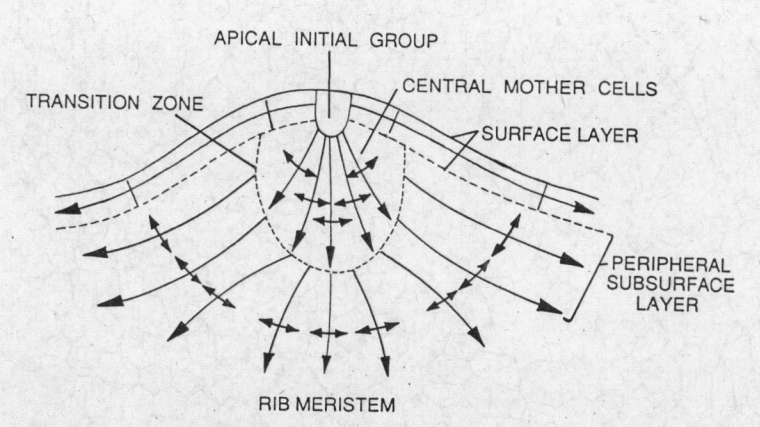

Fig. 6.5. The shoot apex. The zones and their mode of growth seen in the shoot apex of *Ginkgo biloba* in L.S. The apical initial group gives rise to the surface layer by anticlinal divisions, and adds to the central mother cells by periclinal divisions. The outermost products of divisions in the mother-cell zone are shifted toward the transition zone where they divide periclinally. (After Foster).

ontogeny much information supports the view that in the ferns the types of apex with several initials is more primitive that the one with a single apical cell (Bower, 1890, 1935; Wardlaw, 1945).

Gymnosperms. They commonly show several interrelated growth zones which are derived from a group of surface initials. These initials divide periclinally resulting in the formation of a subsurface group of cells, known as *mother cells*. The cell division is quite slow in the interior of this group but is active on its periphery. The derivatives of the divisions along the flanks of mother-cell group combine with products resulting from anticlinal divisions of the apical initials. All together these lateral derivatives form a mantle-like peripheral zone of deeply stainable small cells which are less differentiated (eumeristem) than the mother cells and the cells or initiating zone. The derivatives produced at the base of the mother-cell zone become pith cells, and usually they pass through a rib-meristem form of growth. The part of the pith may arise from the peripheral zone. The peripheral mantle of cells, rich in cytoplasm, is the seat of origin of leaf primordia and of the epidermis, the cortex and the vascular tissues of the axis.

The details of above mentioned pattern vary in the different groups of gymnosperms. The cycads have wide apices and possess large number of cells in the initiating zone. This way the initiating zone occupies a large portion of the surface of the meristem and its periclinal derivatives converge toward the centre of the meristematic mound. This is characteristic of the cycads. The mother-cell zone in cycads is ill-defined. The rib meristem derived from the base of the central zone is conspicuous.

In *Ginkgo* the zones are quite conspicuous. The central mother-cell zone differentiates close to the apical initials. The rib meristem and the peripheral zones are sharply deliminated from the central zone. (see Fig. 6.5).

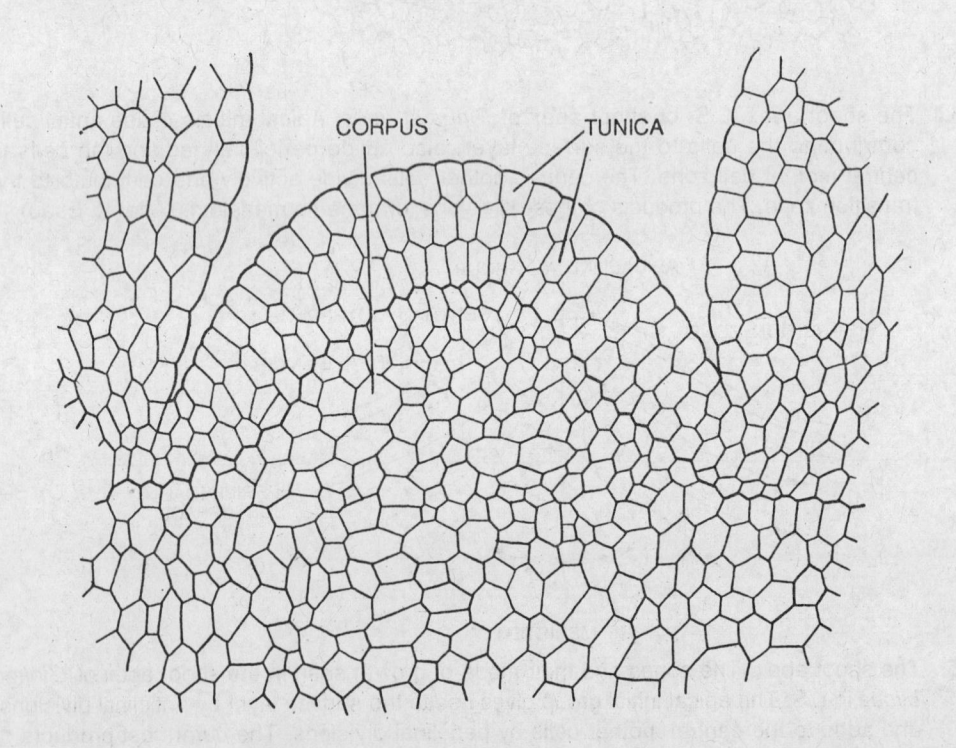

Fig. 6.6. Tunica and Corpus organization.

Apical Meristems

The apical zonation in *Pinus* and other Coniferales is less diversified than in the cycads and less well defined than in *Ginkgo*.

In Gnetales, there is a definite separation into a surface layer and an inner core derived from its own initials. The shoot apices of *Ephedra* and *Gnetum* (Johnson, 1950) have been described as having a tunica-corpus pattern of growth.

Angiosperms. There is *tunica-corpus organization* in the shoot apex of angiosperms. One to five layers tunica have been observed in the dicotyledons, and one to three-layered in the monocotyledons. However, tunica-corpus organization is not found in *Saccharum officinarum*. To draw a clear cut demarcation line in between tunica and corpus is not simple matter. In angiosperms, the number of parallel periclinal layers in the shoot apex may vary during the ontogeny of the plant body and under the influence of seasonal growth changes.

In the angiosperms the segregation of apical-meristem zones is more definite than in lower groups. There are two sets of initials, one above the other, which give rise to *tunica* and *corpus*. The tunica has no or only rare periclinal divisions and ranges in thickness from several layers to one with two or three layers probably most frequent. The number of layers in the tunica may vary even in an individual plant.

OTHER THEORIES OF SHOOT APEX ORGANIZATION

In support of shoot apex organization other theories have also been propounded. Dermen (1947) put forth his *Histogenic layer concent*. According to him there is no distinct layer of apical meristems. He named

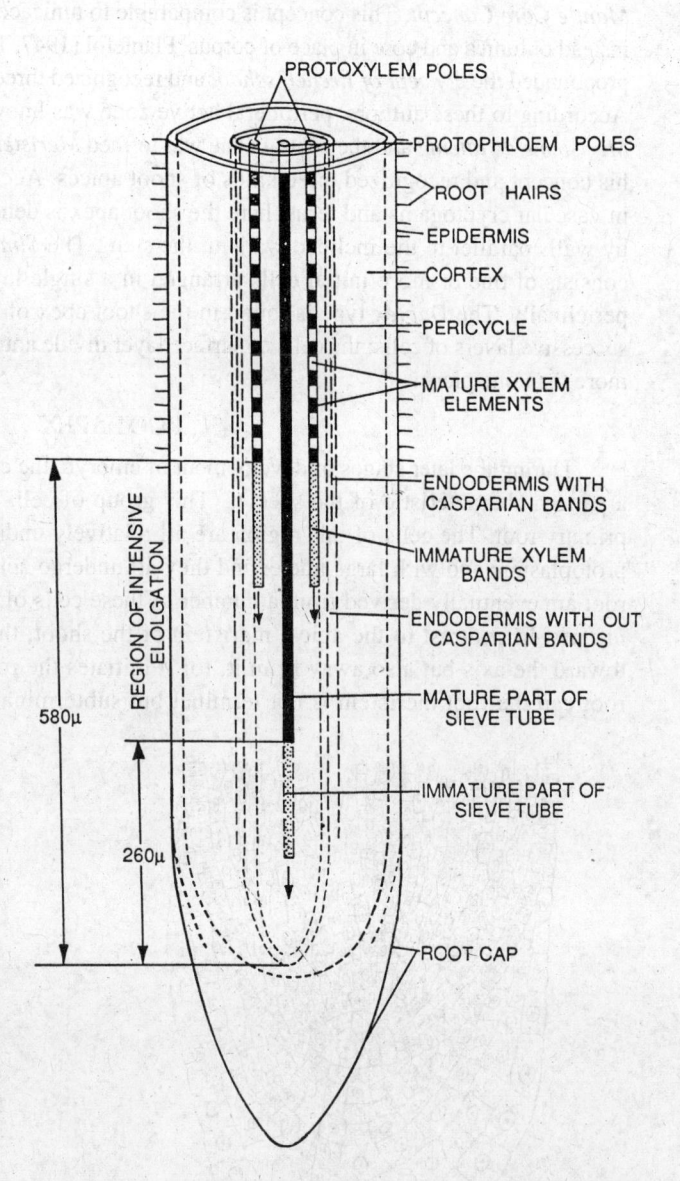

Fig. 6.7. L.S. of root tip of tobacco. The epidermis and the root cap show common origin. The cortex and the vascular cylinder have separate initials in the apical meristem. The pericycle is delimited close to the apical meristem. The sieve tubes mature first within the vascular cylinder. The root hairs develop beyond this region. The Casparian strips develop in the endodermis close to the position of first mature xylem elements. (After K. Esau).

the different layers of apical meristem as L_1, L_2, L_3, etc. He recognized these layers on the basis of their origin. However, this concept did not get any support. Popham and Chan (1950) put forth *Mantle Core Concept*. This concept is comparable to tunicacorpus theory. They used the term mantle instead of tunica and core in place of corpus. Plantefol (1947, 1950), Buvat (1955) and Amefort (1956) propounded the *concept of French school* and recognized three distinct regions in the apical meristem. According to these authors, peripheral active zone was known as *Anneu initial*, the zone next to it *Meristeme de attente* and the central zone was termed *Meristem medullaire*. Newman (1961) put forth his concept and recognized three kinds of shoot apices. According to him, *Monoplex* type is found in vascular cryptogams and ferns; here the shoot apex is denoted by one or more cells which divide by walls parallel to the inclined walls in the stem. The *Simplex* type is found in gymnosperms; it consists of one or more initial cells arranged in a single layer; these cells divide anticlinally and periclinally. The *Duplex* type is found in the shoot apex of angiosperms; it consists of atleast two successive layers of cells; the cells of surface layer divide anticlinally and that of inner layer divide in more than one plane.

ROOT APEX

During the later stages of development of embryo, the cells at the root pole become arranged in a pattern characteristic of the species. This group of cells comprises the *apical meristem* of the primary root. The cells of this region are all relatively undifferentiated and meristematic, densely protoplasmic and with large nuclei and they all undergo active division. The tissues of the mature root are eventually derived from a number of these cells of the apical meristem, which are termed *initials*. In contrast to the apical meristem of the shoot, that of the root produces cells not only toward the axis but also away from it, for it initiates the root cap and because of the presence of root cap the root meristem is not terminal but subterminal in its position, in the sense that it is

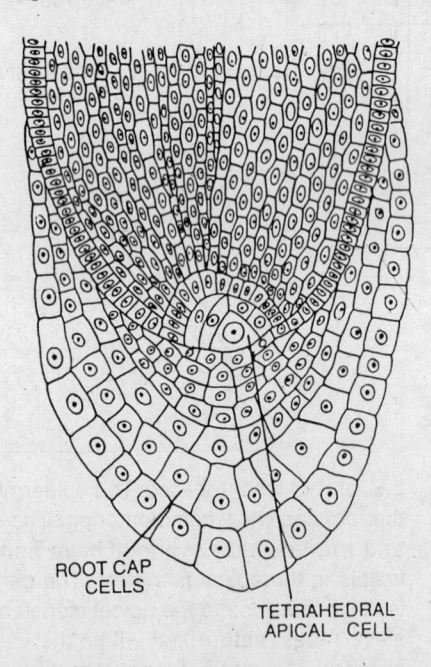

Fig. 6.8. Root apex of *Pteris* (fern) in longitudinal section showing tetrahedral apical cell.

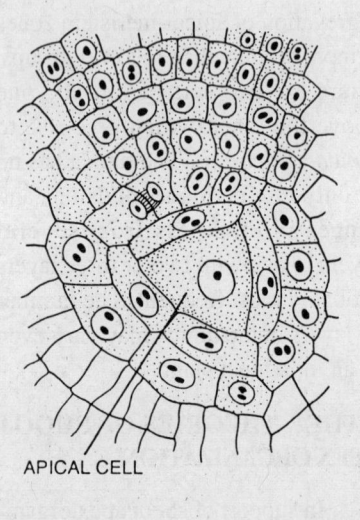

Fig. 6.9. Root apex of *Pteris* (fern) in cross section showing four-sided apical cell, one side being in contact with the root cap.

located beneath the root cap. The root apex also differs from the shoot meristem in that it forms no lateral appendages comparable to the leaves, and no branches. The root branches are usually initiated beyond the region of most active growth, and they arise endogenously. It also produces no nodes, and internodes, and therefore, the root grows more uniformly in length than the shoot, in which the internodes elongate much more than the nodes.

APICAL CELL THEORY

This theory was put forth by Nageli. In the roots of vascular cryptogams (pteridophytes), *e.g.*, *Dryopteris*, a single tetrahedral apical cell is present, it is generally thought that by its division this gives rise to all the tissues of the root. However, the apical cell theory was superceded by the histogen theory.

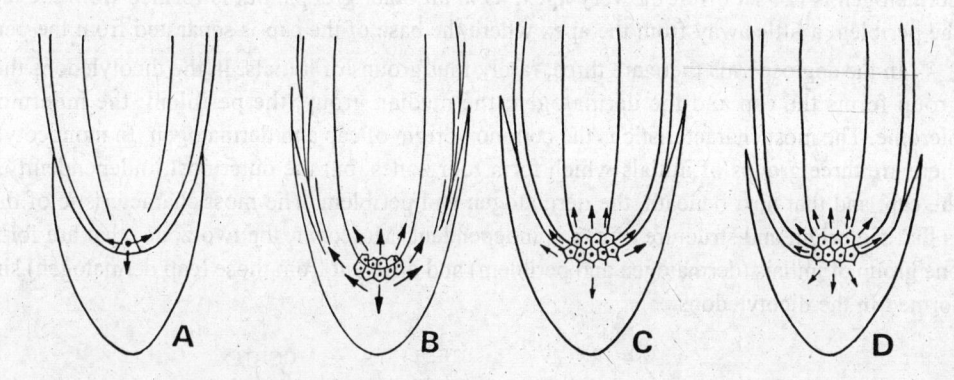

Fig. 6.10. The root apex. Diagrams showing different root apex types, A, initial solitary, cap distinct; B, initials in two groups, cap not distinct structurally; C, initials in three groups, cap not distinct; D, initials in three groups, cap distinct, and independent in origin. (After E. & M.)

In number, the initials range from one to many. Where the initials are more than one, they are arranged in one to four fairly distinct, uniseriate groups. In each group there are one to several initials. Where there is more than one group, the groups lie adjacent to one another on the longitudinal axis of the root. Each of these groups quickly develops one or more growth zones. In many plants these zones appear to represent 'the histogens'. The terms dermatogen, periblem and

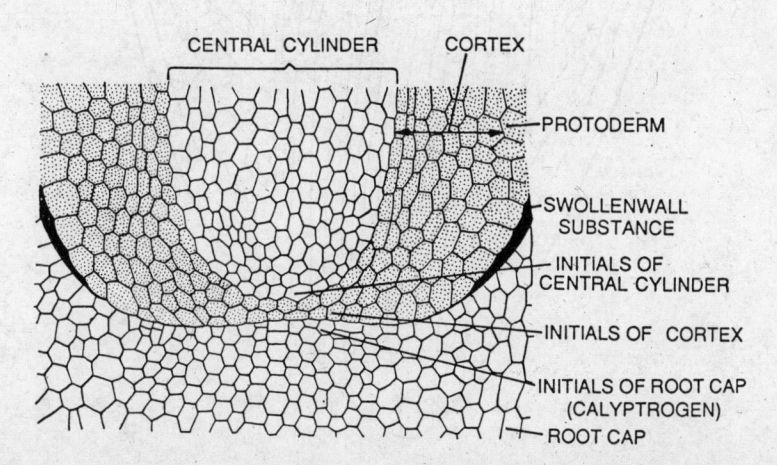

Fig. 6.11. The root apex. Root tip of *Zea mays* in L.S. The swollen wall substance originates through the gelatinization of the wall between the root cap and the protoderm.

plerome are no longer in general use in descriptions of stem ontogeny but they have been continued to indicate general zones in studies of root development. A fourth histogen, the *calyptrogen*, is added where the cap has an independent origin. There are basic patterns for the major plant groups. The pattern is determined by the number of initials, the number of groups of these initials, the zones formed by each group, the morphological nature of the cap, and the degree of independence of the cap.

The vascular cryptogams such as horsetails, most of the ferns, some species of *Selaginella* have a solitary apical cell in the root. This one cell forms the entire root and the cap.

In many gymnosperms there are two groups of initials. The inner forms the plerome; the outer forms the periblem and the cap. The cap appears as a distal proliferation of the periblem. A dermatogen is not set off at the very apex, as in all other groups, but is formed from the layers of the periblem a little away from the apex where the base of the cap is separated from the periblem.

In the angiosperms there are three, rarely four groups of initials. In the dicotyledons the distal group forms the cap and the dermatogen; the median group, the periblem; the innermost, the plerome. The most characteristic is the common origin of cap and dermatogen. In monocotyledons, there are three groups of initials which form four zones, but the outermost, independently, forms the cap, and that next beneath, the dermatogen and periblem. The most characteristic of this type is that the origin and structure of cap is independent. Moreover, the two zones that are formed by one group of initials (dermatogen and periblem) and different from those (cap dermatogen) similarly formed in the dicotyledons.

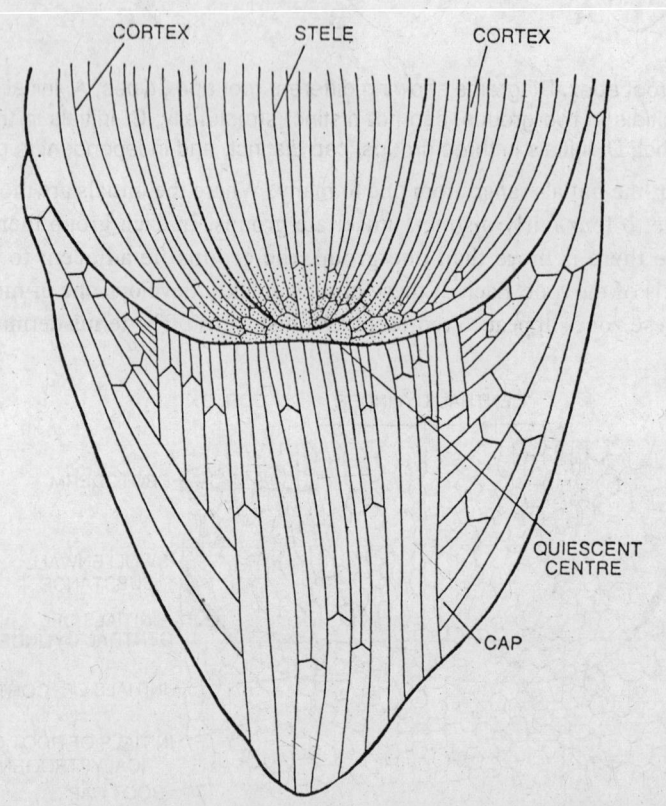

Fig. 6.12. Root apex. Pattern of cell lineages in the root apex of Zea mays. The cortex, vascular cylinder or stele, root cap and quiescent centre are indicated. (After Clowes).

Korper-kappe theory

This theory was put forth by Schuepp in 1917. Since the root changes in diameter during growth, there are various points at which a single longitudinal file of cells has become a double file as a result of cell division. At these points a cell first divides transversely and thereafter one of its daughter cells divides longitudinally. This was known as T division, because the cell walls form a T-shaped structure. In some zones of the root, mainly in the centre, the bar of the T faces the root apex, in other it faces away from the apex ($\perp$). These zones of the root, delimited by the planes of cell division, were called *Korper* (body) and *Kappe* (cap) respectively. This theory may be compared with the tunica-corpus theory in the shoot apex.

Mainly in roots with a very regular arrangement of cells in the apical meristem, such as of *Zea mays*, it is possible to conclude from the study of cell lineages that there is a central region of cells which divide rarely or not at all. The cells on the periphery of this hemispherical or cup-shaped region are meristematic. This inactive or passive region of cells is known as '*quiescent centre*'.

The Quiescent Centre

In the apical meristem of root of *Zea mays*, and other plants with a regular arrangement of cells in the apical meristem, it is possible to conclude from the study of cell lineages that there is a central region of cells which divide, rarely or not at all. These inactive or passive cells constitute the *quiescent centre*. The cells on the periphery of this hemispherical or cupshaped region are meristematic and may be regarded as the constituents of the promeristem (Clowes, 1958). By various techniques, the existence of quiescent centre, has now been demonstrated in the root apices of a considerable number of species. The quiescent centre develops during the ontogeny of the root (Clowes, 1958). A quiescent centre is not found in the roots with a single apical cell. In 1956, Clowes was able to show that there was a central region (quiescent centre) in the roots of *Zea* where the cytoplasm had the lower content of RNA and where the cells had smaller nucleoli. He was also able to demonstrate that the cells in the quiescent centre did not actively synthesize DNA.

The physiological and cytological properties of the cells in the quiescent centre have now been studied in a number of species. The cells in the region have a lower concentration of DNA, RNA and protein than any other cells in the root apex (Clowes, 1958; Jensen, 1958). The cells of quiescent centre also have fewer mitochondria, little endoplasmic reticulum, and the smallest dictyosomes, nuclei and nucleoli (Clowes, 1964). They are less sensitive to radiation damage than other cells of the meristem (Clowes, 1959, 1964).

The function of the quiescent centre may be to provide a reserve block of diploid cells within the root. The quiescent centre may be the site of hormone synthesis.

THE PROMERISTEM

Clowes (1961) has defined the promeristem as that part of the root apex which is capable of giving rise to all the tissues of the root. In the roots of vascular cryptogams, for example, the promeristem would consist of apical cell only, and in angiosperms it would comprise the initials of the histogens. This way there is a tendency to regard the promeristem as a rather small region, situated terminally in the root apical meristem, below the root cap. On the basis of modern work it has been suggested that in many roots the promeristem is broad and consists of a somewhat cupshaped group of cells on the periphery of a central inactive region. This grouping of the initial cells of the promeristem was suggested by Clowes (1950) on the basis of an anatomical study of the root apex of *Fagus sylvatica*.

THEORIES OF STRUCTURAL DEVELOPMENT AND DIFFERENTIATION

As has been discussed by several workers (Foster, 1939, 1941; Sifton, 1944; Wardlaw, 1945), the view regarding the number, the arrangement, and the activity of the initial cells and their derivatives in the apical meristems has undergone many changes since the shoot apex was first recognized by Wolff in 1759 as an undeveloped region from which the growth of the plant proceeded.

The Apical Cell Theory

This theory was put forth by Nageli in 1858. Solitary apical cells occur in many of algae, bryophytes and vascular cryptogams (pteridophytes). The discovery of the apical cell in cryptogams led to the concept that such cells exist in phanerogams (seed plants) as well. The apical cell was interpreted as a constant structural and functional unit of apical meristems governing the whole process of growth. However, this was confirmed by later researches that this theory may hold good for cryptogams but is not applicable to the phanerogams. Further researches have refuted the universal occurrence of apical cells and replaced it by a concept of independent origin of different parts of plant body. The apical cell theory was superceded by the *histogen theory*.

The Histogen Theory

This was introduced in 1870 by Hanstein who considered that the primordial meristem was sharply separable into three distinct zones or *histogens*. According to this theory the apical meristem or growing region of the stem and root are composed of small mass of cells which are all alike and are in a state of division. These meristematic cells constitute promeristem. The cells of the promeristem soon differentiate into three regions — *dermatogen*, *periblem* and *plerome*. Every zone consists of a group of initials and is called a *histogen* or a *tissue builder*. *Dermatogen* — This is the single outermost layer of the cells which later gives rise to the epidermis of the stem. In the root it is also single layered, but at the apex it merges into the periblem and just outside the periblem the dermatogen cuts off many new cells resulting into a small celled tissue, the *calyptrogen*, which is also meristematic and gives rise to the *root cap*. *Periblem* — This region is found internal to the dermatogen, and is the middle region of the apical meristem. It is single layered at the apex but in central part it becomes multilayered. It develops into the cortex of the stem. In the roots it is also single layered at the apex and many layered in the central portion. In the case of root, it also develops into the cortex. *Plerome* — It is the central meristematic region of stem apex and lies internal to the periblem. It is also composed of thin walled isodiametric cells. Ultimately it develops and differentiates into the central stele consisting of primary vascular tissues and ground tissues, such as, pericycle, medullary rays, and medulla. In the roots the function of plerome is practically same as in stem. At a little distance behind the apex certain strands of cells show a tendency to elongate. These strands of elongated cells make the *procambium*. The procambial strands ultimately become differentiated, into vascular bundles. A portion, however, remains undifferentiated, and it forms the cambium of the vascular bundle.

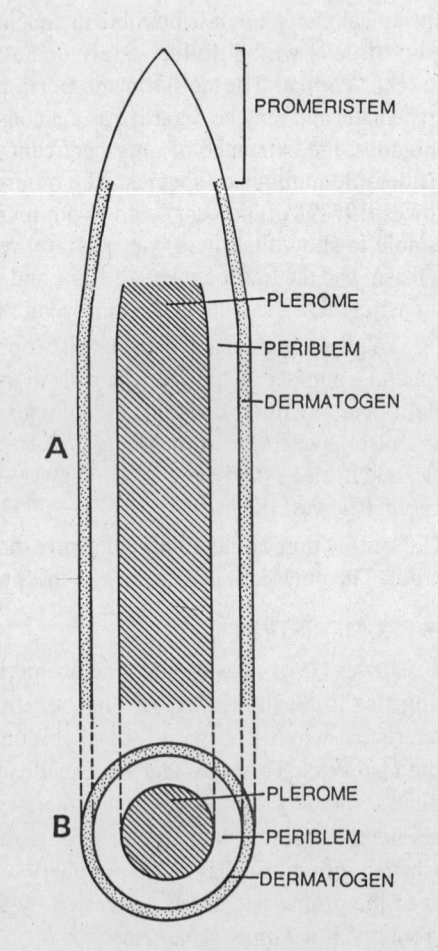

Fig. 6.13. The shoot apex—diagram showing histogen regions. A, longitudinal section; B, transverse section. (After E. & M.).

Apical Meristems

Recent investigations have revealed that there is no strict relationship between the development of the histogens and various regions of plant body and the segmentation and layering of the cells in the apical meristem. However, the distinction of these histogens in an apex cannot be made in some plants, and in others the regions have no morphological significance.

The Tunica-Corpus Theory.

This theory was put forth by Schmidt in 1924. The apical cell theory and the histogen theory were developed with reference to both the root apex and the shoot apex. Later attention became centred largely on shoot apices, and with the result the tunica corpus theory was developed. According to this theory, there are two zones of tissues in the apical meristems — *the tunica* consisting of one or more peripheral layers of cells, and the *corpus*, a mass of cells enclosed by the tunica. According to this theory different rates and method or growth in the apex set apart two regions. The layers of the tunica show predominantly anticlinal divisions, that is, they are undergoing surface growth. In the corpus the cells are large, with arrangement and planes of cell division irregular, and the whole mass grows in volume. Each layer of the tunica arises from a

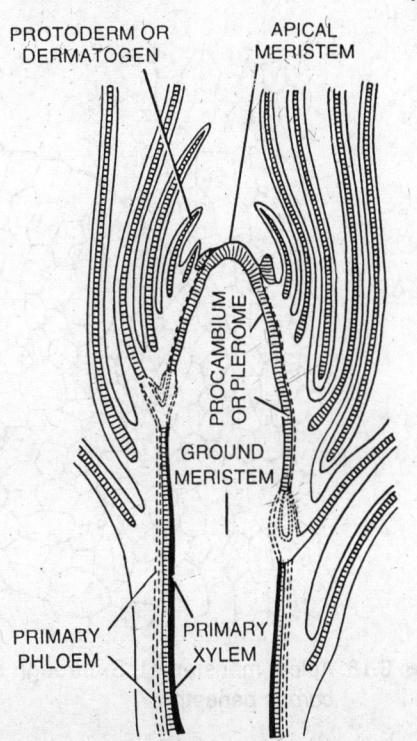

Fig. 6.14. Apical meristems. L.S. of stem tip (diagrammatic) showing the histogens.

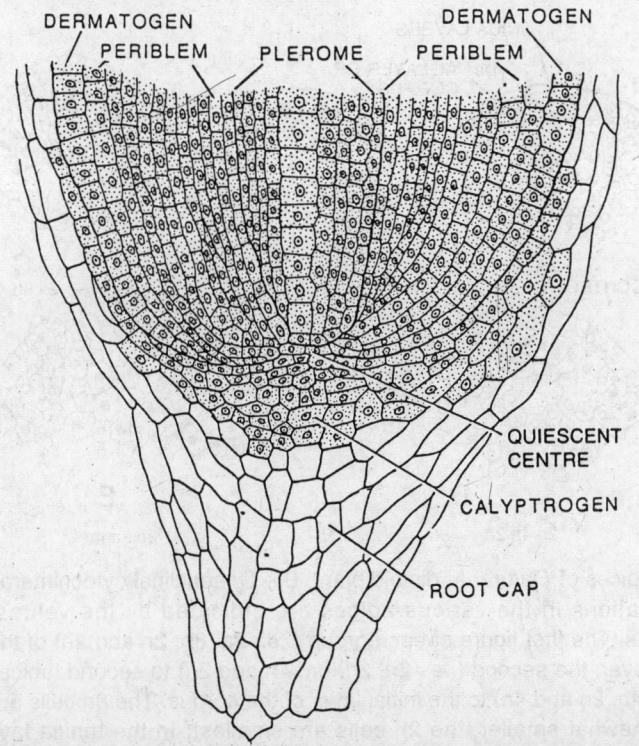

Fig. 6.15. Root apex. L.S. of a tip of onion root showing histogens and root cap.

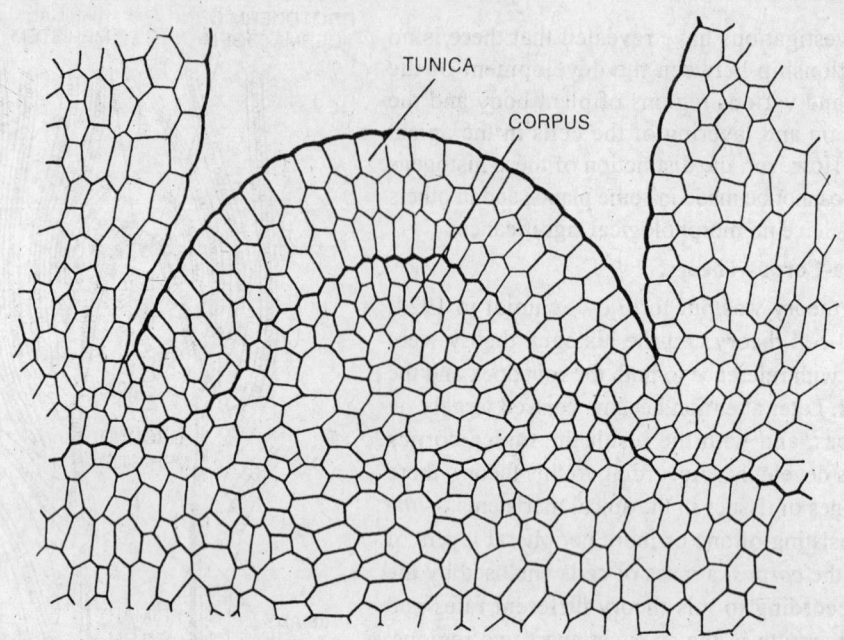

Fig. 6.16. Apical meristem. L.S. through shoot apex of *Vinca*, showing three-layered *tunica* and *corpus* beneath it.

group of separate initials, and the corpus has one layer of such initials. In the tunica the number of layers of initials is equal to the number of layers of tunica, that is, each layer of tunica has its own layer of initials. The corpus arises from a single tier of initials which divide first periclinally to give rise

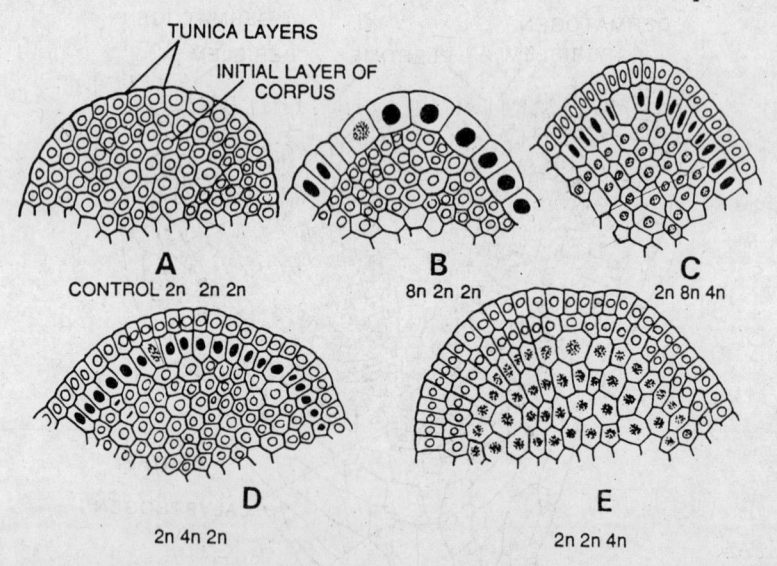

Fig. 6.17. Shoot apices of *Datura*. A, diploid plant; B – C, periclinal cytochimeras. The chromosomal combinations in the various apices are indicated by the values given below each diagrams. The first figure of each group (*i.e.,* 2n, 8n, 2n and 2n) of three refers to the first tunica layer; the second (*i.e.,* 2n, 2n, 8n, 4n and 2n) to second tunica layer; the third (*i.e.,* 2n, 2n, 4n, 2n and 4n) to the initial layer of the corpus. The 8n cells are largest the 4n cells are somewhat smaller, the 2n cells are smallest. In the tunica layer the divisions are anticlinal, in the corpus they occur in various planes. (After Satina *et. al.*, 1940).

to group of derivatives, which divide in various planes resulting in the formation of the inner mass of cells.

The number of initials varies from few to many. For example, in small very slender apices, such as those of grass seedlings, there may be only one or two in the tunica and about two in the corpus.

In vascular plants, the differentiation of the zones of stem-apex follows more or less definite patterns that seem to be characteristic of the major groups. These patterns show increasing complexity from the lower to the higher groups and appear to represent a series in specialization from simplicity to complexity.

As regards the concept of tunica and corpus, there may be several types which are found in the stem apices of several vascular plants. The types may be as follows.

The primitive type of stem apex having no distinction of tunica and corpus. *Lycopodium*, *Isoetes*, *Selaginella* (pteridophytes) and *cycads* (gymnosperms) belong to this group. They have simple apices with surface initials and no distinction of tunica and corpus. In *Lycopodium*, the initiating layer is weakly defined, having uniseriate surface area which divides freely both anticlinally and periclinally. Here all the cells of the layer are morphologically alike. The anticlinal divisions increase the area of the surface layer, whereas the periclinal divisions form the inner core.

The stem apices with weak tunica-corpus demarcation. The demarcation of tunica and corpus layers begins in some of the lower conifers. In *Abies* and *Pinus* (Coniferales), the initials make an apical uniseriate group. These initials further give rise to a central core and an enveloping uniseriate layer by both periclinal and anticlinal divisions. The uniseriate layer that envelops the central core suggests a tunica in appearance, but there is no clearcut demarcation between the tissues of two regions. However, in the apices of *Sequoia sempervirens*, the initials are a small group of surface cells in one tier, with both anticlinal and periclinal divisions. These divisions result in the formation of a dermatogen like layer and a central mass. The outer layer suggests a tunica and the central mass, the corpus. The species of *Cryptomeria* and *Taxodium* (Coniferales) have a dermatogen, in which there are no periclinal divisions. There appears to be structural segregation of tunica and corpus in the apices of many Coniferales, but there is only one tier of initials and no independent meristematic regions are recognized.

The stem apices with distinct tunica and corpus. In angiosperms the demarcation of meristematic zones of apical region is usually more distinct and definite than in lower groups. There are two sets of initials, one above the other, which give rise to tunica and corpus which seem to be completely independent. The tunica has no or only rare periclinal divisions. It ranges in thickness from one to several layers. Usually there occurs two or three layers. The larger numbers of tunica layers occur more frequently in the dicotyledons. A single-layered tunica occurs in the grasses. However, in monocotyledons the number of tunica layers is one to three. In *Zea*, tunica divides periclinally, which shows an exceptional condition. The number of the layers in tunica may vary even in an individual plant. The corpus varies from a large complex type to a slender, simple type.

Significance of the tunica corpus theory. The tunica-corpus theory served well in the establishment of meristematic patterns of the shoot apices of seed plants. The position, number and behaviour of the initiating cells in seed-plant stems, and early stages in the development of primary body of the shoot are now much better understood. The tunica-corpus theory is of topographical value in studies of detailed development. The lateral organs of the stem, *i.e.*, leaves, branches and floral organs, arise near the apex and studies of tunica and corpus have added greatly to a knowledge of origin and early development of these organs.

ORIGIN OF LEAVES

A leaf initiates by periclinal divisions in a small group of cells at the side of an apical meristem. In angiosperms, the tunica and the corpus are responsible for leaf initiation. In the dicotyledonous plants the periclinal divisions initiating the leaves occur, not in the surface layer, but in one or more layers beneath it. If the tunica is single-layered, such divisions take place within the corpus, otherwise they occur both in tunica and corpus or in the tunica only. In certain monocotyledonous plants the superficial tunica layer undergoes periclinal divisions and gives rise to some or most of the tissue. In the case of gymnosperms the leaves initiate from the peripheral tissue zone. In the vascular cryptogams (pteridophytes) the leaves are initiated either from single superficial cells or from groups of such cells.

The periclinal divisions which initiate a leaf primordium are responsible for the formation of a lateral prominence on the side of the shoot apex. This prominence constitutes the leaf base which is also known as *leaf buttress*. Subsequently the leaf grows upwardly from the buttress. As shown in the figure, in *Hypericum* spp., the apical meristem is less prominently elevated above the youngest leaf buttress.

Before the initiation of a new leaf primordium the apical meristem appears as a rounded mound. It gradually widens, and, then leaf buttresses are initiated on its sides. While the new leaf primordia grow upward from the buttresses, the apical meristem again becomes like a small mound.

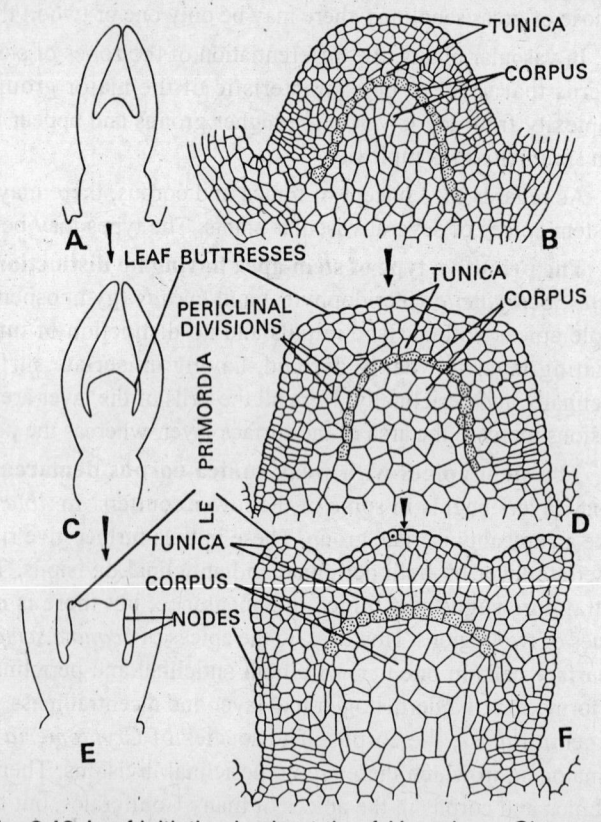

Fig. 6.18. Leaf initiation in shoot tip of *Hypericum*. Change in shape and histology of shoot apex. Longitudinal sections are seen in this figure. Leaves are in pairs at each node, in decussate arrangement. Protrusions in axis below the leaves in A, C and E are leaf bases of next lower pair of leaves; B, D and F, stippling indicates outer-boundary cells of corpus and their immediate derivatives; F, four sided figure indicates the presumptive place of origin of an axillary bud.

ORIGIN OF BRANCHES

In angiosperms, branches commonly are initiated in close association with the leaves — They originate in the axils of the leaves, and in their nascent state they are known as *axillary buds*. The axillary buds commonly initiate somewhat later than the leaves subtending them and therefore, it is not always clear whether the meristem of the axillary bud is derived directly from the apical meristem of the main shoot or whether it originates from partly differentiated tissue of the internode. Both situations may occur because plants vary with regard to time of appearance of axillary buds. On the

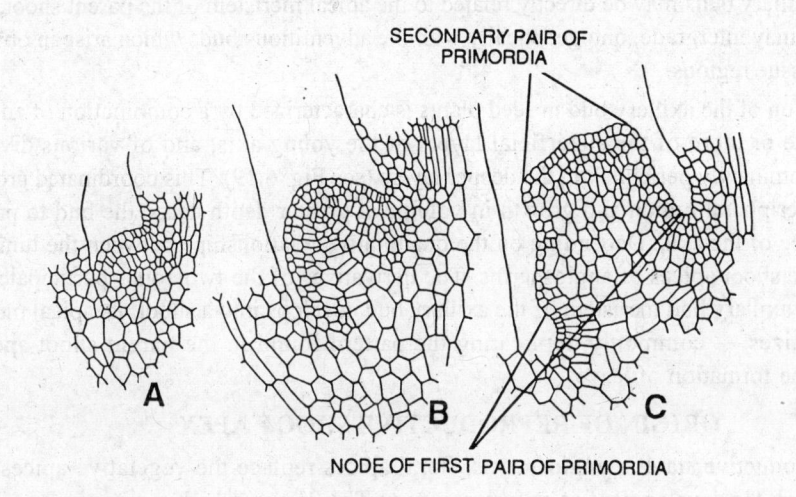

Fig. 6.19. Initiation of axillary bud in *Hypericum*. It is developed by derivatives of three layers of tunica of bud (A – C). Third layer divides periclinally and gives rise to third and fourth layers of tunica and to corpus of bud, C, second pair of leaf primordia is being initiated.

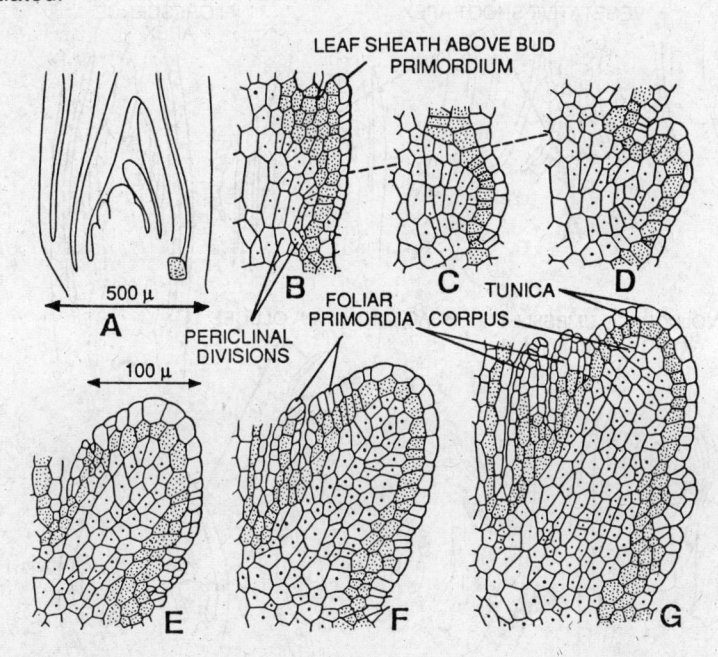

Fig. 6.20. Initiation of lateral bud in *Agropyron*. A, low-power view of shoot tip with several leaf primordia; stippled part indicates position of bud; it is developed by derivatives of the two-layered tunica and the corpus. B – G derivatives of second layer of tunica are stippled, and those of corpus indicated by a single dot in each cell. Bud is initiated by periclinal divisions in corpus derivatives (B, C). Anticlinal divisions occur in tunica derivatives; bud emerges above surface of stem (D). Tunica derivatives remain in a biseriate arrangement at apex of bud and form its two-layered tunica (E, G). Foliar primordia arise on bud (E – G).

one hand, the axillary buds may be directly related to the apical meristem of the parent shoot; on the other hand they may intergrade, ontogenetically with the adventitious buds which arise in obviously differentiated tissue regions.

The initiation of the axillary bud in seed plants is characterized by a combination of anticlinal divisions, in one or more of the superficial layers of the young axis, and of various divisions, sometimes predominantly periclinal, in the deeper layers (see Fig. 6.19). This coordinated growth in surface of the peripheral region and growth in volume at greater depth cause the bud to protrude above the surface of the axis. Depending on the quantitative relationships between the tunica and the corpus in the shoot apices of angiosperms. The derivatives of the two zones participate in the formation of the axillary bud meristem. If the axillary bud develops into a shoot, its apical meristem gradually organizes — commonly duplicating the pattern found in the parent shoot apex and proceeds with the formation of leaves.

ORIGIN OF REPRODUCTIVE SHOOT APEX

In the reproductive state in angiosperms, floral apices replace the vegetative apices either directly or through the development of an inflorescence. The flower, which may occur singly or as part of an inflorescence, is formed during the reproductive phase of growth. It develops from a terminal or lateral vegetative shoot apex and results in the culmination of meristematic activity of

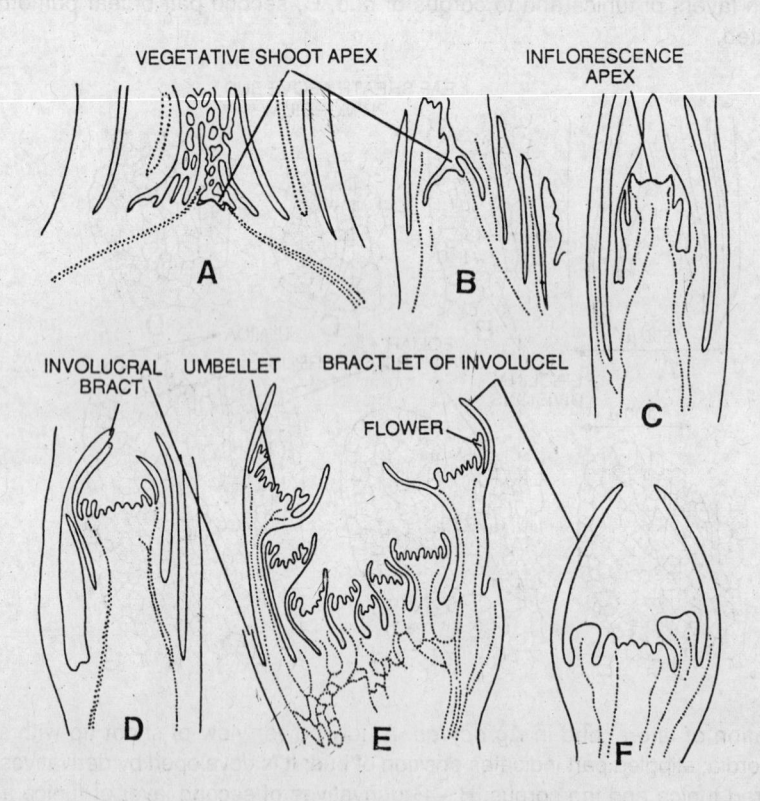

Fig. 6.21. Transformation of apical meristem during shift from vegetative growth to development of floral apices in carrot. The inflorescence is a compound umbel. A, vegetative shoot apex at base of rosette of leaves; B, shoot apex at approach of reproductive stage by internodal elongation. C, D, flattened inflorescence apices producing bracts and umbel primordia; E, compound umbel in young state; F, each flower develops a flat apex and forms floral organs.

that particular meristem. Thus, the floral apex, like the leaf primordium and unlike the vegetative shoot apex, shows determinate growth.

The change to the reproductive stage may be first detected by the modified growth habit of the shoot. When the flowers develop an axillary-branch inflorescence, there appears an acceleration in production of axillary buds, which shows one of the earliest indication of approaching flowering. Simultaneously, the nature of foliar organs subtending the axillary buds also changes. They develop as bracts more or less distinct from the foliage leaves. Here during the reproductive stage the axillary buds appear earlier and grow more vigorously than the subtending bract primordia. The next feature that reveals the beginning of the reproductive state is the sudden increase in the elongation of internodes.

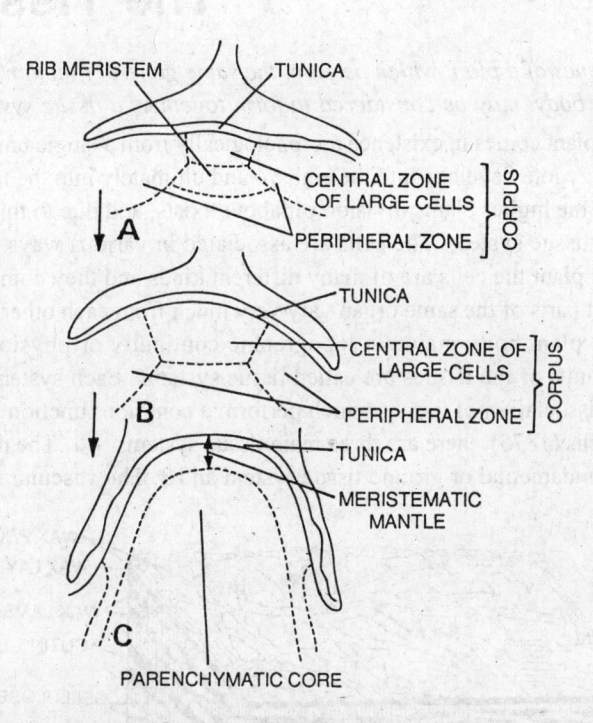

Fig. 6.22. Modification in Zonation of vegetative apex during change into inflorescence in *Succisa*. A, apex when forming foliage leaves, B – C, two stages of development of inflorescence, B, initiation of inflorescence is associated with cessation of growth in length and disappearance of rib-meristem, C, central and peripheral zones are reorganized to form, together with the tunica, a meristematic mantle, overarching a parenchymatic core.

From the viewpoint of histology and cytology the reproductive meristem differs from the vegetative meristem in varying degrees. It may have the same quantitative relationship between the *tunica* and the *corpus* as was present in the vegetative apex, or the number of separate surface layers may be reduced or increased. The most conspicuous change is exhibited in the distribution of the eumeristematic and the more highly vacuolated cells. In many species the apex of the inflorescence or the flower shows a uniform, small-celled mantle-like zone of one or more layers enclosing a large celled core; this type of apex may be flatter and wider than the vegetative one. It is not necessary that the mantle may coincide with the tunica; a part of corpus may be included in it. The cells of the central tissue enlarge and become vacuolate, and the meristematic activity remains restricted to the mantle zone. This activity is concerned only with the production of floral organs.

7

The Tissue System

"All the tissues of a plant which perform the same general function, regardless of position or continuity in the body, may be considered to form together, a tissue system."

A vascular plant comes in existence morphologically from a single unicellular zygote. By further development the zygote develops into the embryo and ultimately into the mature sporophyte or plant body. Usually in the higher plants division of labour exists, and due to this, the tissues are arranged in system called tissue system. The cells are associated in various ways with each other and form tissues. In higher plant the cells are of many different kinds and they combine into tissues in such a way that different parts of the same organ vary very much from each other. The arrangement of cells and tissue in the plant body maintain topographic continuity or physiologic similarity, or both, together. These units of the tissues are called tissue systems. Each system usually consists of only one tissue or an association of tissues which perform a common function and have the same origin. According to Sachs (1875), there are three main tissue systems —1. The dermal or epidermal tissue system. 2. The fundamental or ground tissue system and 3. The vascular tissue system.

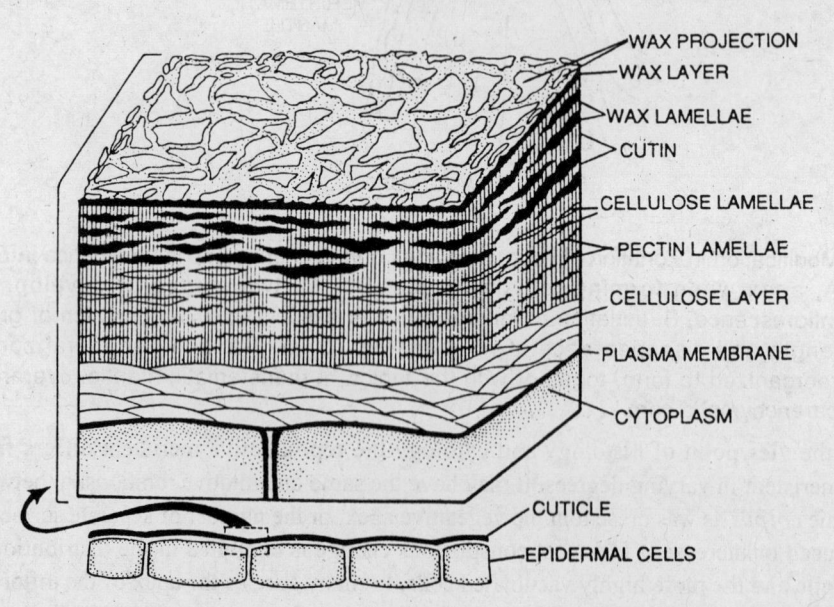

Fig. 7.1. The Epidermis. Schematic representation of the structure and composition of the cuticle and cell wall of foliar epidermal cells.

1. THE DERMAL OR EPIDERMAL TISSUE SYSTEM

The dermal system forms the outer protective covering of the plant and is represented, in the primary plant body, by the *epidermis*. During secondary growth the epidermis may be replaced by another dermal system the *periderm*, with the cork cells forming the new protective tissue.

THE EPIDERMIS

The epidermis usually consists of a single layer of cells which cover the whole outer surface of the plant body. The word is derived from two words of Greek origin, *epi*, upon, and *derma*, skin. It is a continuous layer except for certain small pores, called stomata and lenticels. According to the histogen theory introduced by Hanstein in 1870, it is derived from dermatogen of apical meristem. In meristematic regions it is, of course, undifferentiated, and in older stem and roots it may have been destroyed by secondary growth. Mostly the epidermis is single layered, but in many plants it has been described as bi-or multiseriate. In the leaves of India rubber plant

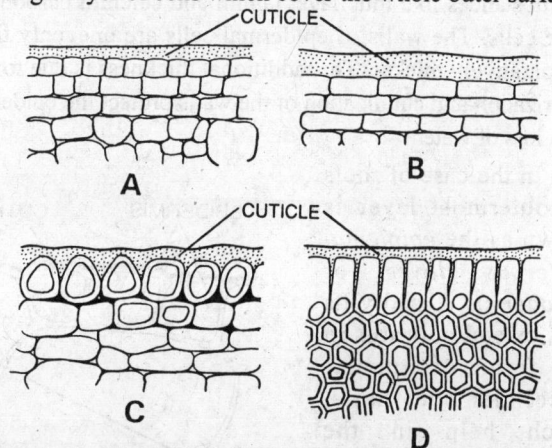

Fig. 7.2. The cuticle. A, in fruit of *Citrus sinensis*; B, in fruit of *Malus pumila*; C, in *Dracaena* stem; D, in leaf of *Dasylirion serratifolium*. (After Eames and Mac Daniels.)

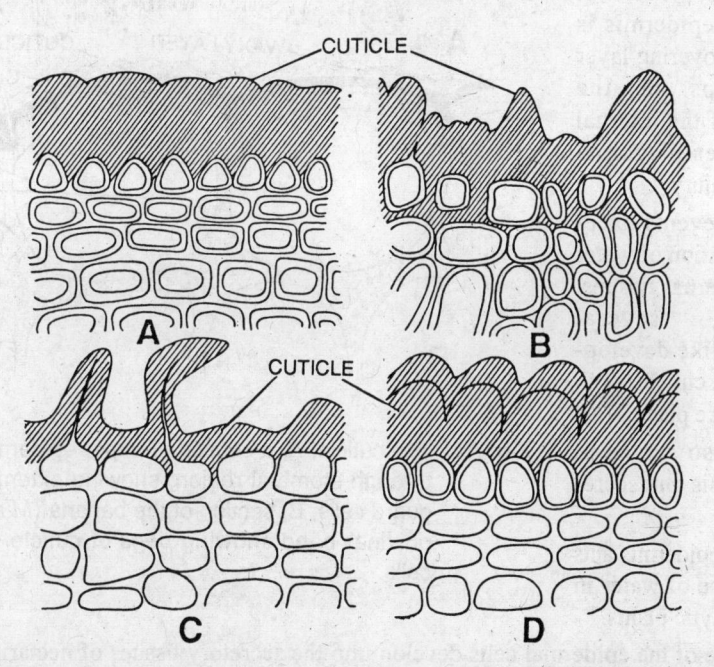

Fig. 7.3. The cuticle. The figure shows the section of the epidermis and adjacent tissues. A, thick cuticle of *Acer* stem, the older outer layer cracking and disintegrating; B, thick cuticle of *Smilax* stem; C, thick cuticle of *Vaccinum* stem; D, thick cuticle of *Cornus* stem, showing two stages, the older, outer layers are replaced by new ones developed below.

(*Ficus elastica*), banyan tree (*Ficus bengalensis*), oleander (*Nerium* spp.), etc. It becomes two to multilayered. The epidermal cells may be somewhat irregular in outline, usually varying in shape and size and arranged very close to each other having no intercellular spaces among them. The cells possess a large central vacuole and thin peripheral cytoplasm. The cells may contain leucoplasts, anthocyanins and chromoplasts, but no chloroplasts except in guard cells. In the epidermal cells of certain aquatic (*e.g., Hydrilla*) and shade loving plants the chloroplasts are also found. Sometimes, the substances like mucilage, tannin and calcium carbonate crystals (cystoliths) are also found in these cells. The walls of epidermal cells are unevenly thickened. The inner and radial walls are comparatively thicker. This additional thickness is due to the impregnation of suberin or cutin. The suberization and cutinization of the walls protect the epidermis from mechanical injuries and prevent from loss of water.

In the case of roots the outermost layer is known as the *epiblema, piliferous layer* or *rhizodermis*. Usually its cells extend outwards in the form of tubular unicellular root hairs, which help in the absorption of water and mineral nutrients from the soil.

Functions of the Epidermis

1. The epidermis is primarily a covering layer which helps in the protection of the internal soft tissues against mechanical injury.

2. It prevents excessive evaporation of water from the internal tissues, for this, several adaptations like development of thick cuticle, wax, hairs, etc., take place.

3. It also serves in photosynthesis and secretion.

4. The epidermis acts as store house of water in many xerophytic plants.

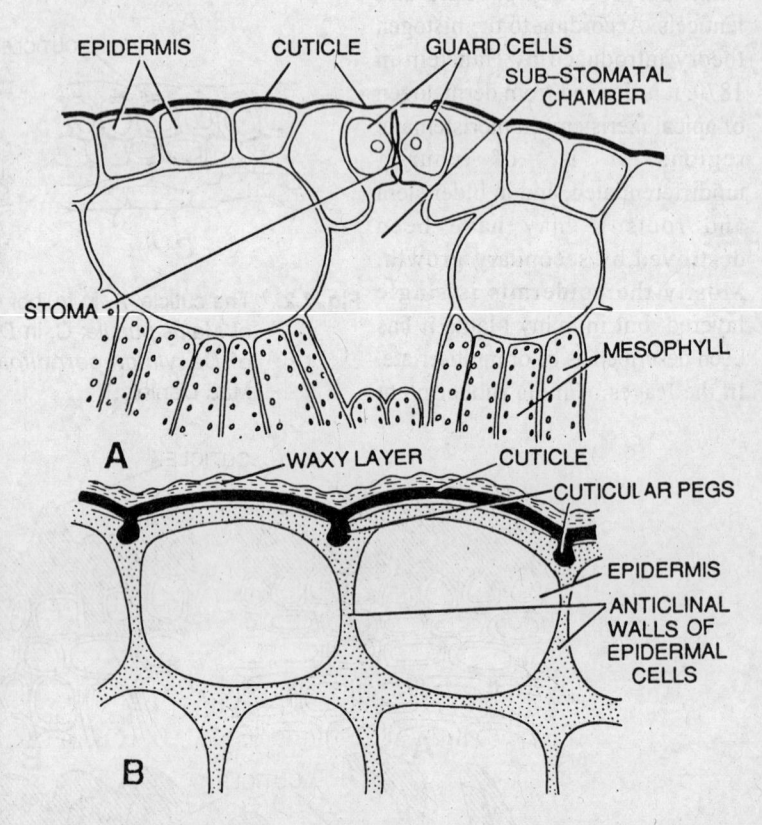

Fig. 7.4. The cuticle. A, a section of upper epidermis of *Musa* leaf through stomatal region, showing extent of cuticle over guard cells; B, section of the banana (*Musa*) leaf through pulvinar band showing *pegs* of cuticle over epidermal cells.

5. Some of the epidermal cells develop into the secretory tissues of nectaries, the stomata of leaves and stems, and the absorbing hairs of roots.

STOMATA

The stomata are minute pores which occur in the epidermis of the plants. Each stoma remains surrounded by two kidney or bean shaped epidermal cells the *guard cells*. The stomata may occur on

The Tissue System

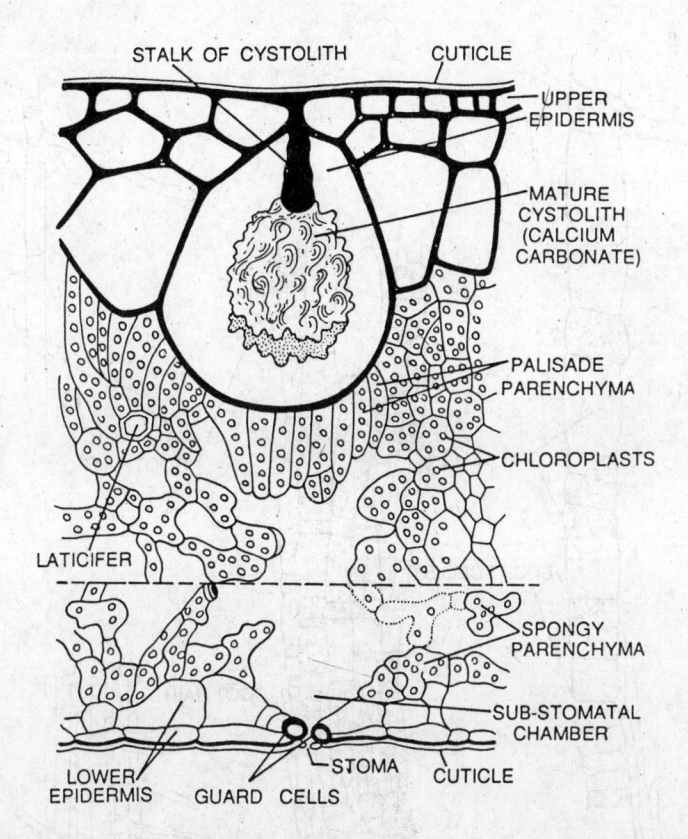

Fig. 7.5. Epidermis. Multiple epidermis in T.S. of *Ficus elastica leaf*. A mature cystolith with calcium carbonate deposited on its stalk in the epidermal cell.

any part of a plant except the roots. The epidermal cells bordering the guard cells are called *accessory cells* or *subsidiary cells*. Generally the term stoma is applied to the stomatal opening and the guard cells. The guard cells are living and contain chloroplasts in them. They also contain a larger proportion of protoplasm than other epidermal cells. Usually in the leaves of dicotyledons the stomata remain scattered whereas in the leaves of monocotyldons they are arranged in parallel rows. The

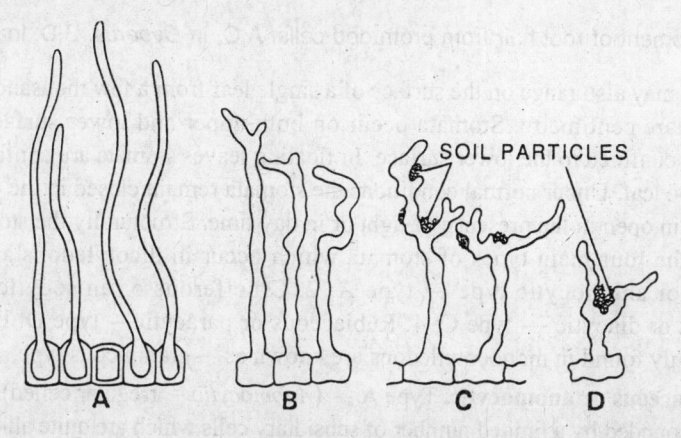

Fig. 7.6. Root hairs, A, grown in water; B, grown in moist soil; C and D, grown in dry soil. (After Schwarz).

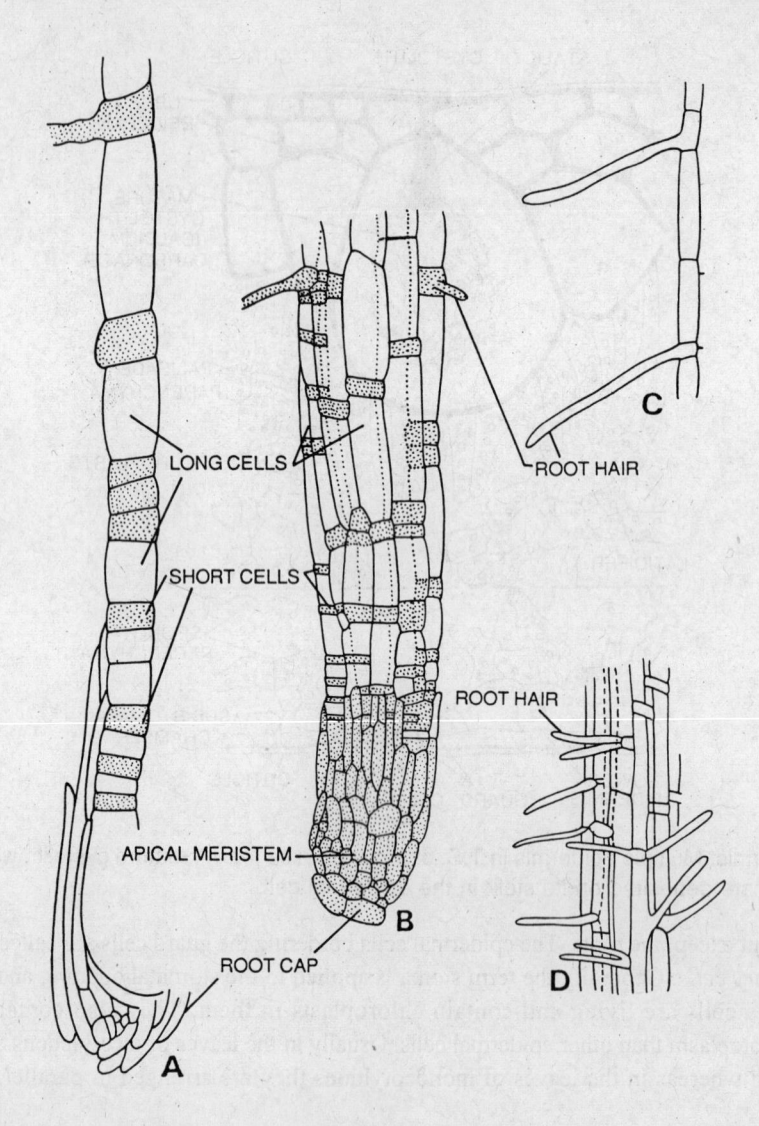

Fig. 7.7. Development of root hair from protruded cells. A,C, in *Cyperus*. B,D, in *Anigozanthos*.

number of stomata may also range on the surface of a single leaf from a few thousand to hundreds of thousands per square centimetre. Stomata occur on both upper and lower surfaces of leaf, but especially they are confined to the lower surface. In floating leaves stomata are confined only on the upper surface of the leaf. Under normal conditions the stomata remain closed in the absence of light or in night or remain open in the presence of light or in day time. Structurally the stomata may be of different types. The four main types of stomata which occur in dicotyledons are known as 1. Ranunculaceous or anomocytic type — type A; 2. Cruciferous or anisocytic — type B; 3. Caryophyllaceous or diacytic — type C; 4. Rubiaceous or paracytic— type D. the fifth type of stomata is commonly found in monocotytledons are known as —gramineous type.

1. Ranunculaceous or anomocytic. Type A — (*Anomocytic* = irregular celled). In this type the stoma remains surrounded by a limited number of subsidiary cells which are quite alike the remaining epidermal cells. The accessory or subsidiary cells are five in number.

The Tissue System

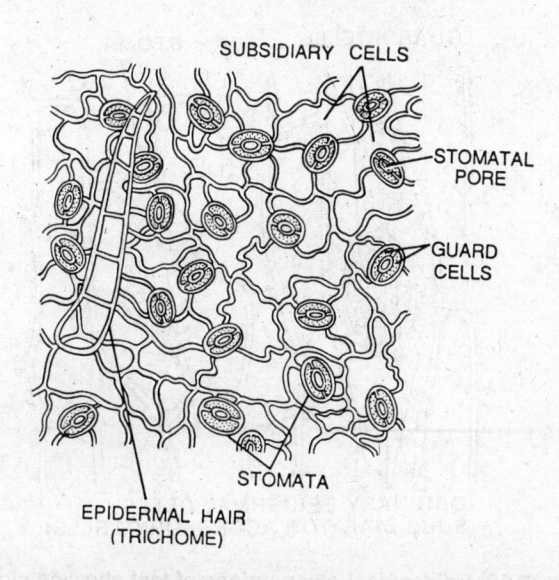

Fig. 7.8. The epidermis—stomata. Surface view of epidermis of a dicotyledonous leaf showing stomata. An epidermal hair is shown at the left.

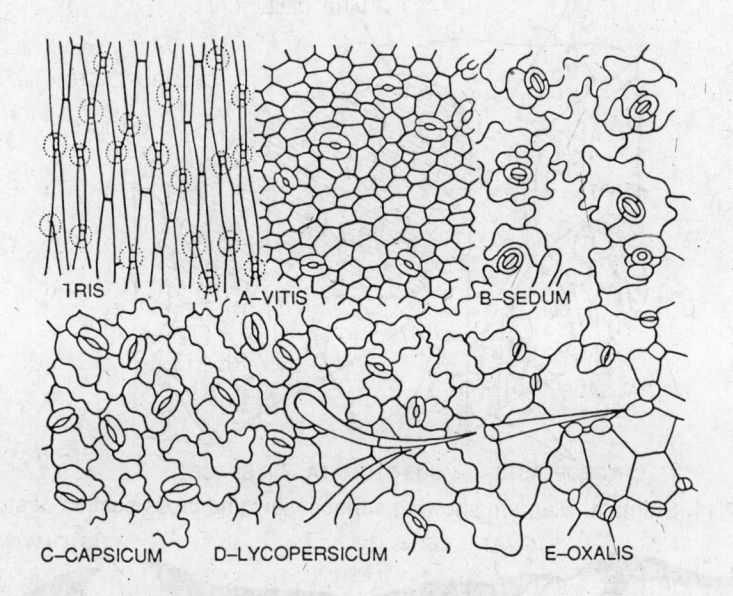

Fig. 7.9. The Epidermis. Surface views of abaxial leaf epidermis. Upper figure on left side of *Iris*, sunken stomata in longitudinal rows. A, dispersed stomata of *Vitis*; B, anisocytic stomata of *Sedum*; C, raised stomata of *Capsicum;* D, raised stomata of *Lycopersicon;* E, sunken stomata of *Oxalis*.

    2. Cruciferous or anisocytic. Type B — (*Anisocytic* = unequal celled). In this type stoma remains surrounded by three accessory or subsidiary cells of which one is distinctly smaller than the other two.

    3. Rubiaceous or paracytic. Type C — (*Paracytic* = parallel celled). In this type, the stoma remains surrounded by two subsidiary or accessory cells which are parallel to the long axis of the pore and guard cells.

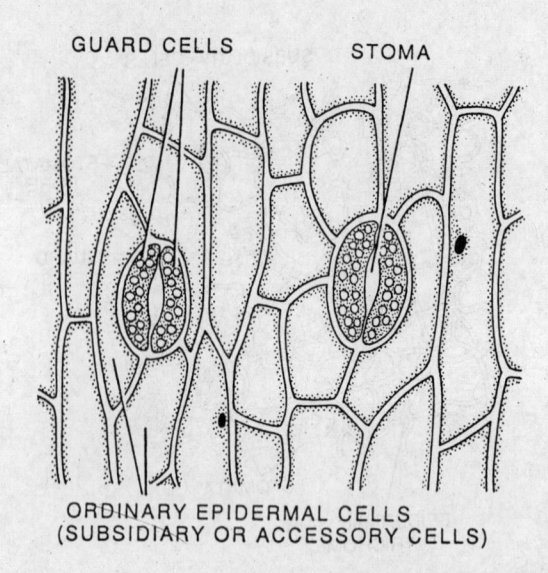

Fig. 7.10. Epidermis. Lower surface of leaf showing stomata.

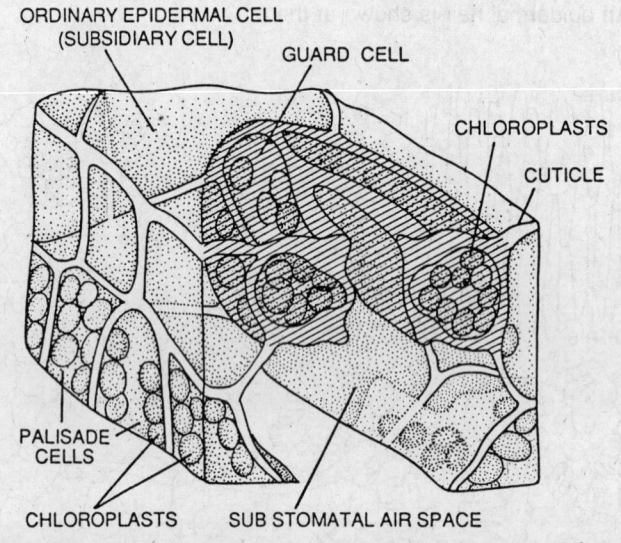

Fig. 7.11. Stomata. Diagram showing surface view and cross section of stoma.

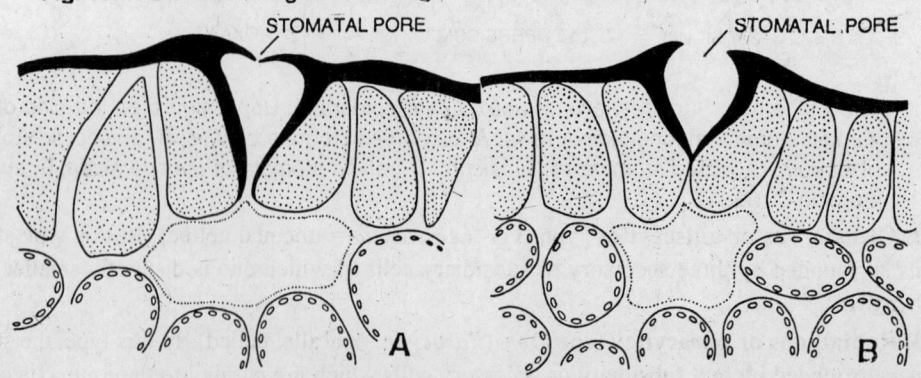

Fig. 7.12. Stomata. A—B, closed and open stomatal pores.

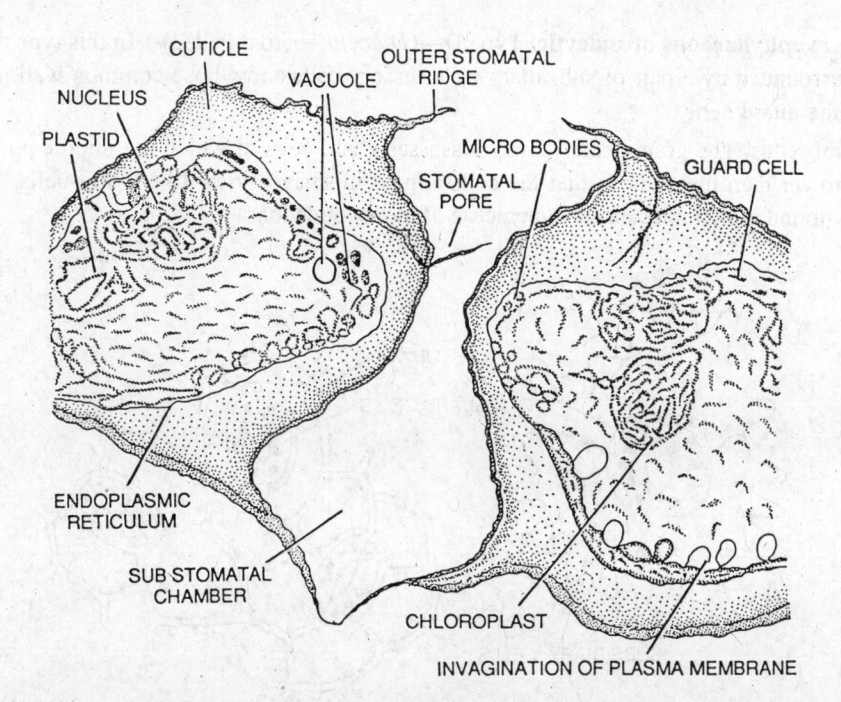

Fig. 7.13. Guard cells of *Vicia faba* as seen in electron microscope.

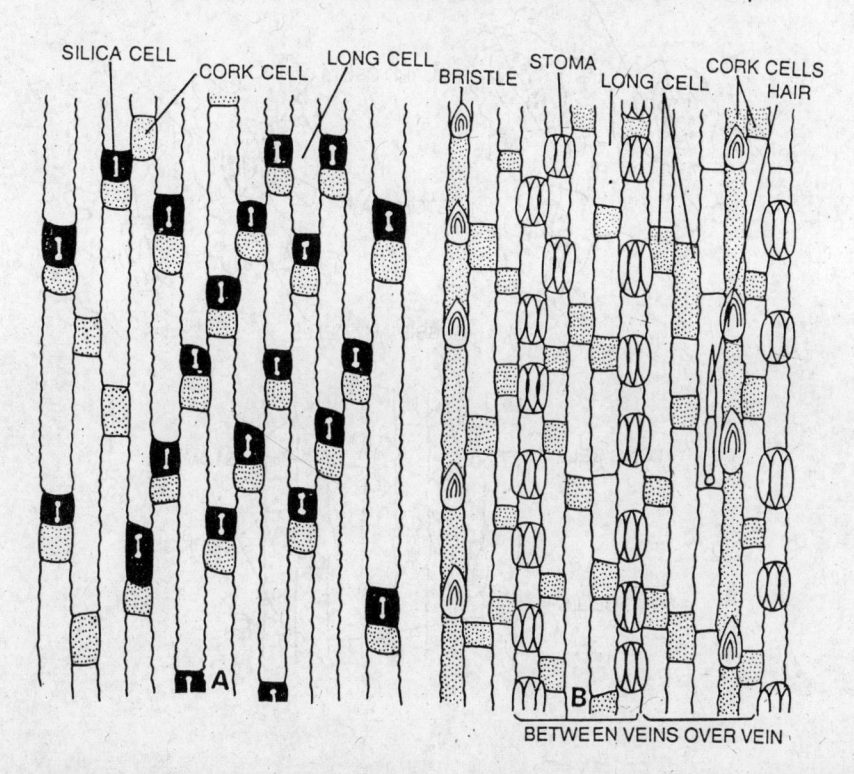

Fig. 7.14. Epidermis. A, epidermis of sugarcane stem showing alternation of long cells with pairs of short cells, the cork cells and silica cells; B, lower epidermis from leaf blade of sugarcane, showing distribution of stomata and their epidermal cells (After Artschwager).

4. Caryophyllaceous or diacytic. Type D—(*Diacytic* = cross celled)—In this type the stoma remains surrounded by a pair of subsidiary or accessory cells and whose common wall is at right angles to the guard cells.

Gramineous. The gramineous stoma possesses guard cells of which the middle portions are much narrower than the ends so that the cells appear in surface view like dump-bells. They are commonly found in Gramineae and Cyperaceae of monocotyledons.

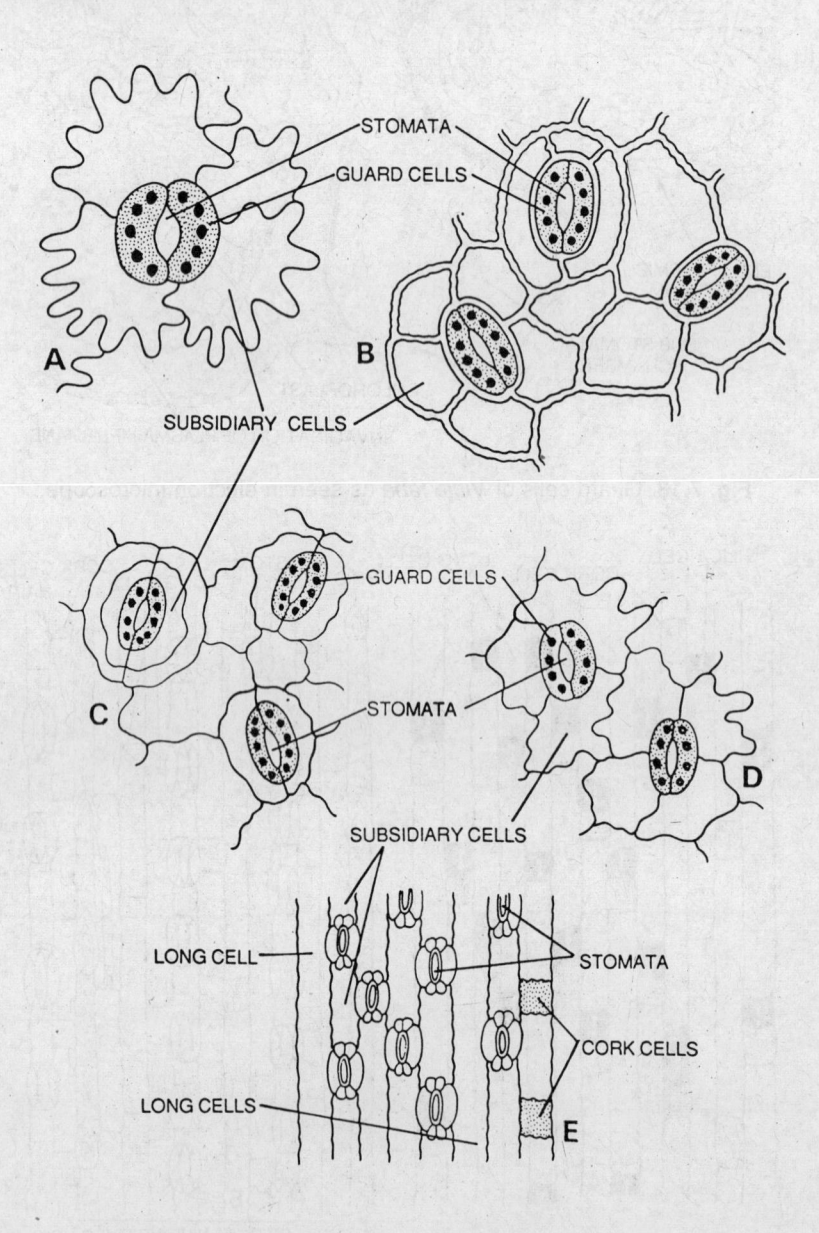

Fig. 7.15. Stomata—types of stomata. A, anomocytic or irregular celled type (ranunculaceous type); B, anisocytic or unequal celled type (cruciferous type); C, paracytic or parallel celled type (rubiaceous type); D, diacytic or cross-celled type (caryophyllaceous type); E, gramineous type.

The Tissue System

Coniferous Stomata

They are sunken and appear as though suspended from the subsidiary cells arching over them. In their median parts the guard cells are elliptical in section and have narrow lumina. At their ends they have wider lumina and are triangular in section. The characteristic of these guard cells is that their walls and those of the subsidiary cells are partly lignified and partly non-lignified.

Function of Stomata

They are used for the exchange of gases in between the plant and atmosphere. To facilitate this function, each stoma opens in a sub-stomatal chamber or respiratory cavity. Evaporation of water also takes place through stomata.

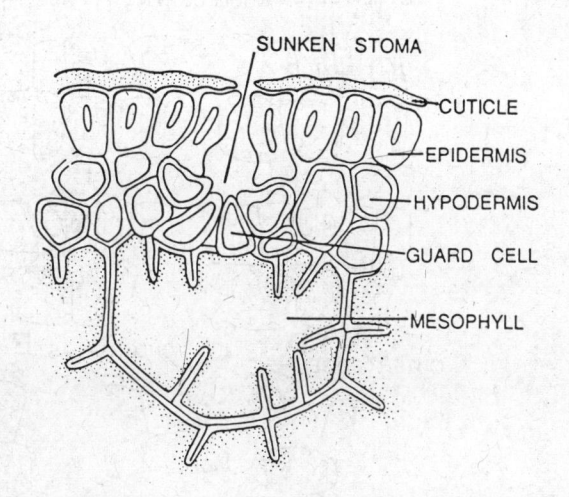

Fig. 7.16. Sunken stoma. Sectional view of a sunken stoma of pine leaf.

HAIRS OR TRICHOMES

Some of the epidermal cells of most plants, grow out in the form of hairs or trichomes. They may be found singly or less frequently in groups. They may be unicellular or multicellular and occur in various forms. They vary from small protuberances of the epidermal cells to complex branched or stellate multicellular structures. The cells of the hairs may be dead or living. Very frequently the hairs lose their protoplasm in their cells.

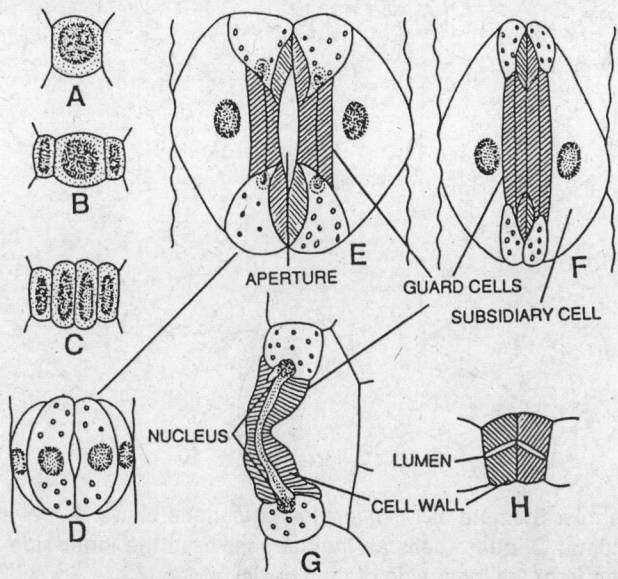

Fig. 7.17. Development (A—D) and structure of stomata (E—H) in sugarcane (a monocot). A, stoma mother cell; B, stoma mother cell with two subsidiary cells derived from two adjoining cells; C, early stage of guard-cell development; D, young stoma with two guard cells; E and F, mature stomata seen from the outer surface in open (E) and closed (F) states; G, L.S. of one guard cell; H, transection through the central portion of two guard cells from a closed stoma.

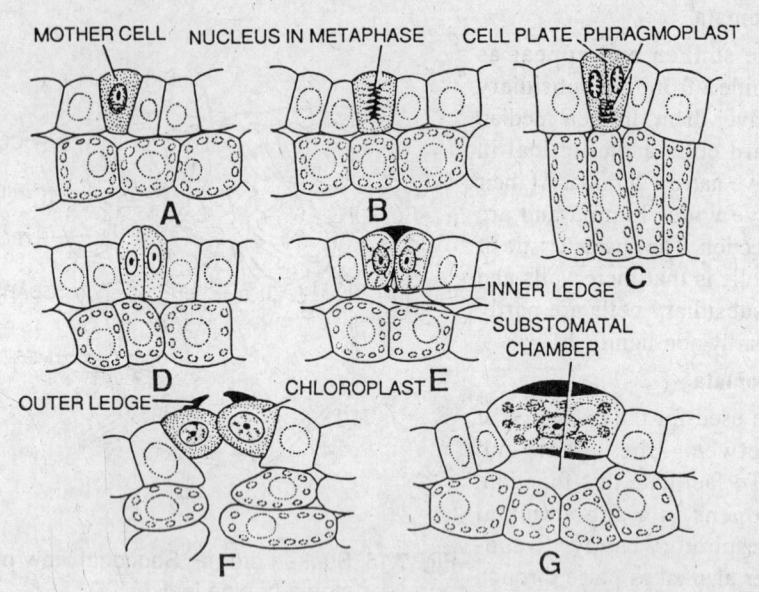

Fig. 7.18. Development of stoma in *Nicotiana* leaf as seen in sections. A–C, stoma mother cell before and during division in two guard cells; D, young guard cells with thin walls; E, thickening of the walls of guard cells begun; F, mature guard cells with unevened thickened walls; G, one mature guard cell.

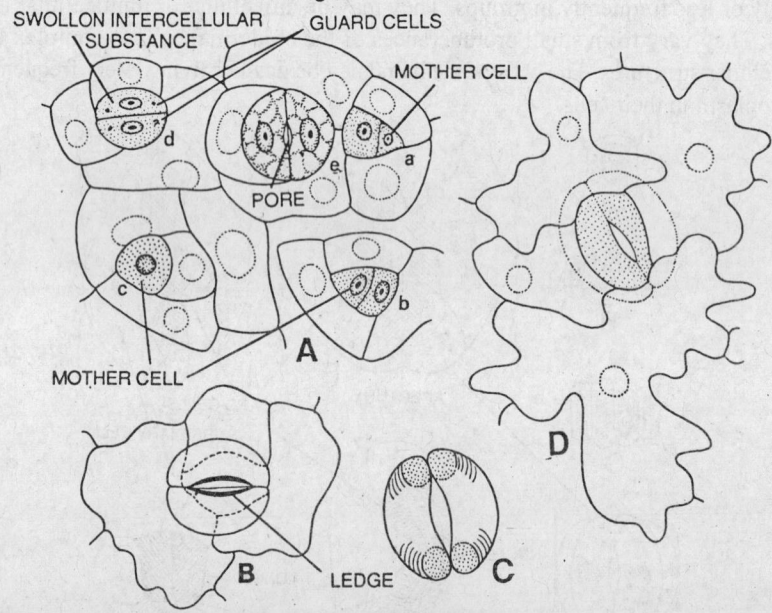

Fig. 7.19. The epidermis. Stomata of *Nicotiana* leaf in surface views. A, developmental stages. B, mature stoma; C, guard cells as they appear from the inner side of the epidermis; D, stoma seen from the inner side of the abaxial surface.

The hairs may be of several types, as—*stinging hairs; laticiferous hairs, bladder like hairs, mucilage hairs, arachnoid hairs, calcified or silicified hairs, non-glandular shaggy hairs, glandular shaggy hairs, non-glandular tufted hairs, two-armed non-glandular hairs, stellate glandular hairs, branched non-glandular hairs, branched glandular hairs, capitate sessile hairs,*

The Tissue System

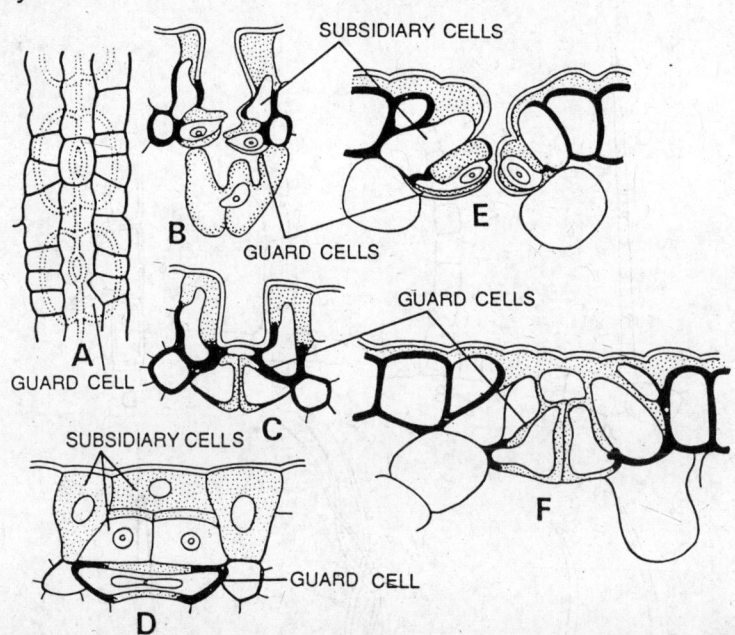

Fig. 7.20. Stomata of conifer leaves. A, surface view of the epidermis of *Pinus merkusii* showing two deeply sunken stomata, guard cells are overarched by subsidiary and other epidermal cells; B—D, stoma and other cells from *Pinus* spp; E and F, stomata and other cells from *Sequoia*.

glandular capitate stalked hairs, non-glandular peltate hairs, glandular peltate hairs and *uniseriate hairs.*

Trichomes may be classified into different morphological categories. One common type is referred to as *hair*. The hairs may be subdivided into (i) *unicellular*; and (ii) *multicellular*. The unicellular hairs may be unbranched or branched. Multicellular hairs may consist of a single row of cells or several layers. Some multicellular hairs are branched in dendroid (tree-like) manner, others have branches oriented largely in one plane (stellate hairs.)

Some important types have been described here.

Stinging hairs. They are one of the most interesting types of the trichomes. it contains a poisonous liquid and consists of a basal bulb like portion from which a stiff, slender and tapering structure is given out. This tapering structure ends in a small knob like or a sharp point. The tip is usually somewhat oblique, and as the body of an animal or human being comes in its contact with some force, the tip is broken off, and the sharp pointed end readily penetrates the skin of the animal, and fluid is being transferred from the basal knob of the hair to the body of the animal.

Glandular hairs. Many plants possess glandular hairs. These hairs may secrete oil, resin or mucilage. A typical glandular hair possesses a stalk and an enlarged terminal portion, which may be referred to as gland. The glandular hairs may be uni- or multicellular. Active secretory cells of glandular trichomes have dense protoplasts and elaborate various substances, such as volatile oils, resins and mucilages, and gums. These substances are excreted and accumulate between the walls and cuticle. Their final removal from the hair occurs by rupture of the cuticle.

Scale or peltate hair. A common type of trichome is the *scale,* also called *peltate hair* (from the latin *peltatus*, target-shaped or shield like, and attached by its lower surface). A scale consists of a discoid plate of cells, often borne on a stalk or attached directly to the foot.

Cell Wall

The cell walls of trichomes are commonly of cellulose and are covered with a cuticle. They may be lignified. Plant hairs often produce thick secondary walls as, for instance, the cotton seed hairs or

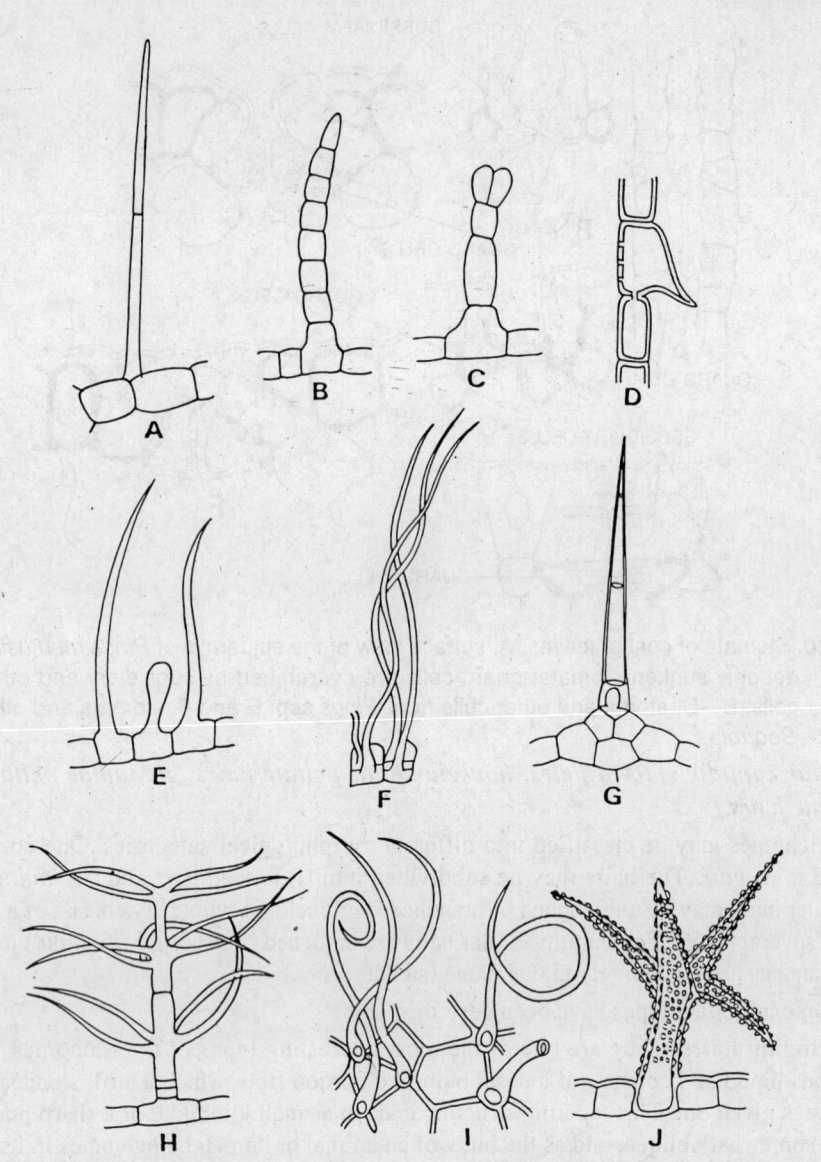

Fig. 7.21. Various types of trichomes (hairs). A, from petal of *Epigaea*; B, from leaf of *Coreopsis*; C, from petal of *Phryma*; D, from leaf of *Avena*; E, from sepal of *Heliotropium*; F, from stem of *Onopordum*; G, from *Cucumis* leaf; H, from *Platanus* leaf; I, from *Rubus* fruit; J, from *Aubrietia* stem. (After E. & M.)

the climber hairs of *Humulus*. The walls or trichomes are sometimes impregnated with silica or calcium carbonate. Their contents are varied in relation to function. Cystoliths and other crystals may develop in hairs.

Development

A trichome is initiated as a protuberance from an epidermal cell. The protuberance elongates, and if it develops into a multicellular structure various divisions may follow the initial elongation.

Trichomes and Taxonomy

The trichome types have been successfully used in the classification of genera and even of species in certain families and in the recognition of interspecific hybrids (Metcalfe and Chalk, 1950;

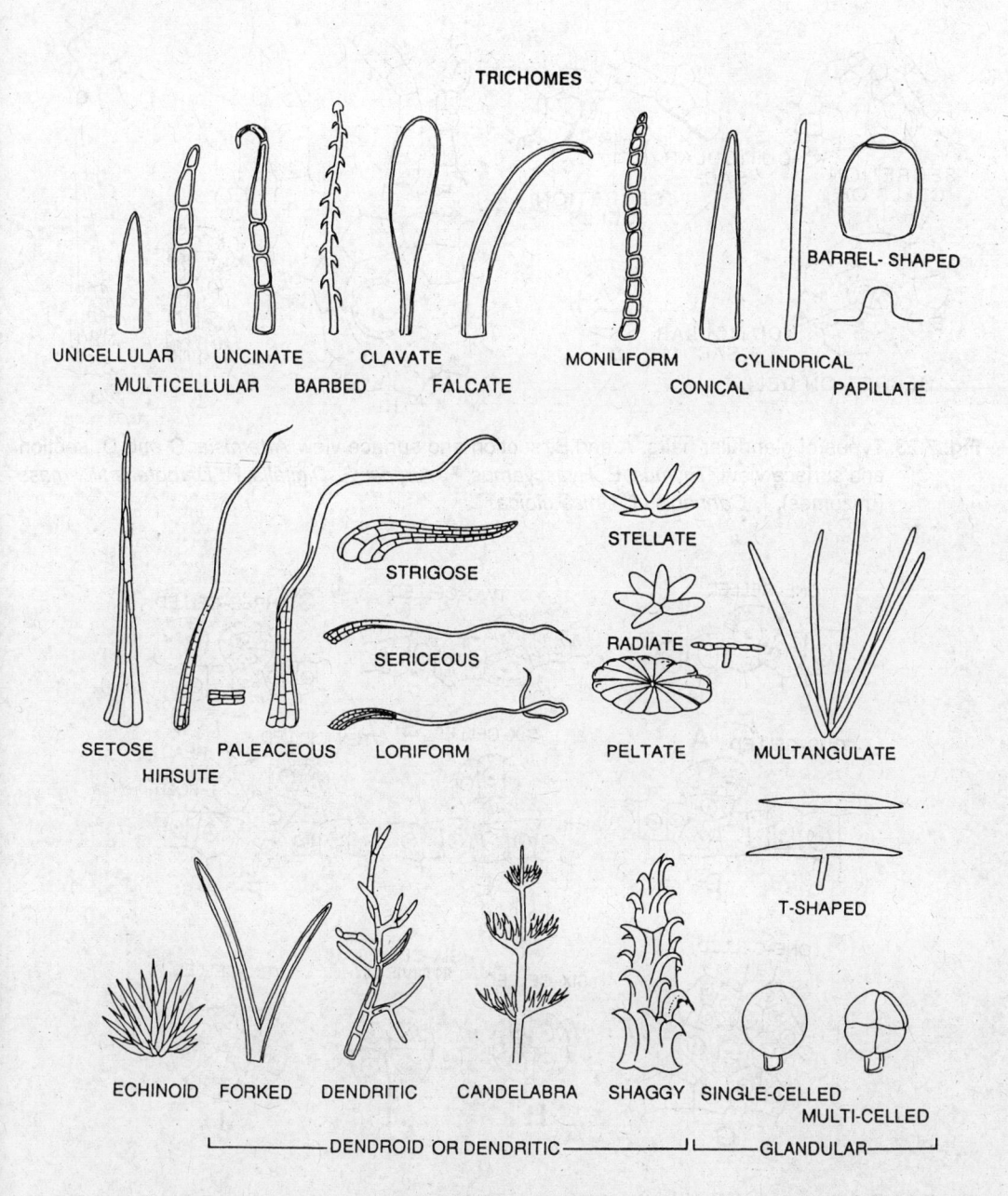

Fig. 7.22. Trichomes of different types.

Hoff, 1950; Sporne, 1956; King and Robinson, 1969, 1970; Ramayya, 1972). King and Robinson (1969, 70) used trichome types to determine genetic limits among Compositae. Ramayya (1972), used the trichome types to establish genetic limits among Magnoliales.

Functions of Trichomes

Generally a dense covering of wooly trichomes controls the rate of transpiration. They also reduce the heating effect of sunlight. They aid in the protection of plant body from outer injurious agencies.

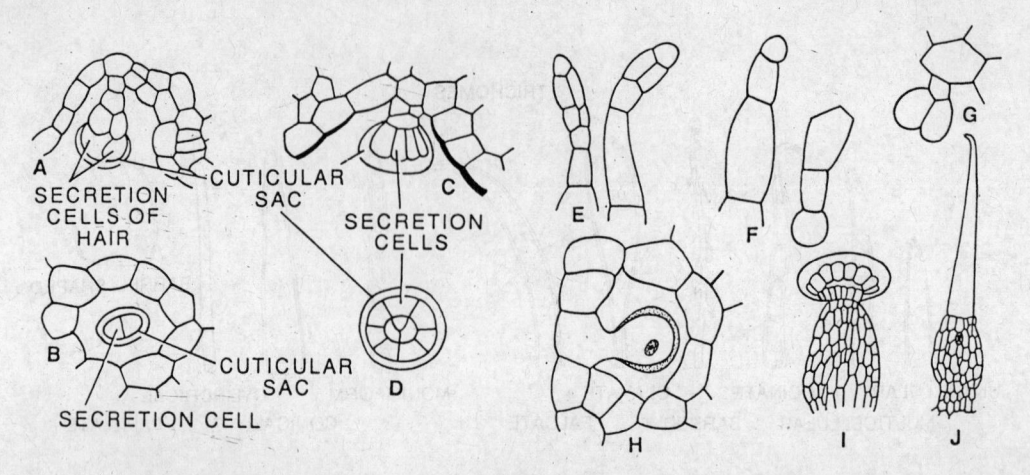

Fig. 7.23. Types of glandular hairs. A and B, section and surface view *Artemisia*; C and D, section and surface view, *Thymus*; E, *Hyascyamus*; F, *Atropa*; G, *Digitalis*; H, *Dryopteris filix-mass* (rhizomes); I, *Cannabis*; J, *Urtica dioica*.

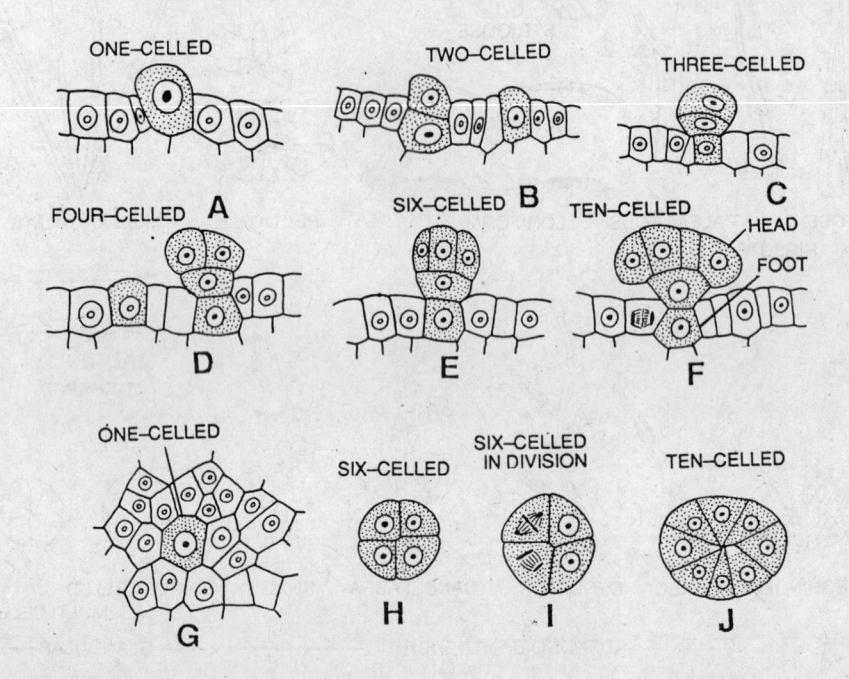

Fig. 7.24. Development of glandular trichomes in *Lingustrum* as seen in sectional (A—F) and surface (G—J) views.

2. THE FUNDAMENTAL OR GROUND TISSUE SYSTEM

Usually in stems which possess the vascular system in the form of solid cylinder, the ground tissue found in between epidermis and vascular cylinder is called *cortex*. In such vascular system the central tissue may be called medulla or pith and the rays of parenchyma cells arising from medulla or pith are termed *medullarry* or *pith rays*. If the vascular bundles are found in dispersed condition throughout in the axis (as in monocotyledonous stems) then there is no delimitation of the ground

The Tissue System

tissue into pith and cortex. The ground or fundamental tissue system consists of main bulk of plant body and extends from below the epidermis to the centre, leaving vascular bundles apart.

CORTEX

The ground tissue found beneath the epidermis which surrounds the central cylinder and is delimited from the cylinder by the endodermis is called the *cortex*. Usually the cortex of stems consists of thin-walled parenchyma cells having sufficiently developed intercellular spaces among them. Usually some of the cortical cells or all of them contain chloroplasts at least in young stems. The cortical cells also contain starch, tannins, crystals and other common secretions in them. The cortex may contain collenchyma, sclerenchyma and sclereids in addition to ordinary parenchyma. Collenchyma is usually arranged as a cylinder or in the form of strands near or beneath the epidermis. In most of the dicotyledonous stems collenchyma is often found in the ridges, in the corners and in other portions to give temporary support to the plant body.

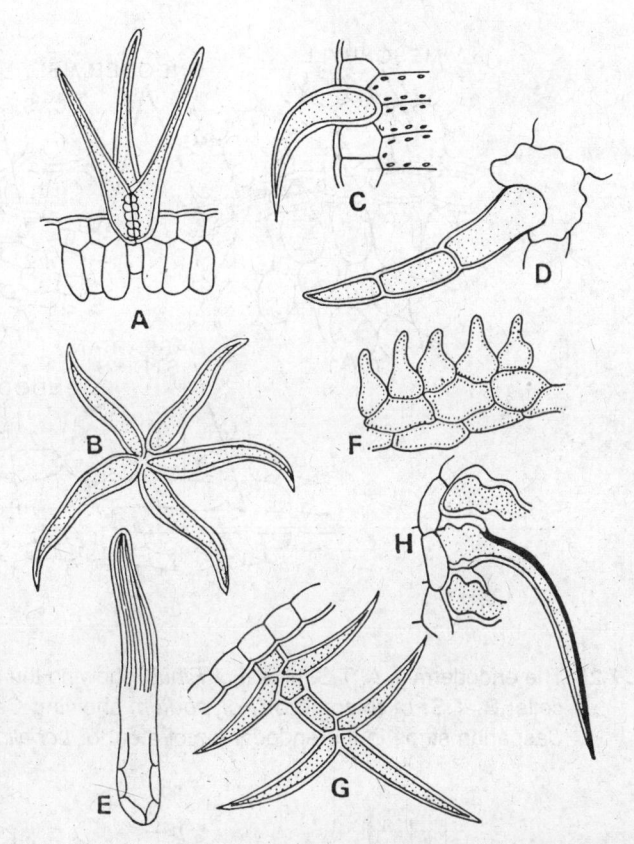

Fig. 7.25. Various kinds of trichomes. A and B, leaf of *Hamamelis*; C, leaf of *Cassia*; D, leaf of *Datura*; E, testa of *Strychnos nuxvomica*; F, petal of *Arnica*; G, leaf of *Verbascum*; H, testa of *Strophanthus*.

Sometimes a few layers of fibres of collenchyma develop just beneath the epidermis forming an outer protective layer called *hypodermis*. Just beneath the hypodermis a few layers of parenchyma and chlorenchyma are found. The innermost layer of the cortex is endodermis which is single-layered and sometimes known as *starch sheath*. The cortex of roots is more homogeneous than that of stems and usually consists of parenchyma only.

The tissues of the cortex are strictly primary and as a whole, mature with the primary tissues of the stele, but there is considerable overlapping of development with secondary-tissue formation within the stele. Collenchyma develops early, but sclerenchymatous cells are usually late in reaching maturity. The cortex of an axis in which marked secondary growth has occurred has tissues crowded and even more or less crushed radially.

Functions of the Cortex

In stems it acts as a protective tissue, but secondarily carbon assimilation, storage of water, storage of food and other functions are also carried on. Collenchyma of the cortical region aids in the temporary mechanical support of plant cody. In roots it is a storage tissue and helps in pumping water from hairs to the xylem.

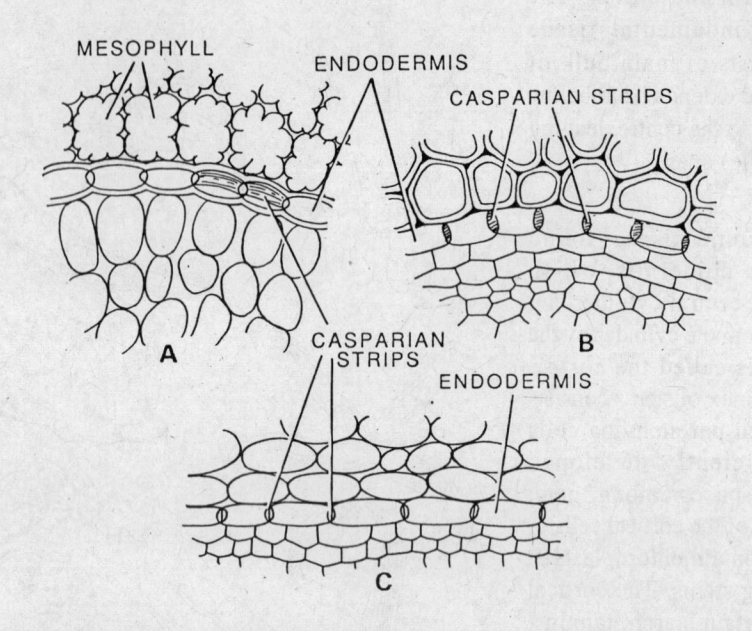

Fig. 7.26. The endodermis. A, T.S. of leaf of *Pinus*, showing the Casparian strips in the endodermal cells; B, T.S. of rhizome of *Polypodium* showing Casparian strips in endodermis; D, Casparian strips in the endodermis of stem of *Lobelia*.

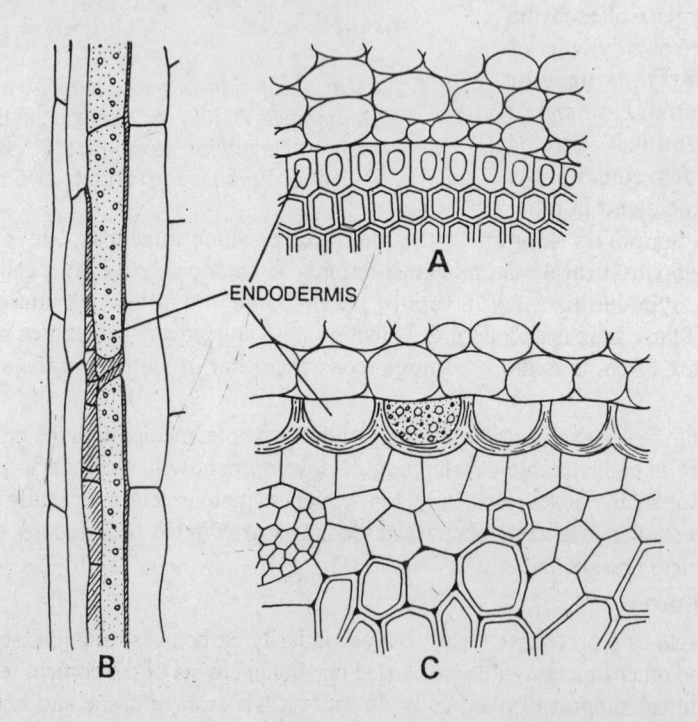

Fig. 7.27. The endodermis, thick walled types. A and B, T.S. and L.S. of *Smilax* root; C, T.S. of *Musa* root.

ENDODERMIS OR STARCH SHEATH

This is a uniseriate layer of cells delimiting the cortex from stele. It consists of barrel-shaped cells arranged quite close to each other having no intercellular spaces among them. The cells of endodermis are elongated and arranged parallel to the long axis of the vascular tissue. They are living cells and the protoplasts are those of typical parenchyma cells. Starch, tannin, mucilage and nuclei are frequently found in endodermal cells. Due to the presence of starch in the endodermal cells, it is also known as *starch sheath*. In stems, it is inconspicuous and found in the form of wavy layer, and in certain cases it becomes altogether obliterated, whereas in roots this layer is well-defined and circular in appearance.

Commonly the endodermal cells are of two types— primary or thin-walled and secondary or thick-walled. In primary or thin-walled cells certain thickenings of suberin are developed in the form of a band or strip which run completely around the cell on the radial walls and end walls, are called *Casparian strips* or *Casparian bands*. These bands range in width from minute threads to broad bands that occupy the entire radial wall. In transverse section, the strips are often called *Casparian dots* or *radial dots*. In secondary or thick-walled endodermal cells the radial and inner walls and sometimes all the walls are thickened by suberin lamellae laid down over the earlier formed wall with its Casparian strips. The thick wall is strongly suberized like the Casparian strips. The Casparian band was first recognized as a wall structure by Caspary (1865–66) and is therefore known as the Casparian strip or band. Among the thick-walled cells of the endodermis, as in many roots, there occur occasionally isolated thin walled cells usually opposite the protoxylem elements, which are known as *passage* or *transfusion cells*. Through these thin walled cells the sap absorbed by root hairs enters the xylary elements.

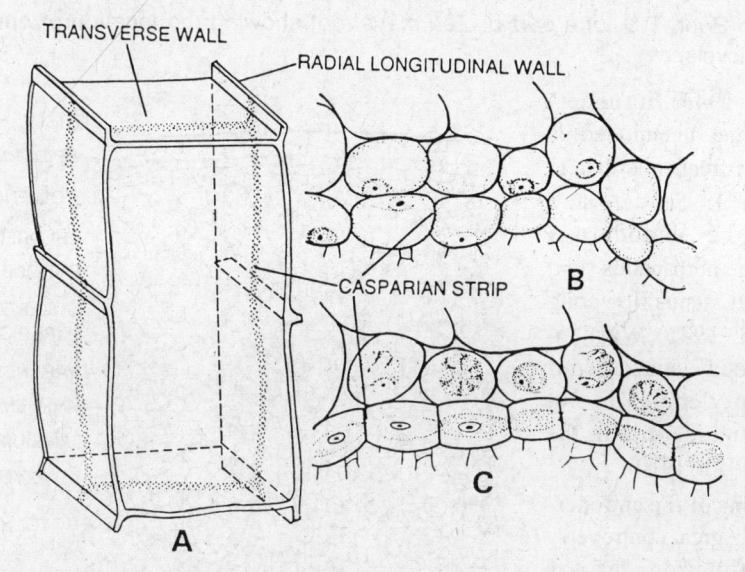

Fig. 7.28. Endodermis. A, diagram of an endodermal cell showing Casparian strip; B and C, the endodermal cells showing Casparian bands.

The endodermis is commonly clearly differentiated in the stems of the vascular cryptogams (pteridohytes) and is found here with Casparian strips and with the additional suberin lamella, but apparently not with the secondary cellulose layer (Guttenburg, 1943). The endodermis occurs in lower vascular plants around the periphery of vascular cylinder, sometimes also between the pith and the vascular tissues. In some ferns the endodermis encloses individual steles. In seed plants the endodermis is quite distinct in the roots, but in a number of herbaceous angiosperms, the stems develop an endodermis with Casparian strips, and also with somewhat thickened walls. According to

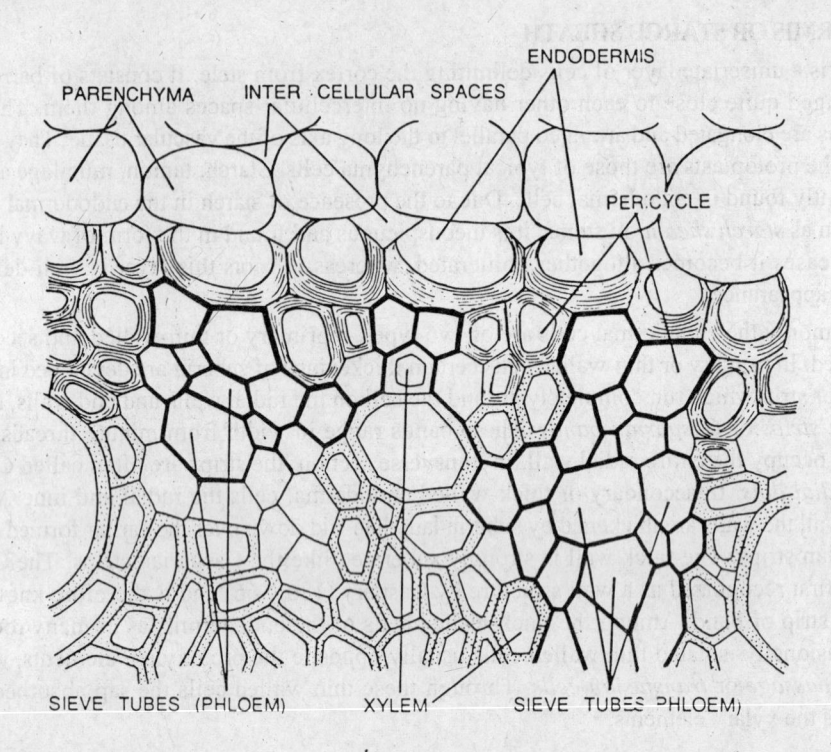

Fig. 7.29. The Root. T.S. of a part of *Zea mays* root showing the location of endodermis and pericycle.

Guttenburg (1943), in underground rhizomes an endodermis develops more frequently than in aerial stems. In *Senecio* and *Leonurus*, the endodermis develops in the herbaceous stem when the plant attains flowering stage (Datta, 1945; Warden, 1935). The woody gymnosperms and the dicotyledons do not possess an endodermis in the aerial stems (Plaut, 1910).

Functions of the endodermis. There is a great controversy about the functions of the endodermis. According to some workers it is a protective layer or sort of accessory inner epidermis. According to others it is connected with the maintenance of root pressure. It is also thought that it acts as an air dam which prevents diffusion of

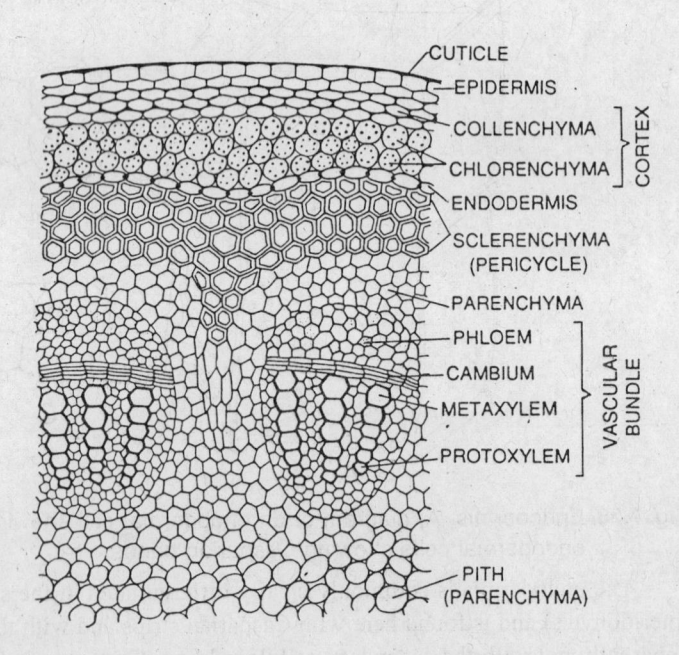

Fig. 7.30. Anatomy of dicot stem. T.S. of a part of stem of *Tinospora* showing complete band of lignified pericycle.

The Tissue System

air into the vessels and thus they escape from closing. It may serve as storage tissue having starch grains in many dicotyledons.

PERICYCLE

In dicotyledonous stems, the pericycle is a multilayered zone found in between the endodermis and the vascular bundles. It always occurs as a thin cylinder of tissue completely encircling the vascular bundles and the pith. Towards inner side this pericyclic zone is limited by the primary phloem, whereas towards outer side it is limited by the endodermis. It may be a complete sclerenchymatous zone as in many cucurbits or it consists of both sclerenchyma and parenchyma cells (*e.g.*, in sun-flower and in many other members of Compositae). Sometimes when endodermis is altogether absent, the pericycle merges with the cortex. Typically the pericycle consists of parenchyma, as in most roots and in the stems of the pteridophytes. Recently it has been shown that the abundant 'pericyclic fibres' of some plants are a part of the primary phloem. Certainly many of them—such of those of *Cannabis* (hemp) and *Linum* (flax)—belong to the phloem and are known as phloem fibres, bast

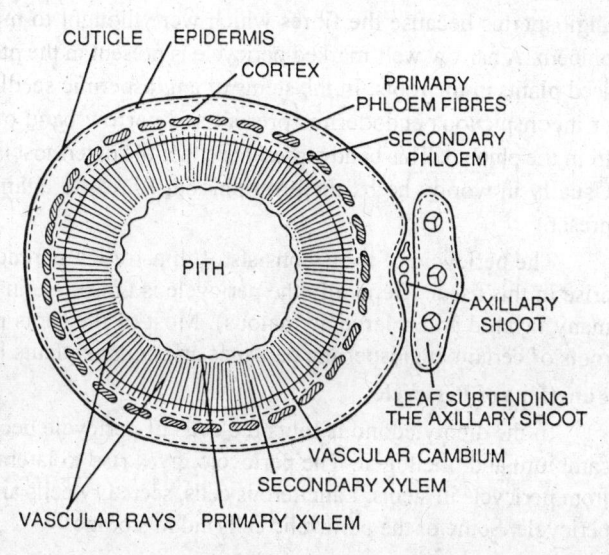

Fig. 7.31. T.S. of flax stem (diagrammatic).

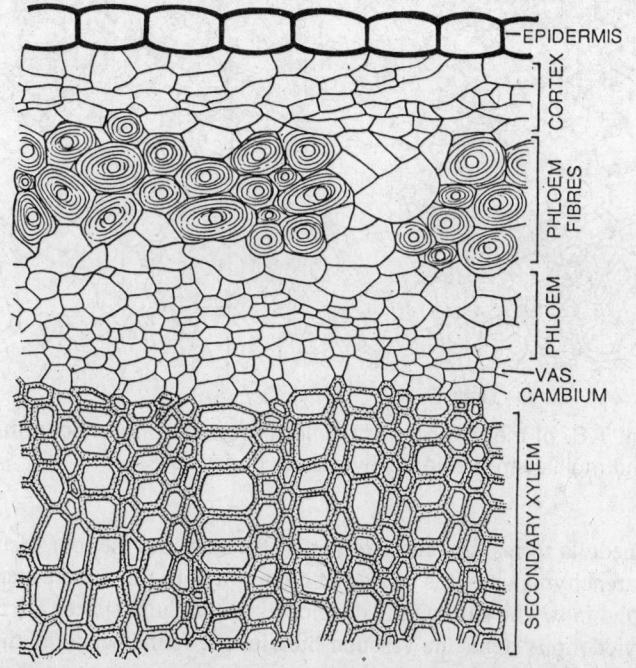

Fig. 7.32. T.S. of stem of *Linum usitatissimum* L. showing phloem fibres and their relation to other tissues of stem.

fibres or hard bast. In many plants such pericyclic fibres are thought to be associated with the phloem of vascular bundles. It is also claimed that no pericycle is present in the stems of many angiosperms because the fibres which were thought to make up much of this layer belong to the phloem. A narrow well-marked pericycle is present in the pteridophytes, in both root and stem, and in seed plants in the roots. In the stems of angiospermic seedlings and herbaceous angiosperms, a true or inconspicuous endodermis present and narrow band of parenchyma separates the endodermis from the phloem. This band represents either the outermost layer of the primary phloem or a pericycle. Usually in woody herbs the protophloem fibres lie against the endodermis, and no pericycle is present.

The pericycle of roots consists of thin-walled parenchyma. The lateral roots of angiosperms arise in this tissue. Generally the pericycle is uniseriate in the roots (*in Smilax* root the pericycle is many layered and sclerenchymatous). Most of the roots possess distinct pericycle. However, the roots of certain angiospermic parasites and aquatic plants lack pericycle.

Functions of Pericycle

In the dicotyledonous roots the cells of pericycle become meristematic and form the vascular cambium and phellogen. The pericycle gives rise to lateral roots. The adventitious roots originate from pericycle in stems. Laticiferous cells, secretory cells and other specialized cells may occur in the pericycle. Some of the pericyclic cells aid in storage.

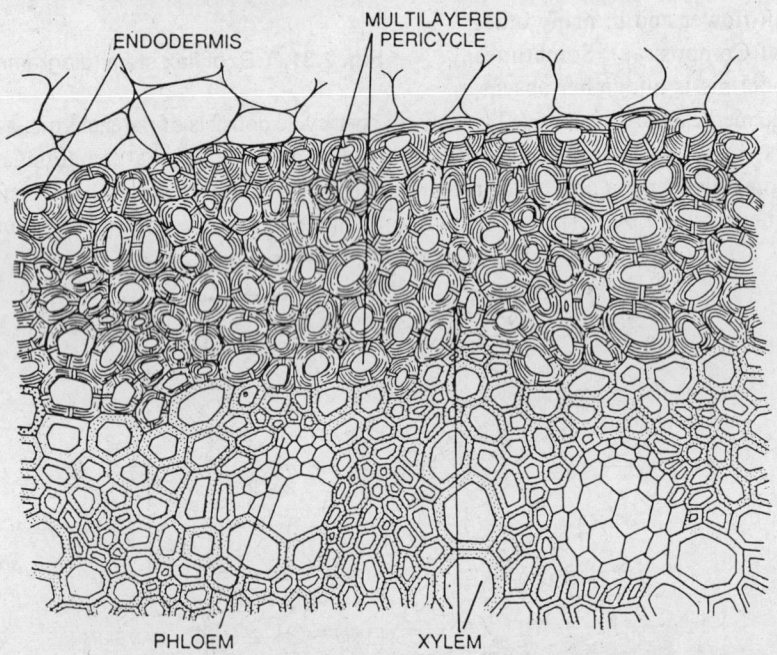

Fig. 7.33. The root. T.S. of the inner part of *Smilax* root showing endodermis (very-thick-walled cells) and multilayered sclerenchymatous pericycle.

PITH

The pith or medulla forms the central region of the stem and the root. Usually the pith of dicot stems is largely parenchymatous. It is devoid of chlorophyll in the mature state but starch forming leucoplasts are found in it. The sufficiently developed intercellular spaces are found among the pith cells. In monocotyledonous stems the vascular bundles are found scattered throughout the ground tissue and pith is not distinguishable. In the dicotyledonous roots the pith is scanty or lacking. However, in monocotyledonous roots it is well developed. In certain monocotyledonous roots (*e.g.*,

The Tissue System

Canna), the pith is sclerenchymatous. The extensions of the pith in the form of narrow parenchymatous strips are called *medullary* or *pith rays*. The pith of many plants is partially obliterated during the growth of the stem and in such cases the stem becomes hollow. However, in such cases the nodes retain their pith. Certain pith cells possess tannin and crystals. Certain specialized structures like laticifers or secretory canals may also occur in the pith cells. In many cases the peripheral portion of the pith is demarcated from its central portion, due to the presence of smaller cells, their contents and sometimes even by the presence of chloroplasts in them (*e.g., Lantana, Anagalis*, etc.). In certain plants (*e.g., Equisetum*) there is an inner endodermis. In such cases the medullary rays are not connected with the central pith region. In some plants belonging to Umbelliferae and Compositae the medullary rays are composed of sclerenchymatous cells or of both sclerenchyma and parenchyma cells. As regards the ontogeny of pith it develops from the ground meristem and may be treated as the inner portion of the ground or fundamental tissue system.

Function of the Pith

The pith cells mainly serve as storage tissue. In case the pith is sclerenchymatous it acts as mechanical tissue and provides mechanical strength to the plant. The medullary ray consisting of parenchyma cells, serve as channels for the transport of food materials and water from the central part (pith) to peripheral region (cortex) of the stem. In dicotyledonous stems some of the parenchyma cells of medullary rays become meristematic and give rise to interfascicular cambium.

3. THE VASCULAR TISSUE SYSTEM

The vascular tissue system consists of a number of vascular bundles which are found to be distributed in the stele. The stele is the central cylindrical portion of the stem and the root, commonly surrounded by the endodermis, and consists of vascular bundles, pericycle, pith and

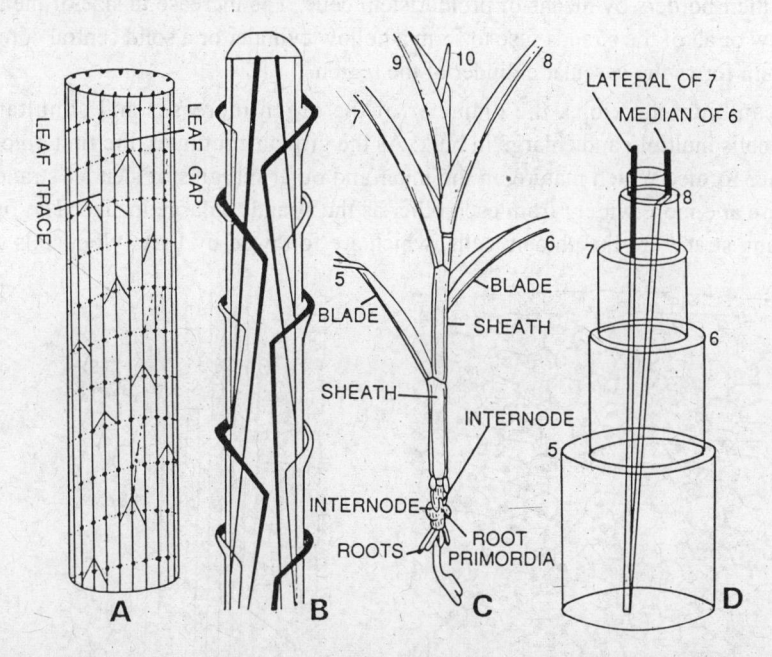

Fig. 7.34. Primary vascular system of angiosperms. A dicotyledon (*Linum* spp.) with the vascular system in the form of a network of leaf traces. The traces are connected with each other; B. monocotyledonous type of a vascular system, one median and one small lateral leaf traces are for each leaf. C, young *Zea mays* plant; D, diagram of part of plant in C showing the median trace of leaf 8 and its connection with the lateral trace of leaf. 7. (A, after Esau; B, after Linsbauer, C and D, after Sharman).

medullary rays. Each vascular bundle consists of xylem and phloem tissues with or without cambium. In roots separate xylem and phloem strands are found. The function of this system is to conduct water and other nutrients from roots to leaves through the xylem and translocation of prepared carbohydrates from leaves to other storage organs and growing regions of plant body through the phloem. The vascular bundle elements are derived from the procambial strands of the primary meristem. The vascular bundles may be arranged in circular ring as in the dicotyledonous stems and the roots, on the other hand, they are found to be scattered throughout the axis in the monocotyledonous stems. In certain plants the vascular bundles remain scattered within the well defined pith, such bundles are called medullary vascular bundles (*e.g., Mirabilis, Boerhaavia, Bougainvillaea, Achyranthes, Amaranthus,* etc.). Vascular bundles may also occur in the cortical region of the stem, such bundles are known as cortical bundles (*e.g., Casuarina, Nyctanthes*).

PRIMARY VASCULAR TISSUE

The Procambium

The first cells to mature, in either leaf or stem, belong to the vascular tissue. In the promeristem, where all cells are isodiametric and alike, continuing longitudinal divisions set apart in some areas strands of elongate, slender cells with dense cytoplasm. Thus, meristematic tissue forms the primary phloem and xylem and is known as the *procambium*. Here, the term procambium is used to indicate the meristematic tissue that gives rise to the morphological vascular units. The first procambium appears as isolated strands very close to the apex in stem and root. It is continuous backward in the older tissues with older promeristem strands and mature vascular tissue. The slender procambium strands increase in diameter by longitudinal cell division within themselves and by the addition of new cells on their borders by means of promeristem cells. The increase in size of the strands is so great that a few or all of the strands fuse to form a hollow cylinder or a solid central core. Ultimately the procambium forms the vascular cylinder of the region.

As procambium develops, the diameter of the organ increases and simultaneously the promeristem cells multiply and enlarge in bulk. As the growth continues, the first phloem and first xylem cells are formed which mature on the inner and outer margins of slender strands. They are separated more and more widely from each other as the strands enlarge in size. The first maturing cells in a young strand are the phloem cells, which are followed by first xylem cells thereafter.

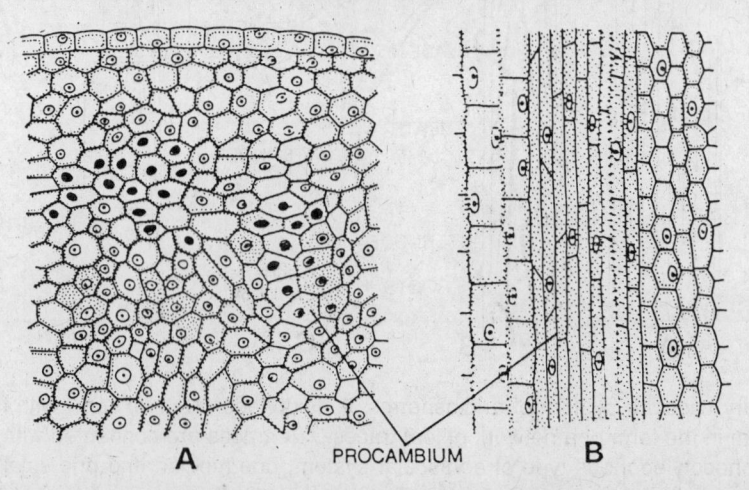

Fig. 7.35. Procambium. A and B, T.S. and L.S. of stem tip of *Linum usitatissimum* showing procambium. The procambium cells possess dense cytoplasm. (After Eames and MacDaniels).

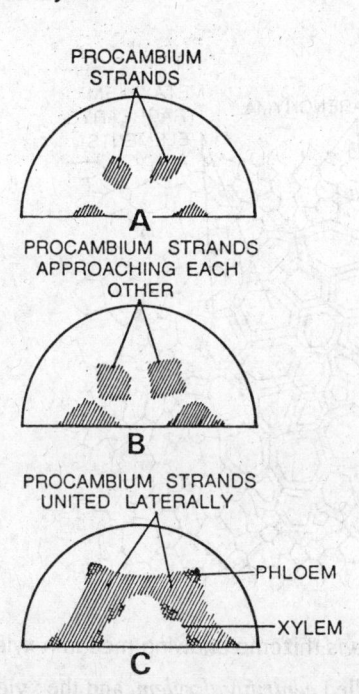

Fig.7.36. Procambium. Stages in the development of a procambium cylinder from promeristem. A, first procambium appeared in the form of strands in promeristem; B, strands of procambium increased in diameter, oldest portions (outer and inner) have become phloem and xylem; C, procambium strands united laterally, more procambium matured as phloem and xylem.

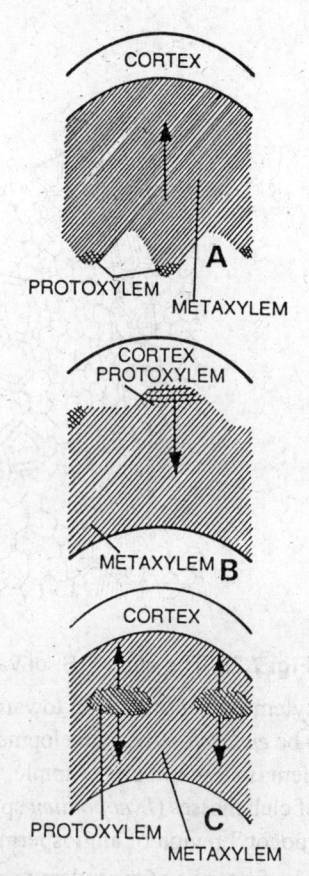

Fig. 7.37. Development of primary vascular tissue. A, centrifugal (endarch); B, centripetal (exarch); C, both centrifugal and centripetal (mesarch).

The fist mature xylem and phloem cells are separated radially by procambium in stems and leaves and tangentially by promeristem in roots.

Constituents of a Vascular Bundle

A vascular bundle of dicotyledonous stem consists of three major zones—(*a*) xylem or wood, (*b*) phloem or bast and (*c*) cambium.

(*a*) **Xylem or wood.** The xylem of a vascular bundle lies towards the centre and is composed of (*i*) vessels or tracheae, (*ii*) tracheids, (*iii*) wood fibres and (*iv*) a patch of xylem or wood parenchyma. The vesslels may possess various kinds of thickenings such as—annular, spiral, scalariform, reticulate and pitted. The tracheids are also found to be associated with the vessels. In the similar way wood fibres and wood parenchyma are also found. The xylem or wood parenchyma of secondary wood usually becomes thick-walled and lignified. The xylem elements, *i.e.,* vessels and tracheids, aid in the conduction of water and mineral salts from the roots of the leaves, whereas wood or xylem parenchyma are living tissues, and aid in the storage. The wood fibres give mechanical support to the plant body.

When the development of xylem takes place towards the centre of the axis, or in other words, the protoxylem develops towards the periphery, it is called *centripetal xylem*, and the xylem strand is said to be *exarch*. If the development of xylem is towards periphery of the axis, or in other words,

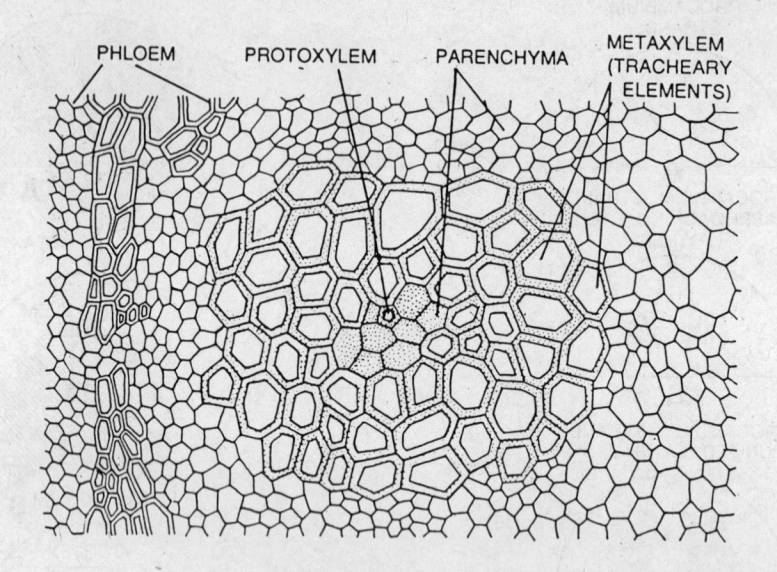

Fig. 7.38. The stem T.S. of vascular bundle of *Pteris* rhizome showing mesarch xylem.

protoxylem elements develop towards the centre, it is called *centrifugal xylem*, and the xylem unit is said to be *endarch*. When development is such that both centripetal and centrifugal xylem are formed, the xylem is *mesarch*. For example, the stem of seed plants in endarch; the root is always exarch; the stem of club mosses (*Lycopodium* spp.) is exarch; mesarch xylem commonly found in the ferns and in the hypocotyl region of angiospermic seedings.

The first cells of the xylem to mature are collectively called the *protoxylem*. The protoxylem is complex tissue made up of tracheids vessels and parenchyma cells. The protoxylem consists of annular spiral and scalariform vessels which may stretch in length very easily. In the stems it lies towards the centre of the axis whereas in the root it lies towards periphery. The vessles of protoxylem have smaller cavities. The xylem which develops afterwards and possesses reticulate and pitted vessels and some tracheids is called *metaxylem*. In the roots towards the centre. The vessels of metaxylem have bigger and wider cavities.

(b) Phloem or bast. Usually in stems, phloem is found away from the centre of the axis towards the periphery and consists of sieve tubes or sieve cells only, or sieve tubes and companion cells only, or sieve tubes, companion cells and phloem parenchyma. In gymnosperms it is represented by sieve cells only. In most of monocotyledons it consists of sieve tubes and companion cells only, whereas in dicotyledons, sieve tubes, companion cells and phloem parenchyma possess simple pits in their walls, particularly which lie against the sieve tubes. Phloem serves for translocation of prepared carbohydrates from leaves to the storage tissue and other growing regions. Sieve tubes translocate proteins and some other carbohydrates, phloem parenchyma conducts amines, amino acids and soluble carbohydrates and companion cells also translocate many soluble food materials. All phloem elements are living and formed of cellulose. In certain cases, the primary phloem is capped by a patch of sclerenchyma called hard bast as in the *Helianthus* stem.

The first cells of the phloem to mature are known as *protophloem*. The protophloem consists of narrow sieve tubes, and is found towards periphery. The inner portion of the phloem consists of bigger sieve tubes called *metaphloem*. The metaphloem is complex tissue and consists of well developed cells of all types such as—sieve tubes, companion cells, phloem parenchyma and sometimes phloem fibres and sclereids.

The Tissue System

(c) Cambium. In between xylem and phloem, a thin strip of primary meristem is found in dicotyledonous stems, called the *cambium*. The cells of cambium are rectangular and thin-walled. The cambium strip may be uniseriate or multilayered. The cambial cells are living, sufficiently elongated and possess oblique ends, but as they become flattened tangentially they look rectangular in cross-section.

Types of Vascular Bundles

A vascular bundle consists of a strand like portion having xylem and phloem of the primary vascular system. According to the arrangement of xylem and phloem in the vascular bundles, they are being arranged in the following main types —(1) radial, (2) conjoint and (3) concentric.

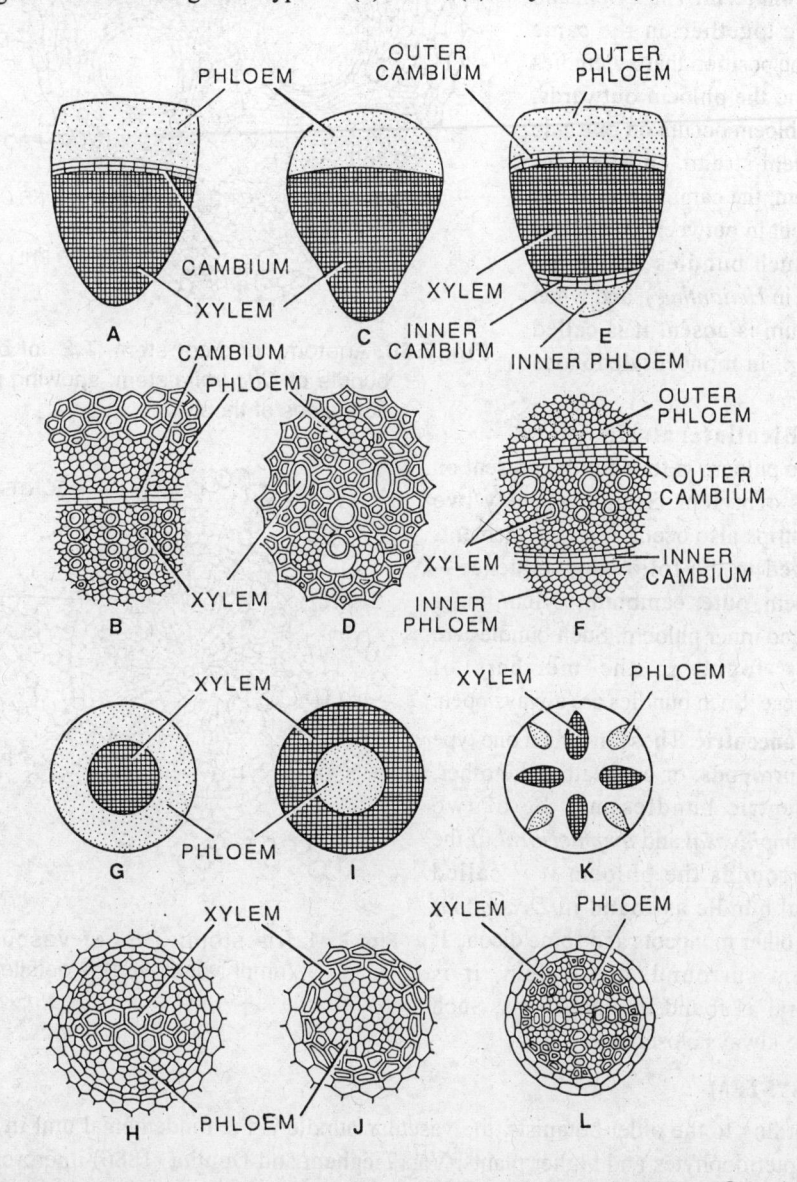

Fig. 7.39. Types of vascular bundles. A and B, conjoint, collateral and open; C and D, conjoint, collateral and closed; E and F, conjoint, bicollateral and open; G and H, concentric and amphicribral; I and J, concentric and amphivasal; K and L, radial.

(1) Radial. Those in which the xylem and the phloem lie radially side by side (*e.g.*, in roots of seed plants). This is most primitive type.

(2) Conjoint. Those in which the two types of tissues are separated from one another. Here, xylem and phloem together form a bundle. The two subtypes are— (*a*) Collateral and (*b*) Bicollateral.

(*a*) Collateral. The xylem and phloem lie together on the same radius in the position that xylem lies inwards and the phloem outwards. Here the phloem occurs on one side of the xylem strand. In dicotyledonous stem, the cambium is found to be present in between xylem and phloem, such bundles are called *open* (*e.g.*, in *Helianthus*), and when the cambium is absent it is called *closed* (*e.g.*, in monocotyledonous stems).

(*b*) Bicollateral. In such bundles the phloem is found to be present on both sides of xylem. Simultaneously two cambium strips also occur. Various elements are arranged in the following sequence— outer phloem, outer cambium, xylem, inner cambium and inner phloem. Such bundles are commonly found in the members of Cucurbitaceae. Such bundles are always open.

(3) Concentric. Those in which one type of tissue surrounds, or ensheaths, the other. The concentric bundles may be of two subtypes, *amphivasal* and *amphicribral*. If the xylem surrounds the phloem it is called amphivasal bundle as found in *Dracaena*, *Yucca* and other monocots and some dicots. If the phloem surrounds the xylem, it is amphicribral as found in many ferns. Such bundles are always closed.

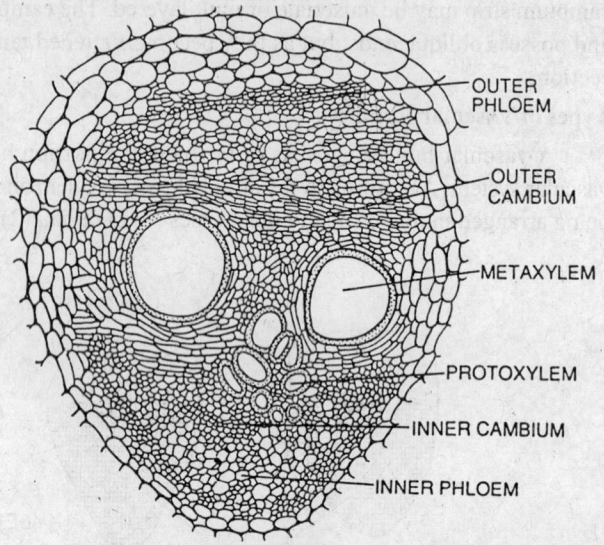

Fig. 7.40. Anatomy of dicot stem. T.S. of bicollateral bundle of *Cucurbita* stem, showing phloem on both sides of the xylem.

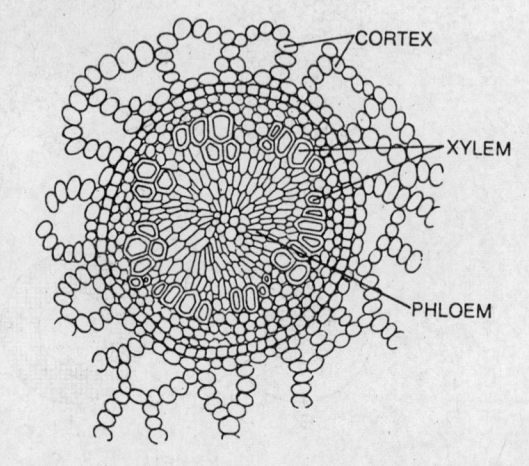

Fig. 7.41. *Iris* stem. T.S. of vascular bundle (amphivasal type)— detailed structure.

STELAR SYSTEM

According to the older botanists, the vascular bundle is the fundamental unit in the vascular system of pteridophytes and higher plants. Van Tieghem and Douliot (1886) interpreted the plant body of vascular plant in the different way. According to them, the fundamental parts of a shoot are the cortex and a central cylinder, is known as *stele*. Thus *the stele is defined as a central vascular cylinder, with or without pith and delimited the cortex by endodermis*. The term stele has been derived from a Greek word meaning pillar. Van Tieghem and Douliot (1886) recognized only three types of steles. They also thought that the monostelic shoot were rare in comparison of polystelic shoots. It is an established fact that all shoots are monostelic and polystelic condition rarely occurs. The stele of the stem remains connected with that of leaf by a vascular connection known as the *leaf supply*.

The steles may be of the following types:

1. Protostele

Jeffrey (1898), for the first time pointed out the stelar theory from the point of view of the phylogeny. According to him, the primitive type of stele is *protostele. In protostele, the vascular tissue is a solid mass and the central core of the xylem is completely sorrounded by the strand of phloem*. This is the most primitive and simplest type of stele.

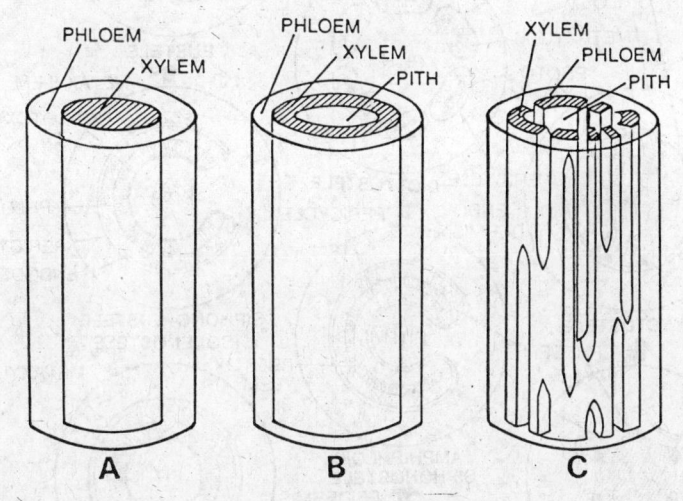

Fig. 7.42. Types of arrangements of vascular tissues in steles. A, protostele; B, siphonostele; C, dictyostele.

There are several forms of protostele:

(*a*) **Haplostele.** This is the most primitive type of protostele. Here the central solid smooth core of xylem remains surrounded by phloem (*e.g.*, in *Selaginella* spp.).

(*b*) **Actinostele.** This is the modification of the haplostele and somewhat more advanced in having the central xylem core with radiating ribs (*e.g.*, in *Psilotum* spp.).

(*c*) **Plectostele.** This is the most advanced type of protostele. Here the central core of xylem is divided into number of plates arranged parallel to each other. The phloem alternates the xylem (*e.g.*, in *Lycopodium*).

(*d*) **Mixed-pith stele.** Here the xylem elements (*i.e.*, tracheids) are mixed with the parenchymatous cells of the pith. This type is found in primitive fossils and living ferns. They are treated to be the transitional types in between true protosteles on the one hand and siphonosteles on the other (*e.g.*, in *Gleichenia* spp. and *Osmunda* spp.).

2. Siphonostele

This is the modification of protostele. A stele in which the protostele is medullated is known as *siphonostele*. Such stele contains a tubular vascular region and a parenchymatous central region. Jeffrey (1898) interpreted that the vascular portion of siphonostele possesses a parenchymatous area known as a *gap* immediately above the branch traces only or immediately above leaf and branch traces. On the basis of these branch and leaf gaps Jeffrey (1910), distinguished two types of

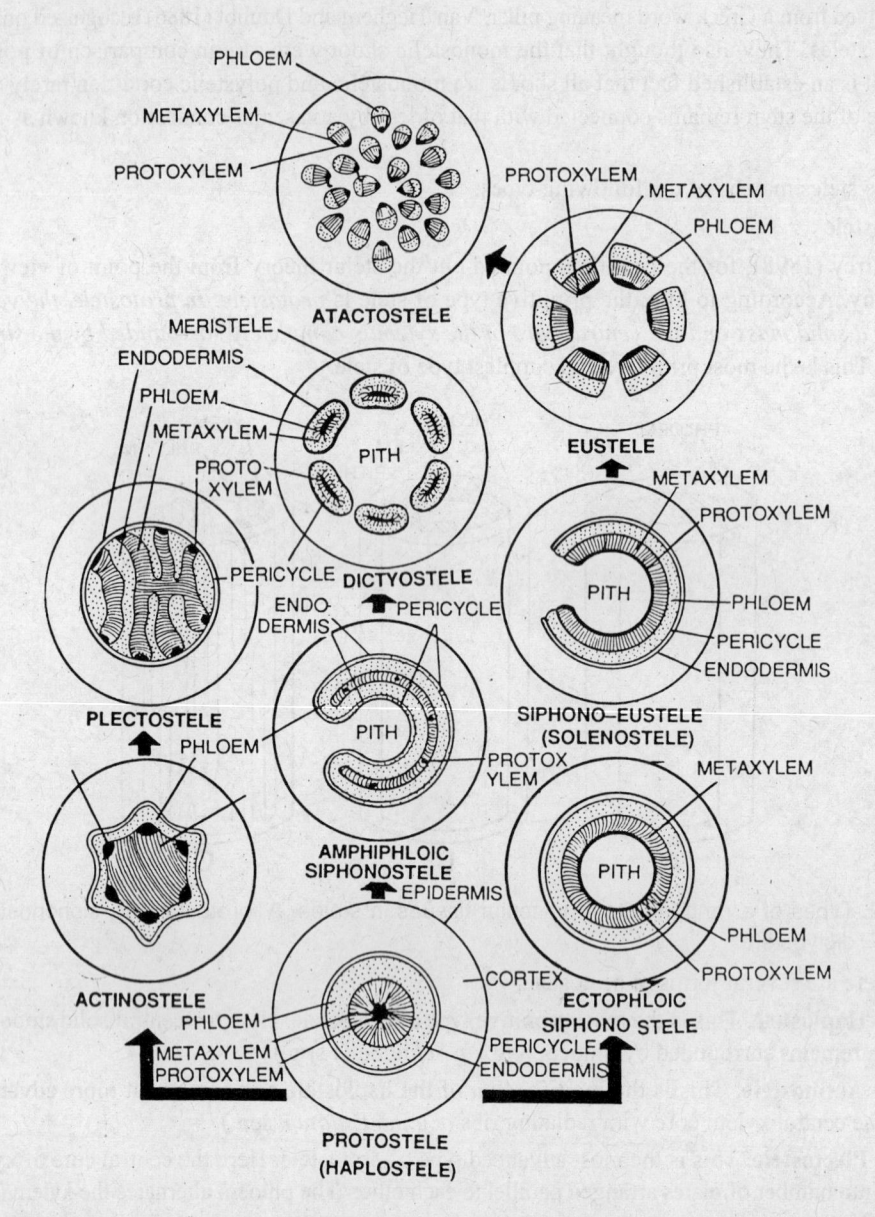

Fig. 7.43. The stelar system. Different types of steles arranged in evolutionary sequence.

siphonosteles. In one type, however, the leaf gaps are not found and they are known as *cladosiphonic siphonosteles*. In the other type both leaf and branch gaps are present and they are known as *phyllosiphonic siphonosteles*.

Jeffrey (1902, 1910, 1917) interpreted the evolution of the siphonostele from the protostele as follows. He supported that the parenchyma found internal to the phloem and xylem has been originated from the cortex. The supporters of this theory believe that the inner endodermis found to the inner face of the vascular tissue and the parenchyma encircled by this endodermis have been originated from the cortex. According to Jeffrey and other supporters of this theory the siphonosteles with internal endodermis are more primitive than those without an internal endodermis.

The siphonosteles which do not possess the inner endodermis are believed to have been originated by disintegration of inner endodermis during evolution.

According to the theory proposed by Boodle (1901), and Gwynne Vaughan, the siphonostele has been evolved from the protostele by a transformation of the inner vascular tissue into parenchyma.

A siphonostele may be of the following types:

(*a*) **Ectophloic.** In this type of siphonostele, the pith is surrounded by concentric xylem cylinder and next to xylem the concentric phloem cylinder.

(*b*) **Amphiphloic.** In this type of siphonostele the pith is surrounded by the vascular tissue. The concentric inner phloem cylinder surrounds the central pith. Next to the inner phloem is the concentric xylem cylinder which is immediately surrounded by outer phloem cylinder (*e.g.,* in *Marsilea*).

3. Solenostele

The vascular plants have been divided into two groups on the basis of the presence or absence of the leaf gaps. These groups are— *Pteropsida* and *Lycopsida*. The ferns, gymnosperms and angtosperms are included in Pteropsida, whereas the lycopods, horse-tails, etc., are included in Lycopsida. The simplest form of siphonostele has no gaps, such as some species of *Selaginella*. However, among the simplest siphonostelic Pteropsida and siphonostelic Lycopsida, the successive leaf gaps in the stele do not overlap each other and are considerably apart from each other.

According to Brebner (1902), Gwynne-Vaughan (1901) such siphonosteles which lack overlapping of gaps are known as *solenosteles*. They may be ectophloic or amphiphloic. Some authors (Bower, 1947; Wardlaw, 1952; Esau, 1953) however, interpret the solenostele as an amphiphloic siphonostele.

4. Dictyostele

In the more advanced siphonosteles of Pteropsida, the successive gaps may overlap each other. Brebner (1902) called the siphonosteles with overlapping gaps as *dictyosteles*. In such cases the intervening portion of the vascular tissue between lateral to such leaf gaps is known as *meristele*. Each meristele is of protostelic type. The dictyostele with many meristeles looks like a cylindrical meshwork.

5. Polycylic stele

This type of stelar organization is the most complex one amongst all vascular cryptogams (pteridophytes). Such type of steles are siphonostelic in structure. Each such stele possesses an internal vascular system connected with an outer siphonostele. Such connections are always found at the node. A typical polycyclic stele possesses two or more concentric rings of vascular tissue. This may be a solenostele or a dictyostele. Two concentric rings of vascular tissue are found in *Pteridium aquilinum* and three in *Matonia pectinata*.

6. Eustele

According to Brebner (1902), there is one more modification of the siphonostele known as *eustele*. Here the vascular system consists of a ring of collateral or bicollateral vascular bundles situated on the periphery of the pith. In such steles, the interfascicular areas and the leaf gaps are not distinguished from each other very clearly. The example of this type is *Equisetum*.

NODAL ANATOMY

LEAF TRACES AND LEAF GAPS

A shoot bears *nodes'* and *internodes*. At each node, portions of the vascular system are deflected into the leaf, which is attached at this node. A vascular bundle located in the stem but directly related to a leaf, to represent the lower part of the vascular supply of this leaf, is termed the

Plant Anatomy

leaf trace. The leaf trace is defined as follows—*The leaf trace is a vascular bundle that connects the vascular system of the leaf with that of the stem.* A leaf trace is extended between the base of a leaf and the point where it is completely merged with other parts of the vascular system in the stem. One or more leaf traces may be associated with each leaf.

In the shoot of a pteropsid (seed plants and ferns) where the leaf trace diverges into a leaf, it appears as though a portion of the vascular cylinder of the stem is deflected to one side. Immediately above the diverging trace, a parenchymatous tissue is being differentiated instead of vascular tissue in the vascular region of the stem for a limited distance. The

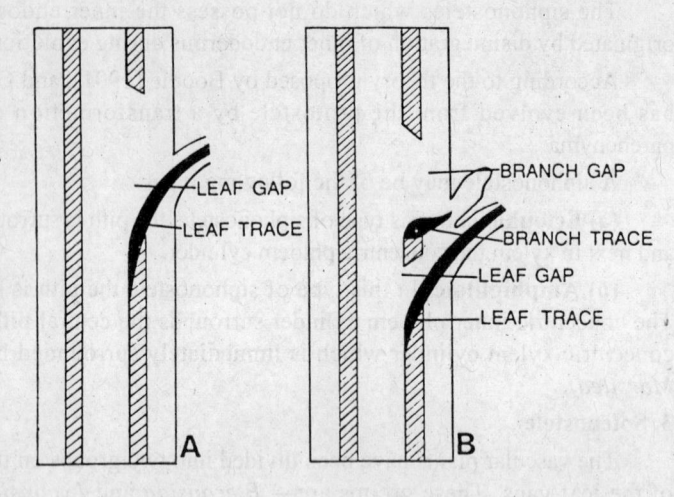

Fig. 7.44. Leaf and branch traces and gaps. A, L.S. of node through leaf trace and gap; B, L.S. of node through branch trace and gap.

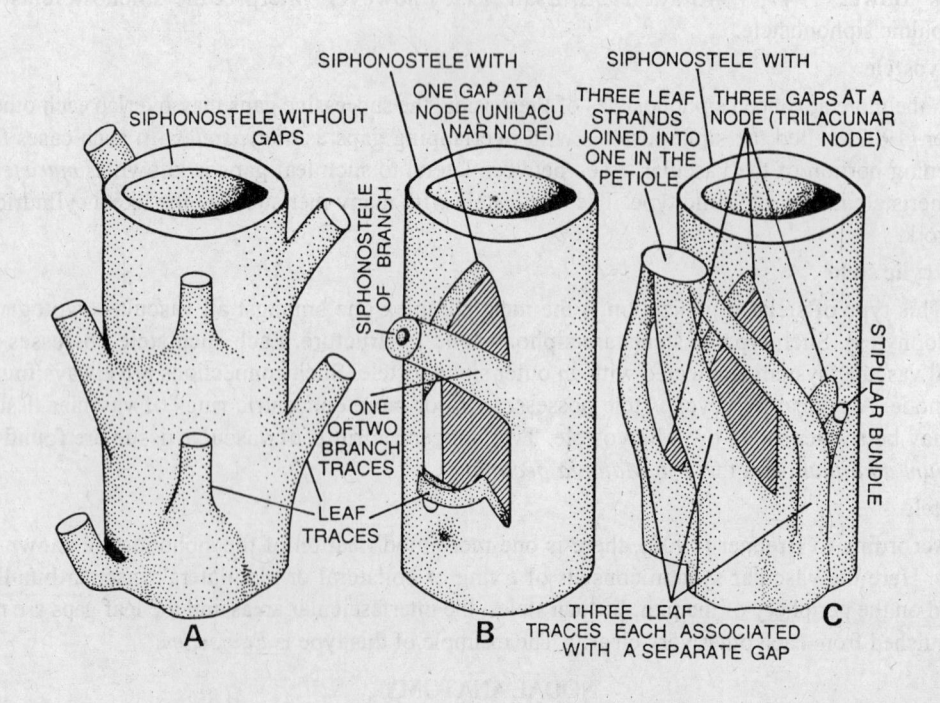

Fig. 7.45. Primary vascular system. A, *Selaginella*—three-dimensional view, siphonostelic type without leaf-gaps; B, *Nicotiana*—unilacunar node, siphonostelic with leaf gap; C, *Salix*—trilacunar node, siphonostelic type with leaf gaps. (After Esau).

The Tissue System

parenchymatous regions in the vascular system of the stem, located adaxially from the diverging leaf traces, are called *leaf gaps* or *lacunae*. Actually these gaps are not breaks in the continuity of the vascular system of the axis. Lateral connections occur between the tissues above and below the gap. In transverse sections of an axis at the level of a leaf gap, the gap resembles an *interfascicular area*.

The gaps are quite conspicuous in the ferns and angiosperms where the vascular system in the internodal parts of the stem forms a more or less continuous cylinder. In some ferns the leaves are so crowded that the gaps formed at the successive nodes overlap one another and the vascular cylinder appears highly dissected. The transverse sections of such stems show a circle of vascular bundles with the parenchymatous leaf gaps. In certain ferns, gymnosperms and most angiosperms

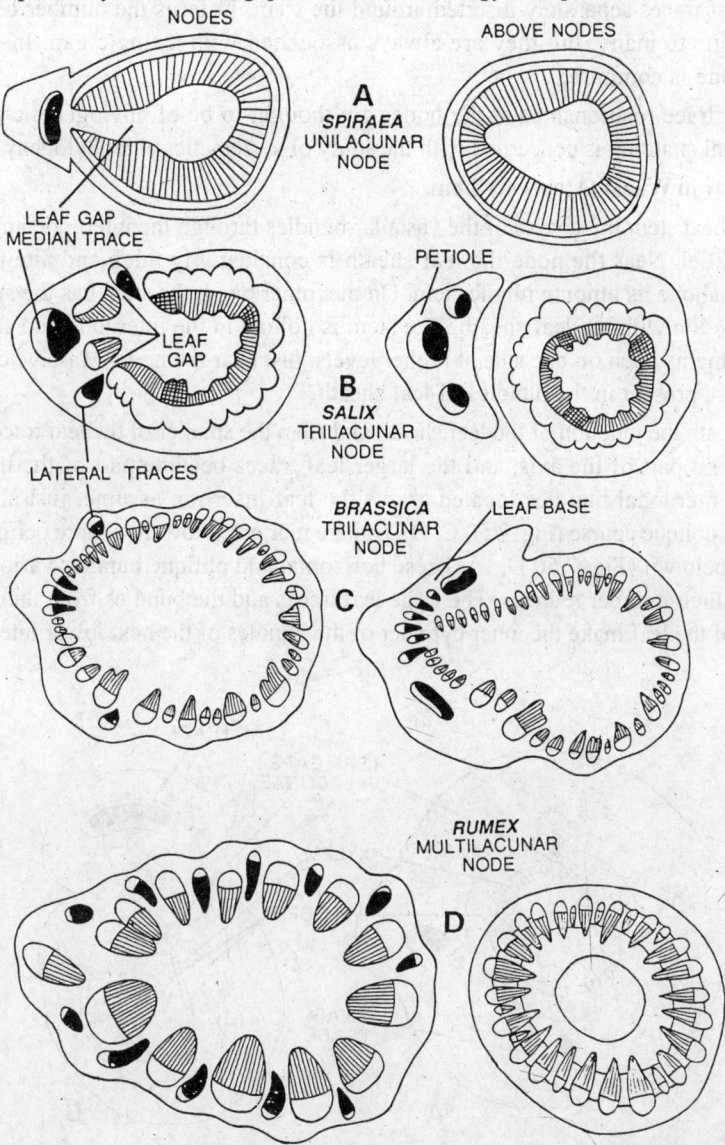

Fig. 7.46. Nodal anatomy of dicotyledons. A. *Spiraea*— each leaf has one leaf trace and one leaf gap (unilacunar node); B, *Salix* — each leaf has three leaf traces and three leaf gaps (trilacunar node); C, *Brassica*—three leaf traces and three leaf gaps per leaf (trilacunar node); D, *Rumex*—many leaf traces and many leaf gaps per leaf (multilacunar node).

the vascular system consists of anastamosing strands. In such cases, the parenchyma that occurs above the diverging leaf trace becomes confluent with the interfascicular areas, thus the recognition of the gaps become uncertain.

There are three common types of nodes in the dicotyledons. The node with a single gap and a single trace to a leaf is known as *unilacunar*; the node with three gaps and three traces to a leaf (one median and two lateral) is known as *trilacunar*; and the node with several to many gaps and traces to a leaf is known as *multilacunar*. The most accepted concept is that the trilacunar condition is primitive in the dicotyledons and that the unilacunar and the multilacunar have been derived from it. Several monocotyledonous plants possess leaves with sheathing bases and nodes with a large number of leaf traces separately inserted around the stem. In ferns the number of traces to a leaf varies from one to many, but they are always associated with a single gap. In gymnosperms a unilacunar node is common.

The leaf trace relationships at the nodes are thought to be of phylogenetic importance, and therefore, nodal anatomy is concerned with the study of systematics and phylogeny of angiosperms.

Nodal Anatomy in Wheat (Monocot) Stem.

In the wheat stem the course of the vascular bundles through the internode and the leaf sheath is almost parallel. Near the node the leaf sheath is considerably thick and attains its maximum thickness just above its union with the stem. On the other hand, the stem has the smallest diameter above the junction with the leaf sheath. The stem is hollow in the internode and solid at the node. The sheath remains open on one side at higher levels, just near the node. Massive collenchymatous bundle caps are present in the bundles of leaf sheath.

Just beneath the junction of the leaf sheath and stem the smaller of the leaf traces are prolonged in the peripheral part of the axis, and the larger leaf traces become part of the inner cylinder of strands. The internodal bundles located above the leaf insertion assume, just above the node a horizontal and oblique course (Fig. 7.50 C, D), and are reoriented toward a more peripheral position in the node and below it (Fig. 7.50 D, E). These horizontal and oblique bundles variously branch and coalesce, and their number reduces. The large leaf traces and the bundles from the internode above the insertion of the leaf make the inner cylinder of the bundles of the next lower internode (Fig. 7.50

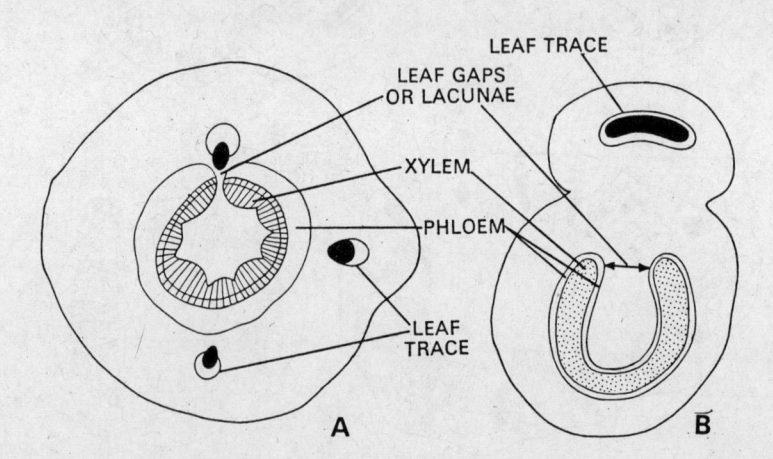

Fig. 7.47. Nodal anatomy. A, nodal anatomy of *Picea* (a conifer) in transection; B, nodal anatomy of *Adiantum* (a fern) in transection. Both possess alternate leaf arrangement, single traces to leaves, and single leaf gaps at the nodes. A, has some secondary growth; B, has phloem on both sides of xylem.

The Tissue System

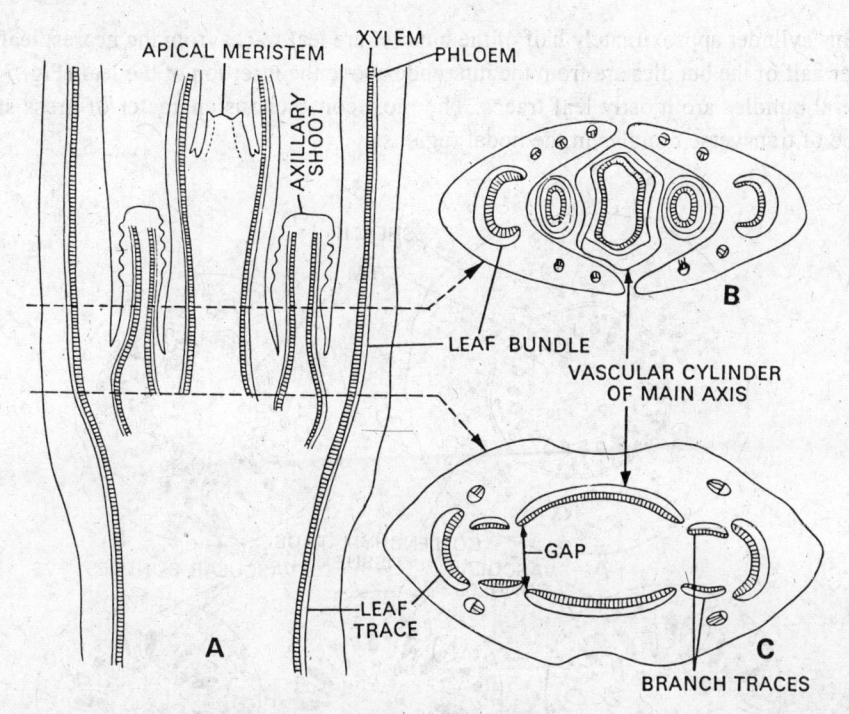

Fig. 7.48. Nodal anatomy. The diagrams depict the primary vascular system of plant with an opposite (decussate) leaf arrangement in L.S. (A) and T.S. (B and C). The branches present in the axils of the leaves are inflorescences. The leaf trace and branch traces in its axil are associated with one common gap. The two branch traces of level C are united into a tubular vascular cylinder in B.

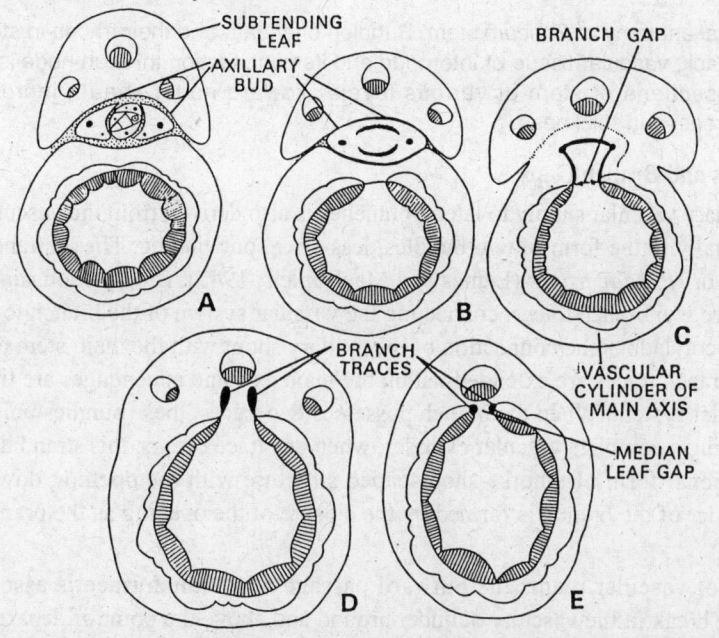

Fig. 7.49. Branch traces and gaps. Diagrams showing vascular connection between an axillary branch (bud) and the main axis in *Salix*. Vascular system of bud is shown in black in the diagram. First two leaves of bud are opposite. The branch gap and the median gap of the subtending leaf are confluent.

E). In this cylinder approximately half of the bundles are leaf traces from the nearest leaf above and the other half of the bundles are from the internode above the insertion of the leaf (Fig. 7.50 E). The peripheral bundles are mostly leaf traces. The most conspicuous character of grass stems is the presence of transverse bundles in the nodal regions.

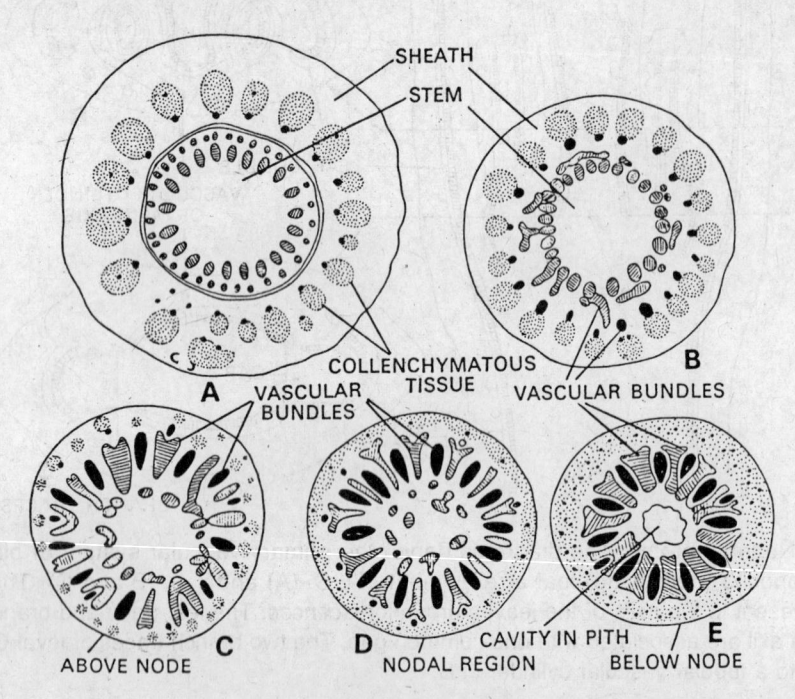

Fig. 7.50. Nodal anatomy of *Triticum* stem. Bundles of sheath and their traces in stem are depicted in black; vascular tissue of internode and its continuation through node is hatched. A—E, transections of stem at various levels. Toward node, sheath increases and stem decreases in thickness.

Branch Traces and Branch Gaps

The primary vascular supply to lateral branches is also derived from the vascular system of the main axis, usually in the form of two bundles, less often, one bundle. These strands are known as *branch traces* or *ramular traces* (Eames and MacDaniels, 1947). Dicotyledons and gymnosperms commonly have two branch traces, connecting the vascular system of the branch to that of the main stem. In monocotyledons the connection of the axillary shoot with the main stem consists of many strands. The branch traces are extended within the main axis and appendages are tied together by a primary vascular system. When the branch possesses two traces, these bundles unite within a short distance, forming a complete vascular cylinder; when one trace occurs, this strand usually possesses the cross-sectional form of a horse-shoe shaped structure with the opening downward, and the vascular cylinder of the branch is formed by the closure of the opening as the branch traces passes out.

In most of vascular plants the outward passage of a branch trace is associated with the formation of a break in the vascular cylinder around and above the point of departure of the trace. This opening is known as *branch gap*, which always accompanies a branch trace. Branch gaps are present in all vascular plants which possess a pith. However, in protosteles the gaps do not occur because there is no pith. Branch gaps are commonly lower than leaf gaps and extend for greater distances in the axis.

The Tissue System

CLOSING OF LEAF GAPS

The features which characterize the nodal structure do not perpetuate in the secondary body. A cambium develops in the parenchyma of the leaf gap and forms vascular tissues in continuity with those bordering the gap. This phenomenon is known as closing of the gap. The parenchyma cells near the margin of the gap are the first to change into cambium and those in the inner portion change later. This process takes place gradually, and the gap parenchyma is maintained as such within the secondary body until and cambium is differentiated throughout the entire tangential width of the gap.

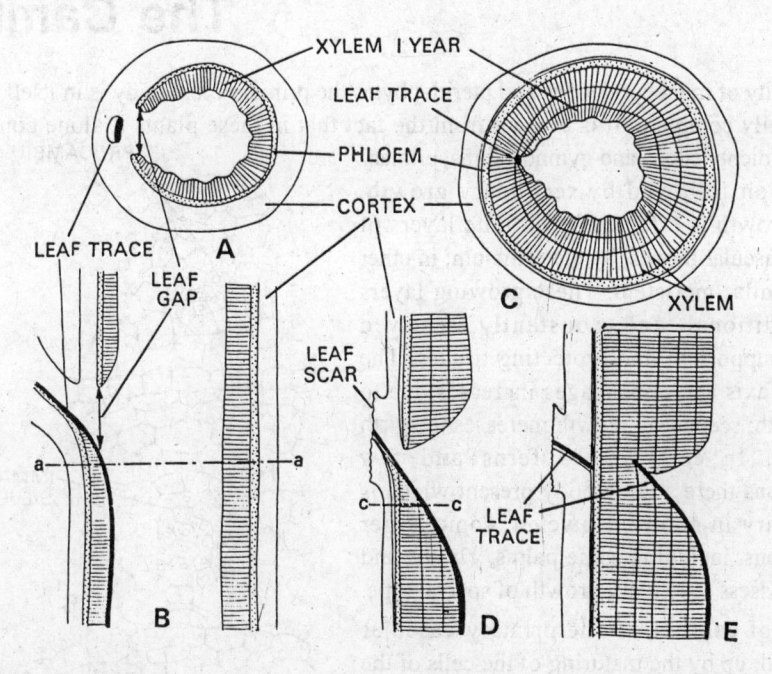

Fig. 7.51. Closing of leaf gaps by secondary growth. A and B, T.S. and L.S.. through nodal region of stems in first year of growth; C—E, T.S. and L.S. through stems several years old; D and E, depict stages in closing of gaps and rupture of leaf trace (E).

In the leaf trace itself complicated changes take place during secondary growth. The primary xylem is buried by the secondary tissues while the phloem is pushed outward. The upper part of the trace diverges outwardly and crosses the plane of the cambium. The part of the cambium that differentiates above the trace in the gap region, produces vascular tissue between the trace and the vascular cylinder. This tissue which increases in amount, exerts a pressure upon the trace and ultimately causes its rupture. The break is filled with parenchyma which is changed into cambium and connects the cambium of the lower part of the trace with that formed in the gap. After this cambium has formed some secondary tissues, the end of the trace below the break becomes embedded in secondary xylem (Fig. 7.51 E). The upper severed end is carried outward, and in time it may be thrown off, together with the cortex, by the activity of the periderm. Since the cambium within the trace itself pushes the trace phloem outward, the buried part of the trace consists of xylem only.

8

The Cambium

In majority of monocotyledons and pteridophytes the primary plant body is in itself structurally and functionally complete. It is evident from the fact that in these plants it alone constitutes the plant. In most dicotyledons and gymnosperms, primary growth is soon followed by secondary growth. Secondary growth is effected by definite layers of initials— in vascular tissues, by the cambium; in other tissues by similar meristem. These growing layers provide additional and constantly renewed conducting, supporting and protecting tissues. The length of the axis and appendages increases due to primary growth; secondary growth increases the girth of the axis. In certain tree ferns and few monocotyledons there a large body present which is wholly primary in nature. However, some larger monocotyledons, including some palms, *Yuccas* and *Dracaenas* possess secondary growth of special type.

Origin of cambium. The primary vascular skeleton is built up by the maturing of the cells of the *procambium* strands to form xylem and phloem. The plants which do not possess secondary growth, all cells of the procambium strands mature and develop into vascular tissue. In the plant which have secondary growth later on, a part of the procambium strand remains meristematic and gives rise to the cambium proper. In roots the formation of cambium differs from that in stems because of the radial arrangement of the alternating xylem and phloem strands. Here the cambium arises as discrete strips of tissue in the procambium strands inside the groups of primary phloem. Later on, the strips of cambium by their lateral extension are joined in the pericycle opposite the rays of primary xylem. The secondary tissue formation is most rapid beneath the groups of phloem so that the cambium, as seen in the transverse section of older roots, soon forms a circle.

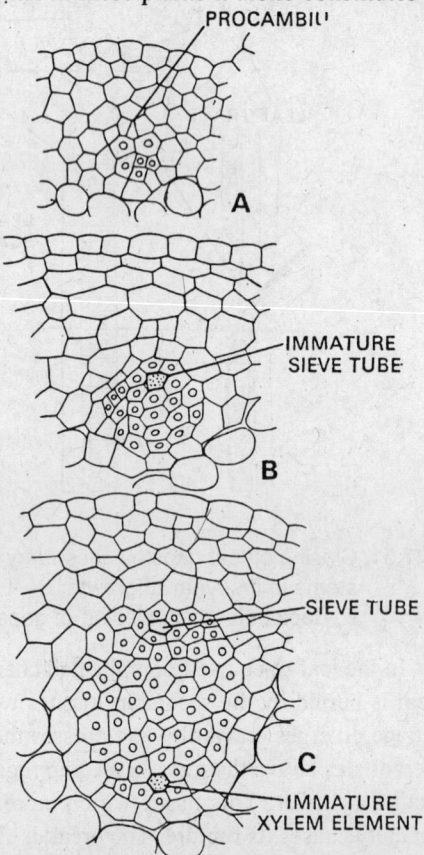

Fig. 8.1. Primary vascular differentiation, A–C, successive stages in the development of the procambium in T.S. of *Linum* stem. The first phloem and xylem elements begin to differentiate before procambial strand completes its increase in diameter.

FASCICULAR AND INTERFASCICULAR CAMBIUM

In stems the first procambium that develops from promeristem is usually found in the form of isolated strands. In some plants these first-formed strands soon become, united laterally by additional similar strands formed between them and by the lateral extention of the first-formed strands. During further development this procambial cylinder gives rise to a cylinder of primary vascular tissue (xylem and phloem) and cambium. Later on, a cylinder of secondary vascular tissue is formed that arises in strands as does the primary cylinder. In *Ranunculus* and some other herbaceous plants, the procambium strands, and the primary vascular tissues, do not fuse laterally but remain as discrete strands. More often in herbaceous stems the cambium extends laterally across the intervening spaces until a complete cylinder is formed. Where such extension occurs, the cambium arises from interfascicular meristematic cells derived from the apical meristem. The strips of cambium that arise within collateral bundles are known as *fascicular cambium*, and the cambial strips found in between the bundles are known as *interfascicular cambium*.

DURATION OF CAMBIUM

The duration of the functional life of the cambium varies greatly in different species and also in different parts of the same plant. In a perennial woody plant the cambium of the main stem lives from the time of its formation until the death of the plant. It is only by the continued activity of the cambium in producing new xylem and phloem that such plants can maintain their existence. In leaves, inflorescenes and other deciduous parts, the functional life of the cambium is short. Here all the

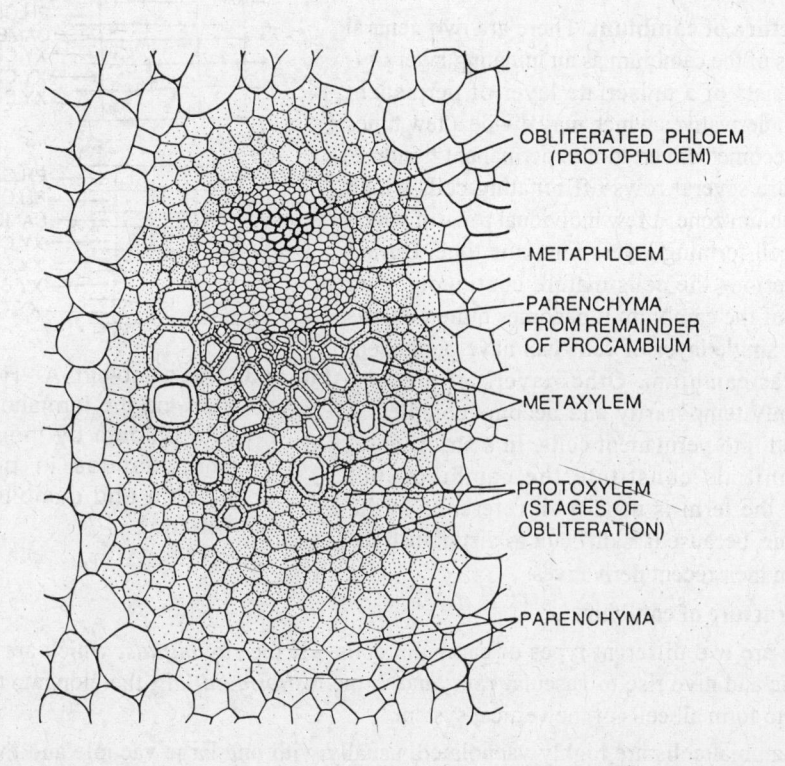

Fig. 8.2. The stem. T.S. of a vascular bundle of *Ranunculus*. This is an example of collateral bundle from a herbaceous dicotyledon lacking secondary growth. (After K. Esau.)

cambium cells mature as vascular tissue. The secondary xylem is directly found upon the secondary phloem in such bundles.

Function of Cambium

The meristem that forms secondary tissues consists of an uniseriate sheet of initials that form new cells usually on both sides. The cambium forms xylem internally and phloem externally. The tangential division of the cambial cell, forms two apparently identical daughter cells. One of the daughter cells remains meristematic, *i.e.* the persistent cambial cell, the other becomes a *xylem mother cell* or a *phloem mother cell* depending upon its position internal or external to the initial. The cambium cell divides continuously in a similar way; one daughter cell always remains meristematic, the cambium cell, whereas the other becomes either a xylem or a phloem mother cell. Probably there is no definite alternation and for brief periods only one kind of tissue is formed. Adjacent cambium cells divide at nearly the dame time, and the daughter cells belong to the same tissue. This way, the tangential continuity of the cambium is maintained.

Structure of cambium. There are two general conceptions of the cambium as an initiating layer— 1. that it consists of a uniseriate layer of permanent initials with derivatives which may divide a few times and soon become converted into permanent tissue; 2. that there are several rows off initating cells which form a cambium zone, a few individual rows of which persist as cell forming layers for some time. During growing periods the cells mature continuously on both sides of the cambium it becomes quite obvious that only a single layer of cells can have permanent existence as cambium. Other layers, if present, function only temporarily and become completely transformed into permanent cells. In a strict sense, only the initials constitute the cambium, but frequently the term is used with reference to the cambial zone, because it is difficult to distinguish the initials from their recent derivatives.

Cellular structure of cambium

There are two different types of cambium cells—1. the *ray initials*, which are more or less isodiametric and give rise to vascular rays; and 2. the *fusiform initials*, the elongate tapering cells that divide to form all cells of the vertical system.

The cambial cells are highly vacuolated, usually with one large vacuole and thin peripheral cytoplasm. The nucleus is large and in the fusiform cells is much enlongated. The walls of cambial cells have primary pit fields with plasmodesmata. The radial walls are thicker than tangential walls, and their primary pit fields are deeply depressed.

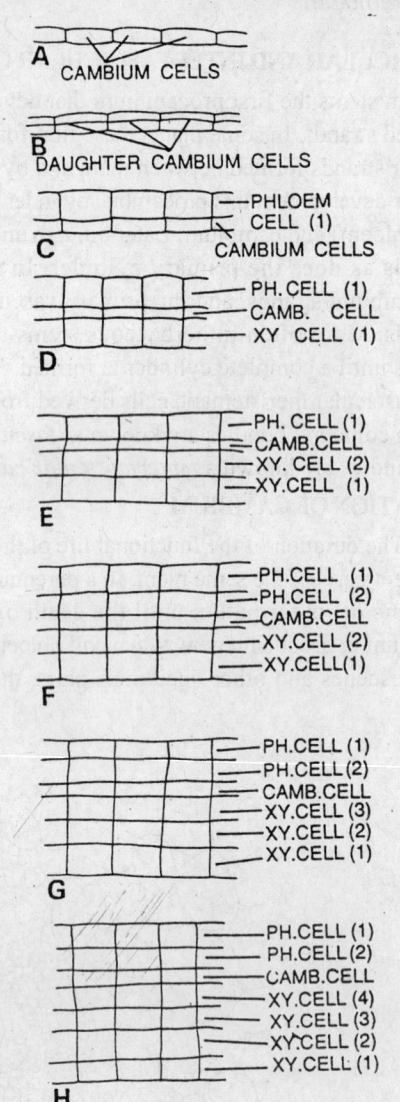

Fig. 8.3. The cambium. A—H, diagrams showing the formation of xylem and phloem by the cambium, and changes in position of phloem and cambium by this activity.

The Cambium

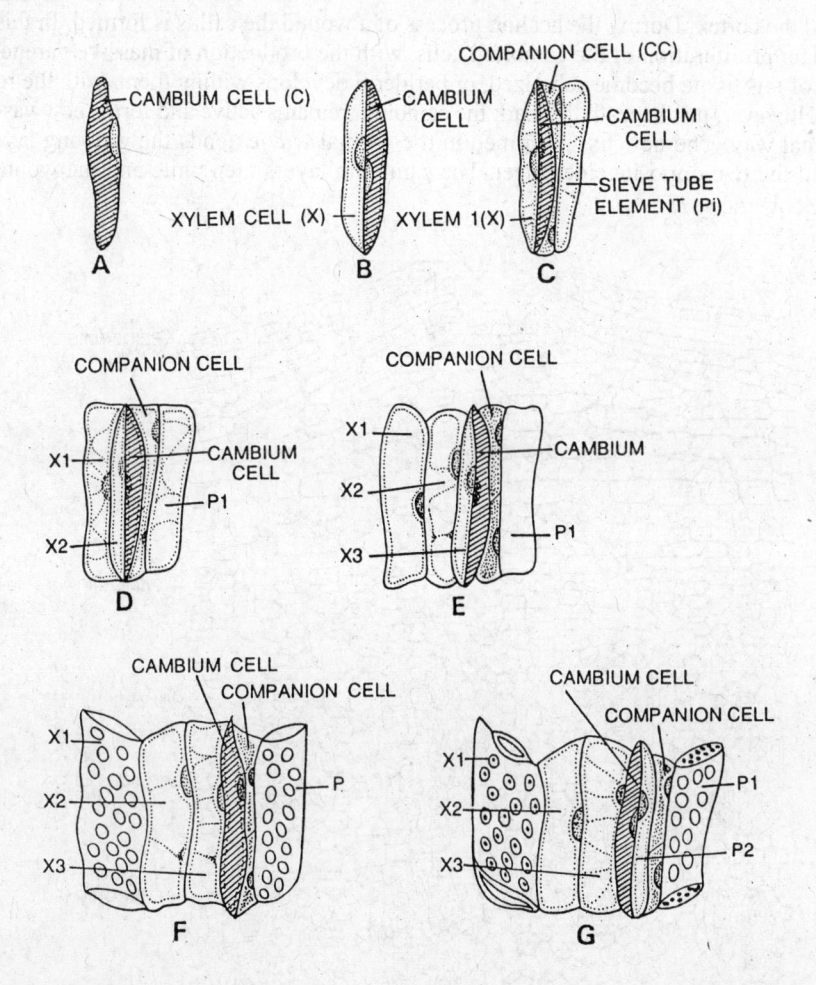

Fig. 8.4. The cambium. Differentiation of phloem and xylem from vascular cambium. A—G, diagrams as seen in radial section, showing stages in differentiation of vascular cambium cells.

Cell Division in Cambium.

With the result of tangential (periclinal) divisions of cambium cells the phloem and the xylem are formed. The vascular tissues are formed in two opposite directions, the xylem cells towards the interior of the axis, the phloem cells toward its periphery. The tangential divisions of the cambium initials during the formation of vascular tissues determine the arrangement of cambial derivatives in radial rows. Since the division is tangential, the daughter cells that persist as cambium initials increase in radial diameter only. The new cambium initials formed by transverse divisions increase greatly in length; those formed by radial divisions do not increase in length.

As the xylem cylinder increases in thickness by secondary growth, the cambial cylinder also grows in circumference. The main cause of this growth is the increase in the number of cells in tangential direction, followed by a tangential expansion of these cells.

Cambium growth about wounds. One of the important functions of the cambium is the formation of *callus* or *wound tissue*, and the healing of the wounds. When wounds occur on plants, a large amount of soft parenchymatous tissue is formed on or below the injured surface; this tissue is known as callus. The callus develops from the cambium and by the division of parenchyma cells in the phloem and the cortex. During the healing process of a wound the callus is formed. In this there is at first abundant proliferation of the cambium cells, with the production of massive parenchyma. The outer cells of this tissue become suberized, or periderm develops within them, with the result a bark is formed. However, just beneath this bark the cambium remains active and forms new vascular tissue in the normal way. The new tissue formed in the normal way extends the growing layer over the wound until the two opposite sides meet. The cambium layers then unite and the wound becomes completely covered.

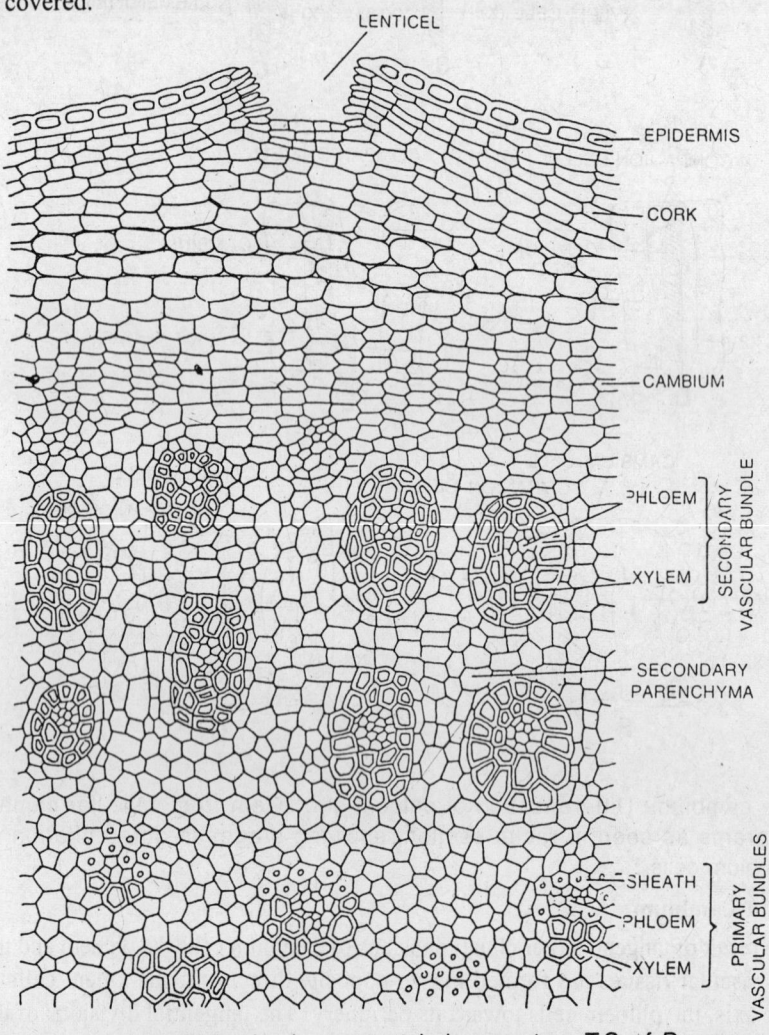

Fig. 8.5. econdary thickening in monocotyledonous stem. T.S. of *Dracaena* stem.

Cambium in budding and grafting. In the practices of budding and grafting, the cambium of both stock and scion gives rise to callus which unites and develops a continuous cambium layer that gives rise to normal conducting tissue. There is an actual union of the cambium of stock and scion of two plants during the practices of budding and grafting and therefore these practices are not commonly found in monocotyledons.

Cambium in monocotyledons. A special type of secondary growth occurs in few monocotyledonous forms, such as *Dracaena, Aloe, Yucca, Veratrum* and some other genera. In these plants the stem increases in diameter forming a cylinder of new bundles embedded in a tissue. Here a cambium layer develops from the meristematic parenchyma of the pericycle or the innermost cells of the cortex. In the case of roots, the cambium of this develops in the endodermis. The initials of cambium strand in tiers to form a storied cambium as found in the normal cambium of some dicotyledons.

9

The Root — Primary and Secondary Structure

The roots are generally divided into two categories: (1) *primary*, normal roots, which originate from the embryo and usually persist throughout life, and (2) *adventitious* roots, which arise secondarily from stem, leaf or other tissues and which may be either permanent or temporary.

The functions of the primary roots are to anchor the plant in the soil, to absorb water and soluble substances and to serve as storehouses of food materials. The functions of adventitious roots are very various. Sometimes they enter the soil and act as primary roots. In other cases they may be modified into climbing organs, stilts or props, thorns, haustoria, etc.

The anatomy of roots is in many respects simpler than that of stems and it is considerably more uniform, perhaps as the result of a rather uniform environment underground, which is in marked contrast with the extreme variability of the conditions affecting aerial shoots. Root anatomy is more ancient in type than that of stems. Roots preserve, in their primary condition, the actinostely of the vascular tissues, which in the lower Pteridophyta is characteristic of both stems and roots.

GENERAL CHARACTERISTICS OF THE ROOTS

1. The roots possess a tendency to grow downwards or sideways rather than upwards.
2. There is absence of chlorophyll in the roots.
3. The roots are not susceptible to the influence of light.
4. There is absence of leaves and hence, of course, absence of buds.
5. The roots possess a root cap over the apex.
6. They possess endogenous origin and branching.
7. In roots, the phloem and xylem are situated on different radii in the primary structure.
8. There is a relatively short zone of growth at the apex.
9. They possess root hairs near the apex.

ANATOMICAL CHARACTERISTICS OF THE ROOT

ROOT CAP

The root cap consists of parenchymatous cells in various stages of differentiation. It is protective in function. Recent experiments have indicated that the root cap has another function that is physiologically of great importance. The root cap is apparently the site of the perception of gravity; thus it appears to be capable of controlling the production in the meristem of the growth-regulating substances involved in *geotropism* or their movement. However, the root cap itself is not the site of synthesis of growth substances. The experiments show that the root cap can control the movement, if not also the synthesis of endogenous auxin in the root apex.

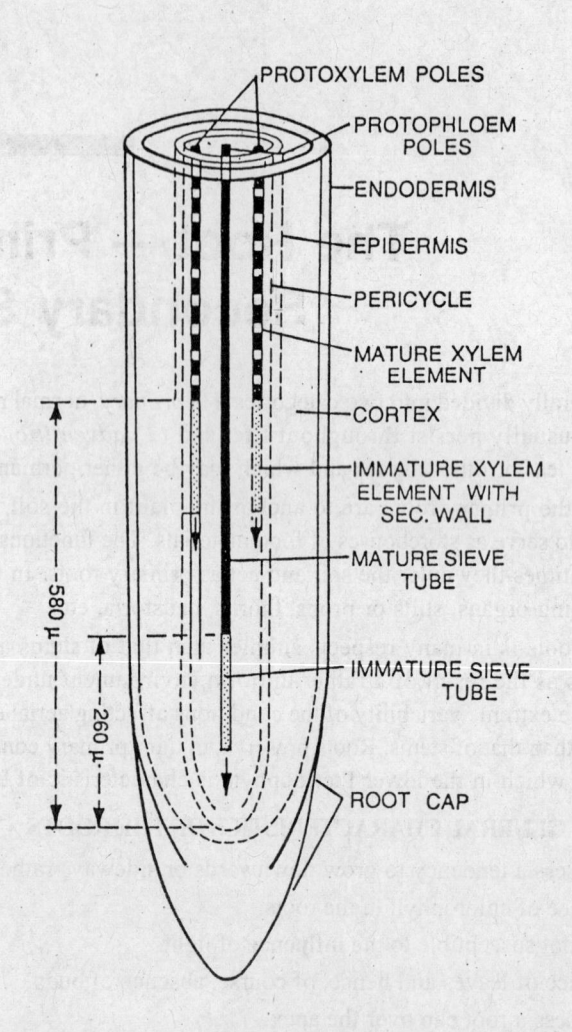

Fig. 9.1. The root. L.S. of a root tip of tobacco. (After Esau).

EPIDERMIS

The epidermis is also known as epiblema or piliferous layer. In most of roots, root hairs develop from some of the epidermal cells at a little distance from the apical meristem. The cells, giving rise to root hairs are known as *trichoblasts*. In the roots of the grass (*Phleum*), trichoblasts are formed by unequal division, of a cell of the immature epidermis, the protoderm. The trichoblast is the distally situated cell of this division. Root hairs are outgrowths of single cells, and function both in the absorption and in anchorage. Root hairs are usually eventually sloughed off, but occasionally they are persistent. Electron microscopy reveals (Leech, Mollenhauer and Whaley, 1963) that the wall of emerging root hair is a continuation of only the inner component of the wall of the epidermal cell which gives rise to it. In some species, root hairs are formed from special cells which are dintinct in size and metabolism from the neighbouring epidermal cells. These cells are known as *trichoblasts*. In other species, root hairs may develop from a whole row of cells, which stain more densely than their neighbours.

The Root—Primary and Secondary Structure

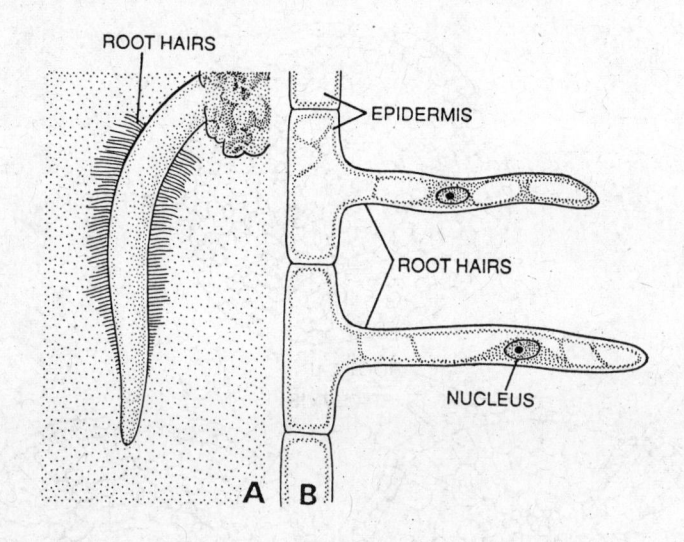

Fig. 9.2. Root hair. A, young seedling of radish with root hairs developing acropetally; B, highly magnified mature root hairs with vacuolated cytoplasm.

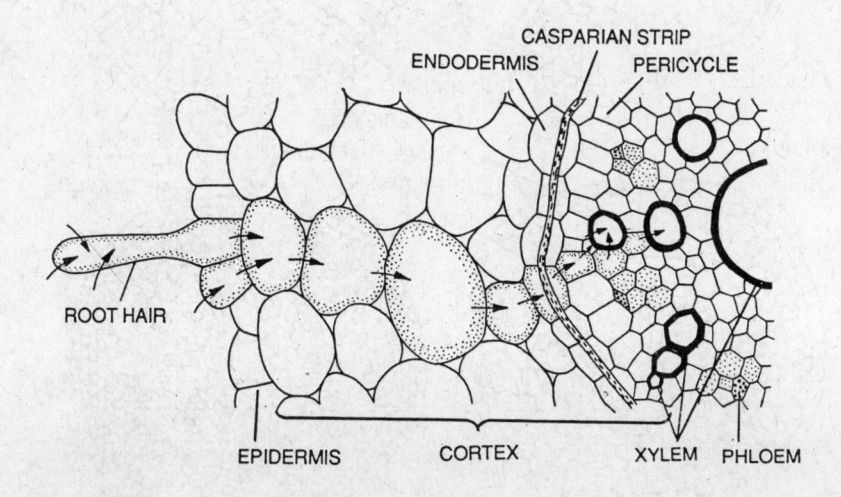

Fig. 9.3. Function of root. In this diagram the arrows indicate the movement of water and salts absorbed from the soil by root hair.

In the aerial roots of certain epiphytic orchids a multiple epidermis, or *velamen* is present. This tissue is derived from the epidermal initials and may be several cells thick. The cells are dead and devoid of contents, the cell walls are strengthened with bands of lignin. On the inner side of the velamen there is a specialized layer of cells which is derived from the periblem and not the dermatogen, and may therefore be considered as the outermost layer of the cortex, the *exodermis*. This layer is composed of alternating long and short cells; the long cells become thick-walled on their radial and outer tangential surface, but the small cells remain thin-walled and are called *passage cells*. The velamen is thought to function as a protective tissue, preventing undue water loss from the delicate cortical cells of the exposed aerial root. Formally this was believed that the cells of the

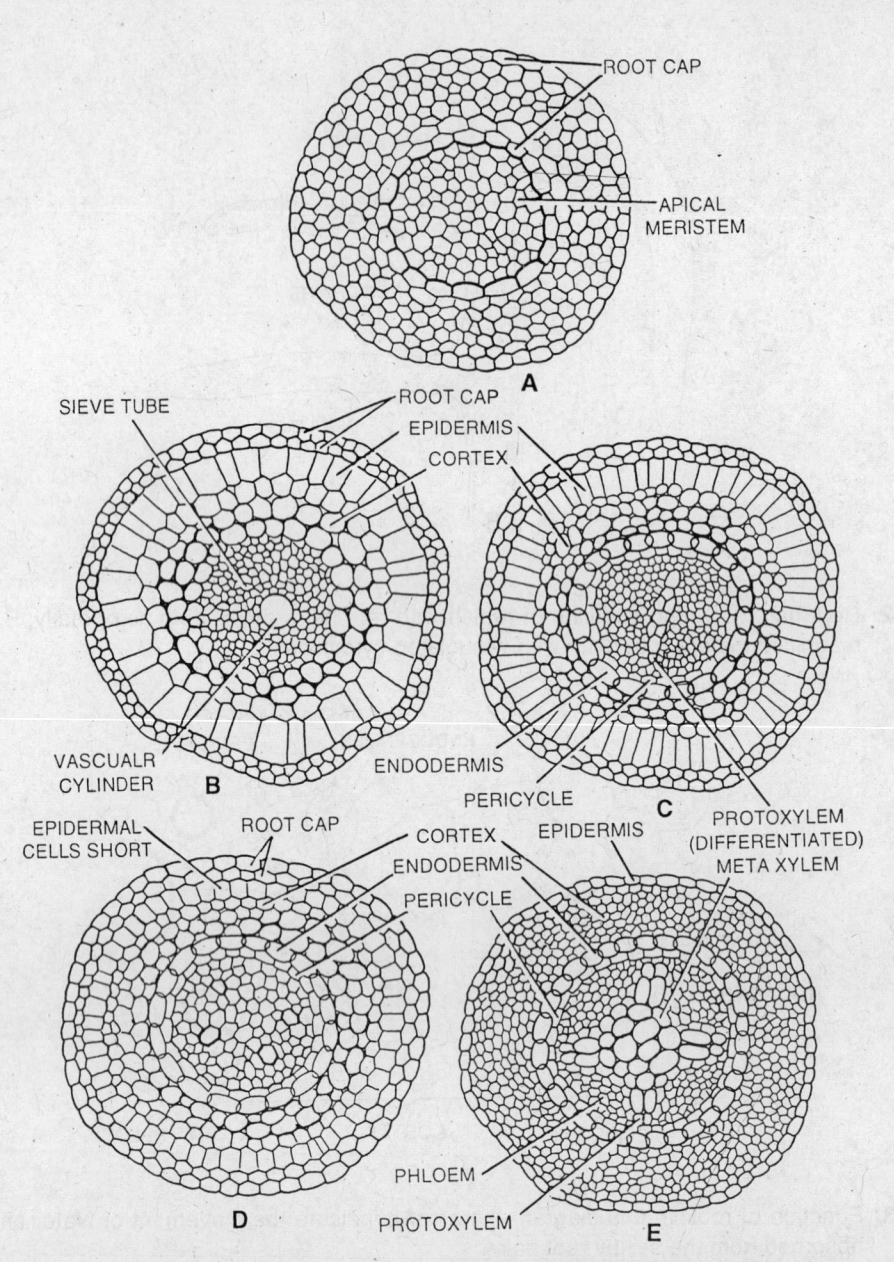

Fig. 9.4. Tissue differentiation in root. Cross sections at various levels of root showing tissue differentiation. A, T.S. root apex showing apical meristem and root tip; B, T.S. showing root cap, epidermis cortex and elements of sieve tubes; C, endodermis, pericycle, sieve elements and xylem elements are seen; D—E, show later stages of development.

velamen also absorbed and conserved water drawn from the atmoshphere, but recent experiments show that the mature velamen and exodermis are nearly impermeable to water and solutes.

CORTEX

In most roots the cortex is parenchymatous. In some roots, the cells of the cortex are very regularly arranged, both radially and in concentric circles. Conspicuous intercellular spaces may be present, and especially evident in aquatic species, where they form a type of aerenchyma. The

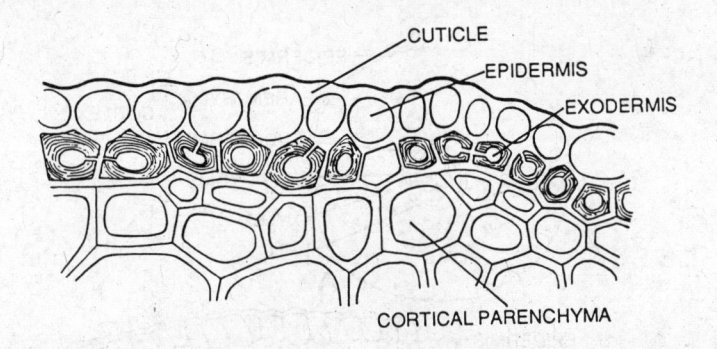

Fig. 9.5. The root. T.S. of the outer part of the *Smilax* root showing a thick-walled exodermis beneath the epidermis. The passage cell of exodermis is thin-walled.

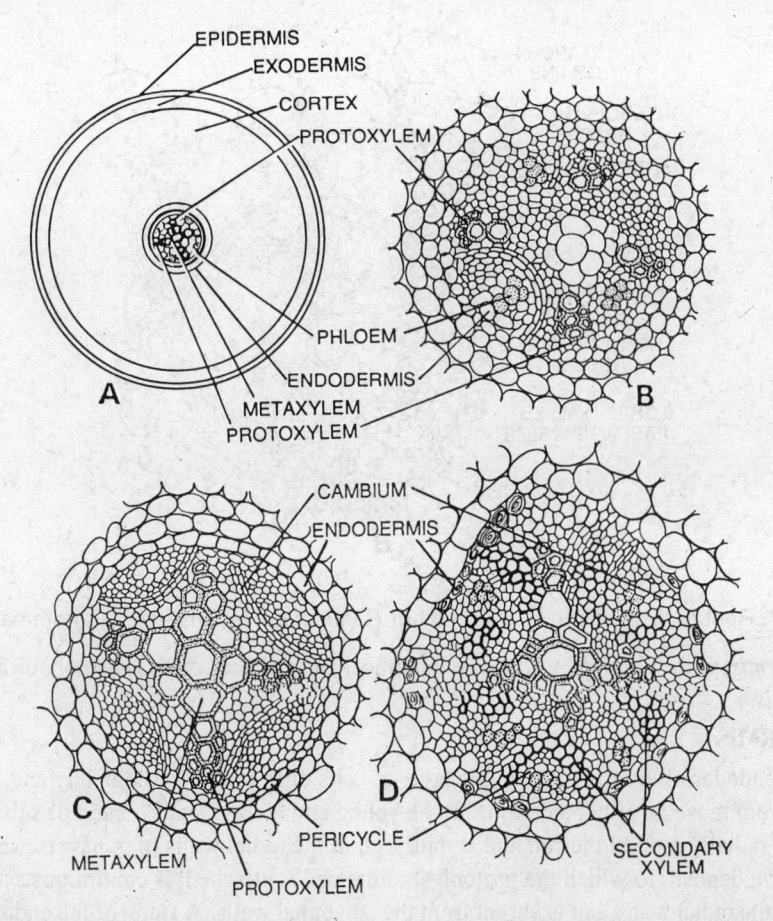

Fig. 9.6. Anatomy of root. A—D, T.S. of tetrarch root of *Ranunculus* in successive stages of development.

cortical cells often contain starch, and sometimes crystals. Sclerenchyma is more common in the roots of monocotyledons than those of dicotyledons. The characteristic trichosclereids are found in the roots of *Monstera*. Collenchyma is occasionally present in roots (*e.g.*, *Monstera*). The outermost layer or layers of the cortex, just beneath the epidermis may be differentiated as an *exodermis*, a kind

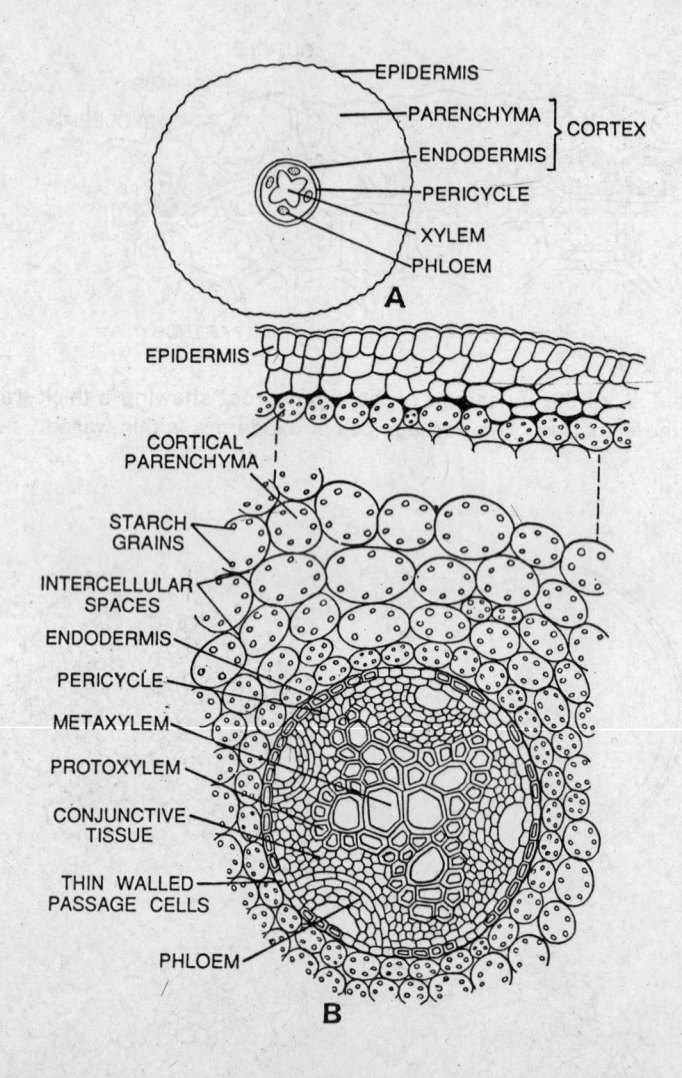

Fig. 9.7. Root of dicotyledon. T.S. of root of *Ranunculus repens*— A, diagrammatic; B, detail.

of *hypodermis*, with suberized walls. The innermost layer of the cortex is usually differentiated as an *endodermis*.

ENDODERMIS

The endodermis comprises a single layer of cells differing physiologically and in structure and function from those on either side of it. In the young endodermal cells a band of suberin, *Casparian strip*, runs radially around the cell and is thus seen in the radial walls in transverse sections of roots. This suberin deposit, to which the protoplast of the cell is attached, is continuous across the middle lamella of the radial walls, but is absent from the tangential walls. A study of the endodermis with the electron microscope (Bonnett, 1968) reveals a thickening of the wall in the region of the Casparian strip. The *plasmalemma* is thicker here and adheres strongly to the cell wall. The thin-walled *passage cells* often remain in the endodermis in positions opposite the protoxylem.

PERICYCLE

The pericycle is usually a single layer of parenchymatous cells lying just within the endodermis and peripheral to the vascular tissues. The pericycle has a capacity for meristematic growth, and gives

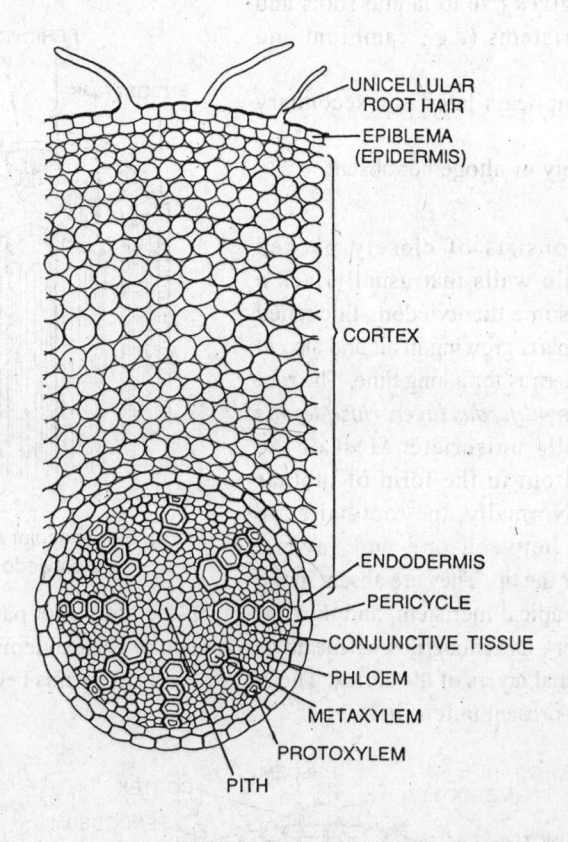

Fig. 9.8. Structure of root. T.S. of a typical monocotyledonous root.

rise to lateral root primordia, parts of the vascular cambium, and usually the meristem which produces cork, the phellogen. The pericycle is sometimes called pericambium.

VASCULAR SYSTEM

The vascular system of the root as seen in transverse section consists of a variable number of triangular rays of thick-walled, lignified tracheary elements, alternating with arcs of thin-walled phloem. In the root, the xylem and phloem do not lie on the same radius. The xylem may form a solid central core, or there may be a parenchymatous or sclerenchymatous pith, as in the roots of many monocotyledons. Roots with 1, 2, 3, 4, 5 and many arcs of xylem are respectively called monarch, diarch, triarch, tetrarch, pentarch and polyarch. The xylem is exarch, *i.e.*, protoxylem lies towards periphery and metaxylem towards the centre. The xylem is always centripetal in its development. The phloem bundle consists of sieve tubes, companion cells and phloem parenchyma. The protoxylem consists of annular and spiral vessels and metaxylem of reticulate and pitted vessels.

The parenchyma found in between xylem and phloem bundles is known as *conjunctive tissue*. The pith may be large, small or altogether absent.

ANATOMY OF DICOTYLEDONOUS ROOTS

The important anatomical characteristics of the dicotyledonous roots are as follows:

1. The xylem bundles vary from two to six number, *i.e.*, they may be diarch, triarch, tetrarch, pentarch or hexarch.

2. The pericycle gives rise to lateral roots and secondary meristems (*e.g.*, cambium and phellogen).
3. The cambium appears later as a secondary meristem.
4. The pith is scanty or altogether absent.

EPIDERMIS

The epidermis consists of closely packed elongated cells with thin walls that usually lack a cuticle and stomata. In some dicotyledons thickened outer walls occur in root parts growing in air and also in roots that retain their epidermis for a long time. The root epidermis (also known as *piliferous layer, rhizodermis* or *epiblema*) is typically uniseriate. Most of the epidermal cells extend out in the form of tubular unicellular root hairs. Normally, the root hairs are confined to a region between one and several centimetres in length near the tip. They are absent in the

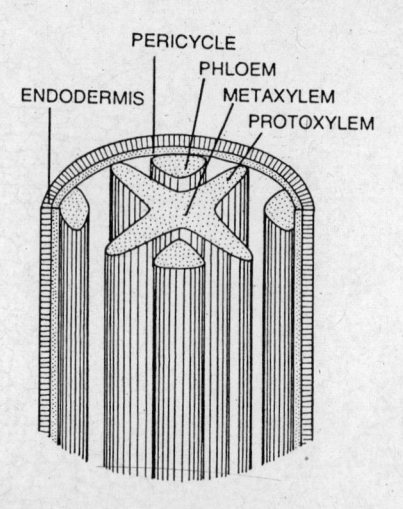

Fig. 9.9. Vascular system—stele of a dicotyledonous root.

nearest proximity of the apical meristem, and they die off in the older root parts. Some roots also develop a specialized layer—the exodermis—beneath the epidermis. The exodermis arises from one or several of the sub-epidermal layers of the cortex. The cell walls of exodermis become suberized. The exodermis is found to be present in few dicots.

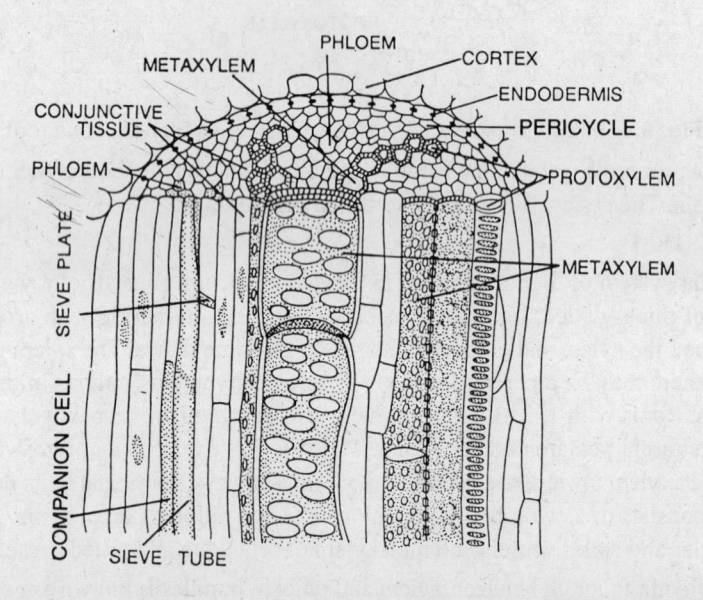

Fig. 9.10. Dicot root—transverse and longitudinal sections of the stele of a typical dicot root.

CORTEX

The cortex is massive and consists of thin-walled rounded or polygonal parenchyma cells having sufficiently developed intercellular spaces among them. The parenchyma cells of the cortex contain abundant starch grains in them. In the roots of dicotyledons which possess secondary growth and shed their cortex early, the cortex consists mainly of parenchyma. As seen in transverse sections, the cortical cells may be arranged in radial rows, or they may alternate with one another in the successive concentric layers. The presence of schizogenous intercellular spaces is typical of the root cortex. In the water plants the intercellular spaces are large and form distinct air spaces. The cortex of roots is usually devoid of chlorophyll. Exceptions are roots of some water plants and aerial roots of many epiphytes (*e.g., Tinospora* spp). Various idioblasts and secretory structures are found in the root cortex. Some dicotyledons (*e.g., Brassica, Pyrus, Prunus, Spiraea*, etc.) may develop prominent reticulate or band-like thickenings in cortical cells outside the endodermis.

ENDODERMIS

The innermost distinct layer of the cortex is known as *endodermis*. The endodermis is uniseriate and almost universally present in the roots. The cells of endodermis are living and characterized by the presence of *Casparian strips* or *Casparian bands* on their anticlinal walls. The strip is formed during the early ontogeny of the cell and is a part of the primary wall. The strip is typically located close to the inner tangential wall. Guttenberg (1943) says, that the suberin-like materials are found in the strips. The cytoplasm of an endodermal cell remains

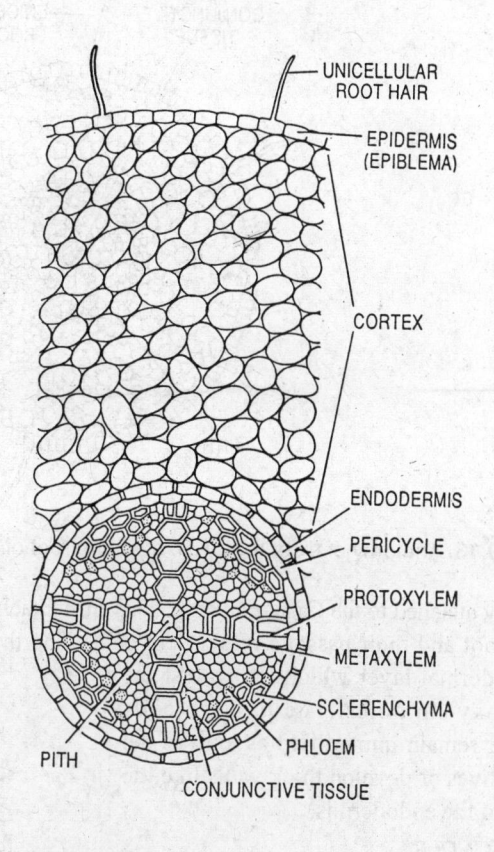

Fig. 9.11. Dicot root—T.S. of gram root.

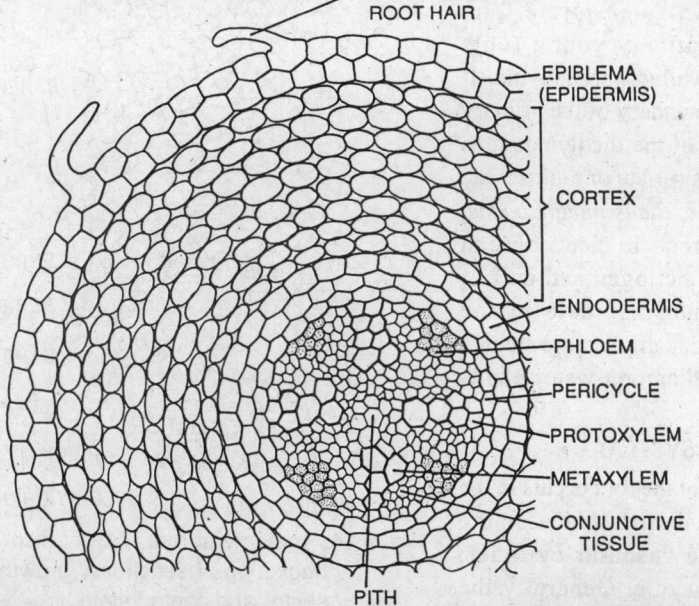

Fig. 9.12. T.S. of young dicotyledonous root. The cambium has not developed.

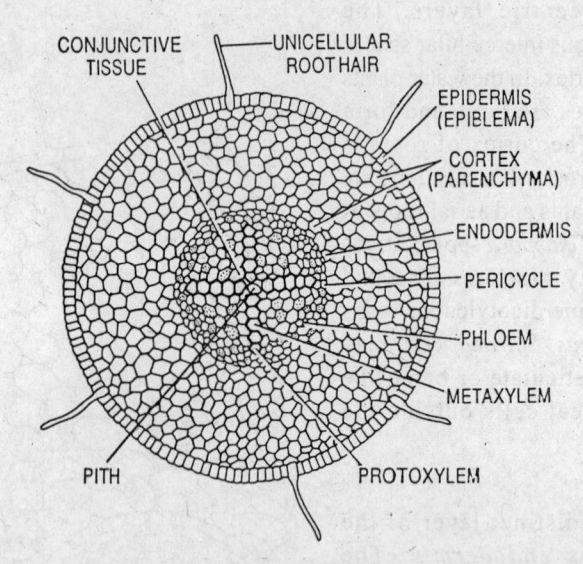

Fig. 9.13. Anatomy of dicot root. T.S. of young root of *Phaseolus radiatus* prior to cambium formation.

firmly attached to the Casparian strip. This firm attachment controls the movement of the materials in the root and their passage into xylem cells. The thin-walled *passage-cells* are also found in the endodermal layer which lie against the protoxylem poles. The passage cells either remain unmodified as long as the root lives or develop thick walls like the rest of the endodermis.

PERICYCLE

The layer next to the endodermis is commonly known as *pericycle*. The pericycle of relatively young roots consists of thin-walled parenchyma. It makes the outer boundary of the primary vascular cylinder of the dicotyledonous roots. It may be uniseriate or multiseriate (*e.g., Morus, Salix, Ficus benghalensis,* etc.). The lateral roots in dicots arise in this tissue. The phellogen and part of vascular cambium originate in the pericycle. Roots without pericycle are rare but may be found among water plants and parasites.

THE VASCULAR SYSTEM

The phloem of the root occurs in the form of strands distributed near the periphery of the vascular cylinder, beneath the pericycle. Generally the xylem forms discrete strands, alternating

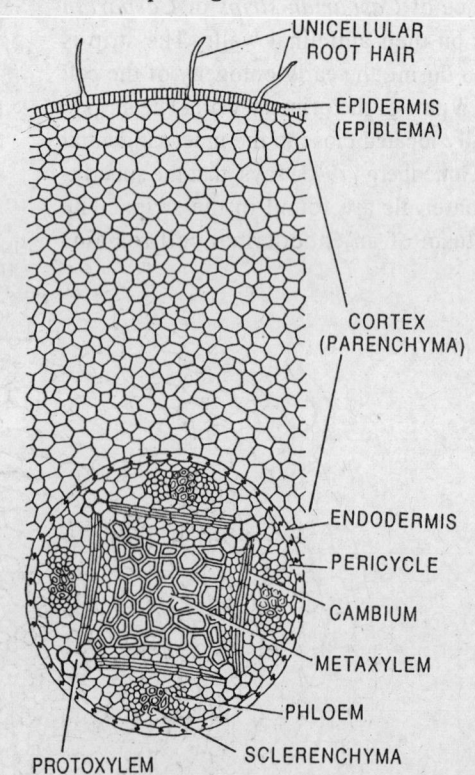

Fig. 9.14. Anatomy of root. T.S. of dicot root showing beginning of secondary growth. Details of a sector and central stele.

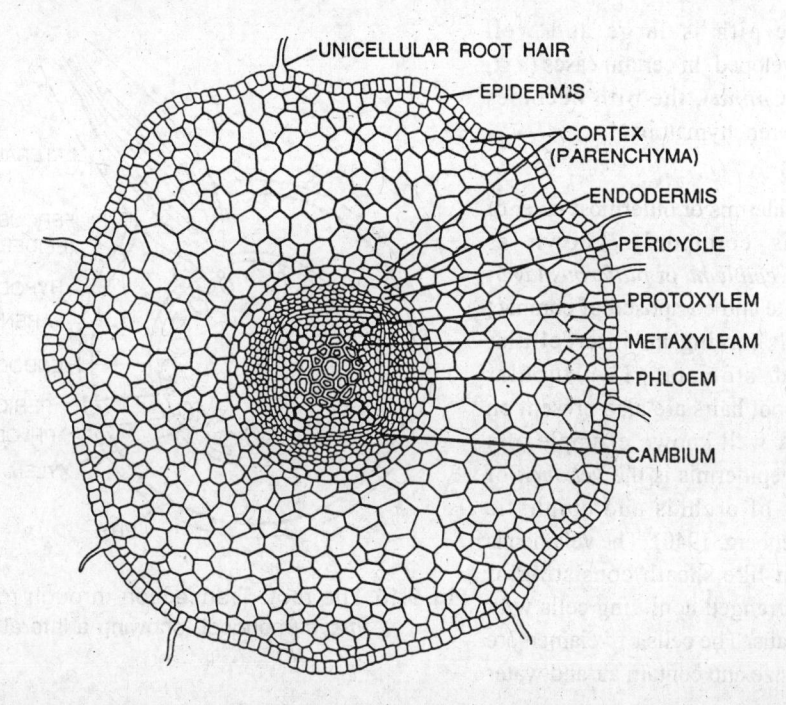

Fig. 9.15. Anatomy of dicot root. T.S. of root of *Phaseolus radiatus* showing beginning of cambium formation.

with the phloem strands. Sometimes the xylem occupies the centre, with the strand-like parts projecting from the central core like ridges. If xylem is not differentiated in the centre, the centre is occupied by a pith. The root typically shows an exarch xylem, *i.e.*, the protoxylem is located near the periphery of the vascular cylinder, the metaxylem farther inward. The phloem is also centripetally differentiated, *i.e.*, the protophloem occurring closer to the periphery than the metaphloem. Most dicotyledons have few xylem strands. The taproot is frequently di-, tri-, or tetrarch, but it may have five to six and even more poles, (*e.g.*, many Amentiferae, *Castanea*). Only one xylem strand occurs in the slender root of the hydrophyte *Trapa natans*. In *Raphanus, Daucus, Linum, Lycopersicon* and *Nicotiana* the roots are diarch. In *Pisum* the root is triarch. In *Cicer, Vicia, Helianthus, Gossypium* and *Ranunculus* the roots are tetrarch. In Certain dicots the root of the same plant may show di-, tri-, and tetrarch xylem. For example, tetrarch and polyarch roots have been reported from *Nymphaea chilensis*. (Wardlaw, 1928). Banerji (1932) reported tri-, tetra-, and pentarch roots in *Enhydra fluctuans*. Such roots are known as *heteroarchic roots*. The protoxylem consists of annular and spiral vessels whereas metaxylem of reticulate and pitted vessels. The phloem strand consists of sieve tubes, companion cells and phloem parenchyma. The parenchymatous conjunctive tissue occurs in between xylem and phloem strands. The pith is scanty or altogether absent.

ANATOMY OF MONOCOTYLEDONOUS ROOTS

The distinctive anatomical characters of the monocotyledonous roots are as follows:
1. The xylem groups are numerous (polyarch condition) and generally vary from twelve to twenty.
2. The pericycle gives rise to lateral roots only.
3. The cambium is altogether absent even in later stages, as there is no secondary thickening in such roots.

4. The pith is large and well developed. In certain cases (*e.g.*, in *Canna*), the pith becomes sclerenchymatous.

EPIDERMIS

The epidermis or outermost layer of the root is commonly known as *rhizodermis, epiblema* or *piliferous layer*. It is uniseriate and composed of compact tabular cells having no intercellular spaces and stomata. The tubular unicellular root hairs are also present on this layer. A well known example of a multiseriate epidermis is the *velamen* of aerial roots of orchids and epiphytic aroids (Guttenberg, 1940). The velamen is a parchment-like sheath consisting of compactly arranged nonliving cells with thickened walls. The cells of velamen are quite big in size and contain air and water

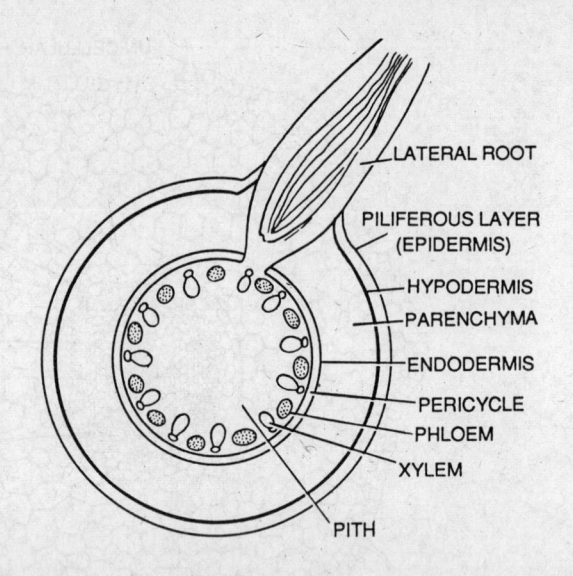

Fig. 9.16. The root. Transection through root of *Zea mays* (monocot) showing a lateral root.

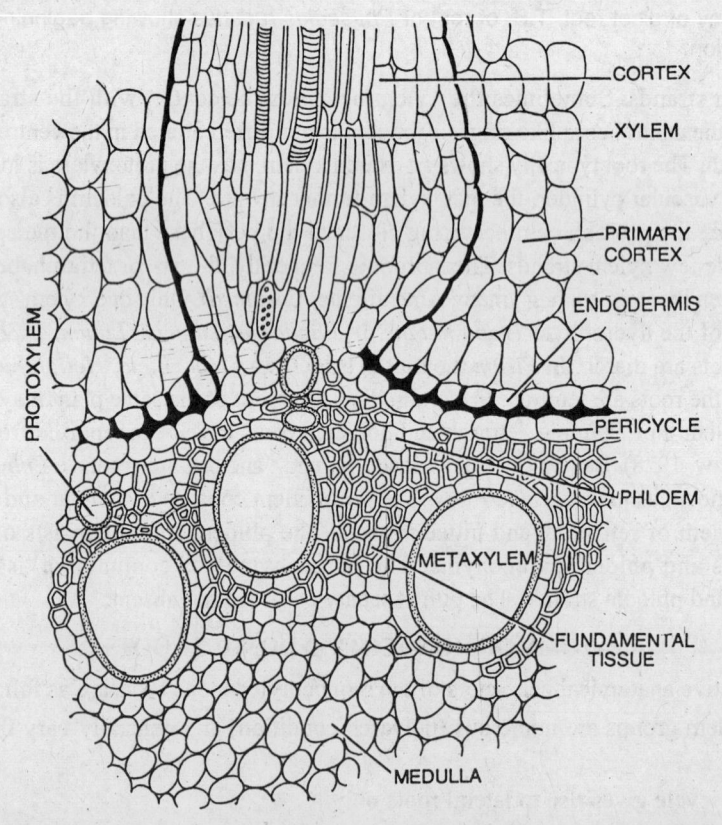

Fig. 9.17. The root. Transection through root of *Zea mays* (monocot) showing a lateral root. The cortex and xylem are clearly visible in the lateral root.

in them. The cell walls develop fibrous thickenings. Generally, beneath the epidermis there are present one or more layers of *exodermis*. Usually the exodermis consists of a single row of cells with thickened outer and lateral walls except certain passage cells which remain thin-walled.

CORTEX

Immediately beneath the epidermis a massive cortex lies consisting of thin-walled parenchyma cells having sufficiently developed intercellular spaces among them. Usually in an old root of *Zea mays* a few layers of cortex immediately beneath the epidermis undergo suberization and give rise to a simple or multilayered zone—the *exodermis*. This is protective layer which protects internal tissues

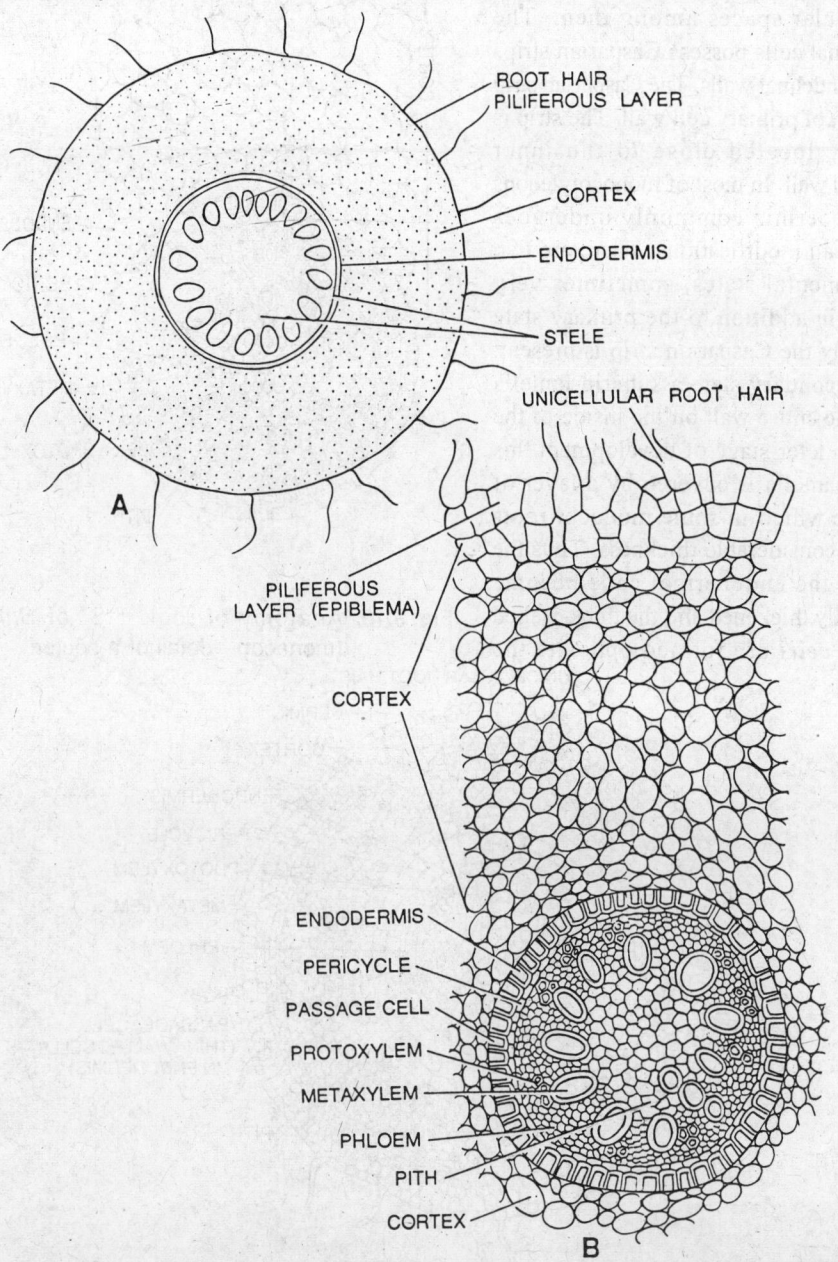

Fig. 9.18. Root. T.S. of monocotyledonous root (*Iris*). A, diagrammatic; B, detailed.

from injurious agencies. The starch grains are abundantly present in the cortical cells. The sclerenchyma cells are commonly found in the cortex of monocotyledons.

ENDODERMIS

The innermost layer of the cortex is called the endodermis. It is composed of barrel-shaped compact cells having no intercellular spaces among them. The endodermal cells possess Casparian strips on their anticlinal walls. The Casparian strip is the part of primary cell wall. The strip is typically located close to the inner tangential wall. In most of monocotyledons the endodermis commonly undergoes certain wall modifications. There are two developmental states, sometimes very distinct, in addition to the primary state when only the Casparian strip is present. In the secondary state a suberin lamella covers the entire wall on the inside of the cell. At a later stage of development this suberin lamella is covered by a layer of cellulose which in some monocot roots attains a considerable thickness. Thus the walls of the endodermal cells become sufficiently thickened and the thick-walled *passage cells* are formed opposite the

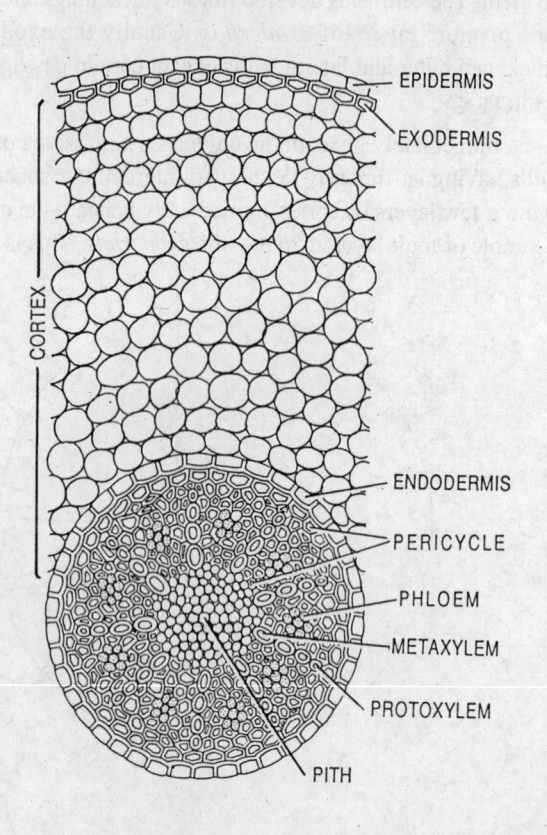

Fig. 9.19. Anatomy of root. T.S. of *Smilax* root (monocot)—detail of a sector.

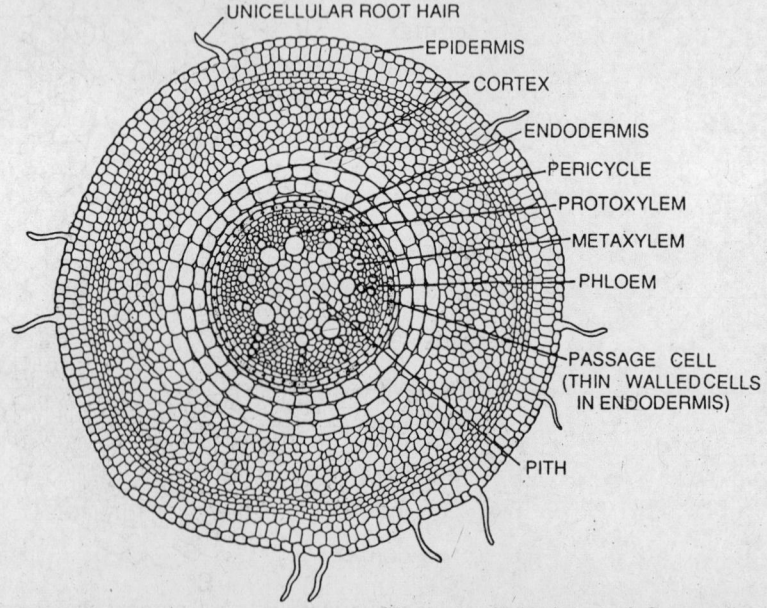

Fig. 9.20. Anatomy of monocot root. Cross section of root of *Commelina*.

protoxylem poles. The passage cells are meant for diffusion and are also called the *transfusion cells*.

PERICYCLE

It is usually uniseriate and composed of thin-walled parenchymatous cells. In the monocotyledons, the pericycle often undergoes sclerification in older roots, partly or entirely. In many monoctyledons (*e.g.*, some Gramineae, *Smilax, Agave, Dracaena,* palms) the pericycle consists of several layers. The pericycle may be interrupted by the differentiation of xylem (many Gramineae and Cyperaceae) or phloem elements (Potamogetonaceae) next to the endodermis (Guttenberg, 1943). Here, the pericycle gives rise to lateral roots only.

VASCULAR TISSUE

The vascular tissue consists of alternating strands of xylem and phloem. The phloem occurs in the form of strands near the periphery of the vascular cylinder, beneath the pericycle. The xylem forms discrete strands, alternating with the phloem strands. The centre is occupied by large pith which may be parenchymatous or sclerenchymatous. Bundles are numerous and referred as *polyarch*. The adventitious roots of Palmae and Pandanaceae have considerably higher number of vascular

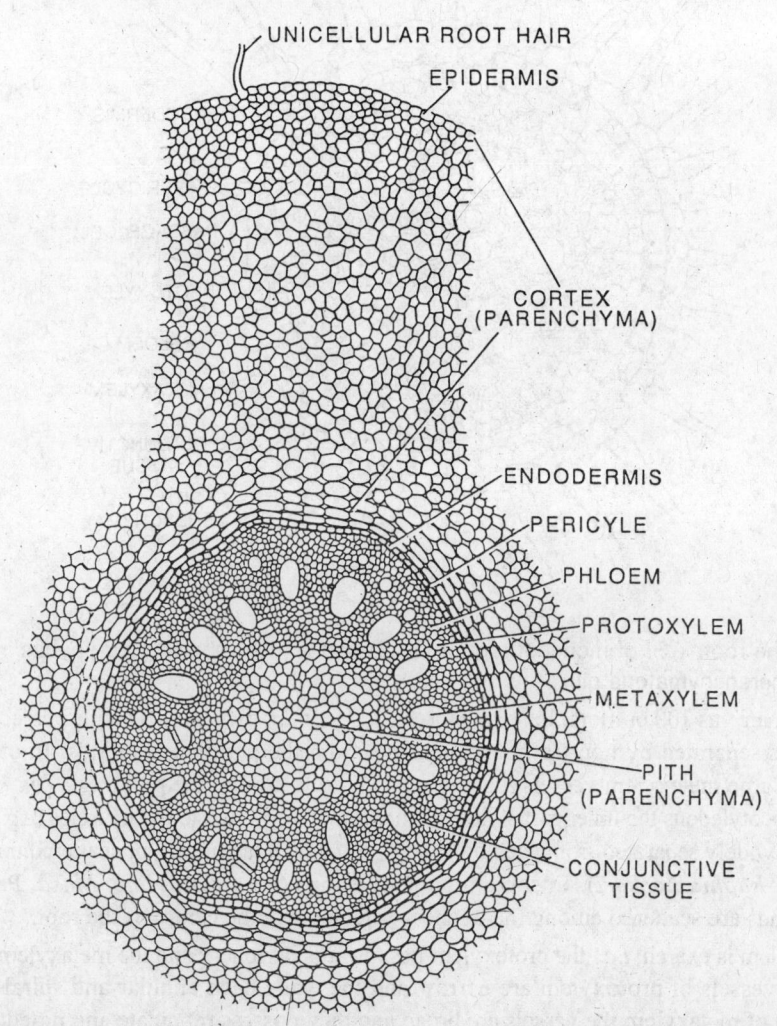

Fig. 9.21. Anatomy of monocot root. T.S. of a portion of root of *Hedychium coronarium*, showing stele and cortex.

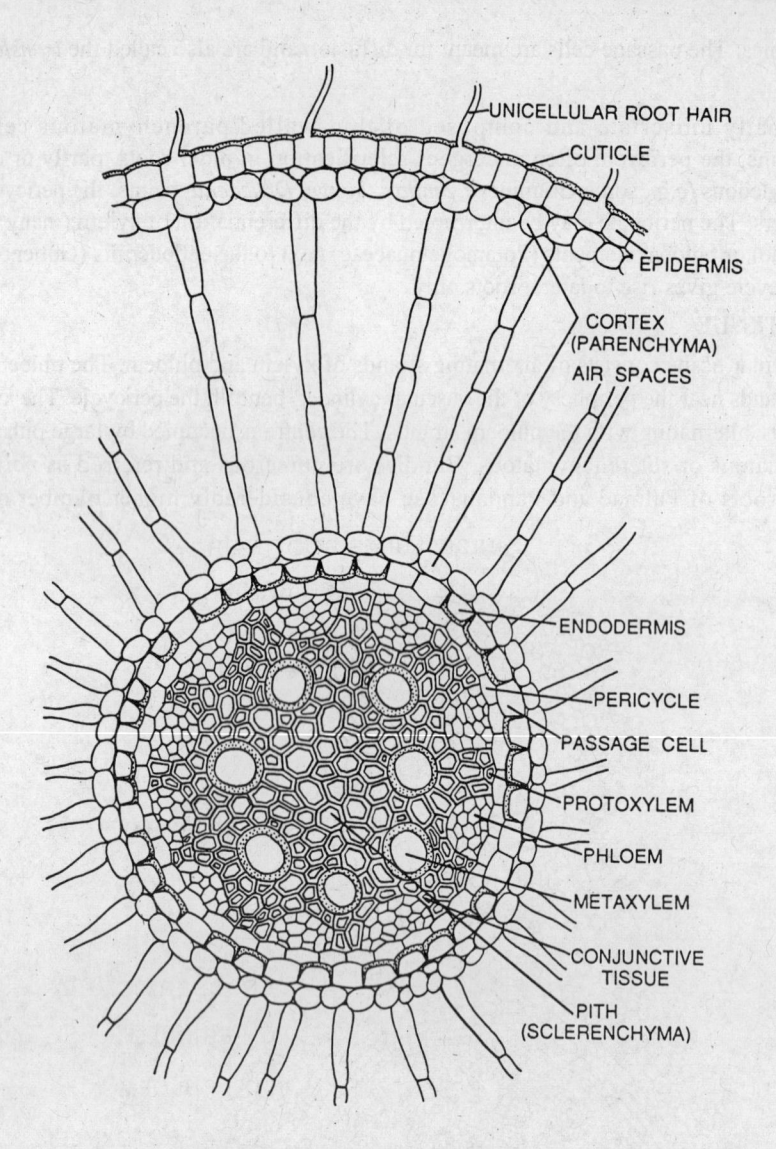

Fig. 9.22. The root. T.S. of monocot root of *Oryza sativa* (rice) showing air-spaces in cortex and sclerenchymatous pith.

bundles, as many as 100 or more. In some roots (*e.g., Hydrilla, Triticum*), a single vessel occupies the centre and is separated by non-tracheary elements from the peripheral strands. In other variable numbers of large metaxylem vessels are arranged in circle around the pith (*e.g., Zea mays*). In the woody monocotyledons the inner metaxylem elements may form two to three circles (*e.g., Latana*), or they may be widely separated from each other (*e.g., Phoenix dactylifera*), or scattered throughout the centre (*e.g., Raphia hookeri*). In some monocotyledons (*e.g., Cordyline, Musa*, Pandanaceae) phloem strands are scattered among the tracheary elements in the centre of the root.

The xylem is exarch, *i.e.,* the protoxylem lies towards periphery and the metaxylem towards the centre. The vessels of protoxylem are narrow and the walls have annular and spiral thickenings whereas that of metaxylem the vessels are broad and they possess reticulate and pitted thickenings.

The phloem strands consist of sieve tubes, companion cells and phloem parenchyma. The phloem strands are also exarch having protophloem towards the periphery and metapholem towards

The Root—Primary and Secondary Structure

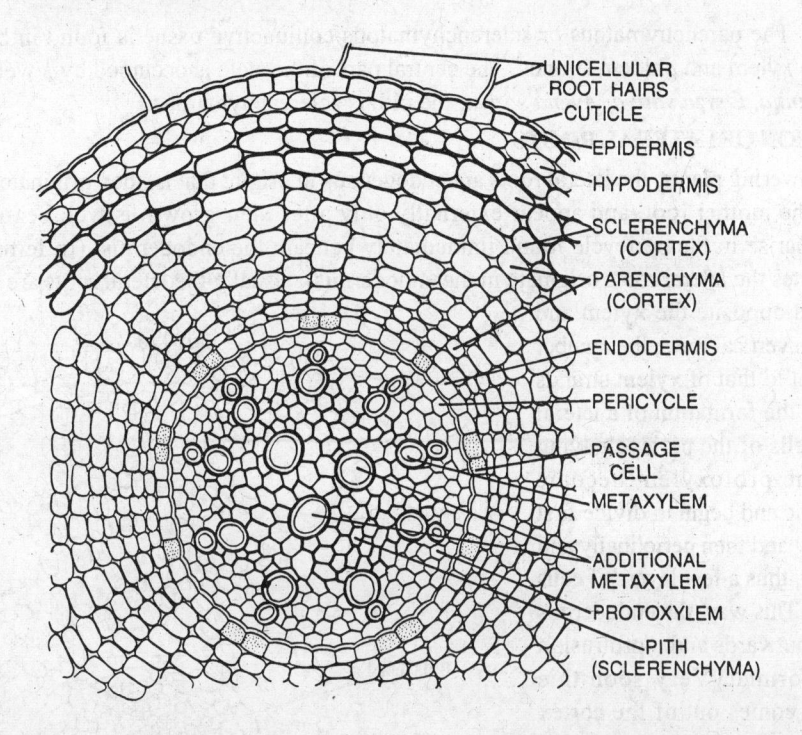

Fig. 9.23. The root. T.S. of monocot root (*Avena sativa*—Oats) showing sclerenchymatous pith and additional metaxylem vessels. The cells of cortical region are being arranged in distinct radial rows.

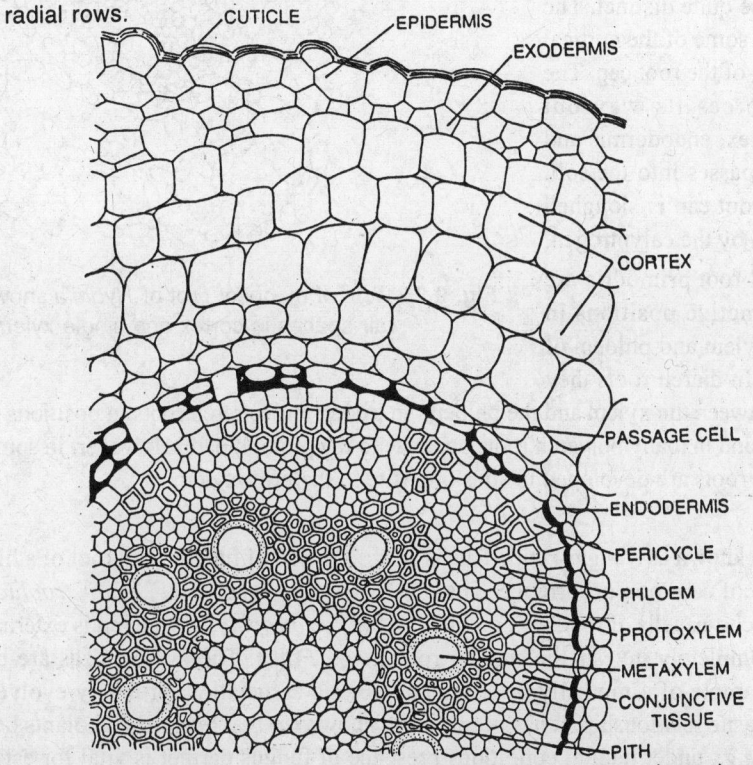

Fig. 9.24. The root. T.S. of monocotyledonous root (grass).

the centre. The parenchymatous or sclerenchymatous conjunctive tissue is found in between and around the xylem and phloem strands. The central part of the stele is occupied by a well developed pith. In *Canna, Oryza sativa, Avena sativa*, the pith is sclerenchymatous.

FORMATION OF LATERAL ROOTS

In flowering plants, the lateral roots are endogenous in origin, that is, they originate in the inner tissue of the mother root, and appear externally only after their growth is well begun. The root meristems arise in the pericycle found immediately beneath the endodermis. (In ferns, and other pteridophytes the lateral roots originate in the endodermis). Usually the lateral roots are restricted to the regions opposite the xylem and come out in vertical rows, the number being equal to that of xylem strands present. In the formation of a lateral root, the cells of the pericycle lying against the protoxylem become meristematic and begin to divide first tangentially and then periclinally and anticlinally, thus a few layers of cells are cut off. This way, the endodermis is pushed outwards and a protrusion is being formed. Very soon this protrusion comes out of the cortex and three regions of root apex, that is, dermatogen, periblem and plerome, become quite distinct. The endodermis and some of the cortical cells form a part of the root cap. The lateral root forces its way out through the cortex, endodermis and epidermis, and passes into the soil. Very soon the root cap is sloughed off and renewed by the calyptrogen.

The lateral root primodria are formed in distinctive positions in relation to the xylem and phloem of the parent root. In diarch roots they usually occur between the xylem and the phloem, in triarch and tetrarch roots in positions opposite to the protoxylem, and in many polyarch roots opposite to the protophloem. However, in some polyarch roots, the lateral roots are developed in sites opposite to the protoxylem.

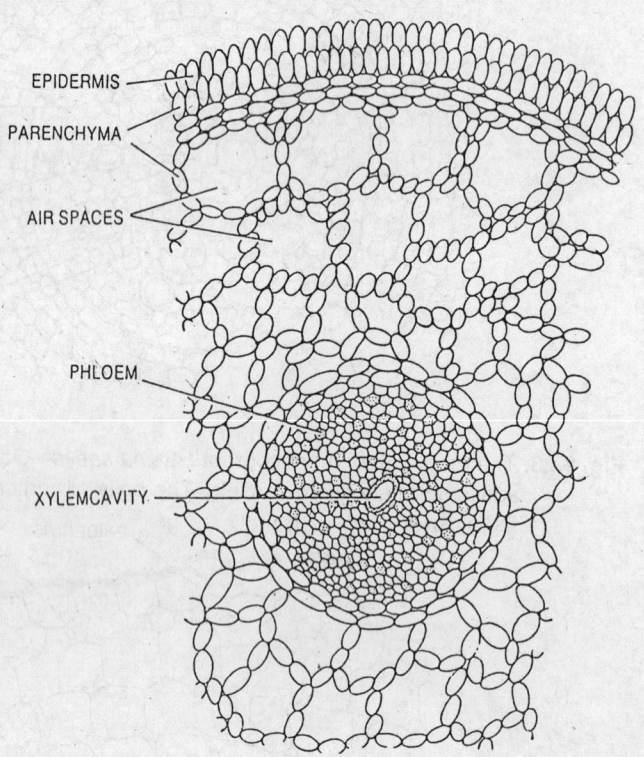

Fig. 9.25. T.S. of monocot root of *Hydrilla* showing large air spaces in cortex and single xylem cavity.

MYCORRHIZA

This is also known as '*fungus root*'. This is an association of fungus with root of a higher plant. Mycorrhizas are of common occurrence. There are two main types —(1) *endo- trophic*, in which fungus is within cortex cells of root, *e.g.*, orchids, and (2) *ectotrophic*, in which it is external, forming a mantle that completely invests the smaller roots, *e.g., Pinus*. The mycorrhizas are believed to constitute an example of a mutually beneficial symbiotic association, probably evolved from an original host parasite relationship. It has been clearly shown (*a*) that mycorrhizal plants benefit from the association, *e.g.*, under natural conditions presence of fungus partner is vital for establishment

The Root—Primary and Secondary Structure

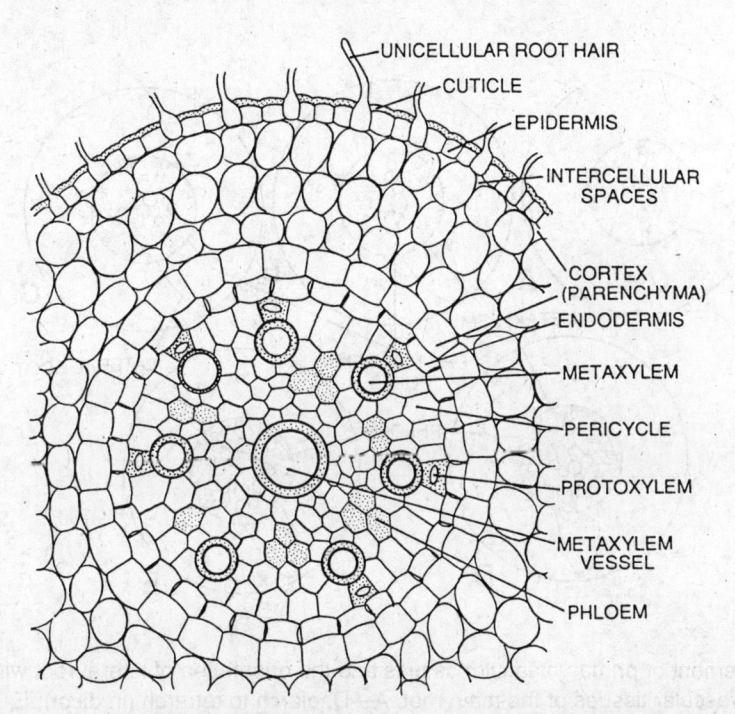

Fig. 9.26. The root. T.S. of monocot root of *Hordeum vulgare* showing the central metaxylem vessel.

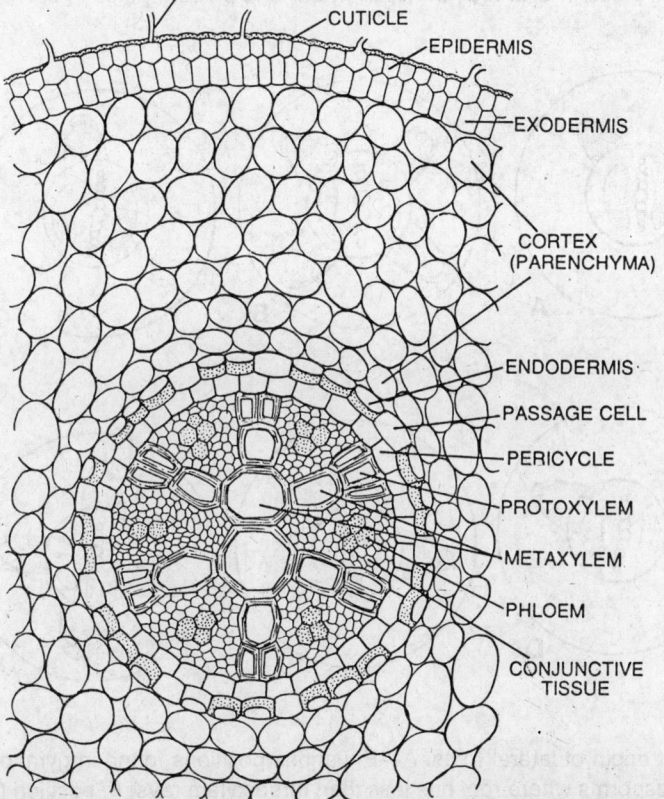

Fig. 9.27. The root. T.S. of monocotyledonous root (*Allium cepa*).

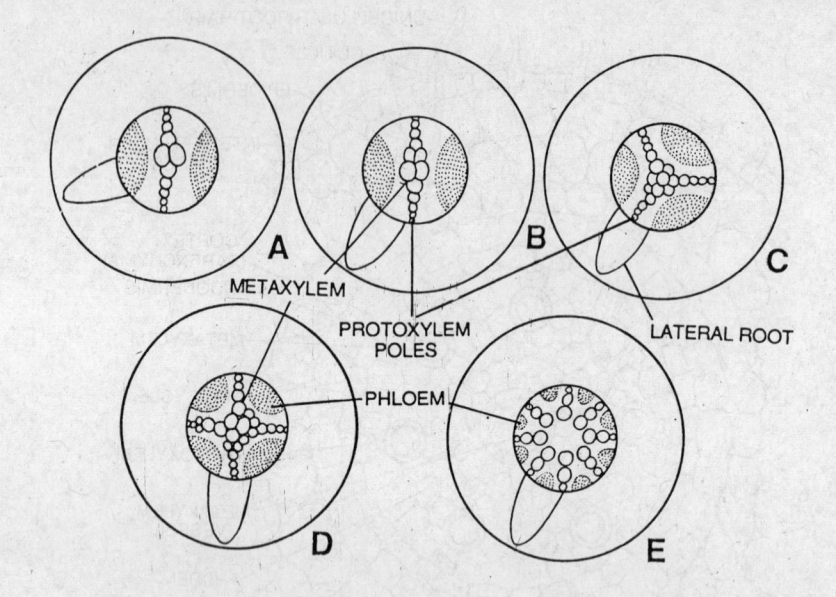

Fig. 9.28. Arrangement of primary vascular tissues and the orientation of lateral root with reference to the vascular tissues of the main root. A—D, diarch to tetrarch (in dicot); E, polyarch (in monocot). Lateral root arises opposite the phloem poles in A and E; opposite the protoxylem poles in C and D; between xylem and phloem poles in B.

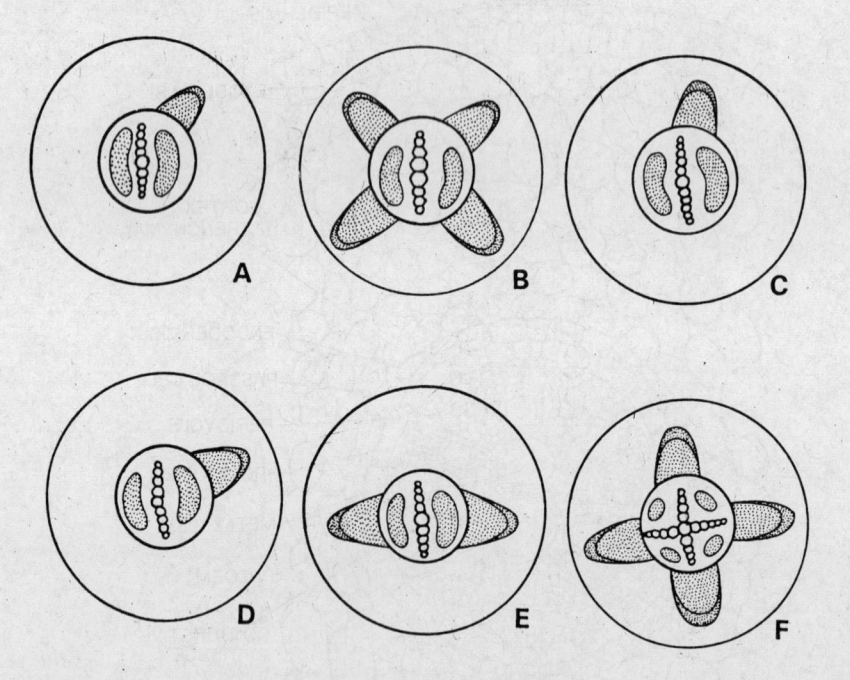

Fig. 9.29. Position of origin of lateral roots. A—E, various positions found in gymnosperms as well as in angiosperms where root has less than three xylem rays; F, position (opposite xylem rays) found in roots of all vascular plants where there are three or more xylem rays.

The Root—Primary and Secondary Structure

and growth of seedling trees of a number of different species, *e.g.*, pines, and (*b*) that association of fungus with tree is necessary for development and reproduction by fungus.

FORMATION OF ADVENTITIOUS ROOTS

The adventitious roots may occur on the hypocotyl of a seedling, at nodes and internodes of stems, and in roots. They may be formed in young organs or in older tissues which are still meristematic. Most adventitious roots arise endogenously. Stem-borne adventitious roots make the main vascular system in vascular cryptogms (pteridophytes), in most monocotyledons, in dicotyledons propagating by means of rhizomes or runners, in water plants, in saprophytes and in parasites. The roots which are developed on cuttings, directly from the stem or from the callus tissue, are also adventitious.

Usually the adventitious roots are initiated in the vicinity of differentiating vascular tissues of the organ which gives rise to them (Datta and Majumdar, 1943). In young organ, the adventitious primordium is initiated by a group of cells near the periphery of the vascular system. In older organ, it is located deeper, near the vascular cambium. In young stems, the cells that form the root primordium are derived from the interfascicular parenchyma, while in older stems from a vascular ray. In certain cases the adventitious roots are initiated by divisions in the cambial zone (Smith, 1936). Usually the seat of the root primordium in the case of stems is known as pericycle. The origin of the adventitious roots in the vascular ray, or in the cambium places the young root close to both the xylem and the phloem of the mother axis and makes the vascular connection between the two organs (see fig. 9.30).

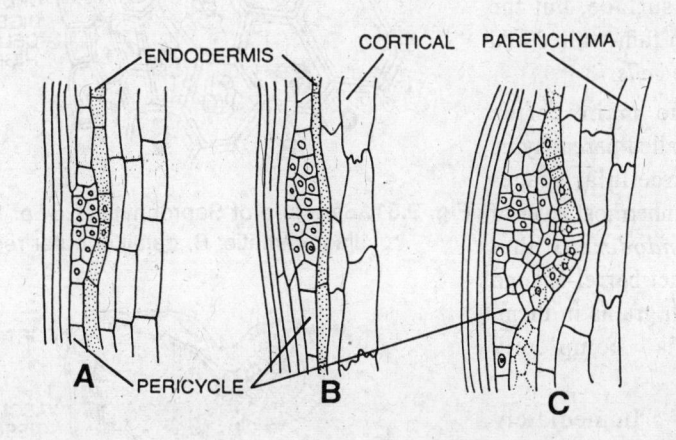

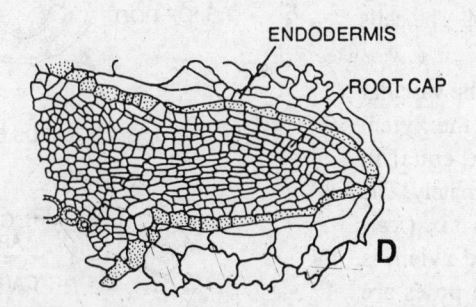

Fig. 9.30. Origin of lateral root. A, initiation of branch of carrot root through formation of meristematic cells in pericycle; B—C, enlargement of meristematic region, D, young root pushing through cortex.

ANATOMY OF EPIPHYTIC ROOTS
ANATOMY OF ORCHID ROOT (MONOCOT)

The orchids (of family Orchidaceae-monocotyledons) are epiphytes. They possess the aerial roots hanging in the air. The anatomy of the root of *Dendrobium* (an orchid) is given here.

Velamen. The velamen consists of several layers of dead cells often with spirally thickened and perforated walls, which act as sponge, soaking up water that runs over it. The velamen is multiple epidermis. The velamen is thought to function as a protective tissue, preventing undue water loss from the delicate cortical cells of the exposed aerial root.

Exodermis. It is the outermost layer of the cortex. This layer is composed of alternating long and short cells; the long cells become thick-walled on their radial and outer tangential surface, but the small cells remain thin-walled and are called *passage cells*.

Cortex. The main cortex consists of thin-walled parenchyma cells having intercellular spaces among them. The innermost layer of the cortex is *endodermis* that consists of compact barrel-shaped cells having starch grains in them. The endodermis completely encircles the stele.

Pericycle. Immediately beneath the endodermis a single-layered pericycle is found. The cells are thin-walled.

Vascular system. The vascular bundles are radial, *i.e.*, the xylem and phloem strands are equal in number and arranged alternately. The bundles are more than six (*i.e.*, polyarch condition). The xylem is exarch. The protoxylem poles are found towards periphery and metaxylem towards centre. The vessels of the protoxylem are narrow

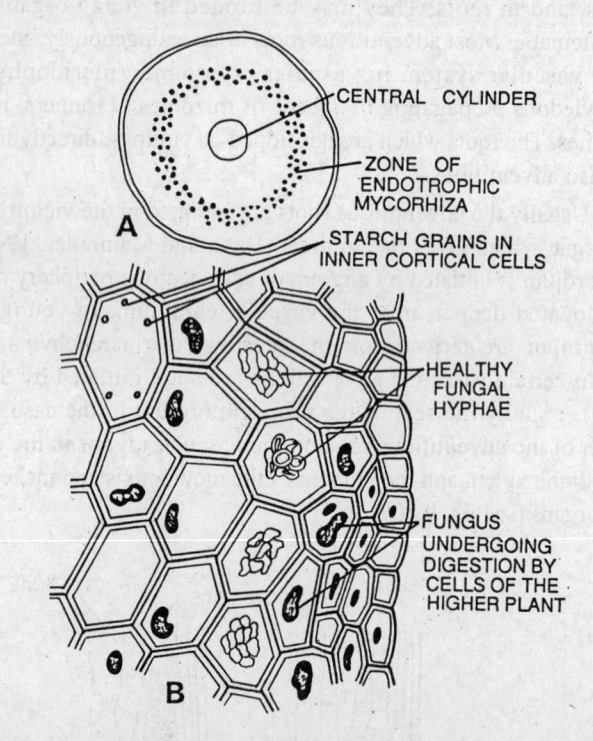

Fig. 9.31. Structure of Saprophyte. T.S. of *Neottia* root; A, diagrammatic; B, detail of outer region.

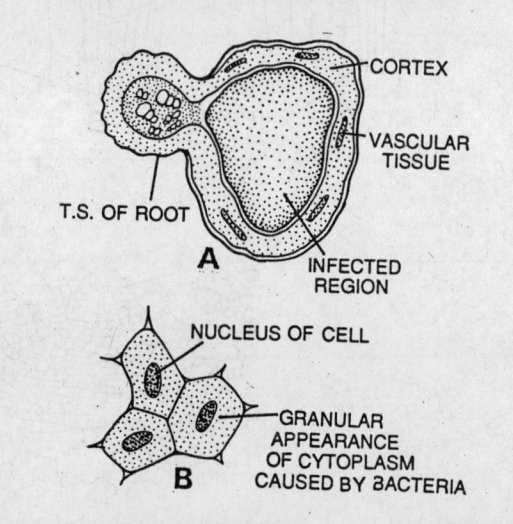

Fig. 9.32. Root. The root nodules. A—B, T.S. of root in region of nodule.

The Root—Primary and Secondary Structure

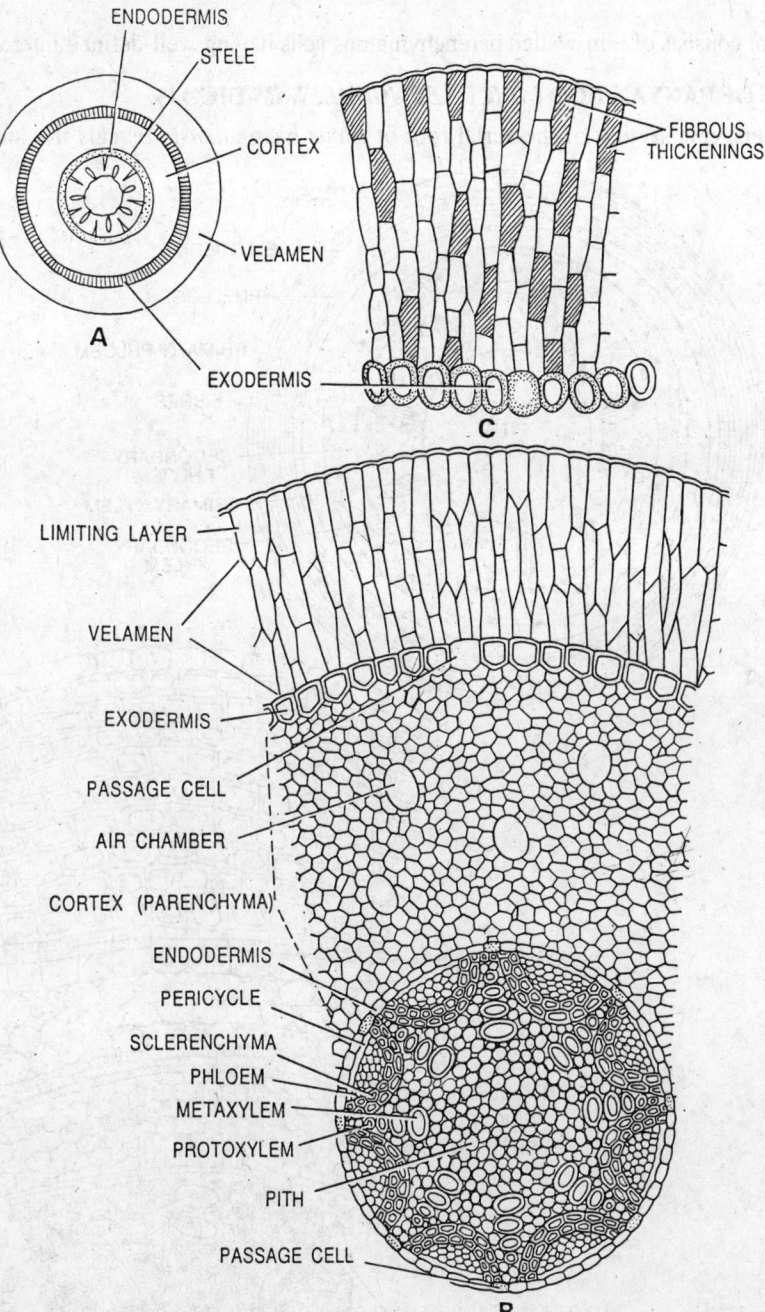

Fig. 9.33. Anatomy of Orchid root (Monocot—Orchidaceae)—A, T.S. of root (diagrammatic); C, velaman and exodermis enlarged; B, T.S. of *Dendrobium* (orchid) root showing detailed structure.

and possessing annular and spiral thickenings whereas the vessels of metaxylem are broad and possessing reticulate and pitted thickenings on their walls. The phloem bundles are made up of sieve tubes, companion cells and phloem parenchyma.

The *conjunctive tissue* is represented by the presence of parenchyma cells in between and around the vascular bundle.

The *pith* consists of thin-walled parenchymatous cells having well-defined intercellular spaces.

ANATOMY OF BANYAN ROOT (*FICUS BENGHALENSIS*-DICOT)

The transverse section of the aerial root of *Ficus benghalensis* reveals the undermentioned structure :

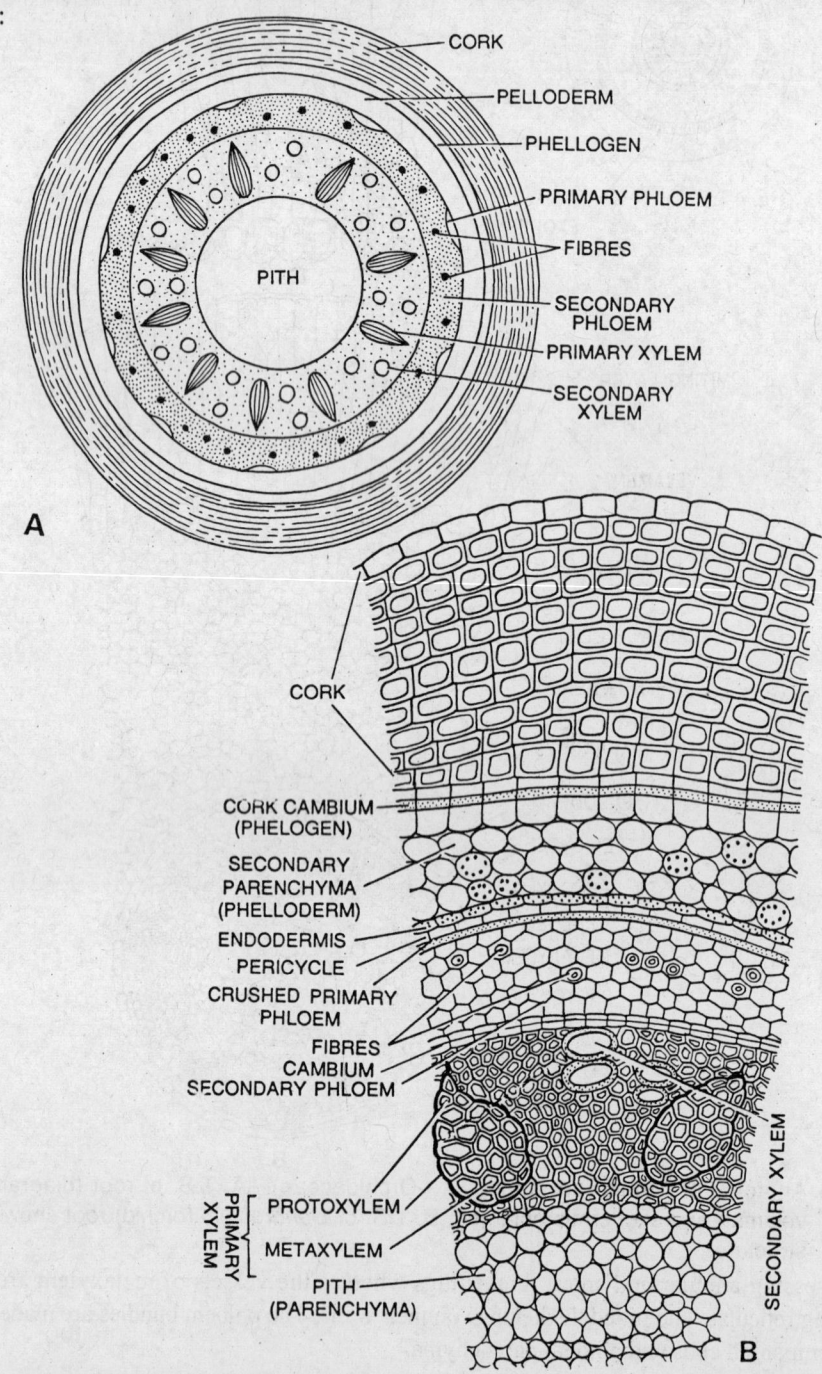

Fig. 9.34. *Ficus benghalensis.* T.S. of dicot aerial root showing periderm, crushed primary phloem, secondary phloem with fibres, cambium, secondary xylem, primary xylem and large pith.

Rhizodermis. The outermost limiting layer of the root is represented by rectangular cells. A thick cuticular layer may also be seen.

Cork. Just beneath the rhizodermis there are several layers of cork cells. These cork cells are somewhat rounded, suberized and having intercellular spaces.

Phellogen. The phellogen or cork cambium is found beneath the cork cells. The rectancular cells of cork cambium divide tangentially forming cork cells towards the outer side and secondary cortex towards the inner side. The secondary cortex is composed of a few layers of parenchyma cells.

Endodermis and pericycle. In young roots, the endodermis and pericycle are distinctly clear, but in older roots they are inconspicuous.

Vascular system. Just beneath the pericycle a crushed layer of primary phloem is visible, which is immediately followed by well developed secondary phloem. The secondary phloem is composed of sieve tubes, companion cells, phloem parenchyma and phloem fibres.

The cambium forms a complete ring in the older roots. Just beneath the cambium there is secondary xylem. The secondary xylem possesses vessels, tracheids and xylem fibres. The primary xylem strands are easily recongnizable having protoxylem poles towards periphery and metaxylem towards centre. The vascular strands are more than six in number.

In the central region there is well developed pith. The pith is composed of thin-walled parenchyma cells having well developed intercellular spaces.

ANATOMY OF THE ROOT OF *TINOSPORA CORDIFOLIA* (DICOT.)

The transverse section of the aerial root of *Tinospora cordifolia* shows the undermentioned structure:

Rhizodermis. The outermost limiting layer of the root is represented by rectangular cells.

Cork. Just beneath the rhizodermis there are few layers of rectangular cork cells. The cork cells are arranged in radial rows.

Phellogen or cork cambium. The phellogen is found beneath the cork. The cells of phellogen divide tangentially forming cork towards the outer side and secondary cortex (of chlorenchyma) towards inner side.

The endodermis and pericycle are inconspicuous.

Vascular system. The crushed primary phloem is found beneath the secondary cortex (chlorenchyma). Next to crushed primary phloem there is secondary phloem. The secondary phloem is composed of sieve tubes, companion cells and phloem parenchyma. In between secondary phloem and secondary xylem there is distinct cambium. Towards the inner side of the cambium there is secondary xylem. The secondary xylem has large vessels. The primary xylem strands are easily recognisable possessing protoxylem poles towards periphery and metaxylem towards centre. There is tetrarch condition. Scanty pith is present in the centre. The meduallary rays of parenchyma are visible. (See figs. 9.36 and 9.37).

ANATOMY OF STORAGE ROOTS

The underground roots may become very much thickened and serve as organs for the storage of food. Such is the case in sweet potatoes, radishes, turnips, carrots and dahlias. In such roots the food may be stored largely in the cortex or xylem region or in both.

In turnips food is stored largely in the xylem, and the phloem and cortex are relatively narrow. In the radish and sweet potato the xylem is also the chief region of food storage, but food is also stored outside the xylem. In the carrot there is a more even distribution between xylem and bark. In beets there are alternate layers of xylem and phloem owing to the formation of successive cambia. The

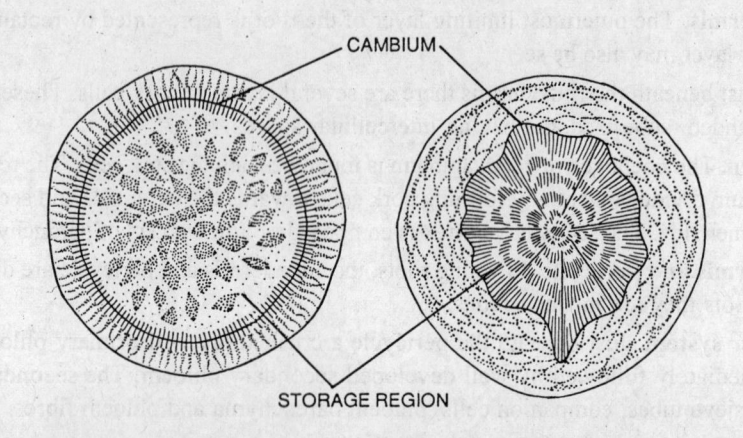

Fig. 9.35. Anatomy of storage roots (dicot). Left—cross section of storage root of *Ipomoea batatas*; right—cross section of root of carrot.

secondary tissues of the root accumulate starch in the same kind of cells as those of the stem, that is various parenchymatous and some sclerenchymatous cells of the xylem and the phloem. In general, roots possess a higher proportion of parenchyma cells than do stems.

***Daucus carota.* Carrot (Umbelliferae-Dicot).** In this case, the hypocotyl and base of taproot form jointly one fleshy structure.

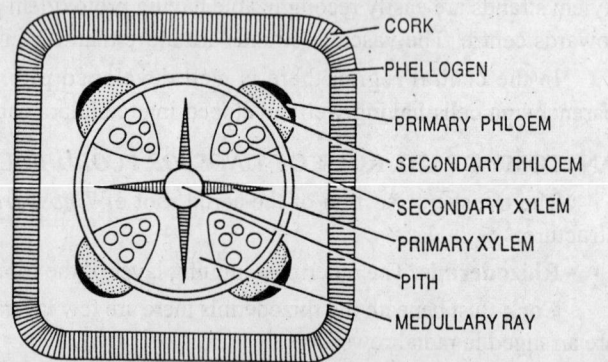

Fig. 9.36. T.S. of the aerial root of *Tinospora cordifolia* (dicot).—diagrammatic.

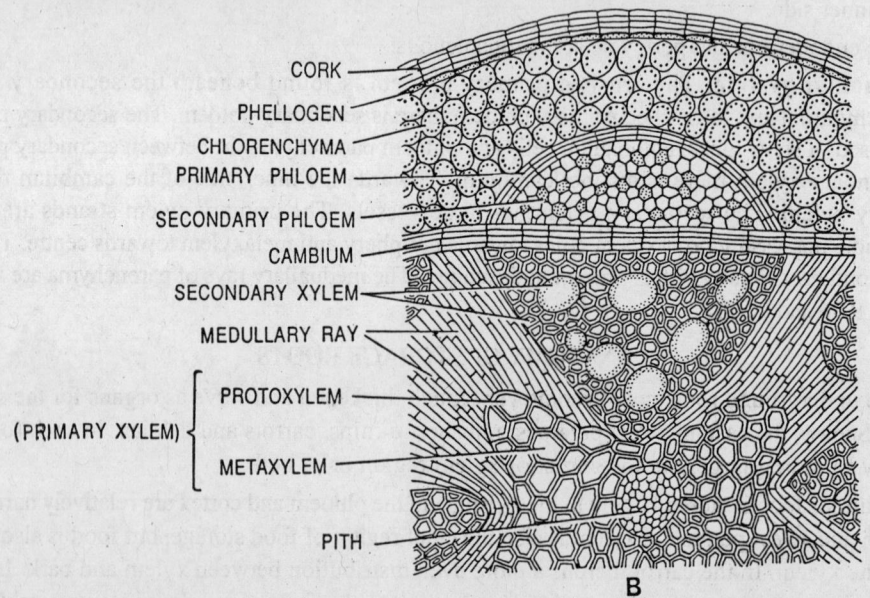

Fig. 9.37. T.S. of the aerial root of *Tinospora cordifolia* (dicot.) showing secondary growth-detail.

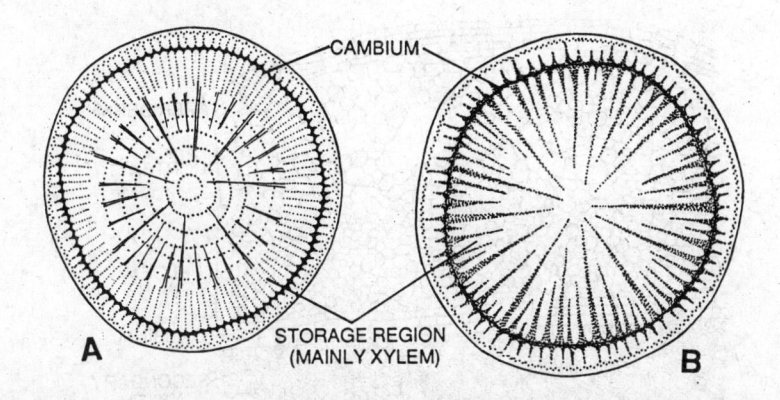

Fig. 9.38. Anatomy of storage root (dicot.) A, cross section of turnip root; B, cross section of radish root.

Here, the fleshy organ has large amount of storage parenchyma associated with ordinary arrangement of tissues. In this type of development, where the hypocotyl and the upper part of the tap root, after sloughing off the cortex, become fleshy through a massive development of parenchyma in the phloem and the xylem. Besides the cambial activity the massive parenchyma, that makes the storage tissue, adds to the thickness of the root.

Raphanus sativus. Radish (Cruciferae-Dicot)

The fleshy roots of radish show a proliferation of parenchyma in the pith and in the secondary xylem, and a differentiation of concentric vascular bundles within this parenchyma. Here, the fleshy roots show a diarch primary xylem. The normal cambium cuts off secondary phloem towards periphery and secondary xylem towards the centre. Several concentric vascular bundles are seen in the transverse section of the fleshy root. The concentric bundles are composed of secondary cambial rings with a few vascular elements in the centre.

Ipomoea batatas. Sweet potato (Convolvulaceae-Dicot.)

It exhibits a complicated type of anomalous secondary thickening. In primary state the root is pentarch or hexarch. The cortex is delimited by a single-layered distinct endodermis from the stelar region. In the normally developed but highly parenchymatous primary and secondary xylem, anomalous cambia arise around individual vessels or vessel groups and produce phloem rich in parenchyma and with some laticifers away from the vessels, and

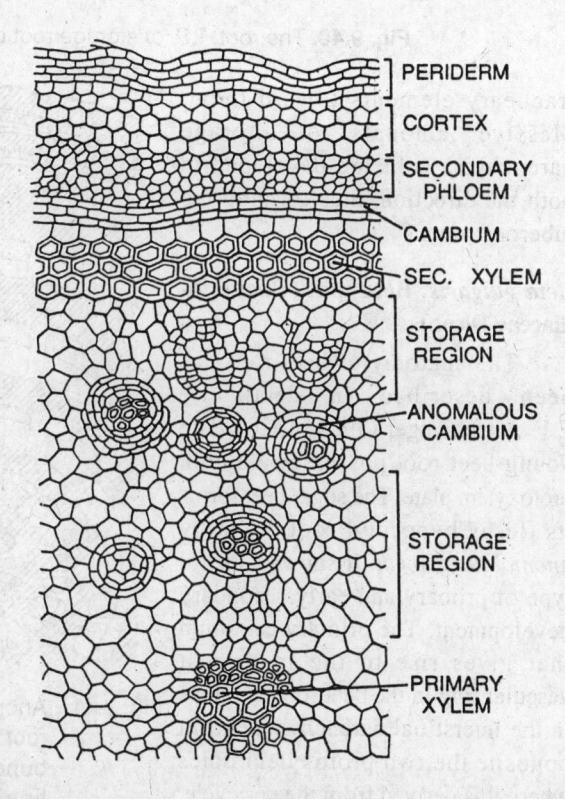

Fig. 9.39. The root. T.S. of a sector of root of *Raphanus sativus* (dicot.)

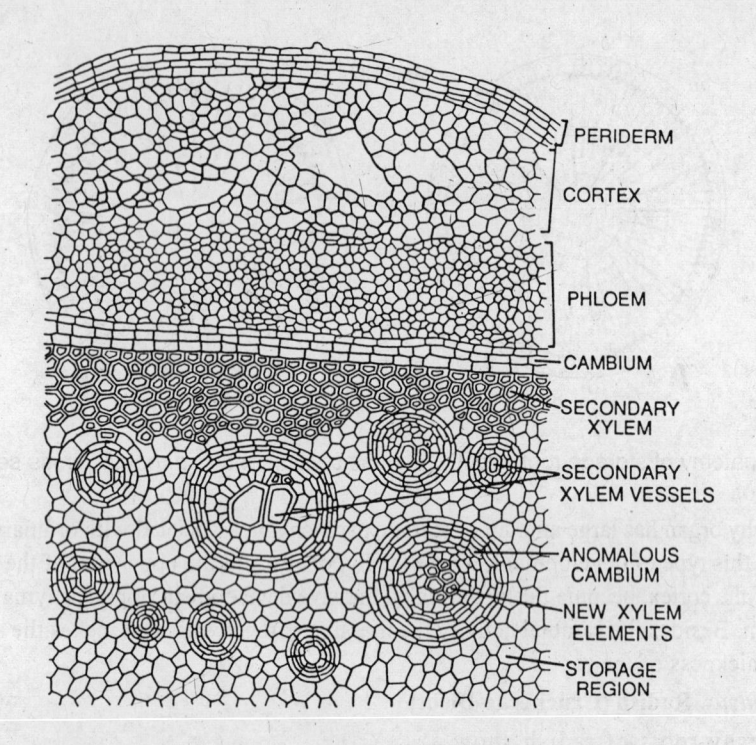

Fig. 9.40. The root. T.S. of storage root of *Ipomoea batatas* a sector.

tracheary elements toward them. Massive amounts of storage parenchyma cells are developed in both the direction, thus forming the tuberous roots.

Beta vulgaris. Beet root (Chenopodiaceae-Dicot.)

The anatomy of beet root has been described in detail by E.F. Artschwager (1924, 1926). The young beet root possesses a diarch protoxylem plate. The sugar beet forms its fleshy hypocotyl root organ by *anomalous growth*. It shows a useful type of primary and early secondary development. The primary cambium that gives rise to the innermost vascular ring in the beet root develops in the interstitial parenchyma except opposite the two protoxylem poles where it is derived from the pericycle. The first secondary cambium in the

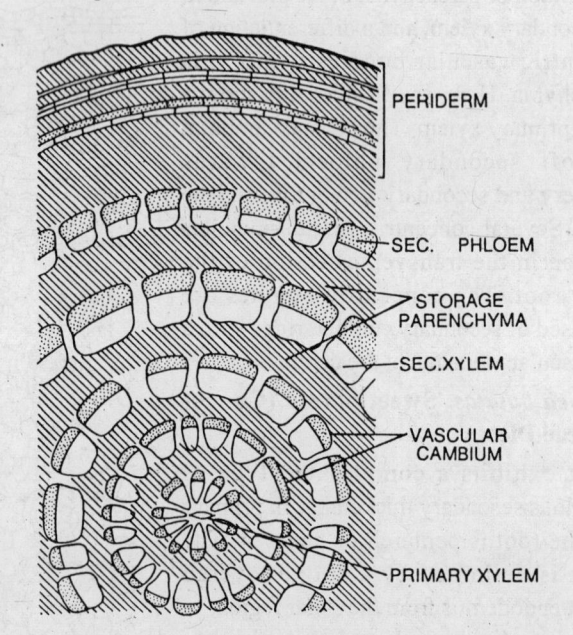

Fig. 9.41. Anomalous secondary growth in *Beta vulgaris* root (T.S.), showing alternate layers of vascular bundles and proliferated pericycle, phloem, lignified xylem cells in radial rows, xylem parenchyma and secondary interfascicular tissue. (After Eames and MacDaniels).

The Root—Primary and Secondary Structure

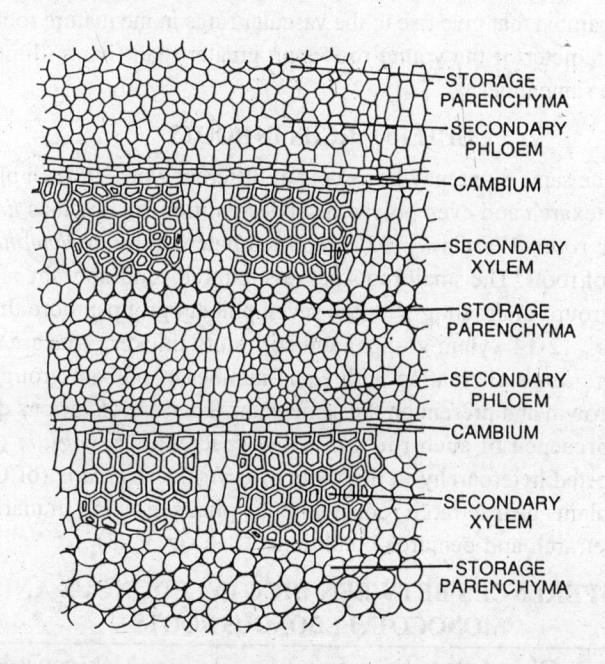

Fig. 9.42. The root. The root of *Beta vulgaris* showing growth layers.

root and lower part of the hypocotyl arises in the phloem parenchyma, whereas in the upper part of the hypocotyl is derived from the pericycle. Later, however, a series of supernumerary cambia arise outside the normal vascular cylinder and produce several increments of vascular tissue, each consisting of a layer of parenchyma, parenchymatous xylem and parenchymatous phloem. Practically

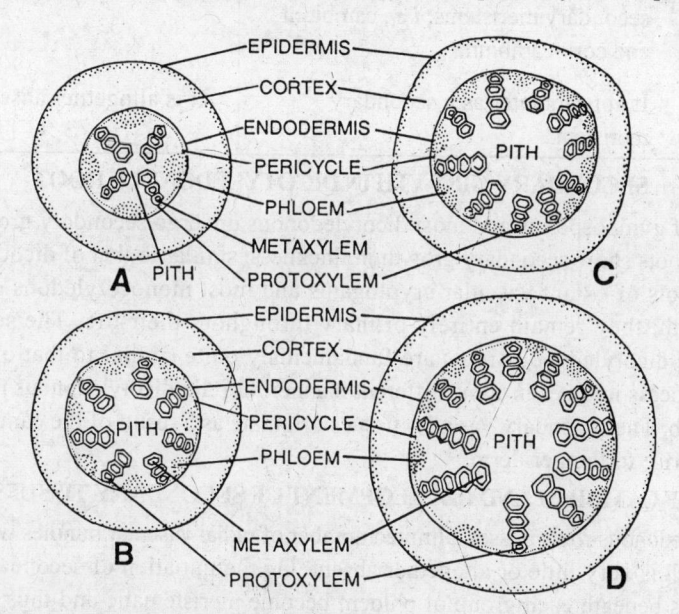

Fig. 9.43. Heteroarchy in roots of *Nymphaea chilensis*. A, root with tetrarch condition; B, root, with seven xylem strands; C, root with ten xylem strands; D, root with twelve xylem strands. (After Wardlaw, 1928).

all the supernumerary cambia that give rise to the vascular rings in the mature root have already been developed when the diameter of the young root is no greater than five millimetres or so. All the cambia are active at the same time.

HETEROARCHY IN ROOTS

In certain cases, the same plant may bear roots of different types. In such plants, the roots may be tetrarch, pentarch, hexarch and even polyarch. This condition is known as *heteroarchy*, and the roots, the *heteroarchic roots*. Wardlaw (1928) reported heteroarchy in *Nymphaea chilensis*. This plant bears four types of roots. The small roots possess tetrarch condition; the slightly thicker roots possess seven xylem groups alternating with the seven phloem groups; more thicker roots possess polyarch condition (*i.e.*, 12-14 xylem groups) and still more thicker roots are also polyarch (*i.e.*, they bear 16-18 xylem groups alternating with equal number of phloem groups). However, these roots lack secondary growth and present on the same plant. They are of various diameters. Wardlaw (1921) reported the presence of such roots in some species of *Eryngium* (of Umbelliferae). Majumdar (1932) reported heteroarchy in the roots of *Enhydra fluctuans* (of Compositae). Here, the normal terrestrial plants bear tetrarch roots while the plants growing in marshy places bear the roots having triarch, tetrarch and pentarch conditions. (See fig. 9.43)

DIFFERENCES BETWEEN DICOTYLEDONOUS AND MONOCOTYLEDONOUS ROOTS

		Dicotyledonous Root	Monocotyledonous Root
1.	Xylem bundles	The number varies from two to six (di- to hexarch), rarely more.	Usually they are numerous, rarely a limited number (*e.g.*, in Onion).
2.	Pith	It is small or absent.	It is large and well developed.
3.	Pericycle	It gives rise to lateral roots and secondary meristems, *i.e.*, cambium and cork-cambium.	It gives rise to lateral roots only.
4.	Cambium	It appears later as a secondary meristem.	It is altogether absent.

SECONDARY GROWTH IN DICOTYLEDONOUS ROOT

The roots of gymnosperms and most dicotyledonous undergo secondary growth. Most of the dicotyledonous roots show secondary growth in thickness, similar to that of dicotyledonous stems. However, the roots of extant vascular cryptogams and most monocotyledons do not show any secondary growth; they remain entirely primary throughout their life. The secondary tissues developed in the dicotyledonous roots are fundamentaly quite similar to that of dicotyledonous stems, but the process initiates in some different manner. Certain dicotyledonous roots do not show secondary growth. The secondary vascular tissues originate as a result of the cambial activity. The phellogen gives rise to the periderm.

FORMATION OF CAMBIUM AND DEVELOPMENT OF SECONDARY TISSUES

The dicotyledonous roots posses a limited number of radial vascular bundles with exarch xylem. Normally the pith is very little or altogether absent. On the initiation of secondary growth, a few parenchyma cells beneath each group of phloem become meristematic and thus as many cambial strips are formed as the number of phloem groups. The cambial cells divide tangentially again and again and produce secondary tissues. Thereafter some of the cells of single layered pericycle become meristematic lying against the protoxylem groups, which divide and form a few layers of cells. The first

The Root—Primary and Secondary Structure

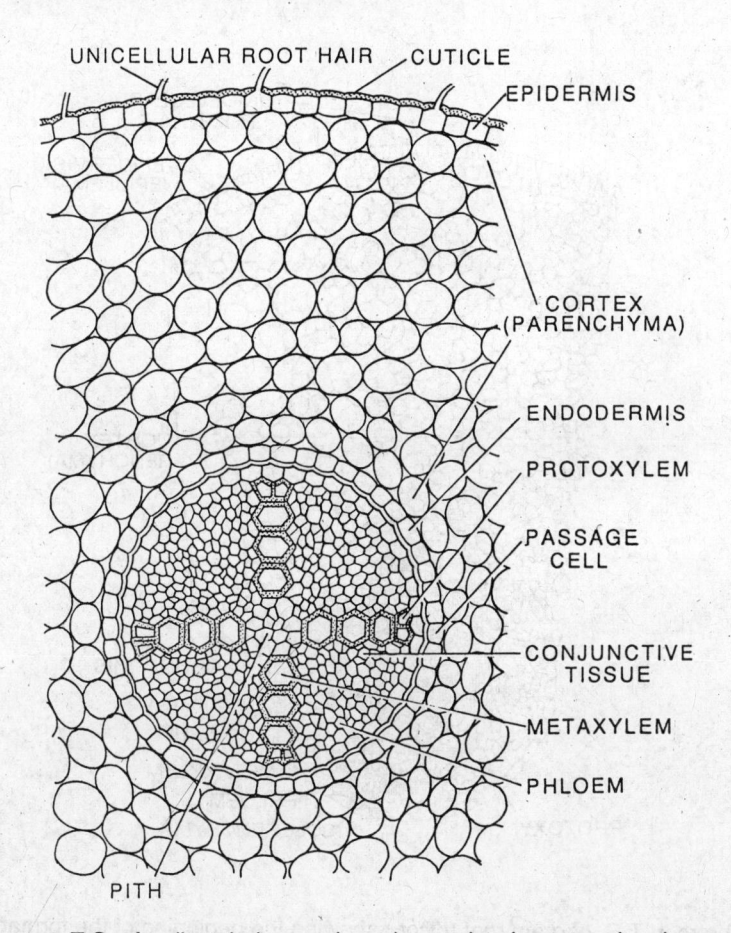

Fig. 9.44. The root. T.S. of a dicotyledonous (gram) root showing tetrarch xylem and small pith.

formed cambium now extends towards both of its edges and reaches the inner most derivatives of the pericycle, thus giving rise to a complete ring of cambium. The cambium ring is wavy in outline, as it passes internal to phloem and external to xylem groups. The cambial cells produce more xylem elements than phloem. The first formed cambium produces secondary xylem much earlier, and the wavy cambium ring ultimately becomes circular. Now whole of the cambium ring becomes actively meristematic, and behaves in the similar way as in the stem, giving rise to secondary xylem on its inner side and secondary phloem towards outside.

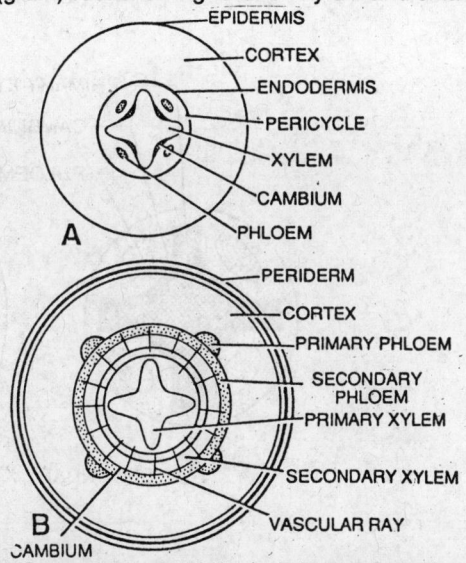

Fig. 9.45. Secondary growth in root A, cross section of a root without secondary growth; B, the same after considerable secondary growth.

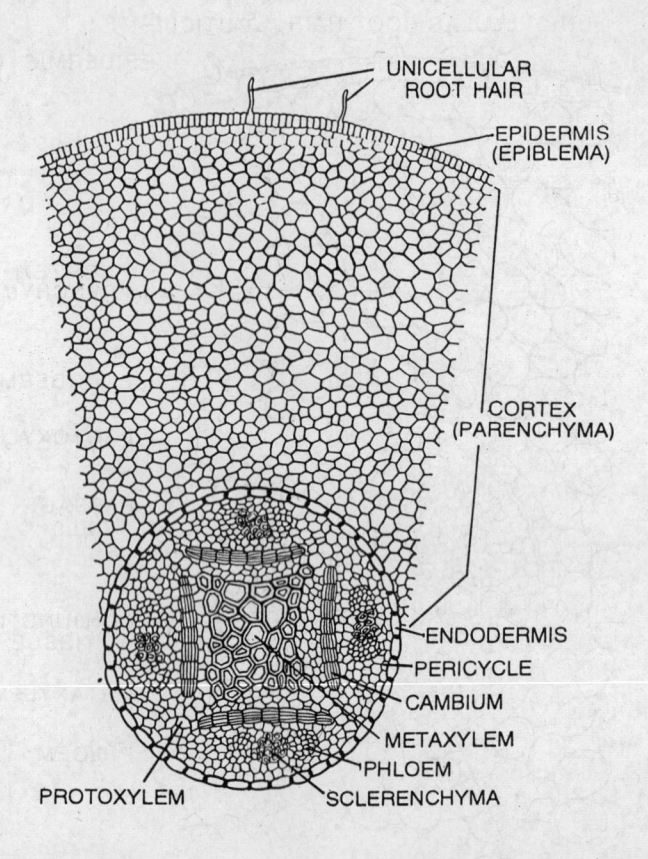

Fig. 9.46. The root. T.S. of gram root (dicot.) showing the beginning of the formation of cambium.

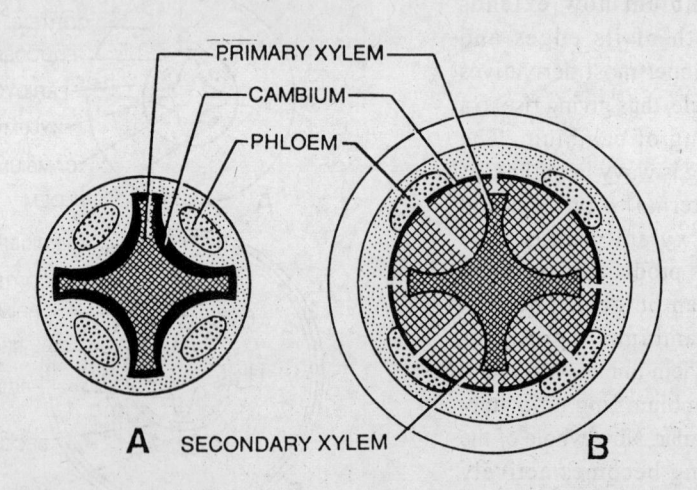

Fig. 9.47. Secondary growth of root. A, diagrammatic cross section of a dicot root showing the cambium; B, a similar section showing secondary phloem and xylem with vascular rays (medullary rays) indicated in white lines.

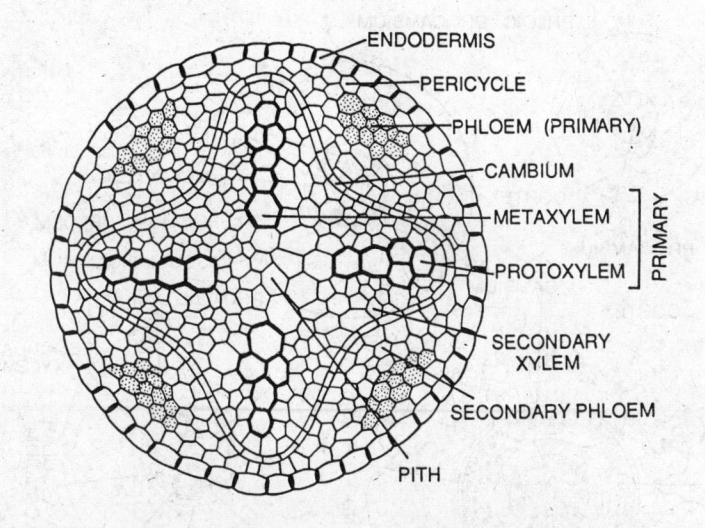

Fig. 9.48. Beginning of secondary growth in dicotyledonous root.

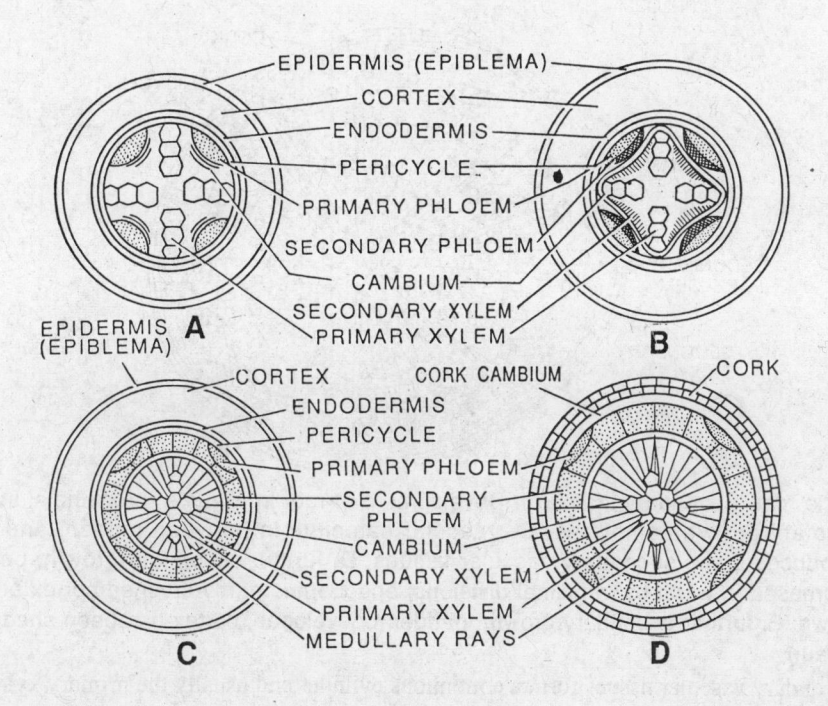

Fig. 9.49. Secondary growth in dicot root. A—D, diagrams showing stages in the secondary growth of a typical dicotyledonous root.

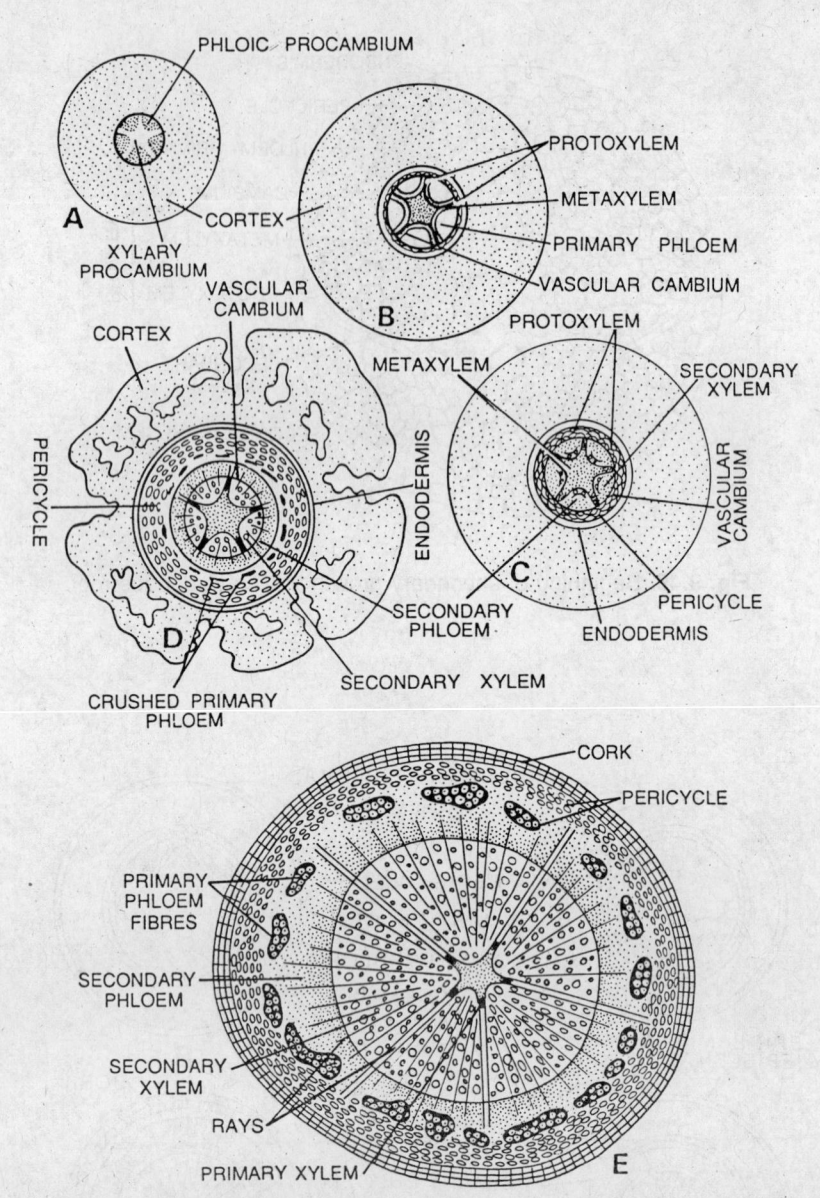

Fig. 9.50. The root. Secondary structure. Development of root in *Pyrus*. A, procambial state; B, primary growth completed; C, vascular cambium in between phloem and xylem produced some secondary vascular tissues; D, further secondary growth, pericycle increased in width by periclinal divisions; endodermis partly crushed cortex breaking down; E, further secondary growth, periderm developed, cortex has been shed. (After Esau.)

The secondary vascular tissues form a continuous cylinder and usually the primary xylem gets embedded in it. At this stage distinction can be made only by exarch primary xylem located in the centre. The primary phloem elements are generally seen in crushed condition. The cambial cells that originate from the pericycle lying against the groups of protoxylem function as ray initials and produce broad vascular rays. These rays are traversed in the xylem and phloem through cambium; this is characteristic feature of the roots. Normally, such rays are called *medullary rays*.

The Root—Primary and Secondary Structure

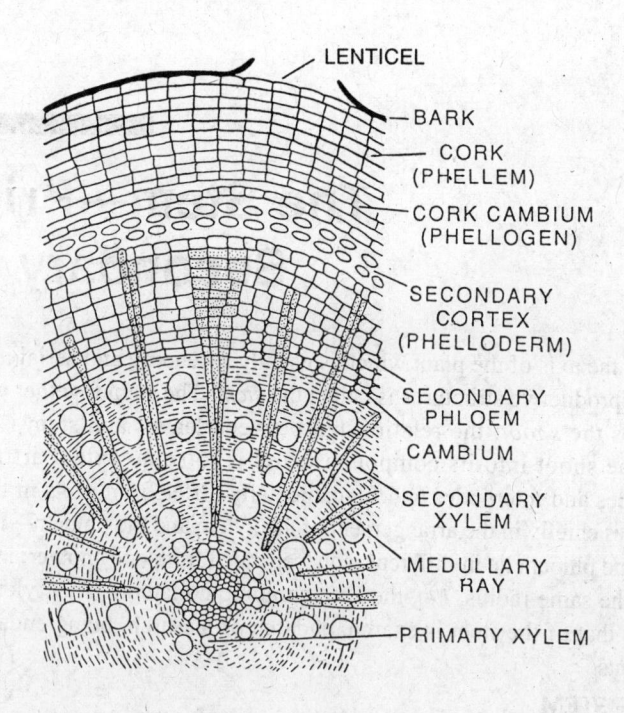

Fig. 9.51. Secondary growth. T.S. of dicotyledonous root showing secondary growth (later stage).

PERIDERM

Simultaneously the periderm develops in the outer region of the root. The single layered pericycle becomes meristematic and divides, giving rise to *cork cambium* or *phellogen*. It produces a few brownish layers of *cork cells* or *phellem* towards outside, and the *phelloderm* on the inside. The phelloderm does not contain chloroplasts. The pressure caused by secondary tissues ruptures the cortex with endodermis, which is ultimately sloughed off. The epiblema dies out earlier. Lenticels may also be formed.

10

The Stem—Primary and Secondary Structure

The part of the axis of the plant which is usually ascending and aerial in nature, and also bears the leaves and reproductive structures is called the *stem*. The stem together with the leaves which it bears constitutes the *shoot*; the relationship between leaves and stem is very close one and a separation of the shoot into its component parts is to some extent artificial. The stem bears conspicuous nodes and internodes and fundamentally differ from roots in their vascular structure. The difference lies chiefly in the arrangement of the xylem and the phloem—in the root the strands of primary xylem and phloem lie in different radii, separated from one another; in the stem the strands lie side by side in the same radius, *i.e.*, they are conjoint, collateral. The xylem of the root is always exarch, whereas that of the stem is exarch, endarch or mesarch, being endarch most commonly in present-day plants.

ORIGIN OF THE STEM

The first stem meristem is organized during the development of the embryo. The fully developed embryo commonly consists of an axis, the *hypocotyl-root axis*. The axis bears at its upper end, one or more cotyledons and the shoot primordium, whereas at its lower end it bears the root primordium covered with a root-cap. The *radicle* (embryonic root) is found at the lower end of the hypocotyl and the embryonic shoot is found above the insertion of the cotyledons. The embryonic shoot is composed of an axis bearing unextended internodes and one or more leaf primordia. This shoot (first bud) is commonly known as *plumule* and its stem part is termed *epicotyl*. The origin of shoot organization is found in the *hypocotyl-cotyledon system* where the hypocotyl is the first stem unit of the plant and the cotyledons are the first leaves. The hypocotyl is located below the cotyledonary node, but not in between nodes. During the germination of the seed, the root

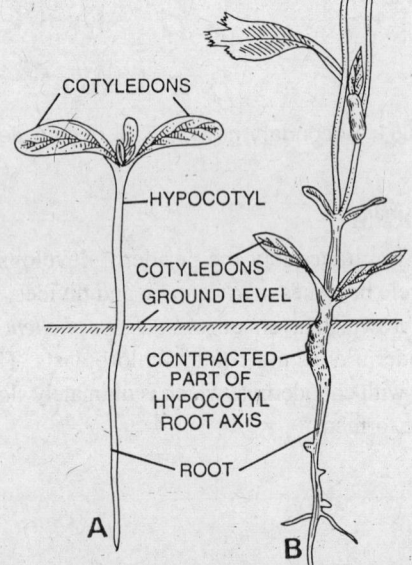

Fig. 10.1. Seedling of *Medicago sativa*. A, the younger seedling bears its cotyledons high above the ground level; B, the older seedling has pulled down the cotyledons close to the ground by contracting the hypocotyl and the upper part of the root.

The Stem—Primary and Secondary Structure

meristem forms the first root, whereas the shoot meristem develops the first shoot by adding new leaves, nodes and internodes to the shoot system formed in the embryo. The lateral stems normally arise by the development of new apical meristems laterally in the terminal meristem of the mother axis. The adventitious branches develop on both stem and roots, by the formation of meristems secondarily in the pericyclic, phloic or cambial regions.

ROOT-STEM TRANSITION

The root and stem make a continuous structure called the *axis* of the plant. The vascular bundles are continuous from the root to the stem. The epidermis, cortex, endodermis, pericycle and secondary vascular tissues are directly continuous in the two organs, root and stem, but the arrangement of vascular bundles is quite different in the two organs. The stems possess collateral bundles with endarch xylem, whereas the roots possess radial bundles with exarch xylem. Of course,

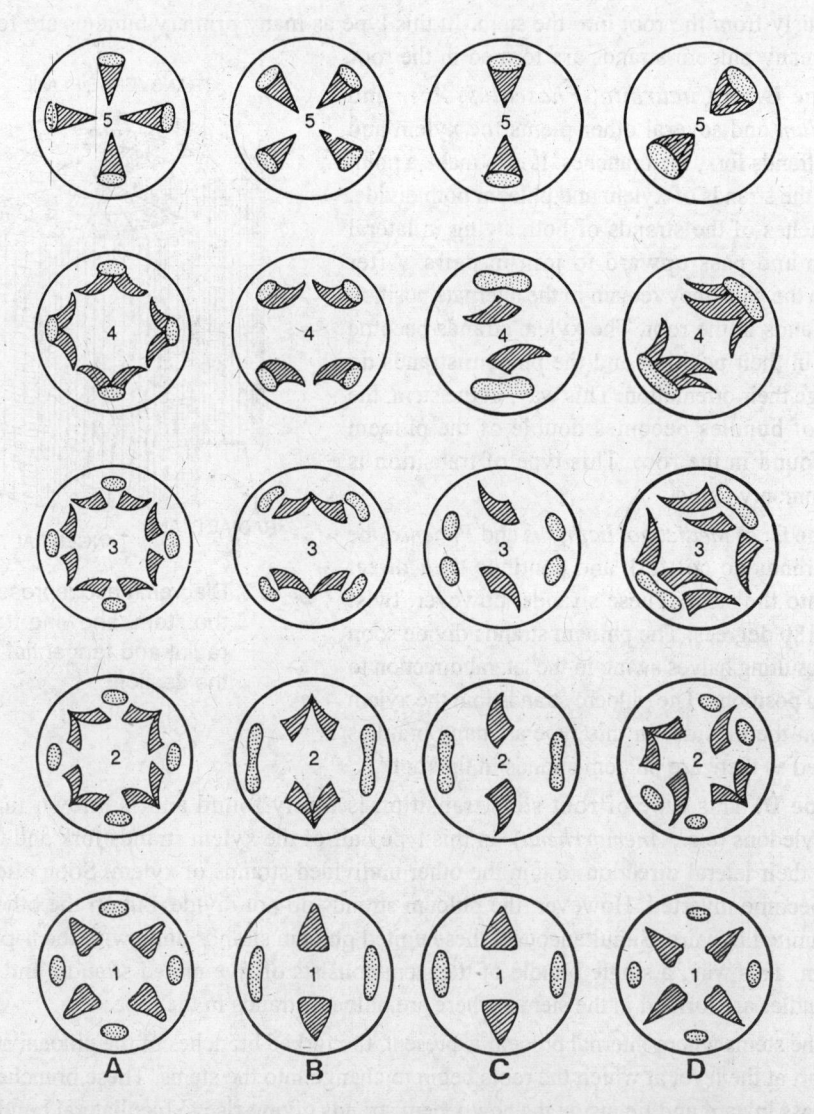

Fig. 10.2. Root-stem transition. Diagrams of four types. A, B, C and D. A, *Fumaria* type; B, *Cucurbtia* type; C, *Lathyrus* type; D, *Anemarrhena* type. (After Fames and MacDaniels).

a region exists where these changes occur and the two different types of vascular tissues maintain their continuity. The change of position involving inversion and twisting of xylem strands from exarch to endarch type is referred to as vascular transition, and the part of the axis where these changes occur is called *transition region*. Commonly this region is quite small and rarely of several centimetres. These changes may be found gradually or abruptly in the top of the radicle or at the base of the hypocotyl, near its middle region, or in the upper part. The phloem bundles remain practically in the same position. The transition may be of four main types which vary from species to species.

Type A. In *Mirabilis*, *Fumaria* and *Dipsacus*, and other plants each xylem strand of the root divides by radial division in the branches. As these branches pass upward, they swing in their lateral direction; one bends towards right and the other goes to the left. Simultaneously these branches join the phloem strands on the inside. The phloem strands, however, do not change their position and also remain unchanged in their orientation. They remain in the form of straight strands continuously from the root into the stem. In this type as many primary bundles are formed in the stem as many phloem strands are formed in the root.

Type B. In *Cucurbita*, *Phaseolus*, *Acer* and *Trapaeolum* and several other plants the xylem and phloem strands fork, the branches. It is to make a point here that the strands of xylem and phloem both divide. The branches of the strands of both swing in lateral direction and pass upward to join in pairs. After joining in the pairs they remain in the alternate position of the strands in the root. The xylem strands become inverted in their position and the phloem strands do not change their orientation. This way, in the stem, the number of bundles becomes double of the phloem strands found in the root. This type of transition is more commonly found.

Type C. In *Medicago*, *Lathyrus* and *Phoenix*, the xylem strands do not fork and continue their direct course into the stem. These strands, however, twist through 180 degrees. The phloem strands divide soon and the resulting halves swing in the lateral direction to the xylem positions. The phloem strands join the xylem strands on the outside. In this type as many bundles are formed as there are phloem strands in the root.

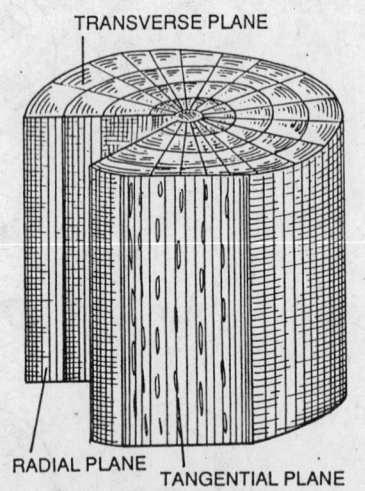

Fig. 10.3. Diagrammatic representation of the stem, showing transverse, radial and tangential planes of the section.

Type D. This type of root-stem transition is rarely found and is known in only a few monocotyledons (*e.g.*, *Anemarrhena*). In this type half of the xylem strands fork and the branches swing in their lateral direction to join the other undivided strands of xylem. Soon after the xylem strands become inverted. However, the phloem strands do not divide, but on the other hand they become united in pairs. Simultaneously these united phloem strands unite with the triple strands of the xylem. This way, a single bundle of the stem consists of five united strands, and thus half as many bundles are formed in the stem as there are phloem strands in the root.

In the stems where internal phloem is present, the forked branches of the phloem strands of the root depart at the level at which the roots begin to change into the stems. These branches of phloem strands pass inward and lie inside the new xylem strands giving rise to bicollateral bundles. In some of the monocotyledonous plants the transitory region is very short and from that too several lateral roots are given out.

The Stem—Primary and Secondary Structure 213

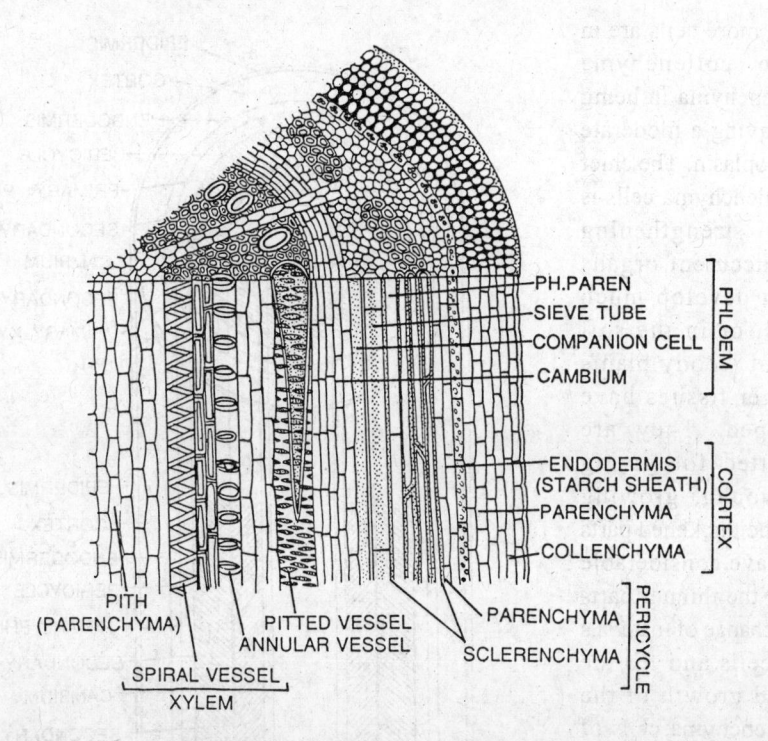

Fig. 10.4. Anatomy of dicot stem. Diagrammatic combined transverse and longitudinal sections of stem.

The seed plants are generally divided into two groups—the *angiosperms* and the *gymnosperms*. The angiosperms are further sub-divided into *dicotyledons* and *monocotyledons*. In a cross section of the stems of dicotyledons and gymnosperms the vascular bundles are found to be arranged in a ring, whereas in most of the monocotyledonous stems the vascular bundles are numerous and scattered.

ANATOMY OF DICOTYLEDONOUS STEMS

In young dicotyledonous stems there are three distinct regions—the epidermis, the cortex and the stele.

Epidermis. The epidermis consists of a single layer of cells and is the outermost layer of the stem. It contains stomata and produces various types of trichomes. The outer cell walls are greatly thickened and heavily cutinized. The cells are compactly arranged and do not possess intercellular spaces. In transverse section the cells appear almost rectangular. It serves mainly for restricting the rate of transpiration and for protecting the underlying tissues from mechanical injury and from disease-producing organisms.

Cortex. The region that lies next to the epidermis is the *cortex*. The innermost layer of the cortex is the *endodermis*, known also as the *starch-sheath*. It consists of a single layer of cells which surrounds the stele and contains numerous starch grains. Frequently it is most easily distinguishable from the surrounding tissue by the presence of these starch grains. The part of the cortex situated between the epidermis and the endodermis is generally divided into two regions, an outer zone of *collenchyma* cells and an inner zone of *parenchyma* cells.

Collenchyma. On the inside of the epidermis there is usually a band of collenchyma. The cells of the collenchyma are modified parenchyma cells with cellulose walls thickened at the angles where three or more cells are in contact. The collenchyma resembles parenchyma in being alive and in having a moderate amount of protoplasm. The chief function of collenchyma cells is to serve as strengthening material in succulent organs which do not develop much woody tissue, or in the soft young parts of woody plants before stronger tissues have been developed. They are especially fitted for giving strength to young, growing organs, since the thickened parts of the walls have considerable rigidity, while the thinner parts allow for an exchange of materials between the cells and for the stretching and growth of the cells. The collenchyma cells of stems sometimes contain chloroplasts and carry on photosynthesis.

Parenchyma. The parenchyma cells are generally regular in shape, have comparatively thin walls, and

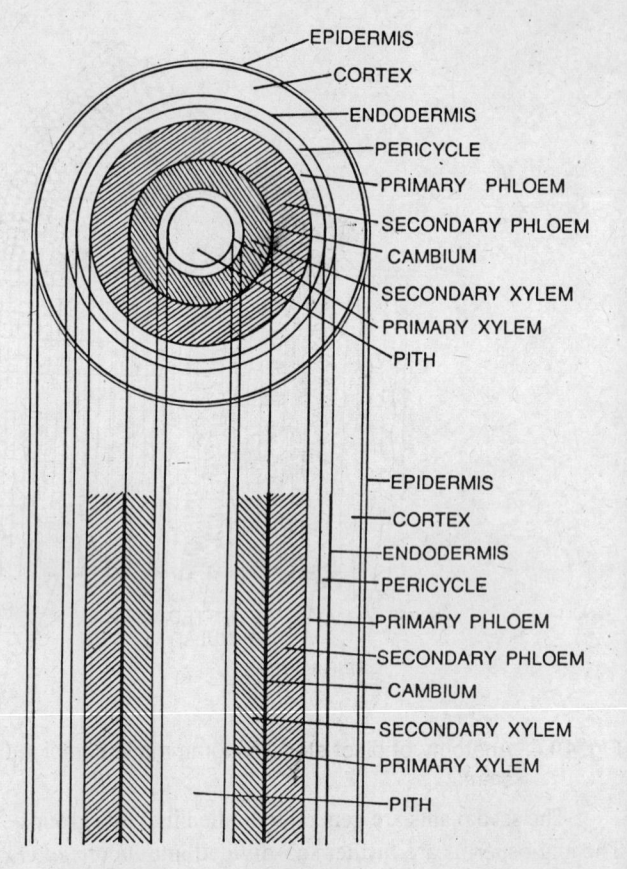

Fig. 10.5. The stem. Diagram showing structure of dicot axis. A, transverse section; B, longitudinal section. (After Eames and MacDaniels).

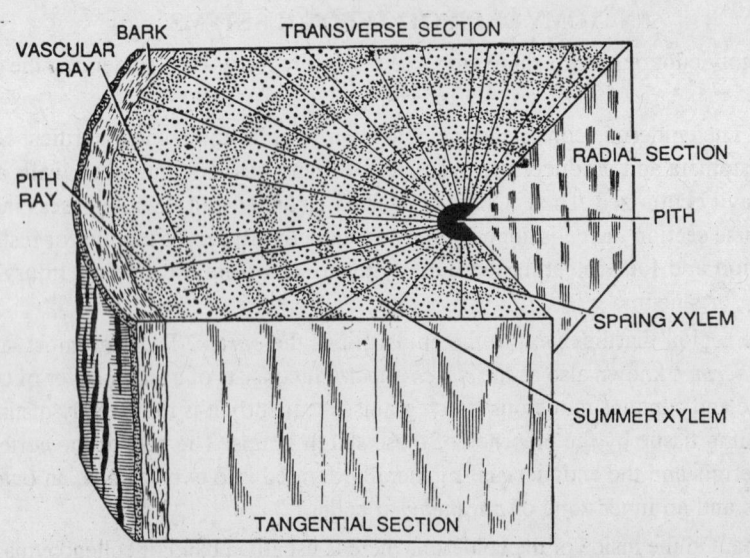

Fig. 10.6. The dicot stem. Three dimensional structure.

The Stem—Primary and Secondary Structure

are not greatly elongated in any direction. They are living cells and contain a moderate amount of protoplasm. When they are exposed to the light they develop chloroplasts and are known as *chlorenchyma cells*. Chlorenchyma cells are thus only a special kind of parenchyma cells. The parenchyma cells in the cortex of a stem are near enough to the light so that some or all of them develop chloroplasts and perform photosynthesis. The turgid parenchyma cells frequently help in giving rigidity to an organ. The function of parenchyma cells is important in succulent stems and in the young parts of the stems and woody plants before strong mechanical tissues have been developed. The parenchyma cells serve for the slow conduction of water and food. In the case of the cortex of stem it becomes evident that the water which is received by the collenchyma and the epidermis must be conducted through the parenchyma. The parenchyma is the special storage tissue of plants.

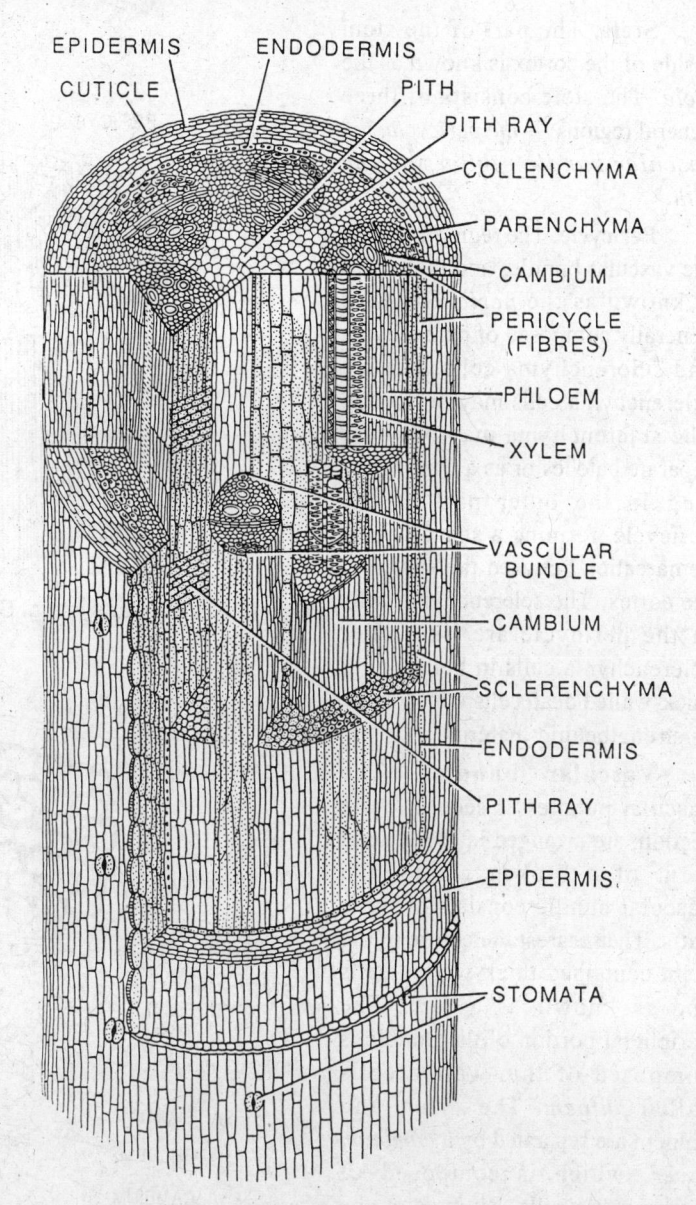

Fig. 10.7. Anatomy of dicot stem. Three dimensional diagram of a portion of an internode of a stem to show transverse, radial and tangential sections.

Sclerenchyma. The sclerenchyma cells are found in the cortex of some stems. There are two varieties of these scelerenchyma cells — short or irregularly shaped cells, known as *stone cells*, and *sclerenchyma fibres*. Sclerenchyma fibres are long, thick-walled dead cells and serve as strengthening material. Stone cells give stiffness to the cortex. The sclereids have been reported from the cortex of many water plants (*e.g., Limnanthemum, Nymphaea,* etc.).

Endodermis. The innermost layer of the cortex is the endodermis consisting of barrel-shaped, elongated, compact cells, having no intercellular spaces among them. Usually the cells contain starch grains and thus the endodermis may be termed at starch sheath.

Stele. The part of the stem inside of the cortex is known as the *stele*. The stele consists of three general regions — *the pericycle, the vascular bundle region* and *the pith*.

Pericycle. The region between the vascular bundles and the cortex is known as the *pericycle*. It is generally composed of parenchyma and sclerenchyma cells, but the sclerenchyma cells may be absent. The sclerenchyma may occur as separate patches or as a continuous ring in the outer part of the pericycle, forming a sharp line of demarcation between the stele and the cortex. The sclerenchyma cells in the pericycle are like other sclerenchyma cells in being long, thick-walled dead cells which serve as strengthening material.

Vascular bundles. The vascular bundles as seen in cross section, are arranged in the general form of a broken ring. Each vascular bundle consists of three parts. That nearest the centre of the stem contained thick-walled cells and is known as *xylem*. The peripheral portion of the bundle is composed of thin-walled cells called *phloem*. The xylem and phloem are separated by a *cambium layer*, which is composed of meristematic cells. By division the cambium layer increases the size of vascular bundles by forming xylem cells on the inner side and phloem cells on the outer side. In some stems the bundles are separate and run the length of the internode. In others they are more or less united and form a hollow cylinder in which the medullary rays occur as radiating plates with slight vertical extension.

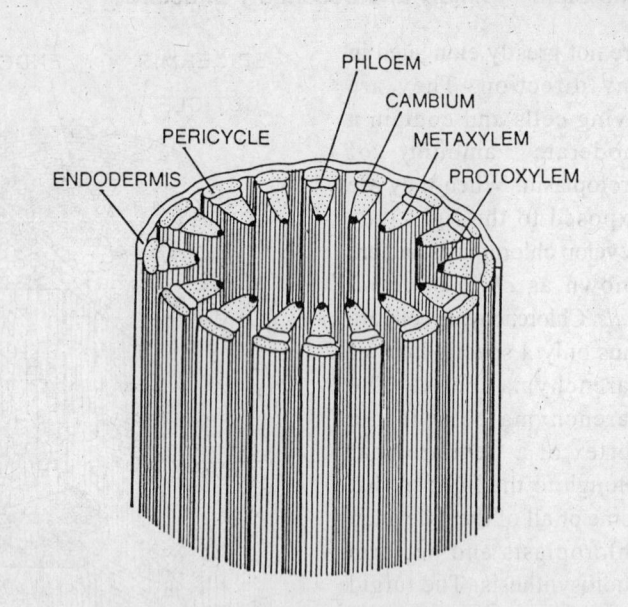

Fig. 10.8. Vascular system. Stele of a dicotyledonous stem.

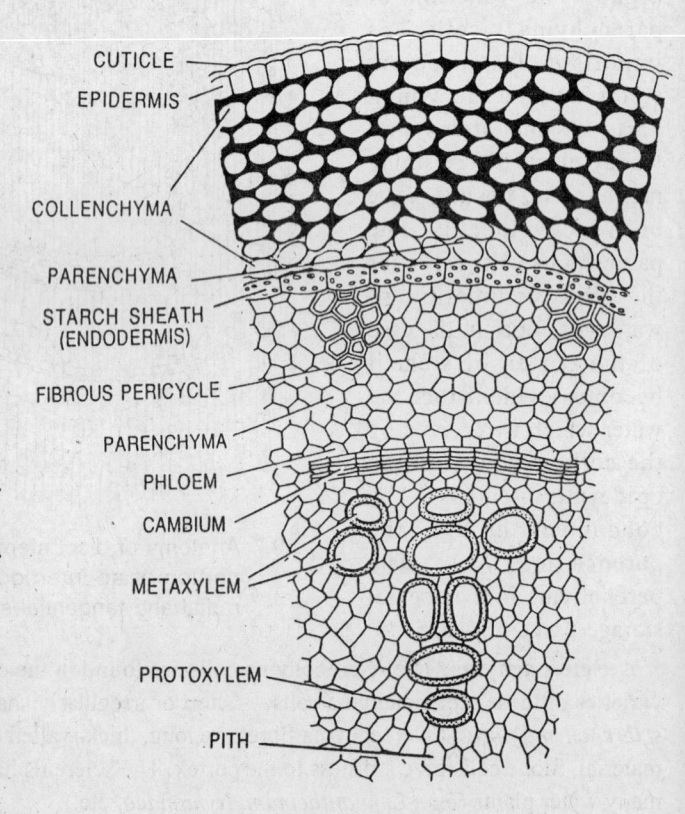

Fig. 10.9. Anatomy of dicot stem. T.S. of a sector of stem of *Ricinus communis*.

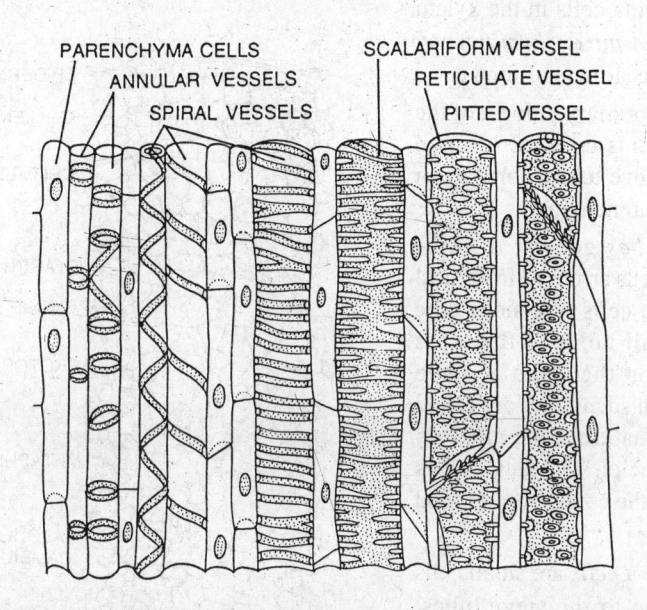

Fig. 10.10. The primary xylem. Types of vessels that occur in primary xylem in an elongating branch. Annular vessels formed first, and are therefore the oldest are more stretched. The pitted vessel formed last is the youngest. Elongation has stopped, so the pitted vessels will not be stretched.

Xylem. The xylem which is formed before the activity of the cambium has begun to produce xylem and phloem cells is called *primary xylem*. It is composed of two parts. The xylem formed first is nearest the centre of the stem and is called *protoxylem*. The more peripheral part of the primary xylem is known as *metaxylem*.

The xylem is composed of three different types of cells— tracheary cells, that include *tracheids* and *vessels*; *wood fibres* and *wood parenchyma*.

The *tracheids* are elongated dead cells, with walls that are thick in some places and thin in others. They serve both as water conducting and as strengthening cells. The walls of the tracheids are heavily impregnated with lignin.

The *vessels* are composed of rows of tracheary cells the cavities of which are connected by the total or partial disappearance of the cross walls. The diameter of vessels is usually much greater than that of tracheids. They form long tubes, and therefore, they constitute the principal water-conducting elements of the dicotyledonous stem.

The tracheary cells may be divided into several types according to the method by which the walls are thickened. *Annular* tracheary cells have thickenings in the form of rings, while *spiral* tracheary cells have spiral thickenings. *Pitted* tracheary cells have walls which are uniformly thickened except for thin places in the form of pits. When ladder-like thickenings are present, the vessel is said to be *scalariform*.

The protoxylem is composed largely of annular and spiral vessels and parenchyma, while the tracheary elements of the secondary xylem are pitted.

The *wood fibres* are long, slender, pointed dead cells with greatly thickened walls and only comparatively few small pits. They serve as strengthening cells. The tracheids that have a structure approaching that of wood fibres are called *fibre tracheids*.

The parenchyma cells in the xylem are known as *wood parenchyma*. They serve mainly for the storage of food.

Phloem. The primary phloem of the dicotyledonous stems consists of three types of cells — sieve tubes, companion cells and phloem parenchyma.

The *sieve tubes* consist of thin-walled, elongated cells arranged in vertical rows. The adjacent cells of a sieve tube are united by small holes in the cross walls. The areas on the walls of sieve tubes which contain such holes are called *sieve plates*. The mature sieve tubes do not contain any nuclei. The sieve tubes serve primarily for the conduction of food material.

The *companion cells* are small cells which are attached to the sieve tubes. Each companion cell is the sister cell of a sieve-tube cell, the two being formed by the division of a mother cell.

The phloem contains parenchyma cells whose structure is very similar to that of other parenchyma cells. These are known as *phloem parenchyma*.

Cambium. There lies a layer of meristematic cells between the xylem and the phloem is known as the *cambium*. The cambium consists of a single layer of cells which, by division gives rise to xylem cells toward the centre of the stem and phloem cells toward the periphery. At first the cambium is confined to the bundles, but later the parenchyma cells of the pith rays which lie between the edges of the cambium in the bundles divide and form a layer of cambium which reaches across the pith rays and connects that in the bundles, so that the cambium becomes a continuous cylinder.

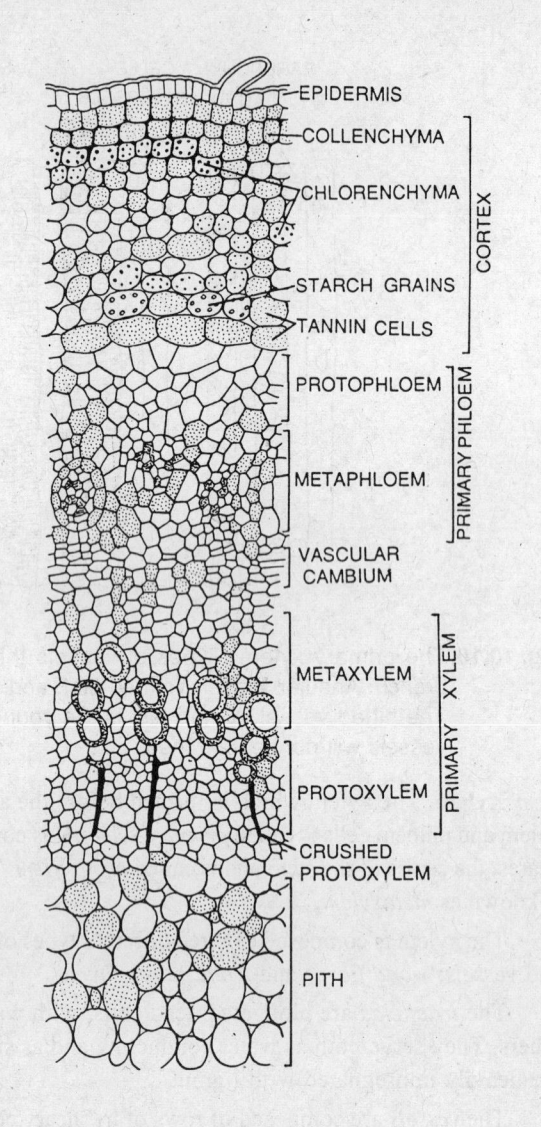

Fig. 10.11. Anatomy of dicotyledonous stem. T.S. of *Prunus* stem showing details.

Pith rays. The vascular bundles are separated from each other by radial rows of parenchyma cells known as pith rays. The pith-ray cells are usually elongated in a radial direction. They serve primarily for the conduction of food and water radially in the stem and for the storage of food.

Pith. In a dicotyledonous plant the centre of the stem is composed of thin-walled parenchyma cells and is known as the *pith*. The cells have distinct intercellular spaces.

VARIATIONS IN STEM STRUCTURE

The above mentioned description of structure of the stems is applicable to the great majority of dicotyledonous plants, but there are a few which show minor variations. The relative development of the various parts, however, varies greatly in different species. In some cases the pith is wide, while in

The Stem—Primary and Secondary Structure

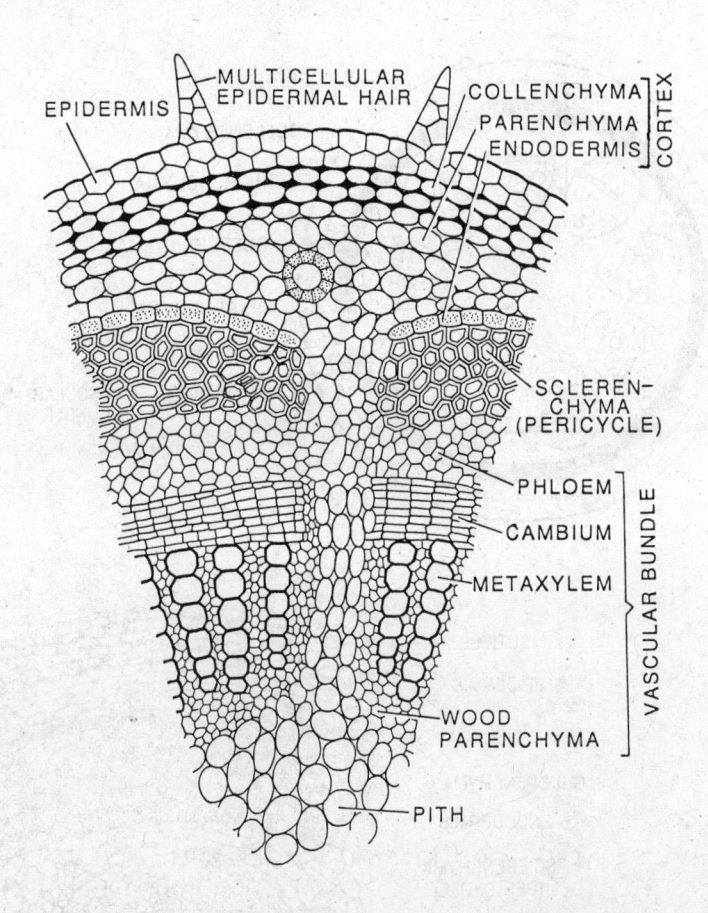

Fig. 10.12. T.S. of a dicotyledonous stem (*Helianthus annuus*).

others it is narrow. It may be wide and transitory and its early disappearance results in a hollow stem. The vascular bundles vary considerably in number and size, while the pith rays and cortex vary in width. Bundles which have the phloem only on the outside of the xylem are called *collateral* bundles. The bundles of some plants have phloem on both the outside and the inside of the xylem (*e.g.*, in the members of Cucurbitaceae) and known as *bicollateral* bundles.

ANATOMY OF *CUCURBITA* STEM

Epidermis. This single outermost layer consists of compact barrel shaped cells having no intercellular spaces. The epidermis remains covered with a thin cuticle. Some of the epidermal cells possess multicellular epidermal hairs.

Cortex. This region consists of external collenchyma, chlorenchyma (photosynthetic tissue) and endodermis.

(a) **Collenchyma.** This lies immediately beneath the epidermis consisting of many layers of the cells in the ridges, whereas in furrows it is only two or three layered or sometimes altogether absent.

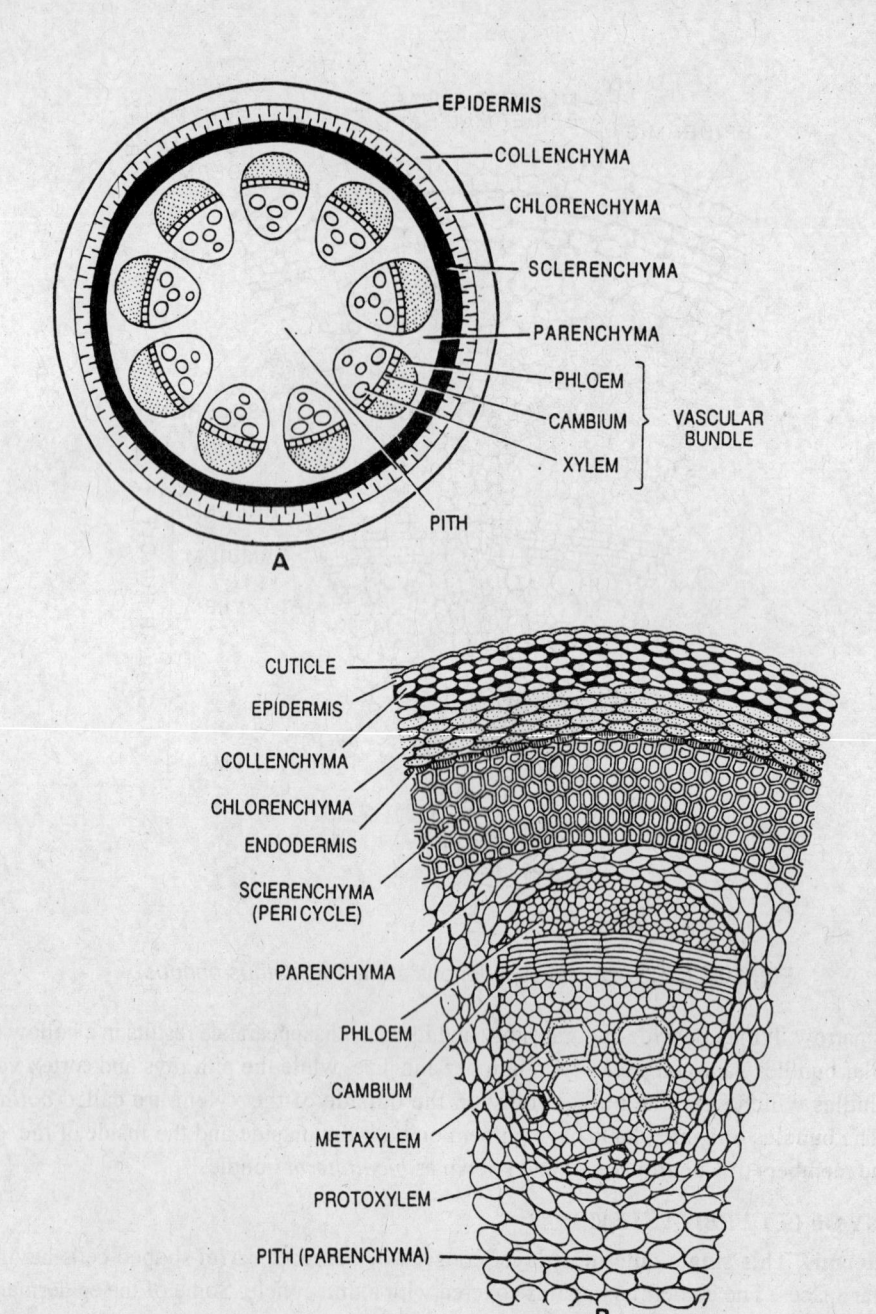

Fig. 10.13. Anatomy of dicot stem. T.S. of young stem of *Aristolochia*. A, diagrammatic; B, detail of a sector.

   (b) Chlorenchyma. Just below the collenchyma two or three layers of parenchyma containing chloroplasts (chlorenchyma—photosynthetic tissue) present which help in the process of assimilation.

The Stem—Primary and Secondary Structure

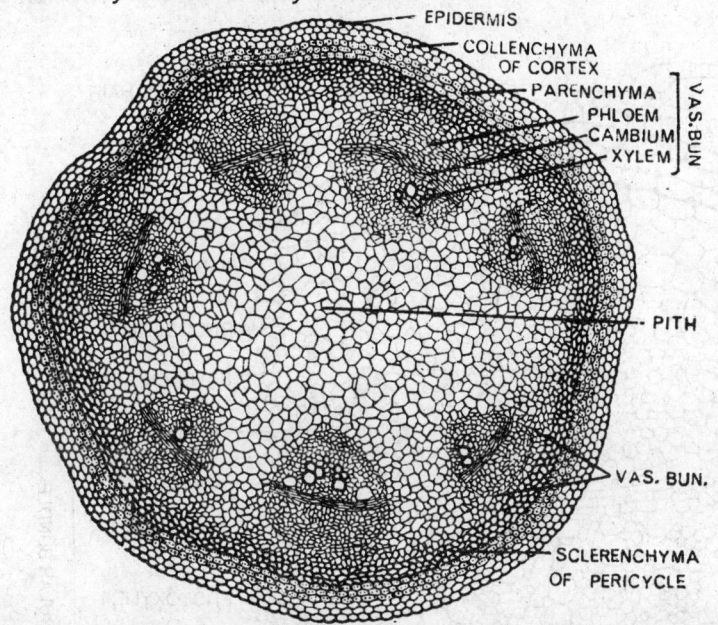

Fig. 10.14. The stem. Cross section of a young stem of *Aristolochia*. Just beneath the uniseriate layer of epidermis a few layers (2 or 3) of collenchyma are present. Chlorenchyma is also visible. The pericyclic region consists of a continuous band of sclerenchyma. Seven bundles are arranged in a ring. Each bundle shows typical dicot characteristics. Pith is large and conspicuous.

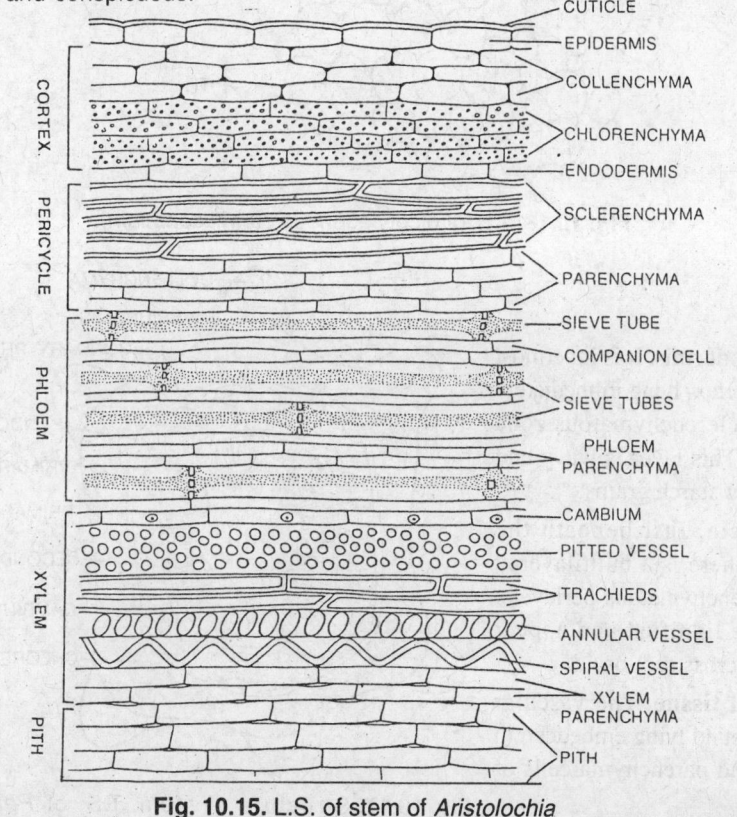

Fig. 10.15. L.S. of stem of *Aristolochia*

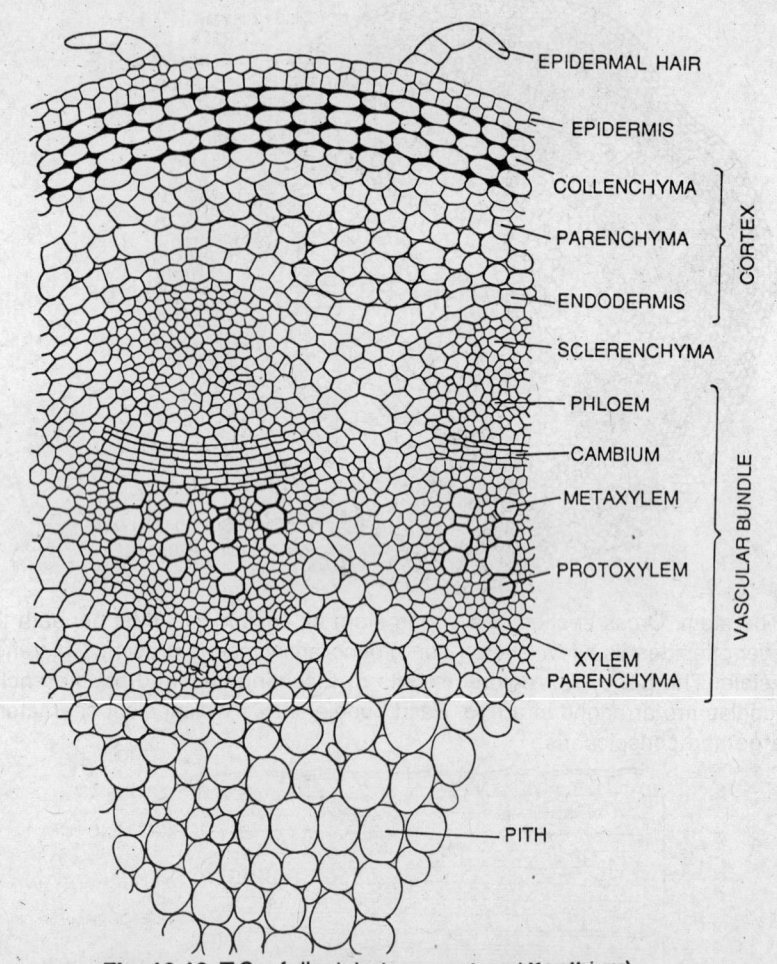

Fig. 10.16. T.S. of dicotyledonous stem (*Xanthium*).

(c) **Endodermis.** It is innermost layer of the cortex, lying immediately outside the sclerenchymatous zone of pericycle. This layer is wavy and contains many starch grains.

Pericycle. Just beneath the endodermis there is a multilayered zone of sclerenchymatous pericycle. The cells are lignified and appear polygonal in cross section.

Ground tissue. The vascular bundles are found lying embedded in the thin walled parenchyma cells of

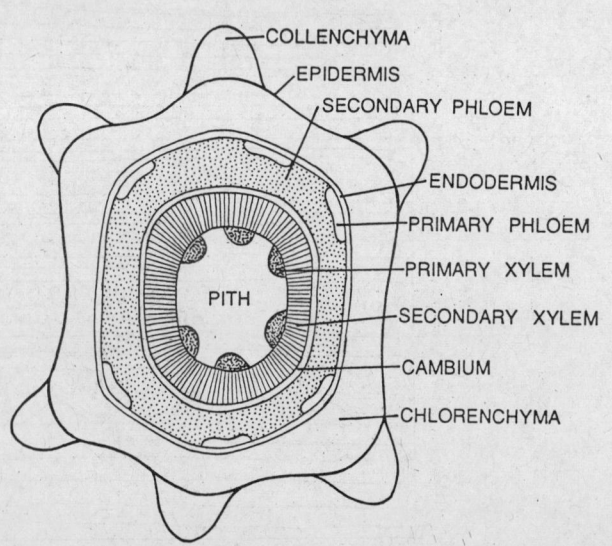

Fig. 10.17. Structure of stem. T.S. of *Peristophe* stem (diagrammatic).

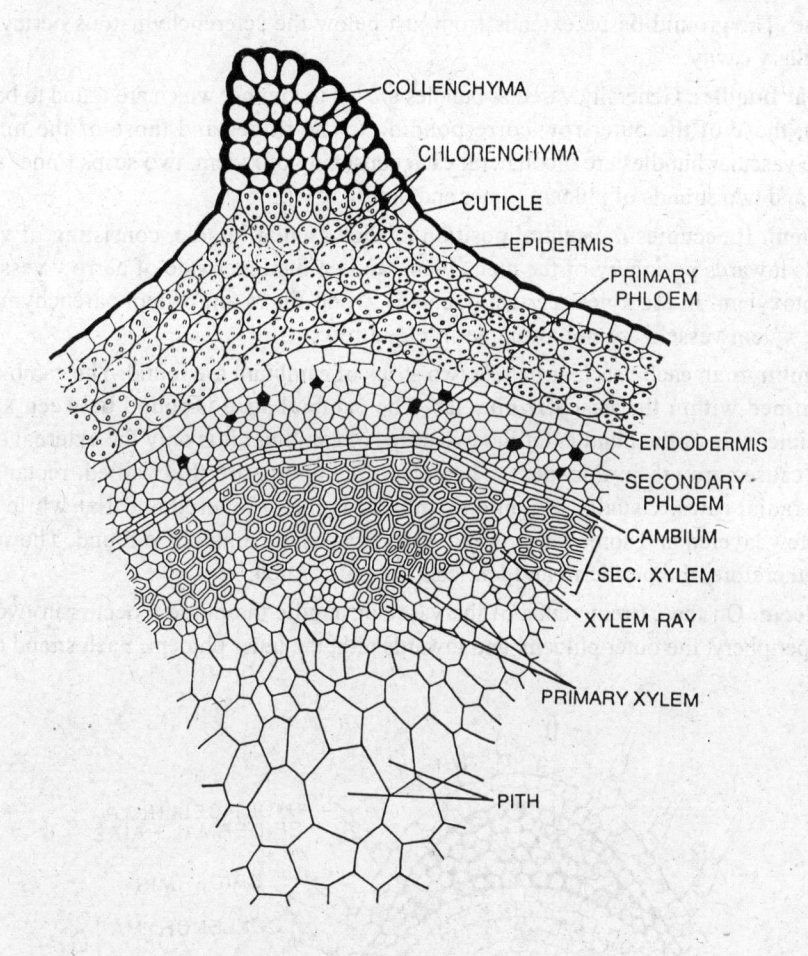

Fig. 10.18. The stem. T.S. of a portion of *Peristrophe* stem (dicot) showing detailed structure.

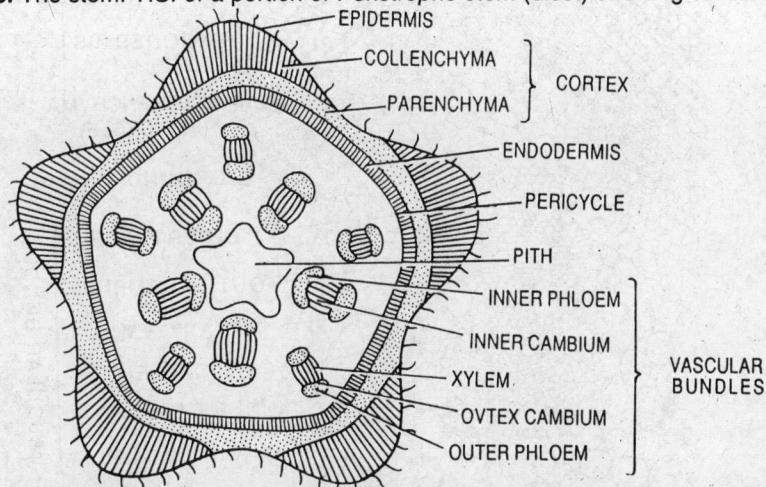

Fig. 10.19. T.S. of *Cucurbita* stem (diagrammatic representation). Each vascular bundle possesses external and internal phloem (bicollateral bundles). Small strands of sieve tubes and companion cells traverse the parenchyma of the vascular region and the tissues of the cortex. (After K. Esau).

ground tissue. The ground tissue extends from just below the sclerenchymatous pericycle to the central medullary cavity.

Vascular bundles. Generally vascular bundles are ten in number which are found to be arranged in two rows, those of the outer row corresponding to the ridges and those of the inner to the furrows. The vascular bundles are bicollateral each consisting of xylem, two strips (inner and outer) of cambium and two strands of phloem (inner and outer).

(a) **Xylem.** It occupies the central position of the vascular bundle, consisting of very wide, pitted vessels towards periphery of the metaxylem, and on the inner side of narrow vessels which form the protoxylem. In the xylem certain tracheids, wood fibres and xylem parenchyma are also present. The xylem vessels are not arranged in radial rows.

(b) **Cambium.** In each vascular bundle two strips of cambium are found. The cambial activity remains confined within the vascular bundles. The cambial strip is found between xylem and phloem on either side of the bundle. Of the two strips of cambium it is only the external one which divides and causes growth in thickness. The cells of cambium are thin walled, rectangular and arranged in radial rows. Usually the outer cambium is many layered and flat while the inner cambium is few layered and somewhat curved. Only fascicular cambium is found. The stem is not woody, and therefore, the periderm and lenticels are not formed.

(c) **Phloem.** On the extreme ends of the vascular bundle the phloem occurs in two patches, towards the periphery, the outer phloem, and towards pith, the inner phloem. Each strand of phloem

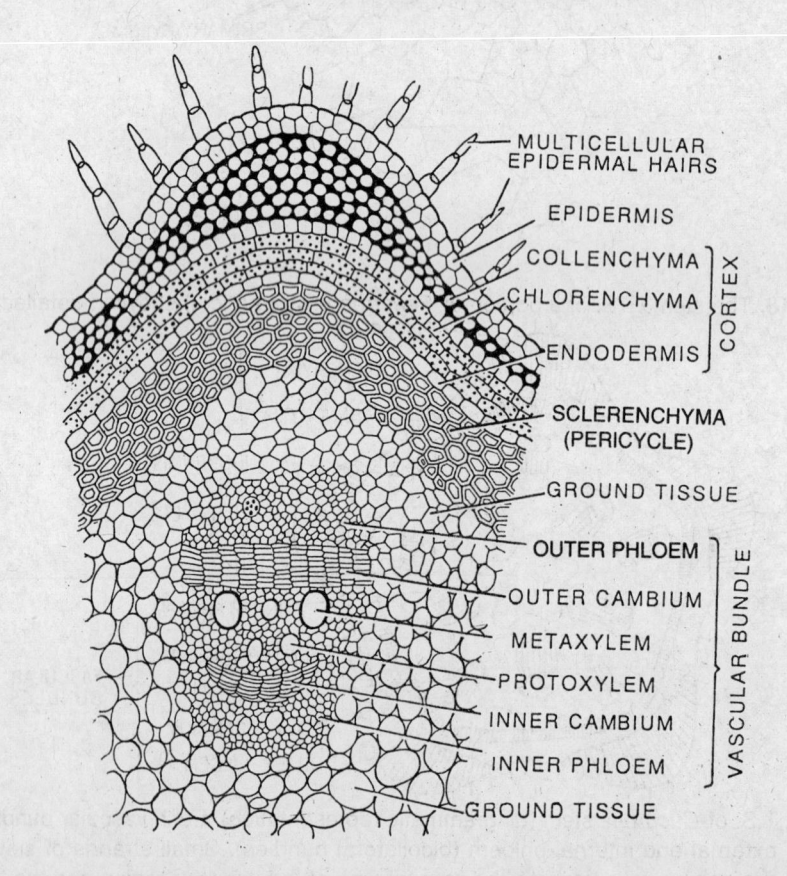

Fig. 10.20. T.S. of young *Cucurbita* stem (dicotyledonous) showing bicollateral vascular bundle.

The Stem—Primary and Secondary Structure 225

consists of sieve tubes, companion cells and phloem parenchyma. Sieve tubes are very well developed. The sieve plates with perforations are also visible. Fibres and ray cells are absent.

SPECIAL STRUCTURE

The bicollateral open vascular bundles are found each consisting of xylem (central position), two strips of cambium (outer and inner) and two patches of phloem (outer and inner). (See figs. 10.19 and 10.20)

ANATOMY OF *BRYONIA* STEM

Epidermis. The single outermost layer consists of compact barrel shaped cells having no intercellular spaces. Usually the epidermis is covered with a thin cuticle. At certain places stomata are also found.

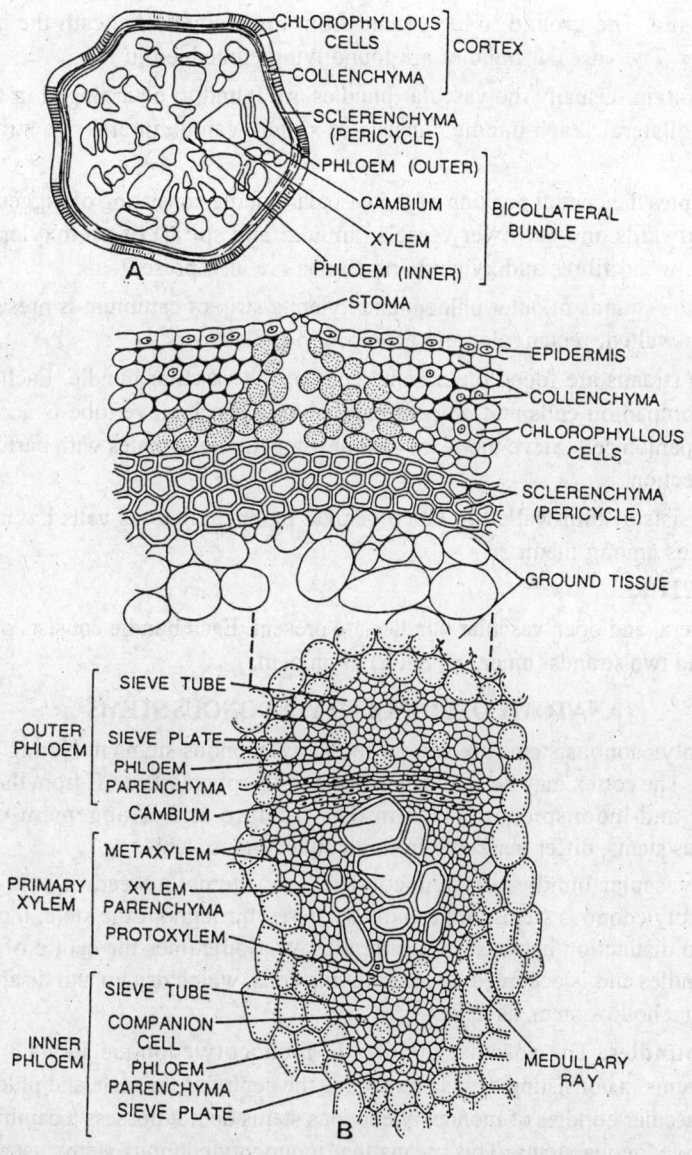

Fig. 10.21. *Bryonia* stem (Cucurbitaceae). Transverse section—A, diagrammatic; B, detailed structure.

Cortex. This consists of collenchyma, loose chlorenchyma and inconspicuous endodermis.

(a) **Collenchyma.** This lies immediately beneath the epidermis consisting of rounded or oval cells (in cross section) usually thickened laterally due to the presence of cellulose. The collenchymatous region is multilayered.

(b) **Chlorenchyma.** Below the collenchyma few layers of chlorenchyma are found. The cells are thin walled, rounded or oval, with chloroplasts and having intercellular spaces among them.

(c) **Endodermis.** The single layered endodermis is inconspicuous. The endodermal cells may contain abundant starch grains in them.

Pericycle. Immediately beneath the endodermis multilayered sclerenchymatous pericycle is found. The cells are lignified and polygonal as seen in cross section.

Ground tissue. The ground tissue extend from immediately beneath the pericycle to the central pith cavity. The vascular bundles are found lying embedded in it.

Vascular system. Usually the vascular bundles are found to be arranged in two rows. They are open and bicollateral. Each bundle consists of xylem, cambium and two strands (inner and outer) of phloem.

Xylem occupies the central position of the vascular bundle consisting of bigger vessels (pitted) of metaxylem outwards and narrower vessels (annular and spiral) of protoxylem towards pith. Certain tracheids, wood fibres and xylem parenchyma are also present.

In between the strands of outer phloem and xylem a strip of cambium is present. The cells of cambium are thin walled, rectangular and arranged in radial rows.

The phloem strands are found on extreme ends of the vascular bundle. Each strand consists of sieve tubes, companion cells and phloem parenchyma. Each sieve tube is accompanied by a conspicuous companion cell. Sieve tubes are well developed. Sieve plates with perforations are also visible in cross section.

Pith. It consists of thin-walled rounded or oval parenchymatous cells having well defined intercellular spaces among them.

SPECIAL STRUCTURE

The bicollateral and open vascular bundles are present. Each bundle consists of central xylem, a cambial strip and two strands (inner and outer) of phloem.

ANATOMY OF MONOCOTYLEDONOUS STEMS

The monocotyledonous stems are similar to dicotyledonous stems in having an *epidermis*, a *cortex* and a *stele*. The cortex may be well developed and sharply marked off from the stele, or it may be quite narrow and inconspicuous. It is in the structure and arrangement of bundles that monocotyledonous stems, differ markedly from dicotyledonous stems.

Stele. The vascular bundles of monocotyledonous stems, instead of being arranged in a cylinder as in dicotyledonous stems, are usually scattered throughout the stele, including the pith, so that there is no distinction between pith and pith rays. Sometimes the centre of the stele is free from vascular bundles and is occupied by parenchyma cells, which dry up and disappear at an early stage, resulting in a hollow stem, as in most grasses.

Vascular bundles. The vascular bundles of monocotyledonous stems are like those of dicotyledonous stems in consisting of xylem towards the centre of the stele and phloem towards the periphery. The vascular bundles of monocotyledonous stems do not possess a cambium layer which is found in dicotyledonous stems. This means that monocotyledonous stems usually do not have secondary thickening. Each bundle remains more or less completely surrounded by a sheath of sclerenchyma cells, the *bundle sheath*, which is particularly well developed on the sides toward the

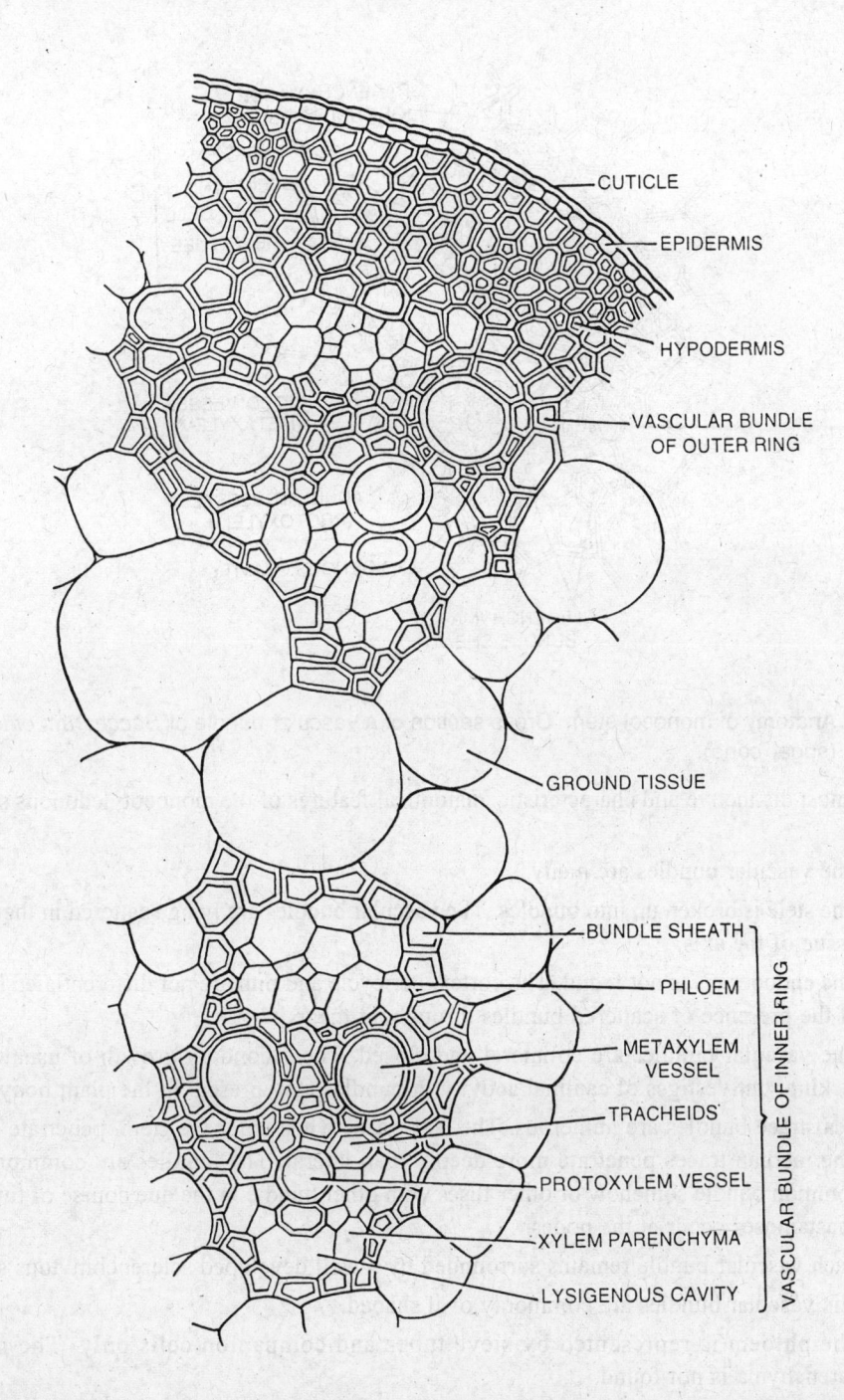

Fig. 10.22. T.S. of a portion of stem of grass (*Cynodon dactylon*) typical monocot stem, showing detailed internal structure.

centre and toward the periphery of the stem. The phloem is made up mostly of sieve tubes and companion cells, and the xylem of vessels and wood parenchyma.

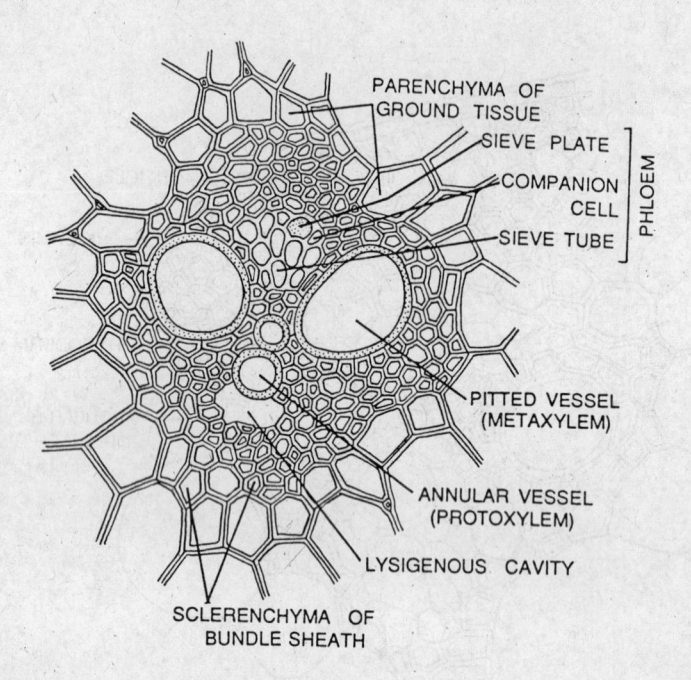

Fig. 10.23. Anatomy of monocot stem. Cross section of a vascular bundle of *Saccharum officinarum* (sugar cane).

The most distinctive and characteristic anatomical features of the monocotyledonous stem are as follows:

1. The vascular bundles are many.
2. The stele is broken up into bundles. The vascular bundles are lying scattered in the ground tissue of the axis.
3. The endodermis is not found. The cortex, pericycle and pith are not differentiated because of the presence of scattered bundles throughout the axis.
4. The vascular bundles are collateral and closed. The secondary growth of usual type is lacking, but vestiges of cambial activity in bundles may present in the plant body.
5. Leaf trace bundles are numerous. The leaf traces when enter the stem, penetrate deeply. The median traces penetrate more deeply than lateral. The bundles are common. Each common bundle somehow or other fuses with other bundle in the due course of time. The anastamoses occur at the nodes.
6. Each vascular bundle remains surrounded by a well developed sclerenchmytous sheath.
7. The vascular bundles are commonly oval shaped.
8. The phloem is represented by sieve tubes and companion cells only. The phloem parenchyma is not found.
9. The pith is not marked out.
10. Usually sclerenchymatous hypodermis is present.
11. Usually epidermal hairs are not present.

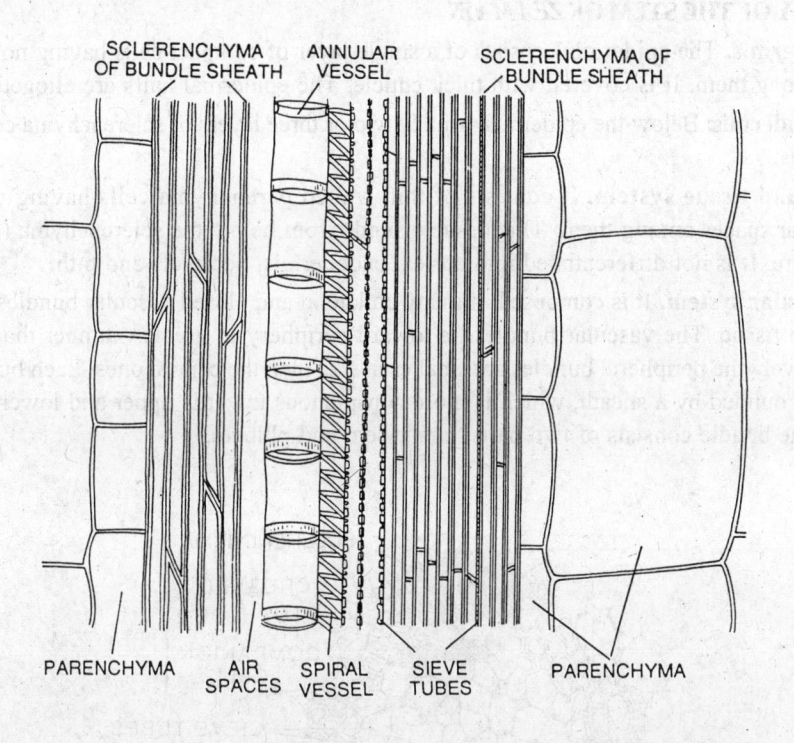

Fig. 10.24. Anatomy of monocot stem. L.S. of a vascular bundle of Saccharum officinarum (sugar cane).

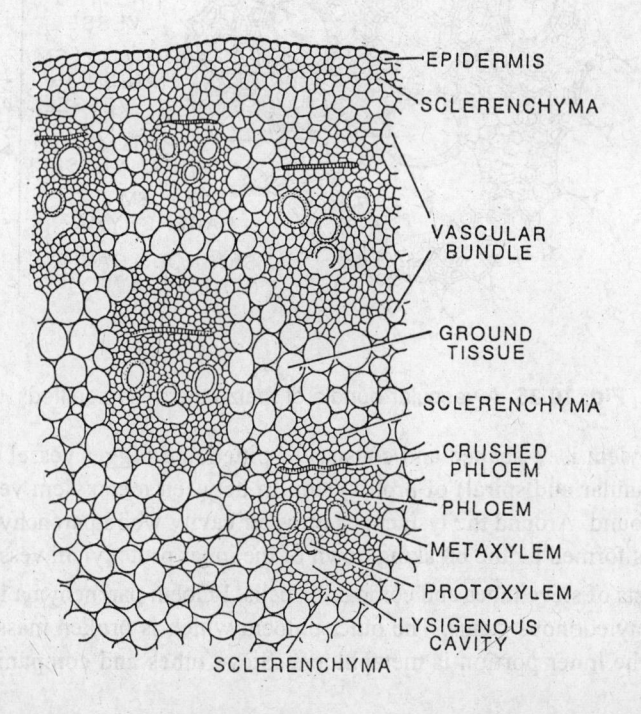

Fig. 10.25. The monocotyledonous stem. T.S. of maize stem showing sclerenchymatous hypodermis, parenchymatous ground tissue and vascular bundles.

ANATOMY OF THE STEM OF ZEA MAYS

Epidermis. The epidermis consists of a single layer of compact cells having no intercellular spaces among them. It is covered with thick cuticle. The epidermal hairs are altogether absent.

Hypodermis. Below the epidermis, usually two or three layers of sclerenchyma cells represent hypodermis.

Ground tissue system. It consists of thin walled parenchyma cells having well-defined intercellular spaces among them. This tissue extends from below the sclerenchyma (hypodermis) to the centre. It is not differentiated into cortex, endodermis, pericycle and pith.

Vascular system. It is composed of many collateral and closed vascular bundles scattered in the ground tissue. The vascular bundles lie toward periphery in greater number than the centre. Comparatively the peripheral bundles are smaller in size than the central ones. Each bundle is more or less surrounded by a sheath, which is more conspicuous towards upper and lower sides of the bundle. The bundle consists of two parts, *i.e.*, xylem and phloem.

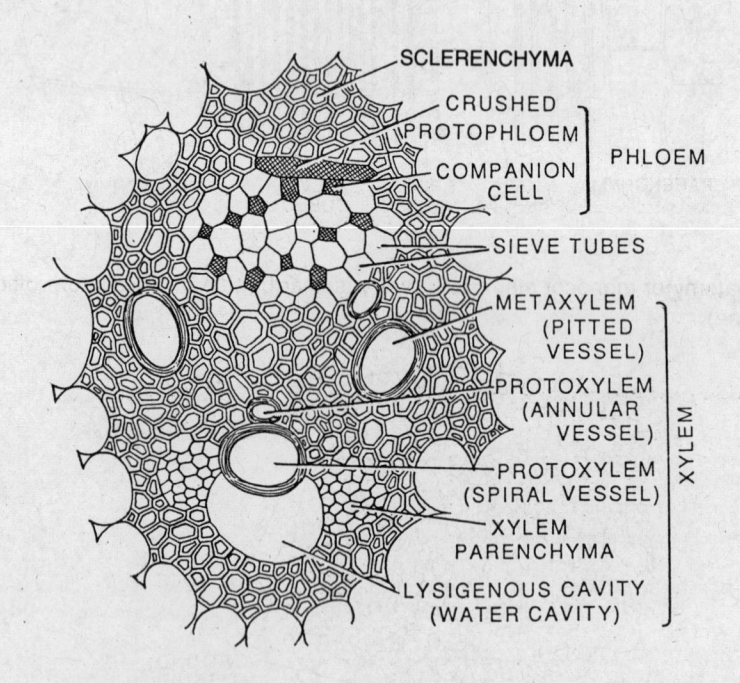

Fig. 10.26. A vascular bundle of maize stem—magnified.

Usually the xylem is Y-shaped and consists of pitted and bigger vessel of metaxylem and smaller vessels (annular and spiral) of protoxylem. In between metaxylem vessels, small pitted tracheids are also found. Around the lysigenous or water cavity wood parenchyma is present. The lysigenous cavity is formed by the breaking down of the inner protoxylem vessel.

Phloem consists of sieve tubes and companion cells. Phloem parenchyma is altogether absent in most of monocotyledonous stems. The outer phloem which is broken mass may be called as protophloem and the inner portion is metaphloem. Sieve tubes and companion cells are quite conspicuous.

The Stem—Primary and Secondary Structure 231

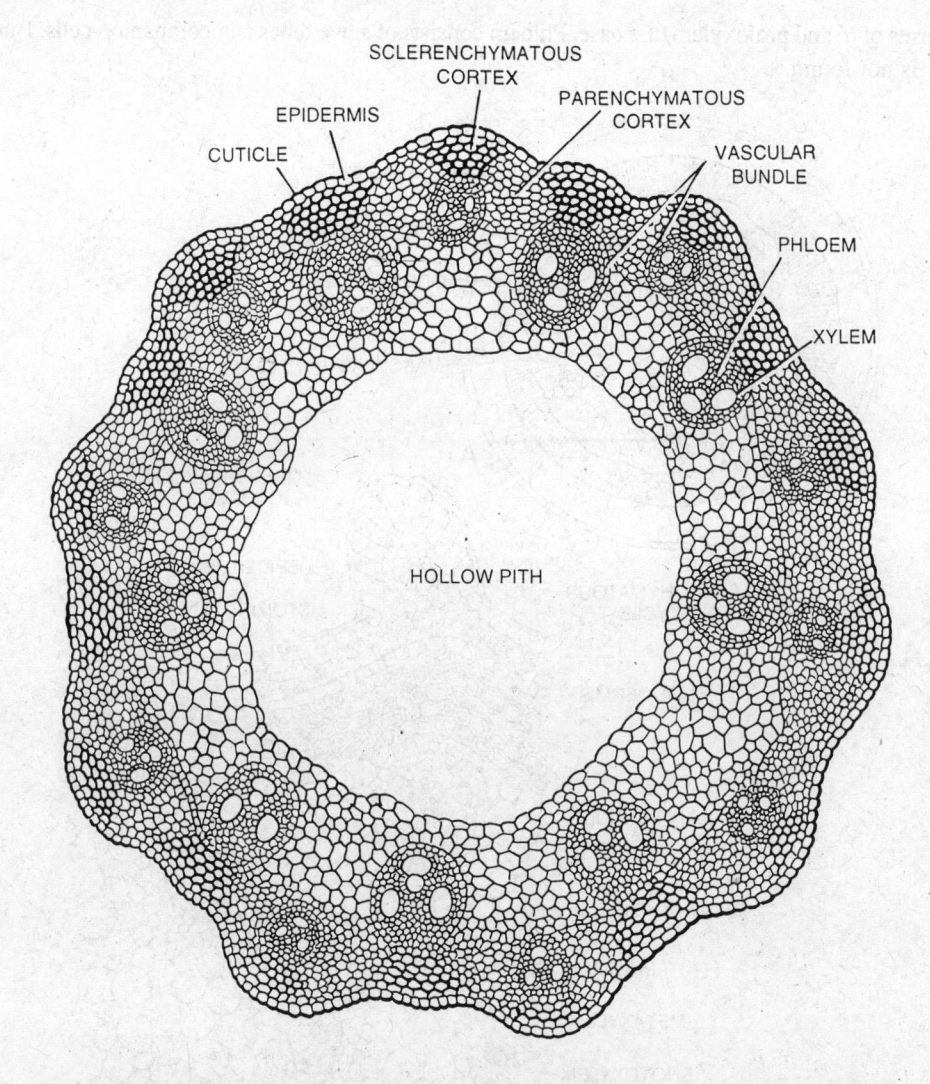

Fig. 10.27. T.S. of stem of *Avena sativa* (oat), showing hollow pith in the centre.

ANATOMY OF THE STEM OF *ASPARAGUS*

Epidermis. It is outermost uniseriate layer composed of approximately rounded cells with cuticularized outer walls.

Ground tissue system. Just beneath the epidermis a few layers of parenchyma are found which contain chloroplasts in them. This may be called *cortex*. The innermost layer of the cortex consists of compact cells and called the *starch sheath*. Below the starch sheath a multilayered complete band of sclerenchyma occurs, which gives mechanical support to the stem. The rest of the portion is ground tissue which consists of thin walled parenchyma cells having well developed intercellular spaces among them. The vascular bundles remain scattered in the ground tissue.

Vascular system. The vascular bundles remain scattered in the ground tissue. The central bundles are comparatively larger than the peripheral ones. They are always collateral and closed. Each vascular bundle consists of xylem and phloem. The xylem is Y-shaped. The metaxylem vessels

232   *Plant Anatomy*

form arms of Y and protoxylem, the base. Phloem consists of sieve tubes and companion cells. Bundle sheath is not found.

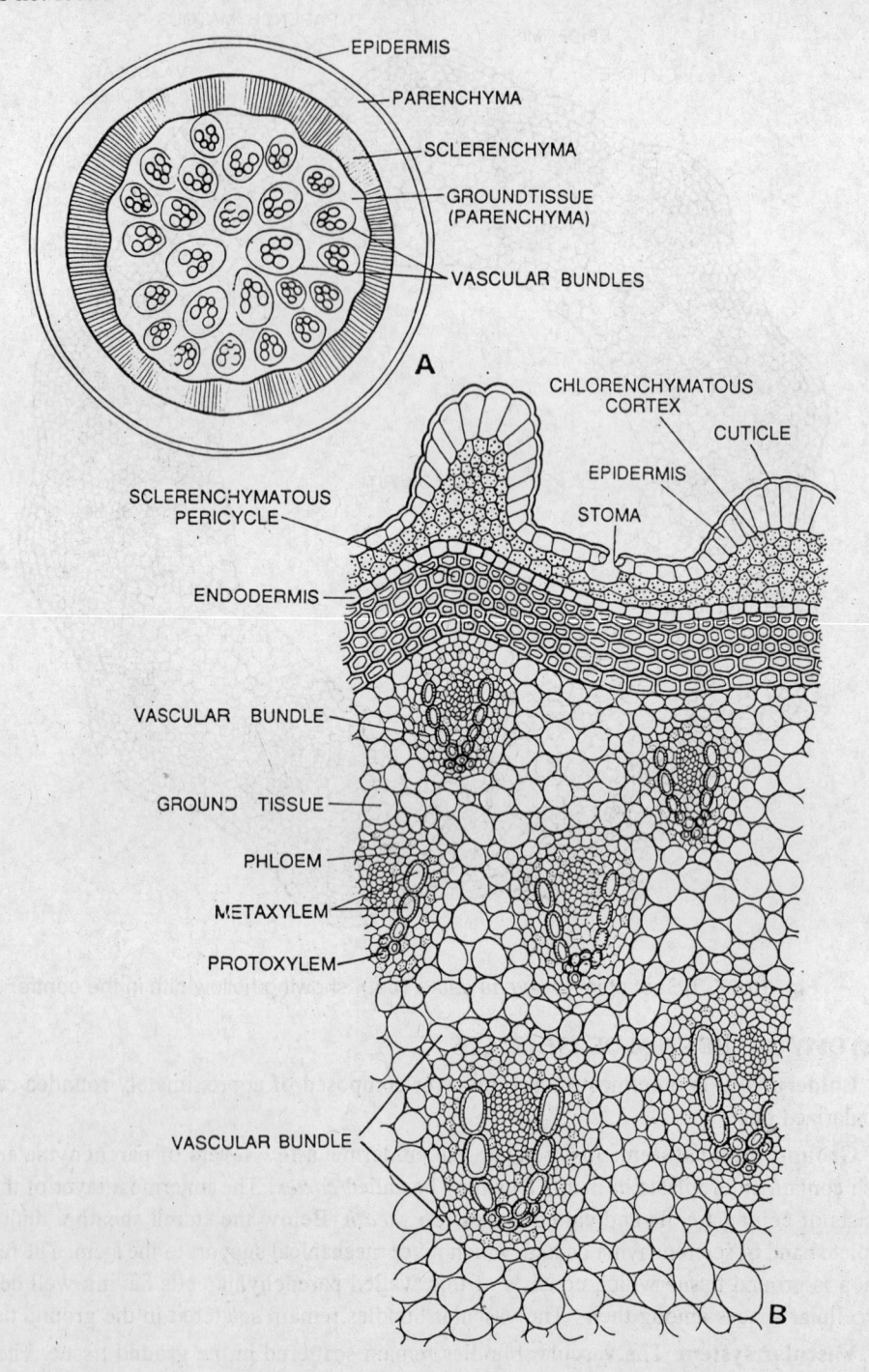

Fig. 10.28. Anatomy of monocotyledonous stem. T.S. of a vascular bundle of *Asparagus* lacking secondary growth. Crushed protophloem and protoxylem are also seen.

ANATOMY OF THE SCAPE OF *CANNA*

Epidermis. It is the outermost uniseriate layer consisting of small, polygonal cells with cuticularized outer walls.

Ground tissue system. Just beneath the epidermis a few layers of parenchyma occur forming small cortical region. The cells of cortex are sufficiently large and polygonal. Immediately below the cortex, a single-layered chlorophyllous tissue is found consisting of chloroplast bearing cells. The sclerenchyma patches also remain attached to the chlorophyllous tissue here and there. The rest of the portion consists of a continuous mass of large, thin walled, parenchymatous cells having sufficiently developed intercellular spaces among them. It is called the ground tissue.

Vascular bundles. They are many and of various sizes, lying scattered in the ground tissue. The bundles are closed and collateral. Each bundle is incompletely surrounded by a sheath of sclerenchyma called *bundle sheath*. The outer sclerenchyma patch of the bundle is more distinct and cap like whereas inner patch is not so developed. Each bundle consists of xylem and phloem. The xylem is situated on the inner side and the phloem towards outer side. The xylem consists of a large spiral vessel with one or two smaller vessels of same nature. The phloem consists of sieve tubes and companion cells.

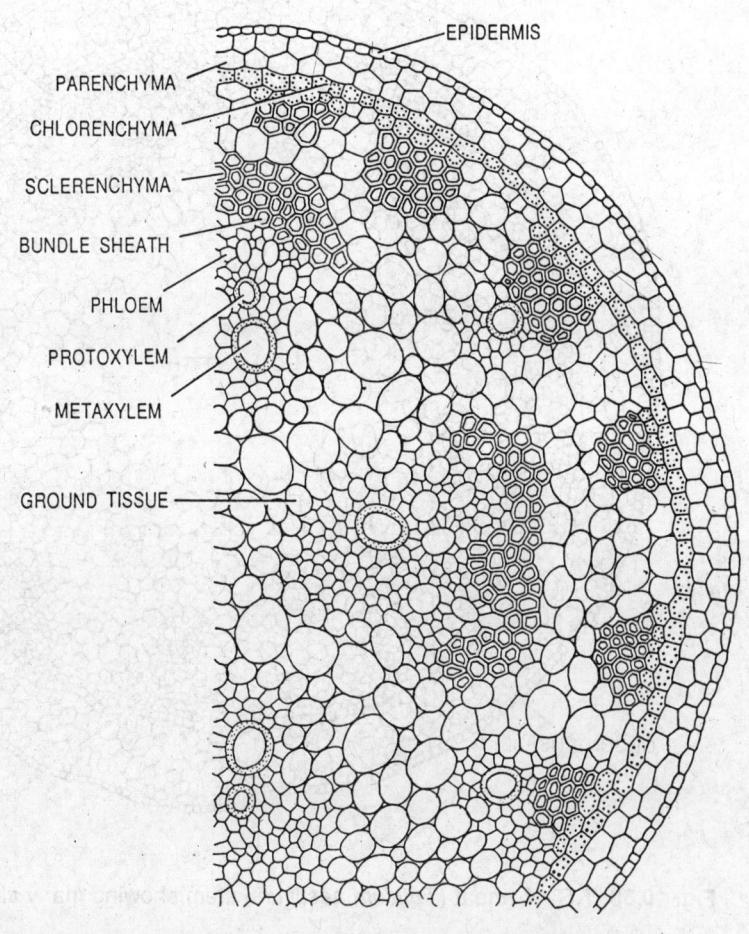

Fig. 10.29. The monocotyledonous stem. T.S. of scape of *Canna*.

ANATOMY OF THE STEM OF WHEAT (*TRITICUM AESTIVUM*)

Epidermis. It is the outermost uniseriate layer, usually composed of compact tabular cells with cuticularized outer walls. The stomata are also seen here and there on the epidermis.

Ground tissue system. Just beneath the epidermis the sclerenchyma cells occur in small patches which are not arranged in a continuous band, but are interrupted by chlorenchyma tissue here and there. The stomata are confined on the epidermis only in chlorenchymatous regions. The rest of the ground tissue consists of thin walled rounded or oval parenchyma cells having sufficiently developed intercellular spaces among them. The central region of the stem is hollow.

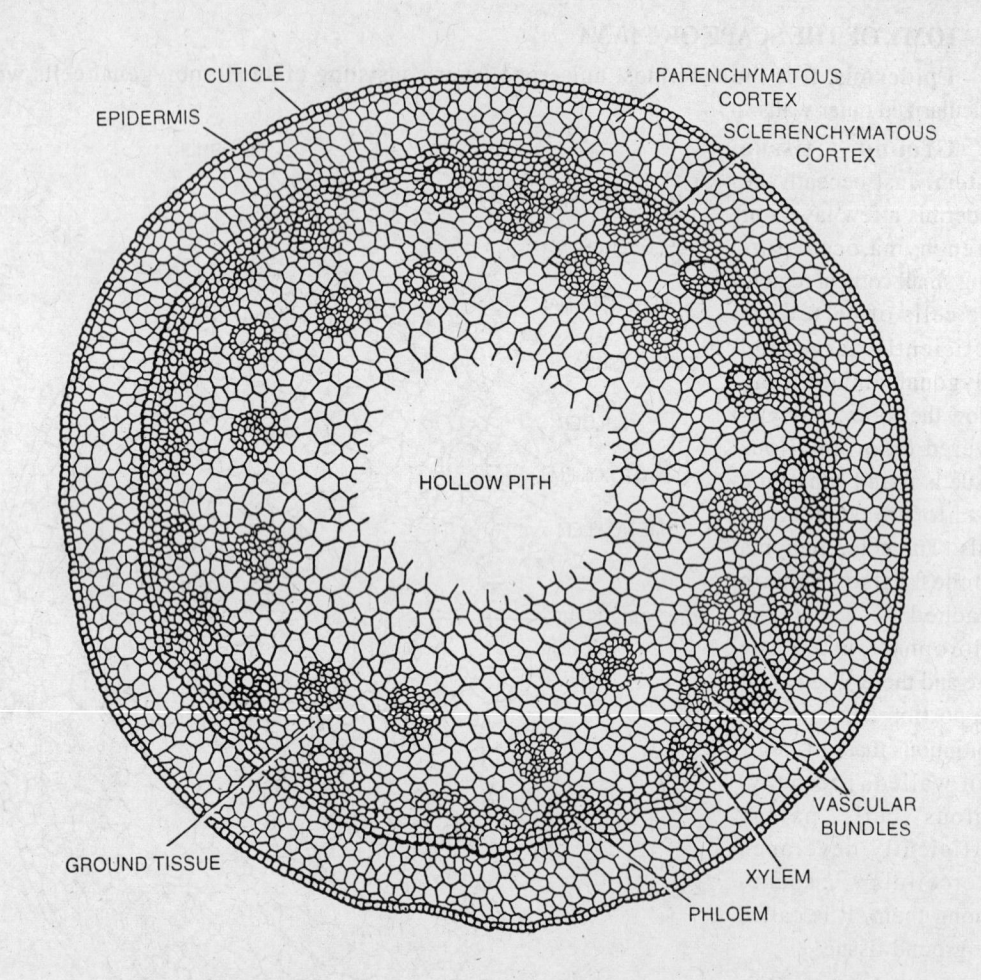

Fig. 10.30. T.S. of wheat (*Triticum aestivum*) stem showing many closed vascular bundles.

Vascular bundles. The closed and collateral vascular bundles occur in two series. The peripheral series consists of smaller bundles, whereas the inner series is of bigger bundles. The vascular bundles of outer series are lying embedded in sclerenchyma band. The bundles of inner series are also surrounded by sclerenchymatous bundle sheath like that of maize stem. The bundle sheath of peripheral bundles actually touches the epidermis.

ANATOMY OF THE PHYLLOCLADE

Numerous stems are specialized for photosynthesis and take the place of leaves in the manufacture of carbohydrates. Some stems which are specialized for photosynthesis are round (*e.g.*, Casuarina, Euphorbia tirucalli,* etc.), others are flattened (*e.g.,* Cocoloba-*Muehlenbeckia platyclada*), and others even have the form of leaves (*e.g., Ruscus, Myrsiphyllum, Phyllocladus protractus,* etc.). Such stems as those of cacti (*e.g., Opuntia, Carnegiea,* etc.) are specialized both for photosynthesis and for water storage. In the following paragraphs, the anatomy of the phylloclade of cocoloba—(*Muehlenbeckia platyclada—Homalocladium platycladum*) and *Ruscus* has been discussed.

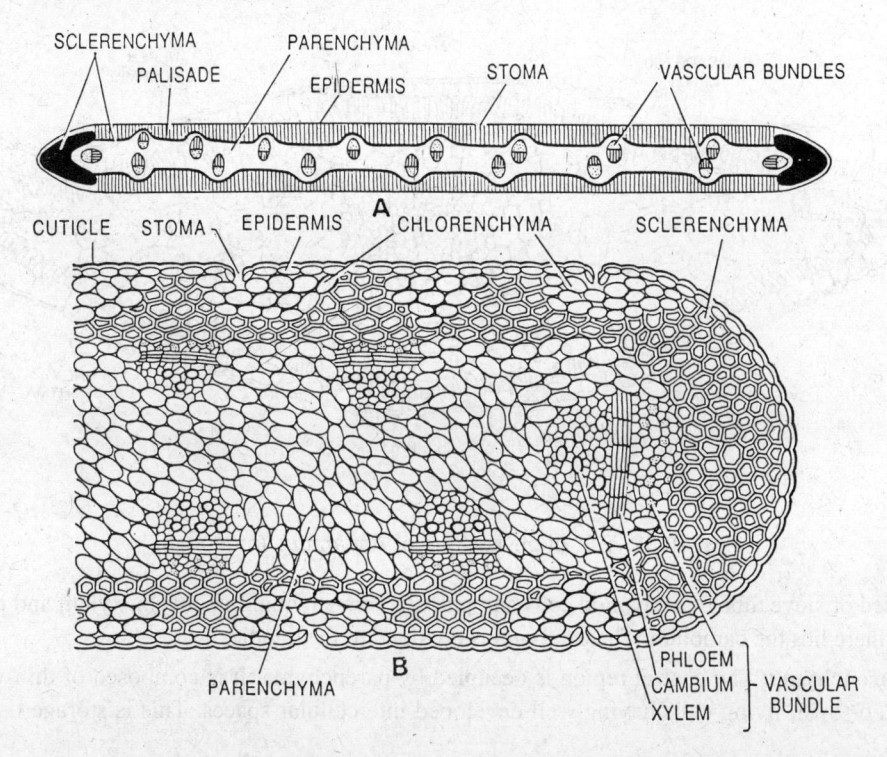

Fig. 10.31. *Homalocladium platycladum* (*Muehlenbeckia platycladus*)—phylloclade (dicot.). A. transection of phylloclade (diagrammatic); B, transaction of phylloclade showing detailed structure. The stomata, the cambium, the vascular bundles, the sclerenchyma and the chlorophyllous cells are clearly visible.

ANATOMY OF THE PHYLLOCLADE OF COCOLOBA (*HOMALOCLADIUM-PLATYCLADUM—MUEHLENBECKIA PLATYCLADA*). DICOT.

Epidermis. The surfaces are bounded on both the sides by upper and lower epidermal layers. A thin cuticle covers the epidermis. The epidermis is interrupted by numerous stomata on both the surfaces. Distinct guard cells of the stomata and sub-stomatal chambers are visible.

Chlorenchyma. Just beneath the epidermis there are few layers of chlorophyllous cells. The chlorophyllous cells are found only underneath the upper and lower epidermal layers. However, the chlorenchyma is not found at the edges of the phylloclade. The stomata are confined to chlorophyllous regions. There are well developed intercellular spaces among these cells. This is assimilatory tissue.

Sclerenchyma. The multilayered sclerenchyma tissue is found at the edges of the phylloclade. This is meant for mechanical strength. Just below the chlorenchyma there is a single layer of sclerenchyma cells, which delimits the central parenchyma and the peripheral chlorenchyma. Usually each vascular bundle is capped by a well developed sclerenchymatous patch.

Vascular bundles. Around the central parenchyma the vascular bundles are found to be arranged in the peripheral region. The vascular bundles of the two corners are bigger in size than the remaining ones. Each vascular bundle is capped by a sclerenchymatous patch, and is composed of xylem, phloem and cambium. The xylem consists of metaxylem and protoxylem groups. The xylem parenchyma is also present. The phloem strand lies towards periphery. The phloem is

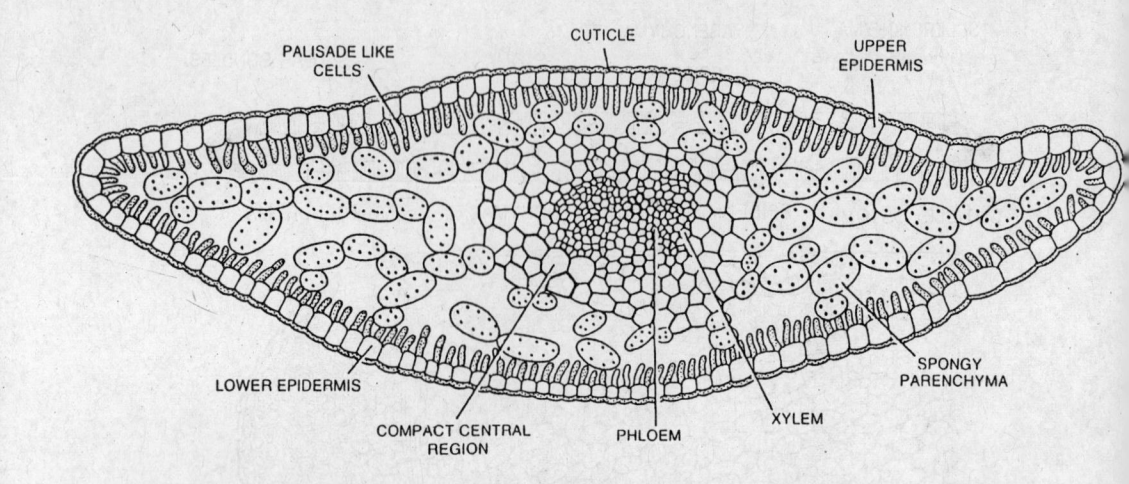

Fig. 10.32. T.S. of phylloclade of *Asparagus* (monocot).

composed of sieve tubes, companion cells and phloem parenchyma. In between xylem and phloem strands there lies the cambium. The cambium is confined to the bundle.

Parenchyma. The central region is occupied by parenchyma. It is composed of thin-walled, rounded or oval, living cells having well developed intercellular spaces. This is storage tissue.

ANATOMY OF THE PHYLLOCLADE OF *RUSCUS* (MONOCOT)

Epidermis. The phylloclade possesses two surfaces (upper and lower). Both the surfaces remain bounded by upper and lower epidermal layers. The upper epidermis consists of a single row of radially elongated epidermal cells. The epidermis is interrupted by stomata at certain places. The substomatal chambers and guard cells with chloroplasts are distinctly seen. The upper surface becomes somewhat bulged in the central region. The lower surface in the central region becomes somewhat angular, otherwise the anatomy of the lower epidermis is quite similar to that of upper epidermis.

Chlorenchyma. Immediately below the upper epidermis few layers of chlorophyllous cells are present. These cells are rounded or oval shaped, containing chloroplasts and having well developed intercellular spaces among them. In between upper and lower epidermis the well developed parenchyma is present.

Vascular system. It is well developed and represented by many amphivasal (phloem surrounded by xylem) vascular bundles. The phloem bundles remain surrounded by a sclerenchymatous sheath. The phloem consists of sieve tubes and companion cells.

SECONDARY GROWTH IN DICOTYLEDONOUS STEMS

The primary body of the plant is developed from the apical meristem. Sometimes as in monocotyledons and pteridophytes, the primary plant body is complete in itself and does not grow in thickness by cambial activity. However, in dicotyledons, the primary permanent tissues make the fundamental parts of the plant, and the further growth in thick mness is completed by cambial activity, called *secondary growth in thickness*. The tissues, formed during secondary growth are called *secondary tissues*. Secondary tissues may be of two types—the vascular tissues that are

The Stem—Primary and Secondary Structure 237

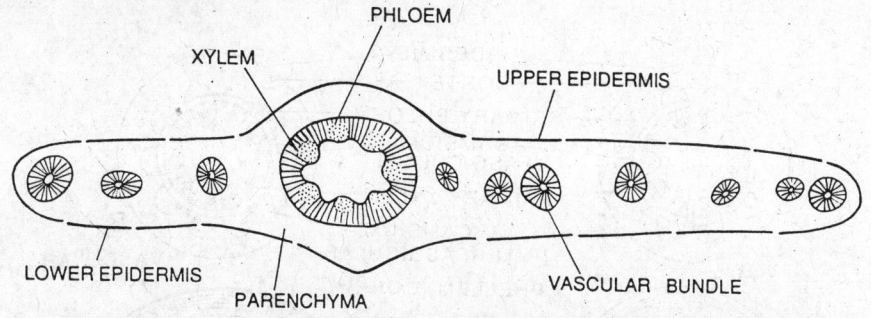

Fig. 10.33. *Ruscus* (monocot). Transection of phylloclade (diagrammatic).

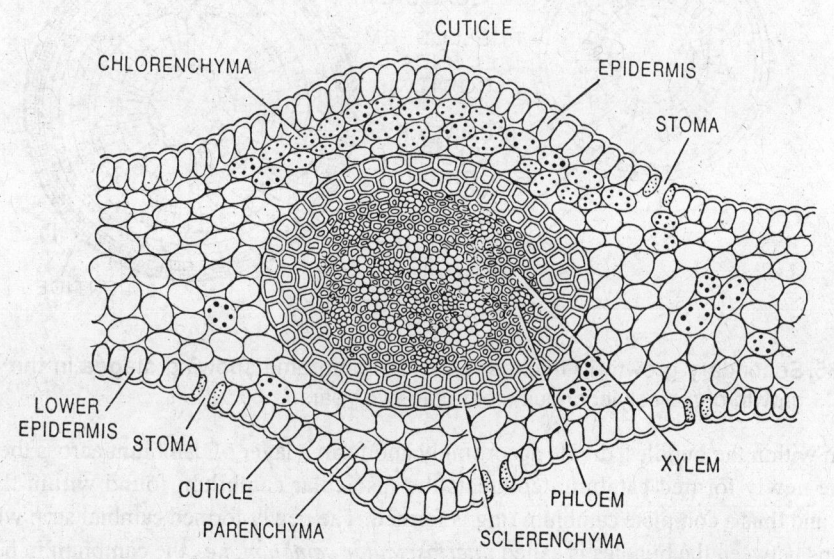

Fig. 10.34. *Ruscus* (monocot). Transection of phylloclade (detail).

called *secondary tissues*. Secondary tissues may be of two types—the vascular tissues that are developed by the true cambium, and cork and phelloderm, which are formed by phellogen or cork-cambium.

In a typical dicotyledonous stem, the secondary growth starts in the intra- and extrastelar regions. The process is as follows.

CAMBIUM

The vascular bundles of dicotoledonous stems are collateral and open, and arranged in a ring. They contain a single layer of cambium cells, which separate the xylem from the phloem, called *fascicular cambium*, i.e., the cambium of the vascular bundle, (*fascicle* = bundle). When the primary xylem and primary phloem are first differentiated there is no cambium across the pith rays or medullary rays to connect the edges of the cambium within vascular bundles. As soon as the differentiation of the first xylem and phloem of the bundles takes place, the cells of the pith or medullary rays which lie in between the edges of the

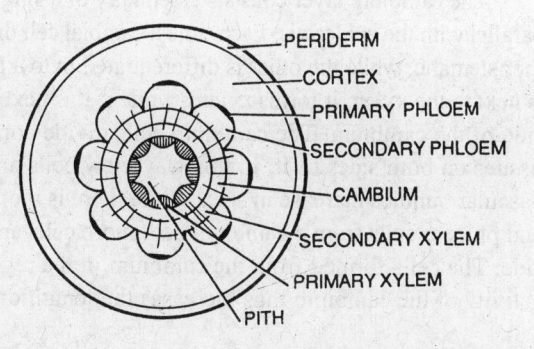

Fig. 10.35. Secondary growth. A herbaceous stem with some secondary growth.

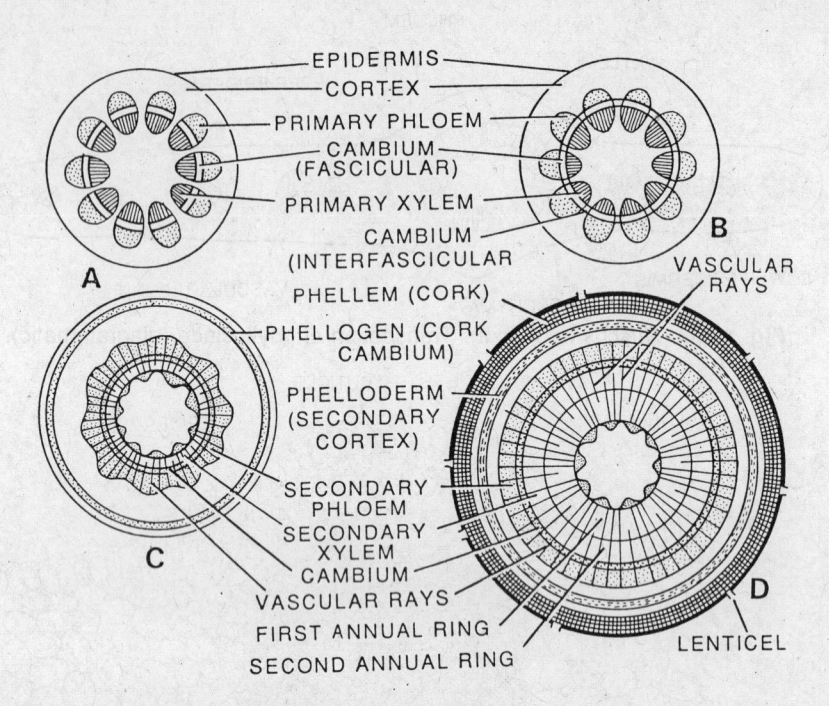

Fig. 10.36. Secondary growth in thickness. A – D, diagrams showing stages in the secondary growth of a dicotyledonous stem upto two years.

cambium within the bundles, divide accordingly and form a layer of cambium across the medullary rays. The newly formed cambium connects the fascicular cambilum found within the vascular bundles, and thus a complete cambium ring is formed. The newly formed cambial strip which occurs in the gaps between the bundles is called *interfascicular cambium*, *i.e.*, the cambium in between two vascular bundles. Thus a complete cambium ring is formed.

The cambium layer consists essentially of a single layer of cells. These cells divide in a direction parallel with the epidermis. Each time a cambial cell divides into two, one of the daughter cells remains meristematic, while the other is differentiated in to a permanent tissue. If the cell that is differentiated is next to the xylem it forms xylem, while if it is next to phloem it becomes phloem towards the outer side of the cambium. The cambium cells divide continuously in this manner producing secondary tissues on both sides of it. In this way, new cells are added to the xylem and the phloem, and the vascular bundles increase in size. While there is more or less alternation in the production of xylem and phloem cells from a cambium cell, more cells are formed on the xylem side than on the phloem side. The cells formed from the cambium in the region of the pith rays become pith-ray cells. The activity of the cambium thus increases the length of the pith rays grow equally. (See Figs. 8.2 and 8.3).

The formation of new cells from the cambium result in an enlargement of the stem that is known as the secondary thickening. The formation of new cells in secondary thickening continues throughout the life of the plant. It is in this way that the trunks of trees continue to grow in diameter. The cambium perpetuates and remains active for a considerable long period of time.

The thin-walled cells of the vascular cambium are highly vacuolate and in this respect are unlike most other meristematic cells. The electron microscopic structure reveals their highly vacuolate nature. Many ribosomes and dictyosomes, and well developed endoplasmic reticulum, are present (Srivastava, L.M., 1966).

The Stem—Primary and Secondary Structure 239

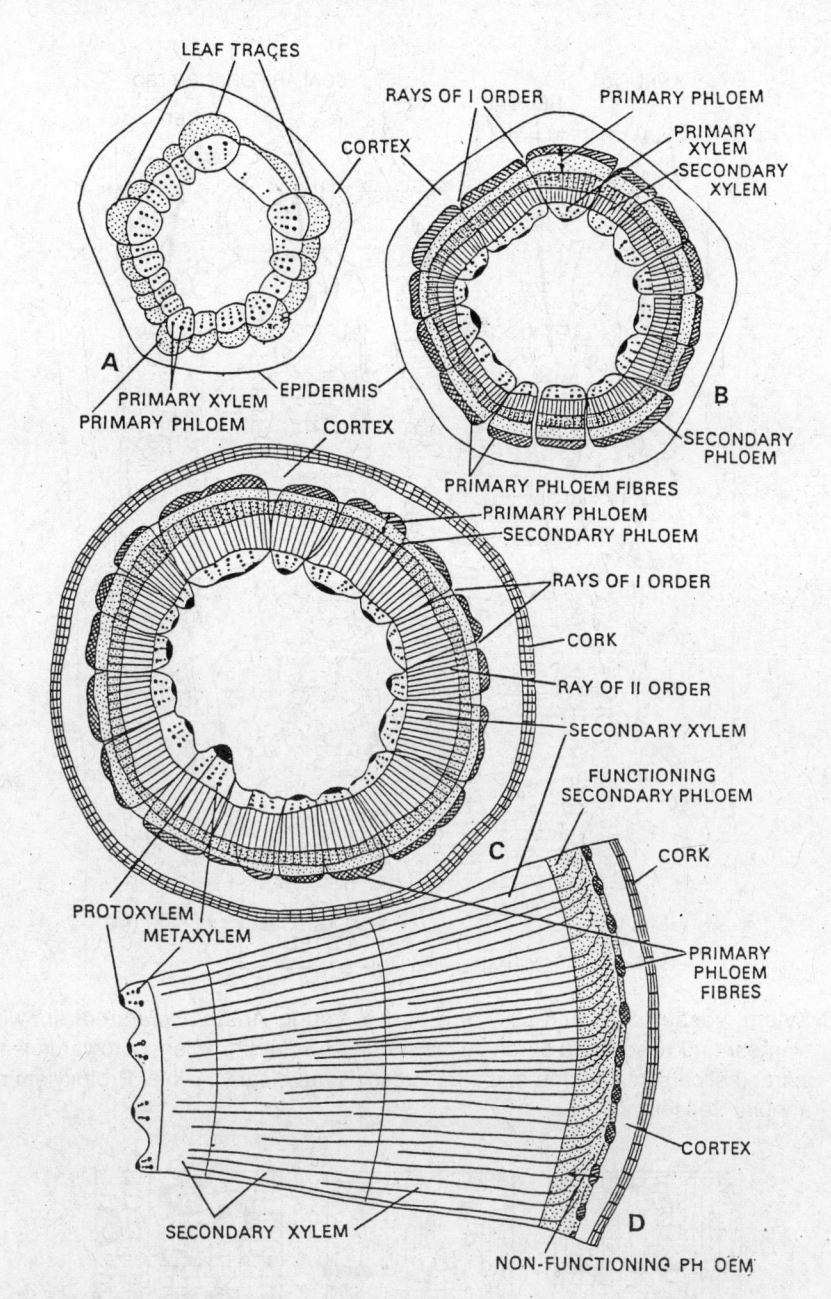

Fig. 10.37. Secondary growth. A – D, primary and secondary structure of *Prunus* stem in transverse sections.

SECONDARY XYLEM

The cambium ring cuts off new cells on its inner side are gradually modified into xylary elements, called the *secondary xylem*. This tissue serves many important functions, such as conduction of water and nutrients, mechanical support, etc. The secondary xylem of tree trunks is of great economic value, since it constitutes the timber and wood of commerce.

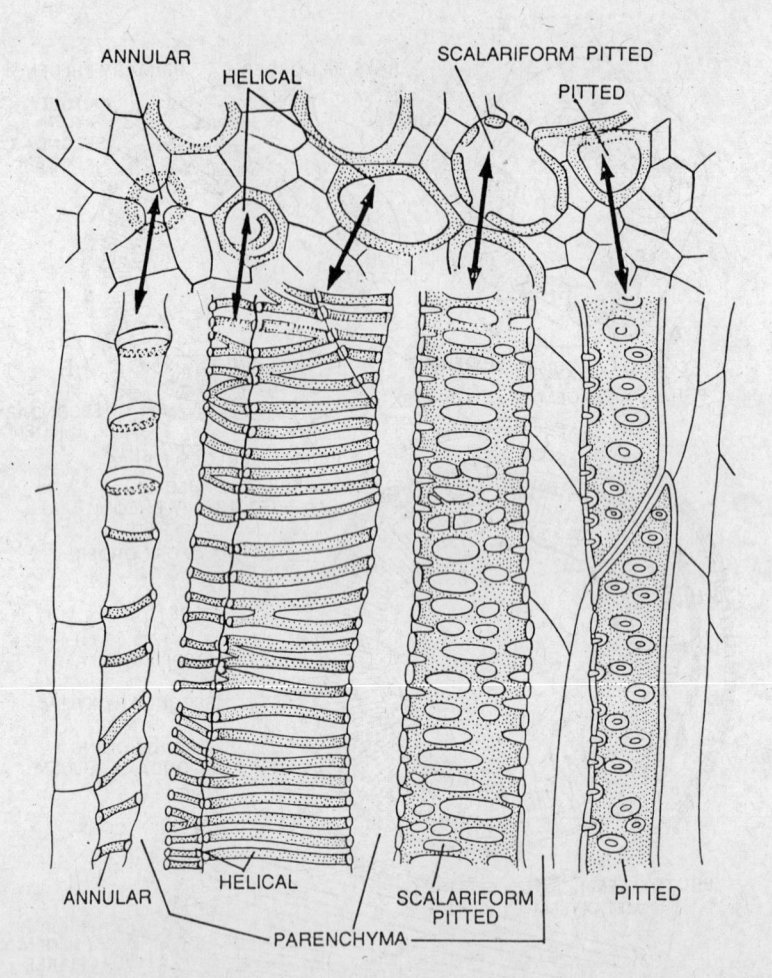

Fig. 10.38. Xylem. vessel. T.S. and L.S. of a part of young *Aristolochia* stem showing vessel element and associated parenchyma. Earlier part of the xylem is towards left. Annular, spiral (helical), scalariform and pitted vessels are clearly visible. Protoxylem consists of annular vessels.

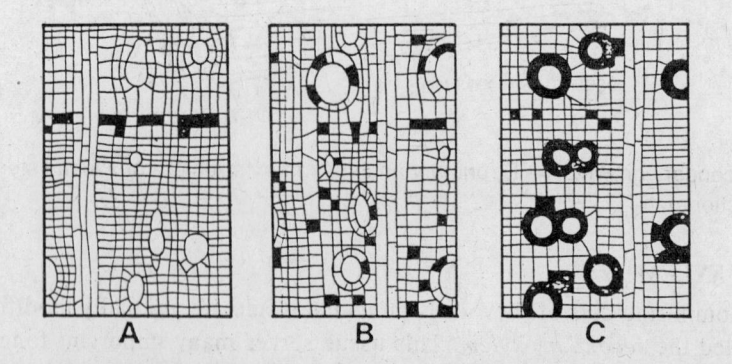

Fig. 10.39. Secondary xylem. Diagrams showing distribution of wood parenchyma, the parenchyma cells are shaded. A, terminal; B, diffuse; C, vasicentric. (After E and M).

The Stem—Primary and Secondary Structure

The secondary xylem consists of a compact mass of thick-walled cells so arranged as to form two systems—a longitudinal (vertical) and a transverse radiating system. The longitudinal system consists of elongate, overlapping and interlocked cells—tracheids, fibres and vessel elements—and longitudinal rows of parenchyma cells. All these cells possess their long axes parallel with the long axis of the organ of which they are a part.

The secondary xylem consists of scalariform and pitted vessels, tracheids, wood fibres and wood parenchyma. These elements of secondary xylem are more or less similar to those occur in primary xylem. Vessels or tracheae are most abundant and are usually shorter than those of primary xylem. Mostly the vessels are pitted. Annular and spiral tracheids and vessels are altogether absent. Xylem parenchyma cells may be long and fusiform, but sometimes they are short. They are living cells and usually meant for storage of food material (starch and fat) in them. Tannins and crystals are frequently found in these cells. Xylem parenchyma may occur either in the association of the vessels or quite independently. The fibres of secondary xylem possess thick walls and bordered pits.

DISTRIBUTION OF WOOD (XYLEM) PARENCHYMA

Wood parenchyma is distributed in *three* ways: (*i*) Terminal wood parenchyma; (*ii*) diffuse or metatracheal wood parenchyma and (*iii*) vasicentric or paratracheal wood parenchyma.

Terminal wood parenchyma. In some gymnosperm woods, wood parenchyma is absent; in other (*e.g.*, *Larix* and *Pseudotsuga*), and in some angiosperm woods (*e.g.*, *Magnolia* and *Salix*), wood parenchyma cells occur only in the last-formed tissue of the annual ring. Such woods have *terminal wood parenchyma.*

Diffuse or metatracheal wood parenchyma. Where parenchyma occurs not only in this region, but also remains scattered throughout the annual ring, some of the cells lying among the tracheids, and fibre-tracheids the plant has *diffuse* or *metatracheal wood parenchyma* (*e.g.*, in *Malus, Quercus, Diospyros* etc.).

Vasicentric or paratracheal wood parenchyma. Where parenchyma occurs at the edge of the annual ring and elsewhere only about vessels and does not occur isolated among tracheids and fibres, the plant possesses *vasicentric* or *paratracheal wood parenchyma* (*e.g.*, in *Acer, Fraxinus* etc.).

Xylem rays. The xylem rays or wood rays, extend radially in the secondary xylem. They are strap or ribbon like. They originate from the ray initials. The xylem rays run as a continuous band to the secondary phloem through the cambium, thus forming a continuous conducting system. All vascular rays are initiated by the cambium and, once formed, are increased in length indefinitely by the cambium. Commonly these rays are known as medullary rays, or pith rays, on the basis of their similarity and parenchymatous nature with the pith rays of herbaceous dicotyledonous stems. These radial rays may be best called *vascular rays*, as these rays are of vascular tissue partly of xylem and partly of phloem.

The xylem rays traverse in the secondary xylem and establish communication with the living cells of the vascular tissue. In gymnosperm wood where no wood parenchyma is present, every tracheid is in direct contact with at least one ray. Vessels also in their longitudinal extent, come into contact with many rays. In herbaceous stems, such as of *Ranunculus*, where vascular bundles are separated by projecting parenchymatous wedges, and in vines, such as *Clematis*, where the bundles are separated by bands of secondary parenchyma, vascular rays are not found. The xylem rays help in the exchange of gases. They also aid in the conduction of water and food from phloem to the cambium and xylem parenchyma.

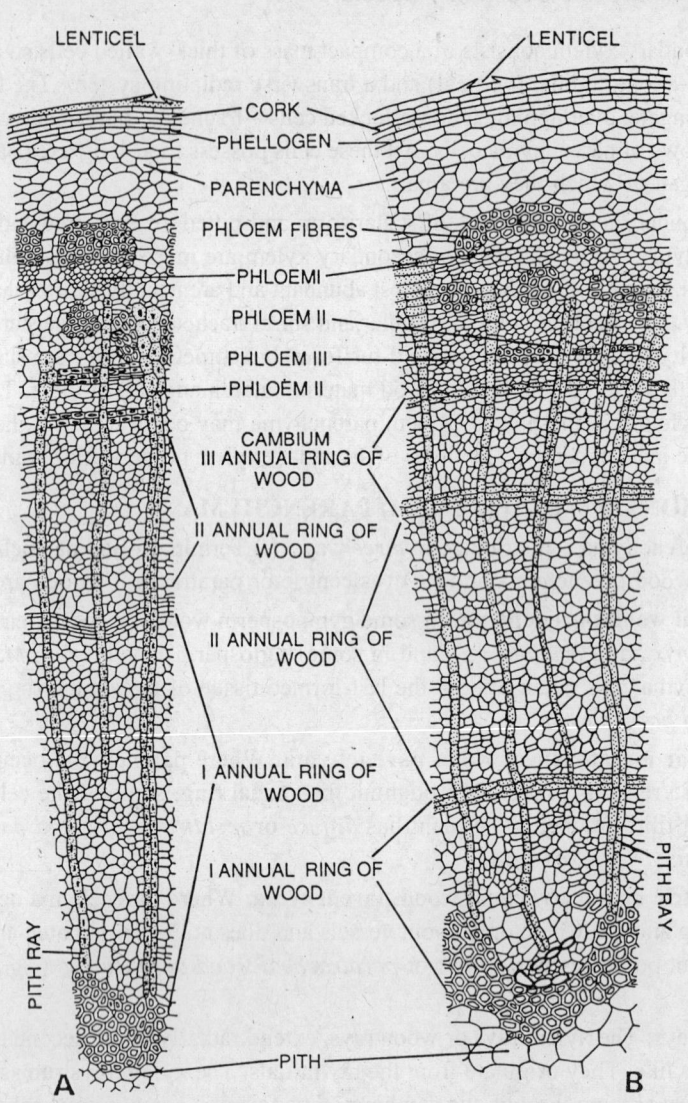

Fig. 10.40. Stem—secondary structure. A, cross section of a portion of two-year old stem of *Liriodendron tulipifera*, B, cross section of a portion of three-year old stem of *Liriodendron tulipifera* (tulip tree) with three annual rings of wood.

ANNUAL RINGS OR GROWTH RINGS.

The secondary xylem in the stems of perennial plants commonly consists of concentric layers, each one of which represents a seasonal increment. In transverse section of the axis, these layers appear as rings, and are called *annual rings* or *growth rings*. They are commonly termed as annual rings because in the woody plants of temperate regions and in those of tropical regions where there is an annual alternation of growing and dormant period, each layer represents the growth of one year. The width of growth rings varies greatly and depends upon the rate of the growth of tree. Unfavourable growing seasons produce narrow rings, and favourable seasons wide ones. Annual or growth rings are characteristic of woody plants of temperate climates. Such rings are weakly developed in tropical forms except where there are marked climate changes such as distinct moist and dry seasons. Annuals and herbaceous stems show, naturally, but one layer.

The Stem—Primary and Secondary Structure

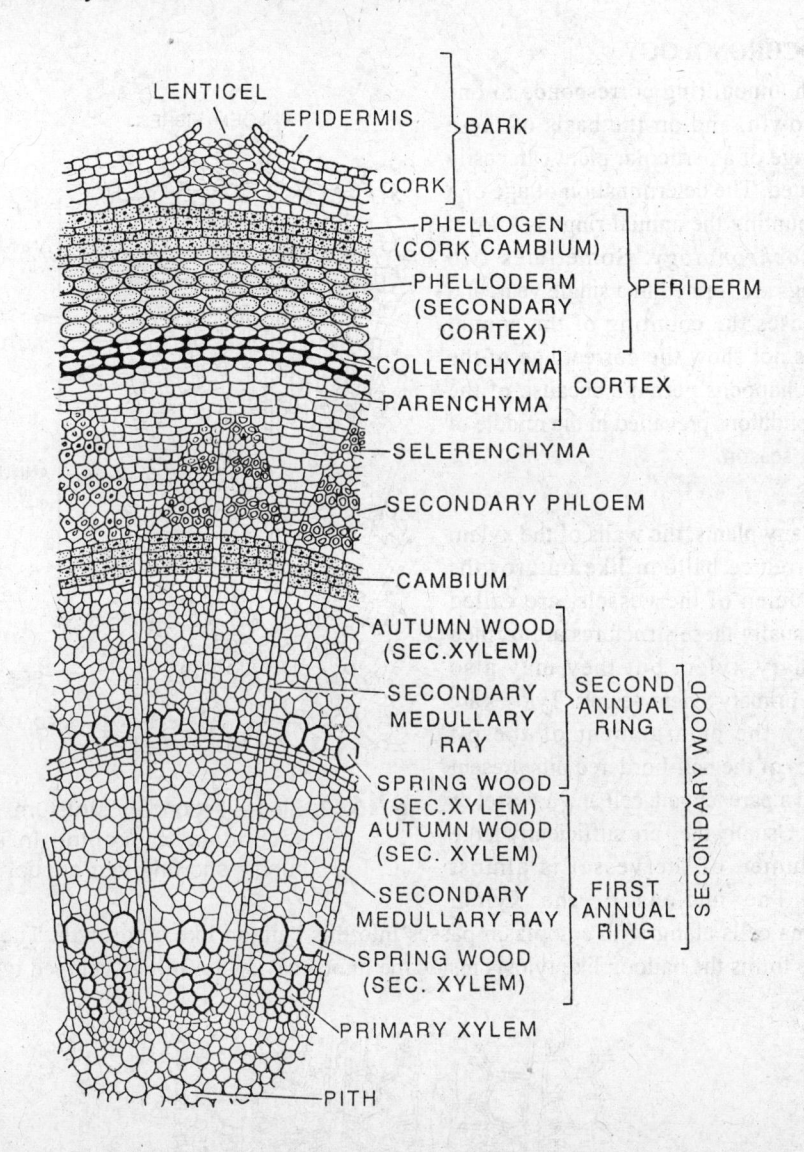

Fig. 10.41. Secondary growth in thickness. T.S. of a two-year old dicotyledonous stem.

In regions with a pronounced cold season, the activity of the cambium takes place only during the spring and summer seasons thus giving rise the growth in diameter of woody plants. The wood of one season is sharply distinct from that of the next season. In spring or summer the cambium is more active and forms a greater number of vessels with wider cavities. As the number of leaves increases in the spring season, additional vessels are needed for the transport of sap at that time to supply the increased leaves. In winter or autumn season, however, there is less need of vessels for sap transport, the cambium is less active and gives rise to narrow pitted vessels, tracheids and wood fibres. The wood developed in the summer or spring season is called *spring wood* or *early wood*, and the wood formed in winter or autumn season is known as *autumn wood* or *late wood*. However, the line of demarcation is quite conspicuous between the late wood of one year and the early wood of next year. An annual ring, therefore, consists of two parts—an inner layer, early wood, and an outer layer late wood.

DENDROCHRONOLOGY

Each annual ring corresponds to one year's growth, and on the basis of these rings the age of a particular plant can easily be calculated. The determination of age of a tree by counting the annual rings is known as *dendrochronology*. Sometimes two annual rings are formed in a single year, and in such cases the counting of the annual rings does not show the correct age of the tree. This happens perhaps because of the drought conditions prevailed in the middle of a growing season.

TYLOSES

In many plants, the walls of the xylem vessels produce balloon like outgrowths into the lumen of the vessels, are called *tyloses*. Usually these structures are formed in secondary xylem but they may also develop in primary xylem vessels. Tyloses are formed by the enlargement of the pit membranes of the half-bordered pits present in between a parenchyma cell and a vessel or a tracheid. Usually they are sufficiently large and the lumen of the vessel is almost blocked. The nucleus of the xylem parenchyma cells along with cytoplasm passes into this balloon like outgrowth. The delicate pit membrane forms the balloon like tylosis inside the lumen cavity. In fully developed tyloses, starch

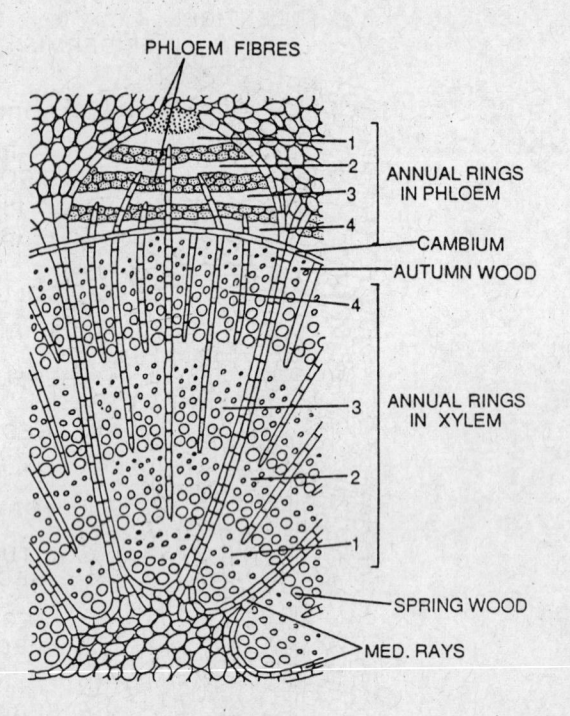

Fig. 10.42. Stem—secondary structure. Diagram of secondary thickening in a vascular bundle showing four annual rings.

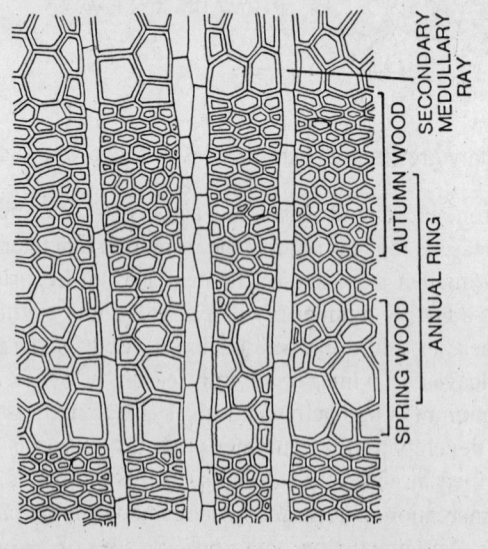

Fig. 10.43. Annual rings. An annual ring in sectional view (magnified).

The Stem—Primary and Secondary Structure 245

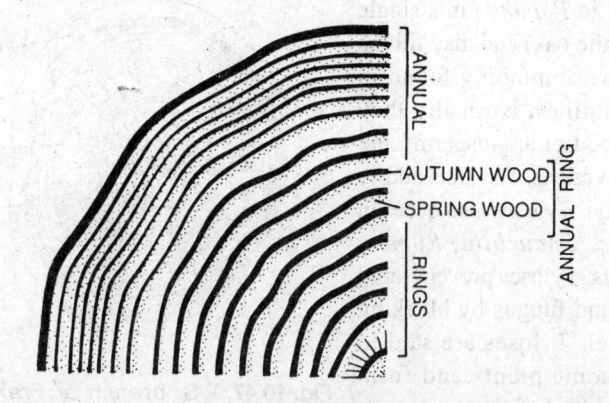

Fig. 10.44. Annual rings (growth rings)—cut surface of a stem showing annual rings.

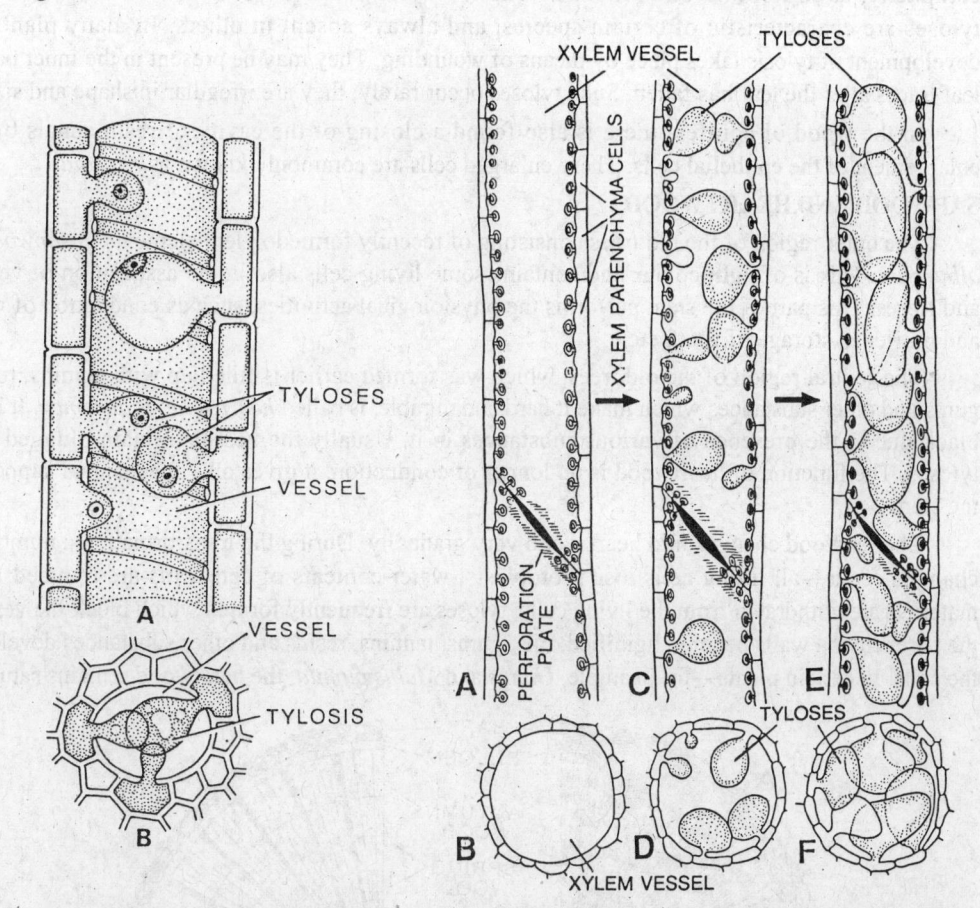

Fig. 10.45. Tyloses. A, L.S. of vessel with tyloses; B, T.S. of vessel with tyloses.

Fig. 10.46. Tyloses. A – F, development of tyloses in xylem vessels depicted in L.S. and T.S.

crystals, resin gums and other substances are found, but they are not found very frequently. The wall of tylosis may remain thin and membranous or very rarely it becomes thick and even lignified. The tylosis may remain very small or sufficiently large in size as the case may be. They may be one or few in number (*e.g.,* in *Populus*) in a single cell or many (*e.g.,* in white oak) and may fill the complete cell. They are commonly found in many angiospermic families. Normally they develop in the heart wood of angiosperms and block the lumen of the vessels, and thus add to the durability of the wood. Tyloses also occur in the vessels of *Coleus, Cucurbita, Rumex, Asarum* and *Convolvulus*. Tyloses prevent rapid entrance of water, air and fungus by blocking the lumen of the vessel. Tyloses are said to undergo division in some plants and form *multicellular tissue*, which fills the lumen compactly, as in *Robinia* and *Maclura*. The tyloses are characteristic of certain species, and always absent in others. In many plants the development of tylosis takes place by means of wounding. They may be present in the inner part of leaf traces after the leaf has fallen. Such tyloses occur rarely; they are irregular in shape and size.

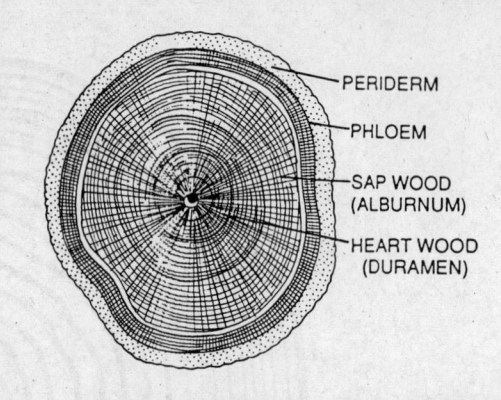

Fig. 10.47. T.S. branch of *Fraxinus*, showing heart and sapwood. (diagrammatic).

In the wood of conifers there is also found a closing of the cavity of resin canals by the enlargement of the epithelial cells. These enlarged cells are commonly known as *tylosoids*.

SAPWOOD AND HEARTWOOD

The outer region of the old trees consisting of recently formed xylem elements in *sapwood* or *alburnum*. This is of light colour and contains some living cells also in the association of vessels and fibres. This part of the stem performs the physiological activities, such as conduction of water and nutrients, storage of food, etc.

The central region of the old trees, which was formed earlier is filled up with tannins, resins, gums and other substances which make it hard and durable, is called *heartwood* or *duramen*. It looks black due to the presence of various substances in it. Usually the vessels remain plugged with tyloses. The function of heartwood is no longer of conduction, it gives only mechanical support to the stem.

The sapwood changes into heart wood very gradually. During the transformation a number of changes occur—all living cells lose protoplasts; water contents of cell walls are reduced; food materials are withdrawn from the living cells; tyloses are frequently formed which block the vessels; the parenchyma walls become lignified; oils, gums, tannins, resins and other substances develop in the cells. In certain plants—for example, *Ulmus* and *Malus pumila*, the heartwood remains saturated

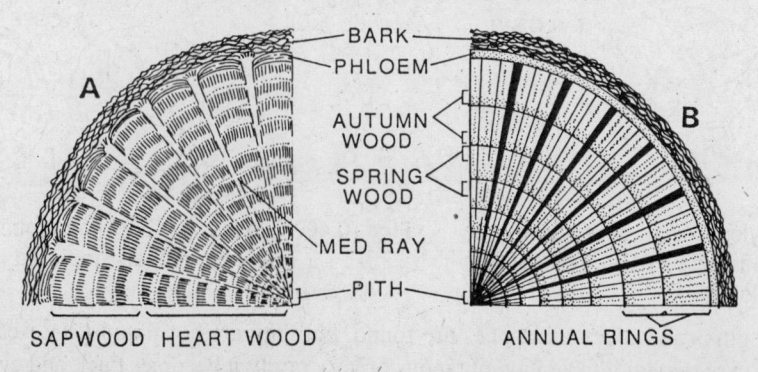

Fig. 10.48. Transections of old dicot stems—left, sector showing heartwood and sapwood; right, sector showing annual rings.

The Stem—Primary and Secondary Structure

with water; in other plants, for example, in *Fraxinus* the heartwood may become very dry. The oils, resins and colouring materials infiltrate the walls, and gums and resins may fill the lumina of the cells. In *Diospyros* and *Swietenia*, the cell cavities are filled with a dark-coloured gummy substance. The colour of heartwood, in general, is the result of the presence of these substances. Generally the heartwood is darker in colour than sapwood. However, in some genera, such as *Betula*, *Populus*, *Picea*, *Agathis* the heartwood is hardly darker in colour than the sapwood.

The proportion of sapwood and heartwood is highly variable in different species. Some trees do not have clearly differentiated heartwood (*e.g.*, *Populus*, *Salix*, *Picea*, *Abies*), others possess thin sapwood (*e.g.*, *Robinia*, *Morus*, *Taxus*), the still others possess a thick sapwood (*e.g.*, *Acer*, *Fraxinus*, *Fagus*).

From economic point of view, heartwood is more useful than sapwood. Heartwood, as timber, is more durable than sapwood, because the reduction of food materials available for pathogens by the absence of protoplasm and starch. The formation of resins, oils and tannins, and the blocking of the vessels by tyloses and gums, render the wood less susceptible to attack by the organisms of decay. The haemotoxylin is obtained from the heartwood of *Haematoxylon campechianum*. Because of the absence of resin, gums and colouring substances, sapwood is preferred for pulpwood, and for wood to be impregnated with preservatives.

SECONDARY PHLOEM

The cambial cells divide tangentially and produce secondary phloem elements towards outside of it. Normally, the amount of secondary phloem is lesser than the amount of secondary xylem. In most of the dicotyledons; usually the primary phloem becomes crushed and functionless and the secondary phloem performs all physiological activities for sufficiently a long period of time.

This is a complex tissue made up of various types of cells having common origin in the cambium. These cells are quite similar to the cells of primary phloem. However, the secondary phloem

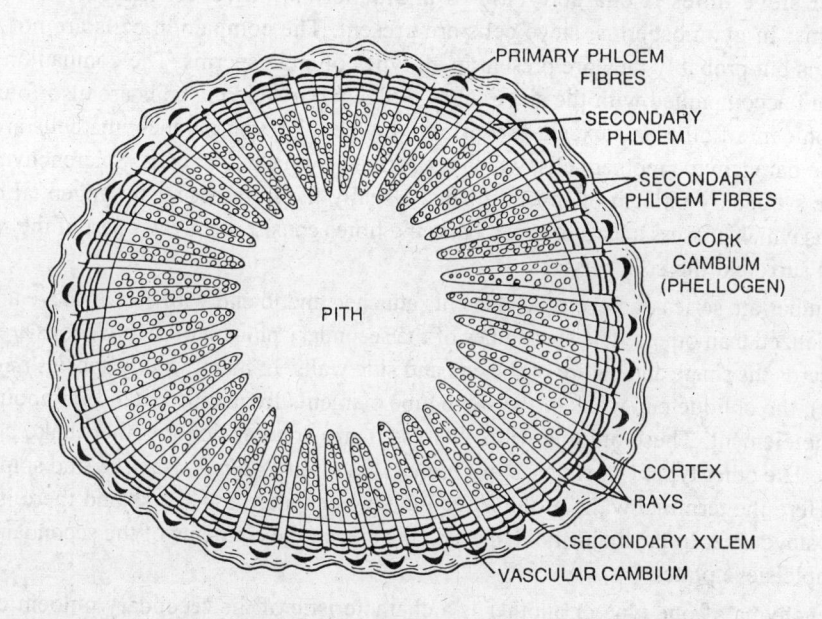

Fig. 10.49. T.S. of grapevine (*Vitis vinifera*) stem, showing arrangement of the vascular tissues. The epidermis, the cortex and the primary phloem were cut off by the activity of the cork cambium which interpolated a layer of cork between primary and secondary phloem. (After K. Esau).

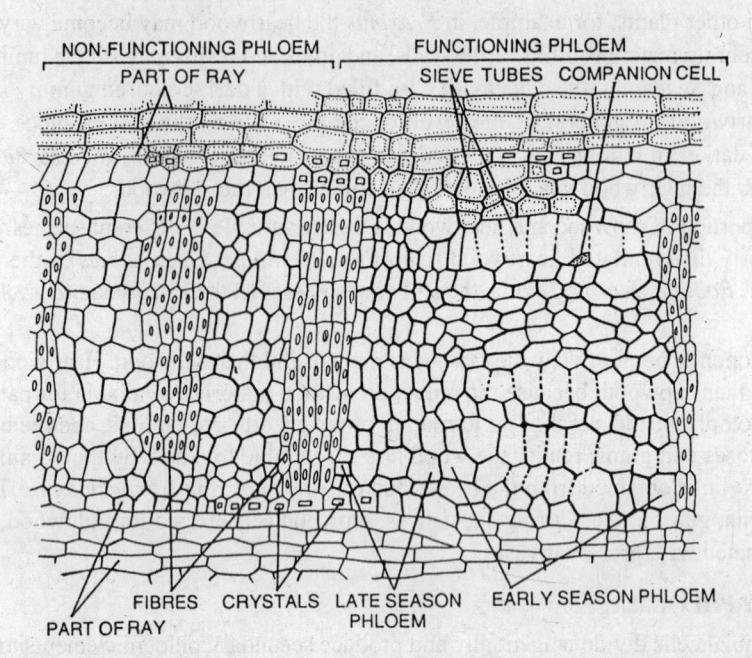

Fig. 10.50. Secondary phloem. T.S. of the secondary phloem of *Vitis vinifera* (grapevine). (After Esau).

possesses a more regular arrangement of the cells in radial rows. The sieve tubes are comparatively larger in number and posses thicker walls. The elements of secondary phloem are sieve tubes, companion cells, phloem parenchyma and phloem ray cells. Sometimes sclerenchyma is also found. Presence of sieve tubes is characteristic of angiosperms, however, they are not found in gymnosperms. In gymnosperms, sieve cells are present. The companion cells are not found in gymnosperms but probably they are present in all types of angiosperms. The companion cells are usually found accompanied with the sieve tubes. Phloem parenchyma cells are also found in the secondary phloem of all plants except few primitive types. Phloem parenchyma cells are formed directly from parenchyma mother cells, which are formed from cambial cells. Sclerenchyma is also found in the secondary phloem of several plants. Usually the fibres occur in tangential bands. In certain plants which possess a hard or tough bark, the fibres consist the greater part of the secondary phloem and surround the softer tissues.

Sieve tubes are series of sieve-tube elements attached end to end with certain sieve areas more highly specialized than others. The sieve tubes of the secondary phloem of dicotyledons are of many types as regards the shape and nature of the end and side walls. In many woody species (*e.g.*, *Carya cordiformis*), the oblique end walls of the sieve tube elements frequently extend for about half the length of the element. These oblique walls possess many areas which together make *compound sieve plates*. The other type, *i.e.*, *simple sieve plate* is found in *Robinia*, *Maclura* and some species of *Ulmus*. Here the terminal walls of the sieve-tube elements are transverse and there is a single specialized sieve area. In the majority of species, the sieve tube elements of the secondary phloem possess simple sieve plates.

Sclerenchyma of one type or another is a characteristic of the secondary phloem of several species. Fibres occur frequently in definite tangential bands (*e.g.*, in *Liriodendron* and *Populus*). In *Cephalanthus*, the fibres are found singly. However, in *Carya cordiformis*, the fibres constitute the greater part of the secondary phloem and surround the groups of softer tissues. All conditions have been reported in gymnosperms. The phloem of *Pinus strobus* lacks sclerenchyma; well developed tangential bands of fibres are found to be present in *Juniperus*, and large masses of sclereids are present in *Tsuga*. In *Thuja occidentalis*, the fibres are arranged in uniseriate tangential rows. These rows of fibres alternate with rows of sieve cells and phloem parenchyma.

In *Platanus* and *Fagus* sclereids are the only type of sclerenchyma present in the phloem. The sclereids are found abundantly in the older, living, but nonconducting phloem of the woody plants.

Phloem rays. The pholem rays are usually present in the vascular tissues developed by the cambium. The vascular rays are formed in the cambium and develop on either side of it with the secondary xylem and secondary phloem of which they are a part. The phloem rays may be one to several cells in width. Normally they are of uniform width throughout their length. They may increase in width outwardly, the increase being due to the multiplication of the cells or to the increase in size of cells toward the outer end of the ray. The phloem rays may be one cell wide (*e.g.*, in *Castanea* and *Salix*), two or three cells wide (*e.g.*, in *Malus pumila*) or many cells wide (*e.g.*, in *Robinia* and *Liriodendron*). However, in oaks there are two types of phloem rays—one very broad and the other uniseriate.

Commonly the phloem ray cells in woody plants, as seen in transverse section, are rectangular and radially elongated. In herbaceous plants, commonly the ray cells are globose. In *Cephalanthus*, *Agrimonia* and *Potentilla* the ray cells closely resemble the phloem parenchyma cells. All phloem ray cells are parenchymatous with active protoplasm, but as they become older many of them become sclereids.

A special type of ray cell known as *albuminous cell* is found in gymnosperms. These albuminous cells are found to be situated at the upper and lower margins of the phloem rays. The albuminous cells differ from the ordinary ray cells both structurally and functionally. They are joined directly with the sieve cells by sieve areas. They do not contain starch, and are of much greater vertical diameter than the normal ray cells. They retain their protoplasts as long as the sieve cells with which they are connected function. It is thought that they function like companion cells of angiosperms.

Seasonal rings in secondary phloem. The tissues of the secondary phloem are generally arranged in definite tangential bands. These layers of tissue have the appearance of annual rings. However, these ring like bands do not possess definite seasonal limits like those of secondary xylem, because there is no sharp distinction between the phloem cells formed in the early and late growing season. Seasonal formation of sclerenchyma bands may exist, but this is not constant feature. In tropical plants new layers of phloem and xylem are formed with each period of new growth.

Function. The functions of secondary phloem are normally the same as that of primary phloem. The various cells of secondary phloem are structurally adapted for the function of translocation of food. The sieve tubes, companion cells and some phloem parenchyma cells are especially adapted for lengthwise conduction, and certain phloem rays help in horizontal conduction to and from the xylem and the cambium. Some of the phloem parenchyma cells in some plants act as storage tissue of starch, crystals and other organic materials.

Economic importance. The secondary phloem of various trees and shrubs of the Malvaceae, Tiliaceae, Moraceae has provided *bast fibres* for economic purposes. The tapa cloth of Pacific islands is composed of mainly of phloem fibres. Tannin obtained from the secondary phloem of various plants is utilized for the preparation of spices and drugs. Secretory canals are abundantly found in the secondary phloem, and the secretions are of much economic value—such as rubber is obtained from the latex of *Hevea brasiliensis*, and resins from various gymnosperms.

PERIDERM

Due to continued formation of secondary tissues, in the older stem, and roots, however, the epidermis gets stretched and ultimately tends to rupture and followed by the death of epidermal cells and outer tissues, and a new protective layer is developed called *periderm*. The formation of periderm is a common phenomenon in stems and roots of dicotyledons and gymnosperms that increase in thickness by secondary growth. Structurally, the periderm consists of three parts—1. a meristem known as *phellogen* or *cork cambium*, 2. the layer of cells cut off by phellogen on the outer side, the *phellem* or *cork*, and 3. the cells cut off by phellogen towards inner side, the *phelloderm*.

The periderm appears on the surface of those plant parts that possess a continuous increase in thickness by secondary growth. Usually the periderm occurs in the roots and stems and their branches in gymnosperms and woody dicotyledons. It occurs in herbaceous dicotyledons, sometimes limited to the oldest parts of stem or root.

Phellogen. In contrast to the vascular cambium, the phellogen is relatively simple in structure, and composed of one type of cells. The cells of phellogen appear rectangular in cross-section, and somewhat flattened radially. Their protoplasts are vacuolated and may contain tannins and chloroplasts. Except in the lenticels, intercellular spaces lacking.

When we consider the place of origin of the meristem forming the periderm, it becomes necessary to distinguish between the first periderm, and the subsequent periderms, which arise beneath the first and replace it as the axis increases in circumference. In most stems the first phellogen arises in the subepidermal layer. In a few plants the phellogen arises in the epidermal cells (*e.g., Nerium, Pyrus*). Sometimes only a part of the phellogen is developed from epidermis while the other part arises in subepidermal cells (*e.g., Pyrus*). In some stems the second or third cortical layer initiates the development of periderm (*e.g., Robinia, Aristolochia, Pinus, Larix,* etc.). In still other plants the phellogen arises near the vascular region or directly in the phloem (*e.g.,* in *Caryophyllaceae, Cupressaceae, Ericaceae, Punica, Vitis,*

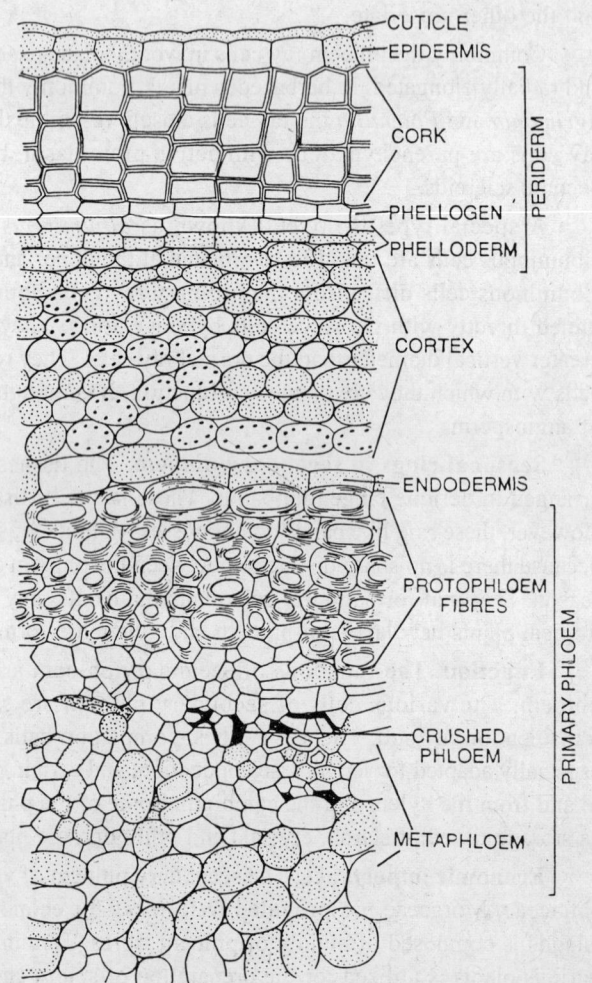

Fig. 10.51. Secondary growth in dicot stem. T.S. of *Prunus* stem showing periderm, cortex and primary phloem details.

The Stem—Primary and Secondary Structure

etc.). If the first periderm is followed by the formation of others, these are formed repeatedly, in successively deeper layers of the cortex or phloem.

At the time of the beginning of the development of a phellogen in epidermal cells, the protoplasts lose their central vacuoles and the cytoplasm increases in amount and becomes more richly granular. As soon as this initial layer develops, it divides tangentially and, to a lesser extent radially, in the similar way as division takes place in true cambium. The derivative cells are normally arranged in radial rows.

Generally, several to many times as many cells are cut off toward the outside (phellem-cork cells) as toward the inside (phelloderm). Phelloderm cells are few or absent; rarely phelloderm is greater in amount than phellem.

Phellem (cork cells). The cells that constitute phellem are commonly known as *cork cells*. They are like the phellogen cells from which they are derived. As seen in tangential section, they are polygonal and uniform in shape, and often radially thin as seen in cross section of the stem. The cells of the commercial cork (*Quercus suber*) are radially elongated as seen in transverse section. In the periderm of *Betula* and *Prunus*, the cork cells are elongated tangentially as seen in cross-section. There are no intercellular spaces among cork cells.

Commercial cork. The development of the periderm layers in the cork oaks (*Quercus suber*) is of special interest. The

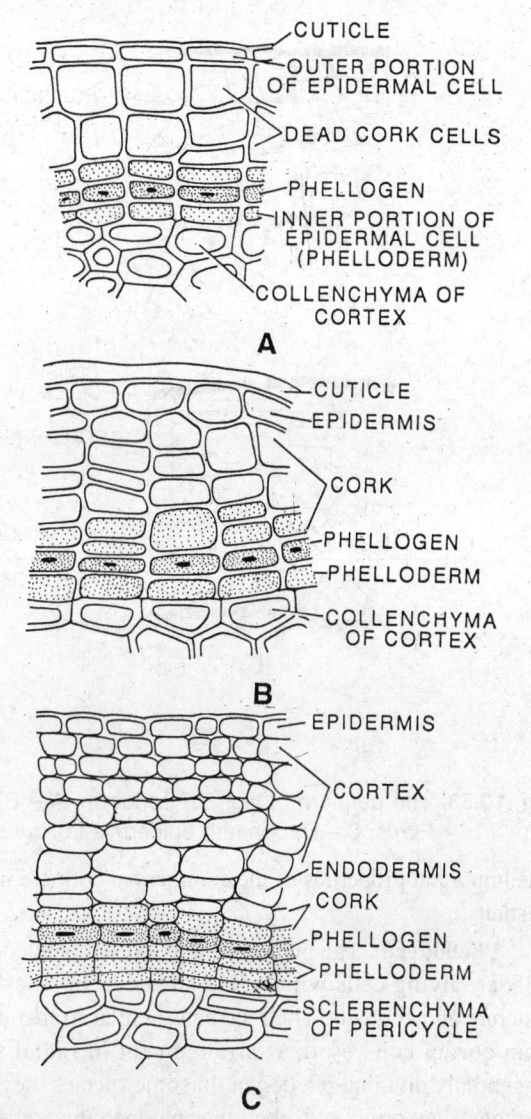

Fig. 10.52. Phellogen. Origin and its development. A, epidermal; B, hypodermal; C, deep-seated.

ability of the plant produce phellogen in deeper layers when the superficial periderm is removed is utilized in the production of commercial cork from the cork oak (*Quercus suber*). At the age of about twenty years, when the tree is about 40 cm in circumference, this outer layer, known as virgin cork is removed by stripping to the phellogen. The exposed tissue dries out to about 1/8 in. in depth. A new phellogen is established beneath the dry layer and rapidly produces a massive cork of a better quality than the first. After nine or ten years the new cork layer has attained sufficient thickness to be commercially valuable and is in turn removed. Of course, this cork is of better quality than the virgin cork, but of inferior quality than the cork obtained at the third and subsequent strippings. These strippings take place at intervals of about nine years until the tree is 150 or more years old. After the successive strippings the new phellogen layers develop at greater depth in the living tissue. The cortex is lost after few strippings and the subsequent cork layers are formed in the secondary phloem.

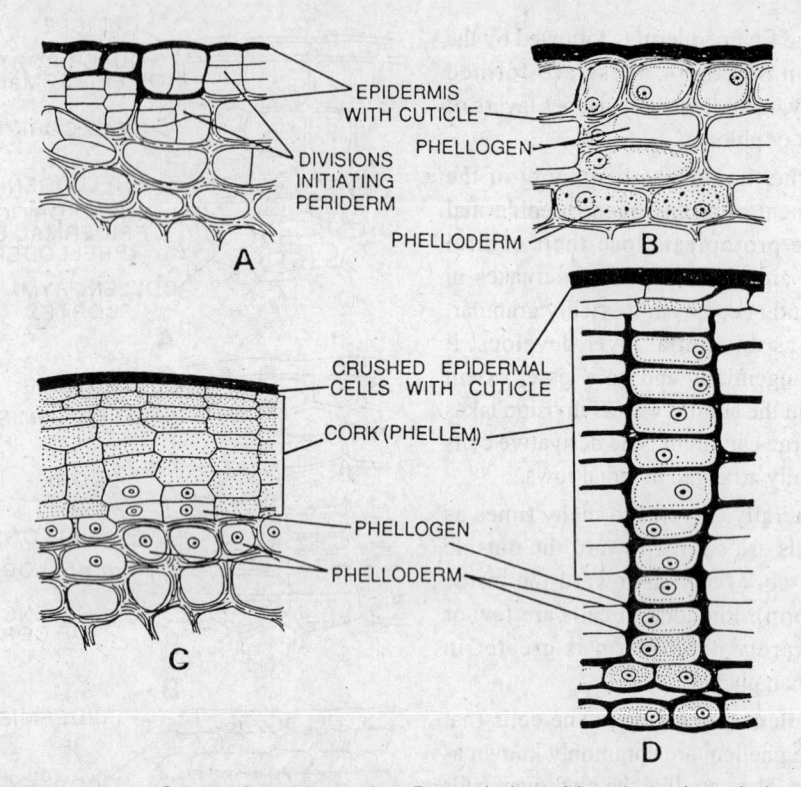

Fig. 10.53. The periderm. Origin of periderm. A – B, partly in epidermis and partly beneath it, in *Pyrus*; C – D, beneath epidermis in *Prunus*.

The important properties of the commercial cork are its imperviousness, its lightness, toughness and elasticity.

Phelloderm. The phellogen cuts off the phelloderm cells towards inner side. The phelloderm cells are living cells with cellulose walls. In most plants, they resemble cortical cells in wall structure and contents. Their shape is similar to that of phellogen cells. They may be distinguished from cortici cells by their arrangement in radial series resulting from their origin from the tangentially dividing phellogen. In some species they act as photosynthetic tissue and aid in starch storage. They are pitted like other parenchyma cells. Occasionally, the sclereids and other such specialized cells occur in phelloderm. The term *secondary cortex* is sometimes applied to phelloderm, which does not seem to be appropriate.

BARK

The term *bark* is commonly applied to all tissues outside the vascular cambium of the stem, in either primary or secondary state of growth. In this way, bark includes primary phloem and cortex in stem with primary tissues only, and primary and secondary phloem, cortex and periderm in stem with secondary tissues. This term is also used to denote the tissue that is accumulated on the surface of the stem as a result of the activity of cork cambium. As the periderm develops, it becomes separated, by a non-living layer of cork cells from the living tissues. The tissue layers thus separated become dead. The term bark in restricted sense is applied to these dead tissues together with the cork layers. In wider sense the term is applied to denote the tissues outside the vascular cambium. However, the term bark is loose and non-technical.

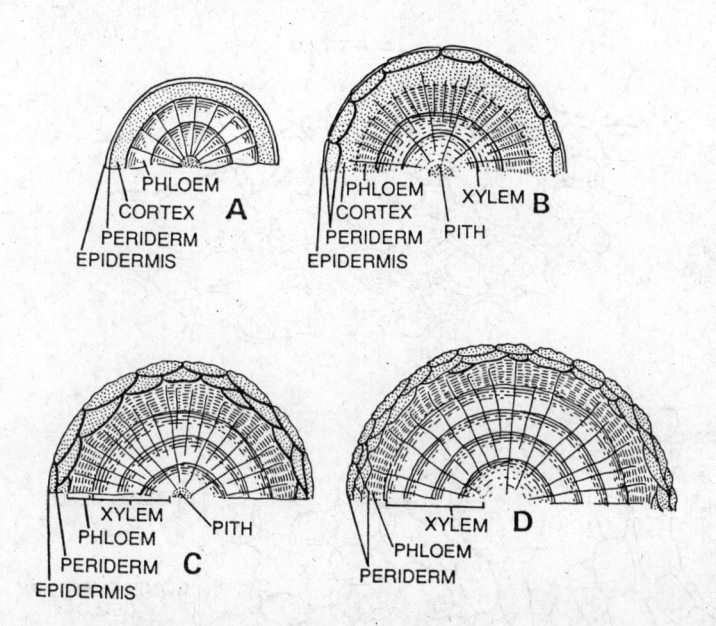

Fig. 10.54. The periderm. Diagrams showing the position and extent of successively formed periderm layers. A, a one-year-old twig, the first periderm layer, a complete cylinder formed beneath the epidermis; B, a two-year-old twig, the epidermis and first periderm ruptured; C, a three-year-old stem, the outer tissues weathered away and more periderm layer formed deeper in the stem; D, a four-year-old stem, the cortex and outer secondary phloem with their periderm layers largely weathered away.

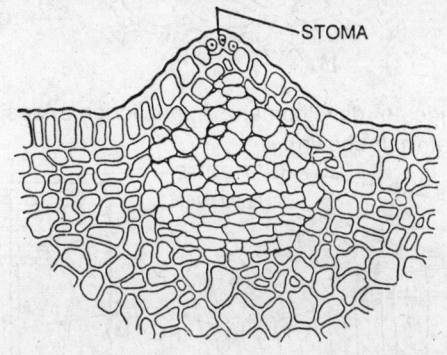

Fig. 10.55. The lenticel. Beginning of lenticel formation under a stoma of *Morus alba* (mulberry).

RHYTIDOME

In most of plants, as soon as the first phellogen ceases to function, second phellogen develops in the tissue below the first one. In this way additional layers of periderm are formed in the progressively deeper regions of the stem, thus new phellogen layers arise in deeper regions of the cortex which may exceed even upto phloem. As the phellogen arises in deeper region and cuts cork cells or phellem towards outside, all the living cells outside the phellogen do not get water supply and nutrients, and become dead. These dead tissues formed outside the phellogen constitute the *rhytidome*.

In some rhytidomes parenchyma and soft cork cells predominate whereas others contain large amounts of fibres usually derived from the phloem. The manners in which the successive layers of

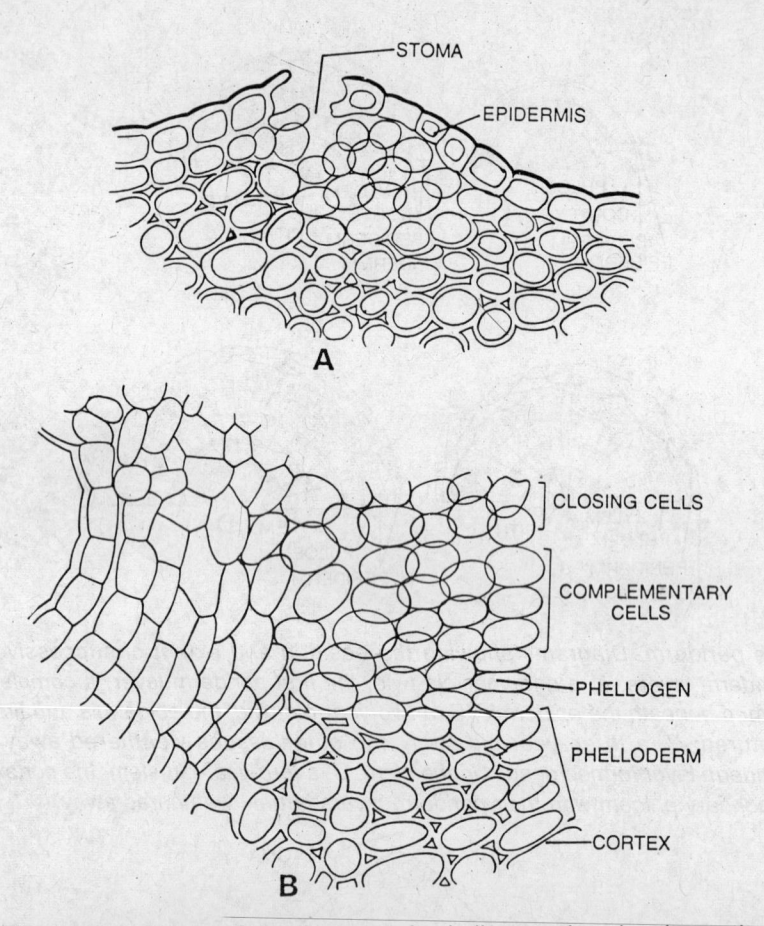

Fig. 10.56. The lenticel. Stages of development. A, phellogen arises just beneath stoma; B, well developed lenticel (half shown).

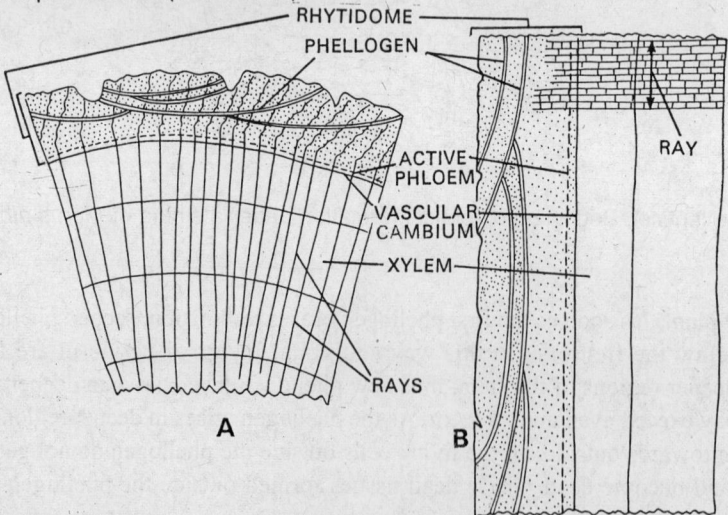

Fig. 10.57. The Periderm. A and B, Transverse and longitudinal sections of part of stem showing rhytidome and its location with reference to vascular tissues; here the rhytidome is composed of periderm and non-living secondary phloem.

The Stem—Primary and Secondary Structure

periderm originate possesses a characteristic effect upon the appearance of the rhytidome. When the sequent periderms develop as overlapping scale-like layers, the outer tissue breaks up into units related to the layers of periderm, and thus formed outer bark is termed scale-bark. On the other hand, if the phellogen arises around the whole circumference of the stem, a ring bark is formed, which shows the separation of hollow cylinders or rings from the stem.

LENTICELS

Usually in the periderm of most plants, certain areas with loosely arranged cells have been found, which possess more or less raised and corky spots where the underneath tissues break through the epidermis. Such areas are universally found on the stems of woody plants. These broken areas are called the *lenticels*. Wutz (1955) defined a *lenticel as a small portion of the periderm where the activity of the phellogen is more than elsewhere, and the cork cells produced by it are loosely arranged and possess numerous intercellular spaces*. These areas are thicker radially than rest of the periderm because of the presence of loose complementary cells. The lenticels perform the function of exchange of gases during night or when the stomata are closed.

Lenticels are first formed immediately beneath the stomata or group of stomata and the number of lenticels, therefore, depends upon the number of stomata or groups of stomata. The lenticels may be scattered on the stems or they may be arranged in vertical or horizontal rows. The lenticels also occur on the roots.

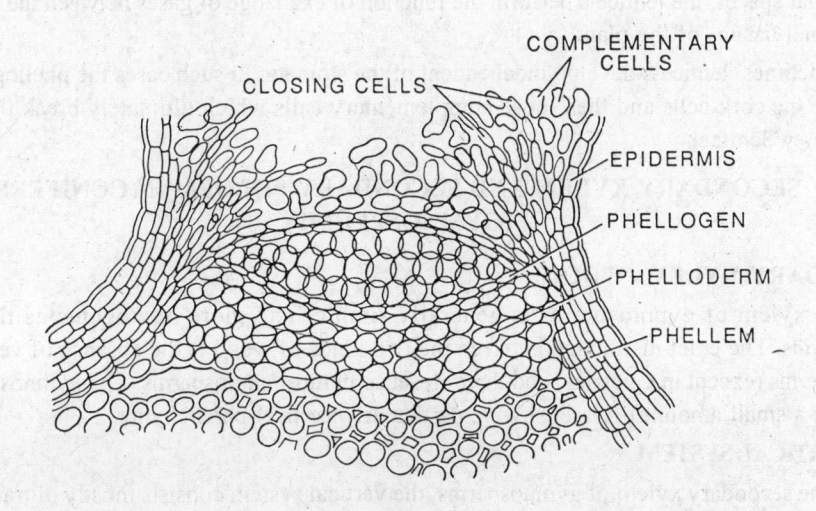

Fig. 10.58. The lenticel. A portion of T.S. of stem of *Prunus* showing lenticel. Successive layers of complementary and closing cells are visible.

The lenticels originate beneath the stomata, either just before, or simultaneously with the initiation of the first layer of the periderm. In most of plants, lenticel formation takes place in the first growing season and sometimes previous to the growth in length has stopped. As the lenticel formation begins, the parenchyma cells found nearabout the sub-stomatal cavity lose their chlorophyll and divide irregularly in different planes giving rise to a mass of colourless, rounded, thin walled, loose cells called *complementary cells*. Such cells are also produced by phellogen towards outside instead of cork cells. As the complementary cells increase in number, pressure is caused against the epidermis and it ruptures. Very often, the outer most cells die due to exposure to outer atmosphere and are replaced by the cells cut off by cork cambium or phellogen. The thin walled loose complementary cells may alternate with masses of more dense and compact cells called

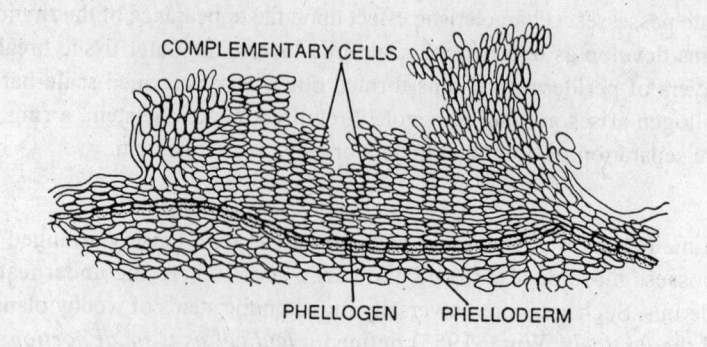

Fig. 10.59. The lenticel of *Mangifera indica* (mango).

the *closing cells*. These cells together form a layer called *closing layer*. With the continuous formation of new loose complementary cells, the closing layers are ruptured. The lenticels are filled up with complementary cells completely in the spring season whereas in the end of the spring season the lenticel becomes closed by the formation of closing layer.

The complementary cells are thin-walled, rounded and loose with sufficiently developed intercellular spaces among them. Their cell walls are not suberized. Due to the presence of profuse intercellular spaces, the lenticels perform the function of exchange of gases between the atmosphere and internal tissues of the plant.

Sometimes, lenticels develop independent of the stomata. In such cases the phellogen cuts for sometime the cork cells and then loose complementary cells which ultimately break the cork and rise to a new lenticel.

SECONDARY XYLEM AND SECONDARY PHLOEM IN CONIFERS (GYMNOSPERMS)

SECONDARY XYLEM (WOOD)

The xylem of gymnosperms is generally simpler and more homogeneous than that of angiosperms. The chief distinction between the two kinds of wood is the absence of vessels in the gymnosperms (except in Gnetales) and their presence in most angiosperms. The gymnosperm wood possesses a small amount of parenchyma, particularly vertical parenchyma.

THE VERTICAL SYSTEM

In the secondary xylem of gymnosperms, the vertical system consists mostly of tracheids. The late wood tracheids possess relatively thick walls and pits with reduced borders, and therefore, they are known as *fibre-tracheids*, but libriform fibres do not occur. The tracheids are long cells (0.5 to 1.1 mm) with their ends overlapping those of other tracheids. The tracheids of existing gymnosperms are interconnected by circular or oval bordered pit-pairs in single, opposite or alternate arrangement. The number of pits on each tracheid may vary from 50 to 300 (Stamm, 1946). The pit-pairs are abundantly present on the ends where the tracheids overlap each other. Generally the pits are confined to the radial facets of the cells. Tori are present on the pit membranes in *Ginkgo*, Gnetales and most Coniferales. The tracheids possess thickenings of intercellular material and primary walls along the upper and lower margins of the pit-pairs. These thickenings are called *crassulae*. Another wall sculpture is represented by the *trabeculae*. They are found in the form of small bars extending across the lumina of the tracheids from one tangential wall to the other. Helical thickenings on pitted walls have been recorded in the tracheids of some conifers. Wherever present, the vertical xylem parenchyma of the Coniferales is found to be distributed throughout the growth ring and occurs in

The Stem—Primary and Secondary Structure

long strands derived from transverse divisions on the mostly long fusiform cambial cells. Some conifers (*Taxus*, *Torreya* and *Araucaria*) do not have parenchyma in the vertical system.

STRUCTURE OF RAYS

The rays of gymnosperms are composed either of parenchyma cells alone, or of parenchyma cells and tracheids. Ray tracheids are distinguished from ray parenchyma cells chiefly by their bordered pits and lack of protoplasts. The ray tracheids possess lignified secondary walls. In some conifers these walls are thick and sculptured, with projections in the form of bands extending across the lumen of the cell. The ray parenchyma cells possess living protoplasts in the sap wood and often dark coloured resinous deposits in the heartwood. The rays of conifers are for the most part only one cell wide and from 1 to 20 or sometimes upto 50 cells high. Ray tracheids may occur singly or in series, at the margins of a ray. The rays serve to transport the assimilation products formed in the leaves and flowing downwards in the phloem in a radial direction into the wood of the stem and roots.

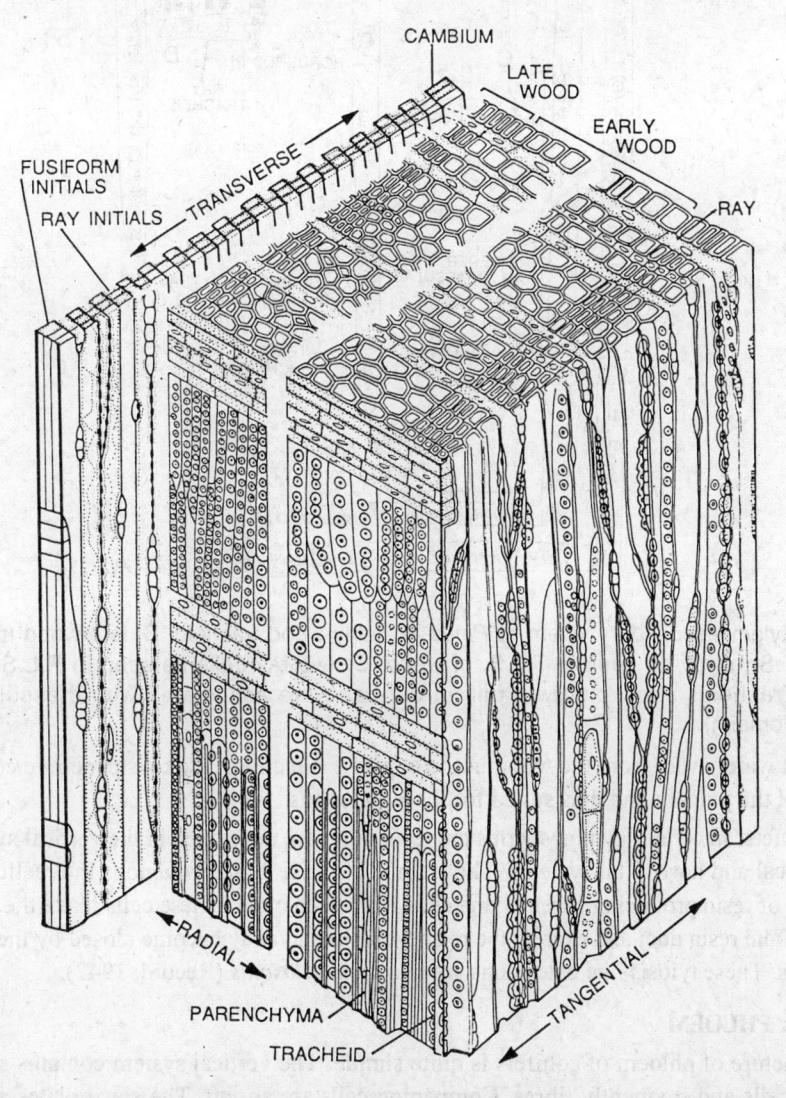

Fig. 10.60. Xylem—wood of conifers. Three-dimensional diagram of the cambium and secondary xylem of *Thuja occidentalis* Linn. (After Bailey).

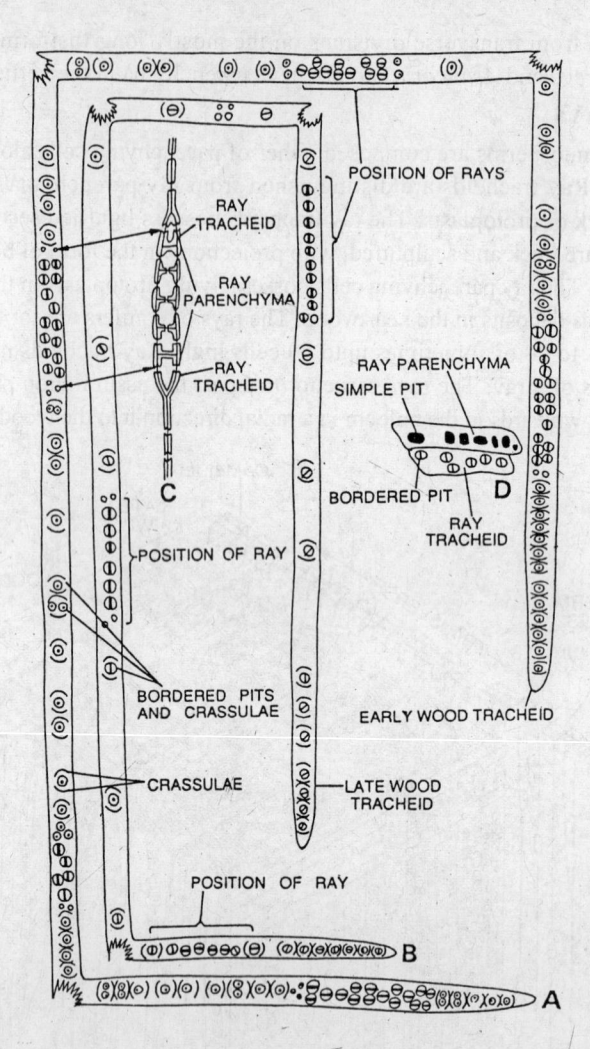

Fig. 10.61. Xylem—secondary xylem of *Pinus*. A, early-wood tracheid; B, late-wood tracheid; C, T.S. of ray as seen in T.L.S. of wood; D, two ray cells as seen in R.L.S. of wood. Tracheids show pits with full borders and are associated with crassulae. (After Forsaith).

They conduct water away from the wood in the opposite direction. The rays penetrate equally into the xylem and the phloem and thus suited for these functions.

Resin ducts. In certain gymnosperms the resin ducts are developed in the vertical system or in both the vertical and horizontal systems. The resin ducts arise as schizogenous intercellular spaces by separation of resin producing parenchyma cells from each other. These cells make the lining, the *epithelium*, of the resin duct and excrete the resin. A resin duct may become closed by the enlarging epithelial cells. These tylosis like extensions are known as *tylosoids* (Record, 1947).

SECONDARY PHLOEM

The structure of phloem of conifers is quite simple. The vertical system contains sieve cells, parenchyma cells and frequently fibres. Companion cells are absent. The sieve plates are present on the lateral walls. The rays are mostly uniseriate and contain parenchyma only or parenchyma and albuminous cells.

The Stem—Primary and Secondary Structure

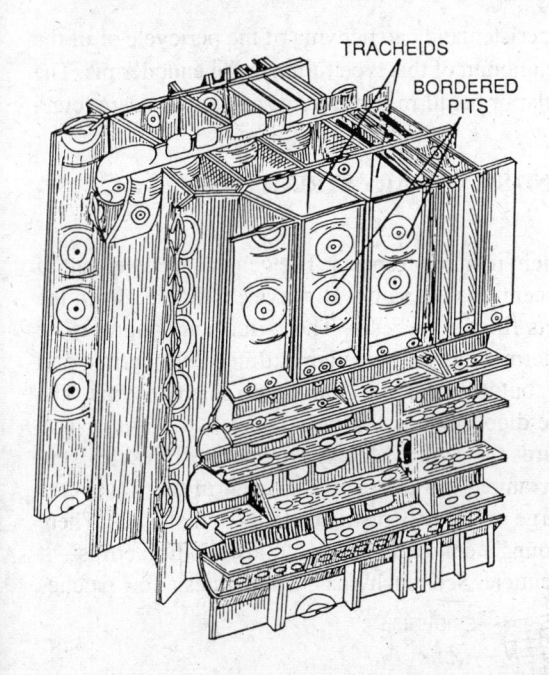

Fig. 10.62. Secondary xylem. Three-dimensional structure of wood of *Pinus strobus*. Tracheids are shown in the figure.

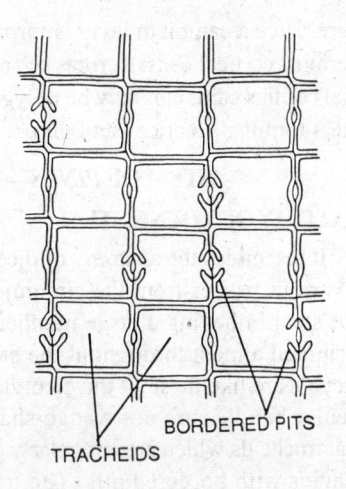

Fig. 10.63. Tracheids with bordered pits of pine stem in T.S.

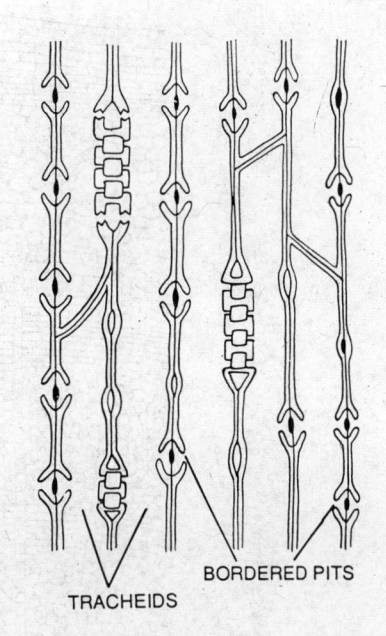

Fig. 10.64. Tracheids with bordered pits of pine stem in T.L.S.

The sieve cells are slender, elongated elements appear like the fusiform initials from which they are derived. They overlap each other at their ends and each sieve cell remains in contact with several rays. The sieve areas are abundantly found on the ends which overlap those of other sieve cells. The connecting strands in the sieve-areas are aggregated into the groups, and the callose associated with the strands in one group fuses into one structure. The phloem parenchyma cells occur in longitudinal strands. They store reserve food material and also contain resins, crystals and tannins. The phloem rays of the trees of advanced age are characteristic in having albuminous cells. The albuminous cells may also occur among the phloem parenchyma cells (in the vertical system) and in still other plants only among the phloem parenchyma cells. The albuminous cells contain dense cytoplasm and distinct nuclei. They are irregular in shape and store food materials. The secondary phloem also contains resin canals. In *Picea canadensis*, they occur in rays and possess cyst-like bulbous expansions.

SECONDARY GROWTH IN THE MONOCOTYLEDONS

Commonly, the vascular bundles of monocotyledons do not possess cambium, and therefore, there is no secondary growth. However, it occurs in some monocotyledons (*e.g., Dracaena, Yucca, Veratrum, Aloe, Sansevieria, Xanthorrhoea, Kingia,* etc.). In these plants the secondary growth occurs by the formation of the cylinder of new bundles embedded in a tissue of less specialized nature. Here a cambium layer is formed from the meristematic parenchyma of the pericycle or of the innermost cortical cells. In roots of some plants, a cambium of this type forms in the endodermis. The initials of this cambium may be polygonal, rectangular or fusiform in different species. They are found in tiers forming a storied cambium.

STEM OF *PINUS*—PRIMARY AND SECONDARY STRUCTURE

ANATOMY OF YOUNG STEM

It resembles the anatomy of dicotyledonous stem in many respects. The general arrangement of the various tissues from the circumference to the centre is the same. However, it differs, from the dicot stem in having a large number of resin ducts filled with resin. These ducts are found to be distributed almost throughout the stem. The epidermis has an irregular outline. Endodermis and pericycle are like those of the dicotyledonous stem, but the pericycle contains no sclerenchyma. The vascular bundles are not wedge-shaped, as in the dicotyledons. Phloem consists of annular and spiral tracheids which are irregularly disposed towards the centre. Metaxylem consists of exclusively tracheids with bordered pits. The tracheids are arranged in radial rows as seen in the transverse section of the stem. The pits of the pine wood are large and mostly restricted to the radial walls. There are no true vessels. The details of the anatomy of young stem of pine are as follows:—Epidermis—It consists of a single layer of cells with a very thick cuticle. Sclerenchyma—Sometimes a few patches

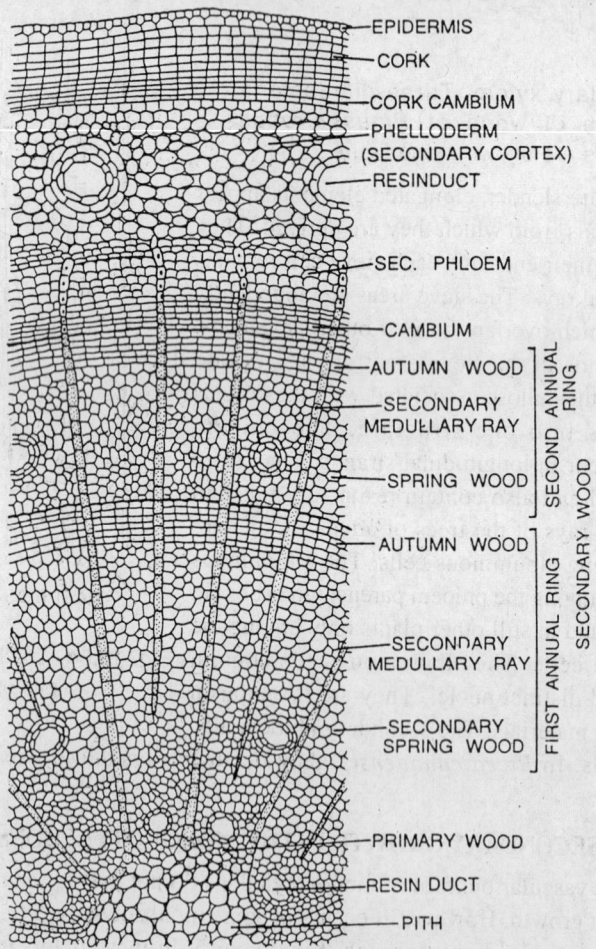

Fig. 10.65. Secondary growth in thickness of the pine stem. T.S. of a two-yearold pine stem.

of sclerenchyma occur here and there below the epidermis. Cortex—Many layers of more or less rounded parenchyma cells, with conspicuous resin ducts lying embedded in the cortex. Endodermis—A single layer lying internal to the cortex; the innermost layer of the cortex is treated as endodermis. Pericycle—It consists of parenchyma cells, there is no sclerenchyma in it. Medullary rays—They run from the pith outwards between the vascular bundles. Pith—There is a well defined pith, consisting of a mass of parenchyma cells. A few resin-ducts are also present in the pith. Vascular bundles—These are collateral and open, and arranged in a ring, as in dicot stem. Each bundle consists of phloem, cambium and xylem. Phloem—The phloem consists of sieve tubes and phloem parenchyma, but no companion cells. It lies on the outer side of the bundle. Cambium—A few layers of thin walled, rectangular cells in between xylem and phloem. Xylem—It consists exclusively of tracheids; there are no true vessels. Resin ducts are also present here. Protoxylem lies towards the centre and consists of a few annular and spiral tracheids which are not disposed in any regular order. Metaxylem lies towards the cambium and consists of tracheids with bordered pits which develop on the radial walls. These tracheids are four sided and are arranged in definite rows.

SECONDARY GROWTH IN THICKNESS OF STEM

The secondary growth in pine stem takes place in exactly the same way as in a dicotyledonous stem. However, the points of differentiation are as follows: The pine stem is characterized by the presence of conspicuous resin-ducts which are distributed throughout the stem. The secondary wood consists exclusively of tracheids with numerous bordered pits on their radial walls. As in the dicotyledonous stem, there are distinct annual rings, consisting of the autumn wood and spring wood. The autumn wood consists of narrow and thick-walled tracheids, and the spring wood of wider and thinner-walled tracheids. The secondary medullary rays are usually one layer of cells in thickness and a few in height. The phloem portion of the medullary ray consists of middle layers of starch—containing cells, called starch cells, and upper and lower layers of protein containing cells, called albuminous cells. The xylem portion of the medullary ray consists of similar starch cells in the middle, and empty cells with bordered pits, called tracheidal cells, in the upper and lower layers. Vessels are absent.

ANATOMICAL DIFFERENCES BETWEEN DICOTYLEDONOUS AND MONOCOTYLEDONOUS STEMS

		Dicotyledonous Stem	Monocotyledonous Stem
1.	Hypodermis	It is collenchymatous.	It is sclerenchymatous.
2.	Cortex	It consists of a few layers of parenchyma.	There is a continuous mass of parenchyma up to the centre. It is commonly known as ground tissue.
3.	Endodermis	It is a wavy layer of compact cells.	
4.	Pericycle	Usually it consists of parenchyma and sclerenchyma.	It is not differentiated into distinct tissues.
5.	Medullary ray	There lies a strip of parenchyma in between vascular bundles.	
6.	Pith	The central cylinder consists of parenchyma.	It is not marked out.
7.	Vascular bundles	(a) They are collateral and open.	They are collateral and closed.
		(b) They are arranged in a ring.	They are scattered.
		(c) They are of uniform size.	They are larger towards the centre.
		(d) Phloem parenchyma present.	Phloem parenchyma absent.
		(e) They are wedgeshaped.	They are usually oval.
		(f) Bundle sheath absent.	Bundle sheath strongly developed.

11

The Stem—Anomalous Structure

Majority of the plants possess stelar structure of the normal type, but several of them have unusual structure. This unusual structure is of many different types. These types differ from the more usual types, and are therefore, sometimes termed *anomalous*. With the result of the combinations of unusual structure, certain anomalous and extremely complex structures are formed. These complex structures are known as *anomalies*. Such anomalies are quite common in angiosperms. These anomalies may be enlisted as follows:

1. Anomalous secondary growth in Dicotyledons.
   (a) Anomalous position of cambium.
   (b) Abnormal behaviour of normal cambium.
   (c) Accessory cambium formation and its activity.
   (d) Extrastelar cambium.
   (e) Interxylary phloem.
2. Absence of vessels in the xylem.
3. Scattered vascular bundles in dicotyledons.
4. Presence of exclusive phloem and xylem bundles.
5. Presence of medullary bundles.
6. Presence of cortical bundles.
7. Intraxylary phloem.
8. Vascular bundles arranged in a ring in Monocots.
9. Secondary growth in Monocotyledons.

ANOMALOUS SECONDARY GROWTH

Several dicotyledons show secondary growth that deviates considerably from the normal secondary growth. The deviating methods of secondary thickening are called *abnormal* or *anomalous*, although the normal and abnormal forms of growth are not sharply separated from one another.

ANOMALOUS POSITION OF CAMBIUM

Stems of several unusual shapes or types are formed by the unusual position of the cambium.

For example in *Thinouia scandens*, while the stem is young, the cambium is thrown into folds or ridges. Here, the tips of the ridges pinched off and after separation develop *steles* in them.

In *Serjania ichthyoctona*, the cambium appears originally in several separate strips, each of which surrounds portions, even individual strands, of the primary xylem and phloem. This type of

The Stem—Anomalous Structure 263

stem appears to be made up of several fused stems. In the older stems, this compound condition becomes more marked, because the parts are separated from each other as the outer layers of each strand die with the result of the development of periderm layers. In this way there is formed a stem composed of strands living together more or less like the strands of a rope.

In *Bauhinia langsdorffiana*, a somewhat similar structure is formed as described above. This structure is brought about by the breaking into strips of the original cambium cylinder, and even of the vascular cylinder formed by this cambium, by the proliferation of xylem parenchyma. The parenchyma in the xylem and phloem increases excessively, thus rupturing the first-formed, original tissues and the cambium sheet which formed them.

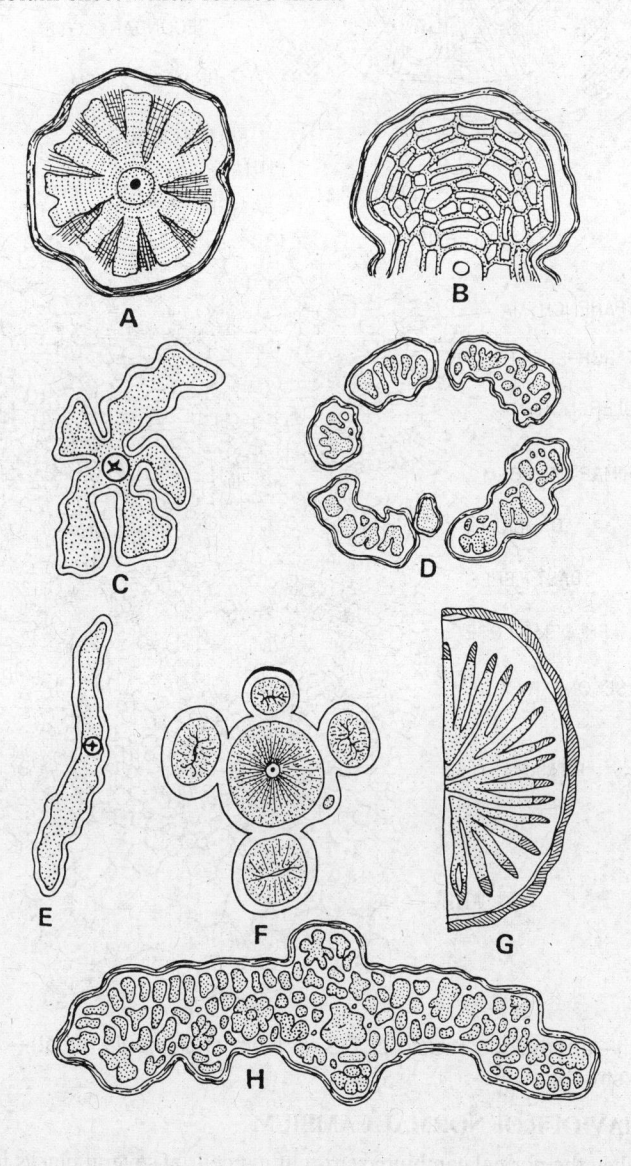

Fig. 11.1. Anomalous structure of stem. A, *Bignonia*; B, *Securidaca lanceolata*, C, *Bauhinia rubiginosa*; D, *Sarjania ichthyoctona*; E, *Bauhinia* spp,. F, *Thinouia scandens*; G, *Aristolochia* spp; H, *Bauhinia langsdorffiana*. (After Schenck).

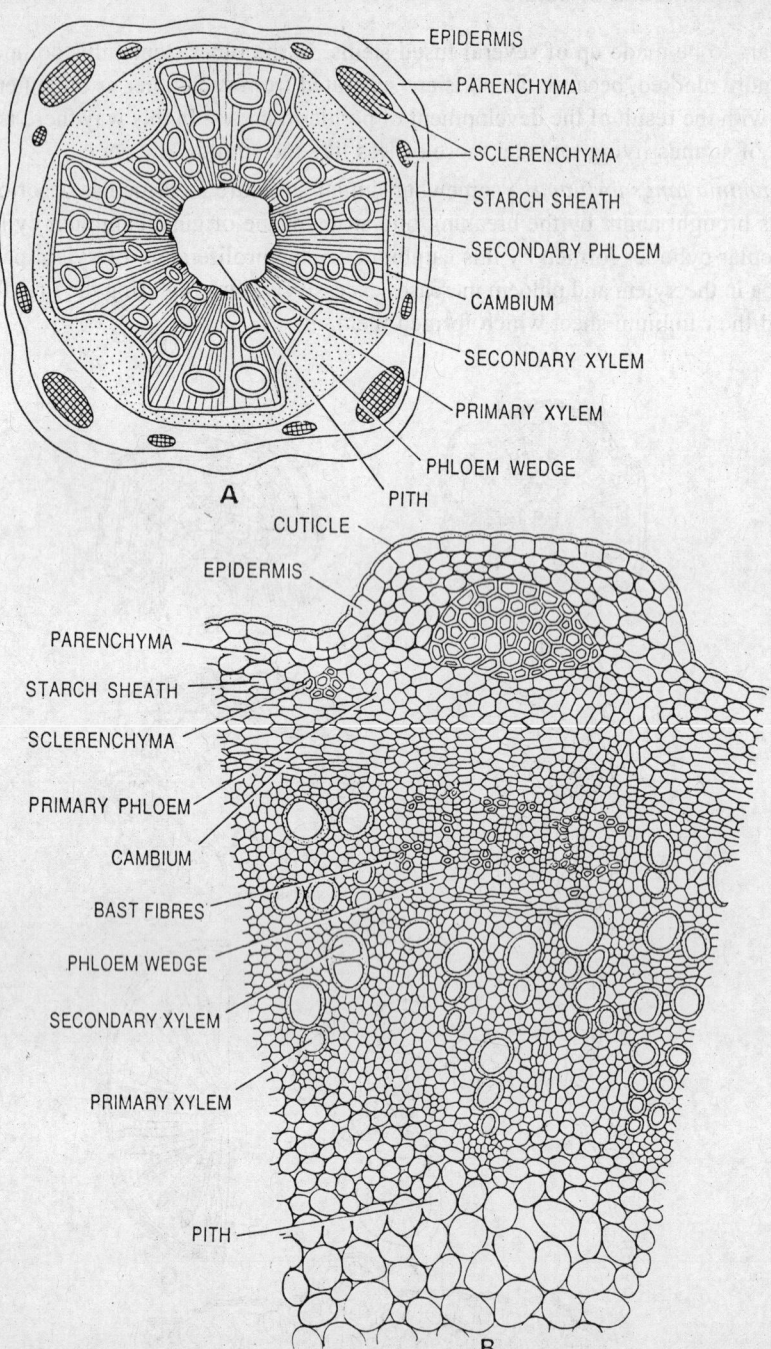

Fig. 11.2. The stem—anomalous structure. T.S. of stem of *Bignonia* (dicot)— A, diagrammatic, B, detailed structure.

ABNORMAL BEHAVIOUR OF NORMAL CAMBIUM

Sometimes when the normal cambium starts cutting cells at several places irregularly, and forms at certain places much larger portions of xylem than of phloem, and at other places more phloem than xylem, and a ridged and furrowed xylem cylinder is produced. This may be of simple structure (*e.g.*, in some Bignoniaceous genus) or very complex (*e.g.*, in *Bignonia* sp.).

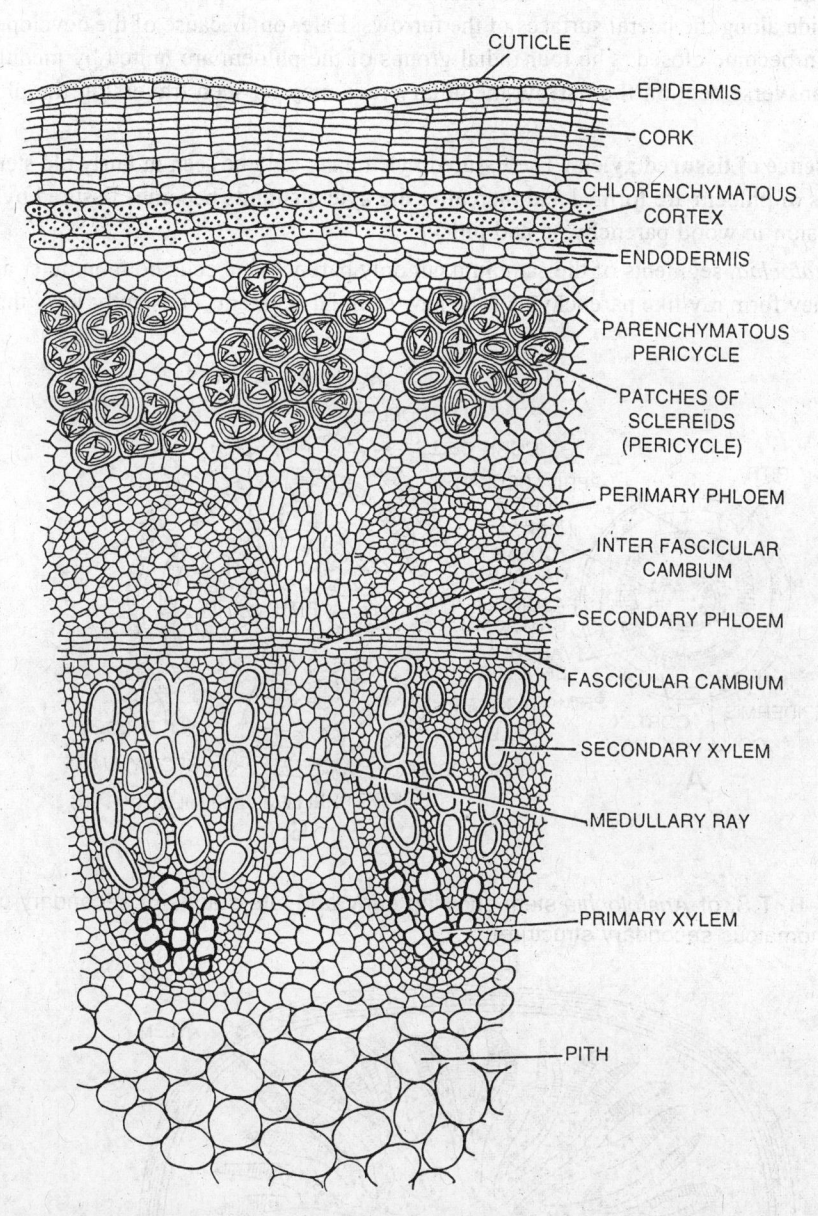

Fig. 11.3. *Aristolochia* (dicot). T.S. of stem showing secondary growth. Beneath the epidermis the cork layers are present; the pericycle splits into groups of sclereids and the interfascicular cambium does not produce secondary vascular tissue.

ANATOMY OF *BIGNONIA* STEM

The most interesting anatomical feature in *Bignonia* of Bignoniceae is the occurrence of anomalous secondary structure. (See fig. 11.2) This is as follows:

A. Presence of phloem wedges in the xylem. The young stems which exhibit this type of structure when mature are provided with a normal ring of vascular bundles. The vessels of the young stem are narrow in diameter. The wood formed in later stages contains wider vessels. As soon as this stage is reached four furrows at four equidistant points appear in the xylem, extending almost to the pith. The cambium is situated on the inside of the furrows. The phloem increases in bulk, and the tissues slide along the lateral surfaces of the furrows. Later on, because of the development the furrows again become closed. The four radial groups of the phloem are united by medullary ray tissue. In transverse section the narrow medullary rays may be seen traversing the phloem of furrows.

B. Presence of fissured xylem. The fissured xylem may only be seen in fairly old stems. First of all wedges of phloem are formed and thereafter the xylem strands becomes fissured by dilation and cell division in wood parenchyma and pith.

In *Aristolochia*, segments of the cambium cut only parenchyma cells both on outer and inner sides, thus they form ray like parenchyma. The new cambial segments constantly form the rays of

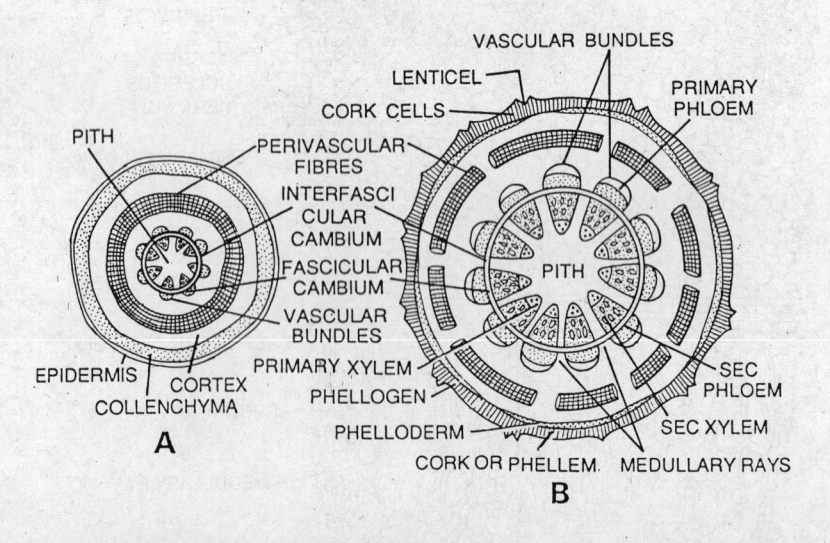

Fig. 11.4. A—B, T.S. of *Aristolochia* stem showing early and late stages of secondary growth—(anomalous secondary structure).

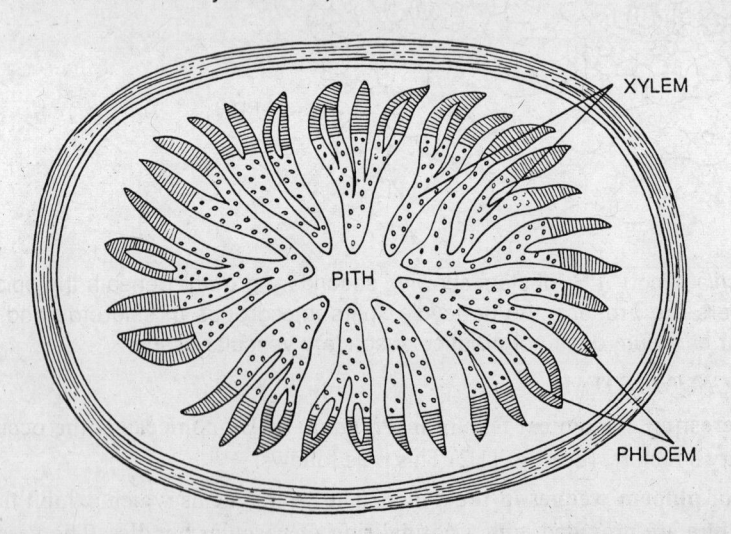

Fig. 11.5. The stem—anomalous structure. T.S. of old *Aristolochia* stem showing anomalous secondary growth.

The Stem—Anomalous Structure

parenchyma thus increasing in diameter. As the vascular cylinder, broken by wide rays, increases in circumference the cylinder of sclerenchyma that encircled the bundles become ruptured and adjacent parenchyma grows intrusively into the gaps. Eventually a very fluted vascular cylinder is formed. Species of *Aristolochia* are woody climbers or *lianes,* which have diverse taxonomic affinities and often show anomalous structural features. Among other characteristics, the vessels are often of unusually wide diameter. (See figs. 11.4 and 11.5).

In *Bauhinia rubiginosa*, there is restriction of the activity of the cambium to certain regions which results in the formation of ridged stems. In other species of *Bauhinia*, the strap like stems are formed because of restricted activity of the cambium in certain regions. In this case the cambium is more active at two opposite sides. (Fig 11.1 E.).

In some climbing plants (*e.g., Vitis, Clematis*), the interfascicular cambium forms only parenchyma, so that the original vascular bundles remain discrete throughout secondary growth.

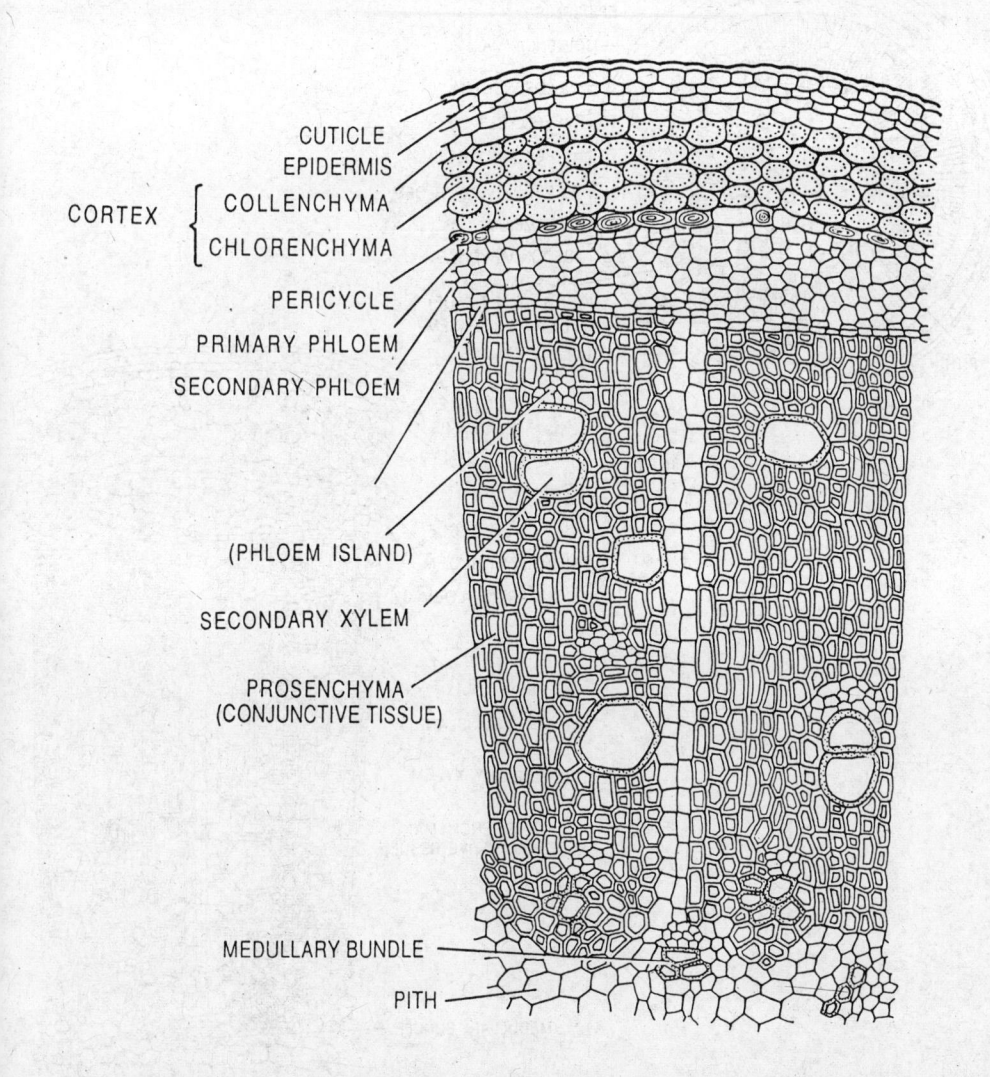

Fig. 11.6. Stem-anomalous structure. T.S. of a portion of stem *Bougainvillaea* (dicot.) showing anomalous secondary growth, periderm, thick-walled conjunctive tissue and medullary bundles.

ACCESSORY CAMBIUM FORMATION AND ITS ACTIVITY

In the stem of *Bougainvillaea*, and other members of the Nyctaginaceae (*e.g., Boerhaavia diffusa; Mirabilis,* etc.) several cambia arise successively in centrifugal direction. Each cambium produces xylem and conjunctive tissue to the inside, and phloem and conjunctive tissue to the outside. The resulting tissue gives the appearance of concentric rings of vascular bundles embedded in conjunctive tissue.

The herbaceous and woody plants of Nyctaginaceae (*e.g., Bougainvillaea, Boerhaavia, Mirabilis*) are noteworthy from the view point of their anatomy that they possess the anomalous secondary growth in thickness of the axis. This anomalous type of secondary growth in thickness takes place by means of development of the successive rings of collateral vascular bundles. In

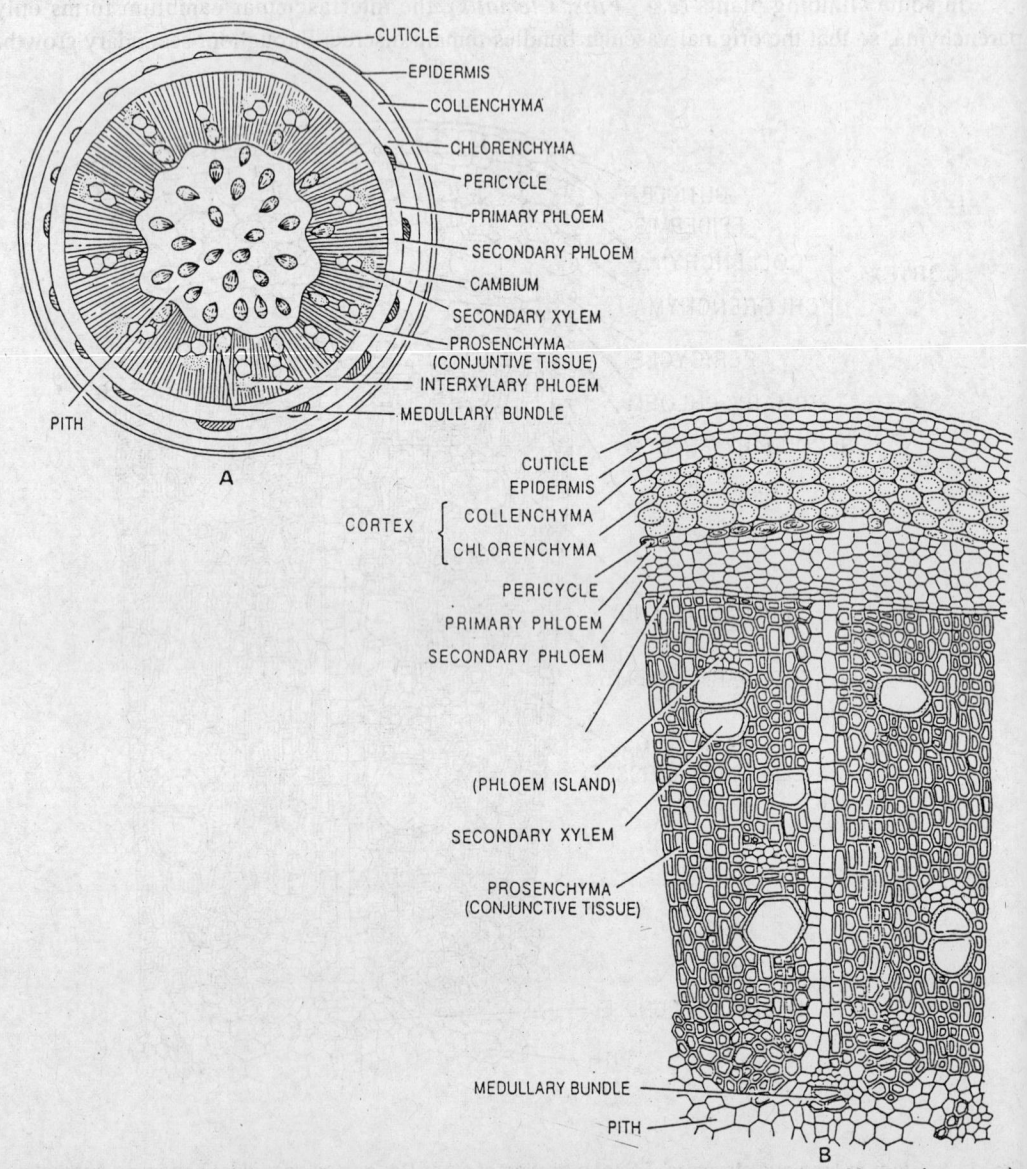

Fig. 11.7. The stem—anomalous structure. T.S. of *Bougainvillaea* stem (dicot-Nyctaginaceae) showing anomalous secondary thickening.

The Stem—Anomalous Structure

herbaceous plants of this family the bundles remain embedded in parenchymatous ground tissue but in the woody species (*e.g., Bougainvillaea*), the ground tissue is somewhat prosenchymatous and lignified. Both types of conjunctive tissue are developed from successive cambia. In the woody species there is no clear differentiation between the xylem and the conjunctive tissue and therefore, sometimes in transverse section of the axis the phloem is seen in the form of island. In certain other cases the strips resembling medullary rays are also seen in the conjunctive tissue.

In *Boerhaavia, Bougainvillaea* and *Mirabilis* the anomalous secondary thickening occurs in the form of succession of rings of vascular bundles. At one time it was believed that the secondary vascular bundles originate in the parenchyma of the pericycle, but according to P. Maheshwari (1930) this inerpretation is not correct for the species of *Boerhaavia* and *Mirabilis*. In the members

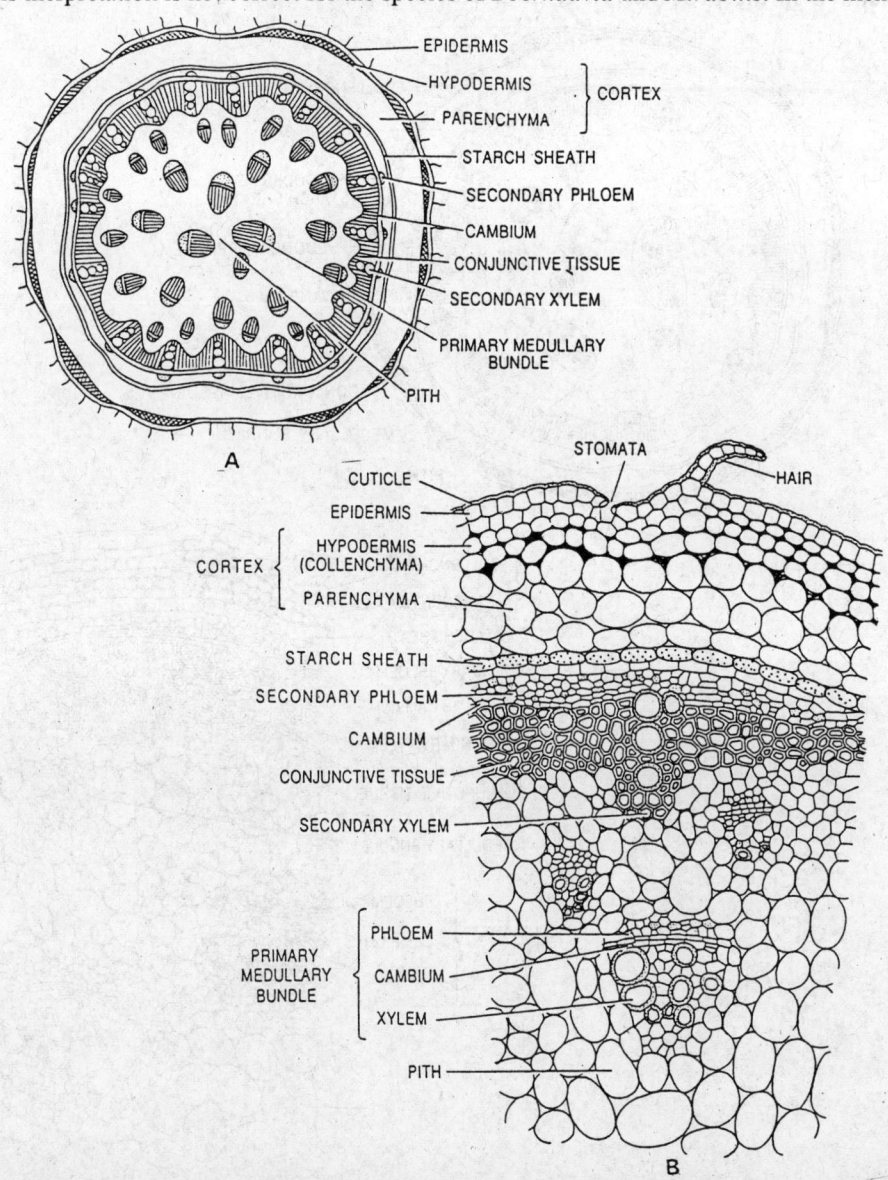

Fig. 11.8. The stem—anomalous structure. T.S. of stem of *Mirabilis* (Nyctaginaceae-dicot)—A, diagrammatic; B, detailed structure.

of family Nyctaginaceae the secondaty vascular bundles remain embedded in parenchymatous, prosenchymatous or lignified conjunctive tissue as the case may be. In *Bougainvillaea glabra*, a robust herbaceous or slightly woody species the inner bundles remain embedded in parenchyma and the outer bundles in the prosenchymatous ground tissue. Sometimes in certain species the narrow radial strips resembling normal medullary rays are also seen traversing in the conjunctive tissue. In *Boerhaavia*, thin-walled lignified groups of parenchyma are associated with the phloem; the phloem groups and adjoining ground parenchyma occasionally appear as concentric annular or band shaped strips of tissue. In the woody species such as *Neea* and *Pisonia*, there is no demarcation between the periphery of the xylem and the surrounding ground tissue, and thus in the bundles embedded in the prosenchyma, the phloem appears in the form of phloem islands. In

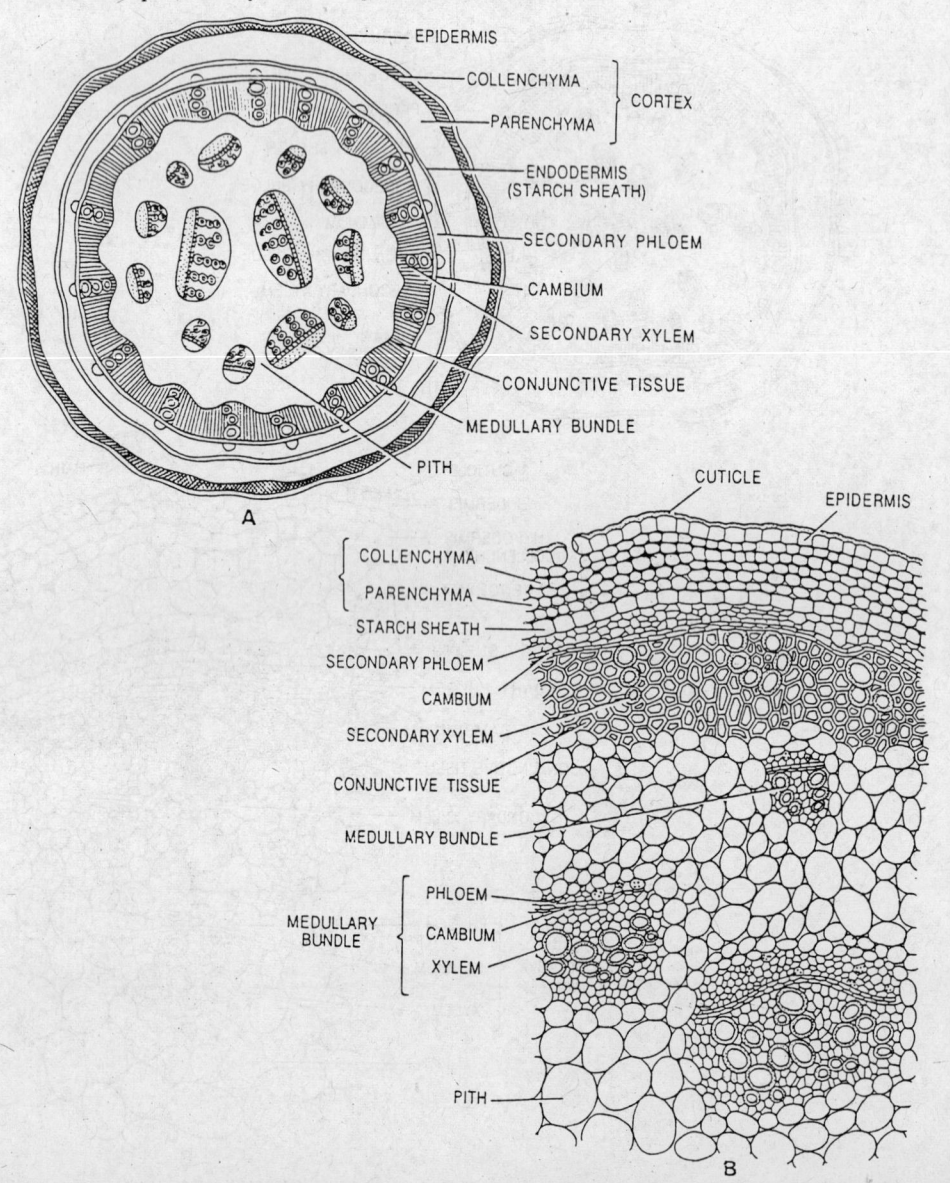

Fig. 11.9. The stem-anomalous structure. T.S. of *Boerhaavia* stem (of Nyctaginaceae). A diagrammatic; B, detailed structure. Description in the text.

The Stem—Anomalous Structure

Bougainvillaea, Mirabilis, Pisonia, the innermost part of the conjunctive tissue, formed by the secondary meristem resembles true pith and thus causes the innermost secondary bundles to appear as if they are medullary in origin.

According to P. Maheshwari (1930), the development of vascular system in *Boerhaavia diffusa* is as follows :

The transverse sections through the young stem of *Boerhaavia diffusa* show two medullary bundles, a middle ring of 6 to 14 bundles and another ring of 15 to 20 or more small bundles. The bundles of the middle ring increase in thickness to a limited extent by a fascicular cambium. The bundles of the outer ring are initially separate, quite small and each provided with its own fascicular cambium. The fascicular cambia of the bundles become interconnected by interfascicular cambia of the bundles become interconnected by interfascicular cambial areas producing a cylindrical meristematic zone which gives rise to interfascicular parenchymatous or slightly prosenchymatous ground tissue between the bundles. Thereafter the cambium ceases to function and a new meristem

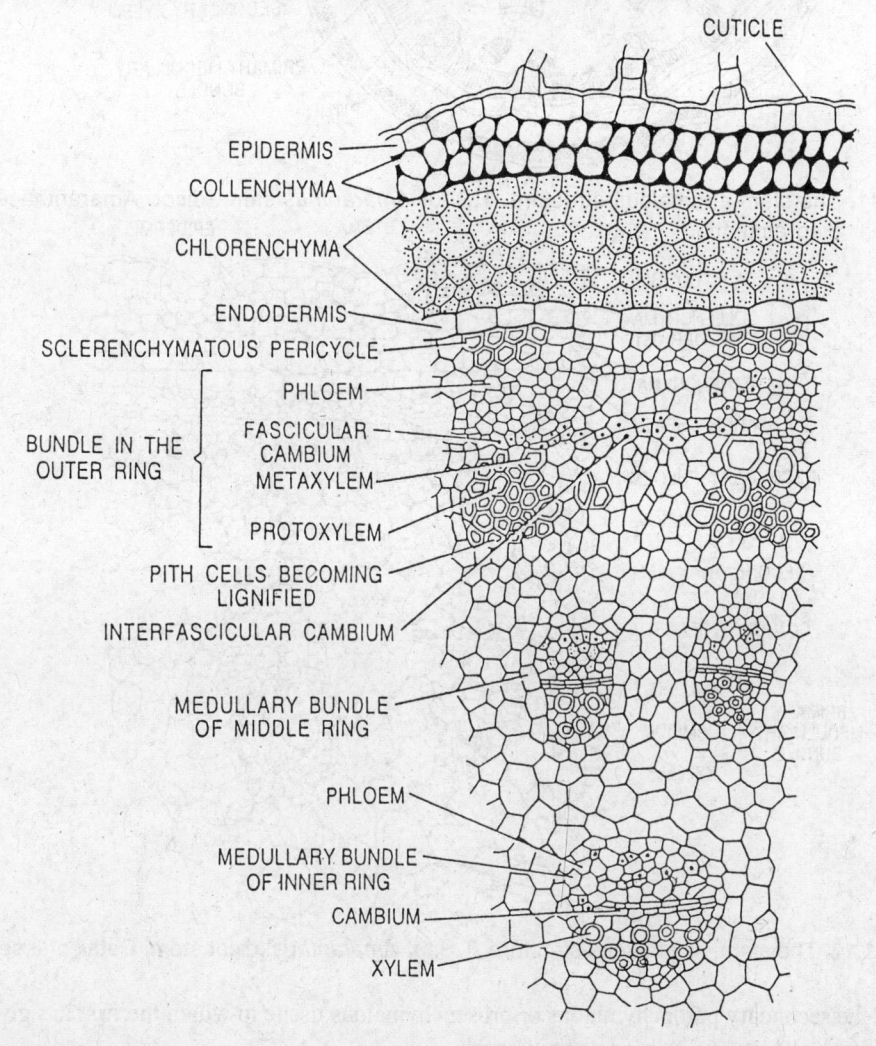

Fig. 11.10. The stem-anomalous structure. T.S. of a sector of stem of *Boerhaavia diffusa* showing periderm, anomalous secondary growth, thick-walled conjunctive tissue, and meduallary bundles.

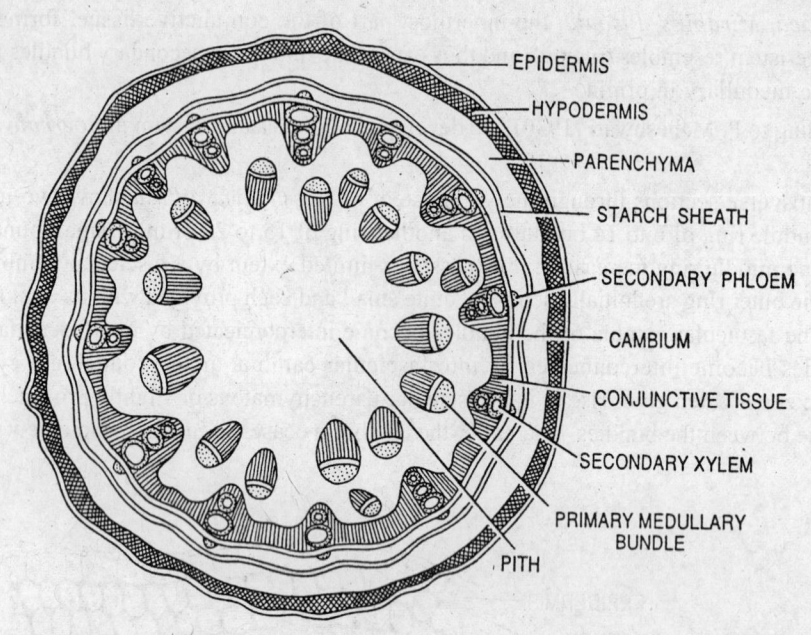

Fig. 11.11. The stem-anomalous structure. T.S. of *Amaranthus* stem (dicot. Amaranthaceae) — diagrammatic.

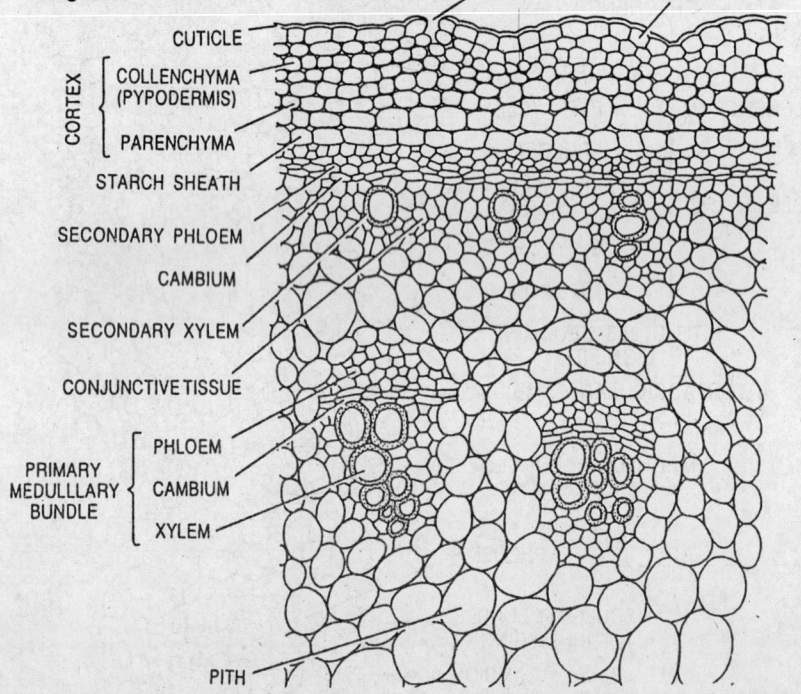

Fig. 11.12. The stem. Anomalous structure. T.S. of *Amaranthus* (dicot stem, Detail of a sector.

arises in the secondary parenchymatous or prosenchymatous tissue to which the first has given rise on the outside. This process repeats at intervals.

Frequently, the successive cambia are ontogenetically interrelated, in that sister cells of one cambium layer become the cambial cells of another layer.

EXTRASTELAR CAMBIUM

In *Amaranthus, Achyranthes* and *Chenopodium*, the extrastelar cambium arises in the pericycle. In *Amaranthus* the cambium is found in the form of a complete ring, whereas in *Achyranthes* it is represented by separate strips or arcs.

Anatomy of *Amaranthus* Stem

The stem of *Amaranthus* of Amaranthaceae shows anomalous secondary structure. The transverse section is somewhat circular in outline and with a single layered epidermis. Immediately beneath the epidermis a multilayered zone of collenchyma is present, which usually remains interrupted by chlorenchyma at several places. The medullary vascular bundles are found, which are many and remain scattered in the pith. The bundles are collateral and open. Cambial activity is found only in the individual bundles, and it ceases soon. Anomalous secondary growth takes place due to the development of a new extrastelar meristem, *i.e.*, the cambium outside the stele in the pericycle region. The cambium cuts off secondary vascular bundles and interfascicular parenchymatous conjunctive tissue. The cambium cuts off cells only towards inner side. At certain places these cells develop into secondary vascular bundles and at other places they make the interfascicular and parenchymatous conjunctive tissue. The secondary bundles remain embedded in conjunctive tissue.

Anatomy of *Achyranthes* Stem

The anomalous secondary thickening in the stem takes place as follows:

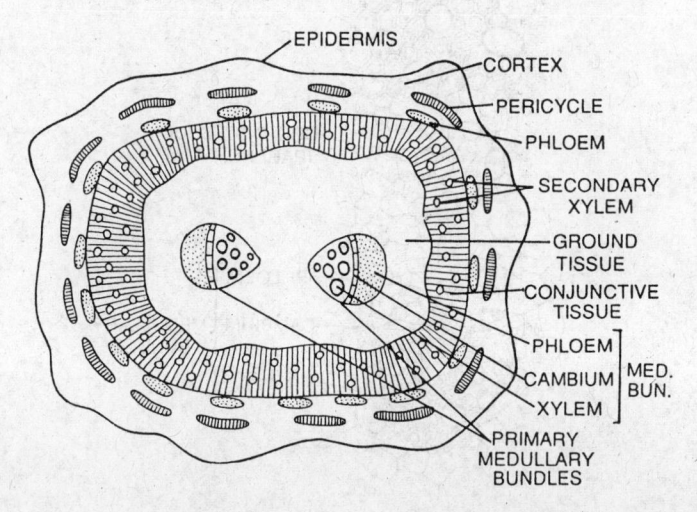

Fig. 11.13. The stem—anomalous structure. T.S. of *Achyranthes* stem (dicot-Amaranthaceae) showing anomalous secondary thickening and medullary bundles (diagrammatic).

The extrastelar secondary arcs or rings of meristematic cells (cambium) appear in the pericycle which give rise to secondary vascular bundles. If the bundles originate from closed ring of cambium they are arranged in concentric circles. When they are formed from cambial arcs, they are irregularly distributed in the ground tissue. The conjunctive tissue in between the bundles consists of parenchyma in some species and of lignified or unlignified prosenchyma in others. When the ground tissue is parenchymatous, the separate individual bundles may not be easily differntiated, because the interfascicular conjunctive tissue and the ground tissue of the true xylem are quite alike in appearance.

In some species of Amaranthaceae, the ground tissue is wholly lignified in some growth zones and parenchymatous in the other growth zones. In still other species the individual vascular bundles

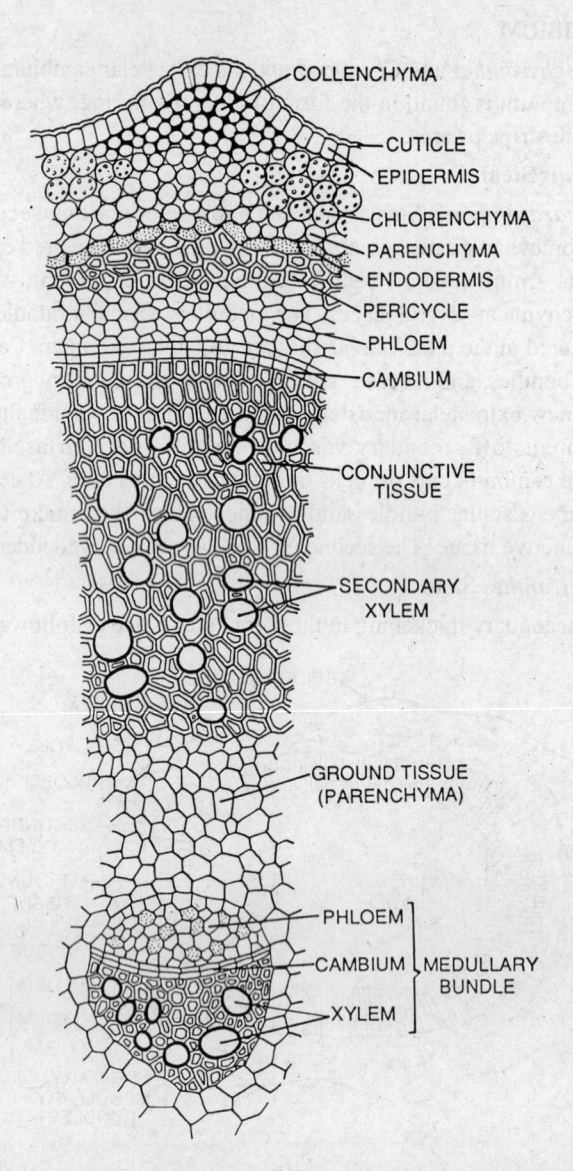

Fig. 11.14. The stem—anomalous structure. T.S. of stem of *Achyranthes* showing anomalous secondary thickening—detail of a sector.

of a single growth ring are connected by sclerenchymatous interfascicular tissue and such growth rings of vascular bundles are separated from each other by parencymatous tissue. Such variations have been recorded from the different species of the family.

Anomalous Secondary growth in *Chenopodium* Stem

Here the extrastelar cambium appears in the pericycle. The first ring of extrastelar cambium is continuous and after the formation of layer of thin walled ground tissue produces secondary vascular bundles. The thin walled ground tissue pushes the primary bundles towards the centre and they look like medullary bundles. Such primary bundles grow in thickness by means of fascicular cambia present within them. Thereafter the first extrastelar cambium produces a number of secondary bundles and prosenchymatous (thick walled) conjunctive tissue.

The Stem—Anomalous Structure

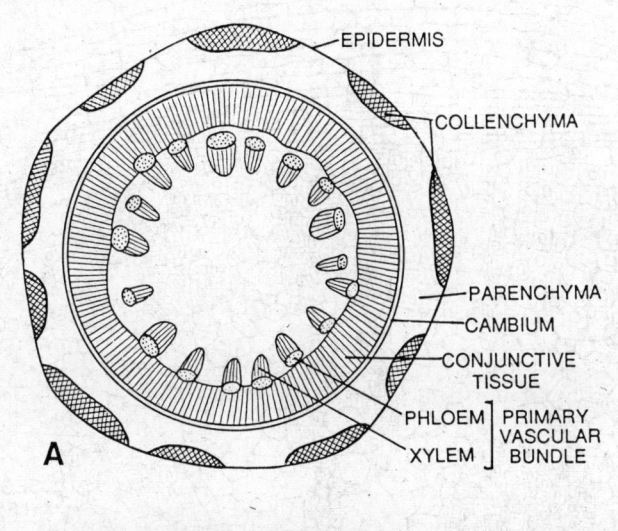

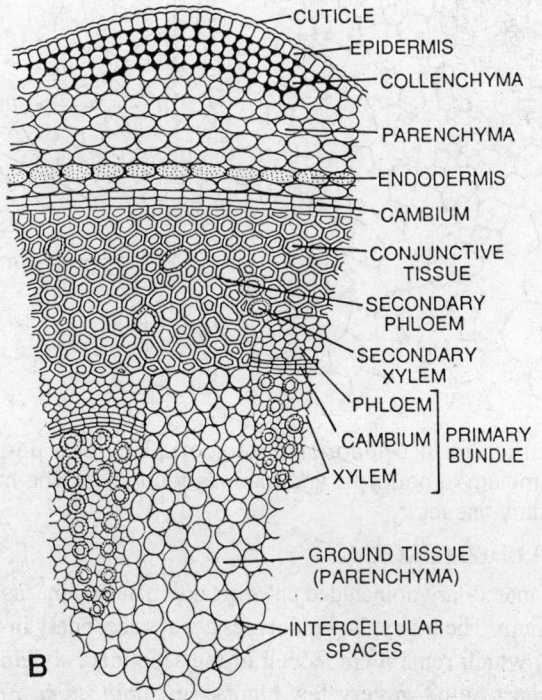

Fig. 11.15. *Chenopodium* (Dicot stem). T.S. of stem showing anomalous secondary thickening. A, diagrammatic; B, detailed structure. Description in text.

Mature stems contain numerous vascular bundles laid down together with the conjunctive tissue around them by a succession of rings or arcs of cambium usually situated in the pericycle and sometimes in the phloem. These secondary meristems also give rise to the conjunctive tissue in which the secondary vascular bundles remain embedded in a concentric, spiral or irregular manner. Sometimes the ground tissue of the bundles and the interfascicular conjunctive tissue are so much alike, that they cannot be differentiated separately from each other.

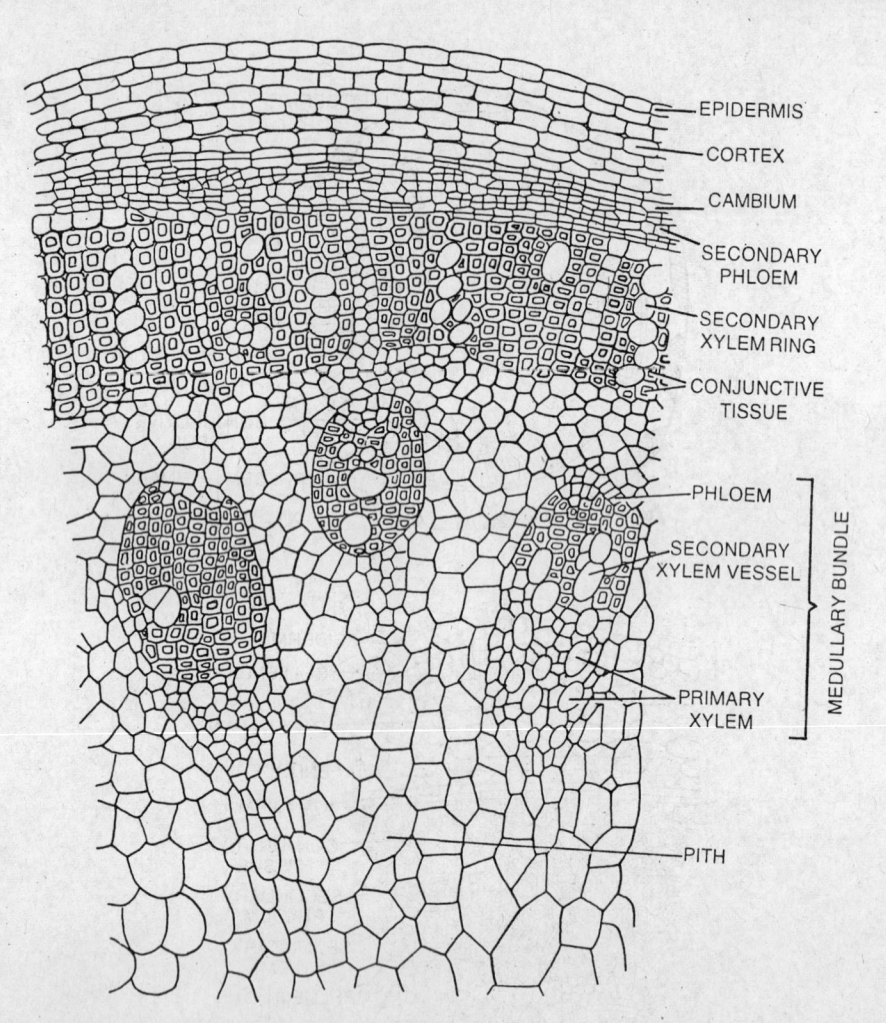

Fig. 11.16. Anomalous structure in *Chenopodium* (dicot). T.S. of a portion of stem showing extrastelar cambium in pericycle and medullary bundles. The medullary bundles also exhibit secondary tissues.

INTERXYLARY OR INCLUDED PHLOEM

The development of interxylary or included phloem takes place by means of the variations found in the activity of the cambium. The interxylary phloem is always secondary in nature and found in the form of strands (islands) which remain embedded in the secondary xylem. The development of included phloem has been studied in very few plants. In *Combretum, Entada, Salvadora* and *Leptadenia*, certain small segments of the cambium cut phloem cells towards the inside of it for a short period of time, instead of cutting xylem cells, which are generally cut in normal conditions. After a short period of time, the cambium once again functions normally, and cut xylem cells inward instead of phloem cells in abnormal conditions, thus inwardly formed phloem becomes embedded in the secondary xylem. This process is repeated several times, and the interxylary phloem patches are developed.

In the other type (*e.g.*, in *Strychnos*) the interxylary phloem patches develop as follows:

At certain points of the general cambium cylinder, the small segments of the cambium cease to function and their cells are being converted into mature conducting tissue. Thereafter, the new

The Stem—Anomalous Structure

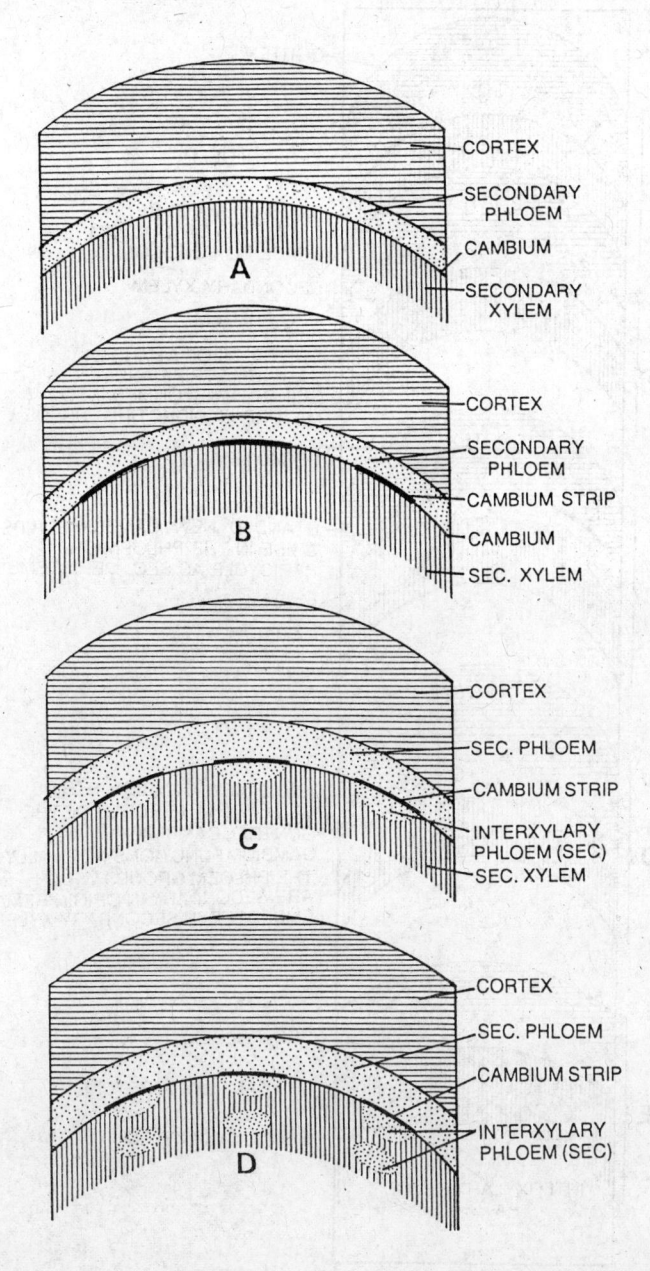

Fig. 11.17. Development of interxylary phloem. *Combretum, Salvadora, Entada, Leptadenia* type.

cambial strips arise as secondary meristems either in the phloem or in the pericycle. Later, these newly formed cambial strips unite with the edges of the segments of the general cambium. Thus, a wavy cambium cylinder is formed. Soon after, this cambial cylinder becomes stretched. The normal activity of the cambium is resumed and thus the phloem cells are engulfed in the secondary xylem. This process is repeated in other parts of the cambium and in this way several phloem patches are engulfed in the secondary xylem. These are interxylary phloem patches which are secondary in origin. Such interxylary or included phloem patches occur in several families (*e.g.*, Asclepiadaceae, Nyctaginaceae, Onagraceae, Salvadoraceae, Loganiaceae, Amaranthaceae and several others).

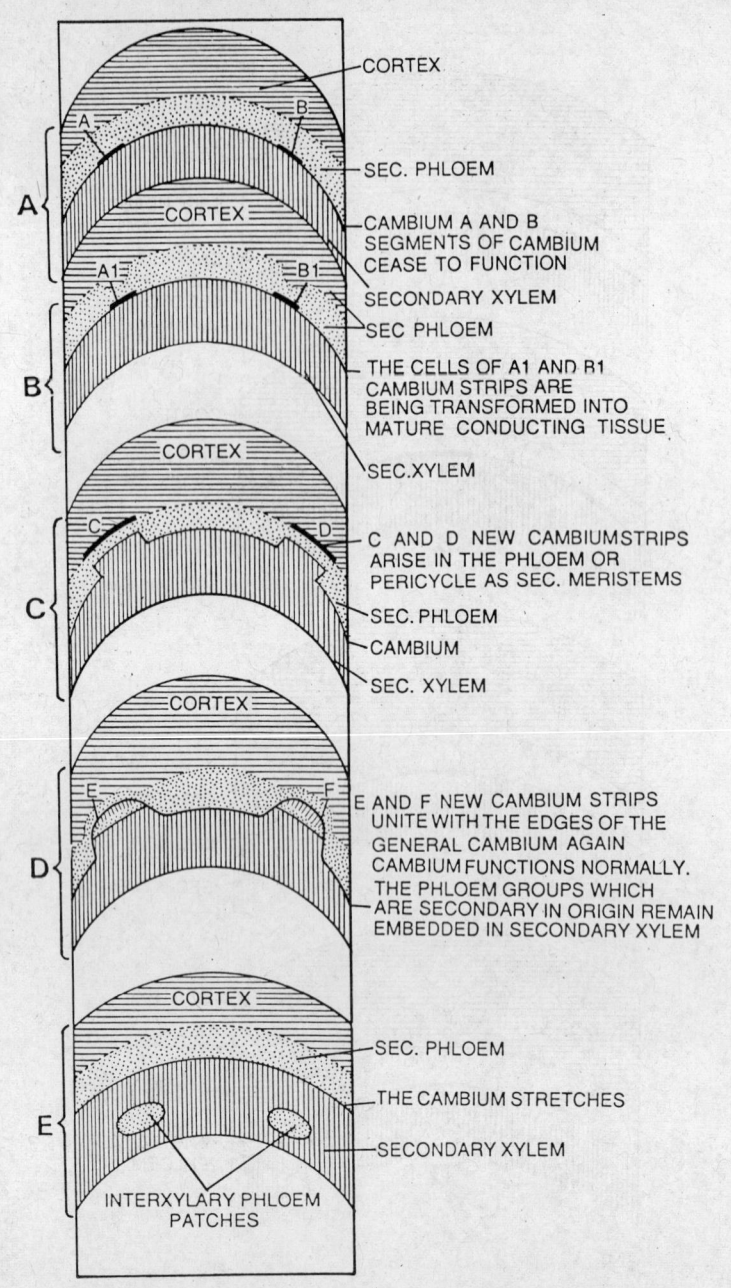

Fig. 11.18. Development of interxylary phloem. *Strychnos* type.

Anatomy of *Leptadenia* Stem.

   Epidermis. The outermost layer epidermis consists of single row of compact cells, and remains covered with a thick cuticle.

   Cortex. This region consists of hypodermis, chlorenchyma and endodermis. The hypodermis consists of one or two layers of thin walled parenchyma cells. Below the hypodermis, a few layers of parenchyma cells are present containing a large number of chloroplasts, and therefore, these cells may be called collectively, the chlorenchyma. The last and very conspicuous layer of the cortex, the

The Stem—Anomalous Structure

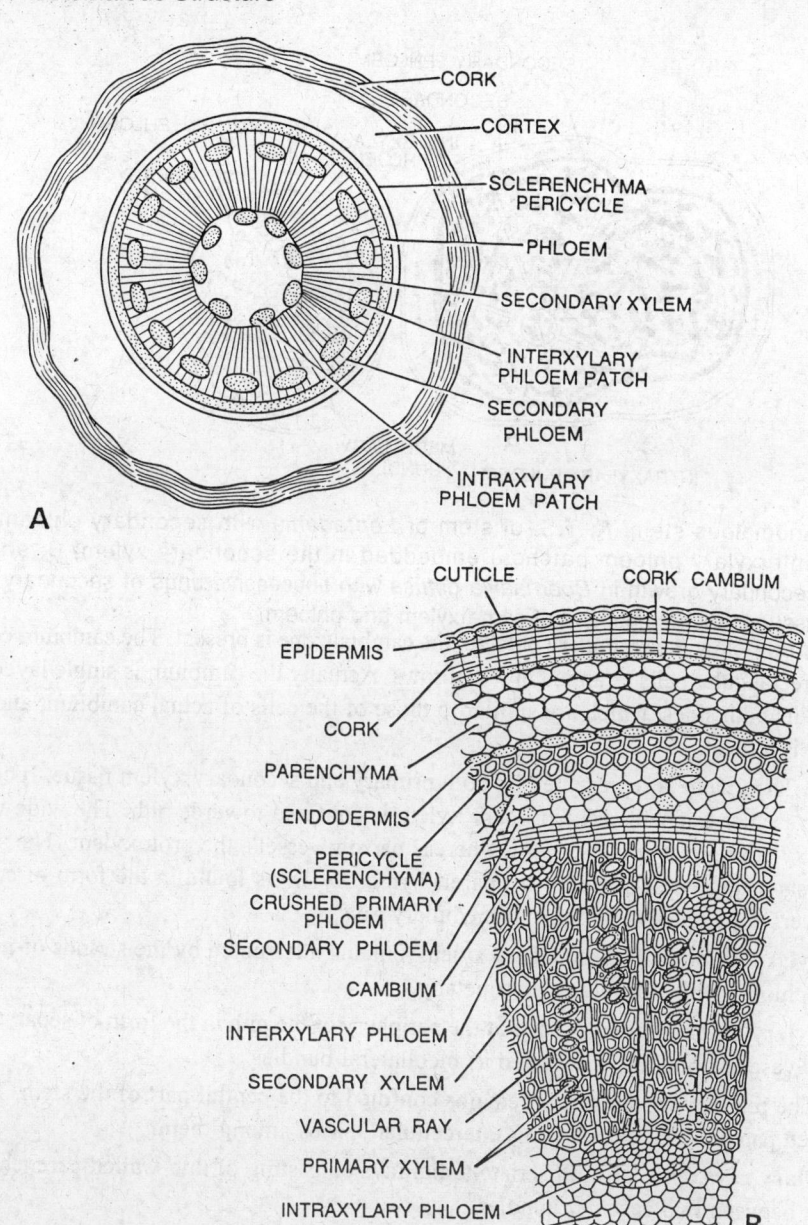

Fig. 11.19. The stem-anomalous structure. T.S. of *Strychnos* stem showing inter-and intraxylary phloem patches. A, diagrammatic; B, detailed.

endodermis consists of barrel shaped and compact cells having no intercellular spaces among them. Usually the endodermal cells contain starch grains.

Pericycle. Below the endodermis, a broad zone of pericycle is present. This pericycle zone consists of thin walled parenchyma cells interrupted by patches of sclerenchyma.

Phloem. The secondary phloem consists of sieve tubes, companion cells and phloem parenchyma. The primary phloem is also seen at certain places opposite the primary xylem vessels in broken or crushed form.

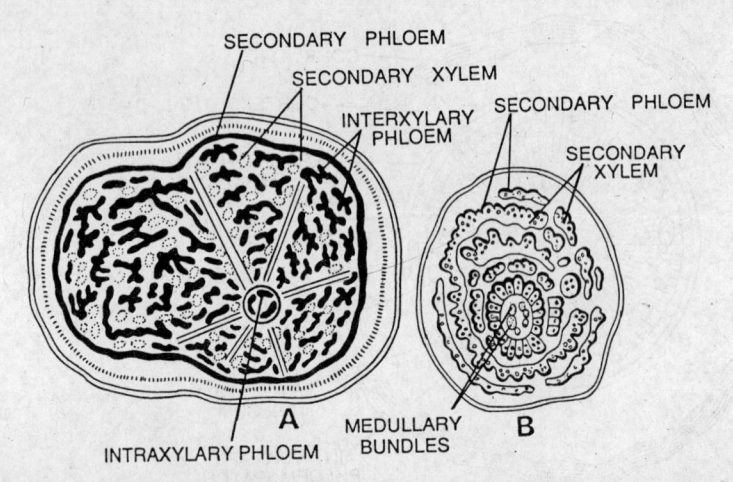

Fig. 11.20. Anomalous stem. A, T.S. of stem of *Leptadenia* with secondary phloem strands (interxylary phloem patches) embedded in the secondary xylem; B, anomalous secondary growth in *Boerhaavia diffusa* with successive rings of secondary vascular tissues, each composed of some xylem and phloem.

Cambium. Below the secondary phloem, the cambial zone is present. The cambium consists of thin walled, rectangular cells arranged in radial rows. Actually the cambium is single layered, but it cuts cells on its both sides, which are similar to those of the cells of actual cambium, and thus the cambial zone is formed.

Xylem. The xylem is represented by both primary and secondary xylem tissue. It consists of vessels and vascular tracheids. The primary xylem is confined towards pith. The wide vessels of the primary xylem represent metaxylem, whereas narrow vessels, the protoxylem. The secondary xylem consists of big vessels and xylem parenchyma. Xylem is found in the form of continuous cylinder traversed by narrow vascular or medullary rays.

Interxylary phloem. The secondary xylem remains interrupted by the strands of interxylary or included phloem which develops centripetally.

Intraxylary phloem. This is found at the periphery of the pith in the form of separate strands, and thus the vascular bundles are treated as bicollateral bundles.

Pith. The pith is very small and remains confined to the central part of the stem. It consists of thin walled parenchyma cells having intercellular spaces among them.

Medullary rays. Uni- or multiseriate radial rays consisting of thin walled parenchyma cells are found in between two vascular bundles.

Latex tubes. They are found among the parenchyma cells.

Anomalous structure. The strands of interxylary phloem patches (secondary) are present in the secondary xylem. The strands of intraxylary phloem are found around the pith region.

Xerophytic characters. A thick cuticle is present over the epidermis. The presence of chlorenchyma in the cortex denotes the xerophytic character. The sclerenchyma patches are well developed in the pericyclic region. The latex tubes are present among the parenchyma cells.

Anatomy of *Salvadora* Stem

Epidermis. The single layered epidermis consists of barrel shaped cells, and remains covered with a very thick cuticle.

Cortex. The cortex consists of thin walled parenchyma cells. Below the epidermis few layers of thin-walled parenchyma cells represent hypodermis. Beneath the hypodermis three or four layers of

The Stem—Anomalous Structure

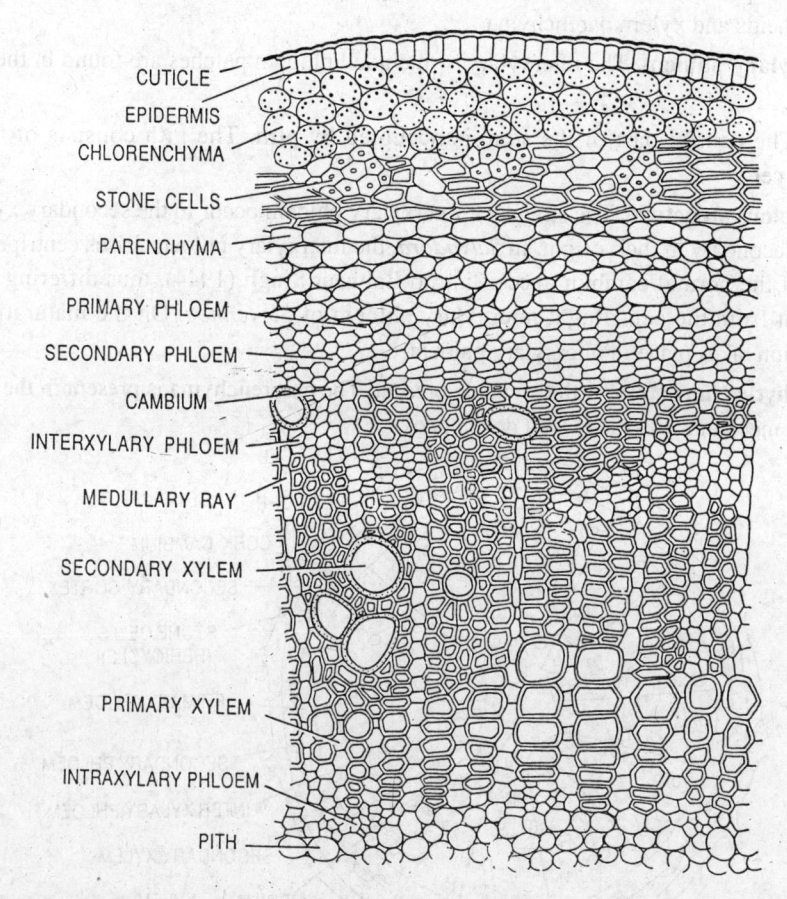

Fig. 11.21. The stem—anomalous structure. T.S. of a sector of stem of *Leptadenia* (Asclepiadaceae—dicot) showing inter- and intraxylary phloem patches.

chlorenchyma are present which contain a large number of chloroplasts. The innermost layer of the cortex is endodermis consisting of barrel shaped closely arranged cells. The endodermal cells contain starch grains in them.

Pericycle. Below the endodermis a very conspicuous pericyclic zone is present which consists of widely spaced strands of thick-walled fibres opposite the vascular bundles.

Phloem. The secondary phloem consists of sieve tubes, companion cells and phloem parenchyma. Sometimes the primary phloem is also seen at certain places, opposite the primary xylem in crushed condition. The phloem is found in the form of complete cylinder usually interrupted by one to three-cells wide medullary rays.

Cambium. In between the cylinders of secondary phloem and secondary xylem, there lies the cambial zone. The cambium consists of thin walled, rectangular cells.

Xylem. Both primary and secondary xylem strands are distinctly seen. The primary xylem is found towards pith. It can be recognized by the presence of wide vessels of metaxylem and narrow vessels of protoxylem. The secondary xylem is found in the form of a circular cylinder, which remains interrupted by parenchymatous medullary rays. The secondary xylem is represented by vessels, tracheids and xylem parenchyma.

Interxylary phloem. The interxylary or included phloem patches are found in the secondary xylem.

Pith. The central part of the stem is occupied by pith. The pith consists of thin walled parenchyma cells.

Anomalous structure. The strands of interxylary phloem occur in the secondary xylem. These patches are secondary in their origin. In *Salvadora*, the interxylary phloem arises centripetally on the inner side of the normal cambium according to Balwant Singh (1944), thus differing in mode of development from the centrifugal interxylary phloem of *Strychnos*. On the maturation cellular disorganization takes place in interxylary phloem.

Xerophytic characters. A thick cuticle present. The chlorenchyma is present in the cortex. The thick-walled mechanical tissue is well developed.

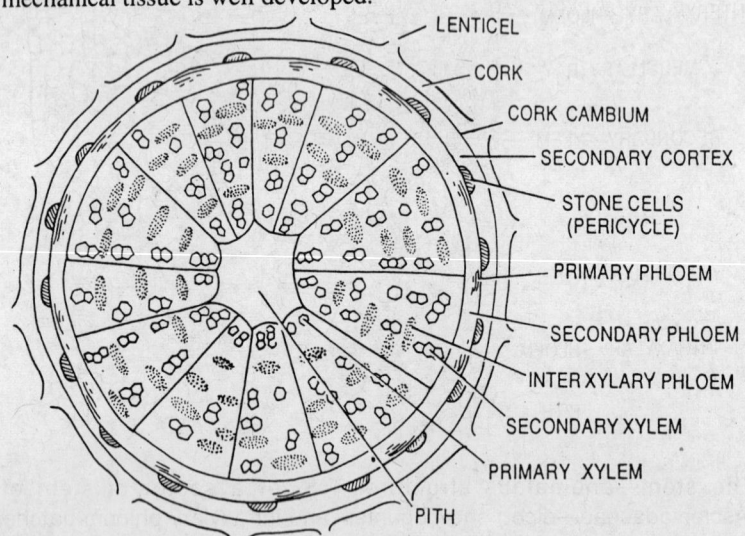

Fig. 11.22. The stem—anomalous structure. T.S. of *Salvadora* stem (dicot—Salvadoraceae)—diagrammatic.

ABSENCE OF VESSELS IN THE XYLEM

Normally the vessels are found in the xylem of angiosperms. However, in some genera belonging to different families of angiosperms, the vessels in the xylem are lacking. These genera and families are as follows: (*Belliolum, Drimys, Exospermum, Pseudowintera* and *Zygogynum* of Winteraceae; *Trochodendron* of Trochodendraceae and *Tetracentron* of Tetracentraceae).

Sometimes in the roots of *Drimys* the vessel like structures appear due to injuries.

The vessels are also not found in some aquatic angiosperms (*e.g., Ceratophyllum, Hydrilla*, etc.).

SCATTERED VASCULAR BUNDLES IN DICOTS

The presence of vascular bundles in a ring is a normal feature of dicotyledonous stems. However, in several members of various families of dicotyledons, the vascular bundles are found

The Stem—Anomalous Structure

to be in scattered condition in the stem (*e.g., Peperomia* and *Piper* of Piperaceae; *Elastostemma sessilis* of Urticaceae; *Thallictrum* and *Anemone* of Ranunculaceae; *Nymphaea* of Nymphaeaceae; *Papaver orientale* of Papaveraceae).

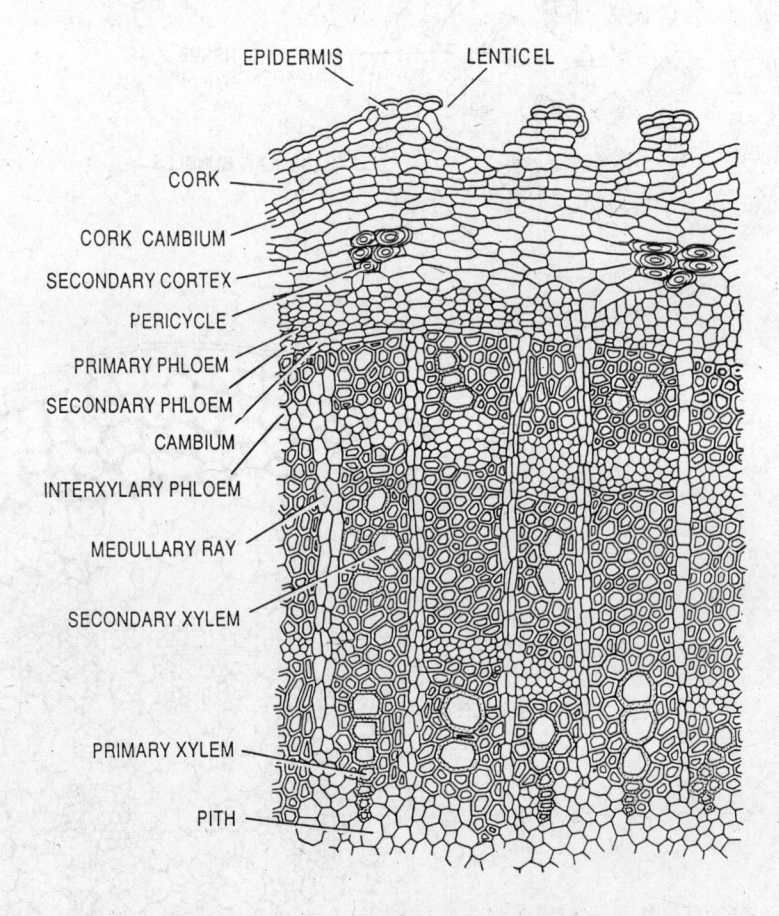

Fig. 11.23. *Salvadora* (dicot). T.S. of stem (detailed) showing interxylary phloem patches.

Anatomy of *Peperomia langsdorfii* Stem

Epidermis. It consists of a single layer of thin walled, compact cells, without intercellular spaces. The cuticle is quite thin.

Cortex. The outer part of the primary cortex consists of a continuous ring of collenchyma. Endodermis is inconspicuous.

Vascular system. It is composed of scattered vascular bundles. The irregular bundles remain embedded in parenchymatous ground tissue. Each bundle consists of phloem, xylem and a strip of cambium.

Anomalous feature. The presence of scattered vascular bundles in parenchymatous ground tissue shows an anomalous feature for dicotyledonous stems.

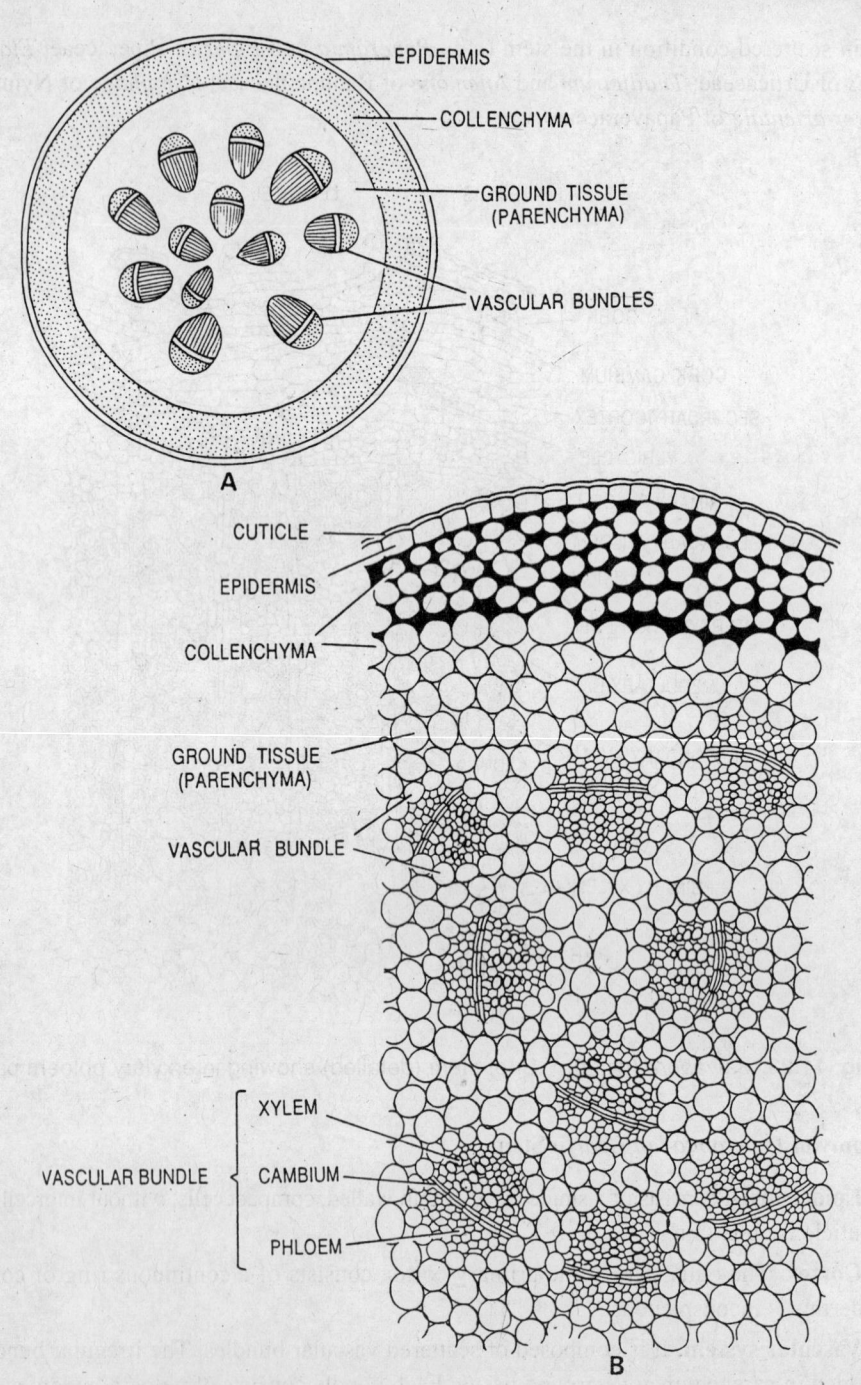

Fig. 11.24. Anomalous structure. T.S. of *Peperomia* (dicot stem). A, diagrammatic; B, detailed, showing scattered vascular bundles.

PRESENCE OF EXCLUSIVE PHLOEM AND XYLEM BUNDLES

Phloem bundles. Sometimes only phloem bundles are found to be developed in between other collateral bundles (*e.g.*, in *Cuscuta*). In *Entada, Combretum, Leptadenia, Salvadora* and others, the phloem patches (interxylary phloem) are developed in the secondary xylem.

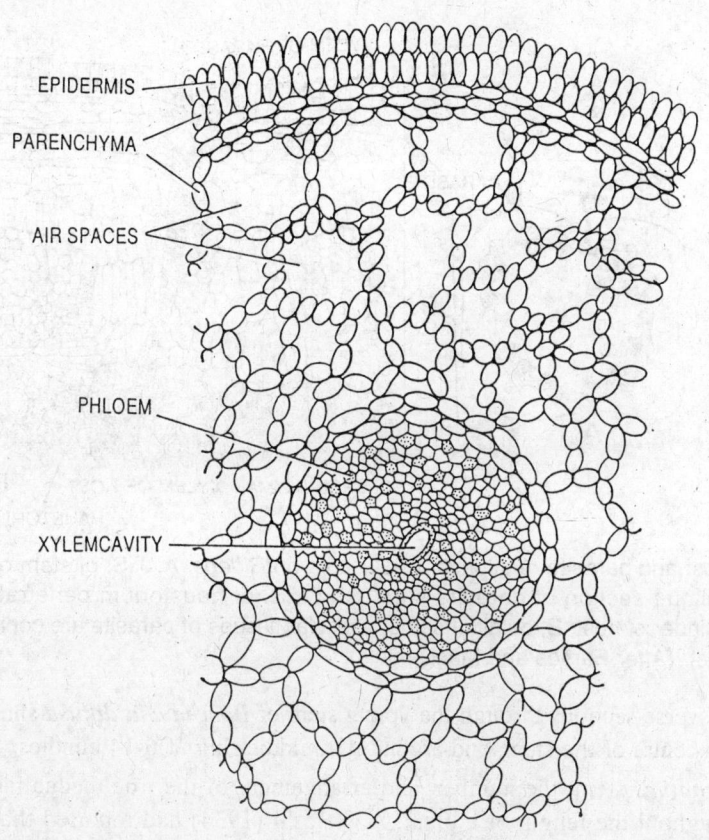

Fig. 11.25. Anatomy of hydrophytic stem. T.S. of an aquatic stem of *Hydrilla* (monocot.)—detailed structure of a sector.

Anatomy of *Cuscuta* Stem

The vascular system of *Cuscuta* consists mainly of primary tissue which may be regarded as reduced form. There are minute groups of vessels accompanied by phloem strands which may be so closely packed that they almost form a closed ring. The phloem is generally more fully developed than the xylem.

Phloem bundles. Usually separate phloem bundles are found in between other collateral groups of xylem and phloem.

Anomalous feature. Normally the collateral bundles are found in the angiospermic stems. Here, the presence of separate phloem bundles denotes anomaly in angiosperms.

Xylem bundles. In *Paeonia* of Ranunculaceae, in addition to normal collateral vascular bundles, exclusive xylem bundles are also found.

PRESENCE OF MEDULLARY BUNDLES

In many dicotyledons, in addition to the normal vascular bundles arranged in a ring, there are found medullary bundles. The medullary bundles may be either scattered or arranged in a ring. In majority of cases the medullary bundles are primary and originate in normal way. The medullary bundles have been recorded in 38 dicot families. Some of them have been enlisted here: (*e.g., Amaranthus, Achyranthes* of Amaranthaceae; *Anemone, Delphinium, Ranunculus chinensis* of Ranunculaceae; *Bougainvillaea, Mirabilis, Boerhaavia* of Nyctaginaceae; *Peperomia, Piper* of Piperaceae; *Begonia menicata* of Begoniaceae; *Chenopodium* of Chenopodiaceae; Cruciferae, Cactaceae, Umbelliferae and others).

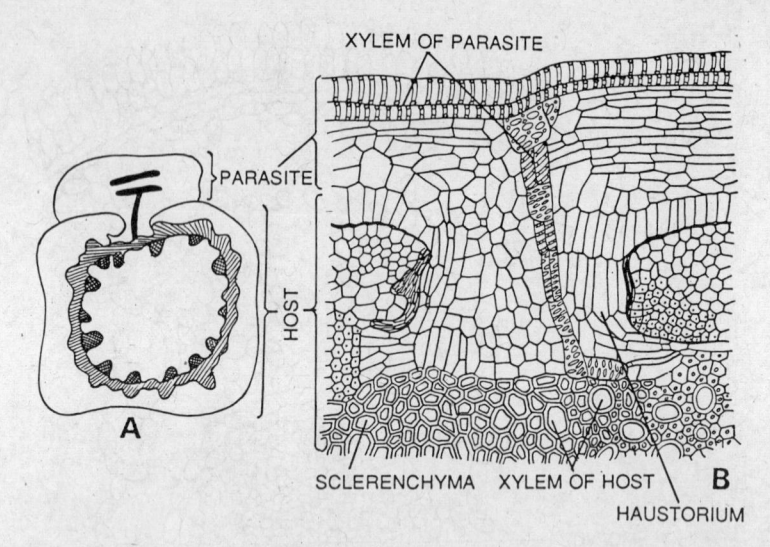

Fig. 11.26. Host and parasite relationship—*Cuscuta* on *Bidens*. A, T.S. of stem of host, somewhat oblique section of parasite stem, and L.S. of haustorium penetrating the vascular cylinder of host; B, detail of same; vascular tissues of parasite are connected to those of host. (After Eames and MacDaniels).

The transverse sections through the young stem of *Boerhaavia diffusa* show two medullary bundles in the centre of the stem, and around it a middle ring of 6-14 bundles.

In *Achyranthes* stem, the number, and arrangement of the true meduallary bundles is not constant throughout the length of a plant. A.C. Joshi (1934) had reported that in *Achyranthes aspera* two separate medullary bundles occur throughout the length of an individual plant. The bundles remain quite separate immediately above or below a node while they unite to form a single *amphixylic bundle* elsewhere in the internode. However, in the inflorescence of the same species there are 4 to 9 medullary bundles.

In *Piper betle*, there are numerous medullary bundles which are found to be scattered in the pith.

In *Piper excelsum*, the medullary bundles are arranged in a ring.

In *Bougainvillaea* and *Mirabilis* of Nyctaginaceae, there are two big medullary bundles in the centre, which may remain surrounded by the smaller bundles.

PRESENCE OF CORTICAL BUNDLES

In several dicots, a ring of vascular bundles is found in cortical region, these bundles may be referred as '*cortical bundles*' of they are to be interpreted as '*leaf traces*'. The presence of such type of bundles has been studied in many families (*e.g.*, Begoniaceae; Cactaceae; Casuarinaceae; Cucurbitaceae, Proteaceae, Oleaceae and several others). A system of inversely oriented vascular bundles has been studied by Majumdar (1941) in *Nyctanthes arbortristis* of Oleaceae.

In the stem of *Nyctanghes arbortristis* (Oleaceae-dicot.), apart from normal vascular bundles which occur in a ring in the central region, there are four inversely oriented vascular bundles at the four ridges of the stem. These cortical bundles are collateral and open. (See fig. 11.28)

INTRAXYLARY PHLOEM

This is also called internal phloem. It occurs usually in the form of strands or as continuous band around the pith. The origin of intraxylary (internal) phloem in most plants is primary. The

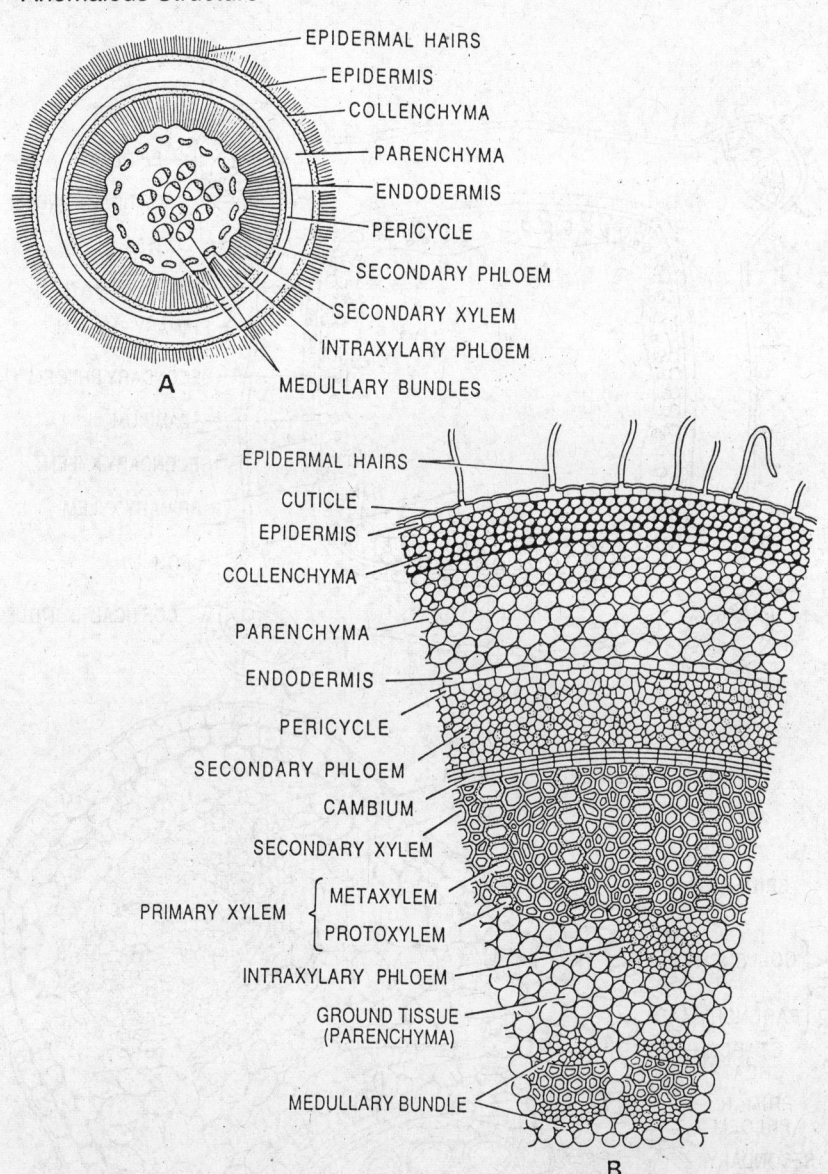

Fig. 11.27. The stem—anomalous structure. T.S. of stem of *Argyereia* (dicot-Convolvulaceae—T.S. showing clothing of epidermal hairs with short stalked cells; cortex of collenchyma and parenchyma, endodermis conspicuous; pericycle scanty; secondary phloem lacking fibres; xylem in the form of cylinder; vessels large and arranged in radial rows; xylem groups occasionally developed in association with the interxylary phloem, thus converting the interxylary phloem with inversely oriented meduallary bundles; pith unlignified; interxylary phloem commonly found in woody species of *Argyereia*. A, diagrammatic; B, detailed.

cells of intraxylary phloem are like those of external phloem except that fibres are few or wanting, and the sieve tubes and companion cells occur in small groups, surrounded by parenchyma. The internal phloem develops after the development of external primary phloem. The bundles are treated as bicollateral, because of the presence of the internal (intraxylary) phloem. The internal phloem is found in 28 dicot families. The names of the important ones are mentioned here (*e.g.,* Asclepiadaceae;

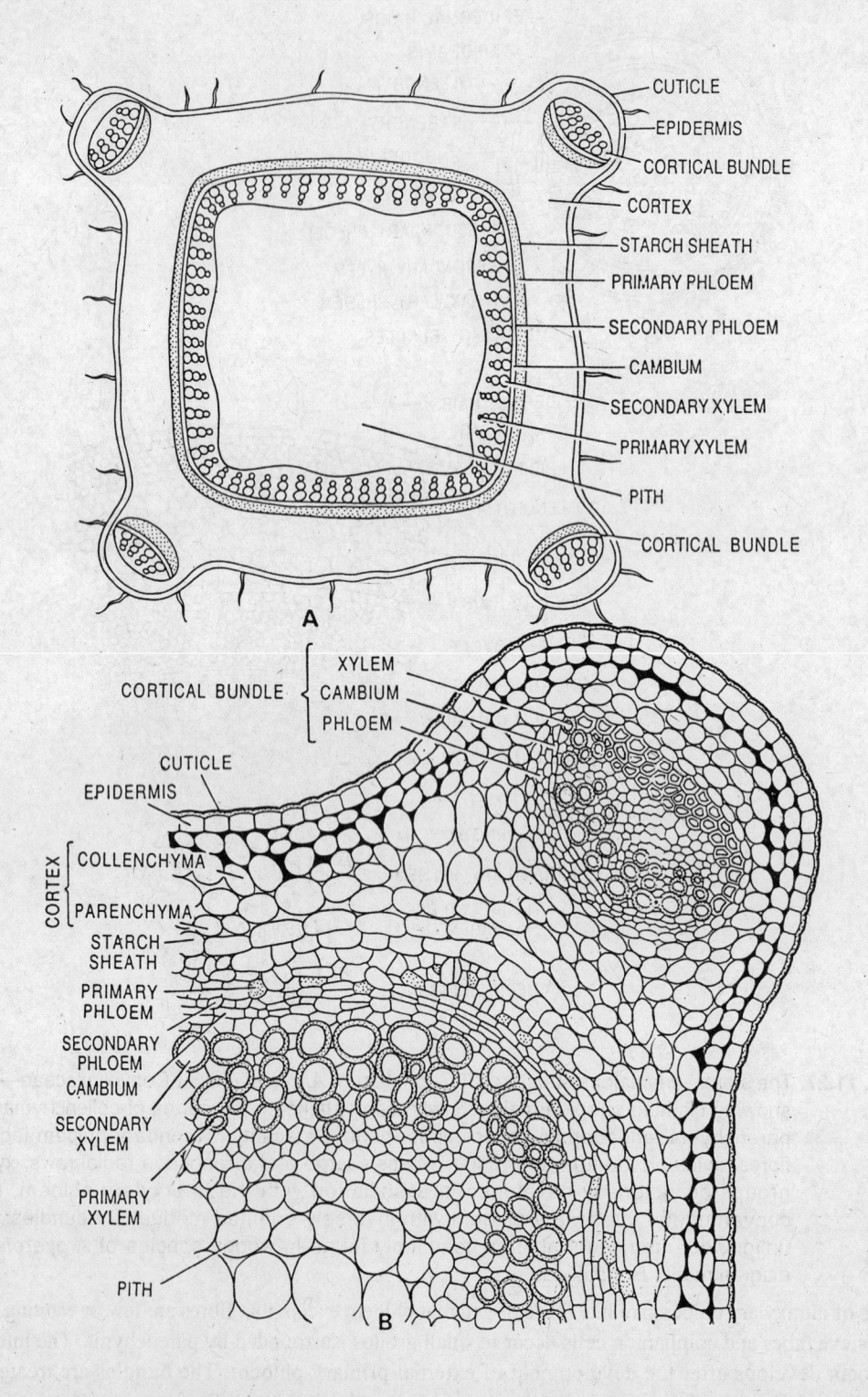

Fig. 11.28. *Nyctanthes arbortristis* (dicot). T.S. of stem showing inversely oriented cortical vascular bundles. A, diagrammatic; B, detailed.

The Stem—Anomalous Structure

Convolvulaceae; Punicaceae; Loganiaceae; Solanaceae, Apocynaceae; Cucurbitaceae; Lythraceae and several others).

Anatomy of *Punica* stem. As seen in cross section, the stem is provided with appendages. Epidermis is single layered. Usually cork arises in the inner part of the pericycle. The pericycle is represented by a ring of fibres along the outer periphery. Phloem and xylem are found in continuous cylinders. Xylem forms a closed cylinder, traversed by narrow medullary rays. The vessels are provided with simple perforations. The pith consists of parenchyma cells. Usually rosette crystals are abundant in the phloem. Secretory cells are also found in cortex and pith.

Special anatomical feature. The most interesting anatomical feature is the occurrence of intraxylary phloem in a continuous strand around the periphery of pith.

Anatomy of the stem of *Convolvulus floridus*. As seen in transverse section, the stem exhibits the following important parts:

Epidermis. This is single outermost layer consisting of thin walled compact cells covered with a thin cuticle.

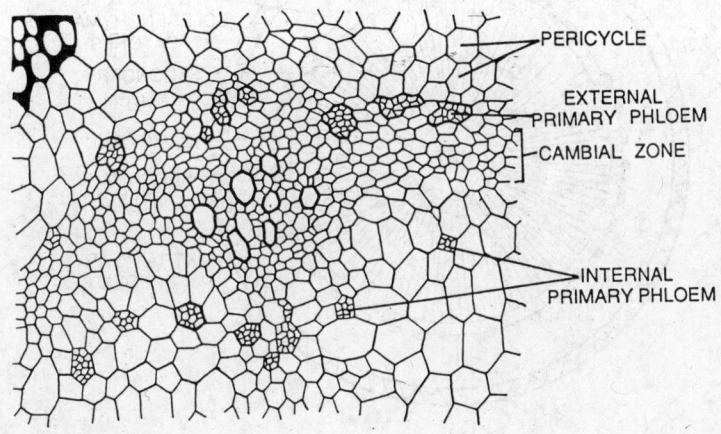

Fig. 11.29. Internal (intraxylary) phloem in T.S. of young stem of *Solanum tuberosum*. Primary phloem cells are seen in small groups external and internal to the cambial zone. (After Artschwager).

Cork and cortex— The cork is usually superficial in origin and consists of thin-walled rectangular cells. Commonly the primary cortex consists of collenchyma. The innermost layer of the cortex is endodermis.

Pericycle—It consists of a thin walled grouped fibres arranged in discontinuous ring. The intervening gaps are filled up with parenchyma cells.

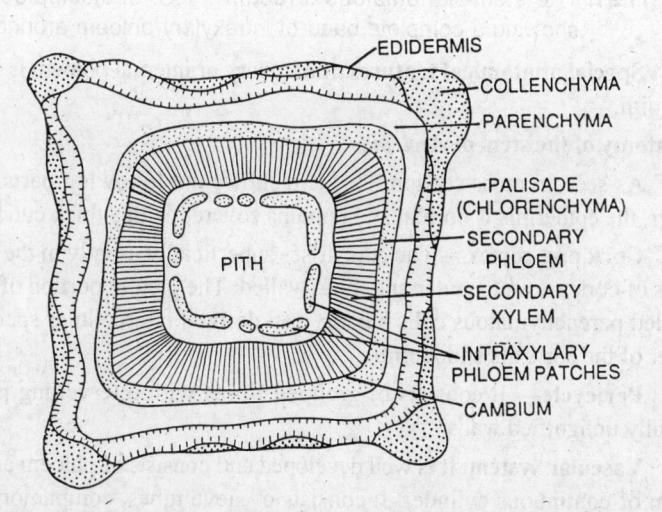

Fig. 11.30. The stem—anomalous structure. T.S. of the stem of *Capsicum* spp. showing intraxylary phloem patches around the pith.

Vascular system. It consists of well developed, collateral and open vascular bundles arranged in a ring. Both primary and secondary tissues are found. Secondary phloem consists of sieve tubes, companion cells and phloem parenchyma and found in the form of continuous cylinder. Xylem is also found in the form of a continuous cylinder traversed by narrow medullary rays. Vessels of primary xylem are comparatively smaller and arranged in radial rows towards pith. Later developed vessels are comparatively larger and confined to certain regions of xylem cylinder.

Laticiferous canals. Commonly the laticiferous canals are found in the cortical region of *Convolvulus*.

Pith— The central region of the axis is occupied by pith, consisting of thin-walled parenchyma cells with intercellular spaces.

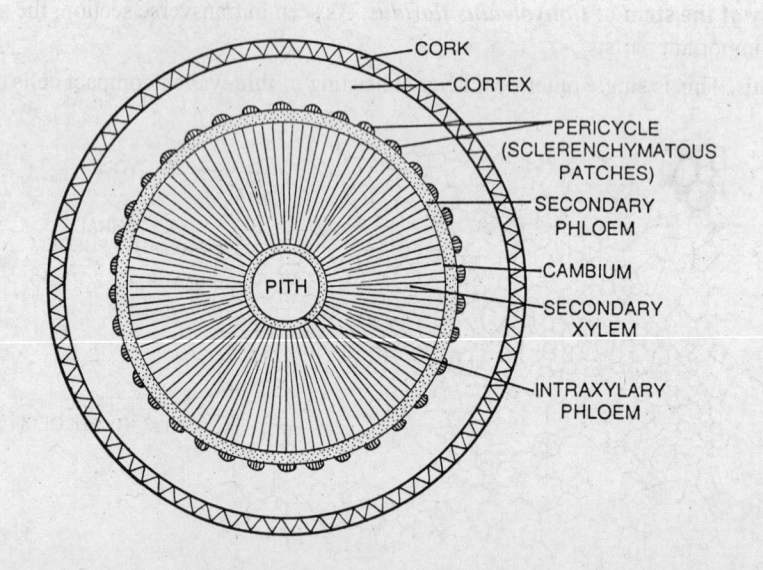

Fig. 11.31. The stem—anomalous structure. T.S. of *Asclepias* stem (dicot-Asclepiadaceae) showing a complete band of intraxylary phloem around the pith.

Special anatomical features. Intraxylary or internal phloem is found in continuous band around the pith.

Anatomy of the stem of *Asclepias*

As seen in cross section, the stem shows the following parts.—**Epidermis**— The outermost layer, the epidermis is normal and ramains covered with a thick cuticle.

Cork and cortex— The cork arises superficially usually in the epidermis or sub-epidermis. The cells of cork are rectangular and thick-walled. The greater portion of the cortex is represented by thin walled parenchymatous cells having well defined intercellular spaces among them. The innermost layer of the cortex is endodermis.

Pericycle— It consists of grouped fibres with intervening parenchyma. The fibres possess usually unlignified walls.

Vascular system. It is well developed and consists of phloem and xylem. Phloem is found in the form of continuous cylinder. It consists of sieve tubes, companion cells and phloem parenchyma. Primary xylem is represented by wide metaxylem vessels and narrow protoxylem vessels. The secondary xylem is found in the form of protoxylem vessels. The secondary xylem is found in the form of continuous cylinder traversed by narrow medullary rays.

The Stem—Anomalous Structure

Special anatomical features. The intraxylary or internal phloem is found in the form of a continuous band around the pith. Laticiferous canals are present among the parenchyma cells.

VASCULAR BUNDLES ARRANGED IN A RING IN MONOCOTS

Normally, the cross section of monocot stem exhibits many scattered bundles. However, in *Tamus communis* of Dioscoreaceae, the vascular bundles are arranged in a ring around large pith. The vascular bundles are also arranged in a ring in several hollow monocots (*e.g., Triticum, Avena, Hordeum, Oryza* and several other members of Gramineae.)

Anatomy of the stem of *Tamus communis*—Epidermis— It consists of a single layer of compact, barrel shaped cells having no intercellular spaces. A thin cuticle covers the epidermis.

Cortex. The presence of cortex is an anomalous feature in monocot stem. The cortex is well developed and consists of thin walled, rounded or oval parenchyma cells having well defined intercellular spaces among them. The innermost layer of the cortex is endodermis. It is conspicuous, consisting of barrel-shaped cells, having no intercellular spaces.

Pericycle. Immediately beneath the epidermis, multilayered sclerenchymatous pericycle is present. This makes a complete cylinder of sclerenchyma.

Ground tissue. It is represented by large, thin walled, rounded or oval parenchyma cells having intercellular spaces.

Vascular system. It consists of many vascular bundles arranged in a ring (anomalous feature). Usually they are arranged in two rings (outer and inner). Each vascular bundle consists of xylem and phloem. The bigger vessels (pitted) towards outside constitute metaxylem, whereas narrower vessels (annular and spiral) towards centre form protoxylem. Phloem consists of sieve tubes and companion cells. Phloem parenchyma is absent. There is no secondary growth or cambium formation.

Pith. The central region of the stem is occupied by pith. It consists of thin walled, rounded parenchyma cells having well developed intercellular spaces among them.

Anomalous features. The vascular bundles are arranged in a ring; well developed cortex is present; endodermis is conspicuous; the sclerenchymatous pericycle is present; the central region of the stem is represented by pith.

SECONDARY GROWTH IN MONOCOTS

The secondary growth accurs in herbaceous and woody Liliflorae (*Aloe, Sansevieria, Yucca, Agave, Dracaena*) and other groups of monocots (Cheadle, 1937). The meristem concerned with this growth is called cambium. The cambium appears to be a direct continuation of a primary thickening meristem (Eckardt, 1941). However, the cambium functions in the part of the axis, that has completed its elongation. The cambium originates in the parenchyma outside the vascular bundles. This part of the axis is sometimes identified as cortex, and sometimes as pericycle. Here, the anatomy of the stem of *Dracaena* has been discussed.

Anatomy of *Dracaena* stem. The cross section of *Dracaena* stem exhibits the following structure: The single layered epidermis remains covered with thick cuticle. The lenticles are also seen on the epidermis. Beneath the epidermis and hypodermis, the cork cambium arises which gives rise to the cork towards outside. Below the cork cambium, well developed parenchyma is present.

Anomalous structure. *Dracaena* shows anomalous secondary growth. The cambium appears in the parenchyma outside the outermost vascular bundles. This region in which the cambium appears, is sometimes identified as cortex, and sometimes as pericycle. The newly formed cambium cuts cells towards outside and inside both. The tissue developed on the inner side of the cambium is usually differentiated into vascular bundles remain separated from each other by lignified tissue, sometimes this tissue remains unlignified and thinwalled. The cells formed on the outer side of the cambium make parenchyma.

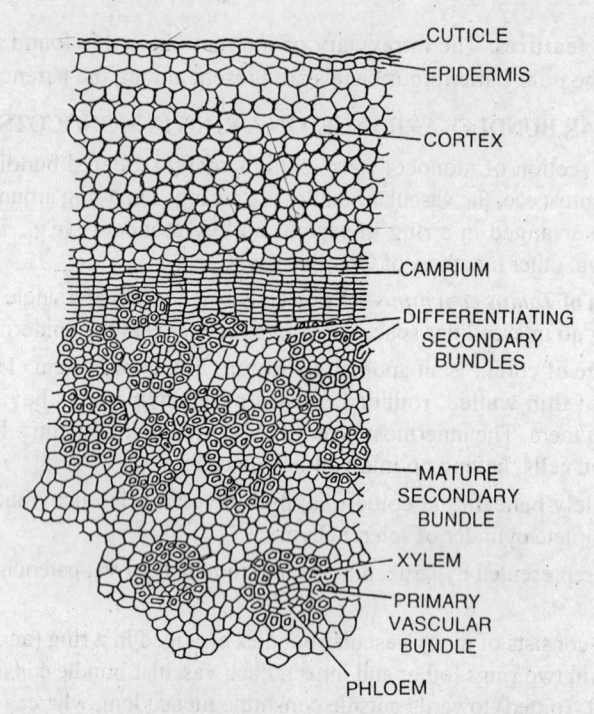

Fig. 11.32. Anomalous secondary growth in *Aloe arborescens* (monocot) stem; T.S. of the stem showing secondary growth.

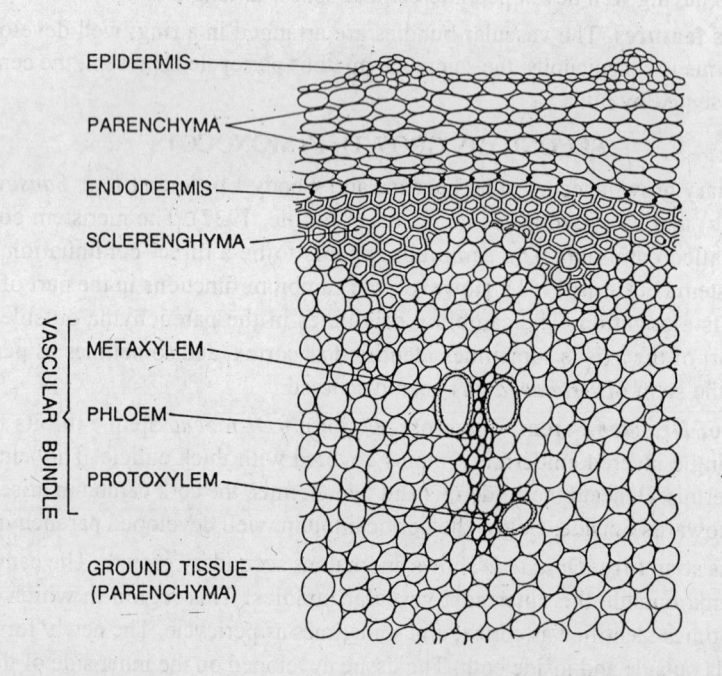

Fig. 11.33. *Tamus communis* (monocot). T.S. of stem (detailed structure) showing vascular bundles in a ring.

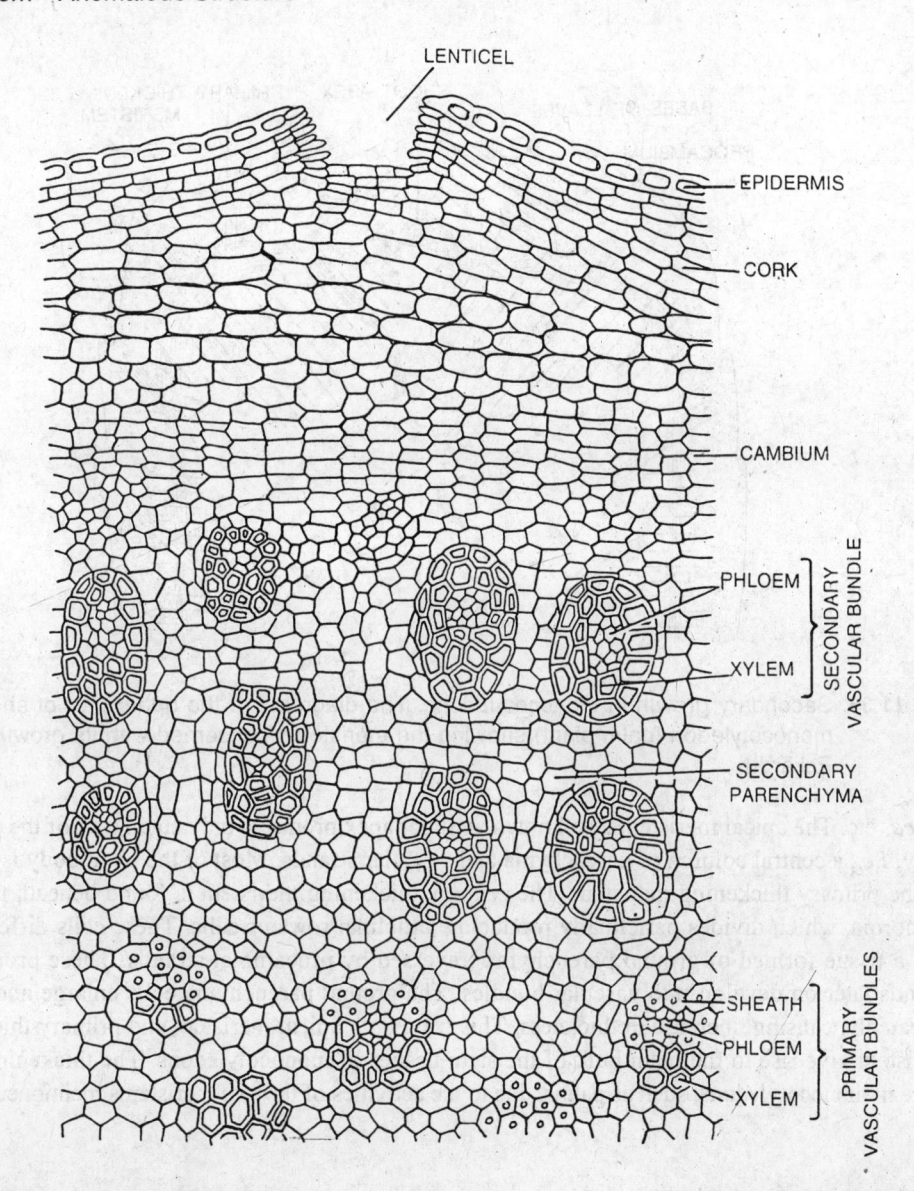

Fig. 11.34. Anomalous secondary growth. T.S. of a portion of stem of *Dracaena* (monocot) showing secondary growth.

Thickening in Palms

The palm stems do not increase in girth, because of any cambial activity but this thickening is the result of gradual increase in size of cells and of intercellular spaces and sometimes of the proliferation of fibre tissues. This is the type of long continuing primary growth. The process is as follows:

Most of the monocotyledons lack secondary growth, but with the result of intense and long continuing primary growth they may produce such large bodies as those of the palms. The monocotyledons often produce a rapid thickening beneath the apical meristem by means of a peripheral primary thickening meristem as shown in figure. The activity of the primary thickening meristem resembles with secondary growth found in certain monocotyledons such as *Dracaena*,

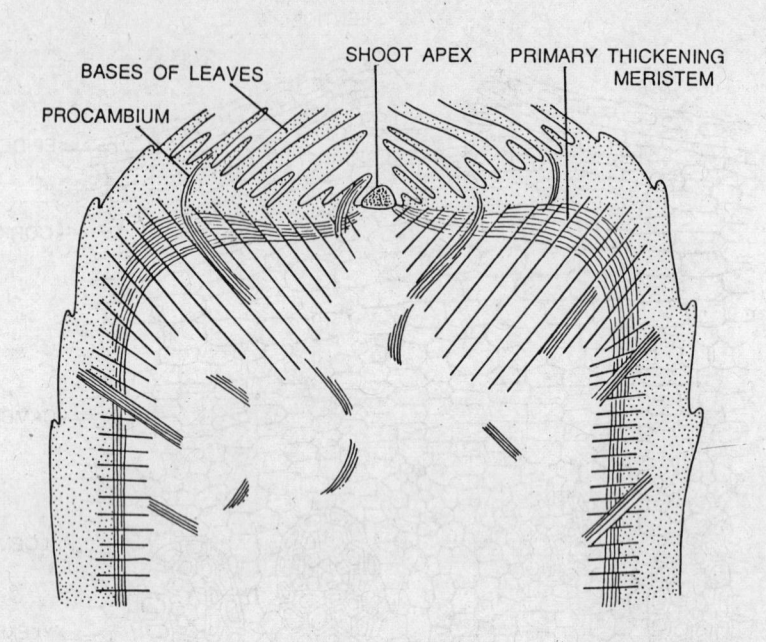

Fig. 11.35. Secondary growth in monocotyledons. The diagrams of the upper part of shoot of a monocotyledon (palm plant) showing the meristems concerned with its growth. (After Eckardt).

Yucca, etc. The apical meristem also known as shoot apex produces only small part of the primary body, *i.e.*, a central column of parenchyma and vascular strands. Most of the plant body is formed by the primary thickening meristem. The primary thickening meristem is found beneath the leaf-primordia, which divides periclinally producing anticlinal rows of cells. These cells differentiate into a tissue formed of ground parenchyma traversed by procambial strands. These procambial strands later on develop into vascular bundles. The ground parenchyma cells enlarge and divide repeatedly, causing increase in thickness. This way, both apical meristem and primary thickening meristem give rise to the main bulk of the stem tissues of monocotyledons. The thickening takes place in monocotyledons, such as palms, due to the activities of the two meristems mentioned above.

12

Anatomy of the Leaf and the Petiole

Commonly there are two types of leaves—1. *dorsiventral leaves* (dicotyledonous) and 2. *isobilateral leaves* (monocotyledonous). The dorsiventral leaves usually grow in a horizontal direction with distinct upper and lower surfaces, the upper being more strongly illuminated than

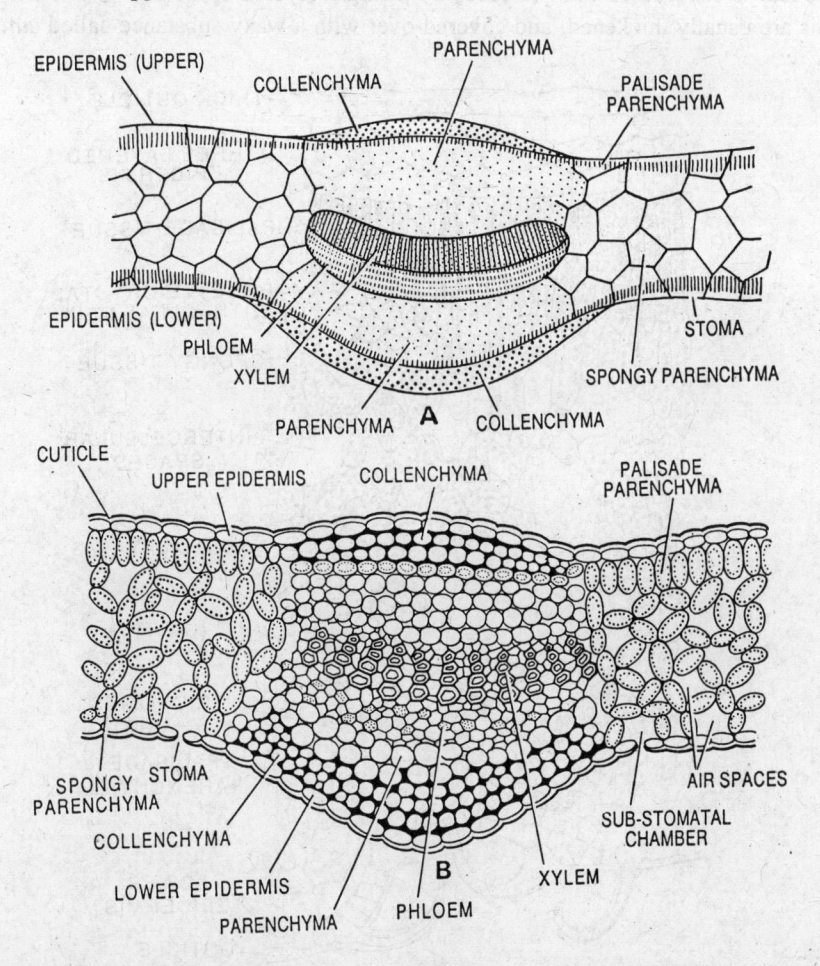

Fig. 12.1. Anatomy of bifacial dicot leaf. A, T.S. of leaf (diagrammatic); B, T.S. of leaf (detailed).

295

the lower. There exists a difference in the internal structure between the upper and lower surfaces of the dorsiventral leaf due to its unequal illumination. Most of the dicotyledonous leaves are dorsiventral. The isobilateral leaves hang vertically so that both surfaces of the leaf receive direct and equal amount of sunlight. The isobilateral leaves possess a uniform structure on both upper and lower surfaces. A very few dicotyledons and most monocotyledons have isobilateral leaves.

Usually the leaf is composed of various tissues, which furnish various functions. In discussions of the form and anatomy of the leaf, it is customary to designate the leaf surface that is continuous with the surface of the part of the stem located above the leaf insertion as *upper, ventral*, or *adaxial* side, the opposite side as the *lower, dorsal* or *abaxial*.

ANATOMY OF DICOTYLEDONOUS LEAF

To study the anatomy of leaf, several vertical sections passing through the mid-rib are required. The internal structure of the dicotyledonous leaf is as follows:

EPIDERMIS

The leaf is covered on both surfaces by a single-layered epidermis. The outer walls of the epidermis are usually thickened, and covered over with a waxy substance called *cutin*. The outer

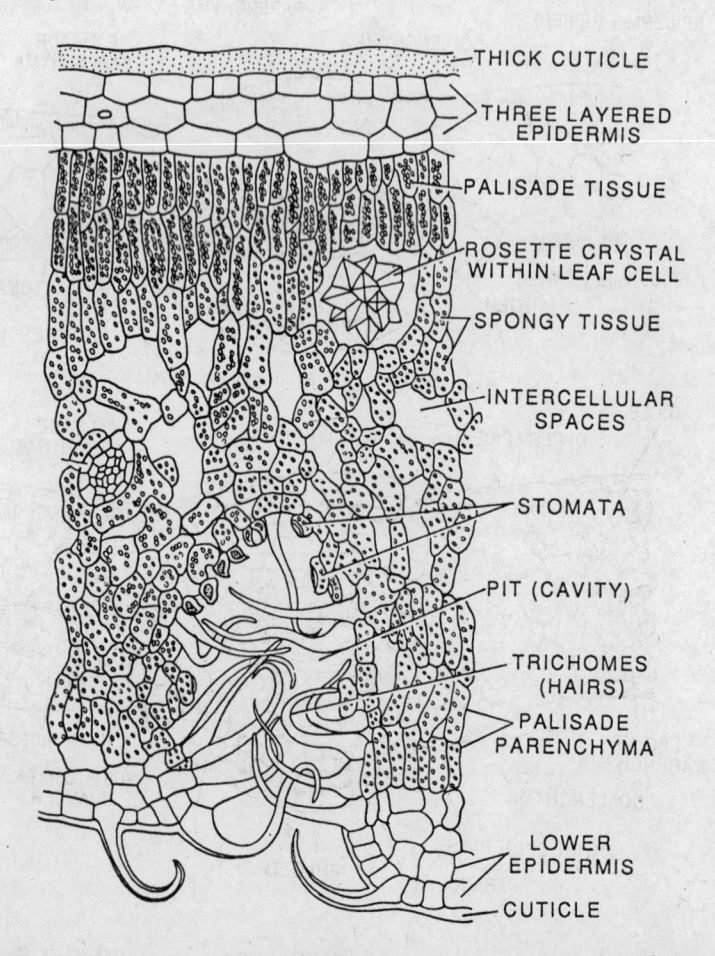

Fig. 12.2. Anatomy of xerophytic leaf. Cross section of a leaf of *Nerium* with upper and lower palisade, three-layered upper epidermis, and stomata in a pit protected by trichomes.

Anatomy of the Leaf and the Petiole

surfaces of the epidermis are frequently covered with a thin or thick cuticle. This cuticular layer is formed of cutin. As the outer walls of the epidermis are thick and cutinized, water does not pass through them rapidly and the transpiration from the surface of the epidermis is greatly reduced, only small quantity of water is evaporated by transpiration. The epidermis checks the transpiration to a great extent. The epidermis also prevents the entrance of pathogens into the interior of the leaf. Another function of the epidermis is the protection of the soft internal tissue of the leaf from the mechanical injuries. Sometimes in the xerophytic leaves the epidermis cells become radially elongated and somewhat lignified. In *Nerium* leaf, the epidermis is multilayerd.

Numerous small openings called stomata, are found in the epidermal layers of the leaves. Stomata are found in most abundance in the lower epidermis of the dorsiventral leaf. They are very few in the upper epidermis and sometimes altogether absent. In the floating leaves, stomata remain confined to the upper epidermis; in the submerged leaves the stomata are absent. In xerophytic leaves either stomata are sunken or situated inside the depressions.

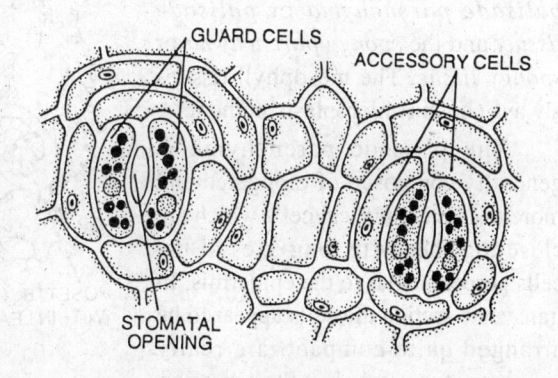

Fig. 12.3. Stomata in epidermal layer—surface view.

Each stoma remains surrounded by two semilunar guard cells. The guard cells are living and contain chloroplasts, they regulate opening and closing of stomata. The guard cells may remain surrounded by two or more *accessory cells* in addition to epidermal cells. The stomata are found in scattered condition.

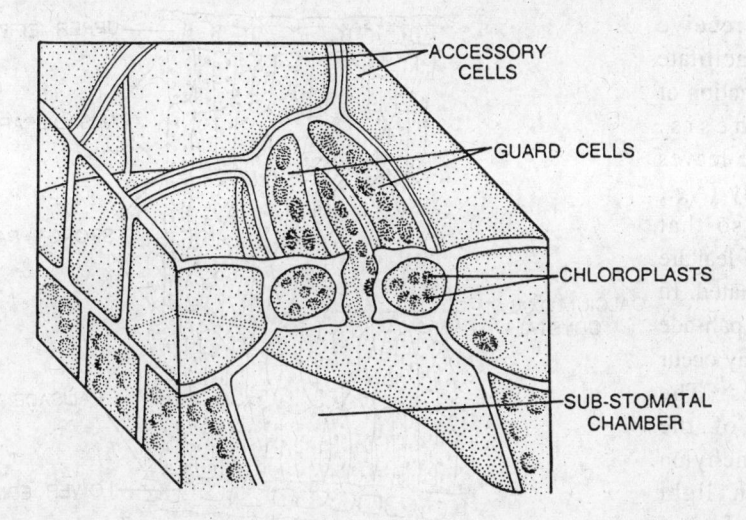

Fig. 12.4. Stomata. Diagrammatic representation of a stoma—surface and sectional views.

Usually the stomata are meant for exchange of gases in between the plant and the atmosphere. To facilitate the diffusion of gases properly, each stoma opens internally into a respiratory cavity or substomatal chamber. The transpiration takes place through the stomata, and the surplus water is being evaporated.

MESOPHYLL TISSUE

The tissue of the leaf that lies between the upper and lower epidermis and between the veins consists of typically thin walled parenchyma is known as *mesophyll*. This tissue forms the major portion of the inner of leaf. Commonly the cells of mesophyll are of two types—the *palisade parenchyma* or *palisade tissue*, and the *spongy parenchyma* or *spongy tissue*. The mesophyll tissues always contain chloroplasts in them.

The palisade parenchyma is generally composed of elongated and more or less cylindrical cells which are close together with long axes of the cells perpendicular to the epidermis. In transverse section the cells appear to be arranged quite compact, are really separate from each other having intercellular spaces among them. The palisade tissue may consist of a single or more layers. These cells are arranged near to the upper surface of the leaf, where they receive sunlight and facilitate to carry the function of photosynthesis. Sometimes the leaves hang vertically (*e.g., Eucalyptus*), so that both surfaces of leaf are equally illuminated. In such leaves the palisade parenchyma may occur on both sides. The compactness of the palisade parenchyma depends upon light intensity. The leaves which receive direct sunlight develop more compact parenchyma in comparison to the leaves which develop in shady places.

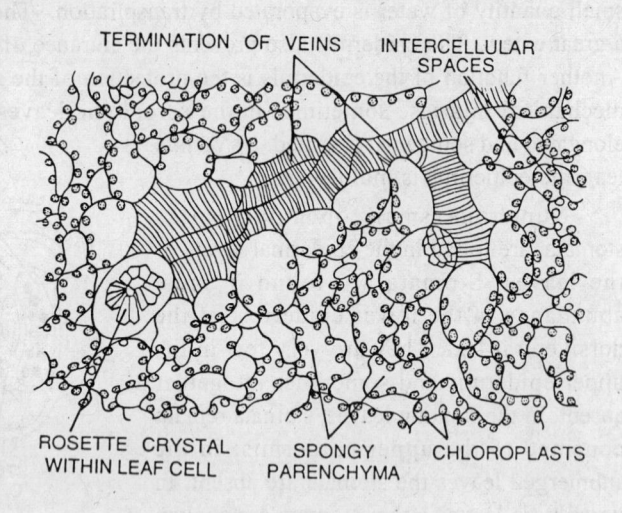

Fig. 12.5. The leaf. Termination of veins in a leaf, as seen in a section cut parallel with the epidermis.

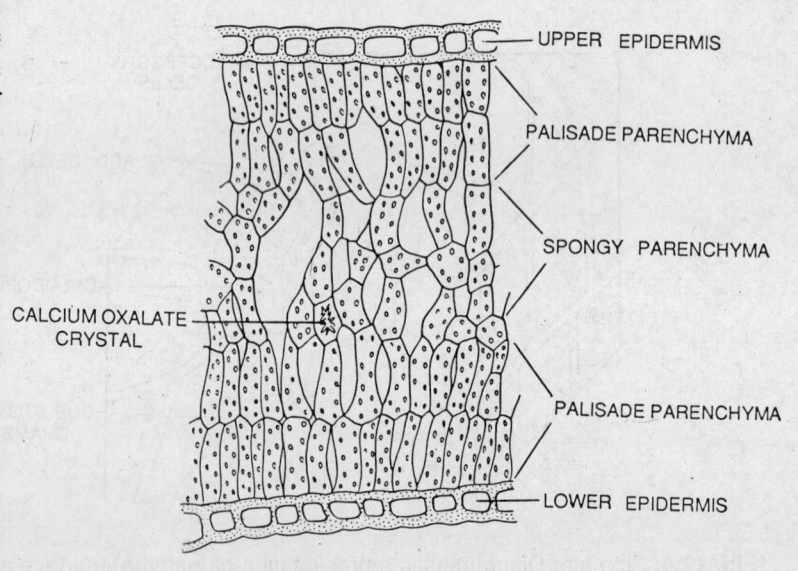

Fig. 12.6. Anatomy of leaf. Cross section of a vertical leaf of *Eucalyptus* showing a palisade layer on each side.

The lower portion of the mesophyll in the leaf is known as spongy parenchyma or spongy tissue. The spongy tissue is usually composed of loose, irregular, thin walled cells having big

Anatomy of the Leaf and the Petiole 299

intercellular spaces (air spaces) among them. The cells of spongy parenchyma also contain chloroplasts and carry on photosynthesis, but in comparison of palisade parenchyma less chloroplasts are developed. Due to the presence of a large air space in the spongy tissue they are more adaptable to the exchange of gases between the cells and the atmosphere.

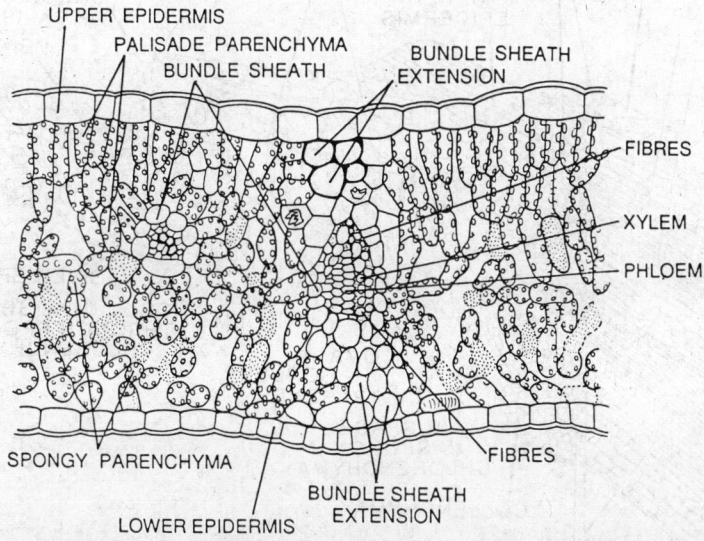

Fig. 12.7. Anatomy of leaf. T.S. of pear leaf. Palisade parenchyma towards upper side consists of 2 or 3 layers. Bundles are enclosed in bundle sheaths. The large vascular bundle has bundle sheath extensions reaching to the epidermis on both sides of the leaf.

The large air spaces that surround the spongy parenchyma cells are near the stomata and directly connected with them. There is therefore a much more free circulation of gases around these cells than around the palisade parenchyma cells, with the result that they are better suited to the exchange of gases between the cells and the surrounding atmosphere. The air spaces of the spongy chlorenchyma are not isolated chambers but a series of intercommunicating passages.

Both spongy and palisade parenchyma contain discoid chloroplasts arranged in parallel rows in the cells. As the chloroplasts are more dense in the palisade tissue than the spongy tissue the upper surface of the leaf appears to be deeper green than the lower surface.

MECHANICAL SUPPORT IN THE LEAF

The functions of the midrib and the lateral veins are to strengthen the leaf. The important tissue giving mechanical strength to leaf are—collenchyma, sclerenchyma, turgid parenchyma and woody xylem.

Collenchyma. In the centre of the upper portion of the midrib, just below the epidermis, there is usually a group of cells which give strength by having thickened walls and by being turgid. A group of the same kind of cells usually occurs also just above the lower epidermis. These cells constitute the *collenchyma*. Collenchyma is composed of living cells with walls which are thickened at the angles where three or more cells come in contact with one another. The thick places in the walls increase the strength of the cells, while the thin places allow for a more rapid transfer of materials from cell to cell than would take place if the cell walls were thickened throughout. These cells are more or less turgid, and so give strength to the leaf in the way also. The weight of the leaf causes it to tend to bend downward, with the result that there is a tendency for the upper portion to be stretched and the lower portion compressed. The collenchyma occurs, therefore, in those parts of the midrib in which there is the greatest need for strengthening material.

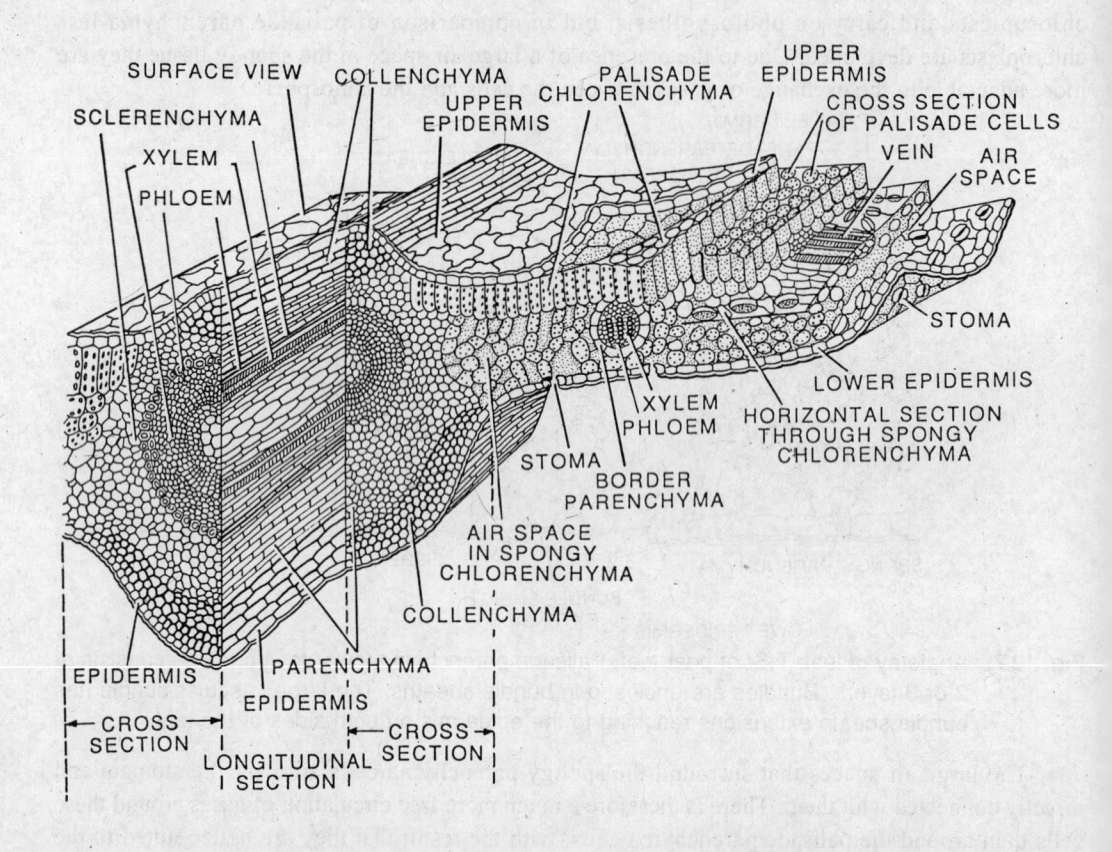

Fig. 12.8. The leaf. A three dimensional view of a section of a midrib and a small part of the leaf blade. On the left side is a cross section of half of the midrib followed by a longitudinal section, and this by a cross section of the remainder of the midrib and a portion of the blade. The leaf on the right side is dissected in various ways to show the arrangement of the tissues. (After Brown).

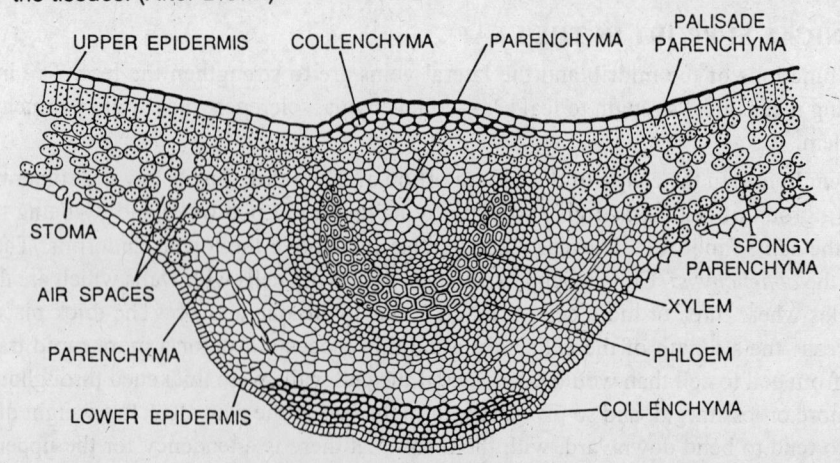

Fig. 12.9. Anatomy of dicotyledonous leaf. Cross section of midrib and leaf blade of leaf of *Erythroxylon*.

Anatomy of the Leaf and the Petiole

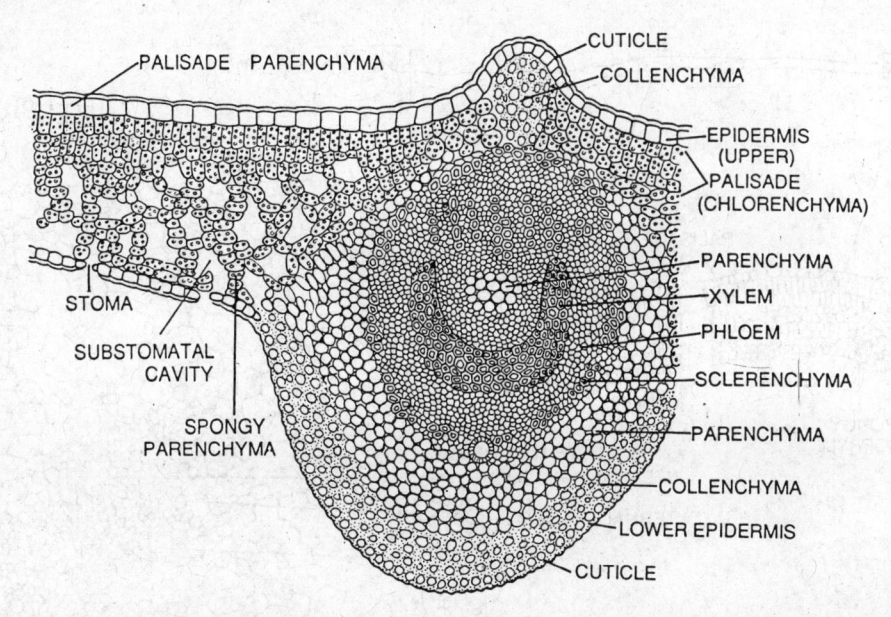

Fig. 12.10. Anatomy of dicotyledonous leaf. Cross section through a midrib and a part of leaf blade of a leaf of *Ixora*.

Sclerenchyma. Usually the sclerenchyma cells or the fibres are associated with the vascular tissues of the leaves. They occur usually as bundle caps adjacent to the phloem. Sometimes the fibres are found on both the sides of large vascular bundle of the leaves. Usually these cells are thick walled, dead and lignified. Their position just exterior to thin-walled phloem affords mechanical protection to the latter. The fibres are greatly elongated in the longitudinal direction of the midrib.

Turgid parenchyma. The regions between the collenchyma cells and the central portion of the midrib are occupied by parenchyma cells. In structure the parenchyma cells are not specially modified for any particular function, but they perform all the general functions of cells to a limited extent. Parenchyma cells have thin walls, but on account of their turgidity they strengthen the midrib.

Xylem. Usually the vessels and tracheids of xylem conduct water, but due to their thick walled nature they also give mechanical support to the leaves. The xylem elements are composed of lignified and dead cells.

ORIENTATION OF VASCULAR TISSUE

In the leaf traces of flowering plants, before they have the stele, the phloem is always found towards the outside of the stem. The leaf traces after their entrance in the petiole and lamina, also maintain the relative position of the xylem and the phloem, *i.e.*, the phloem is always found towards the lower side and the xylem towards the upper side in the leaf. Sometimes the xylem ring remains surrounded by a ring of phloem. The phloem occurs only below the xylem or rarely both above and below it.

Conducting system. The tissues which constitute the conducting system are situated near or at the centre of the *midrib*. This system may have various shapes, *e.g.*, the form of a ring, a crescent shaped ring, a crescent or scattered patches. In the ring shaped conducting system parenchyma cells are usually found in the centre of the ring. The inner part of the ring is composed of xylem (towards upper surface), and phloem (towards lower surface).

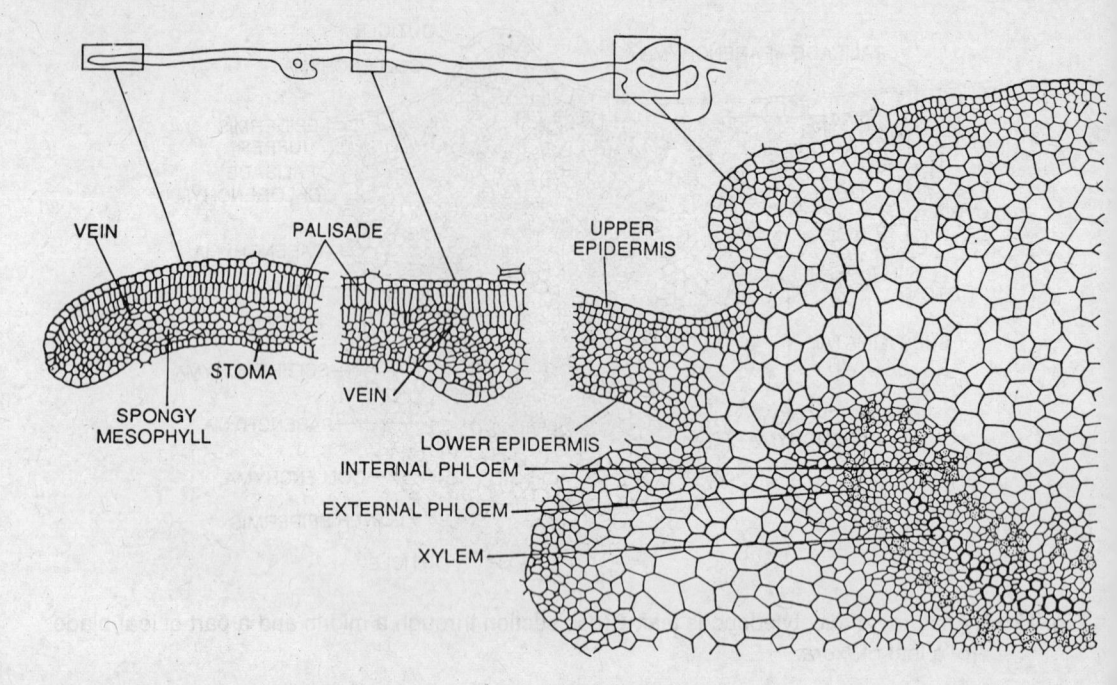

Fig. 12.11. Anatomy of dicotyledonous leaf of *Nicotiana tabacum*. T.S. through a part of leaf. Diagrammatic detail. (After Avery).

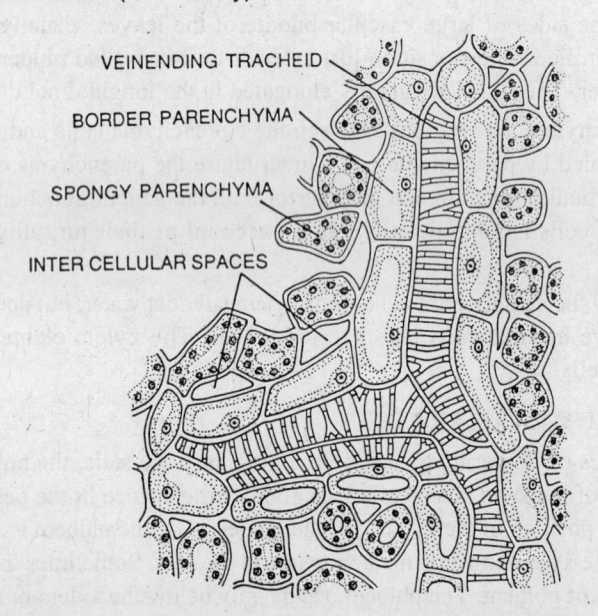

Fig. 12.12. Veins of the leaf. Diagram of vein ending in a leaf. The parenchyma tissue bordering the vein is conspicuous.

Xylem is composed of various kinds of vessels, trachieds wood fibres and wood parenchyma. Specially the vessels are annular and spiral. Xylem conducts water, raw food material and also gives mechanical support to the leaf. The phloem consists of sieve tubes, companion cells and phloem parenchyma. The phloem serves for the translocation of prepared food material from the mesophyll of the leaf.

Anatomy of the Leaf and the Petiole

Veins. The structure of large veins is more or less similar to that of a midrib. As they pass from the base of a leaf blade towards the apex or margin of the leaf, they get reduced in size, and simple in structure. The small veins consist of only of few conducting cells. The xylem is always found towards the upper surface and phloem towards lower even in very small veins. The cells of the mesophyll (chlorenchyma) are usually arranged so that the conduction of materials to and from the veins is facilitated.

The bundle sheath. The larger vascular bundles of dicotyledonous leaves remain surrounded by parenchyma with small number of chloroplasts, whereas the small bundles occur in the mesophyll. However, these small bundles do not remain in contact with intercellular spaces but are commonly enclosed with a layer of compactly arranged parenchyma, the *bundle sheath*. In dicotyledons the bundle-sheath parenhyma is also called *border parenchyma*. The bundle sheaths of dicotyledonous leaves usually consist of cells elongated parallel with the course of the bundle and having walls as thin as those of adjacent mesophyll. In some plants these cells have chloroplasts similar to those the mesophyll (*e.g.*, in *Humulus, Nicotiana tabacum*); in others they have few or no chloroplasts. The bundle sheath cells are in direct contact with the conducting cells of the vascular bundle of parenchyma and on the outer face with the mesophyll tissue. Individual sheath cells may contain crystals.

The parenchymatous bundle sheaths are more common, but in certain dicotyledons bundles of various sizes are enclosed in sclerenchyma, *e.g.*, Winteraceae, Melastomaceae; (Bailey and Nast, 1944; Foster, 1947).

Vertical leaves. The leaves of many species of *Eucalyptus* do not spread out horizontally but hang vertically, so that both surfaces of the leaf receive direct sunlight. In keeping with this fact, palisade chlorenchyma is developed on both sides.

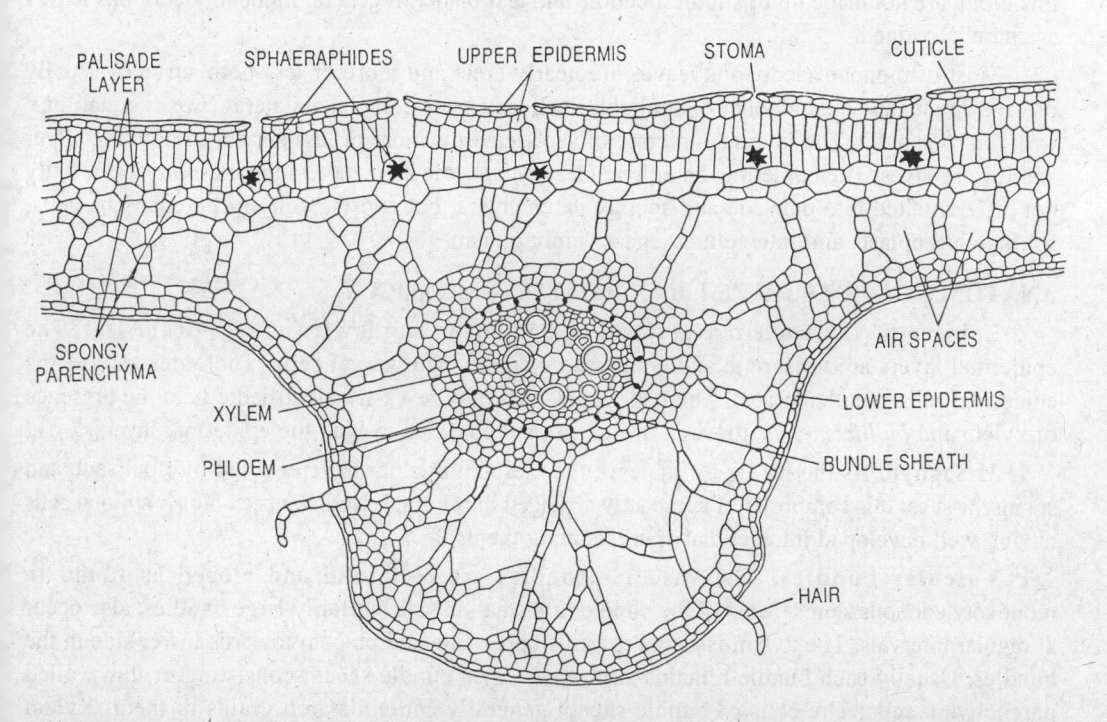

Fig. 12.13. Hydrophytic leaf. T.S. of floating leaf of *Trapa bispinosa*, showing big air spaces, the stomata confined to upper epidermis only.

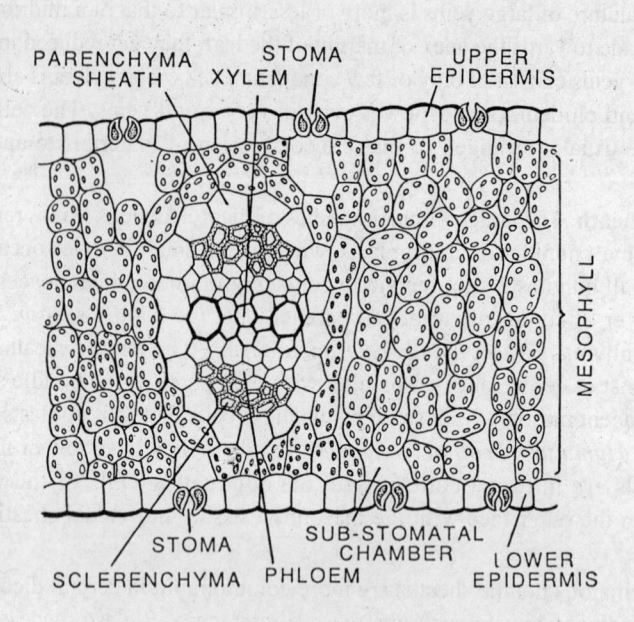

Fig. 12.14. An isobilateral leaf. T.S. of lily leaf.

ANATOMY OF MONOCOTYLEDONOUS LEAF

The monocotyledons as a group show greater diversity of specialized leaf types. The leaves of this group are not made up of stipules, petiole and leaf blade. In general monocotyledonous leaves are parallel-veined.

Most of monocotyledonous leaves are nearly erect and more or less both surfaces usually receive direct and equal amount of sunlight. Such leaves are called isobilateral (*isos* = equal; *bi* = two; *lateris* = side). The internal structure of such leaves is more or less similar in both the upper and lower halves. The epidermis on either sides contains the stomata and the mesophyll is usually not differentiated into palisade and spongy parenchyma, but consists only of parenchyma cells, having chloroplasts and intercellular spaces among them.

ANATOMY OF LEAF OF *ZEA MAYS* (MAIZE)—MONOCOT

Epidermis. The epidermis is found on both upper and lower surfaces of the leaf. The epidermal layers are uniseriate and composed of more or less oval cells. The outer wall of the epidermal cells is cuticularized. The upper epidermis may be easily identified due to the presence of xylem and *bulliform cells* towards it. Stomata are confined to both the epidermal layers.

Mesophyll. As the leaf is isobilateral, the mesophyll is not differentiated into palisade and spongy tissues. It is composed of compactly arranged thin walled, isodiametric chlorophyllous cells having well developed intercellular spaces among them.

Vascular bundles. The vascular bundles are collateral and closed as found in monocotyledonous stems. Most of the bundles are small in size but fairly large bundles, also occur at regular intervals. The xylem is found towards upper side and phloem towards lower side in the bundles. Usually each bundle remains surrounded by a bundle sheath consisting of thin walled parenchyma cells. The cells of bundle sheath generally contain starch grains in them. Xylem consists of vessels and phloem of sieve tubes and companion cells. Sclerenchyma cells occur in patches on both ends of the large vascular bundles which give mechanical support to the leaf.

Anatomy of the Leaf and the Petiole   305

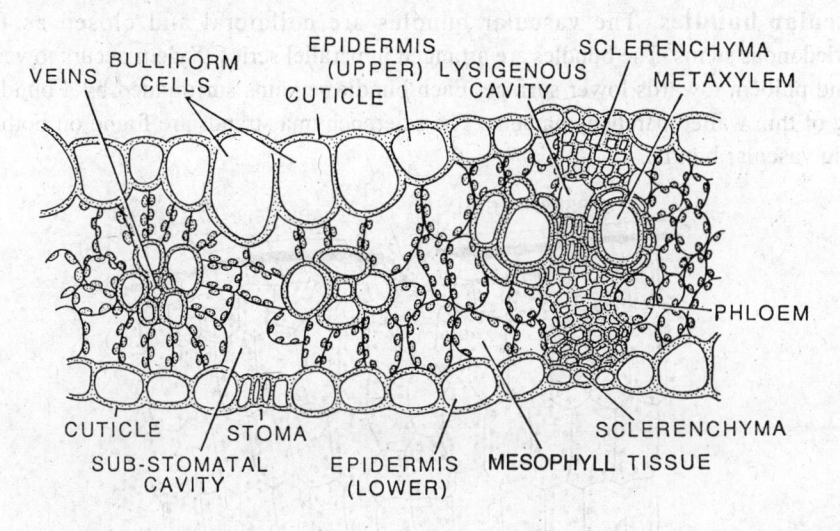

Fig. 12.15. Anatomy of monocotyledonous leaf of *Zea mays* similar structure of the two sides of vertical leaf.

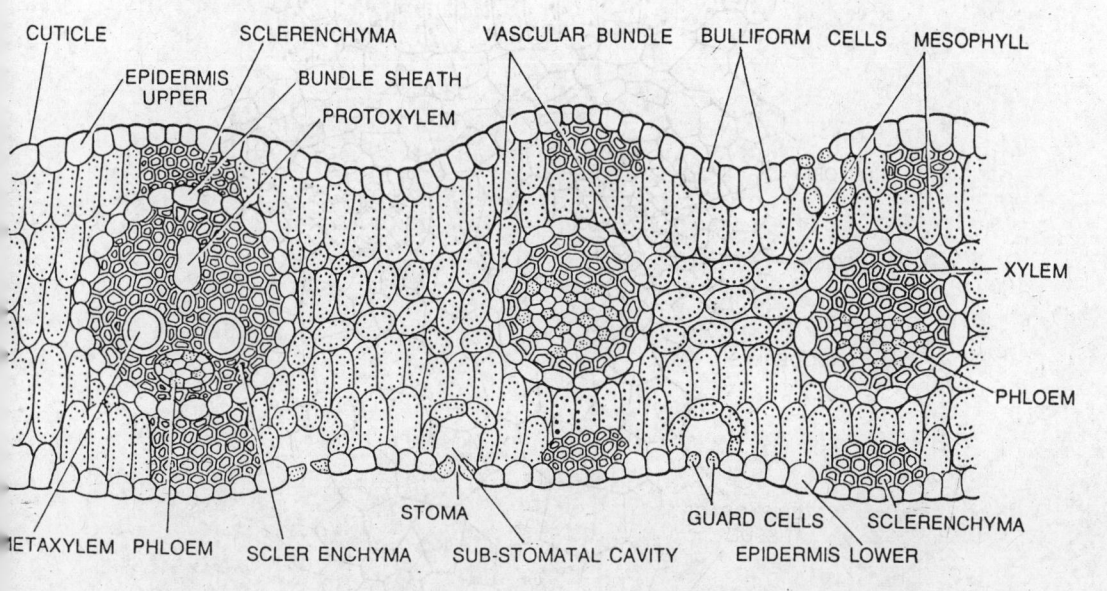

Fig. 12.16. Anatomy of isobilateral leaf. T.S. *Triticum aestivum* (monocot).

ANATOMY OF THE LEAF OF *TRITICUM AESTIVUM* (WHEAT)—MONOCOT

Epidermis. As usual the epidermis layers are found on both upper and lower surfaces of the leaf. The epidermises are uniseriate and composed of more or less oval cells having no intercellular spaces among them. The outer walls of epidermal cells are cuticularized. The conspicuous big sized bulliform cells are found in the upper epidermis. The stomata are confined to both epidermis layers. The sub-stomatal chambers are also seen in vertical section.

Mesophyll. It is composed of more or less oval chlorenchyma cells having intercellular spaces among them. The mesophyll tissue is not clearly differentiated into palisade and spongy parenchyma; however, the cells towards epidermal layers are somewhat elongated and palisade-like. Substomatal chambers are seen beneath the stomata.

Vascular bundles. The vascular bundles are collateral and closed as found in monocotyledonous stems. The bundles are arranged in parallel series. Xylem occurs towards upper surface and phloem towards lower surface. Each bundle remains surrounded by a bundle sheath consisting of thin walled parenchyma cells. The sclerenchyma strands are found on both the ends of each big vascular bundle.

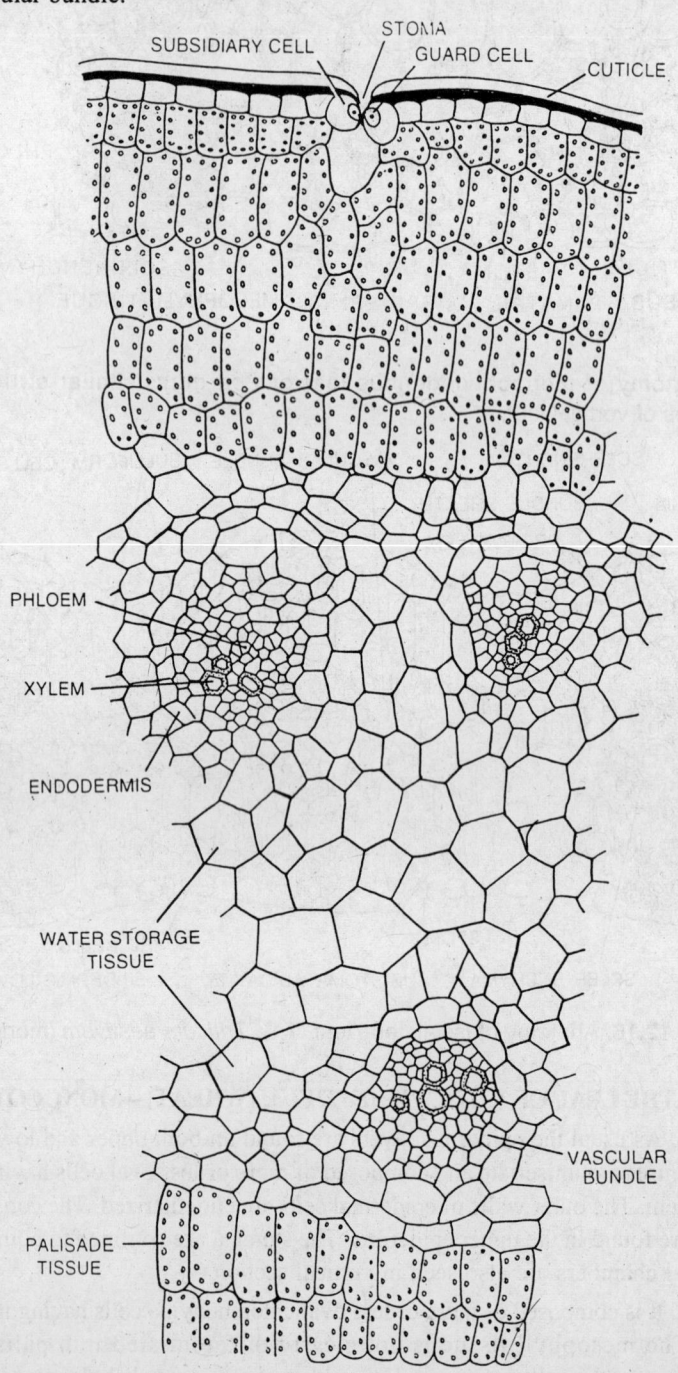

Fig. 12.17. Anatomy of *Aloe* (monocot) leaf. T.S. of a portion of leaf showing detailed internal structure.

Anatomy of the Leaf and the Petiole

ANATOMY OF GYMNOSPERM LEAF

The anatomy of the conifer leaf is described here with reference to the needle of *Pinus* because the leaf of this genus has been investigated in most detail.

The leaf of *Pinus* is xeromorphic. The whole anatomy of the leaf makes it adaptable to withstand the low temperature and the scarcity of water supply.

The outline of the needle (foliage leaf) in a transverse section depends on the number of needles in the dwarf shoot (spur). In *P. monopylla* the spur bears a single needle and therefore, the outline of the needle is circular. In *P. sylvestris* each spur consists of two needles and the outline of each needle is semi-circular. In *Pinus roxburghii* and *P. wallichiana* each spur consists of three needles and therefore, the two flat faces of each needle are towards the inner side and the curved face towards the outside. Here the outline of the needle appears somewhat triangular. The centre of the needle is traversed by one or two vascular bundles surrounded by a peculiar vascular tissue, called *transfusion tissue,* and a thick walled layer the endodermis. Outside the endodermis is the mesophyll. The peripheral layers are the epidermis and the morphologically differentiated hypodermis.

Epidermis. The outermost layer is the epidermis which consists of extremely thick-walled and cuticularized cells. A number of depressions are found over the epidermis. The stomata are developed all over the epidermis in these depressions. The guard cells are sunken in depression below the level of the epidermis.

Hypodermis. Just beneath the epidermis there is a hypodermis which is composed of one or two layers of sclerenchyma cells. The hypodermis is several layered at the corners. The hypodermis is interrupted by air-spaces beneath each stoma.

Mesophyll. The mesophyll is not differentiated into palisade and spongy parenchyma. It consists of thin-walled cells containing a large number of chloroplasts and starch grains. These

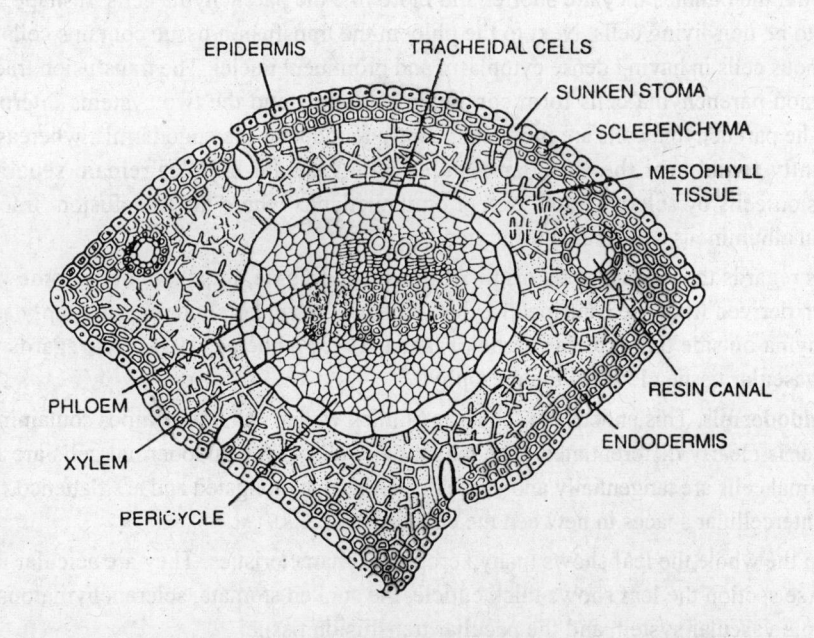

Fig. 12.18. Leaf. T.S. of pine leaf.

thin-walled cells have *peg-like infoldings* of cellulose projecting into their cavities. The presence of these infoldings is probably connected with the development of the air spaces in the leaf. The mesophyll cells are arranged in horizontal layers separated from one another by intercellular spaces. The horizontal strata are not completely detached from each other. These are interconnecting files of cells, making the whole tissue appear like an anastamosing system with a horizontal orientation of spaces (Cross, 1940).

Resin ducts. The leaves have *resin ducts* in the mesophyll. In *Pinus* two lateral ducts occur almost invariably. The resin ducts are lined with thin walled secretory epithelial cells. Outside these cells is a sheath of fibres (hypodermis) with thickened lignified walls. This sclerenchyma is in contact with the hypodermis.

Vascular system. In transverse sections of pine needles the vascular bundles are oriented somewhat obliquely, with the xylem pointing towards the adaxial side, the phloem toward the abaxial side. The xylem is endarch. The protoxylem is partly crushed in mature needles. Outwardly from the crushed elements are some spirally thickened tracheids, probably part of protoxylem, then some metaxylem tracheids with bordered pits. The primary xylem elements are found to be arranged in radial rows, and the rows of tracheary elements are interspersed with rows of parenchyma cells oriented like the rays in a secondary tissue. The individual parenchyma cells are vertically elongated and have transverse end walls.

The sieve cells are also found in radial rows alternating with rows of parenchyma cells. The parenchyma of the phloem is more abundant than that of the xylem; in the phloem some parenchyma cells form starch, whereas others appear to be *albuminous cells* which lack starch but have dense cytoplasm. Some parenchyma cells have crystals.

Transfusion tissue. The transfusion tissue that surrounds the vascular bundles consists mainly of two kinds of cells—living parencnyma cells with non-lignified walls, and thin walled but lignified tracheids with bordered pits. Next to the xylem the transfusion trachieds are somewhat elongated; away from the bundles they are shorter and more like the parenchyma cells in shape. The tracheids appear to be non-living cells. Next to the phloem the transfusion tissue contains cells similar to the albuminous cells in having dense cytoplasm and prominent nuclei. The transfusion tracheids and the transfusion parenchyma cells form continuous systems, and the two systems interpenetrate each other. The parenchyma cells are more abundantly found near the endodermis, whereas tracheids are abundantly found near the vascular bundles. The vascular bundles remain separated from the transfusion cells by sclerenchyma except on their flanks where the transfusion tracheids and the marginal albuminous cells are concentrated.

As regards the phylogenetic origin of transfusion tissue, it is thought by some workers that it has been derived from the centripetal xylem. Other investigators interpret it simply as transformed parenchyma outside the vascular tissue (Takeda, 1913; Abbema, 1934). As regards its function it brings vascular tissue closer to the mesophyll.

Endodermis. This endodermis consists of thick walled cells, sometimes containing starch. This cell layer is clearly differentiated. The tangential walls of the endodermal cells are lignified. The endodermal cells are tangentially and vertically somewhat elongated and are flattened radially. There are no intercellular spaces in between the endodermal cells.

On the whole the leaf shows many xerophytic characteristics. They are acicular in form. In the transverse section the leaf shows thick cuticle, the sunken stomata, sclerenchymatous hypodermis, the simple vascular system and the peculiar transfusion tissue.

ANATOMY OF THE PETIOLE

The tissues of the petiole may easily be compared with the primary tissues of the stem. There is a close similarity between petiole and stem with regard to the structure of epidermis. The ground parenchyma of the petiole is like the stem cortex in arrangement of cells and in number of chloroplasts. The supporting tissue is collenchyma or sclerenchyma. In relation to the arrangement of vascular tissues in the stem, the vascular bundles of the petiole may be collateral (*e.g.*, *Syringa*), bicollateral (*e.g.*, Cucurbitaceae, Solanaceae), or concentric. The primary phloem fibres are differentiated in both the stem and petiole.

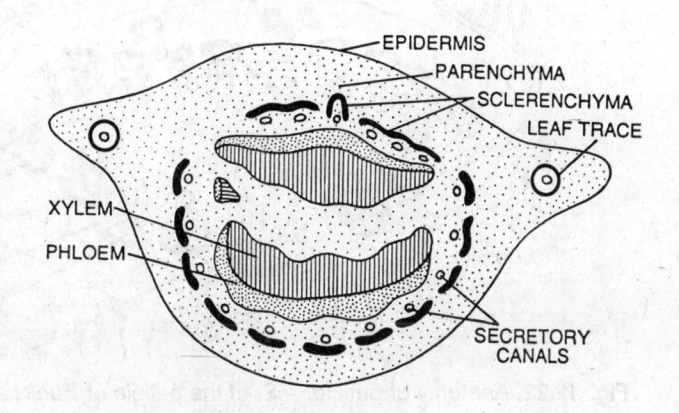

Fig. 12.19. Anatomy of petiole. T.S. of petiole of *Mangifera indica* (dicot)—diagrammatic representation.

The chief anatomical characteristics of the petiole are as follows:

Epidermis. It consists of a single layer of barrel shaped, elongated or radially elongated compact cells having no intercellular spaces among them. The outer walls of the epidermal cells are generally cuticularized.

Hypodermis. Usually a multilayered hypodermis of collenchyma cells is found immediately beneath the epidermis. The sclerenchymatous patches may also be found below the epidermis. Both collenchyma and sclerenchyma make the supporting tissue of the petiole.

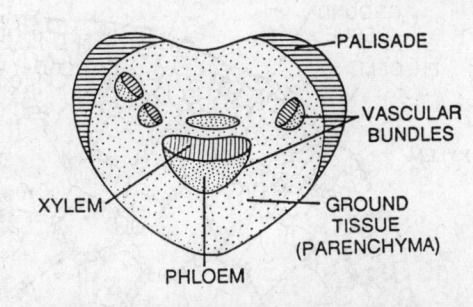

Fig. 12.20. Anatomy of Petiole. T.S. of petiole of *Convolvulus floridus* (dicot-Convolvulaceae)—diagrammatic.

Ground tissue. Just beneath the hypodermis ground tissue is found. It consists of thin walled parenchyma cells having well defined intercellular spaces among them. Usually the vascular bundles are found either arranged in a complete or half ring or scattered in ground tissue.

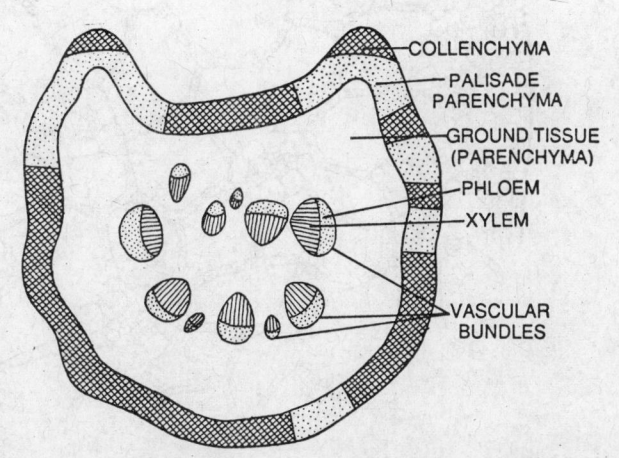

Fig. 12.21. Anatomy of petiole. T.S. of petiole of *Mirabilis nyctaginea* (dicot-Nyctaginaceae)—diagrammatic.

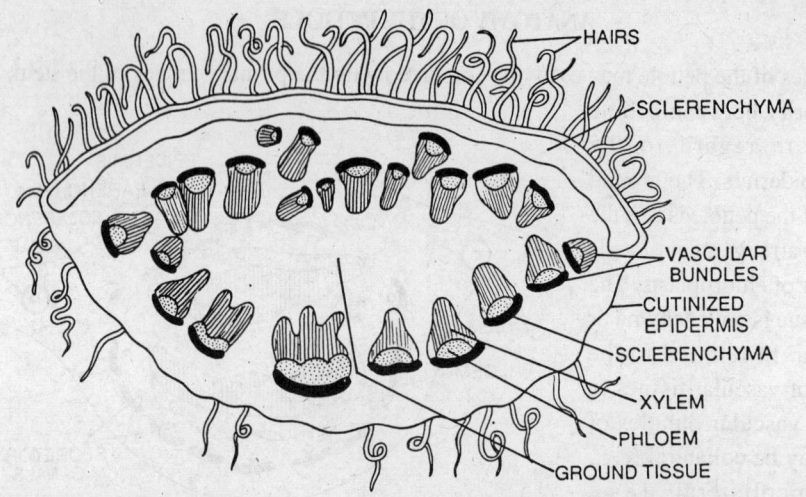

Fig. 12.22. Anatomy of petiole. T.S. of the petiole of *Banksia serrata* (dicot.)—diagrammatic.

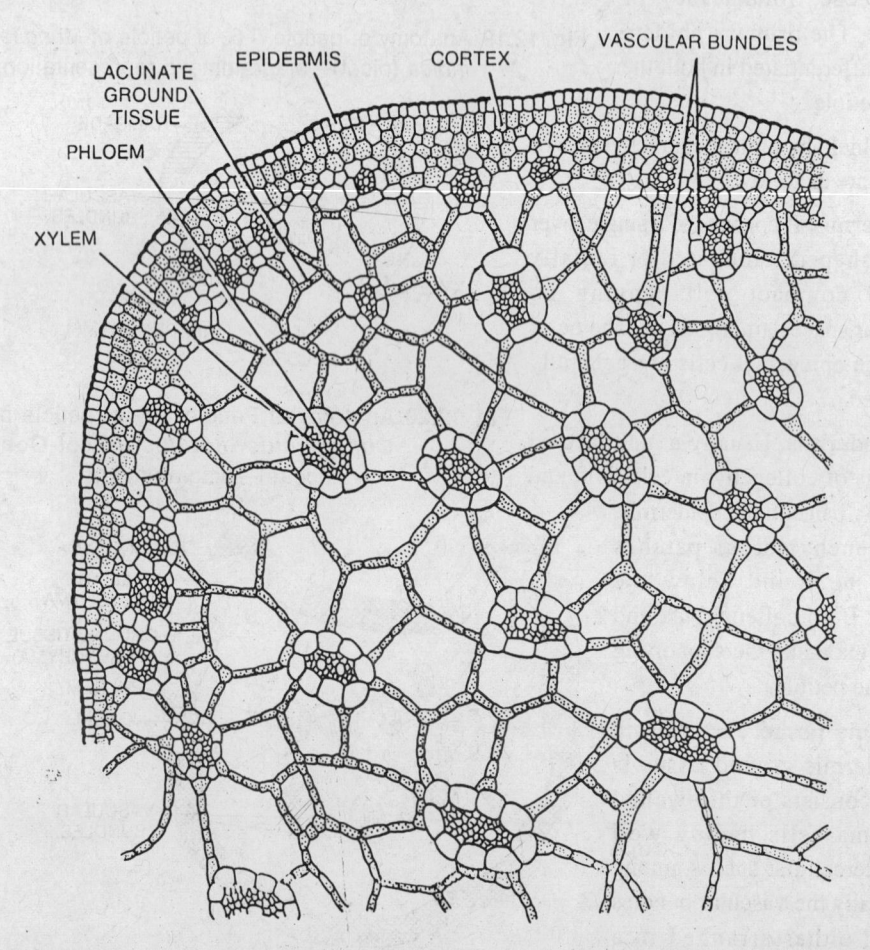

Fig. 12.23. *Eichhornia crassipes* (Water hyacinth monocot). T.S. of petiole, showing the lacunate cortex scattered vascular bundles in the partition walls of lacunae. Every bundle remains surrounded by a bundle sheath. However, xylem and phloem are undifferentiated.

Anatomy of the Leaf and the Petiole

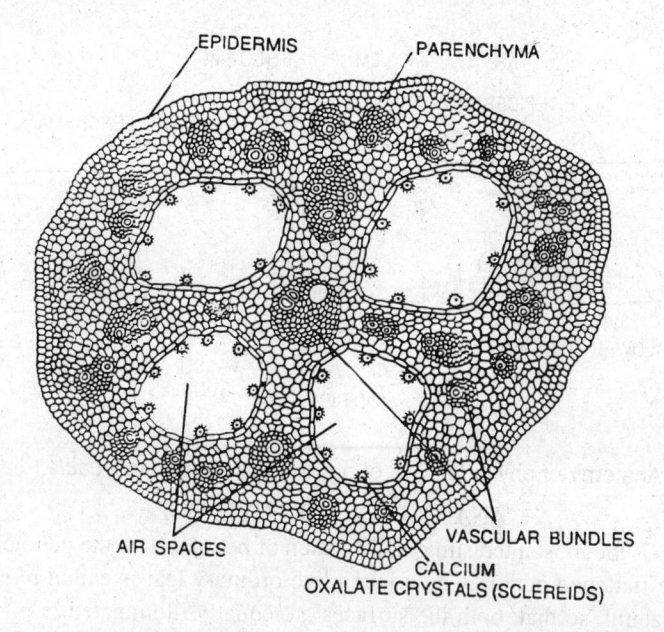

Fig. 12.24. Anatomy of hydrophytic petiole. Cross section of a petiole of *Nymphaea stellata* (dicot) showing large air spaces and sclereids.

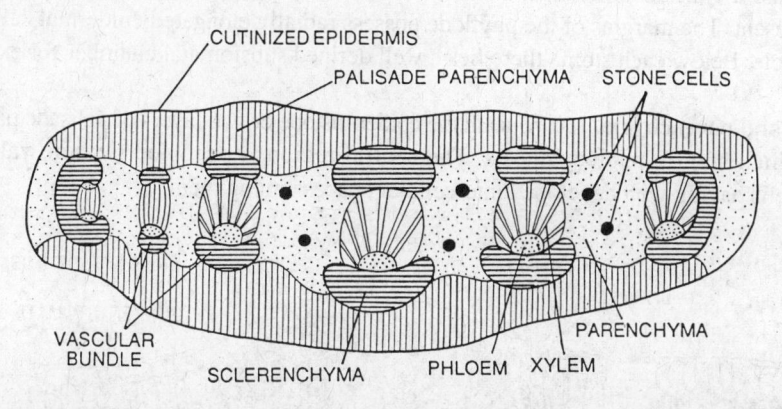

Fig. 12.25. T.S. of petiole of *Hakea dactyloides* (dicot).—diagrammatic.

Vascular bundles. The vascular bundles are of various sizes in the same petiole. In most of the cases, the biggest vascular bundle is found towards lower surface whereas lateral bundles are comparatively smaller in size. Each bundle consists of xylem and phloem. In petiole, the xylem is always found towards upper side whereas phloem towards lower side (as in leaf). Generally the central biggest bundle remains surrounded by single layered endodermal sheath which may or may not be followed by a multilayered pericycle.

Features of Special Interest
1. Usually a groove is present towards upper side.
2. Mostly the vascular bundles are arranged in a semicircle in ground tissue.
3. Mostly the central bundle is biggest and remains encircled by endodermal sheath.
4. The xylem is always found towards upper side and phloem towards lower side.

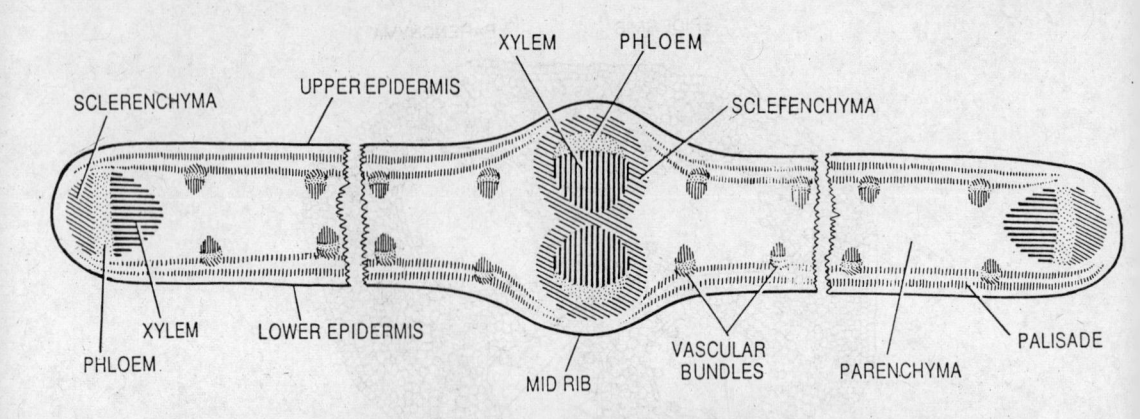

Fig. 12.26. Anatomy of phyllode. T.S. of phyllode of Australian *Acacia* (diagrammatic).

ANATOMY OF THE PHYLLODE

The phyllode is the most interesting modification of petiole. Here the petiole becomes flattened and leaf like. The flattened petiole which looks like ordinary leaf is called *phyllode*. Usually the phyllode is isobilateral, so that, both the surfaces are equally illuminated.

ANATOMY OF THE PHYLLODE OF AUSTRALIAN *ACACIA*

Epidermis. It consists of a single row of cells covered with well developed cuticle. Sunken stomata are present. The margins of the phyllode possess radially elongated epidermal cells covered with thick cuticle. Below each stoma there lies a well defined substomatal chamber for exchange of gases.

Palisade and parenchyma. Just beneath the epidermis one or two layers of palisade parenchyma are present which help in photosynthesis. The central region is occupied by thin walled living parenchyma cells having well developed intercellular spaces among them.

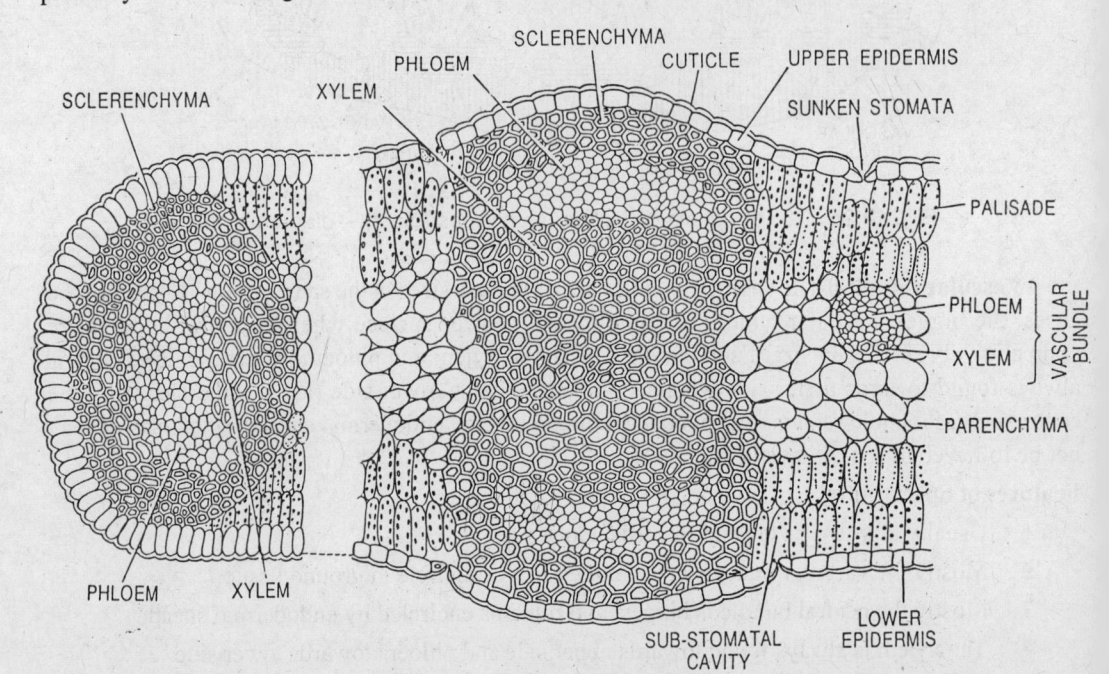

Fig. 12.27. Anatomy of phyllode. T.S. of a phyllode of Australian *Acacia*—detail.

Anatomy of the Leaf and the Petiole

Vascular system. Just like in petiole the vascular bundles form a ring and are arranged below the palisade tissue. The central and marginal vascular bundles are sufficiently big in size. Each vascular bundle consists of xylem and phloem. Around the central and marginal vascular bundles well developed sclerenchyma is found.

Features of Special Interest

1. Vascular bundles arranged in a complete ring. The central and marginal bundles are bigger in size (characteristic of petiole).
2. Palisade tissue is present (characteristic of leaf).
3. Radially elongated epidermal cells are found at the margins (xerophytic character).
4. The cuticle is well developed (xerophytic character).
5. The sunken stomata are present in the epidermis (xerophytic character).
6. The sclerenchyma is well developed (xerophytic character).

ABSCISSION OF LEAVES

The leaves are periodically detached from the perennial plants. The phenomenon of this detachment is complex one. During this phenomenon leaves are separated from the stem without causing any injury to the living tissues in stem and the newly exposed surface is also protected from desiccation and infection. This phenomenon of the separation of leaves from the stem takes place in a particular region of the plant, known as *abscission region* or *abscission zone*. The phenomenon as a whole known as *abscission of leaves* and the separating leaf may be said to abscised. The abscission zone consists of a *separation layer* through which the actual break occurs, and the *protective layer*.

In simple dicotyledonous leaves the abscission zone occurs within the petiole or at its base. In compound leaves the abscission zones occur in the petiole of the leaf as a whole and at the base of individual leaflets.

Within the abscission zone, a few days or even weeks before actual leaf fall, a separation layer develops which aids structurally in the leaf fall. When the tissues, found beneath the separation layer are exposed after the leaf fall, they are protected from desiccation and infection by means of one or more protective layers. At least one of these protective layers lies within the abscission zone. These protective layers are of two types–a *primary protective layer* and the *secondary protective layer* or *periderm*.

From the view point of its structure, the abscission zone is the weakest part of the petiole. As soon as the leaf becomes mature the abscission zone becomes evident. A shallow depression develops

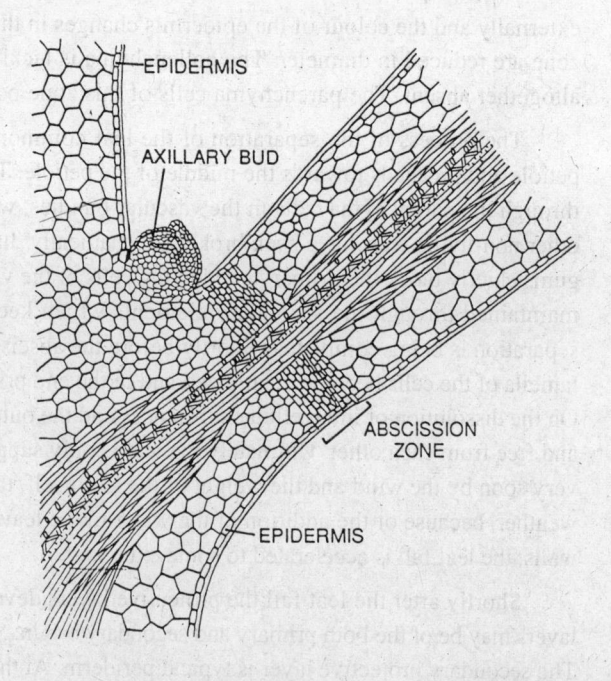

Fig. 12.28. Leaf abscission—formation of the abscission zone.

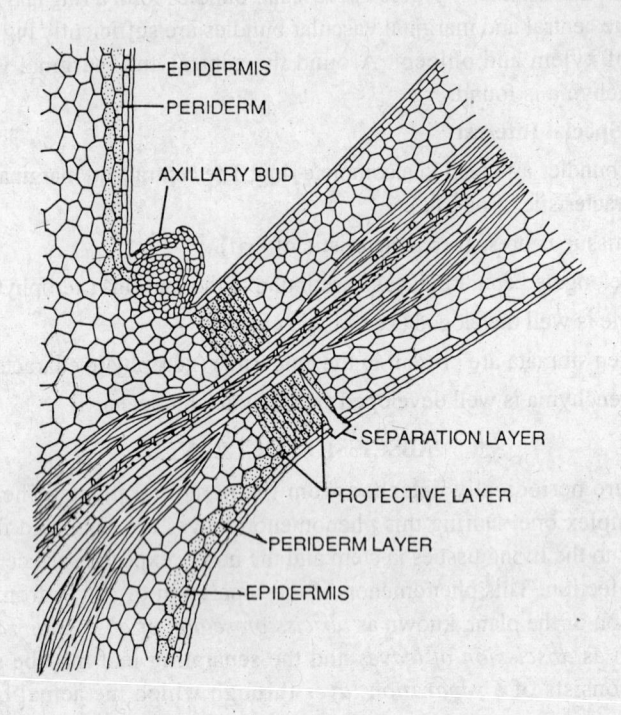

Fig. 12.29. Leaf abscission. Separation within the abscission zone and initiation of periderm.

externally and the colour of the epidermis changes in this zone. The vascular bundles in abscission zone are reduced in diameter. The collenchyma is lacking and the sclerenchyma becomes weak or altogether absent. The parenchyma cells of this zone possess denser cytoplasm.

The process of the separation of the leaf commonly starts from the peripheral region of the petiole and proceeds towards the middle of the petiole. The separation layer remains in continuation through the parenchyma cells in the vascular bundles, whereas the xylem and phloem elements and other non-living cells have been broken mechanically. Just before the actual leaf fall, the tyloses and gums chiefly block the primary conducting cells of the vascular bundles, but sufficient conduction is maintained through the secondary elements which keep the leaf fresh turgid unless and until its separation is being completed. Shortly before the abscise of the leaf, the outer walls and the middle lamella of the cells become gelatinized and in the end prior to leaf fall they break down and dissolve. On the dissolution of intercellular substances and the outer cell walls, the cells become quite separate and free from each other. Ultimately the leaf is only supported by the vascular elements which break very soon by the wind and the weight of the leaf itself, thus separating the leaf from the stem. In wet weather, because of the addition of the water to the leaves, and the hydrolysis of the gelatinous cell walls, the leaf fall is accelerated to some extent.

Shortly after the leaf fall the protective layers develop on the exposed surface. The protective layers may be of the both primary and secondary origin. Sometimes they are only of secondary origin. The secondary protective layer is typical periderm. At the region of separation, a leaf scar is formed. The scar is formed because of the deposition of the substances, which protect the new surface from injuries, infection and loss of water. These substances are found beneath the separation layer in the

Anatomy of the Leaf and the Petiole 315

cells and referred as suberin and lignin. In some plants the periderm develops just after leaf fall. This newly developed periderm beneath the protective layer remains in continuation with the periderm of the stem.

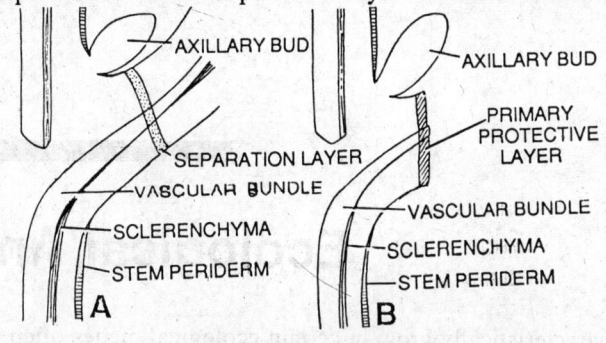

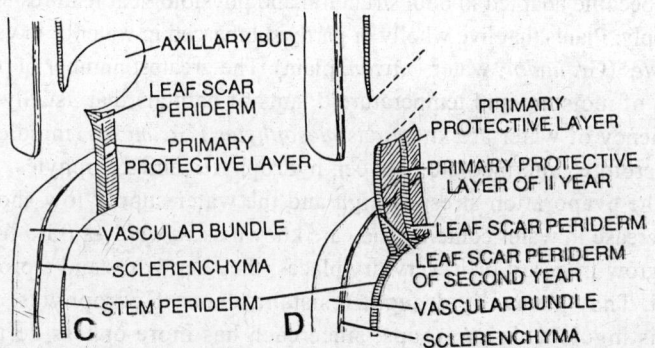

Fig. 12.30. Abscission of leaf in *Castanea* (a woody dicot). A, indicating separation layer extending through vascular bundle which lacks sclerenchyma in the abscission zone; B, soon after the leaf fall, showing primary protective layer below the surface of the scar; C, late autumn stage, showing leaf-scar periderm beneath the primary protective layer; D, second year stage, showing same layers as in C, and in addition a primary protective layer and leaf-scar periderm of the second year, the latter connected with the stem periderm.

13

Ecological Anatomy

The plants which characteristically grow in certain ecological niches often show a type of structure which is believed to be adapted to that particular environment. In the course of evolution, many species have become adapted in both structural and physiological features to habitats with an excessive water supply. Plants that live wholly or partly submerged in water or in very wet places are known as *hydrophytes* (Gr. *hudor*, water; *phyton*, plant). The greatest number of plants grow under average conditions of moisture and temperature. Plants of habitats that usually show neither an excess nor a deficiency of water are known as *mesophytes* (Gr. *mesos*, middle; *phyton*, plant). Mesophytes are therefore, intermediate between hydrophytes and xerophytes. Plants that grow habitually where the evaporation stress is high and the water supply low show characteristic adaptations to a decrease in water content. They are known as *xerophytes* (Gr. *Xeros*, dry; *phyton*, plant). The plants grow in desert or in very dry places; they can withstand a prolonged period or drought uninjured. They are really drought-resistant plants. Hydrophytes, xerophytes, and mesophytes are distinguishable as groups, since each has more or less definite habitat and characteristic appearance. Between hydrophytes and mesophytes on the one hand, and extreme xerophytes and mesophytes on the other, there are found all gradations of form and all degrees of structural adaptations due to intermediate habitats. Plants that grow upon other plants, but do not absorb food from them are known as *epiphytes* (Gr. *epi*, upon; *phyton*, plant), such as many orchids (*e.g., Vanda*). Plants that grow in places rich in decaying organic substances, and derive their nutrition from them, are known as *saprophytes* (Gr. *sapros*, rotten; *phyton*, plant). (*e.g., Montropa*). Plants that grow upon other living plants and absorb their food material from them, are known as *parasites*. For the purpose of the absorption of food the parasites develop haustoria, which penetrate into the tissue of the host plant, and absorb nutrition from it (*e.g., Cuscuta*). Some of the anatomical features of above mentioned groups of plants are described below:

HYDROPHYTES

Plants that grow in water or very wet places. They may be *submerged* or *partly submerged, floating* or *amphibious*. Their structural adaptations are mainly due to the high water content and the deficient supply of oxygen. The various adaptations are as follows:

(*i*) The reduction of protective tissue (epidermis here is meant for absorption and not for protection).

(*ii*) The reduction of supporting or mechanical tissue (*i.e.,* lack of sclerenchyma).

(*iii*) The reduction of conducting tissue (*i.e.,* minimum development of vascular tissue).

(*iv*) The reduction of absorbing tissue (roots chiefly act as anchors, and root hairs are lacking).

(*v*) There is special development of air-chambers (aerenchyma) for aeration of internal tissues.

Ecological Anatomy 317

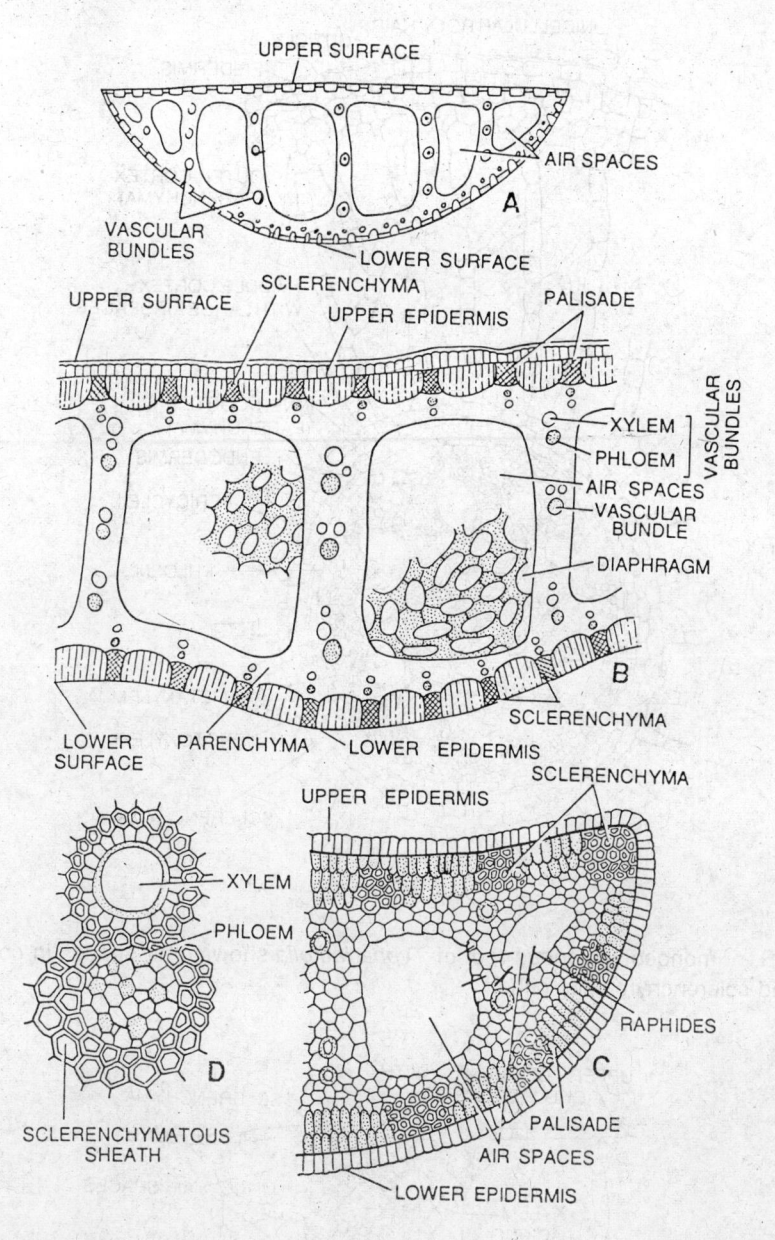

Fig. 13.1. Anatomy of hydrophytic leaf of *Typha* (monocot,). A, T.S. of leaf (diagrammatic); B, detail of a part of A; C, detail of a corner of leaf; D, single vascular bundle.

EPIDERMIS

In aquatic plants, the epidermis is not protective but absorbs gases and nutrients directly from the water. The epidermis in the typical hydrophyte has an extremely thin cuticle, and the thin cellulose walls permit ready absorption from the surrounding water. Commonly the chloroplasts are found in epidermal cells of leaves, especially when the leaves are very thin; these chloroplasts utilize the weak light under water for photosynthesis. In submerged plant, stomata are not present, and exchange of gases takes place directly through cell walls. The floating leaves of aquatic plants have abundant stomata on the upper surface.

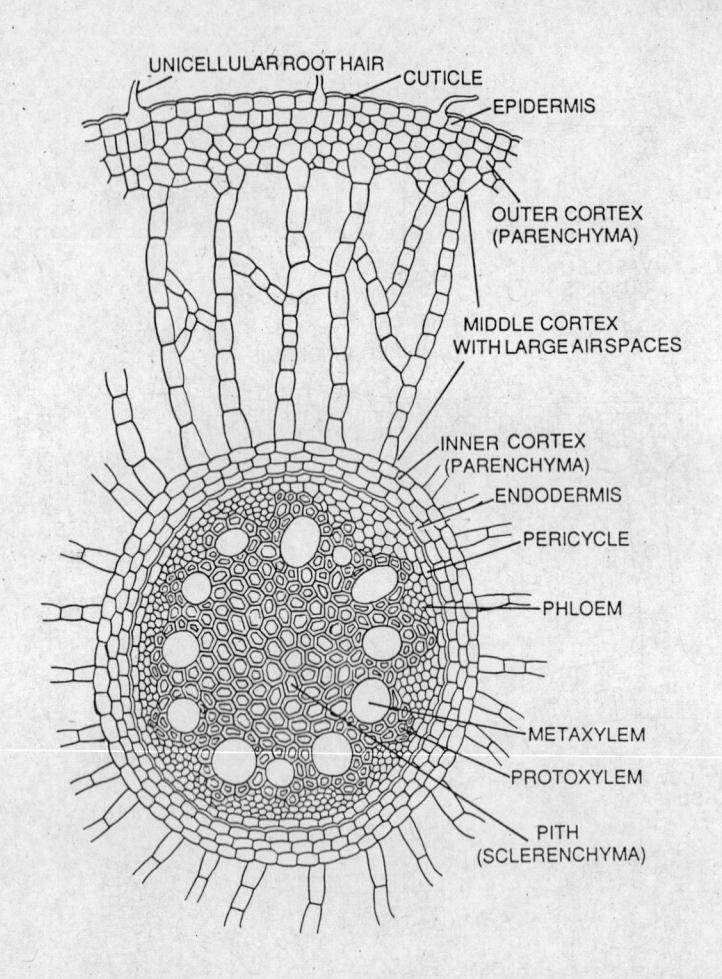

Fig. 13.2. T.S. of monocotyledonous root of *Typha latifolia* showing air spaces in cortical region and sclerenchymatous pith.

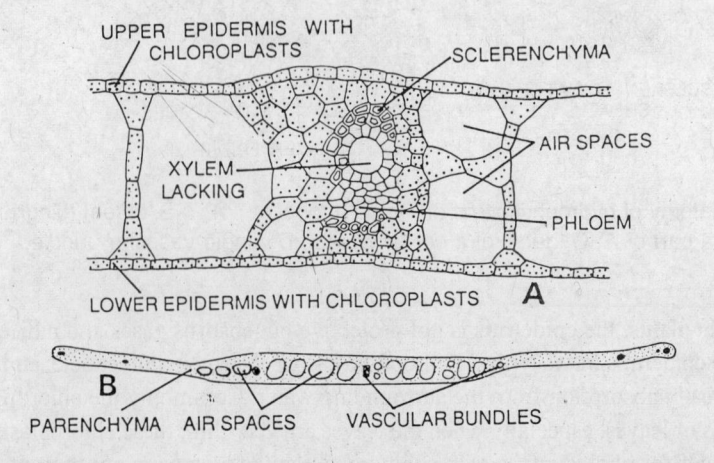

Fig. 13.3. Anatomy of hydrophytes. A, cross section of submerged leaf of *Potamogeton epihydrus* (detail); B, diagrammatic representation of the same.

Ecological Anatomy   319

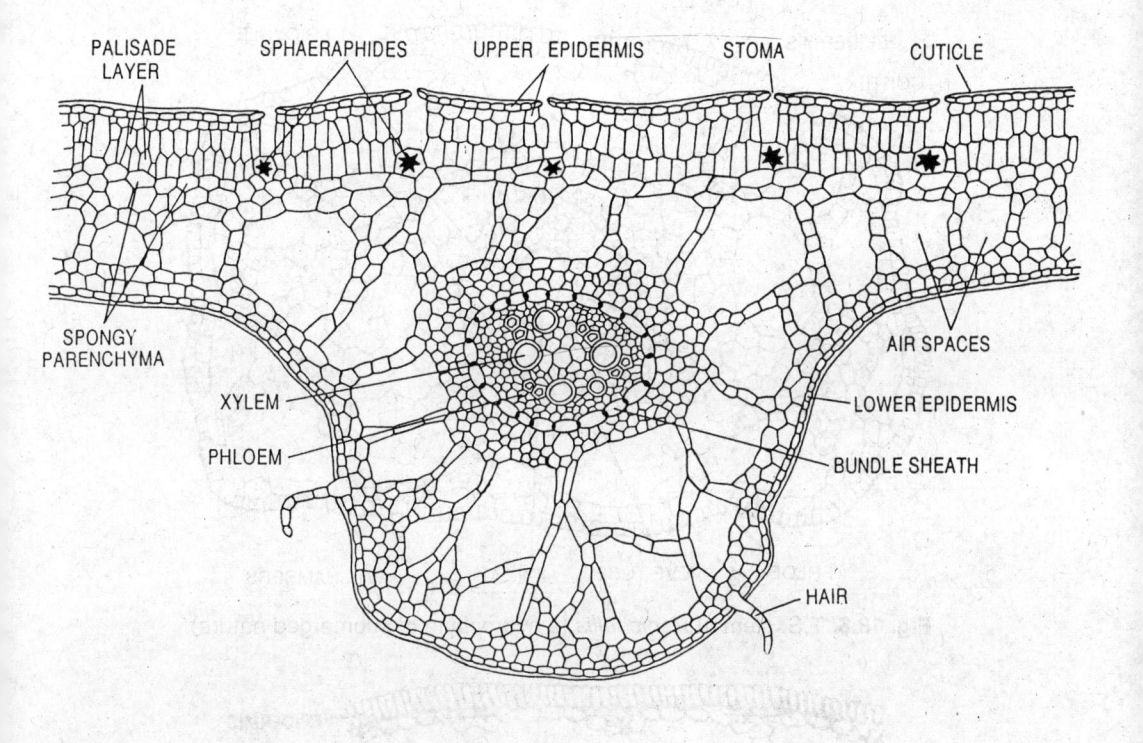

Fig. 13.4. Hydrophytes. T.S. floating leaf of *Trapa bispinosa*, showing big air spaces; the stomata confined to upper epidermis only.

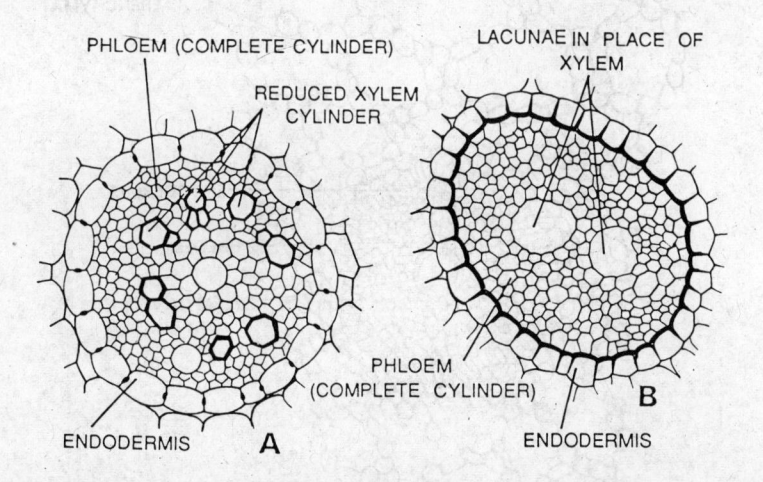

Fig. 13.5. Structure of hydrophytes. A, T.S. of stele of the stem of *Elatine alsinastrum*; B, T.S. of stele of the stem of *Potamogeton pectinatus*.

Lack of sclerenchyma. Submerged plants usually have few or no sclerenchymatous tissue and cells. The water itself gives support to the plant, and protects it to some extent from injury. The thick walls of tissues, their density and the presence of collenchyma in some plants give some

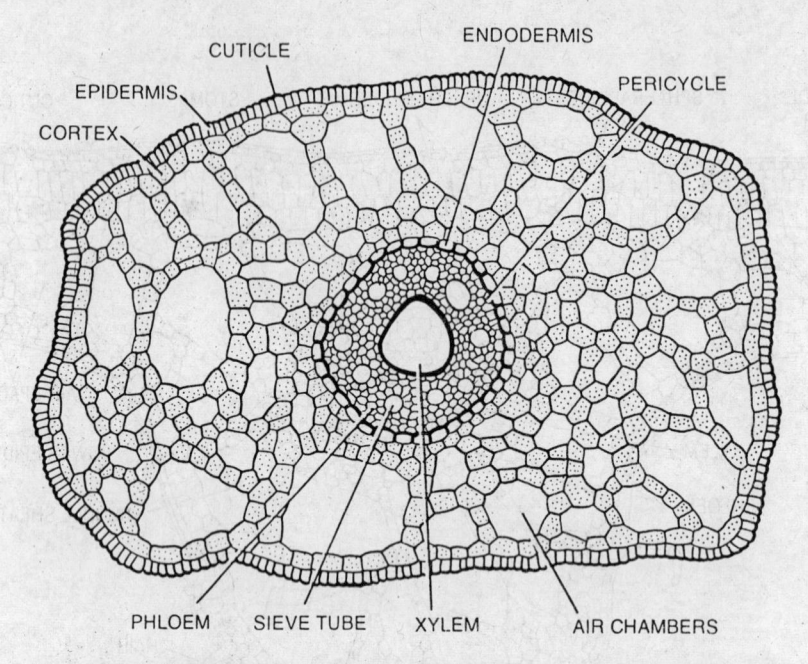

Fig. 13.6. T.S. stem of *Zanichellia* (a hydrophyte of submerged nature).

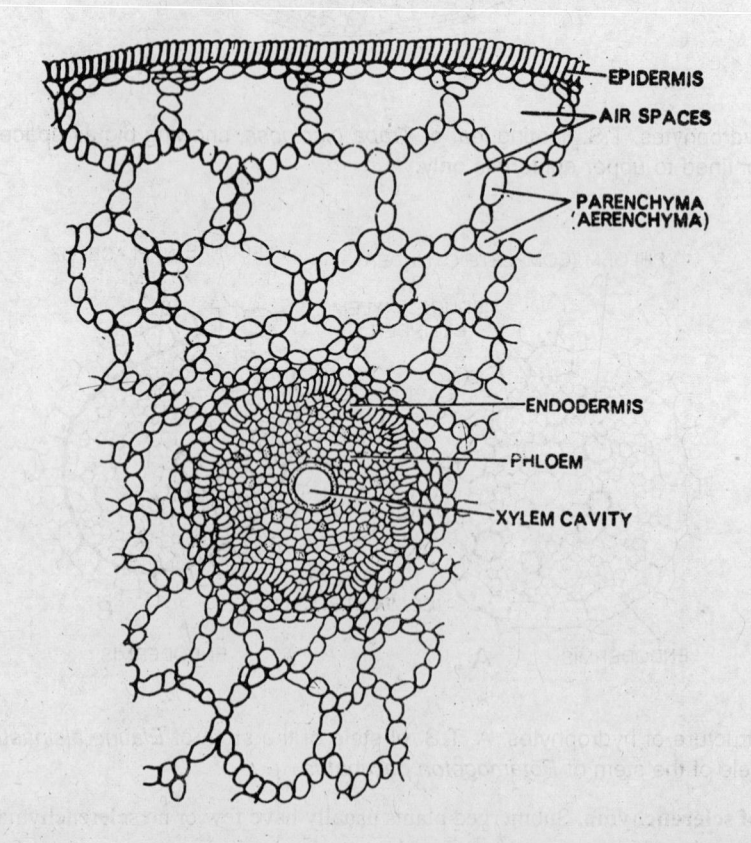

Fig. 13.7. Anatomy of hydrophytic stem. T.S. of a sector of the stem of *Potamogeton* (Monocot.)—detail.

Ecological Anatomy 321

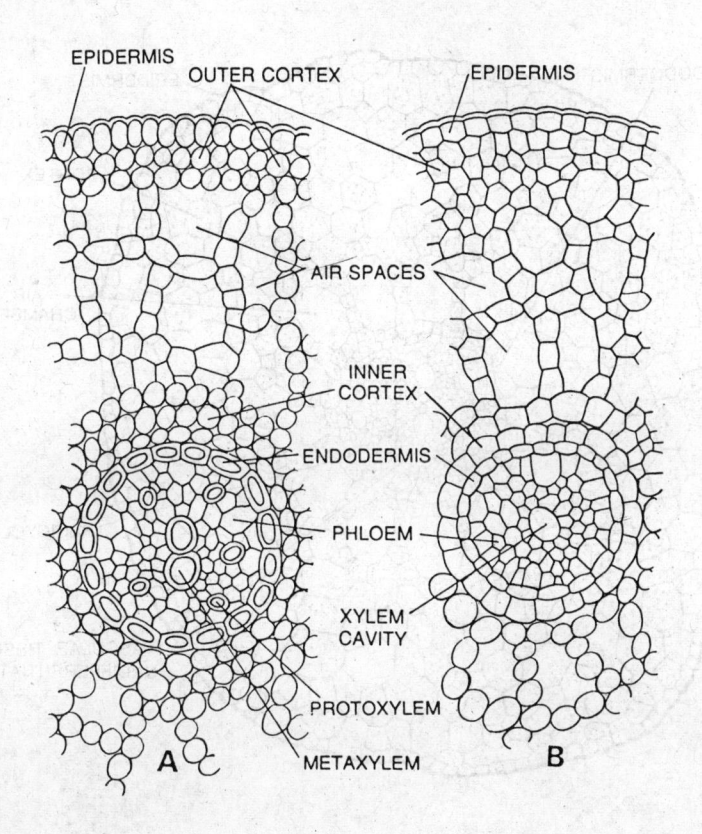

Fig. 13.8. Structure of roots of hydrophytes. A and B transverse sections of the roots of two species of *Potamogeton*.

rigidity. The strands of sclerenchyma occasionally occur, especially along the leaf margins, and increase tensile strength. A few star shaped idoblasts or sclereids are present, which give mechanical support to the body of aquatic plant (*e.g., Limnanthemum, Nymphaea*).

Minimum development of vascular tissue. In the vascular tissues, the xylem shows the greatest reduction and in many aquatic plants consists of only a few elements, even in the stele and main vascular bundles (*e.g., Elatine alsinastrum*). In some aquatic plants (*e.g., Potamogeton pectinatus, P. epihydrus*), in the stele and large bundles, and frequently in the small bundles, xylem elements are lacking. In these plants there is well developed *xylem lacuna* in the position of xylem. These lacunae resemble typical air-chambers (air spaces). In many aquatic plants, the phloem is fairly well developed as compared with the xylem. The endodermis is usually present around the stele, but it is weakly developed.

Reduction of absorbing tissue. The root-system in hydrophytes is feebly developed and root hairs and root cap are absent. In some floating plants such as *Utricularia, Ceratophyllum,* etc., no roots are developed, and in submerged plants such as *Vallisneria, Hydrilla, Naias,* etc., water dissolved mineral salts and gases are absorbed by their whole surface. In plants like *Pistia, Eichhornia, Lemna,* etc., no root cap develops, but root pocket is formed instead. An aquatic plant is, in reality, submerged in or floating upon a nutrient solution. In hydrophytes the root system is functioning chiefly as holdfasts or anchors, and a large part of the absorption takes place through the leaves and stems.

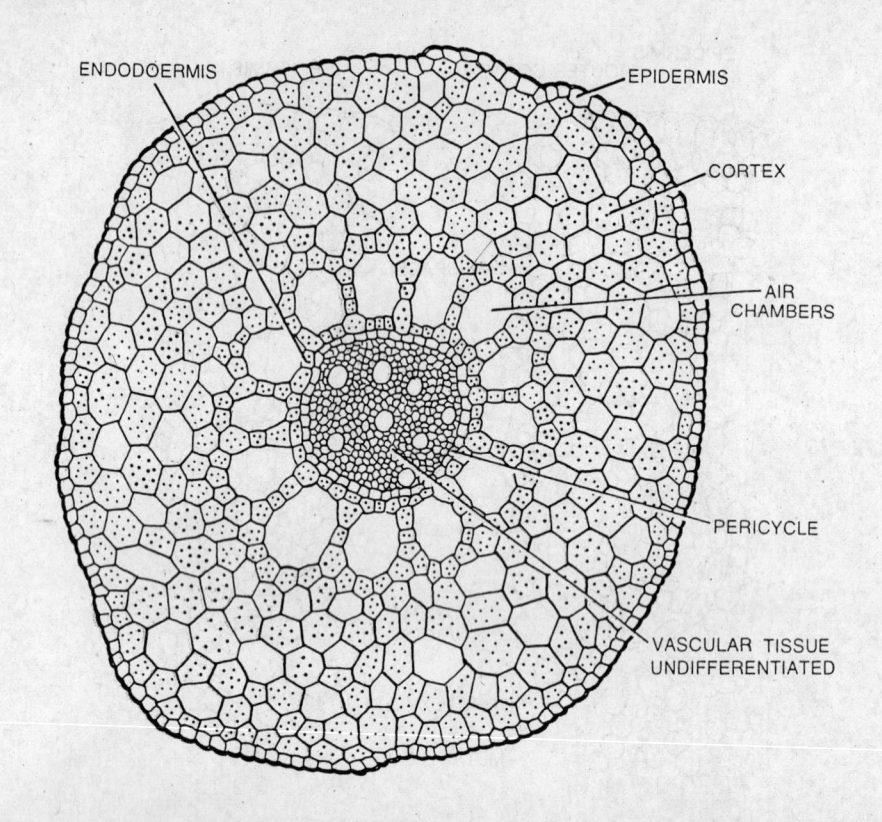

Fig. 13.9. T.S. of stem of *Ceratophyllum* (hydrophyte submerged). The vascular region is mixed protostele. The endodermis and pericycle are quite conspicuous.

Development of air-chambers. Chambers and passages filled with gases are commonly found in the leaves and stems of hydrophytes. The air chambers are large, usually regular, intercellular spaces extending through the leaf and often for long distances through the stem (*e.g., Potamogeton, Pontederia*). The spaces are usually separated by partitions of photosynthetic tissue only one or two cells thick. The chambers prepare an internal atmosphere for the plant. These air-chambers on the one hand give buoyancy to the plant for floating and on the other they serve to store up air (oxygen and carbon dioxide). The carbon dioxide that is given off in respiration is stored in these cavities for photosynthesis, and again the oxygen that is given off in photosynthesis during the daytime is similarly stored in them for respiration. The cross partitions of air passages, known as diaphragms, prevent flooding. The diaphragms are provided with minute perforations through which gases but not water may pass. Another specialized tissue frequently found in aquatic plants that gives buoyancy to the plant parts on which it occurs is *aerenchyma*. The presence of this type of aerenchyma is characteristic of species of *Decodon* and *Lythrum*. Here, very thin partitions enclose air spaces and the whole structure consists of very feeble tissue. Aerenchyma is phellem formed by a typical phellogen of epidermal or cortical origin. At regular intervals individual cells of each layer of phellem elongate greatly in the radial direction while the other cells of this layer remain small. However, the term aerenchyma is applied to any tissue with many large intercellular spaces, but such aerenchyma is quite distinct from the typical aeranchyma mentioned above (of *Decodon*) which is of secondary origin.

Ecological Anatomy

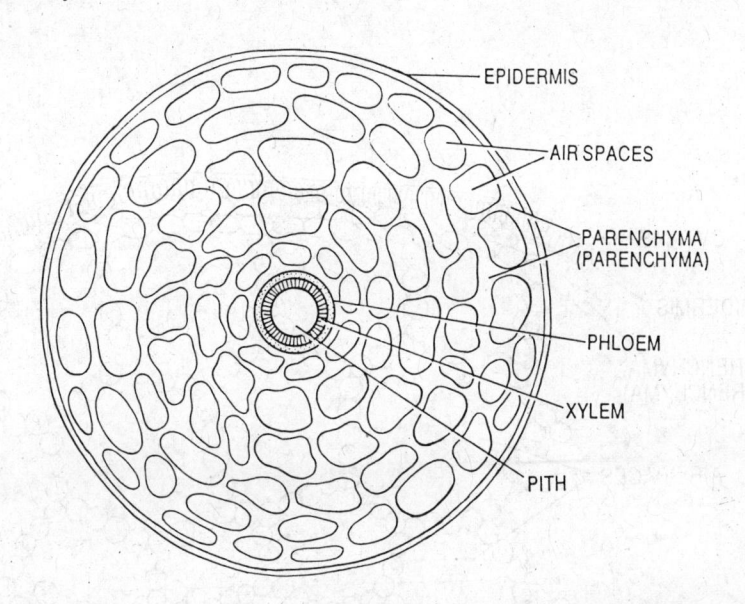

Fig. 13.10. Anatomy of hydrophytic stem. T.S. of the stem of *Hippuris vulgaris* (dicot)—diagrammatic.

List of Common Aquatic Plants.

Submerged— *Vallisneria, Potamogeton, Hydrilla.* **Floating**—*Ceratophyllum, Pistia, Utricularia, Eichhornia, Trapa, Neptunia.*

Plants with floating leaves—*Nymphaea, Nelumbium, Victoria regia, Euryale, Limanthemum.*

Amphibious plants—*Ranunculus aquatilis, Sagittaria, Limnophylla, Cardenthera, Typha, Alisma, Myriophyllum heterophyllum,* etc., (See figs. 13.1 to 13.20)

MESOPHYTES

They grow in habitats that are neither extremely dry nor wet. These are plants that grow under average conditions of temperature and moisture. They are supposed to be intermediate between hydrophytes and xerophytes. In mesophytes the root system is well developed with the tap-root and its branches in dicotyledons, and a cluster of fibrous roots in monocotyledons; root hairs are abundantly produced for the absorption of water from the soil. The stem is solid, erect and normally branched. All the different kinds of tissues, particularly the mechanical and conducting tissues, have reached their full development in the mesophytes. The aerial parts of plants such as the leaves and the branches are provided with cuticle. In dorsiventral leaves the lower epidermis is provided with numerous stomata; there are few stomata or none at all on the upper surface. In erect leaves, as in most monocotyledons, stomata are more or less equally distributed on both surfaces. The stomata are relatively uniform in structure and the guard cells show a maximum capacity for movement. The anatomy of mesophytic plants is quite normal and no special adaptations are found in them. In the foregoing chapters, much emphasis has been given on the anatomical features of the plants that develop in regions of average or optimum water supply, *i.e.*, mesophytes.

XEROPHYTES

They grow in deserts or in very dry places; they can withstand a prolonged period of drought uninjured, for this purpose they have certain peculiar adaptations. The xerophytic plants have to guard against excessive evaporation of water; this they do by reducing evaporating surfaces. Plants produce a long tap root which goes deep into the sub-soil in search of moisture. To retain the water absorbed by the roots, the leaves and stems of some plants become very thick and fleshy (*e.g., Aloe,*

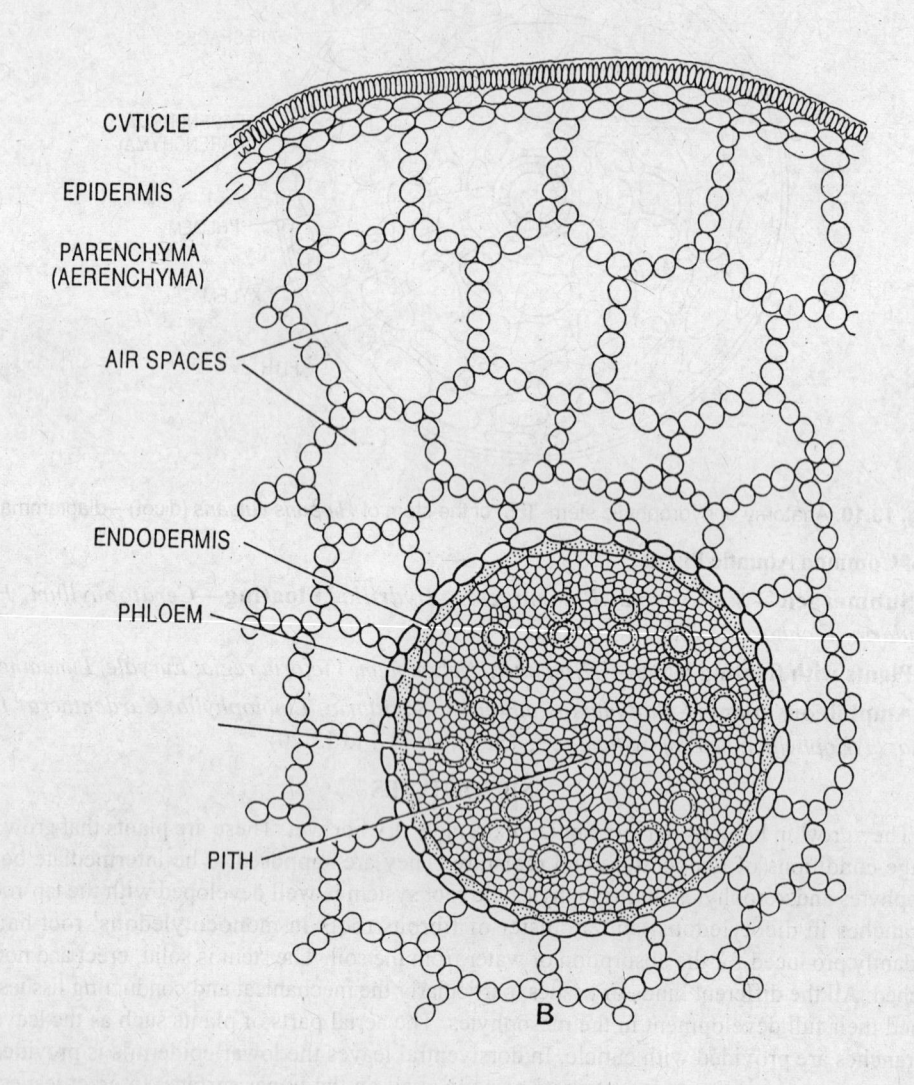

Fig. 13.11. Anatomy of hydrophytic stem. T.S. of stem of *Hippuris vulgaris* (dicot)—detailed structure of a part.

Agave). Water tissue develops in them for storing up water; this is further facilitated by the abundance of mucilage contained in them. Multiple epidermis sometimes develops in the leaf (*e.g., Nerium*). Modification of the stem into phylloclade for storing water and food and at the same time performing functions of leaves is characteristic of many desert plants (*e.g., Opuntia* and other cacti).

In xerophytes certain structural features are also common. Leaves are thick and leathery, with a well developed cuticle and abundant hairs. Well differentiated mesophyll is also present, and there is often more than one layer of palisade tissue (*e.g., Nerium, Hakea*). The walls of epidermal and sub-epidermal cells are frequently lignified, and a distinct hypodermis, may be present. They have

Ecological Anatomy

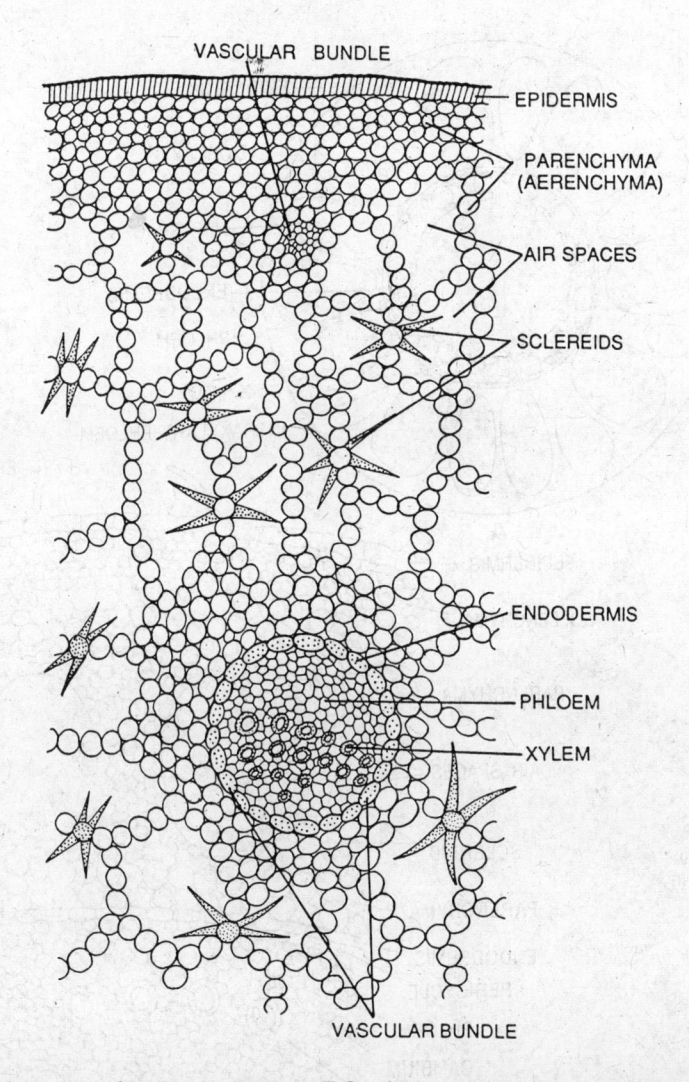

Fig. 13.12. Anatomy of hydrophytic stem. T.S. of *Limnanthemum* stem (dicot.)—detail.

a well developed vascular system and often an abundance of sclerenchyma, either in the form of sclereids or fibres (*Hakea, Ammophila*). The leaf is sometimes cylindrical or rolled. This organization is to protect the stomata, which may occur in furrows. Certain fleshy leaves (*e.g., Sedum*), contain abundant thin-walled cells, the water storage tissue. The characteristic anatomical features of the xerophytes are as follows:

Epidermis and Thick Cuticle

Heavy cuticularization and extreme cutinization of the epidermis and even of sub-epidermal cells are common in xerophytes. The thickness of the cuticle shows various gradations. In certain cases the thickness of cuticle is only slightly greater than normal, like that of plants of semi-xerophytic habitats. In extreme xerophytes the cuticle may be as thick as, or thicker than, the diameter of the epidermal cells. In addition to the presence of thick cuticle, the walls of epidermal cells become cutinized, and sometimes also those of underlying cells. Along with well-developed cutinized layers the epidermal and sub-epidermal cells also become lignified. For example, in the leaflets of *Cycas*, the

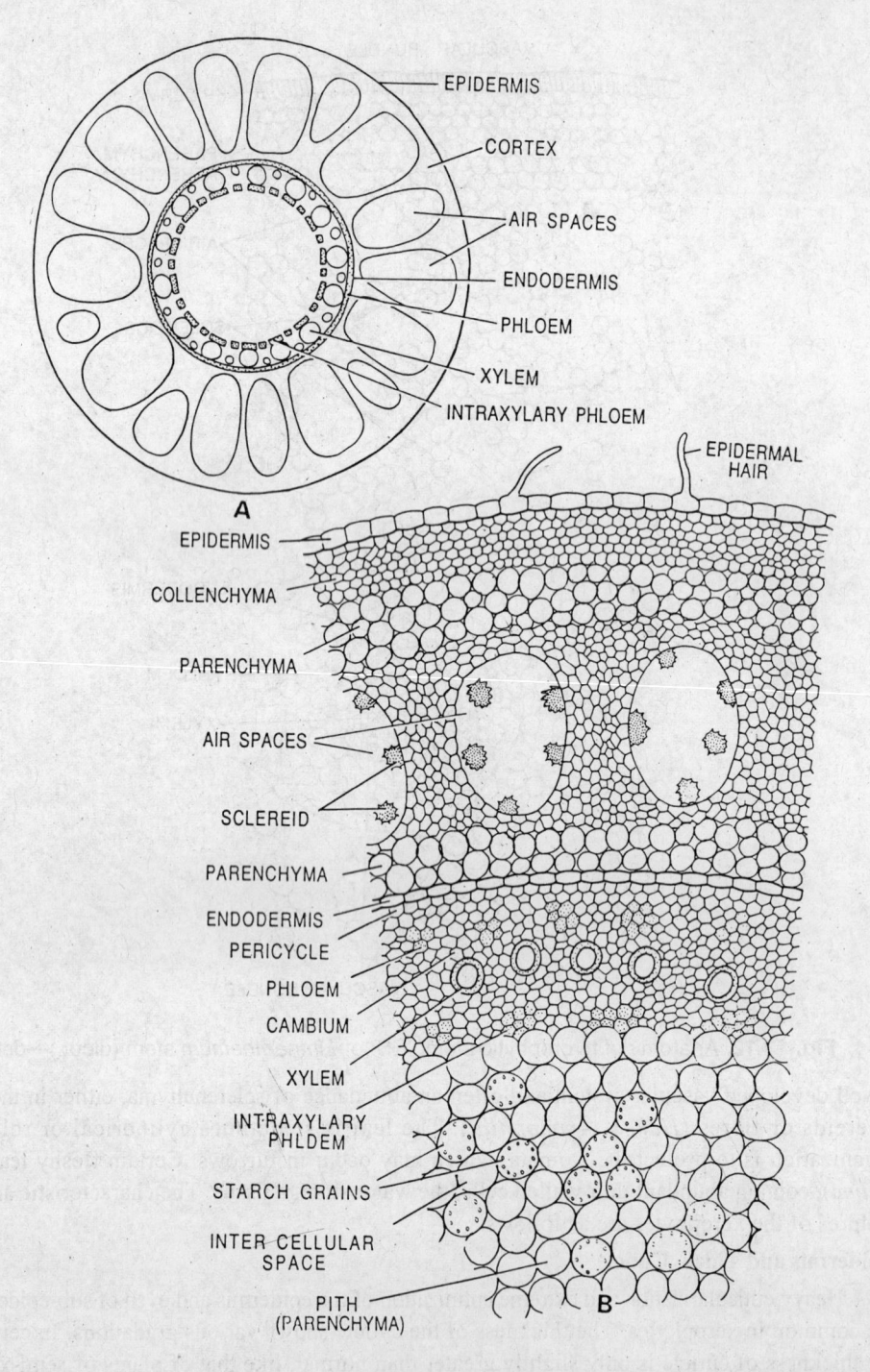

Fig. 13.13. Anatomy of hydrophytic stem. *Trapa bispinosa* (dicot.) A, diagrammatic; B, detail—T.S. showing epidermis, cortex, vascular cylinder and large pith. Cortex consists of collenchyma and parenchyma; big air spaces in the cortical region. Calcium oxalate sclereids present in the cortex. The endodermis, pericycle, phloem, cambium, xylem are present. Intraxylary phloem patches present around the pith. Large intercellular spaces among pith cells.

Ecological Anatomy 327

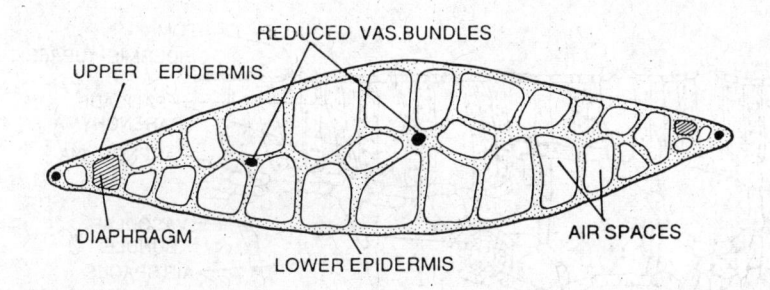

Fig. 13.14. Structure of hydrophyte. Cross section of submerged leaf of *Pontederia* showing diaphragm and air spaces. The air-chambers on the one hand give buoyancy to the plant for floating and on the other they serve to store up air. The cross partitions of air passages, known as diaphragms prevent flooding.

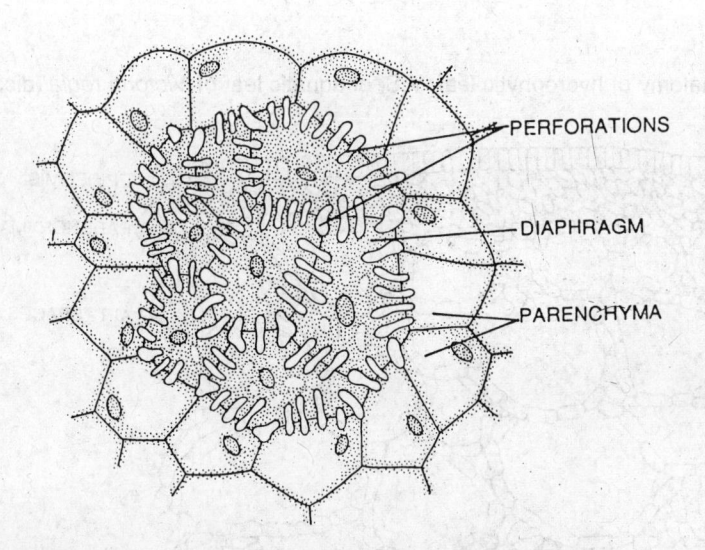

Fig. 13.15. Structure of hydrophyte. Detailed structure of C.S. of diaphragm of *Pontederia*. The diaphragms are provided with minute perforations through which gases but not water may pass.

lignification may extend even to the palisade parenchyma cells. In certain cases the covering of wax is formed on the epidermis (*e.g., Calotropis*). The epidermal cells are usually radially elongated. In the leaves of *Nerium* and *Ficus*, the epidermis becomes multilayered.

In many xerophytes, in addition to a cutinized epidermis, single to multilayered *hypodermis* is also present. In most plants, the hypodermis of leaves is morphologically mesophyll and may be in the form of sheet of fibrous tissue or a layer of sclereids. The hypodermis of the stems seems to be a part of the cortex. The hypodermis of stem and leaves may be cutinized to lignified. In many plants, the mucilage, gums and tannins are commonly found in the hypodermis.

Hairs. In many xerophytic plants, especially those of alpine regions exposed to strong winds, a covering of matted epidermal hairs on the underside of the leaves prevents water loss. Hairs may also be abundant over the entire aerial part of the plant. The thick matting of hairs also prevents rapid

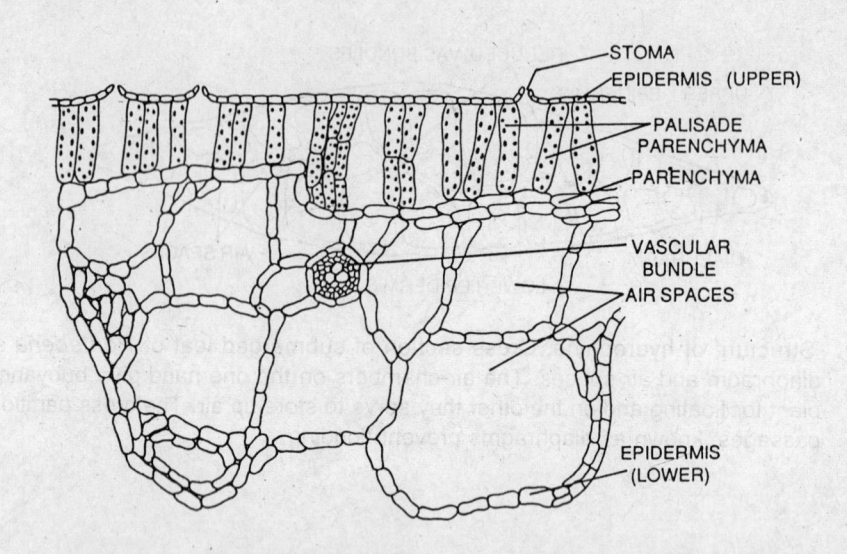

Fig. 13.16. Anatomy of hydrophytic leaf. V.S. of aquatic leaf of *Victoria regia* (dicot.)—detail.

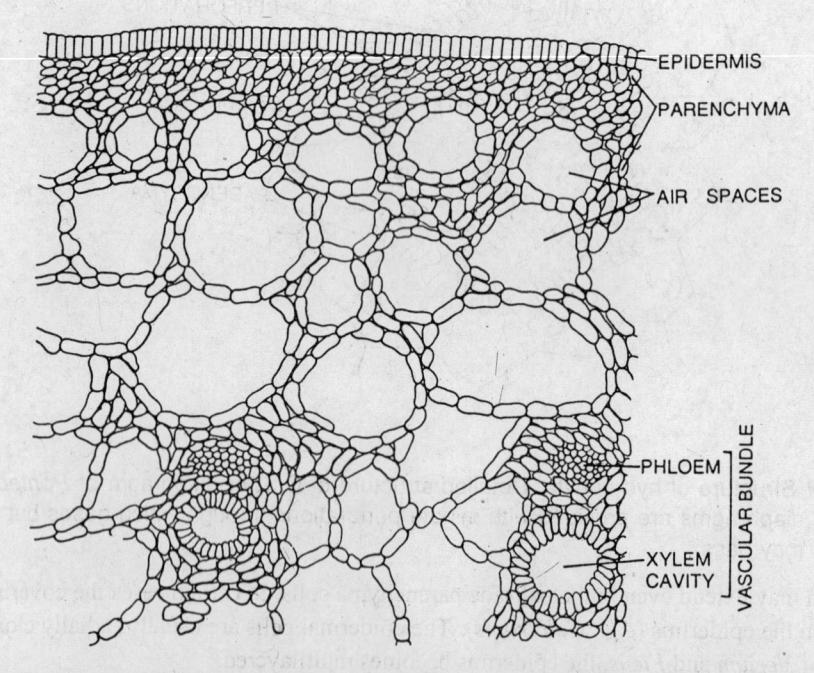

Fig. 13.17. *Nuphar luteum* (dicot.). T.S. of a sector of aquatic petiole showing big air spaces.

evaporation through stomata. The xerophytes that possess abundant hairs, on their leaves and stems, are commonly called *trichophyllous*.

Structure of Stomata.

The stomata are very minute openings formed in the epidermal layer in green aerial parts of the plants.

The stomata are essential for intake of carbon dioxide and oxygen and for the passage inward and outward of other gases. The evaporation of the surplus water takes place through the stomata.

Ecological Anatomy

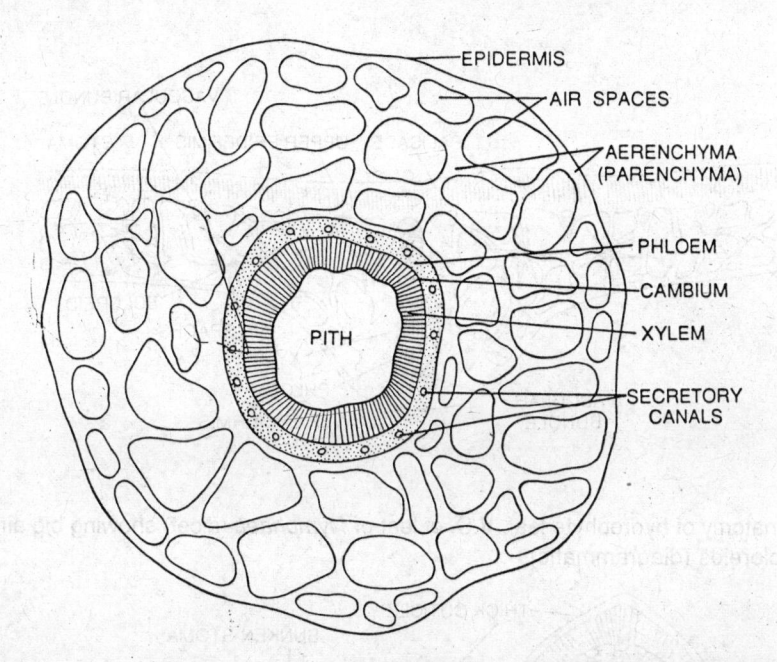

Fig. 13.18. Anatomy of hydrophytes. T.S. of an aquatic dicot stem *Hypericum* spp.—diagrammatic.

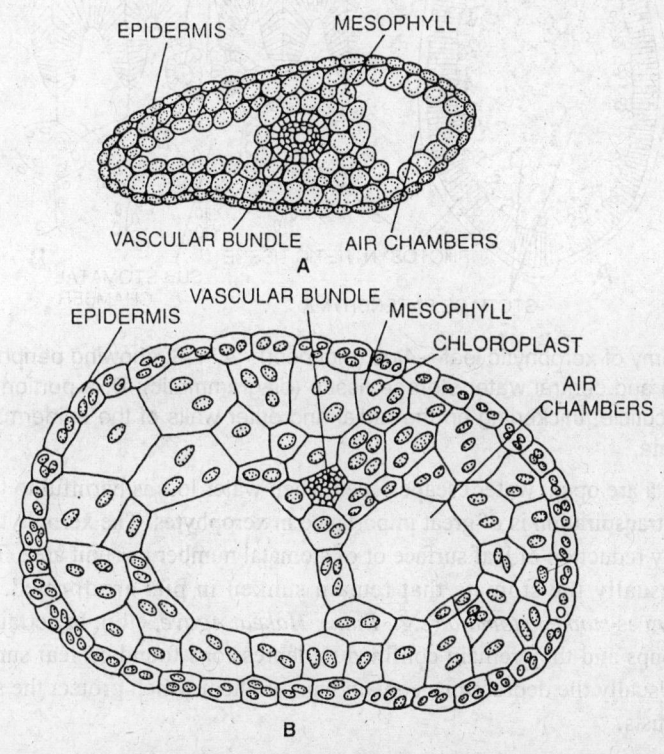

Fig. 13.19. Hydrophytes. A, T.S. through the leaf of submerged plants *Zanichellia palustris*; B, transection of a leaf of submerged hydrophyte *Ceratophyllum* showing the simplest form of vascular bundle. Mesophyll is undifferentiated and the chloroplasts are scattered in the epidermis and other cells.

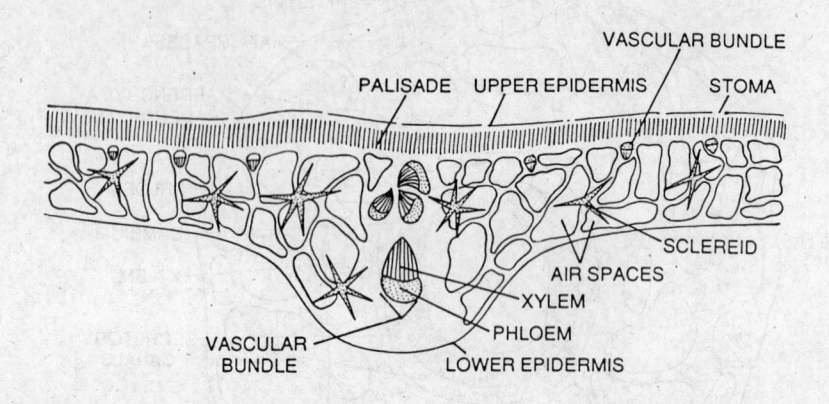

Fig. 13.20. Anatomy of hydrophyte leaf. V.S. of leaf of *Nymphaea* (dicot) showing big air spaces and sclereids (diagrammatic).

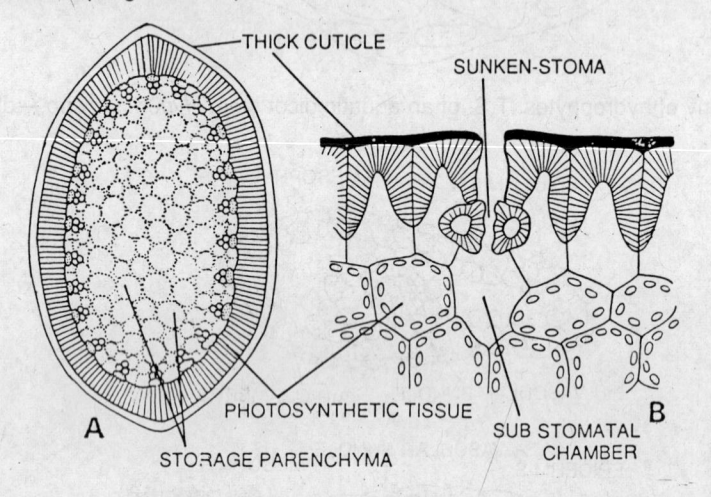

Fig 13.21. Anatomy of xerophytic leaf—*Aloe* (monocot). A, T.S. showing peripheral photosynthetic tissue and central water storage tissue (diagrammatic); B, a portion (detailed) showing thick cuticle, thickening on the radial and outer walls of the epidermal cells and sunken stomata.

When the stomata are open, water escapes even when water loss is harmful to the plant. This way, the reduction of transpiration is of great importance in xerophytes. The xerophytes may possess less stomata, either by reduction of leaf surface or of stomatal number per unit area. To reduce excessive transpiration, usually the stomata that remain sunken in pits are formed. Such stomata are commonly known as *sunken stomata* (*e.g., Pinus, Hakea, Agave*, etc,). In certain cases the stomata are found in groups and they remain confined to depressions found on leaf surface (*e.g., Nerium, Banksia*, etc.). Usually the depressions contain hairs in them which protect the stomata from direct attack of wind gusts.

Sclerenchyma

The xerophytes commonly have a larger proportion of sclerenchyma in their leaf structure than is found normally in mesophytes. The sclerenchyma is either found in groups or in continuous sheets. For example, in *Banksia*, there is a continuous thin sheet of sclerenchyma between the

Ecological Anatomy

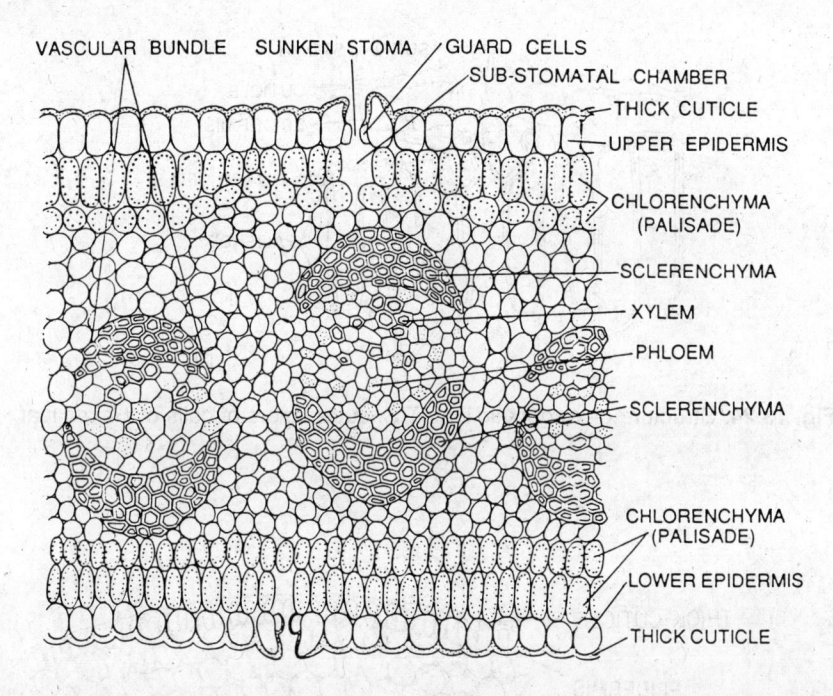

Fig. 13.22. Anatomy of xerophytic leaf. V.S. of leaf of *Agave* (monocot) showing detailed structure.

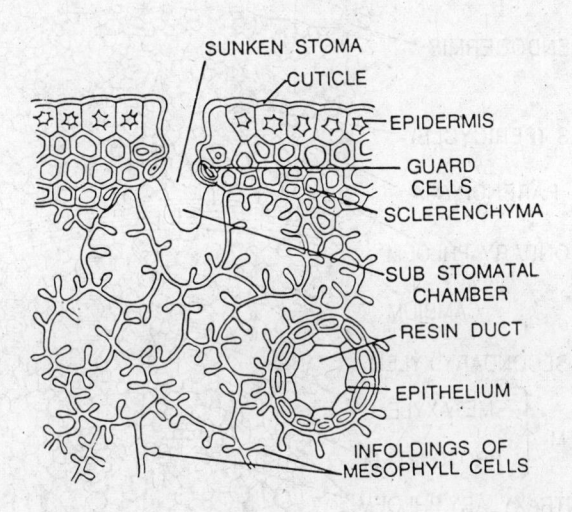

Fig. 13.23. Structure of xerophytic leaf (microphyllous type). C.S. of a marginal part of leaf of *Pinus nigra*.

hypodermis and the mesophyll. In *Dasylirion*, there is well-developed sclerenchyma below the epidermis. The sheets of sclerenchyma check excessive transpiration to some extent and also give mechanical support to the plant body. The xerophytes that possess heavy sclerification of the leaves are known as *sclerophyllous*.

Rolling of leaves. The leaves of many xerophytic grasses, roll tightly under dry conditions. In these grasses, the stomata are confined to the ventral surface of the leaf, so that when the leaf edges

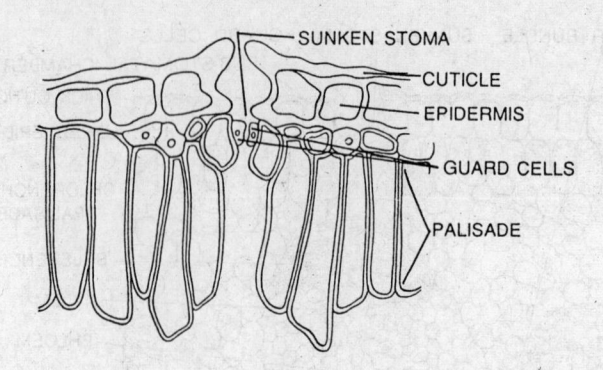

Fig. 13.24. Structure of xerophytic leaf. T.S. outer layers of cells of *Hakea* leaf.

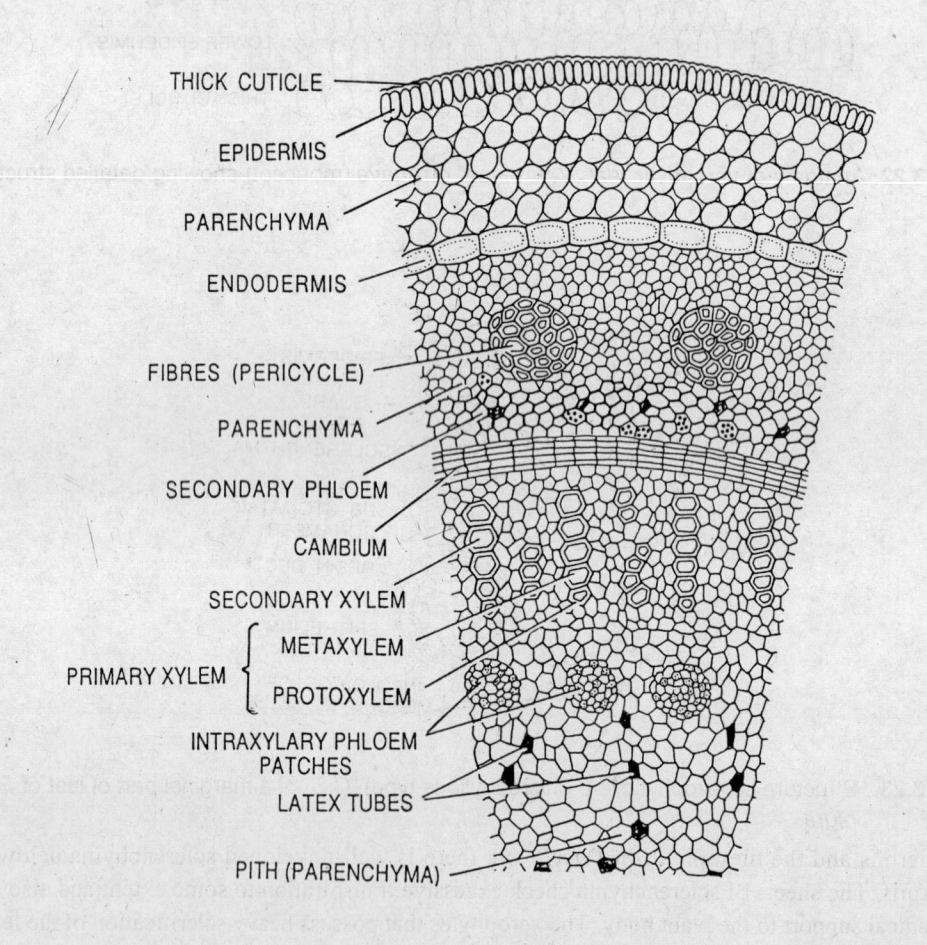

Fig. 13.25. Anatomy of xerophytic stem. T.S. of a sector of *Calotropis* stem (dicot.)—detailed structure.

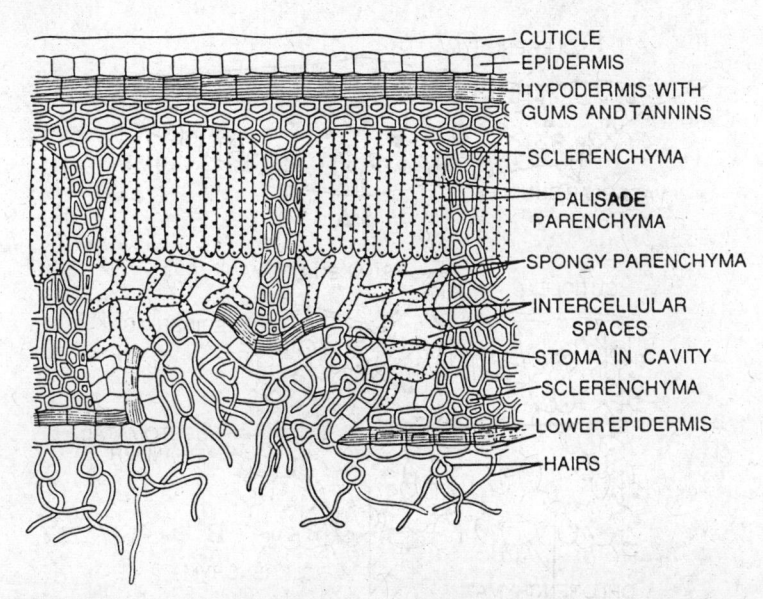

Fig. 13.26. Anatomy of xerophytic leaf. T.S. of *Banksia* leaf (sclerophyllous type).

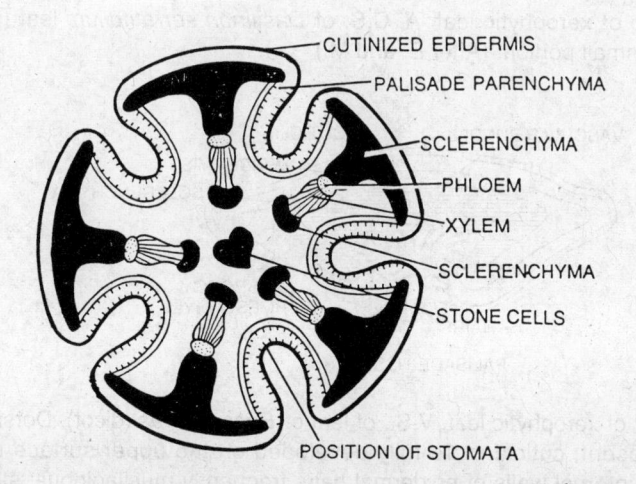

Fig. 13.27. Anatomy of xerophytic leaf. V.S. of leaf of *Hakea sulcata* (sclerophyllous-dicot)—diagrammatic.

roll inward, the stomata are effectively shut away from the outside air. As the stomata are located on the inner surface of the leaf, the air enclosed by the rolled leaf soon becomes saturated with water under the outward water diffusion stops. In *Ammophila arenaria*, there is tight upward folding of the leaf and also the sheltered location of the stomata in furrows, greatly reduce air movement over stomatal areas. Special motor cells (hinge) on the upper surface of the leaf are responsible for the inward rolling of leaves. In the xerophytic grasses, the motor cells are well developed.

Reduced leaf surface. In many xerophytes, the reduction of the leaf surface partly prevents water loss because the total exposed surface of the plant body is relatively small as compared with that of normal mesophytes (*e.g., Equisetum, Pinus, Casuarina, Asparagus,* etc.). In such xerophytes

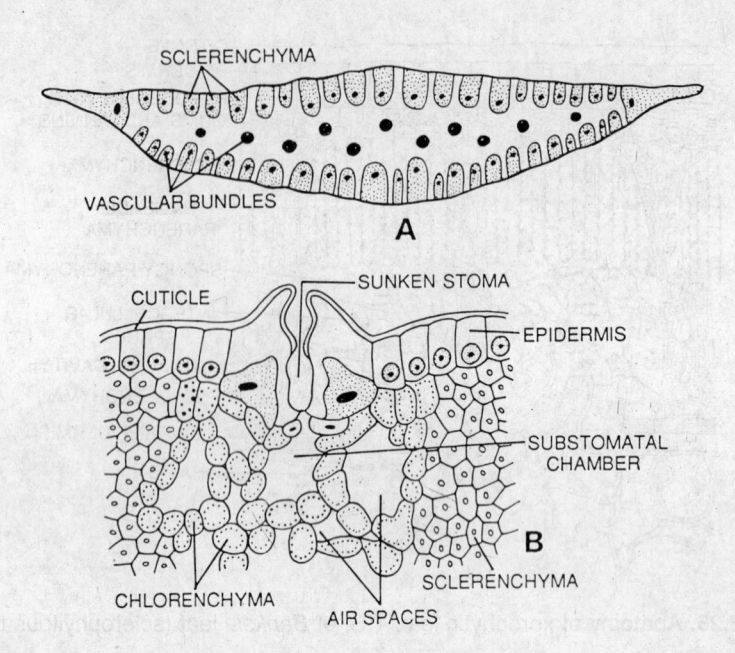

Fig. 13.28. Structure of xerophytic leaf. A, C.S. of *Dasilirion serratifolium* leaf (diagrammatic) B, detail of small portion. (After E. and M.)

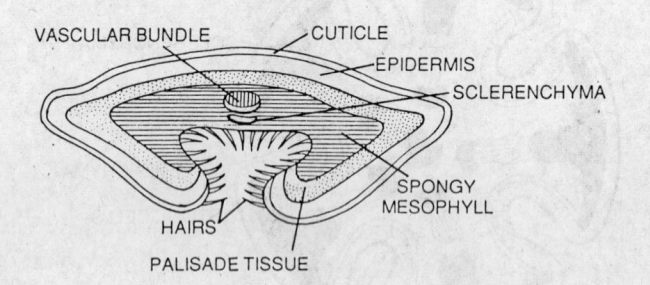

Fig. 13.29. Anatomy of xerophytic leaf. V.S.. of leaf of *Erica cinerea* (dicot). Dorsiventral rolled leaf; hairs present, cuticle more fully developed on the upper surface than on the lower surface; internal walls of epidermal hairs frequently mucilaginous; stomata confined to the grooves on the lower surface mesophyll situated at the leaf margin; vascular bundles mostly accompanied by mechanical tissue.

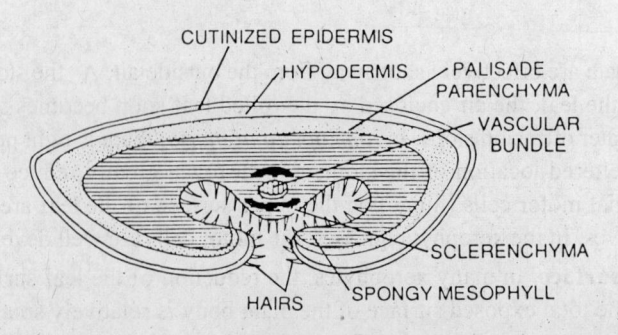

Fig. 13.30. Anatomy of xerophytic leaf. T.S. of dorsiventrally rolled leaf of *Erica tetralix*.

Ecological Anatomy

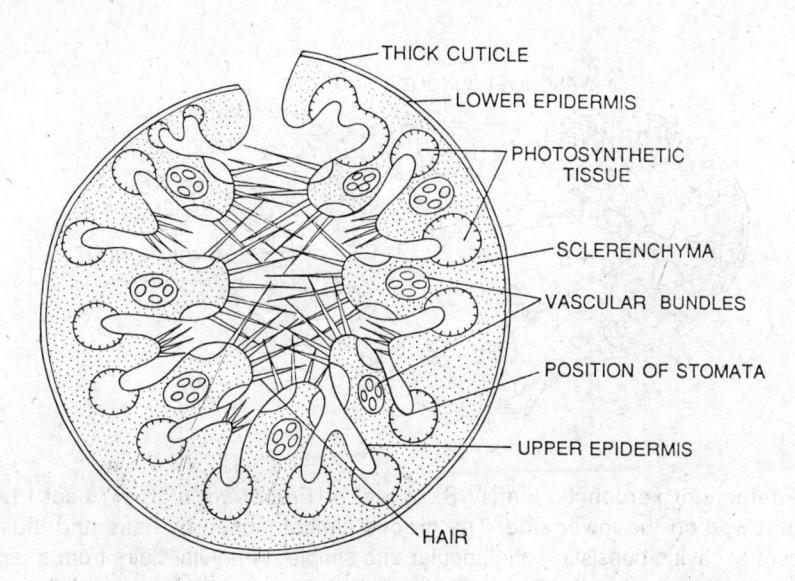

Fig. 13.31. Structure of xerophytic leaf. T.S. (diagrammatic) of *Ammophila arenaria* showing protected stomata.

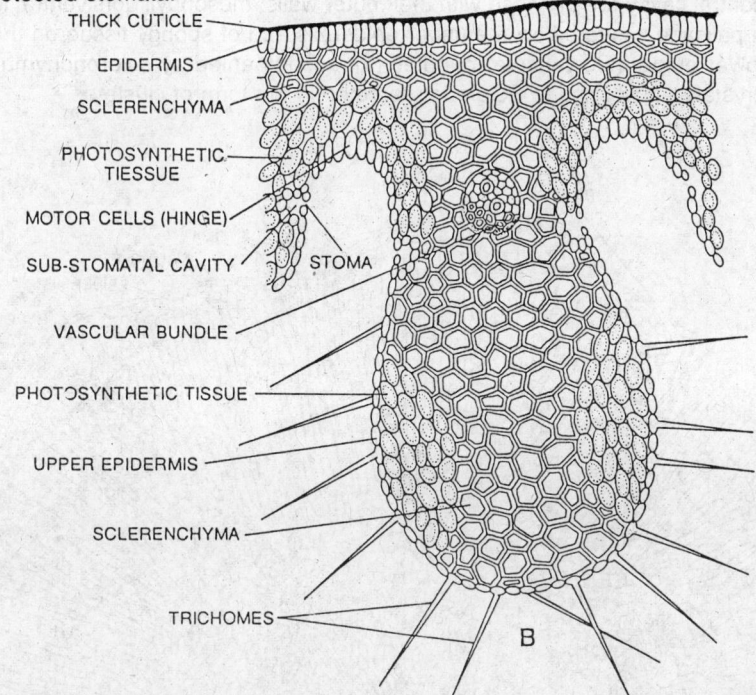

Fig. 13.32. Structure of xerophytic leaf. Part of T.S. of leaf of *Ammophila arenaria*, between two ridges (detailed).

the leaves are either scale-like or very small in size. Usually they are not found in the mature plant, or they persist as small scales or bracts. In some plants (*e.g.*, *Equisetum*, *Polygonella*, etc.) the photosynthesis takes place in the stem where assimilatory tissues are well developed. The reduction of leaf surface is usually accompanied by well-developed sclerenchyma water storage tissue and sunken stomata. Xerophytes, with reduced leaves are known as *microphyllous*.

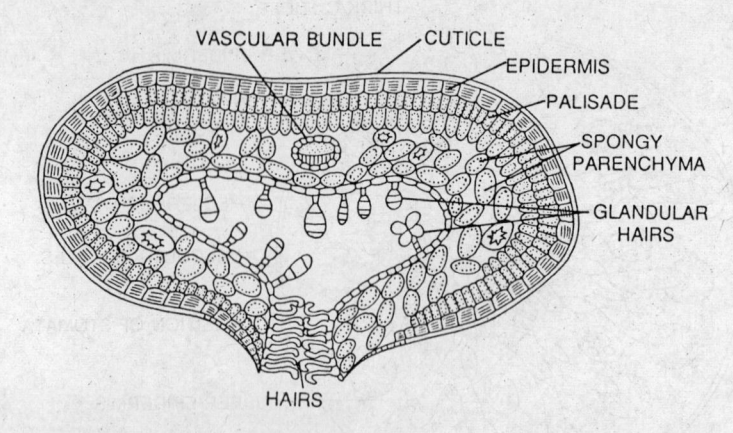

Fig. 13.33. Anatomy of xerophytic leaf. V.S. of leaf of *Empetrum nigrum* (dicot.) Leaf deeply furrowed on the lower side. The groove almost closed by hairs and thus forming a central cavity; consisting of glandular and simple unicellular hairs from a tangled mass closing the entrance to the central cavity; epidermis of the outer surface composed of cells with very thick outer and mucilaginous inner walls; epidermal cells around the central cavity are provided with their outer walls; mesophyll dorsiventral leaf consists of a palisade tissue below the upper epidermis and of spongy tissue on the inside of the lower epidermis, vascular bundles not accompanied by sclerenchymatous element; crystals—according to Solereder secreted in the form of clusters.

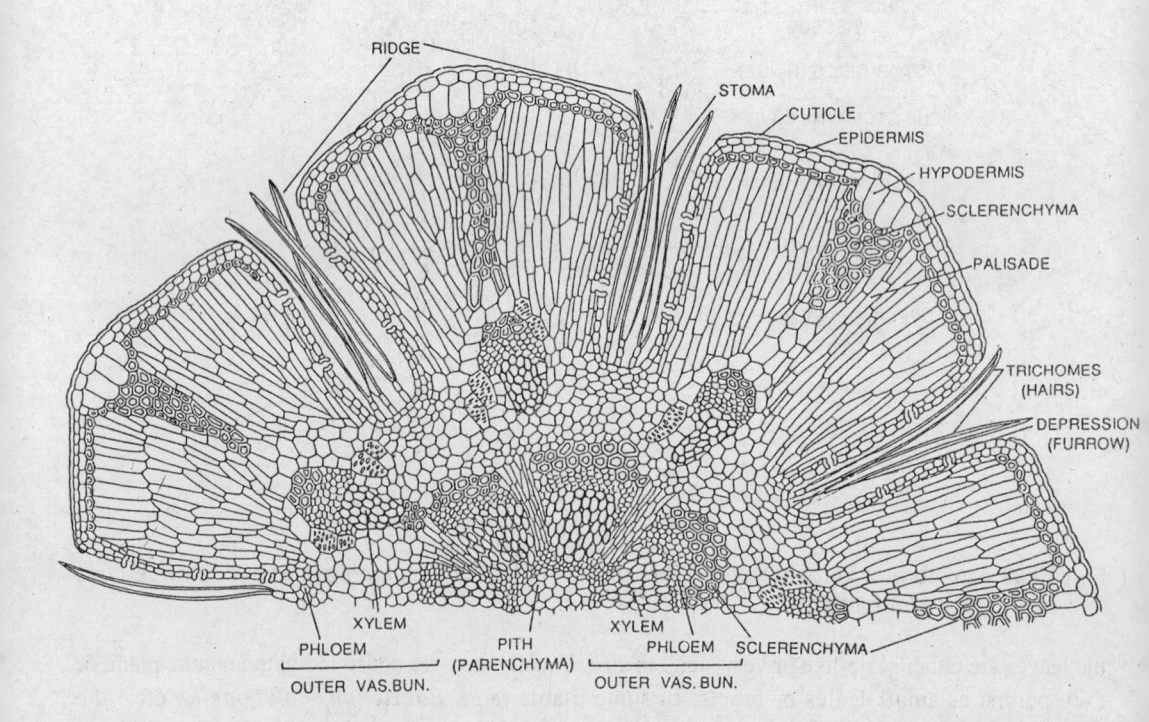

Fig. 13.34. Xerophytes. T.S. of xerophytic stem (*Casuarina equisetifolia*-dicot).

Ecological Anatomy

Needle leaves of gymnosperms. The needle leaves of gymnosperms are of microphyllous type. In the leaves of many conifers, in addition to reduced leaf surface there are heavy cutinization and sunken stomata. A characteristic illustration of this type is the needle of *Pinus*. The outer walls of the epidermal cells are cutinized. The wall is so much thickened that the lumen of the cells is nearly obliterated. The stomata are sunken and arranged in definite longitudinal rows. Beneath the epidermis there is well developed hypodermis, consisting of elongate sclerenchyma cells.

Water storage tissue. Many fleshy xerophytes possess water storage tissue and mucilaginous substance in them. In leaves such tissues are located beneath the upper of the lower epidermis, or upon both sides of the leaf and sometimes in the centre too. The storage cells are usually large and often thin-walled, as in *Begonia* The storage tissue may actually serve as a source or reserve water during drought. The xerophytes, that posses fleshy leaves or stems are known as *malacophyllous*.

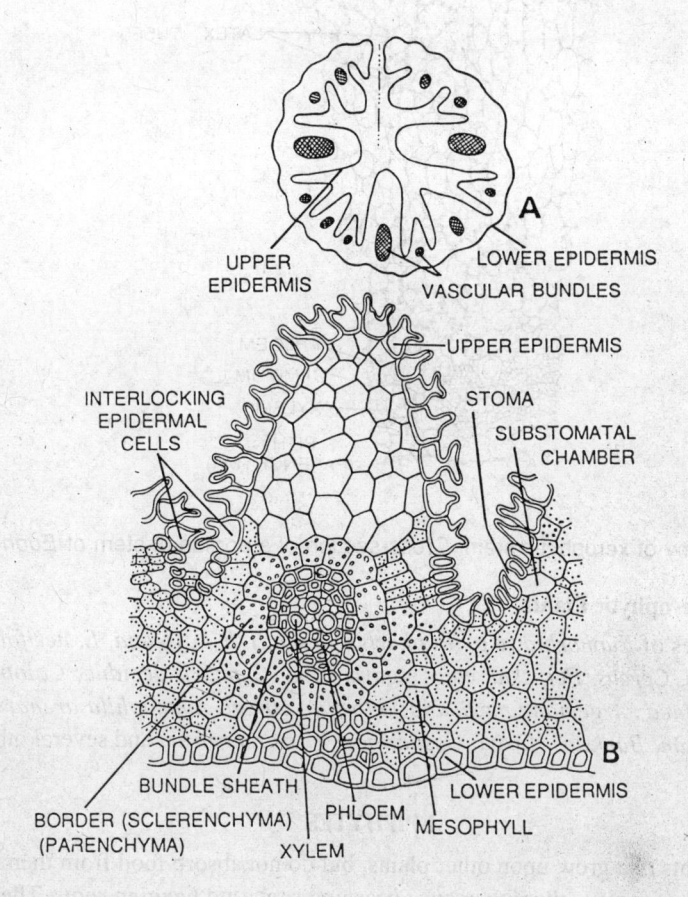

Fig. 13.35. Structure of xerophytic leaf. A, C.S. of rolling leaf of *Spartina*; B, detailed structure of same.

Abundant palisade parenchyma. In the stems of many xerophytes, the palisade tissue is present (*e.g., Capparis decidua*). In the xerophytic leaves the palisade is abundant and compactly arranged.

Latex tubes. In many xerophytic stems and leaves the laticiferous canals are present (*e.g., Calotropis, Euphorbia, Asclepias,* etc.). Due to viscosity of latex the transpiration is reduced to some extent.

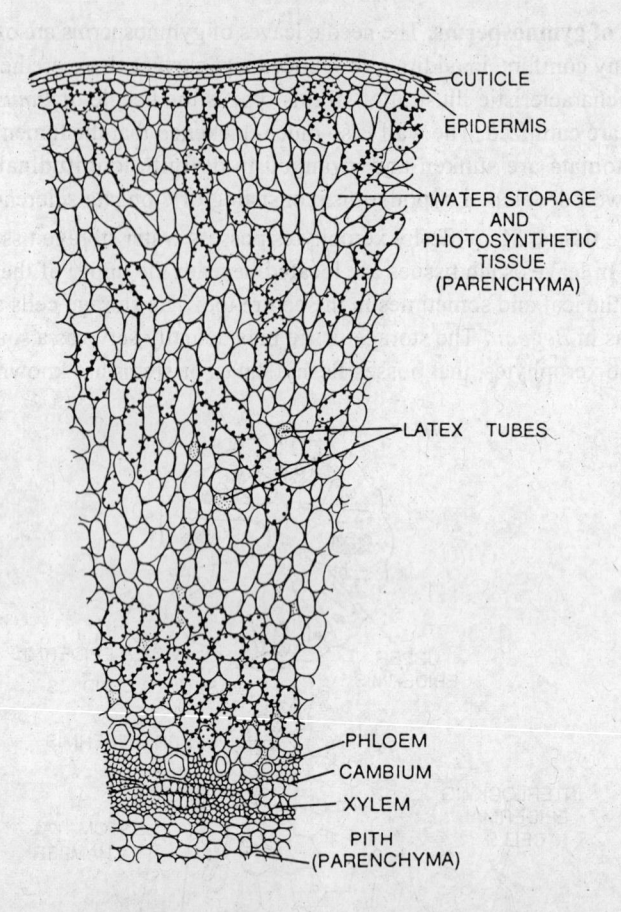

Fig. 13.36. Anatomy of xerophytic stem. Cross-section of a portion of stem of *Euphorbia tirucalli*.

List of Common Xerophytic Plants.

Several species of *Euphorbia* (e.g., *Euphorbia tirucalli, E. royleana, E. nerifolia* etc.); many cacti (e.g., *Opuntia, Cereus, Pereskia*, etc.); *Agave, Aloe; Capparis decidua; Calotropis procera; Carthamus; Portulaca; Argemone mexicana; Salsola; Alhagi, Ammophila arenaria; Dasylirion serratifolium; Nerium; Banksia; Hakea; Casuarina; Erica; Empetrum* and several others. (See figs. 13.21 to 13.36)

EPIPHYTES

These are plants that grow upon other plants, but do not absorb food from them. They usually develop three kinds of roots—*clinging roots, absorbing roots* and *hanging roots*. The clinging roots grow into cracks and crevices in the bark of the supporting plant and fix the epiphyte in proper position on the branch; besides they act as reservoirs of humus which accumulates in the network formed by such roots. The absorbing roots developing from the clinging roots project into the humus and draw food from it. The hanging roots are provided with an outer covering of a special absorptive tissue, called *velamen*, which usually consists of four or five layers of oblong-polygonal cells. The cells are dead containing air or water only, and their walls develop fibrous thickenings. There are also minute pits in the walls. The velamen acts as a sort of sponge and absorbs moisture from the surrounding air and also water trickling down the root (e.g., *Vanda, Dendrobium* and other orchids).

Ecological Anatomy

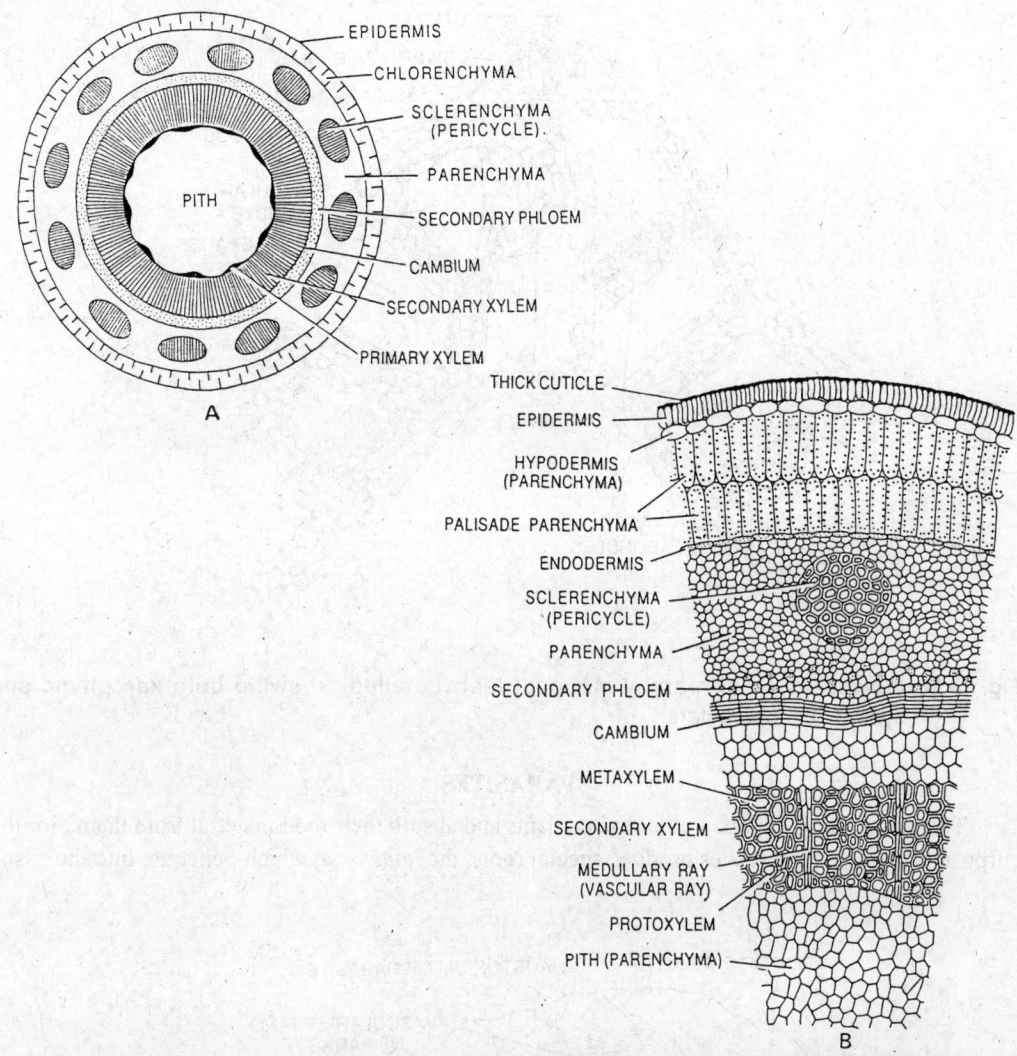

Fig. 13.37. Anatomy of xerophytic stem. T.S. of stem of *Capparis decidua* (dicot).

SAPROPHYTES

These are plants that grow in places rich in decaying organic substances, and derive their nutrition from them. Among the angiosperms *Monotropa* (Indian pipe) and some orchids afford good examples of saprophytes. Total saprophytes are colourless; and the partial ones are green in colour. Their roots become associated with a filamentous mass of a fungus which takes the place of and acts as the root-hairs, absorbing food material from the decomposed organic substances present in the soil. The association of a fungus with the root of a higher plant is known as *mycorrhiza*. Mocorrhizal fungi commonly associated with the roots of many forest trees, orchid seedlings, pine seedlings, etc. Two types of mycorrhiza are seen— *endotrophic*, in which the fungus is internal, usually living within the cortical cells of the root, as in many orchids, and *ectotrophic*, in which the fungus is external, growing attached to the surface of the root, as in conifers and other plants.

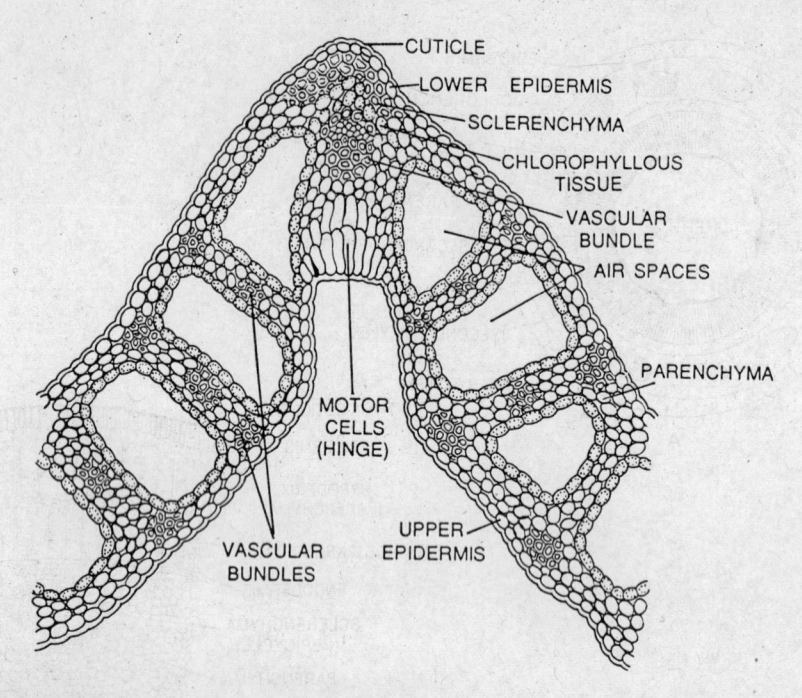

Fig. 13.38. *Molinia*—V.S. of marshy monocot leaf (detailed) showing both xerophytic and hydrophytic characters.

PARASITES

The plants that grow upon other living plants and absorb their food material from them. For the purpose of absorption parasites produce special roots, the *haustoria*, which penetrate into the tissue

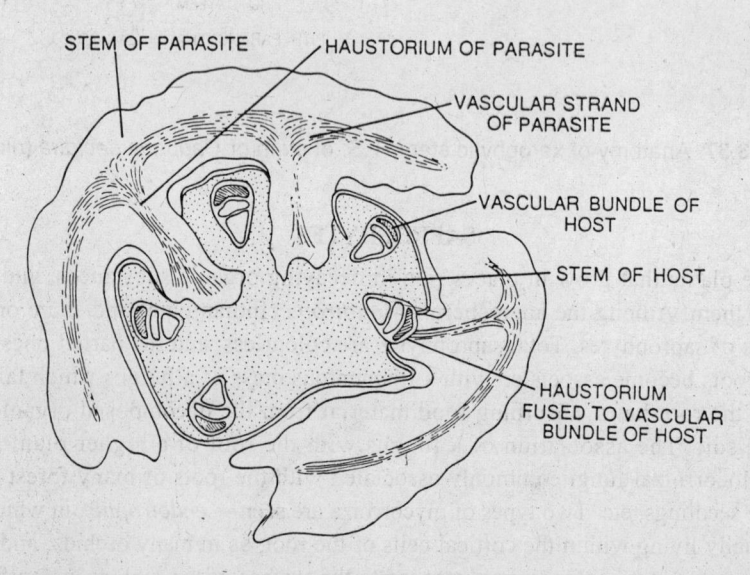

Fig. 13.39. Structure of total stem parasite. Section through haustoria and host of *Cuscuta* (diagrammatic).

Ecological Anatomy

of the host plant and absorb nutrition from it. There are differential degrees of parasitism. Some are *total parasites* and other *partial parasites*. The total parasites are never green in colour as they obtain all their food from the host plant; while the partial parasites develop chlorophyll and are in a position to manufacture food to a greater or less extent. They may be parasitic on the stem and branches or on roots. Accordingly they are said to be *stem-parasites* or *root-parasites* (*e.g.*, total stem parasite—*Cuscuta*; partial stem parasite—*Viscum*; total root parasite *Orobanche*; and partial root parasite, *Striga*). (See figs. 13.39 and 13.40)

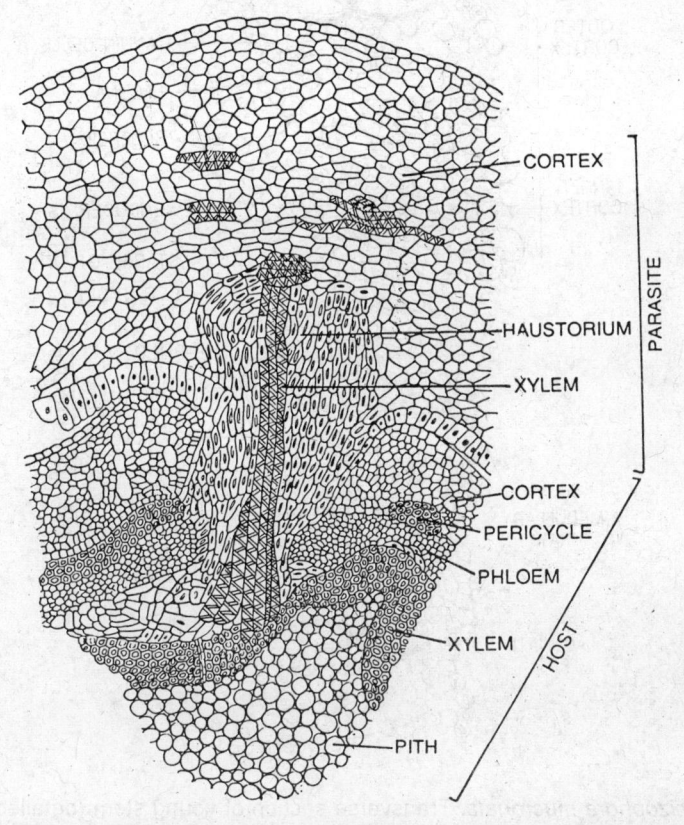

Fig. 13.40. Anatomy of host and parasite. Section through haustoria and portion of host of *Cuscuta*. The cells of the haustorium are between xylem and phloem of the host. The xylem of parasite is in contact with that of the host.

HALOPHYTES

These plants grow in saline soil or water, where there is abundance of salt in the soil; hence halophytes show some special characters. Most of halophytes have succulent leaves and some possess succulent stems. Leaves may be modified into or provided with spines. Typical examples of halophytes are *Suaeda maritima, Salsola, Acanthus ilicifolius, Chenopodium, Basella* and some species of Asclepiadaceae.

Halophytes growing in marshy places near the seashore, as in Sundarbans (West Bengal), form a special vegetation known as the **mangrove**. Mangrove plants produce a large number of stilt roots from the main stem and the branches. In many cases, in addition to the stilt roots special roots, called **respiratory roots** or **pneumatophores**, are also produced in large numbers, such roots develop from underground roots, and projecting beyond the water level they look like so many conical spikes distributed all round the trunk of the tree. They are provided with numerous pores or

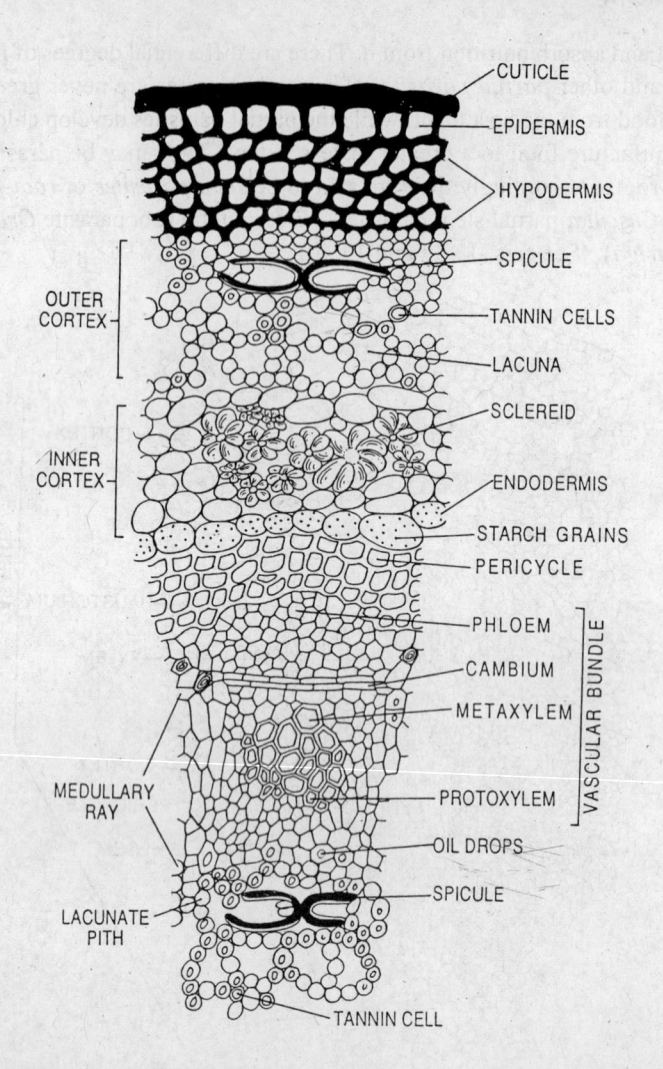

Fig. 13.41. *Rhizophora mucronata.* Transverse section of young stem (detailed structure).

respiratory spaces in the upper part, through which exchange of gases for respiration takes place. Mangrove species also show a peculiar type of germination. The seed germinates inside the fruit while it is still on the parent tree and is nourished by the same. Germination is almost immediate without any period of rest. The radicle elongates to a certain length and swells at the lower part. Ultimately the seedling separates from the parent tree and falls vertically down. The radicle presses into the soft mud, keeping the plumule and cotyledons clear above the saline water. This kind of germination of the seed inside the fruit while the latter still remains attached to the plant is known as **vivipary**. Typical examples of mangrove plants are—*Rhizophora, Sonneratia, Ceriops, Heritiera Excoecaria,* etc. (See figs. 13.41 to 13.48)

The mangrove plants exhibit several xeromorphic characters:

The leaves are succulent. Sometimes the stems also become succelent. Many species are covered with hairs (*e.g., Avicennia*). The epidermis is thick-walled and strongly cutinized. The sunken stomata are found beneath the epidermis. Usually aqueous tissue is present. The mesophyll is almost devoid of intercellular spaces and the palisade tissue in the main chlorenchyma

Ecological Anatomy

(*e.g., Sonneratia*). The nerve ends dilate into water storing tracheids (*e.g., Avicennia*). Long stone cells or bast like mechanical cells are lodged in between the palisade cells (*e.g., Rhizophora, Sonneratia*, etc.). Mucilage cells also occur (*e.g.,* in *Rhizophora, Sonneratia*, etc.).

GENUS *RHIZOPHORA*

T.S. OF STEM

Epidermis. The epidermis is composed of variously shaped cells appearing conical in transverse sections. The epidermis frequently consists of more than one layer, but a true hypodermis of 3–7 layers is also common. The cork in young stems generally arises superficially, usually in the hypodermis. The young stem has a very thick cuticle.

Cortex. The primary cortex is lacunar. H-shaped sclerenchymatous idioblasts are present. The cells of cortex possess pitted walls and are full of tannin and oil. Calcium oxalate crystals are also present. The inner cortex has groups of branched sclereids which give mechanical strength to the lacunate cortex. Te sclereids are lignified thick walled cells with narrow lumina. The endodermis is conspicuous. The endodermal cells possess starch grains.

Pericycle. It consists of a sub-continuous composite ring of sclerenchyma consisting of 3–4 layers of cells.

Vascular bundles. The vascular bundles are conjoint, collateral, endarch and open.

Xylem. The xylem is traversed by rays 2–3 cells wide in *Rhizophora mucronata*. The vessels possess scalariform perforation plates.

Crystals. The crystals are generally clustered.

Secretory elements. Vertically elongated secretory cells containing tannin and/or oil present in the cortex and the pith.

T.S. OF PETIOLE

Transverse sections of petiole exhibit a ring of bundles surrounding additional medullary strands in *Rhizophora mucronata* Lam. Mostly clustered crystals present. **Secretory elements**–cells containing tannin have been recorded in the mesophyll, cortex and pith. The inner cortex possesses H-shaped idioblasts.

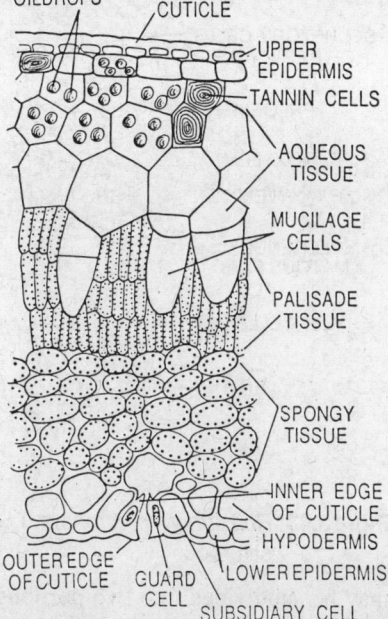

Fig. 13.42. *Rhizophora mucronata.* V.S. of leaf showing detailed structure.

Anatomy of Leaf

The leaves are usually dorsiventral. **Hairs** are mostly unicellular with thick or thin walls. The cuticle is well-developed and often quite thick on both the leaf surfaces. **Cork** warts occur as small black spots on the lower side of the leaf. The epidermis is single layered and consists of rectangular cells. The cells of upper epidermis possess some rod-shaped and cubical crystals of calcium oxalate. **The hypodermis** towards the upper surface is 2 or more layered. The **stomata** are confined to the lower surface. They are depressed and often provided with a front cavity. The mesophyll consists of palisade and spongy tissue. Palisade tissue consists of 1–4 layers. Spongy tissue usually possess large intercellular spaces. Aqueous tissue and mucilage cells are also present beneath the upper epidermis. H-shaped sclerenchymatous idioblasts occur in the palisade tissue, and variously branched ones in the spongy mesophyll.

Anatomy of Root

The well-known breathing roots (pneumatophores) of members of the Rhizophoraceae inhabiting mangrove swamps have large intercellular spaces of the spongy cortex which facilitate gaseous exchange in the special habitat in which the plants grow. The **cork** consists of suberized cells. The cork layer remains interrupted by lenticels at several places. In some cases the cork consists of alternating layers of suberized and ordinary parenchymatous cells in the aerial portion. Next to the cork there lies a thick cortex containing abundant, large, intercellular spaces, arranged radially around the stele. Intercellular spaces are larger in the subterranean than in the aerial parts of the roots. The cortex may be subdivided into two portions–the secondary cortex and the primary cortex. The **secondary cortex** consisting of few layers is found just beneath the cork. The

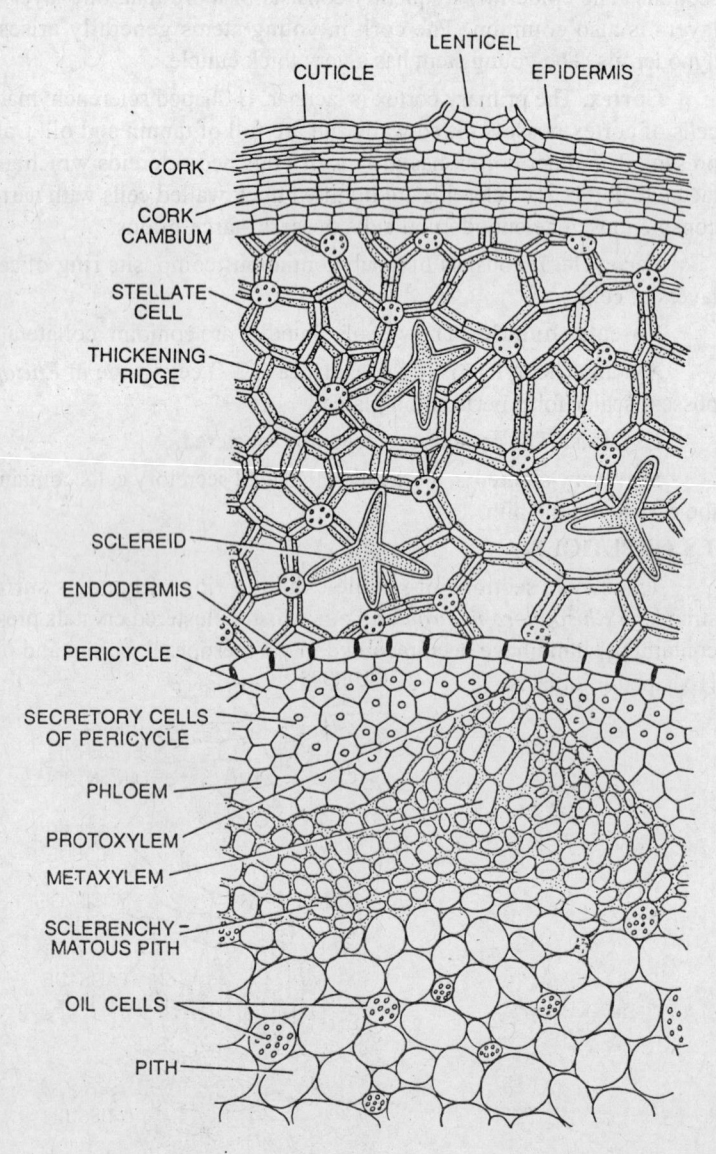

Fig. 13.43. *Rhizophora mucronata*. T.S. of a portion of subterranean stilt root.

Ecological Anatomy

primary cortex lies next to secondary cortex. The primary cortex is broad and lacunar. The primary cortex in the terrestrial part of the root is composed of cells of two kinds–(a) radially elongated cells connected with one another tangentially by short lateral arms; (b) rows of vertically elongated cells which show small, circular lumina in transverse sections. Solereder has recorded the ridges of thickening which are supposed to give mechanical support to the radially elongated cells where they are found. However, according to Bowman these are really cells filled up with mucilaginous sap. Mullan (1933) states that the cortex in the aerial part of the root system of *Rhizophora mucronata* Lam. is much reduced and the lacunae small. Sclerenchymatous **idioblasts** also occur in the cortex. H-shaped sclerenchymatous idioblasts project into the cortical intercellular spaces in the aerial part of the root, but are less numerous below ground. Mullan (1933) has also recorded the vertically elongated tubular cells, filled with tannin and oil, situated at the junctions between the branched cells in the terrestrial part or the root of *Rhizophora mucronata*.

The endodermis is conspicuous. Numerous secretory cells are found in the pericyclic region. The xylem is well formed and strongly lignified. The outer pith is sclerenchymatous whereas the central region of pith is parenchymatous. Several oil cells are present here and there in the pith region.

GENUS *SONNERATIA*

The members of family Sonneratiaceae are generally trees. Some of the species of *Sonneratia* inhabiting Mangrove Swamps from Africa to Australia are provided with **vertical branches of the root** system projecting into the air or water above the mud. The occurrence of **intraxylary phloem** in *Sonneratia* is also noteworthy.

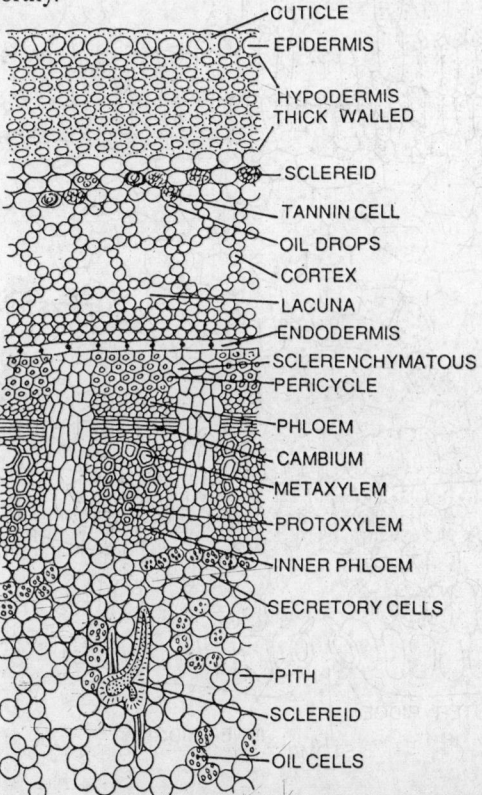

Fig. 13.44. *Sonneratia apetala*. T.S. of stem showing thick-walled hypodermal tissue with sclereids and tannin cells, well developed vascular system, and sclereids in the pith region.

T.S. OF STEM

The young stem of *Sonneratia* is provided with collenchymatous wings. The **cork** arises in the sub-epidermis in *Sonneratia apetala* Ham. Primary **Cortex** consists of spongy parenchyma. The outer part of cortex contains branched sclerenchymatous idioblasts. **Xylem** is found in the form of a continuous cylinder traversed by narrow rays. The vessels possess simple perforations. The **phoem** includes sclerenchyma. The **endodermis** is inconspicuous. The **pericycle** is found in the form of patches over phloem. The **intraxylary** or **inner** phloem is found next to xylem around the pith. The **pith** remains supported by sclerenchymatous elements. The cells of the pith and cortex contain tannin and oil drops. The **secretory cells** occur in the phloem as well as in intraxylary phloem, of *Sonneratia apetala*. According to Mullan (1933), **crystals** in vertical rows of special cells in the secondary phloem of *Sonneratia apetala*. The **secondary cortex** is composed of tangentially elongated cells. Sclereids are also found in the secondary cortex. These cells contain oil drops and tannin.

ANATOMY OF LEAF

The leaves of *Sonneratia apetala* Ham. are isobilateral. The upper **epidermis** possesses cuticular ridges. **Cork** warts are found in some species of *Sonneratia*. The stomata are deeply sunken

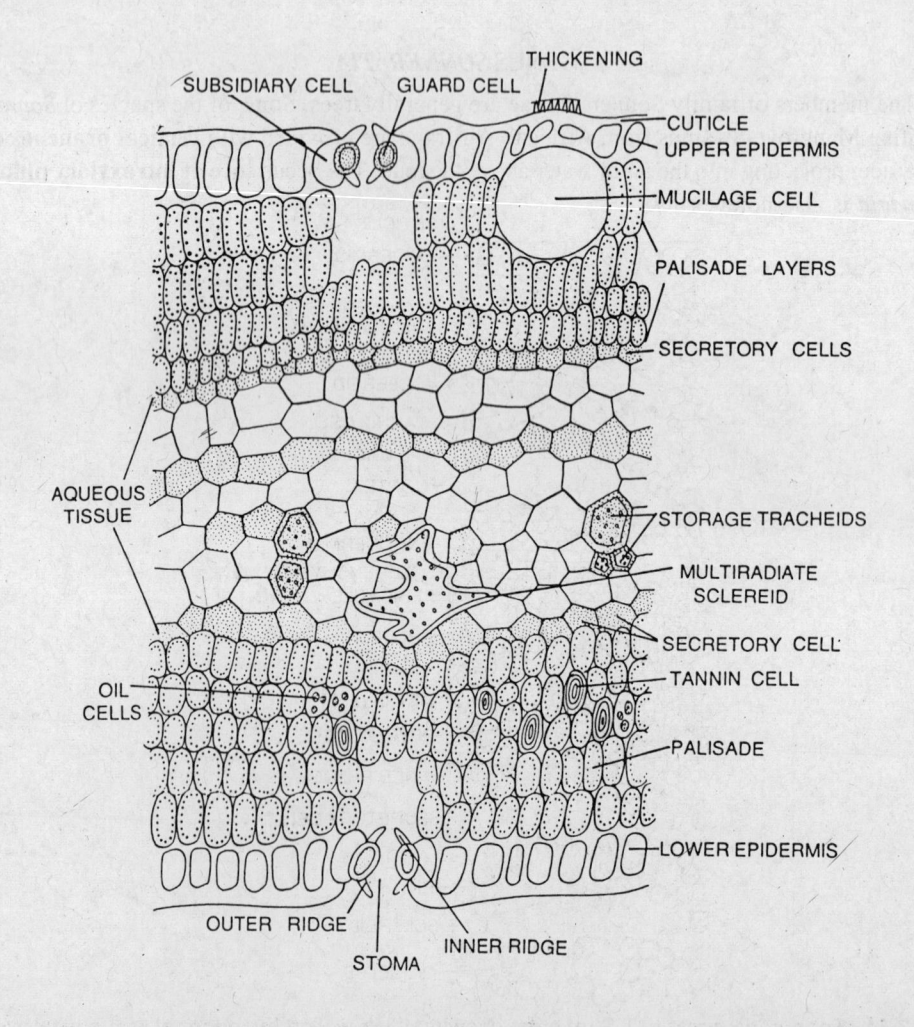

Fig. 13.45. *Sonneratia apetala.* V.S. of leaf showing centrally located aqueous tissue with sclereids and storage tracheids.

Ecological Anatomy 347

and present on both surfaces in *Sonneratia acida* Linn. The stomata are of rubiaceous type and they are equally numerous on both surfaces in *Sonneratia apetala* Ham. Large and rounded mucilage cells are present below the upper epidermis in the mesophyll. The central region consists of aqueous tissue and contains branched sclerenchymatous idioblasts. The oil and tannin cells are also present in the mesophyll tissue.

ANATOMY OF ROOT

The vertical, negatively geotropic portions of the root system are known as **'breathing'** or **'pneumatophores'** The **terrestrial roots** are also present.

The lenticels are present on the pneumatophores. The cork consists of a succession of lamellae each composed of three layers of cells. The cells of the outer layer of the cork are rounded externally and not suberized. The cells of the middle layer of the cork are tabular and suberized and those of

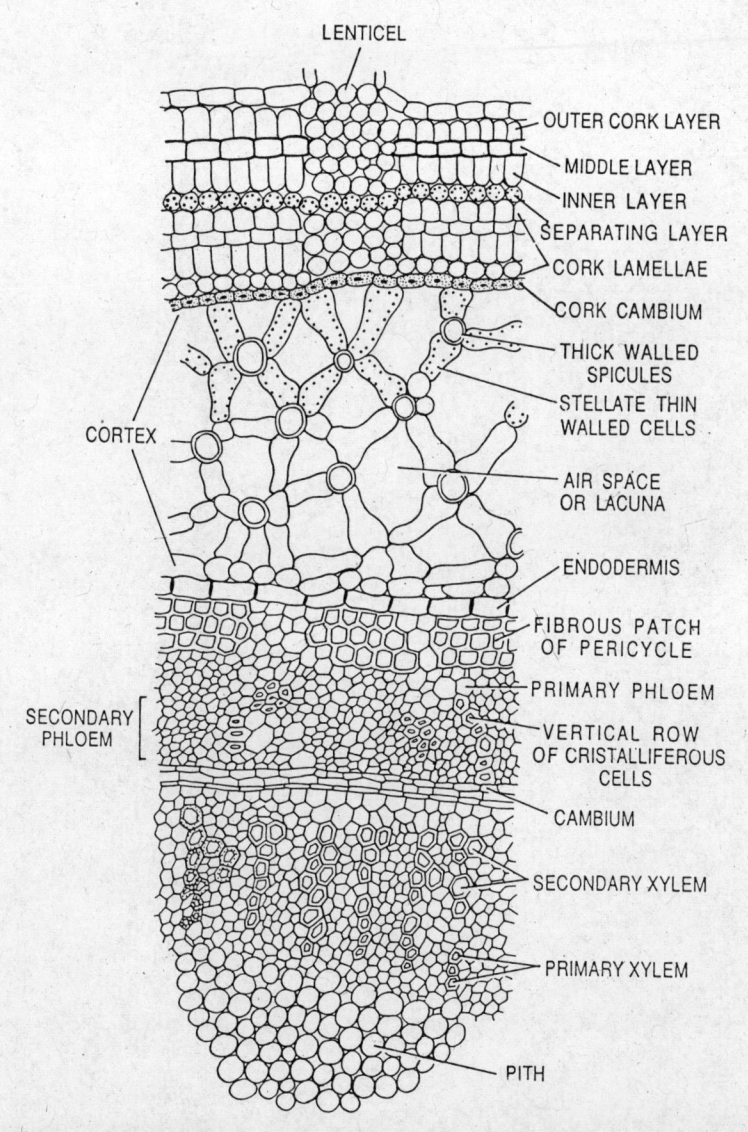

Fig. 13.46. *Sonneratia apetala.* T.S. of a portion of pneumatophore (respiratory root) showing detailed structure.

inner layer are radially elongated, rounded on the inner side and suberized. Two layers of rounded cells, present between each succeeding cork lamella, facilitate separation at these points. The **cortex** is composed of rounded cells interspersed with schizogenous, intercellular spaces. The **cortex** also contains cells with crystals and sclerenchymatous idioblasts. The cortex at the proximal end of the pneumatophore is supported by curved, lignified spicules. The **phloem** includes isolated strands of fibre and vertical rows of crystalliferous cells. The **xylem** is found in the form of a broad feebly lignified ring. The **pith** contains sclerenchymatous idioblasts at the distal end but not at the proximal end.

The terrestrial roots of *Sonneratia* show the following characteristics.

Cortex is lacunar. It is composed of cells mostly triradiate, and supported by thickening ridges and by vertically elongated cells with thick, pitted but lignified walls. Cork arises superficially.

14

Anatomy of the Floral Parts

The large majority of botanists, regard the flower as a modified shoot and its parts homologous of leaves. *Anatomically, the flower is a determinate stem with crowded appendages, with internodes much shortened or obliterated.* The appendages are of leaf rank but differ from those of the vegetative stem in function and shape. In the present text the flower is treated on the basis of the concept of homology between the flower and the shoot in their phylogeny and ontogeny.

PARTS OF THE FLOWER AND THEIR ARRANGEMENT

The flower consists of an axis, also known as receptacle and lateral appendages. The appendages are known as *floral parts* or *floral organs*. They are sterile and reproductive. The sepals and petals which constitute the *calyx* and *corolla* respectively are the sterile parts. The stamens and the carpels are the reproductive parts. The stamens compose the *androecium*, whereas the free or united carpels compose the *gynoecium*.

The vegetative shoot shows unlimited growth, whereas the flower shows the limited growth. In flower, the apical meristem ceases to be active after the formation of floral parts. In more specialized flowers there is a shorter growth period and they produce a small and more definite number of floral parts than the more primitive flowers. In still more advance flowers there are specialized characters, such as, whorled arrangement of parts instead of spiral, adnation of parts of two or more different whorls, cohesion of parts within a whorl, zygomorphic instead of actinomorphic condition, and epigynous condition instead of hypogynous condition.

Sepals. The sepals resemble leaves in their anatomy. Each sepal consists of ground parenchyma, a branched vascular system and an epidermis. The chloroplasts are found in the green sepals but usually there is no differentiation in the palisade and spongy parenchyma. They may contain crystal—containing cells, laticifers, tannin cells and other idioblasts. The epidermis of sepals may possess stomata and trichomes.

The traces are similar in origin and number. From the evidence of vascular system, sepals are clearly, in nearly every case morphologically bracts—that is, they have been derived directly from leaves and are not sterile sporophylls.

Petals. The petals also resemble leaves in their internal structure. They contain ground parenchyma, a more or less branched vascular system, and an epidermis. They may also contain crystal containing cells, tannin cells, laticifers and certain other idioblasts. They contain pigments—containing chromoplasts. Very often, the epidermal cells of the petals contain volatile oils which emit the characteristic fragrance of the flowers. In certain flowers the anticlinal epidermal walls of the petals are wavy or internally ridged, whereas the outer walls may be convex or papillate. The epidermis may also possess stomata and trichomes.

Stamen. Commonly the stamen consists of a two-lobed four-loculed anther. The anther is found to be situated on a slender filament which bears single vascular bundle. In certain primitive dicotyledonous families the stamens are leaf-like and possess three veins, whereas in advance types they are single-veined.

The structure of filament is simple. The vascular bundle is amphicribral and remains surrounded by parenchyma. The epidermis is cutinized and bears trichomes. The stomata may also be present on the epidermis of both anther and filament. The vascular bundle is found throughout the filament and culminates blindly in the connective tissue situated in between the two anther-lobes.

The outermost wall of the anther is the epidermis. Just beneath the epidermis there is endothecium which usually possesses strips or ridges of secondary wall material mainly on those walls which do not remain in contact with the epidermis. The innermost layer is composed of multinucleate cells; this is nutritive in function and known as tapetum. The wall layers which are located in between the endothecium and tapetum are often destroyed during the development of the pollen sacs. On the maturation of the pollen the tapetum disintegrates and the outer wall of the pollen sac now consists of only the epidermis and endothecium. At the time of dehiscence of the anthers the pollen are released out through stomium.

Gynoecium. The unit of gynoecium is called the carpel. A flower may possess one carpel or more than one. If two or more carpels are present they may be united or free from one another. When the carpels are united the gynoecium is known as *syncarpous*; when they are free the gynoecium is said to be *apocarpous*. A gynoecium with single carpel is also classified as apocarpous. The apocarpous gynoecium is termed *simple pistil*, whereas the syncarpous gynoecium is termed *compound pistil*. The carpel is commonly interpreted as foliar structure. The carpel of an apocarpous or syncarpous gynoecium is being differentiated into the ovary and the style. The upper part of the style is differentiated as a stigma. The stigma is sessile.

The ovary consists of the ovary wall, the locule or locules and in a multilocular ovary, the partitions. The ovules are found to be situated on the inner or adaxial (ventral) side of the ovary wall. The ovule-bearing region forms the *placenta*. According to Puri (1952) the position of the placentae is related to the method of union of carpels. In a carpel the placenta occurs close to the margin. Since there are two margins, the placenta is double in nature. The two halves may be united or separate. The number of double placenta in compound ovaries is equal to number of carpels. When the carpels are folded, the ovary is multilocular and the placentae occur in the centre of the ovary where the margins of the carpels meet. This is *axile placentation*. When the partitions of the ovary disappear, it becomes *free-central placentation*. When the carpels are joined margin to margin and the placentae are found to be situated on the ovary placentation is *parietal*.

Most commonly the carpels has three veins, one dorsal or median and two ventral or lateral, and the vascular supply of the ovules has been derived from the ventral bundles. The vascular bundles of the ovary, possessing axile placentation appear in the center of the ovary, with the phloem turned inward and the xylem wall, the outward.

The ovary and style are composed of epidermis, ground tissue of parenchyma, and vascular bundles. The outer epidermis is cuticularized and may have stomata. The ovule consists of a nucellus which encircles the sporogenous tissue. There are two integuments of epidermal origin, and a stalk, *funiculus*. The ovule consists of parenchyma and contains a more or less dominant vascular system.

Anatomy of the Floral Parts

VASCULAR ANATOMY

The study of the vascular anatomy has helped in solving many intricate problems of floral morphology. It has shown that many structures are not what they appear to be or what they are commonly taken to be. The fundamental vascular plan remains more or less unaltered and can always be of some help (Puri, 1952).

Morphologically the flower is a determined shoot with appendages, and these appendages are homologous with leaves. This commonly accepted view is sustained by the anatomy of the flowers. Flowers, in their vascular skeletons, differ in no essential way from leaf stems. They are often more complex than most stems. Taxonomy and comparative morphology have in large measure determined the structural nature of the flower. Anatomy of the flower has aided in the solution of certain puzzling conditions.

Pedicel. The Pedicel and the receptacle have typical structure, with a normal vascular cylinder. The cylinder may be unbroken or it may contain a ring of vascular bundles. In the region where floral organs are borne, the pedicel expand into the *receptacle*. The vascular cylinder also expands and the vascular bundles increase somewhat in number, and finally traces begin to diverge. In the

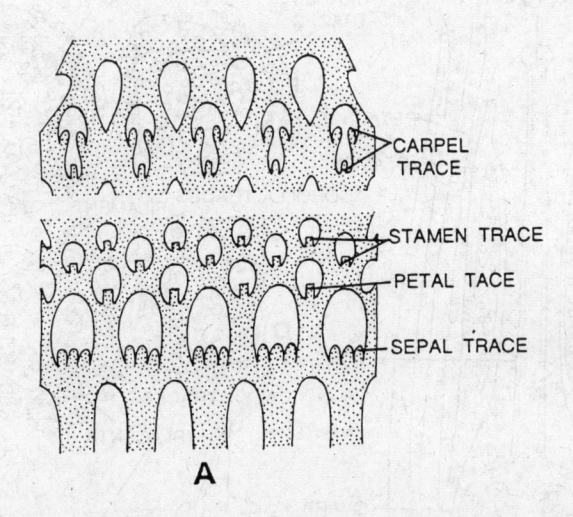

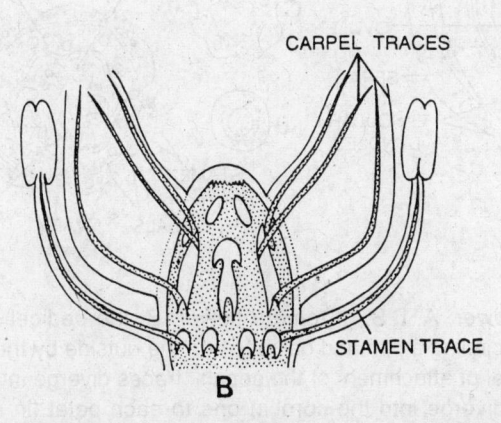

Fig. 14.1. Anatomy of the floral parts. A and B. diagams showing vascular structure of *Aquilegia*.

simplest cases vascular traces for different organs and whorls of organs arise quite independently (*e.g.*, in *Aquilegia*). In other cases various degrees of fusion may take place between bundles situated more or less in the same sectors.

The appendage traces are derived from the receptacular stele exactly as leaf traces are derived in typical stems. When the floral organs are numerous and closely placed, the gap of traces break the receptacular stele into a meshwork.

Sepals. The sepals are with very few exceptions, anatomically like the leaves of the plant in question. A sepal usually receives three traces derived from the same or different sources. As regards the morphological nature of the sepals, they have often been considered as equivalent to bracts and foliage leaves. Such a view is born out by a study of vascular anatomy which reveals practically the same vascular pattern as exists in foliage leaves and bracts of the same plant.

Petals. In their vascular supply the petals are sometimes leaf like, but much more often they are like stamens. The petals may have one, three or several traces. Very commonly there is but one

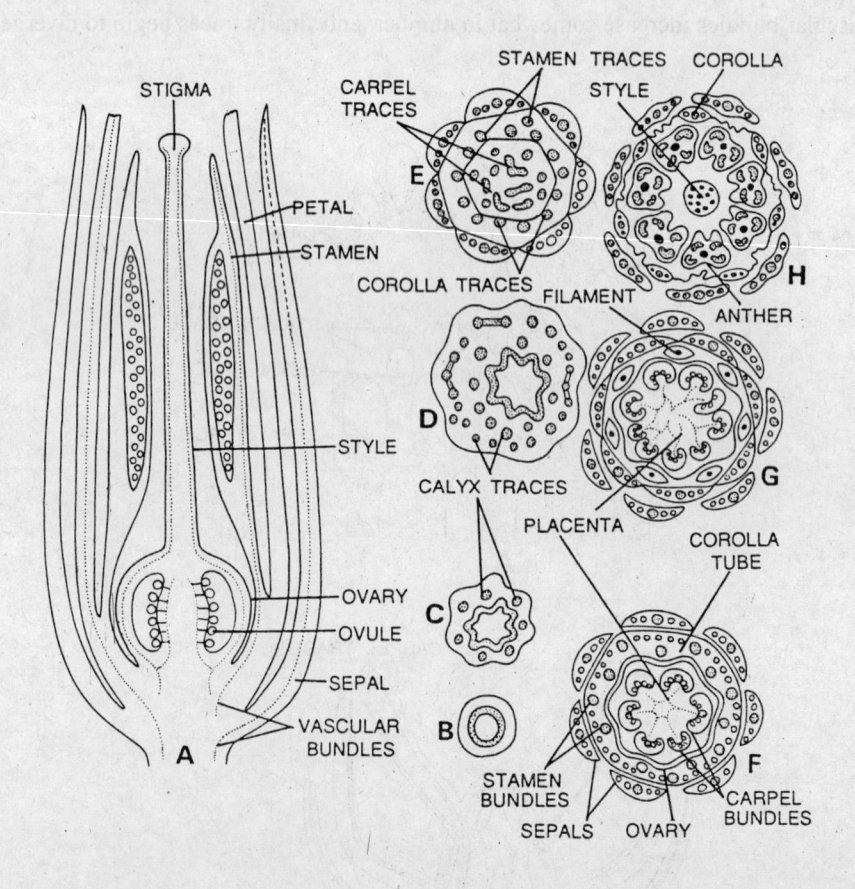

Fig. 14.2. Anatomy of the flower. A, L.S. of tomato flower; B, the pedicel shows a cylinderical vascular region encircling a pith and delimited on the outside by the cortex; C—D, in the receptacle and level of attachment of the sepals, traces diverge into these appendages; D—E, the traces diverge into the corolla, one to each petal (in most dicotyledonous flowers); E—G, traces to the stamens, typically one to each stamen; the carpellary supply of one, three, five or more traces to each carpel; H, a prolongation of the vascular supply into the style (After Esau).

Anatomy of the Floral Parts

trace. The petals appear to be sometimes modified leaves, like the sepals. but in the great majority of families they are sterile stamens. However, since stamens are the homologous of the leaves, it is not always possible to determine from anatomical evidence along whether one trace petals in certain families are modified stamens or whether they have come more directly from leaf-like structures.

Stamens. A stamen generally receives a single trace which remians almost unbranched throughout its course in the filament. In the anther region it may undergo some branching. In a few Ranalian families and rarely elsewhere as in some members of the Lauraceae and Musaceae, three traces are present in each stamen. In *Ravenala* (Musaceae) each filament is traversed by 25 to 28 small vascular bundles. Most of these disappear as the anther is approached, and the system of central bundles consisting of three or four bundles, continues into the connective. From other evidence the above mentioned families appear to be fairly primitive, it seems highly probable that the single trace condition is one of reduction from three.

In the simple flower of *Aquilegia* the stamen traces pass off, one to each organ in several whorls. Above the supermost whorl of stamens the vascular cylinder becomes complete again.

Carpels. The carpel is commonly looked upon as a leaf-like organ folded upward, *i.e.*, ventrally with its margins more or less completely fused and bearing the ovules. This conception has been supported by the anatomy. The details of origin, number and course of the bundles forming the vascular supply are exactly like those of leaves the carpel has one, three, five or several traces. The three trace carpel is most common. The five-trace, carpel is nearly as common as the three traces, and carpels with seven, nine and more traces are increasingly less and less common. The evidence that the one-trace carpel (nearly always an achene) has been derived by reduction from the three-trace type. The median trace which leaves the stele below the other carpel traces, is known as the *dorsal trace* because it becomes the dorsal (midrib) bundle of the folded organ. The outermost traces are known as *ventral* or *marginal traces* because they become the bundles that run along the ventral edge of the carpel, *i.e.*, along or near the margins of the organ if it were unfolded. The upward and inward folding of the sides of the carpel brings about the inversion of these ventral bundles. The phloem remains on

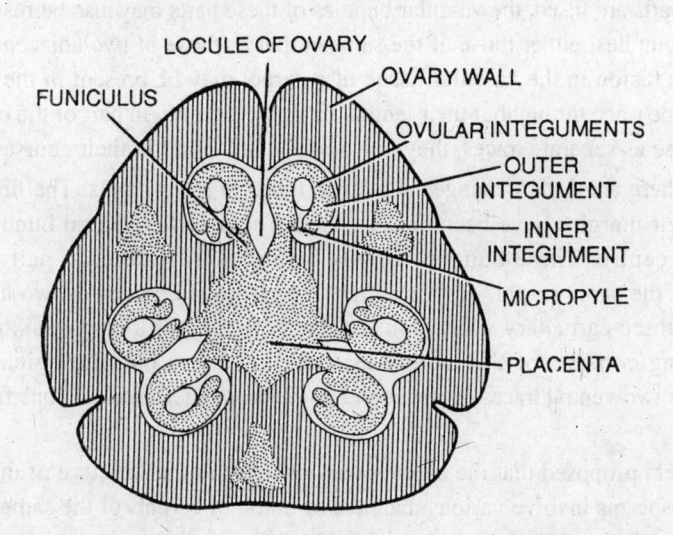

Fig. 14.3. The flower. T.S. of the ovary of *Lilium*, showing six anatropous ovules and the manner of placentation.

the ventral side in the carpel, whereas it is on the dorsal side in the midrib (dorsal) bundle. This important condition may be easily understood when it is remembered that the carpel is leaf-like, with its margins folded upward. The ovule traces are derived from the ventral bundles.

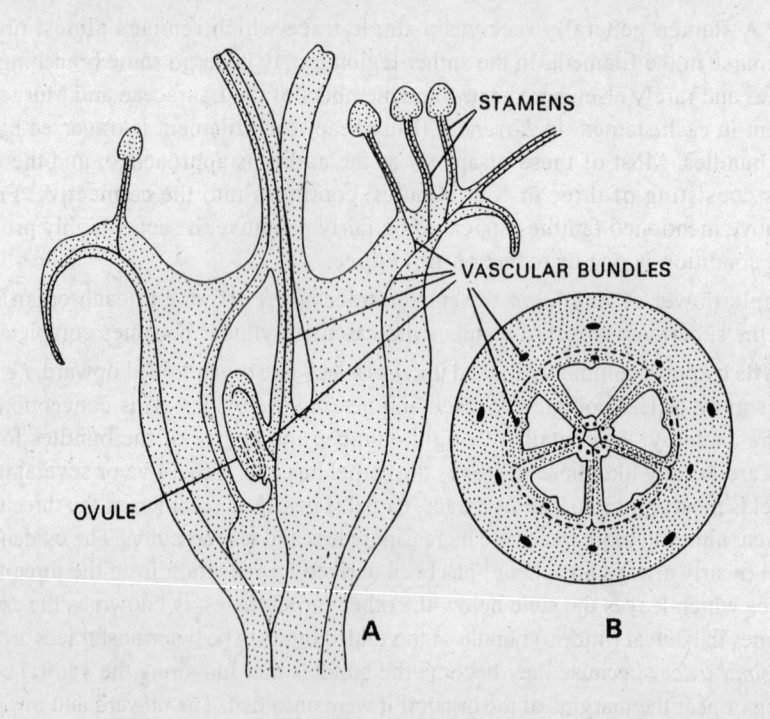

Fig. 14.4. The vascular system of the flower in *Malus pumila*. A, vertical section; B, transverse section.

When floral parts are fused, the vascular bundles of these parts may also be fused. If carpels are united, the lateral bundles, either those of the same carpel or those of two adjacent carpel, may be fused in pairs. The fusion in the vascular tissue of a carpel may be present in the ventral bundles from an origin as one trace throughout their length, or may exist only in part of the carpel; where the ventral bundles arise as separate traces, they may unite at any point in their course.

In syncarpy there are fusion changes similar to those in free carpels. The lines that separate the carpels and their margins have been disorganised. The inverted ventral bundles, form a ring of bundles in the centre. These bundles usually lie in pairs. Here each pair consists of the ventral bundles, of the same carpel, or more often of bundles from each of two adjacent carpels. In the centre of a three carpellary syncarpous ovary there may be a ring of six or three ventral bundles. If the ring consists of three bundles, each bundle is morphologically double and represents either the two ventral traces of one carpel or one from one carpel and one from the adjacent carpel.

Several workers proposed that the evolutionary changes in the structure of the gynoecium of the flower of angiosperms involve various manners of union of carpels of the same flower. In such angiospermous flower the carpel may become joined by their margins to the receptacle (Fig. 14.6 B), or they may grow together laterally in a closed folded condition (Fig. 14.6 C), or they may become laterally united in an open folded condition (Fig. 14.6. A). The junction of carpels in an

Anatomy of the Floral Parts

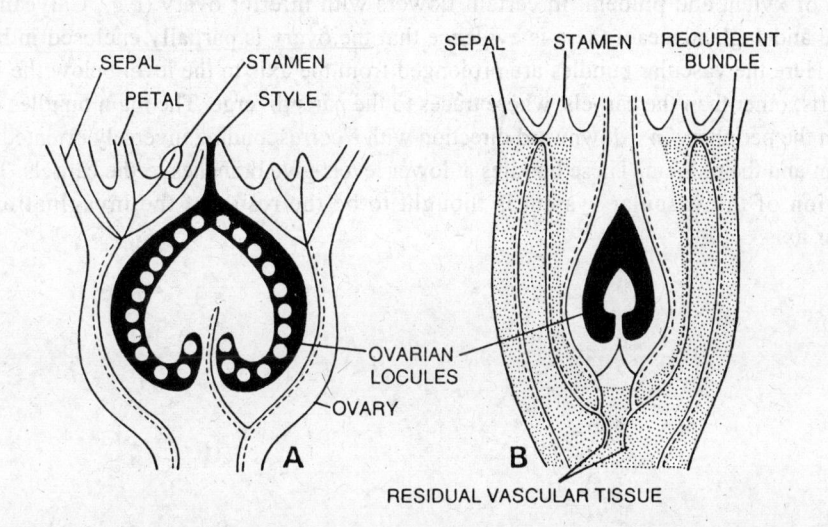

Fig. 14.5. Anatomy of the flower—the inferior ovary, The xylem is represented by solid lines ad phloem by broken lines. A, the ovary embedded within the floral tube (*Samolus floribundus*); B, In *Darbya* the ovary remains embedded in the invaginated receptacle (shown by dots). Here the receptacular nature of the outer tissue is indicated by the presence of recurrent bundles with inversely oriented xylem and phloem (After Douglas, 1944).

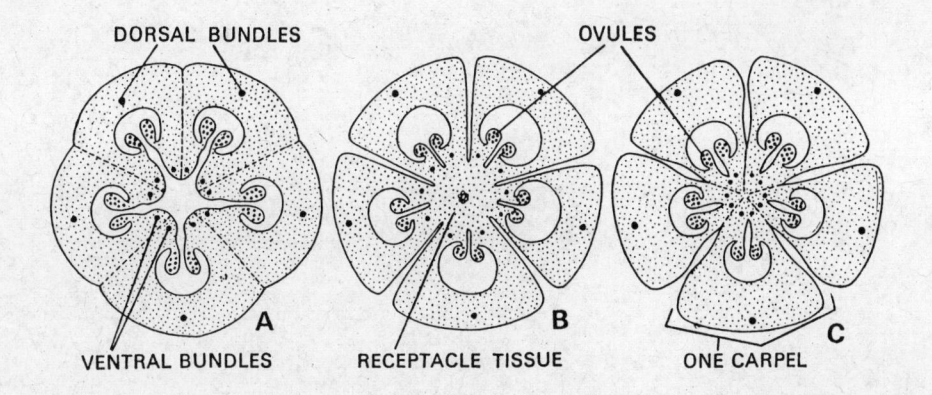

Fig. 14.6. The flower. Gynoecia in transections showing syncarpous tendencies.

open condition may result in a unilocular ovary showing parietal placentation as shown in fig. 14.6 A. Folding combined with union of carpels with each other may form an ovary with as many locules as there are carpels. In such cases the ovules are borne on the central column of tissue where the carpels come together showing, axile placentation (Fig. 14.6 B, C).

The inferior ovary. The inferior ovary is formed by the adnation of the sepals, petals and stamens to the carpels or by the sinking of the gynoecium in a hollowed receptacle with fusion of the receptacle walls about the carpels. The vascular system is thought to show this structure in that the bundles found in the appendages of different whorls are variously fused but all show the usual orientation of xylem and phloem. In certain flowers with inferior ovary (*e.g.*, Calycanthaceae, Santalaceae and Juglandaceae) there is evidence that the ovary is partially enclosed in hollowed receptacle. Here the vascular bundles are prolonged from the axis to the level below the insertion of floral parts, other than the carpels, where traces to the parts diverge. The main bundles continue farther from the periphery in a downward direction with a corresponding inversely oriented position of the xylem and the phloem. These bundles at lower levels give branches to the carpels. This type of orientation of the vascular system is thought to be the result of the invagination of the receptacular axis.

15
Anatomy of the Embryo and Young Seedling

In most embryos, the procambium is commonly found in some isolated areas and a few mature protoxylem and protophloem cells may be present. The anatomy of the embryo proper is the histology of meristems.

ANATOMY OF THE COTYLEDON

The traces of the cotyledon vary in number. Commonly only two traces are found in the lower taxa of both monocotyledons and dicotyledons. Four is a frequent number, with transition to three by the fusion of the middle pair. Three is also common with the strong central bundle sometimes double. The large number of traces are probable uncommon (*e.g.*, *Canna* and some *Araceae*). When four traces are present, the two median traces lie close together. The Iridaceae have a median trace, simple or double, with one or two pairs of laterals. The Araceae have one to many traces. The Liliaceae have all stages in the origin of an odd number of traces—two traces, four traces, a double traces, a median trace with one or more pairs of laterals. Odd numbers result from the fusion of two or more traces. Number, position and fusion of traces coincide with those of leaf traces. A single trace is characteristic of highly specialized cotyledons. In many taxa, commonly, the lateral bundles are free distally and with the median bundle, formed a three veined cotyledon. The two free laterals become major bundles in the anatomy of many specialized cotyledons.

The characteristic feature of many monocotyledons is the presence of two strong lateral traces, that supply the sides of the sheathing leaf base. The median vein that supplies the cotyledon tip, bends downward to the scutellum. The lateral veins continue upward in the sheathing wings, or rarely they also continue downward to the scutellum. The tip of the median bundle, remained unbranched in small and cylindrical forms, but becomes branched in large forms.

In most cotyledons which possess two or three free traces, all the vascular bundles continue as veins towards the tip. In the downbent cotyledons where the tip has been transformed into a scutellum, all the bundles may continue in the scutellum, but usually only the median bundle extends to it; the lateral bundles are shorter and their distal parts recurve below the point of down bending. The down bent cotyledons which show the prominent sheath as a part of the cotyledon, the lateral bundles are also prominent.

An even number of traces in the cotyledons of angiosperms denotes the primitive condition. The primitive taxa (*e.g.*, Ranales, Liliales, Helobiales) have cotyledons with two strong traces, often with pairs of lateral traces. In most gymnosperms even numbered traces are characteristic. In angiosperms the odd-numbered traces are derived from even-numbered traces.

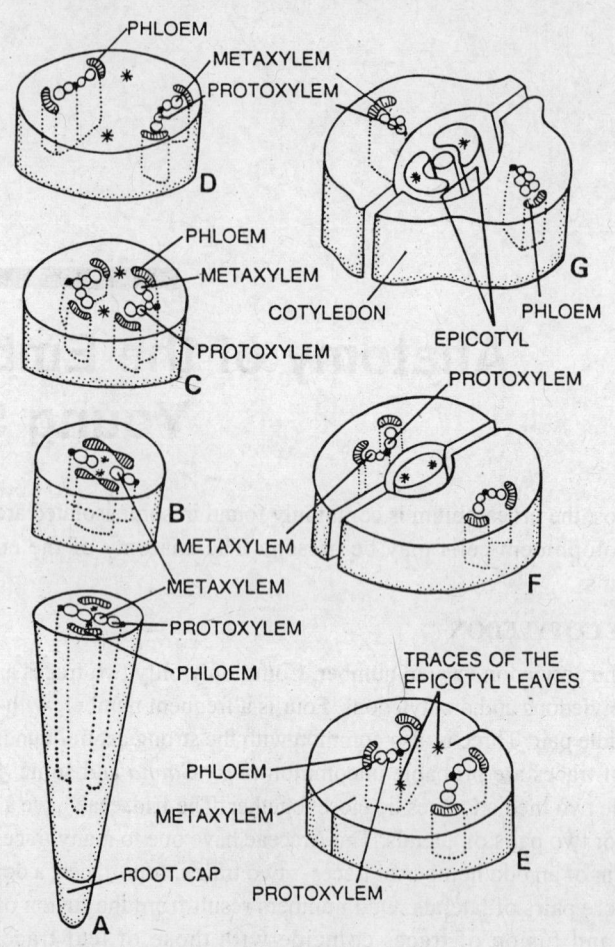

Fig. 15.1. Anatomy of seedling of dicotyledon (*Beta vulgaris*). A – E, diagram showing the connection between the root and the cotyledons in the transition region. Description in text. (After Esau).

The number of cotyledonary traces is fairly constant in families. For example, two traces are commonly found in most of the lower monocotyledonous families. The Liliaceae have two in the lower tribes; two, four and six in others. The Zingiberaceae have two traces. The Amaryllidaceae have two; the Iridaceae, one, two or three. The Araceae have one to several and the Cannaceae have several. Among dicotyledons the woody Ranales have two traces. The vascular system of the cotyledon of monocots has two major bundles or two with additional vascular bundles. The cotyledons of dicot have commonly three major bundles.

ANATOMY OF MESOCOTYL

The mesocotyl is a compound structure which consists of the hypocotyl and the adnate part of the cotyledon. It contains the vascular tissues of the stele of the hypocotyl and one or more vascular bundles of the cotyledon. The external and internal effect of adnation of the cotyledon neck to the hypocotyl is frequently found. External evidence of the fusion is the presence, in some genera, of a longitudinal ridge on the axis of embryo; internal evidence is the presence of a vascular bundle running longitudinal in the cortex of hypocotyl. The superfluous bundle belongs to the down bent adnate neck of the cotyledon that makes the vascular supply of the scutellum, the median bundle of the cotyledon continued downward from the point of downbending of the cotyledon neck. The vascular supply of the scutellum thus follows a roundabout

Anatomy of the Embryo and Young Seedling

course, like an inverted V. With the result of the bending downward of the tip of the cotyledon, the adnate vascular bundles, as seen in cross sections of the mesocotyl, are inverted, the xylem external to the phloem.

When the downrunning bundle of the cotyledon is merged with the vascular tissue of the hypocotyl, its free tip, which supplies the scutellum, is attached at the base of hypocotyl. In a monocotyledonous embryo (e.g., *Zea mays*) the scutellum represents the cotyledon; the mesocotyl represents the first internode of stem; the coleoptile represents the first leaf; and the hypocotyl represents a mere plate of tissue.

ANATOMY OF THE SHEATH OF COTYLEDON

The term sheath of the cotyledon or cotyledonary sheath has been applied to the cap-like or sheathlike structure, the coleoptile, that encloses the plumule, and to the basal part of the cotyledon. The anamotical study of the cotyledonary sheath supports that the sheath is the basal part of the cotyledon.

ANATOMY OF THE HYPOCOTYL

Very occasionally the statement is given that the monocots have no hypocotyl. The basis for this statement is the shortness of the hypocotyl of many grasses that are commonly used to illustrate monocotyledonous embryos. Very short plate like hypocotyls of monocots have received little attention anatomically, because of the complex structure where the transition occurs abruptly and the vascular strands lie almost horizontally. The hypocotyl of some monocots has been described as 'often platelike hardly existent'. Sometimes, in certain monocots it is long and prominent in the embryo, especially where it is a part of mesocotyl. The hypocotyls of dicotyledons are generally, longer than those of the monocotyledons. The cylinderical or platelike hypocotyl is greatly distorted and it tends to be ovoid or spherical in seedlings of *Crocus* and *Raphanus* and in poorly differentiated

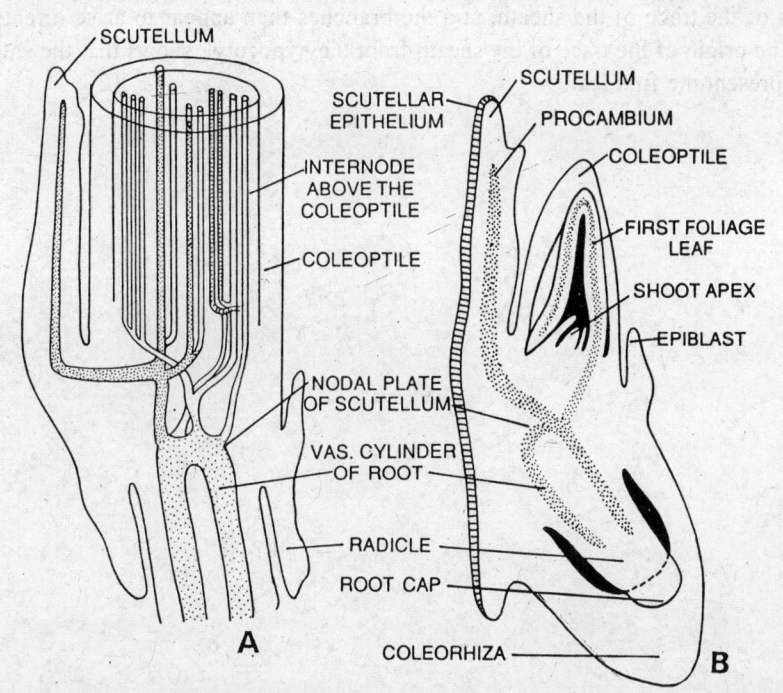

Fig. 15.2. Anatomy of seedling of monocotyledon (wheat). A, three dimensional view of seedling of wheat; B, L.S. of an entire wheat embryo.

embryos of orchids, saprophytes and parasites. The types of transition in vascular structure between root and stem—from radial and exarch to collateral and endarch have been discussed under the head root—stem transition.

ANATOMY OF THE SEEDLING ROOT

The vascular cylinder of the primary root of seedlings is commonly diarch or tetrach. Monarchy is rarely found in primary roots. Polyarchy is frequently found in monocotyledons. It is believed that tetrarchy is basic type since it is associated with arborescent taxa. The diarchy is associated with herbaceous taxa. The Ranunculaceae have diarch condition, whereas the woody families have tetrarch primary roots. In the dicots there are few variations from diarchy or tetrarchy. However, in the monocots there are many variations, but polyarchy is common, the neck, the median part, differentiated as the coleoptile, and the apex, the scutellum.

In cotyledons where there is no downbending of the distal part, all the vascular bundles including those of the sides of the sheath, usually continue into the scutellum. These cotyledons with two or four traces are simple and of primitive type. In cotyledons of advanced type, with one trace which divides to form a median and two lateral strong bundles, only the median bundle continues to the scutellum. The branches extent first upward and laterally and then downward, on the opposite of the sheath. The course of the bundles has been considered remarkable and evidence that the coleoptile is an independent organ, not a part of the cotyledon. The position and course of the veins of sheath support the interpretation that the upper part of sheath is stipular.

Where there is adnation of the cotyledon neck laterally to its lower part, the midvein is bent back upon itself. The two parts are fused and a bicollateral bundle is formed. Under continued specialization of this vascular fusion, the bundle is shortered distally and may become united with vascular stele of the hypocotyl. The shortening and adnation may extend to the point of origin of the branches of the trace of the sheath, and the branches then appear to arise directly from the hypocotyl. The origin of the trace of the sheath from the hypocotyl, shows that the sheath and the coleoptile represent the first leaf.

16

The Mechanical Tissue

No plant can flourish, or even maintain its existence for long, unless it is provided with arrangements which ensure that the plant body as a whole is firmly knit together, and that each of its organs is possessed of the requisite degree of mechanical strength. The plants which are made up of very numerous and varied parts, are subject to all kinds of mechanical injury. An insufficiently strengthened organ is liable, to be broken across, to be torn into pieces, to be bruised or crushed, and so forth, Every plant must safeguard itself again all the possible forms of injury by which its different organs are threatened. The plant organs must therefore be constructed so as to withstand, in some cases a transverse or bending stress, in others a longitudinal pull, in others again, longitudinal compressions, or radial pressures, or shearing stresses. Such mechanical arrangements are indispensable to the welfare of all plants, from the most insignificant of Algae to the tallest or bulkiest of trees. The trunk of a tree has to support the weight of the massive crown, with its large branches and foliage, it must therefore be built like a pillar or column so as to be capable of resisting longitudinal compression. The branches also have to bear a heavy load; these, however, on account of their oblique or horizontal position, are chiefly exposed to bending stresses. In a high wind the mechanical demands are greater in intensity and varied in quality. Trunk as well as branches must that withstand transverse stresses. In other, further that the leaves may not be quickly torn to shreds, the cells and tissues of which they are composed must be firmly united; in addition, the leaf margins are specially strengthened so as to prevent laceration.

In order to maintain the mechanical stability necessary to their welfare, large plants are forced to provide themselves with more reliable mechanical arrangements. This can be done only by applying the principle of division of labour, or in other words, by assigning the task of maintaining stability to special tissues. Such mechanical tissue must of course be more or less perfectly adapted to their special function. The most important types of mechanical tissue have been discussed in the following paragraphs.

COLLENCHYMA

Collenchyma is a living tissue composed of more or less elongated cells with thick primary nonlignified walls. The primary function of collenchyma tissue is support. From the stand-point of morphology, collenchyma is a permanent and simple tissue as it consists of one type of cells. Schleiden in 1839, for the first time introduced the term collenchyma to designate the thick-walled hypodermal tissue of cacti. The term collenchyma is derived from the Greek word *colla*, glue, refers to the thick glistening wall characteristic of collenchyma. The collenchyma is commonly interpreted as thick-walled kind of parenchyma structurally specialized as a supporting tissue (De Bary, 1884; Hayward, 1938).

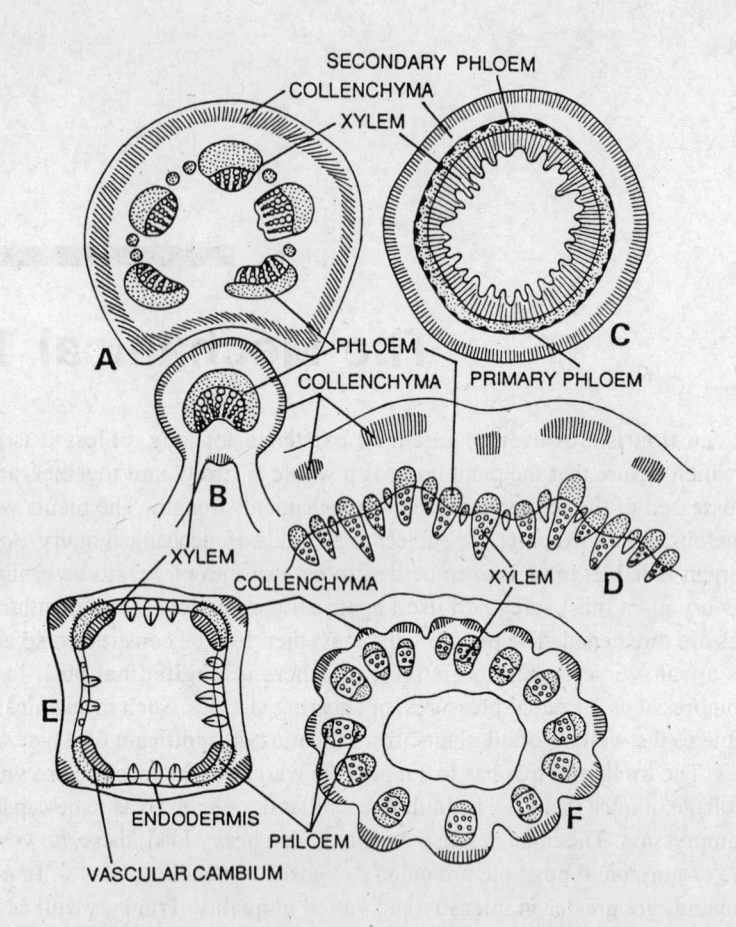

Fig. 16.1. Collenchyma, Position of collenchyma in various plant parts. A, in *Humulus* petiole; B, in the midvein of *Humulus*; C, in the stem of *Sambucus*; D, in *Pastinaca* stem; E, in *Mentha* stem and F, in *Cucurbita* petiole.

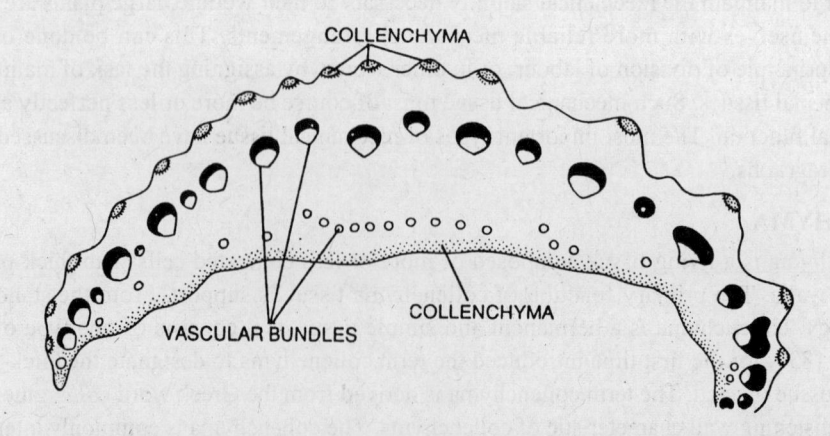

Fig. 16.2. Collenchyma. Distribution of collenchyma and vascular bundles in Celery petiole. The collenchyma is present in the ribs on the abaxial side and in the form of continuous strand on the adaxial side.

POSITION IN THE PLANT BODY

As already stated the collenchyma is a mechanical tissue which gives support to the plant body. It gives support to the growing organs and thereafter to the herbaceous parts of the plant body. It acts as first supporting tissue in stems, petioles, leaves and floral parts. It occurs chiefly in the peripheral parts of the stems, petioles and leaf mid-ribs. It is very commonly found in the ridges and angles of the plant organs. It may also occur in the roots particularly exposed to the light. However, it is absent from the leaves and stems of majority of the monocotyledons. Usually the collenchyma occurs just beneath the epidermis but sometimes one or more layers of parenchyma may develop in between the epidermis and the tissue. Haberlandt (1914), has reported that sometimes the complete epidermis becomes collenchymatous. The collenchyma may occur in the form of continuous or discontinuous cylinder beneath the epidermis. In many stems and leaves of herbaceous plants the collenchyma may form bundle caps and other similar isolated strands. In stems and petioles with bulging ribs, collenchyma is particularly well developed in the ribs. In leaves it may differentiate on one or both sides of the veins and along the margins of the leaf blade.

STRUCTURE

The cells of collenchyma vary in their lengths. The shortest cells are like parenchymatous cells whereas the longest ones are fibre-like in appearance. According to Haberlandt (1914), Majumdar (1941) and others the collenchyma cells may be of about 2 mm. length. The longer cells are tapering at their ends whereas the shorter ones are prismatic in their appearance. In transverse sections both kinds of these cells appear polygonal in their structure. The thickened walls of collenchymatous cells have a high refractive index, and are hence particularly conspicuous in a transverse section on account of their brilliant appearance. Sometimes, the cells of the same strand may vary in shape and size. The strand of the collenchymatous cells results by repeated longitudinal divisions. The cells are living with persistent protoplasts. According to J. Cohn collenchymatous walls contain the remarkably high proportion of 60–70 per cent of water. The cell walls consist of cellulose and pectin. The walls as typical collenchymatous cells are thickened in a highly characteristic manner. The deposition of thickening layers, is restricted to the edges of the cells, or much more pronounced along the edges than in any other part of the wall. This peculiarity is closely connected with the fact that collenchyma serves as the mechanical tissue of growing organs. The wall materials are usually found to be deposited in the angles where many cells join together (*e.g.*, *Ficus, Rumex, Polygonum, Boehmeria, Begonia* etc.). The degree of the wall thickenings to the angles is related to the amount of wall thickenings found on other wall parts. When the general wall thickening becomes too much, the thickening in the corners becomes obscure and the lumen of the cell appears to be circular, in transverse section instead of becoming angular (*e.g.*, in members of Umbelliferae). In the other form of collenchyma the thickening chiefly occurs on the tangential walls (*e.g.*, *Eupatorium, Rheum,* etc.) Here the collenchymatous thickenings are to be present on the walls facing the intercellular spaces (*e.g.*, members of Compositae, *Salvia, Malva, Althaea,* etc.). Muller (1890) had described the above mentioned three forms of collenchyma as angular (*Eckencollenchym*), lamellar (*Plattenocollencym*) and tubular or lacunate (*Luckencollenchym*) respectively. In longitudinal sections the collenchyma shows thin and thick wall portions. According to Majumdar (1941) the end walls of the collenchymatous cells are usually thin and the pointed ends appear thick due to the deposition of wall material. The simple, large or small pits with rounded or slit like apertures are found in collenchyma cells. Such pits are found both in thick and thin walls of collenchyma. In certain cases the collenchyma walls become modified in older parts of plant. For example, in *Tilia, Acer* etc., the collenchyma cells become enlarged and simultaneously their walls become thinner. According to Mullenders (1947), and others the collenchyma cells may become lignified and sometimes develop secondary walls thus changing themselves into sclerenchyma cells. In most of cases the collenchyma is a compact tissue having no intercellular spaces. As Majumdar (1941) reported such spaces are filled up with intercellular materials. The collenchyma cells contain living protoplasts. Chloroplasts are found in numerous collenchyma. Tannins may also be present in these cells.

FUNCTION

Collenchyma is a mechanical tissue which gives support to the growing organs of the plant body. The compactness and the thick walls make it a strong tissue. The collenchyma cells may increase the surface and the thickness of the walls and this way, may develop thick walls during the elongation period of the organ. The collenchyma tissue is flexible and plastic in nature. It possesses tensile strength. The collenchyma cells may be compared with fibres. The fibres are elastic and the collenchyma cells are plastic. The plastic nature of collenchyma walls is important from the standpoint that much of the elongation of the internodes occurs after the walls of collenchyma have been thickened. The plastic nature of collenchyma changes with age. Curtis (1938) reported that the old tissue is harder and more brittle than young. The parts of the plant which have ceased to elongate possess hard collenchyma. According to Esau (1936) in celery, the collenchyma strand is much stronger than the vascular tissue (See Fig. 16.2).

ONTOGENY

From the standpoint of its origin, the collenchyma has been reported to be originated from elongate procambium-like cells which appear very early in the differentiating meristem. It has been thought that the cortical collenchyma and the procambium originate in a common meristem. They possess small intercellular spaces, but such intercellular spaces disappear in angular and lamelar types. As the cells enlarge these intercellular spaces are either closed by enlarging cells or filled up by intercellular substance.

SCLERENCHYMA

The sclerenchyma tissue consists of thick-walled cells which are very often lignified. The principal function of this tissue is mechanical and, therefore, this is known as one of the mechanical tissues. The sclerenchyma cells enable the plant organs to withstand various strains which result from stretching and bending without any obvious damage to the parenchyma tissue. This word has been derived from the Greek and is a combination of *sclerous*, hard and *enchyma*, an infusion; collectively sclerenchyma cells form sclerenchyma tissue. The cells of this tissue are hard, lignified and with low quantity of water. The sclerenchyma cells possess elastic and hard, secondary walls whereas collenchyma cells possess plastic and highly hydrated primary walls. The sclerenchyma cells do not possess living protoplasts at maturity. The walls of sclerenchyma cells are uniformly and strongly thickened.

As regards their classification the sclerenchyma cells are grouped into *fibres* and *sclereids*. The long cells are called fibres whereas the short cells are known as sclereids. Sometimes even the sclereids may be elongated. The cases have also been reported where the fibres may be short contrary to their nature. The pits are less

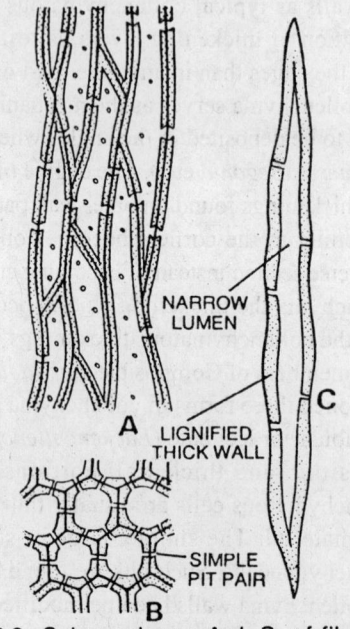

Fig. 16.3. Sclerenchyma A, L.S. of fibres; B, T.S. of fibres; C, a single fibre as seen in L.S.

The Mechanical Tissue

conspicuous in the fibre walls than in the walls of sclereids. As regards their origin, it is thought that the sclereids originate from parenchyma cells through secondary sclerosis and on the other hand the fibres originate from the fibre meristematic cells.

FIBRES

The fibres are the most important mechanical cells which occur in vascular plant. They are recognized for their great tensile strength, flexibility and elasticity and because of these characteristics they enable the plant organs to withstand various types of strains and tensions which result from the action of wind, gravity, etc. In India and abroad many plants are cultivated for the commercial fibres. More important fibre producing plants are — *Cannabis sativa* (true hemp), *Corchorus* spp. (jute), *Boehmeria nivea* (ramie), *Crotolaria juncea* (sann hemp), *Linum usitatissimum* (flax), *Agave fourcroydes* (henequen), *Agave sisalana* (sisal), *Musa textilis* (Manila hemp), *Hibiscus cannabinus* (Madras hemp), *Sansevieria* spp. etc.

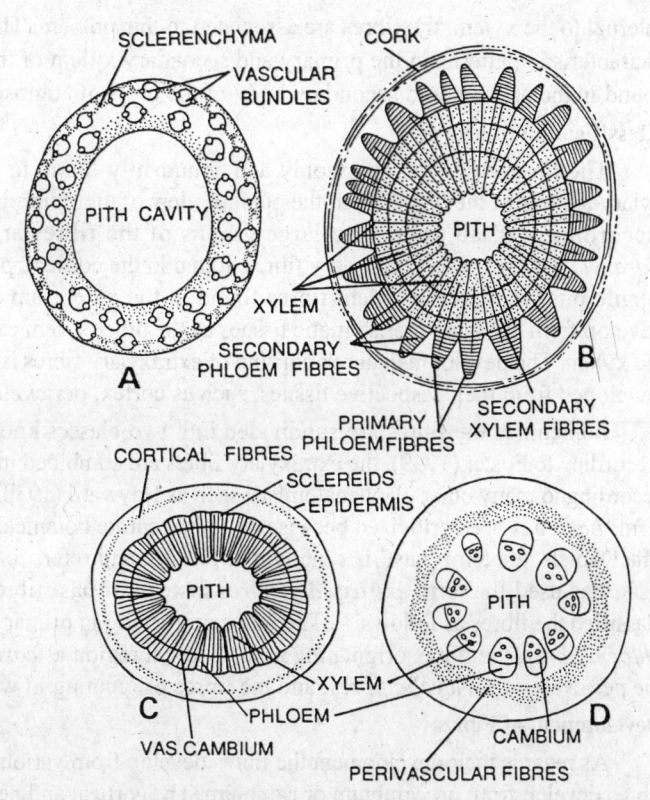

Fig. 16.4. Distribution of the sclerenchyma in various plant organs. A, T.S. of *Triticum* stem with sclerenchyma surrounding the vascular bundles and forming layers in the peripheral part of the stem; B, T.S. of *Tilia* stem with fibres in the primary and secondary phloem and in the secondary xylem; C, T.S. of *Gnetum* stem with fibres in the cortical region and sclereids in the perivascular position; D, T.S. of *Aristolochia* stem with a cylinder of fibres inside the endodermis in perivascular region. (After Esau).

Occurrence of Fibres

The fibres may occur in patches, in continuous bands in the cortical region and the phloem, as bundle sheaths or bundle caps with vascular bundles, and sometimes singly among other cells. They also occur in the xylem and the phloem either in groups or scattered. The fibres are found to be arranged in monocotyledonous and dicotyledonous stems in many characteristic patterns. In many hollow stems of Gramineae (*e.g.*, *Triticum* spp.), the fibres are found to be arranged in the form of a peripheral ribbed cylinder. In the solid stems of *Zea*, *Saccharum*, *Sorghum*, etc., the vascular bundles possess prominent fibre sheaths. These fibre sheaths are more conspicuous in the bundles of peripheral region. The fibres are also common in the leaves of monocot plants. Here they form the bundle caps and bundle sheaths.

In many dicotyledonous stems the fibres form the tangential plates in the outermost part of the primary phloem. In some plants, *Nicotiana*, *Boehmeria*, *Clematis*, *Magnolia*, etc., the fibres develop in the secondary phloem. In many dicotyledonous plants such as *Pelargonium*, *Aristolochia*, *Cucurbita*, etc., the complete cylinders of the fibres are found. In *Polygonum* spp., the fibre strands are found on both inner and outer side of the vascular bundles. In *Nicotiana*, where the phloem is internal to the xylem, the fibres are associated to this phloem. The fibres are found to be arranged in characteristic patterns in the primary and secondary xylem of the angiosperms. The fibres are also found in the primary and secondary body of the roots of angiosperms.

Classification of Fibres

The fibres are most commonly and abundantly found in the cortical, pericyclic, phloic and xylary regions of the plant. From the point of view of their morphology, there are two types of fibres. According to Esau, these two large groups of the fibres are known as *xylem fibres* and the *extraxylary fibres*. The extraxylary fibres, found in the cortical, pericyclic and phloic regions possess simple pits whereas the xylem fibres found in the xylem have bordered pits. The xylem fibres develop from the same meristematic tissues as the other xylem cells and constitute an integral part of the xylem. The developmental relationship of extraxylary fibres is less clear and they are thought to be developed from their respective tissues, such as cortex, pericycle and phloem.

Sometimes the fibres are subdivided into two classes known as *bast fibres* and *wood fibres*. According to Foster (1949), the extraxylary fibres are combined into a group known as bast fibres. But according to many other plant anatomists, such as Hayward (1938), Eames and MacDaniels (1947) the term '*bast*' has been criticized because of its inaccurate botanical meaning. According to Eames and MacDaniels, the term 'bast' is synonym of phloem and refers to the fibres of the secondary phloem. Esau has used the term *extraxylary fibres* instead of bast fibres. Most of the workers, however, classified the fibres as follows — The fibres originating in primary or secondary phloem are known as *phloem fibres*; the fibres originating in the cortical region as *cortical fibres*; the fibres originating in the pericycle as *pericyclic fibres*, and the fibres originating in wood as *wood fibres*.

Development of Fibres

As regards their development the fibres develop from various meristems. The xylem and phloem fibres develop from procambium or cambium. The cortical and pericyclic fibres arise from the ground meristem. From the viewpoint of their development the primary and secondary fibres show pronounced differences. The primary fibres initiate before the elongation of the organ. These fibres may become sufficiently long by elongation while the associated cells are still dividing. On the other hand, the secondary fibres arise in the part of the organ which has already stopped to elongate and they increase in length by intensive growth. This way, the primary phloem fibres attain greater lengths than the secondary ones. For example, in *Cannabis* the average length of the primary phloem fibres is 12.7 mm. whereas of secondary ones is 2.2 mm. According to Esau (1938) and Kundu (1942), in the same plants the longer primary phloem fibres are multinucleate and the shorter secondary phloem fibres are uninucleate. In flax the phloem fibres grow at both apices, upward and downward. In flax and ramie the secondary wall of the fibres develops in the form of distinct lamellae. Each such lamella is tubular in shape and grows from the base upward. Tammes (1907) and Kundu (1942) have reported that the phloem fibres of flax and hemp possess living protoplasts.

FUNCTION

The fibres of various plants have been used economically since ancient time. The flax is known to have been cultivated as early as 3,000 years B.C. in Europe and Egypt. The plants, such as flax, hemp, ramie and jute, where the commercial fibres develop in their phloem, the term fibre stands for a fibre strand.

The commercial fibres have been differentiated into hard and soft fibres. The hard fibres possess heavily lignified wall and found in the leaves of monocotyledonous plants, such as *Agave* spp., *Musa textilis*, *Yucca* and *Phormium tenax*. On the other hand, the soft fibres are with or without lignin. They are soft and flexible. These include the phloem fibres, which originate in the phloem of the plants, such as *Cannabis* (hemp), *Boehmeria nivea* (ramie), *Corchorus capsularis* (jute), *Hibiscus cannabinus* (kenaf), *Linum usitatissimum* (flax) and many others.

The Mechanical Tissue

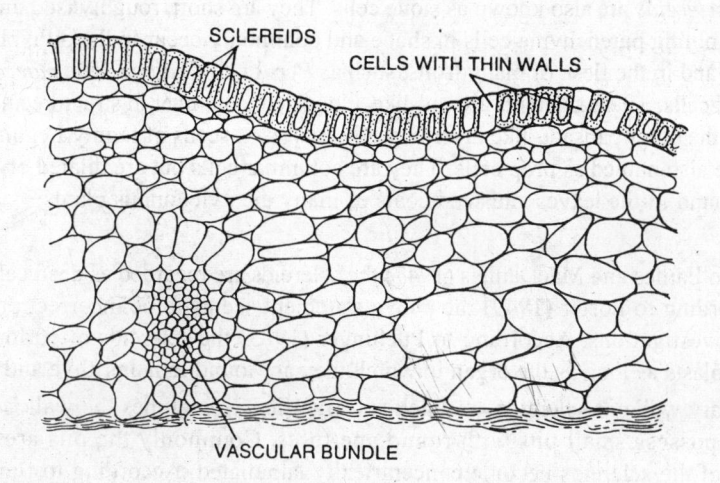

Fig. 16.5. Sclerenchyma—sclereids. Sclereids in epidermal cells of the bulb scale of *Allium sativum* (garlic).

The commercial fibres are separated from the plants by means of a process known as *retting*. In this process the plants are kept under water for a considerable time for bacterial and fungal activity. Because of this process the fibres become soft, and separated from the plant organs mechanically. During this process the pectic enzymes affect the intercellular material and the fibres become loose and softened.

SCLEREIDS

The sclereids occur in a wide range of positions in the plant body. They are very frequently found, either single or in groups in the cortical and pith regions of many dicotyledonous plants. They are also found in the xylem and phloem. Some regions or the tissues of the plant body are almost exclusively composed of the sclereids, such as the hard shells of many fruits and hard coats of many seeds. In many plants the parenchyma cells found in between the primary phloem strands convert into the sclereids by the development of the lignified secondary walls, and thus they form a continuous sclerenchyma cylinder together with the fibres. In *Aristolochia* the gaps of parenchyma are formed in the sclerenchyma cylinder, which later convert into the sclereids by the lignification of the cell walls. In many plants the sclereids develop in the leaves. The sclereids also develop in the epidermis of some protective scales, e.g., *Allium sativum*.

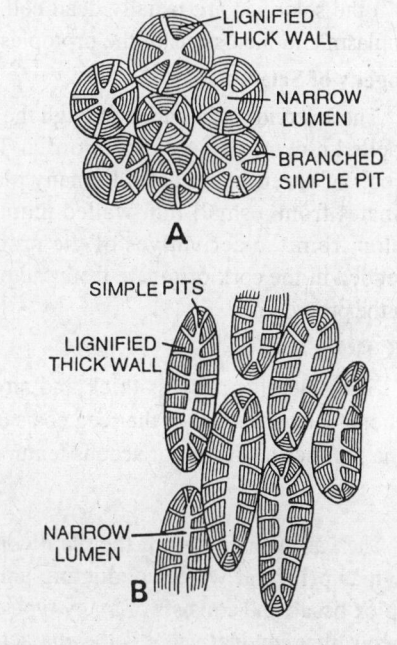

Fig. 16.6. Sclerenchyma. A and B, T.S. and L.S. of sclereids.

Classification of Sclereids

According to the form and structure of the cell four main categories of sclereids have been proposed. They are — (1) *brachysclereids*, (2) *macrosclereids*, (3) *osteosclereids* and (4) *astrosclereids*.

The *brachysclereids* are also known as stone cells. They are short, roughly isodiametric in form more or less resembling parenchyma cells in shape and commonly occur in the pith, phloem, cortex and bast of stems and in the flesh of many fruits, such as *Pyrus* and *Cydonia*. The *macrosclereids* are also known as rod cells. These sclereids are rod-like, elongated, columnar in structure and often found to be arranged in the outer palisade-like layer in seed coats (*e.g.*, seeds of family Leguminosae). The *osteosclereids* are also named as prop cells. They are columnar cells but are dilated or lobed at their ends. They are found in the leaves and seed coats of many dicotyledonous plants.

Structure

According to Eames and MacDaniels (1947) the sclereids are regarded as dead cells when fully mature. But according to Foster (1949) the above statement seems to be incorrect on the basis of several careful investigations. According to Puchinger (1923) the sclereids of various types may retain their protoplasts as long as the organ in which they are found remains alive and functional.

The secondary walls of sclereids are thick and lignified. Sometimes the wall is suberized or cutinized. They possess small pits with round apertures. Commonly the pits are simple. The secondary walls of the sclereids become concentrically lamellated. According to Bailey and Nast (1948) the secondary walls of the sclereids of some species contain crystals. The lumen of the cell is very much reduced and almost filled up with wall deposits. According to Eames and MacDaniels (1947) the sclereids are usually dead cells, and sometimes they contain the shrivelled remains of protoplasm and inclusions of the protoplast, *e.g.*, tannin and mucilage.

Ontogeny of Sclereids

The sclereids either arise through the process of sclerosis of parenchyma cells, or directly from the cells known as sclereid primordia. The typical brachysclereids develop by the process of sclerosis of parenchyma cells. In many plants such as *Camellia, Trochodendron,* etc., the sclereid originates from a small thin-walled uninucleate initial. Within the vascular tissues the sclereids develop from the derivatives of the procambial and cambial cells. The brachysclereids found embedded in the cork originate from the phellogen. The macrosclereids of the seed coats originate from the protoderm.

FUNCTION

The sclereids possess thick and strong walls and therefore, they act as mechanical tissue, producing a hard texture to the seed coats and endocarp of fruits. The idioblastic sclereids in the leaf lamina produce a stiff or coriaceous texture to the leaf.

XYLEM

The vascular system of the plant consists of xylem and phloem. The xylem and phloem are known as principal water conducting and food conducting tissues respectively. The xylem is a complex tissue and consists of many types of living and non-living cells. The presence of tracheary elements, that conduct water is the characteristic of the xylem. Some of the tracheary elements also help in the support. The xylem contains many specialized supporting elements known as fibres and because of its mechanical function, the tissue is included in mechanical tissue. The sclereids formed by the sclerosis of parenchyma tissue may also be present in the xylem. The primary xylem originates from the procambium. The secondary xylem develops by the meristematic activity of the vascular cambium. The primary and secondary xylem tissues may easily be differentiated from each other, but in many characters the xylem of two respective types intergrades with each other. The xylem consists of tracheids and vessels.

TRACHEIDS

The fundamental cell type in the composition of xylem is the *tracheid*. The mature tracheid is non-living and without a protoplast. The tracheids possess walls. In transverse section the tracheids appear to be polygonal or sometimes rounded. The end of the tracheid of secondary wood is chisel-

The Mechanical Tissue

like. They possess pit pairs on their common walls. The lumen of a tracheid is large and free of contents. The tracheids are primarily meant for water conduction and secondarily for mechanical support. The tracheids consist of long, empty thick-walled tubes running parallel to the long axis of the organ of the plant body. In most of the plants the tracheids possess bordered pits with characteristic shape, torus and border. In the tracheids, the secondary thickenings are deposited as rings, continuous spirals as helices of a ladder, giving the wall a ladder-like appearance. Such thickenings are called *annular*, *spiral* and *scalariform* respectively. The secondary wall of the tracheid with reticulate thickenings appears like a net, and sometimes the meshes of the net are elongated transversely. Such thickenings are known as *reticulate* and *scalariform-reticulate* respectively. Besides, the secondary walls are interrupted by means of pits and the thickening is known as *pitted*.

Function of the Tracheid

As already stated the main function of the tracheid is the conduction of water. As they possess thick and firm walls, they also aid in mechanical support. When the plant organs do not have fibres or other mechanical tissue, they play an important role in the support of the plant organ. The tracheids are overlapped and interlocked and united into strands and cylinders. These characters make the tracheids more fit for mechanical support to the plant organ.

Wood fibres and fibre-tracheids

The fibres have been evolved from the tracheids. Here the thickness of the wall increases and the diameter of the lumen cavity decreases. The number and size of the pits is also reduced. In comparison to the tracheids, the fibres possess thicker walls, and reduced pit borders. When the lumen becomes very much narrow and the pits become very small, the typical fibres are formed. The transitional forms occur in between tracheids on one hand and the fibres on the other. Such transitional forms are known as *fibre-tracheids*. The fibre tracheids possess smaller pits with less developed borders than tracheids whereas the libriform fibres possess simple pits. The fibres are most highly specialized supporting elements in those woods which possess specialized vessel members. The fibres with very thick walls and with simple reduced pits, are called *libriform wood fibres* because of their resemblance to the phloem fibres. Such fibres occur in woody dicotyledonous plants (*e.g.*, many members of Leguminosae). In the walls of fibre tracheids and fibres of many genera, the gelatinous layers occur, and therefore, they are known as *gelatinous-tracheids*, *fibre-tracheids* and *fibres*. Sometimes, the tracheids with protoplast divide, so that two or more protoplasts are formed. Such fibre-tracheids are known as *septate fibre-tracheids*. These septa, thus formed are true walls, but they lack secondary layers. These septa remain unlignified. The septate fibre-tracheids occur in many shrubs, woody herbs, vines and tropical trees. The fibres and fibre tracheids aid to the mechanical strength of the xylem and various organs of plant body.

Mechanical Strength of Wood

The mechanical strength, physical properties and suitability for commercial uses of the wood are determined by the composition of xylem tissue and the structures and arrangement of its component elements. In the dicotyledonous woods the length, and thickness of walls of xylem elements, distribution of pits, libriform fibres, and fibre tracheids are mainly responsible for the strength of the wood. These cells have been proved to be more influential when found in dense masses. The wood fibres are very important from the view point of mechanical support and strength to the various plant organs. In dicotyledonous woods the vessel-elements are relatively weak, because their diameters are large and the walls are thin. The woods with aggregated vessels are less resistant to certain stresses than the woods with less and evenly distributed vessels. The late wood is generally stronger than the early wood.

PHLOEM

The phloem as a whole is not mechanical tissue. It is meant for the translocation of solutes. However, the phloem fibres are thick-walled and aid in mechanical support. These fibres are considered as mechanical tissue. In phloem the basic cell type is the *sieve element*. There are two forms of sieve element, *i.e.*, the *sieve cell* (found in gymnosperms and vascular cryptogams) and *sieve tube* (found in angiosperms). The phloem is a complex tissue. It consists of sieve cells and phloem parenchyma only, *e.g.*, in pteridophytes and many gymnosperms; of sieve cells, parenchyma and phloem fibres, *e.g.*, in gymnosperms and sieve tubes, companion cells, phloem parenchyma, phloem fibres, sclereids and secretory cells in angiosperms. Here only those phloem elements have been considered which aid in mechanical support and are treated as mechanical tissue.

PHLOEM FIBRES

In many angiosperms the fibres form a main part of both the primary and the secondary phloem. As compared to xylem fibres, the phloem fibres possess the simple pits with small rounded or linear apertures. The walls of these fibres are lignified. In their development the long tapering ends of the fibres are being interlocked forming the strong fibre-strands. The tangential sheets or cylinders of the fibres are being formed to protect the inner tissues. In certain plants, *e.g.*, *Dirca palustris*, they form more important supporting tissue to give mechanical support to the stem than the xylem cylinder. The fibres of the protophloem prominently aid in the mechanical support, especially in the early stages of the development of the stem. The fibres are variously arranged in the stems of different species. They may be arranged in continuous, uniform or irregular bands. In other species they may be arranged as scattered or isolated strands. In certain species the fibres form the caps of the primary phloem strands. The fibres may be lignified (*e.g.*, *Cannabis*) or they may be non-lignified and composed of cellulose as in *Linum*. These fibres of primary phloem are similar in structure to the fibres of cortical region and of the secondary phloem. In certain species, the secondary phloem fibres mature early and function as mechanical elements (*e.g.*, *Tilia*). In other species, the phloem fibres mature only after the sieve elements stop to function (*e.g.*, *Prunus*). According to Holdheide (1951) these fibres are supposed to be sclerotic phloem parenchyma cells. These fibres, other fibres and vascular bundles form the commercial '*bast*' and therefore, sometimes the phloem fibres are very strong and because of this strength, they are employed in the making of ropes, cords, cloth and mats. The term 'bast' is not appropriately used here. The 'bast' is used to any fibres from the outer parts of the plants which may be even from the cortex or pericycle. The bast includes phloem, phloem fibres, cortical fibres and pericyclic fibres.

Function of Phloem Fibres

The function of fibres and sclereids of phloem is mechanical support. They did in support of plant organs and protect the thin-walled soft tissues. Flax fibres are remarkable for their great tensile strength. They are used in manufacture of linen cloth and the thread. Hemp fibre is a bast fibre which develops in the pericyclic region. As already mentioned, the fibres of hemp are lignified and therefore, they lack elasticity and flexibility as formed in the flax fibres.

PRINCIPLES GOVERNING THE CONSTRUCTION OF MECHANICAL SYSTEM

One of the most important of the considerations that influences an engineer or architect, in deciding upon the plan of construction of a bridge or a roof, is the desire to economise the material. In other words, a design is not satisfactory unless it ensures the maximum of strength and solidity with the minimum expenditure of material. To this end the resistant elements of the structure must be arranged in a particular manner in accordance with approved mechanical principles. An organism (or plant) is confronted with a similar problem when it is called upon to provide itself with a sufficient degree of mechanical strength. Here also the desired result must be obtained with the smallest possible expenditure of material; *hence the principles of construction to which the engineer or architect adheres are precisely those which control the morphogenetic activities of*

The Mechanical Tissue

organisms. The very same mechanical principles, were expressed perhaps with even greater perfection hundreds and thousands of years ago in the skeletal systems of the plants of former geological periods. The most important of the principles that govern the construction of the mechanical system are — inflexibility, inextensibility and incompressibility.

INFLEXIBILITY

If a straight girder be supported at both ends and weighted in the middle, it will bend to a greater or less extent according to the magnitude of the load; as a result of this curvature the upper side of the girder must be slightly shortened, while lower side is correspondingly lengthened. The shortening produces a state of *compression*, the lengthening a state of tension, in the corresponding halves of the girder; the effect in either case, is most pronounced at the upper and lower surfaces, while on approaching the centre of the cross section from either side the corresponding tension gradually diminishes, then falls to zero and finally possess over into the tension of opposite sign. The layer of zero tension is known as the neutral surface. In order, therefore, that a girder may possess maximum inflexibility the available material must be concentrated in the regions of greatest tension, that is to say, near to upper and lower faces. A typical girder thus comes to consist of upper and lower *flanges*, which are firmly joined together by a connecting piece, or web. The cross section of such a girder usually resembles I, the horizontal strokes representing the flanges and vertical line the web.

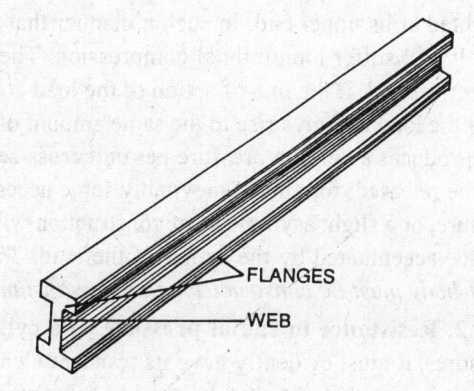

Fig. 16.7. The mechanical tissue—Girder. The wide portions at the top and bottoms are the flanges; the narrow connection is the web.

The strength of girder depends, upon the strength of its flanges. It also increases as the distance between two flanges becomes greater, because the tension due to load varies inversely as the distance between the flanges. Since the web has to bear only a small portion of the total strain, it may be considerably lighter in construction than the flanges which it links together. If a girder is composed of more than one kind of material, the inferior materials used in the construction of the web. *Similarly, in the plant the flanges of a girder are always composed of mechanical cells, whereas the web may consist of vascular tissues or of parenchyma.*

INEXTENSIBILITY

The degree of inextensibility of a structure depends entirely upon the cross-sectional area of its resistant elements. The disposition of these elements is therefore theoretically a matter of indifference. The theory, however, assumes that the stretching force is uniformly distributed, a condition which is not likely to be fulfilled unless the area over which the resistant elements are distributed is very small. If therefore, the resistant elements were scattered, an unequal distribution of tensions would be very liable to occur, and individual strands might break, with the result that the strength of the whole structure would be seriously impaired. The more closely aggregated on the other hand resistant elements are, the more uniform the distribution of tensions is likely to be; *hence the aggregation of the mechanical cells into a single compact and solid mass is the most advantageous arrangement for an organ which has to withstand longitudinal tension.*

INCOMPRESSIBILITY

1. Resistance to longitudinal compression. If an erect prismatic or cylindrical body, with a longitudinal axis greatly exceeding its transverse diameter, be fixed at its lower extremity and weighted at its upper end, in such a manner that the load acts in the direction of the longitudinal axis, it will suffer longitudinal compression. The middle line of such a body can, however, only remain vertical, if the line of action of the load coincides exactly with the longitudinal axis. In this event the load will give rise to the same amount of compression at every point within the body, and thus produces a definite pressure per unit cross-section. In nature this ideal condition can scarcely ever be released; for almost inevitably some accessory circumstance, such as the effect of a lateral pressure, or a slight asymmetry of construction will produce a small deflection, which is there upon at once accentuated by the action of the load. *Thus any columnar structures which occur in the plant-body must be constructed so as to withstand bending.*

2. Resistance to radial pressure. If a cylindrical body is to withstand radial or crushing pressures, it must evidently have its resistant elements arranged in the form of a strong peripheral shell. It is seen that all subterranean and submerged organs have been protected in this way against the radial pressure of the surrounding soil on water. The pressure resisting hollow cylinder or tube must not of course be confused with the similar structure which affords protection against bending strains.

ARRANGEMENT OR DISTRIBUTION OF MECHANICAL TISSUES IN DIFFERENT PLANT ORGANS

As already mentioned the plant body obtains strength from — *parenchyma cells*, *collenchyma cells* and *thick-walled dead cells* (*e.g.*, sclerenchyma, wood fibres, etc.).

The arrangement of the strengthening material is different in leaves, in stems and in roots, and is suited to the special stresses which these various organs have to withstand. In order to understand this arrangement it will be convenient to consider the stresses occurring in a girder, beam. This principle that governs the construction of the mechanical system is known as *inflexibility* and has already been discussed in the sub-head — Principles Governing the Construction of the Mechanical System — of this chapter.

Mechanical tissues in leaves. The leaf is supported at only one end. The weight of the leaf gives it a tendency to bend downward so that its upper surface is stretched, or under tension, while its lower surface is under compression. The greatest stresses are at the upper and lower surfaces and the least stress is in the centre. This arrangement, therefore, calls for longitudinal girders in which are strongest material is near the outer surfaces. The midrib and larger veins of the leaf represent the girders. The principal strengthening material in these is usually collenchyma, and this is generally found as a broad band near the upper and lower surface just within the epidermis. The collenchyma therefore, represents the flanges on an I—beam, while the tissues between the two bands of collenchyma represent the web. The larger veins of grasses usually have sclerenchyma near the upper and lower surfaces, and so act as girder.

Mechanical tissues in stems. If a column, such as a tree trunk, were supporting an evenly distributed weight, the manner in which the strengthening material was arranged would, theoretically, make comparatively little difference. The only stress would be a downward pressure, or compression, the resistance to which would depend more on the cross-sectional area than on the arrangement of the strengthening material. Actually, however, such a condition is seldom attained. When in addition to the vertical pressure of gravity there is a sidewise pressure, as that of wind or of an animal, the side toward the pressure tends to become stretched, or develops tension, and the opposite side is subjected to compression. When the stem of a plant becomes inclined, as by the action of the wind or

The Mechanical Tissue 373

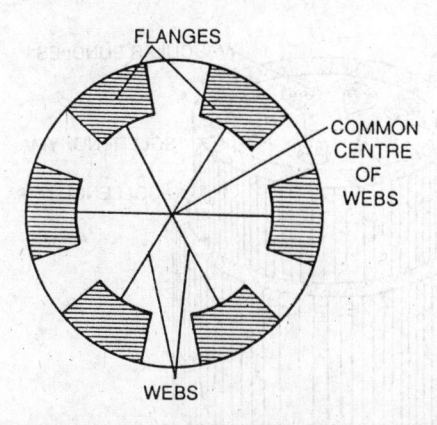

Fig. 16.8. The mechanical tissue. Diagram of cross-section of a compound girder composed of I—beams the webs of which have a common centre.

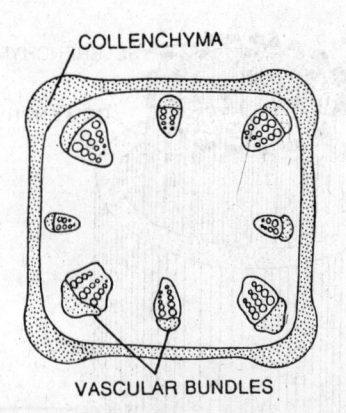

Fig.16.9. The mechanical tissue—collenchyma. Diagram showing arrangement of vascular bundle and collenchyma in a four-cornered stem.

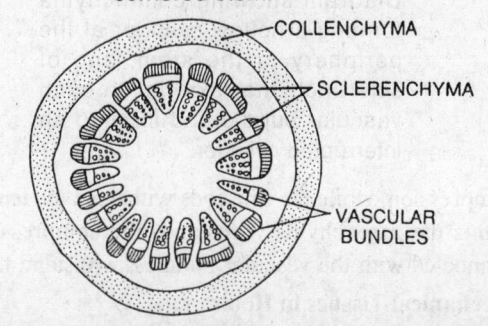

Fig. 16.10. The mechanical tissue—collenchyma and sclerenchyma. Diagram of dicot stem showing sclerenchyma just outside of each bundle, and collenchyma forming a hollow cylinder within the epidermis.

by the weight of the branches, the side that is uppermost is under tension and the lower side under compression. This way, it becomes important to have the strengthening material distributed near the upper and lower surfaces, or, in other words, in the form of an I—beam. As the plant is likely to bend in any direction, however, and thus may develop stress on any side, it is advantageous to have a number of these girders, with the webs crossing each other and the centre of each at the centre of stem. In the four-cornered stems of such plants as the mints or coleus the corners are occupied by a conspicuous development of collenchyma (Fig. 16.9), which thus forms the flanges of two I—beams, the webs of which are crossed. The large vascular bundles are near the corners, so that the phloem is protected by being between the xylem and the collenchyma. In many plants there are strands of sclerenchyma outside of each vascular bundle (Fig. 16.10) and in such cases two strands on opposite sides of the stem represent the flanges of an I—beam. This sclerenchyma which is near a bundle not only serves as one of the flanges of an I—beam but also is in a position to protect the delicate compound girder as that shown in Fig. 16.8 are connected, there is no necessity for the webs; and if they are absent a hollow cylinder results. This type of construction is very frequently found in plants and is particularly evident in hollow stems, such as those of most grasses. In many plants the sclerenchyma is arranged in the form of a hollow cylinder. Such hollow cylinders are, as a rule, near the outer surface of the stem. The strengthening material of the growing part of a stem is usually collenchyma; and this is generally arranged in the form of a hollow cylinder just within the epidermis (Figs. 16.11 and 16.12).

Mechanical Tissues in Monocotyledonous stems

In many monocotyledonous plants the arrangement of the strengthening material is very similar in principle to the reinforcing of concrete in a concrete structure. The concrete withstands

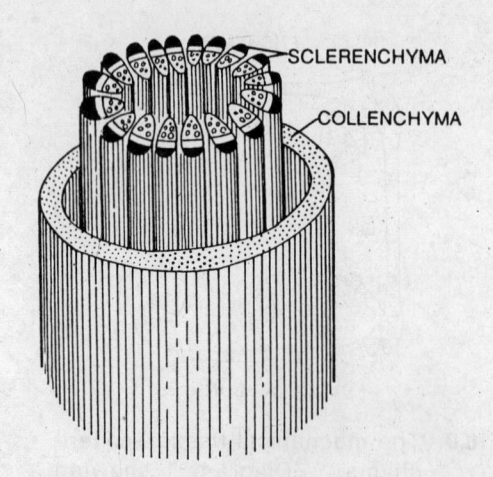

Fig. 16.11. The mechanical tissue—collenchyma and sclerenchyma. Diagram showing collenchyma forming a hollow cylinder at the periphery of the stem, and of sclerenchyma, exterior to each vascular bundle and forming an interrupted cylinder.

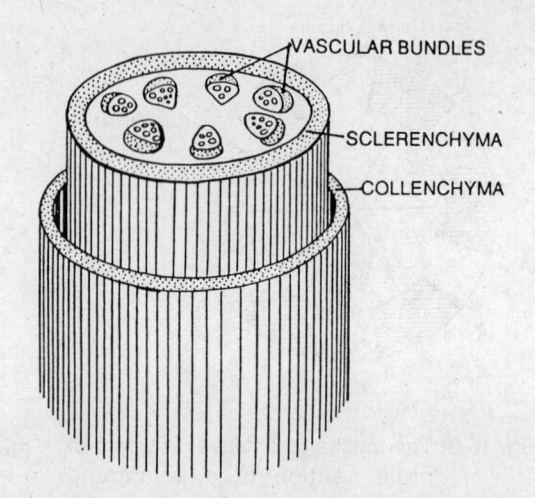

Fig. 16.12. The mechanical tissue—collenchyma and sclerenchyma. Diagram showing arrangement of collenchyma, forming a hollow cylinder at the periphery of the stem, and sclerenchyma forming one exterior to the bundles.

compression, while the iron rods withstand the tension due to movement, etc. In monocotyledonous plants the parenchyma withstands the compression, while the sclerenchyma strands, which are connected with the vascular bundles, withstand the tension.

Mechanical Tissues in Roots

The roots of a plant serve to anchor it in the ground. They act like cables, and the principal stress to which they are subjected is longitudinal tension, or pull. It is evident that when a plant is blown by the wind the roots on the side from which the wind is coming are subjected to longitudinal tension, and if they are not strong enough to withstand this they break and the plant is blown over. The best arrangement of material to withstand this type of stress is in the form of a cord. In roots the vascular bundle and strengthening materials are usually much more centrally located than they are in the stem, the centre being frequently occupied by thick walled elements.

17
Anatomy in Relation to Taxonomy

Anatomy provides evidence concerning the interrelationships of larger groups such as families, or in helping to establish the real affinities of genera of uncertain taxonomic status. Anatomy is of restricted value for distinguishing species or groups of less than specific rank, because the differences between them are usually quantitative rather than qualitative.

Anatomy sometimes proves very helpful for individual identifications. The anatomical methods are of great value in identifying the herbarium specimens which do not bear flowers or fruits. In many cases it becomes possible to assign sterile specimens to a family or a genus. The anatomical methods are also used in the identification of commercial samples of medicinal plants, timbers, fibres, etc. These methods play an important part in checking adulteration, substitution and fraud and sometimes even in detecting the criminals.

It is, however, most important to rely upon anatomical characters of which the taxonomic value has become well established. These characteristics are as follows:

1. Hairs. All taxonomists are familiar with the diversity of external hairs. There are glandular and non-glandular categories, each of which may be sub-divided according to the number of component cells, degree of branching, etc. Whole families may frequently be recognized by the occurrence of one or more distinctive types of hair. In other cases, species and genera can also be recognized on the basis of the structure of hair. Smaller variations in size and density should be treated as a basis for the separation of closely related genera and species only after exhaustive investigation of a wide range of material.

2. Stomata. The term stoma means the pair of guard cells together with the aperture between them. The cells surrounding a stoma differing from the remaining epidermal cells are known as subsidiary cells. There are four main types of stomata found in the dicotyledons. These types are — *'ranunculaceous'*, *'cruciferous'*, *'caryophyllaceous'* and *'rubiaceous'*. These names are given to them after the names of the families in which they were first observed. Now this has been an established fact that the above mentioned types are found in many other families besides those after which they were originally named. The new terms have also been coined for the above mentioned types.

Type A. Stoma surrounded by a limited number of cells that are indistinguishable in size, shape or form from those of the remainder of the epidermis — the *'ranunculaceous'* or *anomocytic* (irregular-celled) type.

Type B. Stoma surrounded by three cells of which one is distinctly smaller than the other two — the *'cruciferous'* or *anisocytic* (unequal celled) type.

Type C. Stoma accompanied on either side by one or more subsidiary cells parallel to the long axis of the pore and guard cells — the *'rubiaceous'* or *paracytic* (parallel celled) type.

Type D. Stoma enclosed by a pair of subsidiary cells whose common wall is at right angles to the guard cells — the *'caryophyllaceous'* or *diacytic* (cross-celled) type.

The *'gramineous'* type of stoma is believed to be confined to the monocotyledons where it is specially characteristic of the Gramineae and Cyperaceae. The 'gramineous' stoma possesses guard cells of which the middle portions are much narrower than the ends so that the cells appear, in surface view, dumb-bell shaped. The subsidiary cells are found lying parallel to the long axis of the pore. (See figs. in Chapter 7).

3. Epidermal cells and hypoderm. Epidermal cells differ considerably in size, shape and outline in different plants. Characters such as a partly or wholly crystalliferous epidermis or the cells with specific chemical contents, are of more importance because of their restricted occurrence. Epidermal cells with vertical or horizontal partitions are of specific or almost generic value, since these features are of restricted occurrence.

4. Veins. The comparative structure of the veins of two leaves, and in particular the structure of the vascular bundles and their relationship to the surrounding tissues, it is important to ensure that the veins of the same order are being examined. It is believed, that certain characters of the veins are of confirmatory value in the identification of species and genera. Veins in which the vascular bundles are accompanied on either side by parenchymatous tissue devoid of chlorophyll or by sclerenchymatous or collenchymatous tissue occupying the whole of the space between the bundle and the upper and lower epidermis are described as vertically transcurrent. The possession of these characters is like-wise of specific and generic value. (See figs. in Chapter 12).

5. Petiole. The petiole is of considerable taxonomic importance, since its structure is little affected by environment change. To obtain a complete picture of the vascular system of the petiole a series of sections must be cut, and to compare them with sections of the petioles in other leaves.

6. Microchemistry. Several types of chemical deposits occur in plant tissues. These deposits are supposed to be particularly valuable as indicators of taxonomic affinity. The chief types of secretory products are as follows:

(a) Crystals. The most common crystals consist of variously shaped deposits of calcium oxalate. The crystalline secretion such as raphides and crystal-sand are more restricted in distribution and therefore of greater taxonomic interest. Sometimes a whole family is characterized by the secretion of a highly distinctive material. For example, myrosin is quite common in Cruciferae. (See figs. in Chapter 2).

(b) Starch. The size, shape and other characters of starch grains are highly distinctive and of considerable taxonomic value. (See figs. in Chapter 2).

(c) Cystoliths. The presence of calcium carbonate crystals, the cystoliths is highly characteristic of certain families *e.g.*, Acanthaceae, Urticaceae. (See figs. in Chapter 2)

(d) Laticiferous tissue. There is a well developed system of tubes or cells in which variously coloured colloidal fluids known as *latex* are secreted. The laticiferous elements are of diagnostic value. Since latex is secreted in quite unrelated families its presence does not indicate taxonomic affinity unless this is also supported by other characters. (See figs. in Chapter 5).

(e) Secretory elements. Several secretory elements such as resins, oils, mucilage, tannins, etc., are of considerable taxonomic value. These substances are deposited in cells, intercellular cavities, elongated sacs and canals. The canals are often being lined with definite epithelium. (See figs. in Chapter 2).

7. Cork. According to Solereder the position in which the cork originates in a young stem is of diagnostic value within limits. In an individual species, the first cork to be formed is often more superficial than that which arises later on. Then again, in a single family there are species which possess either superficial or deep-seated cork respectively, whereas in other families the position of origin of the cork seems to be more constant. (See fig. in chapter 10).

8. Endodermis. The presence of a distinct endodermis in stems is of diagnostic value because of its restricted occurrence. In most dicotyledonous stems the endodermis is inconspicuous. In some the endodermis consists of distinct layer of cells which are differentiated from the neighbouring cells in containing starch. In another group the endodermis consists of cells with well-marked Casparian thickenings. In a few species the endodermis becomes wholly suberized. (See figs. in Chapter 7).

9. Sclerenchyma of pericycle. The presence or absence and nature of the sclerenchyma of pericycle is of great diagnostic and taxonomic value. The most common types seen in the transverse sections of the stem are — (i) An interrupted ring of fibres. (ii) A continuous ring of fibres. (iii) An interrupted ring of mixed fibres and stone cells. (iv) A continuous ring of mixed fibres and stone cells, which is commonly known as 'composite, continuous ring of sclerenchyma. (v) Stone cells present, but no fibres. (vi) Sclerenchymatous elements entirely lacking. The nature of pericyclic sclerenchyma plays an important role in separating species or genera in some families, but in a few instances the arrangement may be typical of a whole family. For example, the pericyclic sclerenchyma is not found in all investigated members of Pittosporaceae, whereas a ring of sclerenchyma of a very characteristic type is found in near about all the members of Geraniaceae. (See figs. in Chapter 7)

10. Width of medullary rays. The transverse sections of the internodes of young stems of many plants show the widely separated vascular bundles by means of broad parenchymatous tissue, the medullary rays. In Ranunculaceae, these medullary rays are quite broad. In extreme cases the vascular bundles are scattered, *e.g.*, Piperaceae and Berberidaceae. In the transverse sections of other plants, the xylem appears as a closed ring traversed by very narrow medullary rays.

11. Bicollateral bundles. The occurrence of bicollateral vascular bundles in the axis of Cucurbitaceae is of much taxonomic value. In certain families, *e.g.*, Solanaceae, Asclepiadaceae the presence of internal phloem makes a good diagnostic feature. (See figs. in Chapter 10)

12. Cortical and medullary bundles, and anomalous secondary thickening. Occurrence of cortical and medullary vascular bundles in various families, genera and species is of diagnostic value on account of their restricted appearance, but the families in which they are found are not closely related to one another. (See figs. in Chapter 12).

13. Wood. The characteristics of the wood (xylem) that possess greatest taxonomic value and diagnostic features are categorized as follows:

(*a*) **Vessels.** The distribution, pattern diameter and frequency as seen in transverse sections, type of perforation and thickening of vessels are of diagnostic value. The presence of tyloses and the diameter of vessel are of much taxonomic importance. (See figs. in Chapter 10).

(*b*) **Wood parenchyma.** The types of wood parenchyma, *i.e.*, *apotracheal* and *paratracheal* are of much diagnostic value. In apotracheal type distribution is independent of the vessels whereas in paratracheal type distribution is determined primarily by the vessels.

The distribution of wood parenchyma in angiosperms is of two main types, *i.e.*, (*i*) Apotracheal type and (*ii*) Paratracheal type.

Apotracheal Type

In such cases the parenchyma cells are not in contact with vessels. This type may be sub-divided into three sub types.

(*a*) **Diffuse apotracheal type.** In this case the parenchyma cells occur singly among the fibres and tracheids.

(*b*) **Banded apotracheal type.** In this case the parenchyma cells occur in bands.

(*c*) **Terminal apotracheal type.** In this case the bands of parenchyma cells are confined to the ends of the growth rings (*e.g.*, in *Michelia, Acer*, etc.).

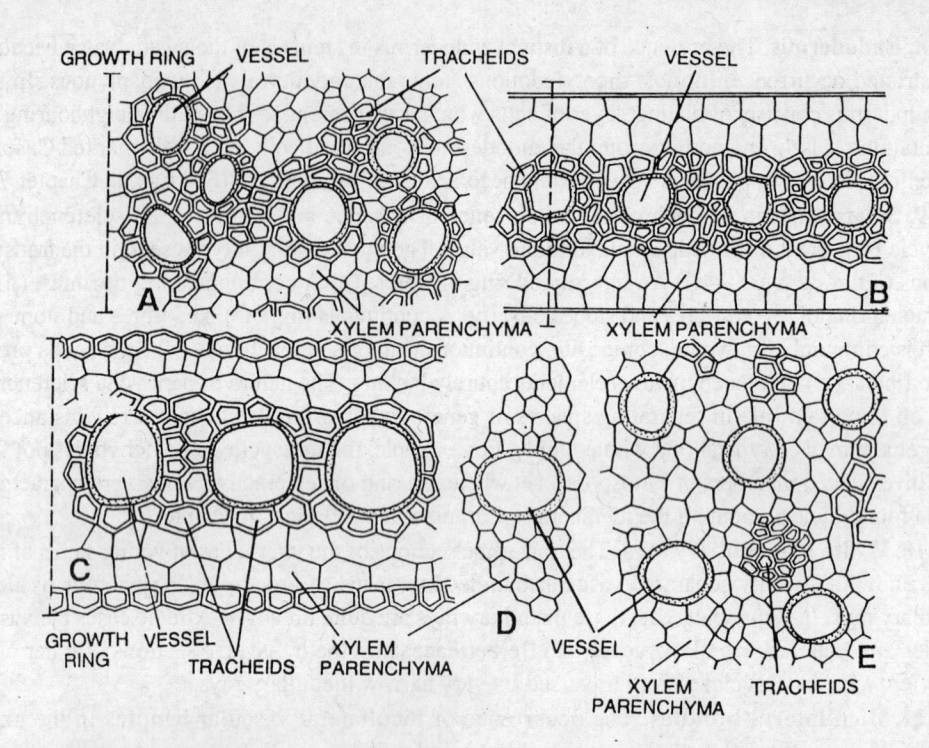

Fig. 17.1. Xylem parenchyma of different types. A, diffuse apotracheal type; B, terminal apotracheal type; C, banded apotracheal type; D, vasicentric ailiform type (also called paratracheal type); E, paratracheal-vasicentric confluent ailiform type.

Paratracheal Type

In such cases the parenchyma cells remain associated with the vessels. This type may be divided into two sub types.

(a) **Abaxial type.** Here the parenchyma cells are found in contact with abaxial surfaces of the vessels (*i.e.*, the surfaces away from the centre of the vessel).

(b) **Vasicentric type.** Here the parenchyma cells completely surround the vessel. In certain cases, the surrounding parenchyma cells make wing-like appendages. This type is termed *ailiform, vasicentric type* (*e.g.*, in some leguminous plants). In still other cases (*e.g.*, in *Terminalia*). The ring-like extensions around many vessels coalesce and form irregular bands. This type is known as *confluent ailiform vasicentric type*.

(c) **Rays.** The width and height of rays are of taxonomic value. Exclusively uniseriate rays is another feature of great value for identification. The frequency of rays is also a diagnostic character.

(d) **Fibres.** The presence of 'libriform fibres' with simple pits and 'fibre tracheids' with bordered pits is of great taxonomic value.

(e) **Storied structure.** This consists of the arrangement of the cells or tissues in horizontal series as seen in tangential section.

(f) **Growth rings.** This character is a standard feature of most descriptions of species.

(g) **Included or interxylary phloem.** The presence of included phloem in the transverse section of the axis of plant is of diagnostic value.

(h) **Intercellular canals.** The presence of radial and vertical intercellular canals is of great taxonomic value.

18
Wood Anatomy in Relation to Phylogeny

There is a great controversy about the origin of vessel in the Gnetales. It was believed that this group owing to its possession of vessels and the occurrence of broad angiosperm like rays in some species, might prove to be a link between other gymnosperms and angiosperms. The controversy centered round the problem of whether the simple perforations in the end-walls of the vessel elements of the Gnetales and the angiosperms respectively have been developed phylogentically along the same lines. It appeared unlikely that the vessel had precisely the same origin in same groups, because the less advanced perforation plates in the Gnetales are typically *foraminate* that is pierced by irregularly arranged, circular holes, whereas the corresponding primitive multi-perforate plates of the dicotyledons are scalariform, with regularly arranged, elongated holes.

It is a common thought that the ancestral tracheal element was the *tracheid,* a long cell with bordered pits, and that vessels have been derived from the tracheids by the disappearance of the pit membrane from some of the pitting in the overlapping walls, thus forming a vertical series of cells that gives an uninterrupted passage for water. Thompson (1918) postulated that the simple perforation of the Gnetales must have been derived from tracheids with sclariform pitting and that therefore there can be no genetic connection of two groups furnish a remarkable instance of the independent development of similar structures.

Bailey (1944), has carried out the ontogenetic studies of the primary and secondary xylem. He considers that the vessels arose first in the secondary xylem and later in the last-formed part of the primary xylem. He points out that Dicotyledons whose vessels in the secondary wood are very unspecialized often possess vessels in the primary xylem that are distinguishable from scalariform pitted tracheids. He states further that 'the fact that structurally primitive angiosperms had sclariformly pitted tracheids, from which vessels originated, rules out any possibility of deriving the angiosperms from the Gnetales or other representatives of the higher gymnosperms'. According to him the development of the vessels in the Gnetales from circular pitted tracheids is of unique type, and entirely different from the derivation of vessel members from sclariformly bordered pitted tracheids in angiosperms. According to Bailey, the vessels have originated independently in the Gnetales and Angiosperms.

Frost (1930) correlated vessel element length with various other characters, of which the most important was the nature of the end-wall. He concluded that the scalariform perforation plate with many bars set in a very oblique end wall is the most primitive type because it is associated with the longest vessel members. He also concluded that subsequent specialization has taken the form of a progressive reduction in the number of bars and perforations and in the steepness of the end-wall to a single, simple perforation in an almost horizontal end-wall. Frost also demonstrated a series with scalariform pitting as the most primitive, progressing through transitional types to opposite and finally to alternate.

Using the series of vessel characters established by Frost (1930) as phylogenetic indicators, Kribs (1935) has defined certain types of ray, which are thought to show degrees of specialization. His conclusions are as follows: The most primitive ray type consists of a combination of multiseriate rays with high uniseriate wings, and numerous and high uniseriate rays leading to a type with multiseriate rays only, or of the multiseriate rays, leading to type with wholly uniseriate rays. Wholly uniseriate rays are never primitive, being derived from multiseriate types, those derived from less advanced types tending to be heterogeneous with large cells.

Type of perforation plate is a much more convenient index of specialization than vessel member length, as the latter can be measured only in sides of macerated materials. The relation between the two characters appears to be quite close to justify the uses of type of perforation alone

According to Beijer (1927) and others the storied structure is related to the length of the cambial initial, and can be used as an index of a very high level of specialization.

According to Metcalfe and Chalk (1950) the homogeneity in rays is an advanced character. It is also apparent that paratracheal parenchyma in some form or other is more commonly associated with advanced than with primitive woods. Of the paratracheal forms the ailiform and confluent appear to be most advanced. This is in general agreement with the conclusions of Kribs (1937).

It is suggested that minute pitting represents a special development peculiar to particular groups rather than a form of specialization towards which the vessels of all moderately advanced woods tend. This feature is likely, therefore, to prove of value as an index of affinity rather than of level of specialization.

Septate fibres are of common occurrence throughout large natural groups (*e.g.*, Meliaceae and Burseraceae), and are a most useful index of affinity. Chalk (1937) suggested that the septate fibres should be regarded as more primitive than the non-septate libriform fibre, but Tippo (1938) reached the opposite conclusion from his study of the woods of the Moraceae and allied families. (See fig. in Chapter 5).

Ring-porousness, or the development of a marked zone of large vessels at the beginning of the growth ring, appears to be accompanied by an increase in the length of the complete vessels in the pore zone, and has been shown to increase the rate of the flow of water through the wood. Gilbert (1940) has stated that there is abundant evidence to indicate that the ring porous type of vessel arrangement represents an evolutionary advance from the diffuse distribution.

Spiral thickening of the secondary wall of the vessel has been regarded by Frost (1931) as an advanced character. Spiral thickening is much more closely linked with ring porousness.

If two or more taxonomic groups have but one important character in common, this should be regarded as suggestive rather than as proof of affinity, unless supported by other evidence.

VESSELS

The most valuable taxonomic vessel characters are the distribution and the pattern of the pores as seen in cross section. The majority of woods have solitary vessels, mixed with some multiples of two or three cells. Divergence in either direction — to exclusively solitary vessels or to more numerous or larger groups and multiples — is useful for identification, but the possession of exclusively solitary vessels, being a common unspecialized type, cannot be used along as indicating affinity. In some groups there is a marked tendency for multiples to be more common or for the vessels to be arranged in radial, oblique or tangential lines, and these characters can be used positively as indicating affinity. The nature of the pitting between vessels and ray or wood parenchyma cells also appears to have some taxonomic significance. The feature known as a *vestured pit* has been shown by Bailey (1933) to be characteristic of a limited number of families (*e.g.*, Apocynaceae, Asclepiadaceae, Capparidaceae, Leguminosae, Rubiaceae and several others). Vasicentric tracheids tend to be associated with particular taxonomic groups (*e.g.*, Apocynaceae, Asclepiadaceae, Casuarinaceae, Linaceae, Myrtaceae, Rutaceae and several others).

PARENCHYMA

This is the most important tissue indicating relationship. The most fundamental distinction is that between *apotracheal* and *paratracheal* parenchyma and in most families the parenchyma is wholly or predominantly apotracheal or paratracheal. In Malvales and in some Dipterocarpaceae, both types may be common in the same wood, but this is rare for it to constitute a distinctive character.

Single crystals of calcium oxalate are found to be scattered in the ordinary cells of wood of ray parenchyma are too common to be of much value, but chambered crystalliferous cells and special forms of crystal, such as raphides and druces, are of significance, as is the occurrence of silica and of less common inclusions such as lapachol and aluminium.

RAYS

Most ray characters, such as heterogeneity and the number of uniseriate rays, indicate specialization rather than affinity. Size is sometimes characteristic or taxonomic groups, *e.g.*, the large rays of the Proteaceae and the Rhizophoraceae and the small rays of Salicaceae. However, it must also be remembered that exclusively uniseriate rays are a form of specialization that has occurred independently in occasional genera of many families. One of the standard features used in wood descriptions in 'rays of two sizes'.

FIBRES

As already mentioned the septate fibres afford a valuable indication of affinity, and that they occur commonly throughout large and small groups that are considered to be closely related on other grounds (*e.g.*, Acanthaceae, Anacardiaceae, Apocynaceae, Cactaceae, Caesalpiniaceae, Compositae, Euphorbiaceae, Labiateae, Rosaceae, Rubiaceae, Rutaceae and several others).

INTERCELLULAR CANALS

Normal intercellular canals are of very limited occurrence and tend to be characteristic of particular groups. The Dipterocarpaceae typically have vertical canals and the Anacardiaceae and Burseraceae commonly have radial canals.

The main emphasis in recent years has been on the anatomy of the secondary xylem and in particular on the evolutionary trends of its elements. The work on anatomical specialization in secondary wood suggests that the application of this knowledge and of similar techniques to the xylem and other tissues of the non-woody plants may ultimately make such a comparison feasible.

The work of Cheadle (1941) on the systematic anatomy of the Monocotyledons indicates other lines along which further advance are likely to be made. There seems to be a considerable evidence in favour of regarding the Magnoliaceae and allied families as retaining many primitive characters. This is supported especially by the lack of vessels in the allies, the homoxylous angiosperms *Drimys*, *Trochodendron* and *Tetracentron*. This is to be regarded as the retention of a truly primitive character rather than as a reduction.

The main general conclusions have been drawn by Bailey (1944) from the studies of the vessel are as follows:

'The independent origins and specializations of vessels in monocotyledons and dicotyledons clearly indicate that, if the angiosperms are monophyletic, the monocotyledons must have diverged from the dicotyledons before the acquisition of vessels by their common ancestors. This renders untenable all suggestions for deriving monocotyledons from vessel-bearing dicotyledons or *vice versa*. Furthermore, the highly specialized structure of the xylem throughout both stems and roots of herbaceous dicotyledons, not only affords conclusive supplementary evidence of the derivation of herbaceous from arboreal or fruticose dicotyledons, but also is an insuperable barrier to the derivation of monocotyledons from herbaceous dicotyledons.'

19

Development of Plant Anatomy in India

The last so many years have witnessed tremendous strides in all branches of Botany, parallel with those in the other biological sciences. Marked activity has characterized the disciplines of Microbiology, Cytology and Genetics, Plant Physiology, Palynology, Tissue culture, Developmental Anatomy, Mycology and Environmental studies.

Main centres of plant anatomy research are Allahabad, Aligarh, Bombay, Dehradun, Delhi, Hyderabad, Chennai, Meerut, Pilani and Vallabh Vidyanagar. The anatomical work done in India on various species is summarized here.

ANGIOSPERMS

Seed and seedling. The structure and development of seed have been investigated in *Plantago* (Misra, R.C. 1964) and *Gossypium* (Ramchandani, S. et. al., 1966). The seedling anatomy has been studied in Umbelliferae and Cucurbitaceae (Desphande, B.S. and Joneja, P. 1962; Deshpande and Kasat, M.L., 1966) and *Plantago* (Misra, R.C. 1966).

Shoot-apex. In *Cuscuta reflexa* photoperiodic induction of vegetative shoots causes in increase of RNA but a decrease in the histone content of the shoot apex (Maheshwari, S.C. and Gupta, S. 1966), J.J. Shah and J.D. Patel (1970 a) studied the ontogeny of the shoot apex in brinjal and chilli. B.C. Kundu and B. Gupta (1964) studied shoot apex of *Xanthium stromarium*.

Root apex. S.K. Pillai and A. Pillai investigated the organization and structure of the root apex of a number of dicotyledons, monocotyledons and gymnosperms. In dicotyledons, mainly the work done in Proteaceae, Cruciferae, Piperaceae, Amaranthaceae, Onagraceae, Gentianaceae, Scrophulariaceae and Compositae. In the evolutionary trend of root apical structure in Scitamineae, Musaceae exhibits primitive structure with a common group of initials. Zingiberaceae, Cannaceae and Maranthaceae show the next stage of specialization with discrete initials for the stele and root cap and a single tier of initials of the protoderm-periblem complex. *Maranta arundinacea* exhibits the next advanced stage of discrete initials for the stele, the root cap and a two-tiered protoderm-periblem complex (Pillai, et al., 1965 a; 1965 b; 1969; Pillai, S.K., 1963; Pillai, A., 1963; 1964; 1966). A cup shaped generative centre is reported in *Cyamopsis* root (S.K. Pillai and K. Sukumaran, 1969).

Nodal anatomy. As reported by N.P. Saxena (1964) the node of *Saxifraga diversifolia* is unilacunar and one-traced. Pant, D.D. and Mehra, B. (1963 a, b) described the nodal anatomy of *Mirabilis jalapa, Oxybaphus nyctagineus, O. viscosus, Bougainvillaea glabra, B. spectabilis* and *Abronia elliptica*. In retrospect of nodal anatomy, they (1964) pointed out that on the basis of existing evidences it is difficult to regard the unilacunar double trace node as primitive for all

Pteropsida. Shah, J.J. (1969) on the basis of observations on nodal anatomy of *Clerodendrum* pointed out that all unilacunar nodes do not appear to be homologous. Ashok Kumar (1976) reported bilacunar two-traced condition, a nodal type new to the angiosperms in *Geranium nepalense* Sweet.

General anatomy. Kapil, R.N. and Rustagi, P.N. (1966) studied the anatomy of the aerial and terrestrial roots of *Ficus bengalensis*. Kapoor, L.D. (1969) studied laticifers in some Papaveraceae. Roy, J.K. (1967) reported that the seminal and adventitious roots of rice have different structure and the adventitious roots arise from the cortex. Ramayya, N. (1966) studied the development of the coenocytic pericyclic fibre of the Apocynaceae and Asclepiadaceae. Pattanath, P.G. and Ramesh Rao, K. (1969) studied the structure of internode of bamboos. Patel R.C. and Shah, J.J. (1971) studied the pattern of internodal elongation in brinjal and chilli.

Sclereids. A.R. Rao (1964) observed multiseriate trichosclereids in seven species of *Scinoplapsus*. Inamdar, J.A. and Patel, R.C. (1971) reported that subsidiary cells of the stomata are transformed into sclereids in *Ipomoea quamoclit*. Bendre, A.M. (1967) recognized four types of floral sclereids in *Fagraea*. A.K.M. Ghouse, M.I.H. Khan and (Mrs.) Shahnaz Khan (1977) studied the structure and distribution of sclereids in *Tectona grandis* L.

Stipules. Y.P.S. Pundir (1976) studied the structure and development of the stipules in *Ficus glomerata* Roxb.

Stomata. K.A. Ahmad (1964 a), J.A. Inamdar (1969, 1970, 1971), G.S. Paliwal (1966, 1969 a, b), D.D. Pant (1963, 1965, 1971), N. Ramayya (1968, 1970), G.L. Shah (1969, 1970, 1971) and their associates have studied the ontogeny and structure of stomata in a number of angiosperms and in some pteridophytes. D.D. Pant (1965) classified the stomata on the basis of their origin; mesogenous, mesoperigenous and perigenous. Pant and R. Banerji (1965) reported transitional forms between paracytic and anisocytic and paracytic and diacytic stomata. Pant and P.K. Kidwai (1966) described triacytic stomata in monocotyledons.

Trichomes and nectaries. M. Farooq and S.A. Siddiqui (1966) described the trichomes on the floats of *Utricularia*. J.A. Inamdar (1967, 1968) studied the ontogeny and structure of trichomes in *Oleaceae* and *Ipomoea*. Ramayya, N. (1969) studied the ontogeny and structure of trichomes in Compositae. Saxena, N.P. (1972) studied the development of glandular hair in *Saxifraga sarmentosa*. Maheshwari, J.K. and Chakraborty, B. (1966), Chavan, A.R. and Deshmukh, Y.S. (1964), Inamdar, J.A. (1968) and Venkata Rao (1967) have studied ontogeny and structure of extrafloral and floral nectaries.

Axillary bud. J.J. Shah and his associates (1969, 1970, 1971) have shown in several angiosperms that the bud meristem originates from the apical meristem. They have also studied the ontogeny of axillary buds, the development of their early primary vascular system and their vascular relationship with the axillary leaf and the axis. Shah, J.J. and Patel, J.D. (1970 b) showed that the extra-axillary position of the flower or inflorescence in brinjal is due to a specialized type of internodal differentiation. Shah, J.J. Patel, J.D. and Kothari, I.L. (1972) surveyed the vascular supply of axillary buds in 44 species of 22 families and discussed relationship of the bud traces with the various types of nodal structure.

Tendril. J.J. Shah and Y.S. Dave (1970 a) reported that tendrils in Vitaceae are modified extra-axillary branches and not modified axillary buds displaced to extra-axillary position by specialized internodal growth. Shah and Dave (1970 b, 1971 a) also reported that the tendril in *Passiflora* originates from a complex of meristem, differentiated from the main axillary bud. They (1971 b) also pointed out that the main tendrillar branch in *Antigonon leptopus* is a modified axillary inflorescence axis and the lateral and terminal tendrils of the axis are modified bracts. In *Cardiospermum halicacabum*, the tendril arises from the axil of the bract (Dave and Shah, 1971).

Cambium, wood and xylem. K.A. Chowdhary (1964) discussed the anatomical structure that brings about the growth rings in tropics. Chowdhury (1969) compared the cambial activity in temperate and tropical trees. J.J. Shah and his associates (1966, 1967) studied the vessel member length-diameter and perforation plate-vessel member length relationships and ontogeny of vessel members in the stem of *Dioscorea alata*. Shah et al. (1968) studied the leaf metaxylem vessels in the genus *Saccharum* and allied genera. Purkayastha, S.K. and Ramesh Rao, K. (1969) reported that there is no definite relationship between specific gravity and proportion of secondary xylem tissues in teak, G.S. Paliwal and N.V.S.R.K. Prasad (1970) studied the seasonal activity of cambium in *Dalbergia sissoo*. P.G. Pattanath (1972) studied the trend of variations in fibre length in bamboos.

Phloem. A.S. Mehta (1964) described the primary phloem in the petiole of *Nymphoides peltatum*, Kundu, B.C. and Shah, B. (1968) reported that types of sieve elements in *Mimosa pudica* but their observations were contradicted by K. Esau (1970). J.J. Shah and R. Jacob recorded specialized phloem parenchyma cells in *Luffa cylindrica* (1969 a). They (1967) also reported slime in the companion cells in *Cordia sebastena*. The peripheral slime and central cavity in sieve tube elements in *Lageneria sicsraria*. (1969 b) and *Luffa cylindrica* (1969 a) were reported by same investigators.

Floral anatomy. The research in floral anatomy in the fifteen years (1963–1977) has been active at Meerut (V. Puri) and Bombay (V.S. Rao). V. Singh (1966) in a series of papers has described the vascular anatomy of members of Helobiae. He (1966 a, b, c, d) has studied the vascular anatomy of the flower of Alismaceae, Butomaceae and Hydrocharitaceae. V. Puri (1977) had reviewed the opinions concerning the nature of ovule in angiosperms and suggested that the angiospermic ovule has possibly arisen independently of the gymnosperms and the two due to lack of any direct link, should not be homologised. Vascular anatomy of the flower in Asclepiadaceae has been studied by V. Puri and R. Shiam (1966). N. Chandra (1962 a, b), from a detailed study of the morphology of the flower of Gramineae, supported the view that it is a branch in the axial of bract, the lemma. R. Shiam (1962–1963) has described the anatomy of the flower of *Cyperus esculentus* and interpreted the single basal ovule as having been derived from free-central ancestry. M.R. Sharma (1962–1965) suggested that the apparent spike inflorescence in *Artocarpus* has been formed by fusion and coalescence of several fertile branches. Saxena, N.P. (1964, 1966, 1969, 1970 a, b) has worked out the floral anatomy of 37 species of Saxifragaceae. According to him there is no obdiplostemony in the family. Y.D. Tiagi (1961, 1963 a, b, 1968), on the basis of evidence derived from floral morphology, anatomy, embryology and Cytology in Cactaceae, considers the family closely related to Calycanthaceae. A.M. Bendre (1967) has described the structure and distribution of sclereids in the flower of three species of *Fagraea*. B. Tiagi and G. Dixit (1966) have studied the vascular anatomy of the flower of a few species of Apocynaceae. T.S. Chauhan (1967), from a detailed anatomical study of some Indian Plumerias, regarded the corolla tube of as a compound structure, the lower portion being receptacular and upper appendicular. C.M. Govil (1963, 1968) on the basis of floral anatomy of Convolvulaceae, has interpreted the nectary in the family as receptacular. The placentation in the family has been regarded as parietal. B. Tiagi and P. Gupta (1963) have also interpreted the placentation in *Evolvulus alsinoides* in the similar way. The floral anatomy of two species of *Cuscuta* has been examined by B. Tiagi (1966) and its retention in the family Convolvulaceae has been suggested by C.M. Govil (1968). The studies on floral anatomy by V.S. Rao deal with monocotyledons (1966, 1969 b, c), placentation (1968 b), dicotyledons (1969 a) and inferior ovary (1968 a). Rao and his associates (1963, 1964) also studied the floral anatomy of Apocynaceae and Rubiaceae. R.L.N. Sastry (1965, 1969) contributed to the floral morphlogy and anatomy of Lauraceae and Berberidaceae. K. Prasad (1976) has reported the vasculature of the flower of *Malcolmia africana* R. Br. of Cruciferae. According to him the traces for sepals and petals exhibit cohesion. Except dorsals the other carpellary traces are not distinct. The carpellary ventrals and the conjoint median laterals show complete fusion and along with parenchyma organize into a 'medullary placental complex'. Such a feature is unique one and quite unlike other investigated members of this family studied so far.

V. Singh (1977) has studied the primary pattern of floral development in all the taxa of the Alismatales as trimerous. Antipetalous stamen pairs are characteristic of the alismatalean flowers with the exception of *Limnocharis*. The trimery becomes more or less obscured in the mature flowers and often been misinterpreted.

N.P. Saxena (1977), described the floral anatomy of seven species of sub family Hydrangeoideae and five of Escallonioideae (Saxifragaceae). On the basis of floral anatomy non-occurrence of obdiplostemony has been suggested. With the exception to tribe Hydrangeae, nectary having diverse position has been observed in all the species studied. The placental ridge of *Deutzia*, *Hydrangea*, *Escallonia* and *Quintinia* has been interpreted; the septum being formed by the fusion of the carpel by their dorsal surfaces and the ridge by swelling of the carpellary margins. Establishment of Hydrangeaceae and Escalloniaceae as separate families has been supported on anatomical and embryological grounds.

Anatomical delimitation of plant taxa. B.C. Kundu and B. Gupta (1963, 1964 a, b) have suggested the use of quantitative microscopy of leaves in distinguishing taxa which is found useful in Solanaceae.

Floral anatomy for delimiting taxa has been utilized by B. Tiagi and G. Dixit (1965) in Asclepiadaceae, Y.D. Tiagi (1964), B. Tiagi (1965, 1966) in *Opuntia*, *Cuscuta* and *Evolvulus*, R.K. Basak (1962) in *Citrullus vulgaris*, Shivaramiah, G. (1967) in *Utricularia*, J.S. Jos (1967) in *Nicotiana*, B.D. Deshpande and A. Singh (1967) in *Withania somnifera* and P.C. Datta and A. Deb (1968 b) in tracing phylogeny of *Rumex*.

Anatomy of axis has widely been used for comparing taxa, in *Corchorus* (R.M. Datta and K. Roy, 1963) in *Ipomoea* (C.M. Govil, 1971), *Phaseolus* (P.C. Datta and A. Saha, 1970). Epidermal and cuticular anatomy has also been used for taxonomic delimitation of medicinal plants (P. Singh and B.C. Kundu, 1962; Y.N. Pandey, 1970; K.J. Ahmad, 1964 a, b). Structure of sclereids has been found very characteristic in the stem of *Saraca indica* (M. Malaviya, 1967), and in seed coats of pulses (P.C. Datta and R.K. Maiti, 1968 e).

GYMNOSPERMS

Xylem. R.J. Rodin (1969) studied xylem elements of Gnetales. A.K.M. Ghouse (1969) prepared a key for the identification of Indian coniferous wood based on microscopic structure.

Sclereids. A.R. Rao and his associates reported brachysclereids and osteoscelereids in *Cephalotaxus drupaceae* (1964) and *Taxus baccata* (1965).

Nodal anatomy. B.S. Deshpande and his associates (1963, 1967) reported unilacunar double trace condition in *Ephedra foliata*.

Abnormal cone. Vandana Sharma and N. Chandra (1977) have studied the vasculature of the abnormal cone of *Gnetum gnemon* L. where in the axil of a collar, an ovule has been replaced by a daughter inflorescence. The vascular supply of the ovule and the daughter inflorescence is similar. However, it is concluded that the daughter inflorescence is not equivalent to an ovule.

Epidermis. A comparative light and scanning microscopic study of the epidermis in Gnetopsida has been done by G.S. Paliwal, G.C. Rampal and Nilima Paliwal (1974). A study has been carried out of the stem and leaf epidermis of *Ephedra distachya, Gnetum gnemon* and *Welwitschia bainesii* with the help of light and scanning electron microscope. The stomata of all the three taxa are invariably sunken. The mature stomata are mostly paracytic but a few are also anomocytic. The number of stomata per unit area is highest in *G. gnemon*, lowest in *W. bainesii*, and *E. distachya* occupies intermediate position. The epidermal cells are enclosed by walls which are straight in *Ephedra* and *Welwitschia*, but sinuous in *Gnetum*. The scanning microphotographs show that *Ephedra distachya* stem has ridges and furrous and stomata occur exclusively on ridges.

Root apex. In the cycads and the conifers the root apical structures are of the open type (Pillai, A. 1963, 1964). In the evolutionary specialization of root apical structures in gymnosperms, the cycads are primitive with the origin of all tissues from a cup-shaped promeristem. The conifers show the next advanced stage, culminating in a type exhibited by the root apex of *Ephedra*.

Seed. R.J. Rodin and R.N. Kapil (1969) studied the seed anatomy of four species of *Gnetum*, the seed coat consists of three layers.

Shoot apex. S.K. Pillai (1963 b, 1964) investigated the structure and seasonal variations in the shoot apex of species of *Podocarpus, Cupressus* and *Araucaria*. Tunica-corpus organization was described for the species of *Araucaria*.

Abnormal vegetative complex. An abnormal vegetative complex was observed in *Cycas circinalis*, with forked rachis and leaflets (P.K. Dublish, O.P. Sharma and P.C. Pande, 1976). The rachis of the abnormal leaf was slightly lobed on the adaxial side and forked at its tip. Anatomy of the rachis below the fork revealed a composite corrugated ring of vascular bundles formed by the fusion of two Omega shaped (ω) rings. When the rachis was distinctly forked, each lobe had its own vascular bundles. The vascular bundles were conjoint, collateral, open and diploxylic.

PTERIDOPHYTES

The ontogeny and structure of stomata in some pteridophytes have been investigated by K.A. Ahmad (1964 a), J.A. Inamdar (1969, 1970, 1971), G.S. Paliwal (1966, 1969 a, b), D.D. Pant (1963, 1965, 1971), N. Ramayya (1968, 1970), G.L. Shah (1969, 1970, 1971) and their associates. Stomata of *Equisetum, Ophioglossum, Psilotum, Tmesipteris* and some leptosporangiate ferns have been studied.

D.S. Loyal and R.K. Garewal (1967) presented their observations on anatomy of *Salvinia auriculata* and *S. natans*, A.R. Rao (1971) on three hymenophyllaceous ferns of India and U. Sen (1966, 1968 a, b) on *Cystodum, Culcita macrocarpa* and *Ophioglossum reticulatum*.

S. Bhambie (1962) has studied the anatomy of the rhizomorph in *Isoetes panchanani*. S. Bhambhie and V. Puri (1963) have studied the shoot apex organization in *Selaginella, Lycopodium* and *Isoetes*. They found that there is a group of meristematic cells at the apex of *Isoetes*, prismatic tissue of secondary phloem and some pitted tracheids included in the protostelic stem of *Isoetes coromandelina*.

P.N. Mehra and S.L. Soni (1971) studied the morphology of vessel elements in four species of *Marsilea* and *Pteridium*.

20

The Fruit, The Fruit Wall And The Seed Coat

THE FRUIT

A fruit is a developed and ripened ovary or ovaries together, often, with adjacent floral organs and other plant parts. Formation of a fruit may occur also without seed development and without fertilization, a phenomenon known as *parthenocarpy*. Fruits range in complexity of their morphology from a single carpel (*e.g.*, legume) to a compound fruit (*e.g.*, pineapple), where an entire inflorescence (including the ovaries, floral parts, bracts, inflorescence axis, etc.) converts into a single succulent mass. According to Winkler (1939) the fruit is the product of the entire gynoecium and any floral parts associated with the gynoecium in the fruiting stage.

The body of the fruit developed from the ovary wall, which covers the seed, is termed the *pericarp*. When the pericarp is not homogeneous histologically and distinguished into outer, inner and median parts then these parts are known as *exocarp*, *endocarp* and *mesocarp* respectively. In strict sense the term *pericarp* refers only to modified ovary wall.

THE FRUIT WALL

Structurally there are two types of fruit walls:

(1) The parenchymatous fleshy, often succulent fruit walls and (2) the sclerenchymatous dry fruit wall.

As regards the structure of the fruit wall, the fruits may be referred as *fleshy* or *dry*. In the dry dehiscent fruits the fruit wall splits open at maturity, whereas in dry indehiscent fruit, the fruit wall remains closed. Dry or fleshy, dehiscent or indehiscent fruit walls occur in fruits derived from both superior and inferior ovaries.

DRY FRUIT WALL

Dehiscent fruit wall. The ovary differentiating into a dry fruit when contains several ovules, it commonly dehisces at maturity. Such type of fruit may develop from a single carpel (follicle, legume) or from several united carpels (capsule). In *follicle* the pericarp usually possesses a relatively simple structure. It may have a narrow exocarp of thick-walled cells and a thin-walled parenchymatic mesocarp and endocarp. The three main longitudinal vascular bundles (one median and two lateral) and the transversely oriented branches from the main bundles become enclosed in sclerenchymatous sheaths. As the fruit matures, the pericarp dries up. The differentiate drying of the parenchymatic and sclerenchymatic tissues of the pericarp creates tensions which results in the splitting of the follicle along the line where the margins of the carpel were fused during the flower development.

The Fruit, The Fruit Wall and The Seed Coat

In comparison to follicle the *legume* has a more complicated structure. In certain legumes, the ovary wall shows a considerable increase in the number of cells after fertilization and then matures into a pericarp with a thick-walled exocarp, a thin-walled parenchymatous mesocarp, and a highly sclerified endocarp. The sclerenchymatous endocarp is made up of several rows of thick-walled cells oriented at an angle to the long axis of the fruit and remains covered internally by a thin-walled epidermis. The thick-walled part of the endocarp is generally differentiated into two distinct layers.

An ovary wall maturing into the pericarp of a *capsule* have a little increase in the number of cells (*e.g.*, in tobacco). The pericarps of capsules possess both sclerenchymatous and parenchymatous tissues in variable distributions. The pericarp of linseed (*Linum usitatissimum*) for example, possesses an exocarp of sclerenchymatous cells and a mesocarp and an endocarp of parenchymatous cells. In contrast, the pericarp of tobacco (*Nicotiana tabacum*) has a thick-walled endocarp, two or three cells in thickness and parenchymatous exocarp and mesocarp.

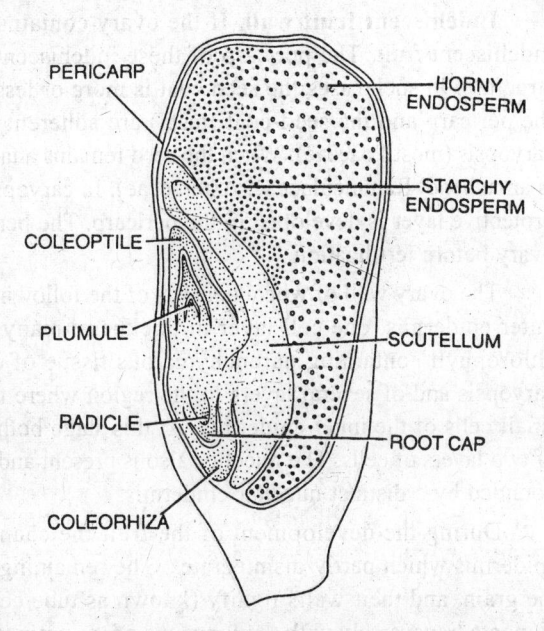

Fig. 20.1. Maize grain in L.S.

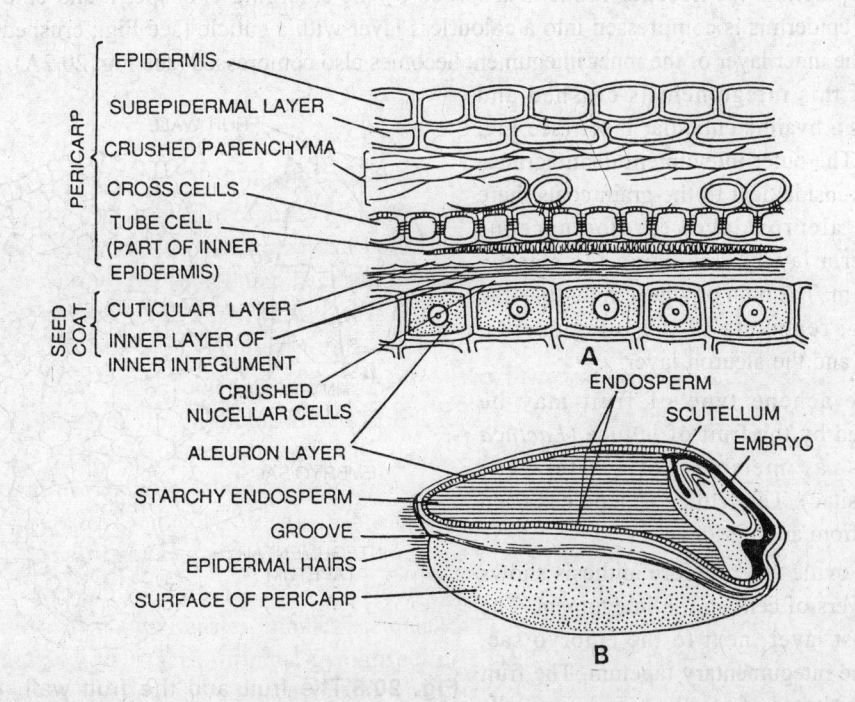

Fig. 20.2. The fruit and its wall. The caryopsis (B) of wheat and its pericarp and seed coat (A).

Indehiscent fruit wall. If the ovary contains a single ovule, it usually develops into an indehiscent fruit. The pericarp of these indehiscent one-seeded fruits resembles a seed-coat in structure. In such cases the seed coat is more or less obliterated during fruit development. When the pericarp and the seed-coat (testa) are adherent in the fruit, the fruit is known as a grain or caryopsis (most Poaceae). When the seed remains attached to the pericarp at one point only, the fruit is an achene (Ranunculaceae, Asteraceae). In caryopsis fruits (*e.g.*, in *Triticum* and *Hordeum*), the protective layer is developed in the pericarp. The pericarp and the seed-coat remain distinct in the ovary before fertilization.

The ovary wall of wheat consists of the following cell layers, beginning from the outside: the outer epidermis, one cell layer in thickness; many layers of colourless parenchymatous cells; chlorophyll containing parenchymatous tissue of one or two layers of cells over most of the caryopsis and of several layers in the region where the caryopsis (grain) is grooved; one layer of small cells of the inner epidermis. At this stage both integuments are intact and each is made up of two layers of cells. The nucellus also is present and consists of several layers of thin-walled cells bounded by a distinct nucellar epidermis.

During the development of the fruit the changes in the ovary wall begin with the inner epidermis which partly disintegrates. The remaining cells elongate parallel with the long axis of the grain, and their walls lignify (known as tube cell, see Fig. 20.2A). The chlorenchyma cells elongate transversely with the long axis of the grain, their chlorophyll disintegrates and their walls lignify; they are cross cells (see Fig. 20.2A). The parenchyma outside the chlorenchyma crushes and remains in the form of crushed parenchyma where the remaining spaces are filled with air (see Fig. 20.2A). One to four layers of this parenchyma persist in the mature grain in the form of compressed subepidermal layer (see Fig. 20.2A). The outer epidermis is compressed also and remain covered with a thin cuticle. The nucellus and integuments also undergo profound changes. Except epidermis the nucellar tissue is absorbed by the enlarging endosperm and embryo. The nucellar epidermis is compressed into a colourless layer with a cuticle (see Fig., crushed nucellar cells). The inner layer of the inner integument becomes also compressed (see Fig. 20.2A). the outer layer of this integument is crushed and becomes a hyaline cuticular layer (see Fig. 20.2A). The outer integument disintegrates. Towards inside next to the grain coats there lies the aleuron layer (protoeinaceous endosperm layer), enclosing the starchy endosperm. The bran of wheat includes—the pericarp, remnants of the integuments, nucellus and the aleuron layer.

The achene type of fruit may be illustrated by the fruit of lettuce (*Lactuca sativa*), a member of Asteraceae (Compositae). The lettuce achene has been derived from an inferior ovary.

An ovule taken before anthesis shows many layers of cells in the integument. The innermost layer, next to the embryo sac, makes the integumentary tapetum. The fruit wall, constituted of small parenchyma cells remain in contact with the ovule. At this

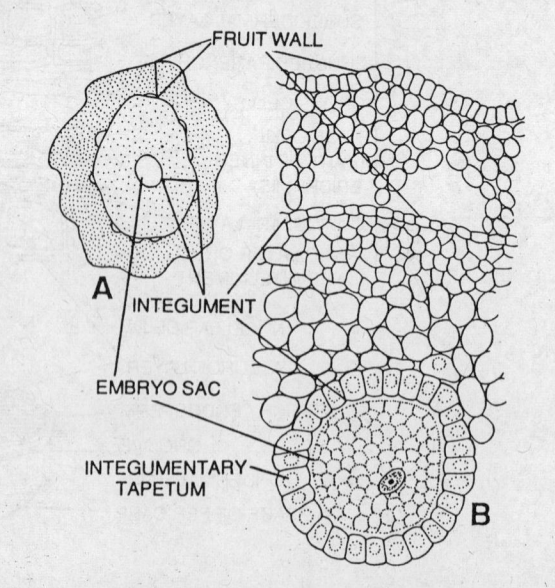

Fig. 20.3. The fruit and the fruit wall. A – B, development of fruit (achene) in *Lactuca*.

stage some cells of the inner fruit wall are disorganized and have left cavities (see Fig. 20.3 A, B). After anthesis where the achene enlarges the integument increases in thickness but also disorganizes next to the integumentary tapetum (see Fig. 20.4 A, B). Eventually the tapetum and all the parenchyma of the integument are disintegrated (see Fig. 20.5 B, C). However, the outer epidermis of the integument persists and develops thick walls (see Fig. 20.5. D). The outer layer of the endosperm develops into a compact layer. This layer and another beneath it are retained in the mature fruit and develop thick walls (see Fig. 20.5 D). There remains a conspicuous cuticle between the endosperm and the remains of the integument (see Fig. 20.5 D). The inner layers of the fruit wall become completely disorganised, but the outer layers persist. Certain parts of the remaining layers project out as ribs and develop sclerenchyma in them. The cells of fruit wall between the ribs are large and possess thin and slightly lignified walls (see Fig. 20.5). In the mature achene of *Lactuca sativa* all the persisting layers become compressed close together and their identity becomes obscured (see Fig. 20.5 D).

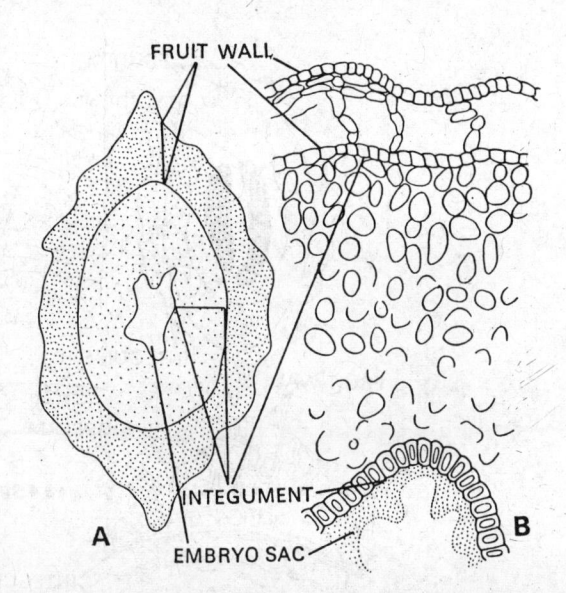

Fig. 20.4. The fruit and the fruit wall. A – B, further stages in the development of fruit (achene) in *Lactuca*.

FLESHY FRUIT WALL

Many monocarpellary or multicarpellary ovaries develop into indehiscent fruits with fleshy fruit walls. The fleshy fruit wall may consist either of the ovary wall (pericarp) or of such a wall (pericarp) fused with the noncarpellary tissue in which it is imbedded. Here, the entire ovary wall or the external part of it differentiates into a parenchymatous tissue whose cells retain their protoplasts in the mature fruit. An immature fleshy fruit wall which has a firm texture becomes softer as the fruit ripens. This softening comes from the chemical changes in the cell contents and in the structure of the walls.

While the fruit wall ripens its colour changes. Immature fruits are green in colour because of the presence of numerous chloroplasts in the outermost cells. The chlorophyll disappears and the carotenoid pigments develop simultaneously. These pigments induce a change to a yellow, orange or red colour. Ripening fruits may form anthocyanins which impart the tissue a red, purple or blue colour. The tannins are frequently accumulated in the outer epidermis.

During ripening of the fruit the composition of carbohydrates also changes. In the fruits such as apple, pear, banana, etc., starch accumulates during ripening but later disappears, whereas sucrose (sugar) increases in amount. In fruits without reserve starch, such as plum, peach, citrus, etc., the fruit ripening is characterised by a decrease in acid content and an increase in sucrose.

In berry fruit, the pericarp possesses a cutinized epidermis and sub-epidermal collenchyma. The inner tissue is parenchymatous, and the inner epidermis is thin-walled.

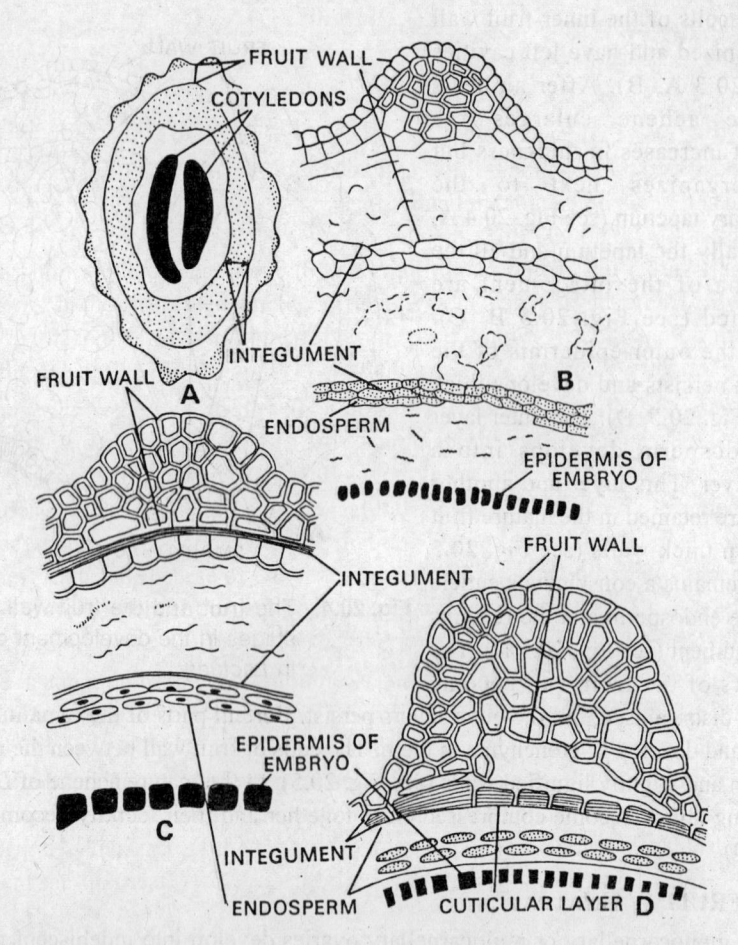

Fig. 20.5. The Fruit and the fruit wall. A – D, further stages in the development of achene of *Lactuca*. D, a part of mature achene.

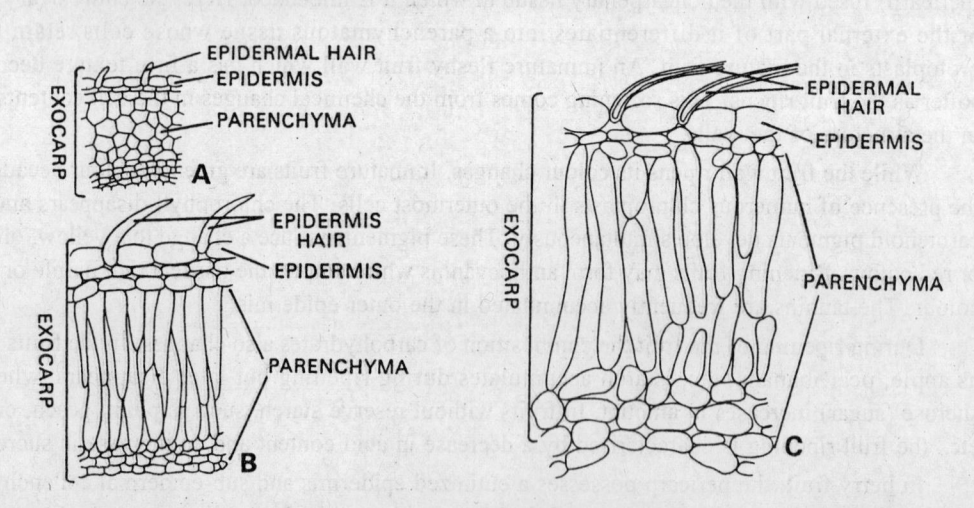

Fig. 20.6. The fruit wall. Fleshy pericarp. A – C, successive stages in the development of exocarp of *Rubus*. A, green fruit soon after fertilization; B, half ripe fruit; C, ripe fruit.

The Fruit, The Fruit Wall and The Seed Coat 393

When the ovary wall matures into a pericarp with a conspicuous stony endocarp and a fleshy mesocarp, the fruit is known as a *drupe* (*e.g.*, peach). In mature peach fruit (*Prunus persica*), the pericarp consists of three distinct parts—the thin exocarp, the thick fleshy mesocarp and the stony endocarp. The exocarp consists of the epidermis and several layers of collenchyma beneath it. A cuticle and numerous unicellular hairs are found on the epidermis (see Fig. 20.7). The fleshy mesocarp possesses loosely packed parenchyma cells which increase in size from the periphery toward the interior (see Fig. 20.7). The smaller cells near the periphery contain most of the chloroplasts in the immature fruit (see Fig. 20.7). The endocarp is composed of tightly packed sclereids and makes the pit or stone of the fruit (see Fig. 20.7).

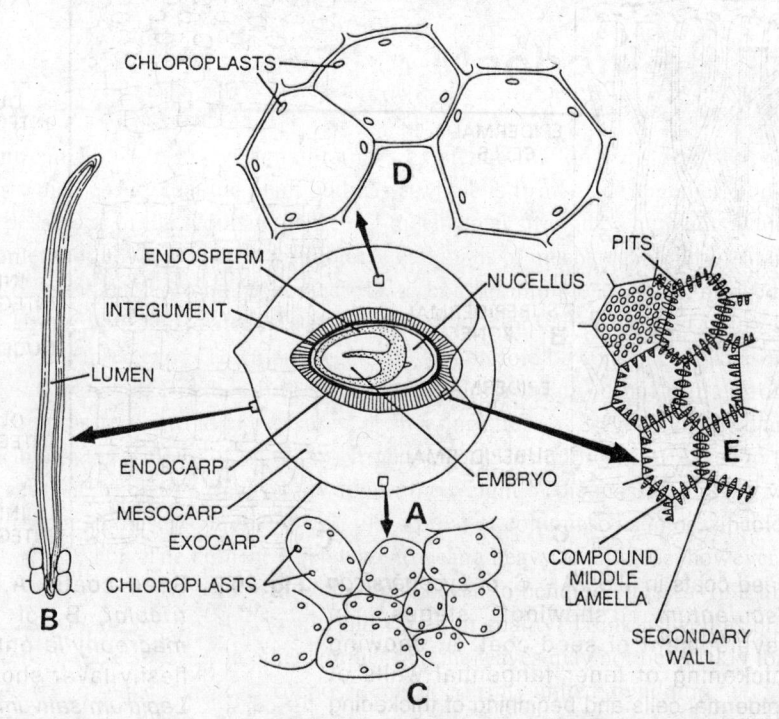

Fig. 20.7. The fruit wall of *Prunus persica* (peach). Description in the text.

THE SEED COAT

The seed coat or testa which develops from the integument or integuments consists of more or less vacuolate thin-walled cells. During the process of maturation of the seed, the seed coat undergoes varied degrees of structural changes. There may be a change in contents and wall structure as well as the disintegration of the original integumentary layers.

The epidermal layer of the seed frequently develops very thick walls and is filled with colouring matter. For example, in the order Leguminales the protodermal cells elongate at right angles to the surface and differentiate into macrosclereids; in *Gossypium* the epidermal cells elongate into hairs, the cotton fibres; in *Linum* and some others the epidermal walls are highly hygroscopic and become mucilaginous on contact with moisture. In *Asparagus*, the mechanically protective tissue is represented by the outer epidermis of the outer integument. The seeds remain also protected by cuticles that originate in the ovule.

The seeds of the order Leguminales have a peculiar wall formation, the light line, in the outer epidermis of the testa. The light line, found somewhat above the middle of the cells, is a compact

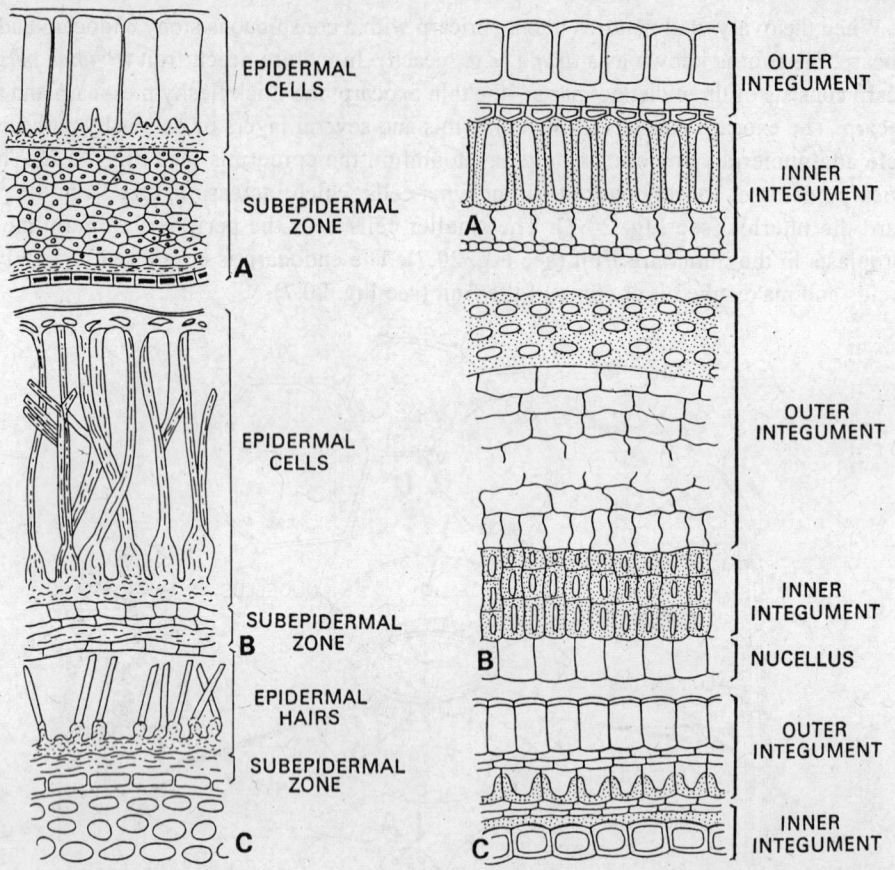

Fig. 20.8. Seed coats in T.S., A – C, of *Lycopersicon esculentum*, showing stages in development of seed coat. A, showing thickening of inner tangential walls of epidermal cells and beginning of thickening in bands on radial walls, numerous cells in sub-epidermal zone; B, showing epidermal cells much elongated, thickening of bands on radial walls greatly increased and extending over parts of outer tangential walls, sub-epidermal zone mostly absorbed; C, showing epidermal hairs—the seed coat consists of the epidermal hairs, a noncellular membranous layer and cellular inner layer one cell in thickness.

Fig. 20.9. Seed coats. A, of *Viola tricolor*; B, of *Magnolia macrophylla* only part of fleshy layer shown; C, of *Lepidium sativum*.

wall region more transparent than the rest of the wall. The hard leguminous seeds attain and maintain a high degree of desiccation.

The structure of the seed coat can be understood well when studied developmentally. Here the development of the testa has been described in the following kinds of seed: (1) the seed derived from an ovule with two integuments and having a mechanically strong seed coat (*e.g.*, in *Asparagus officinalis*): (2) the seed derived from an ovule with two integuments and having a mechanically week seed coat (*e.g.*, in *Beta vulgaris*).

The Fruit, The Fruit Wall and The Seed Coat

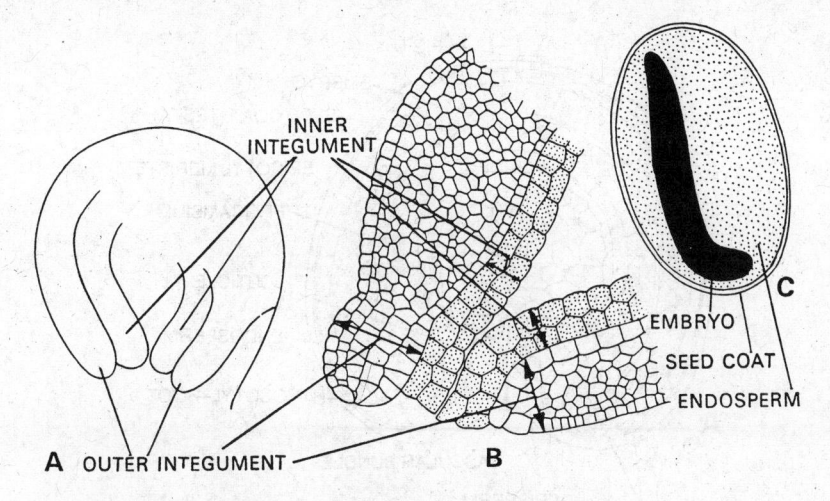

Fig. 20.10. The seed coat. Development of seed coat in *Asparagus* (A – C).

In *Asparagus*, the anatropous ovule has two integuments and a large nucellus. In the mature seed the integuments are being transformed into a black, finely rugose, somewhat brittle seed coat. The nucellus is completely absorbed during the enlargement of the embryo sac. The mature embryo

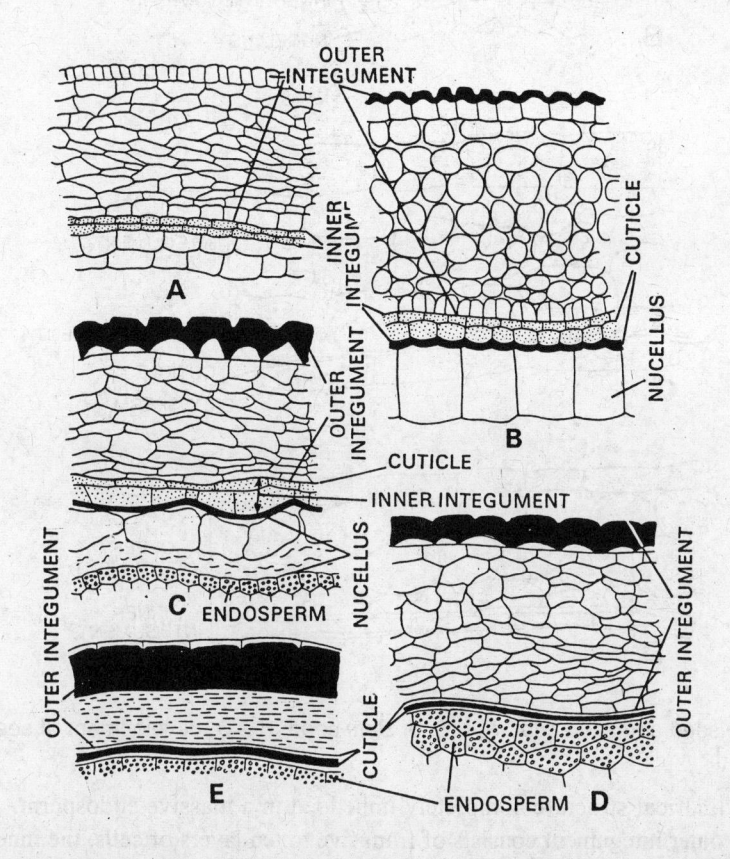

Fig. 20.11. The seed coat. Further stages in the development of seed coat in *Asparagus* (A – E); E, a part of mature seed.

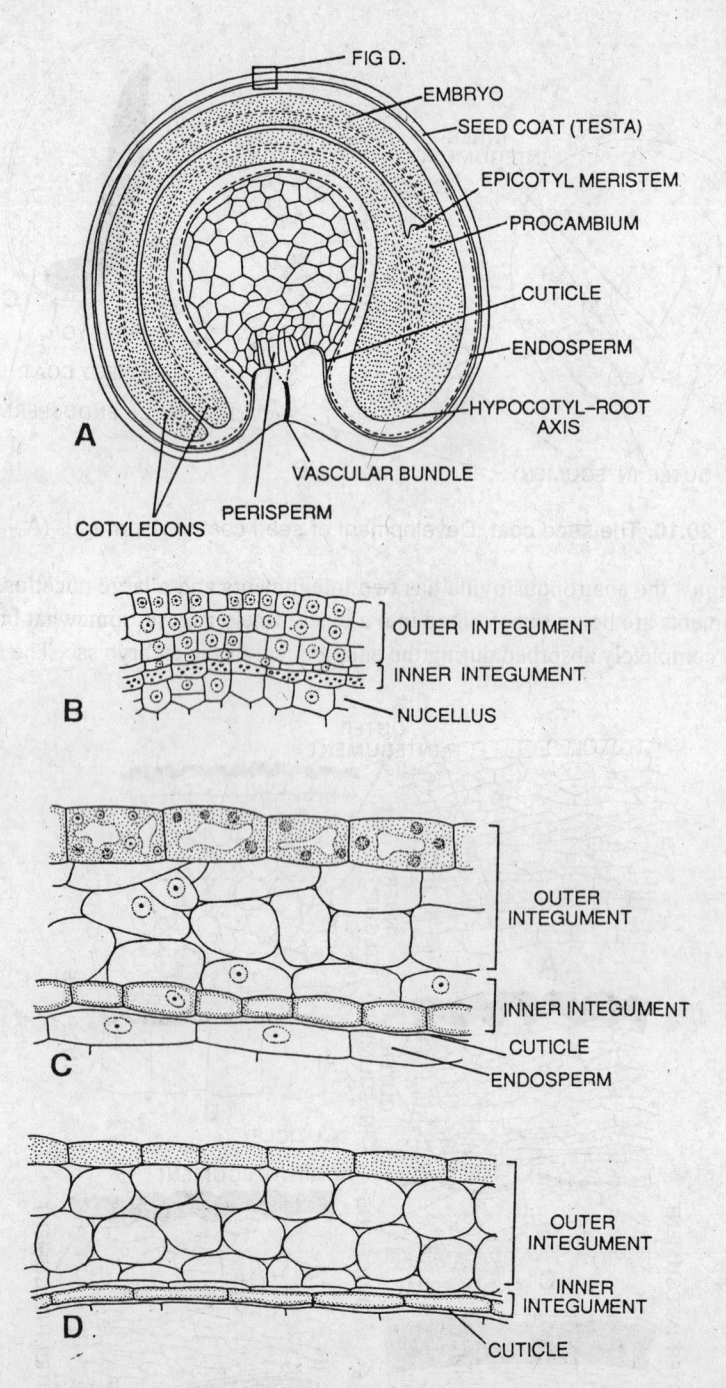

Fig. 20.12. The seed and seed coat. A, L.S. of *Beta* seed; B – D, development of seed coat in *Beta* seed.

is a slender cylindrical structure completely imbedded in a massive endosperm. At the time of pollination the outer integument consists of from five to ten layers of cells, the inner of only two. During the first fortnight after pollination the seed coat attains its maximum thickness as a result of cell enlargement. When the seed coat attains maximum thickness, two cuticles appear, one located between the two integuments, the other, a thicker one, between the inner integument and the nucellus. Later on the seed coat desiccates and shrinks and compresses gradually by the enlarging endosperm. About a month after pollination the cells of the inner integument are disintegrated and compressed so that the two fatty membranes closely approach each other. The walls of the outer epidermal cells continue to thicken until, at maturity, the cell lumina are completely filled up with dark-brown wall material. The outer surface remains covered with a thin transparent pectic membrane. Thus, the principal structural features of the mature seed coat are — a thick-walled epidermis giving mechanical protection and bearing a surface membrane which readily absorbs water and a thick cuticular membrane enclosing the endosperm and embryo.

In *Beta vulgaris*, the campylotropous ovule has two integuments, each two cells in thickness and a large nucellus. During the seed development, a curved sac is formed by breakdown of nucellar cells in continuity with the embryo sac at its chalazal end. The remaining part of the nucellus becomes a storage tissue, the perisperm. The endosperm reduces to a single layer at the micropylar end of the embryo sac. The mature seed is a shiny lenticular structure having a thin seed coat.

In this case the seed coat develops from the two integuments. The protoplasts of the outer layer of the outer integument disintegrate, and the cells become filled with brown resinous material. However, the inner of the outer integument increases in thickness by cell division, and remains thin-walled and parenchymatous. The outer layer of the inner integument becomes disintegrated. The inner layer of the inner integument becomes somewhat thickened, delicately sculptured walls. The outer surface of the seed remains covered with a cuticle. There develops a conspicuous cuticle on the inner side of the inner integument. A tightly packed layer of cells, rich in tannin, is found between the vascular tissues and the perisperm. In mature seed, the walls of these tannin cells give a positive fat reaction. Though the mature beet seed coat is mechanically weak, yet the seed is well protected because it is retained within the fruit, which develops a very hard wall.

21

Embryology of Angiosperms

In a strict sense, embryology is confined to a study of the embryo, but most botanists include under it the study of stamen, carpel, male gametophyte, female gametophyte, fertilization, endosperm formation and embryo formation (embryogenesis).

INTRODUCTION

The *phanerogams* or *spermatophytes* are known as 'flowering' or 'seed-bearing' plants. They constitute the highest division of the plant kingdom and have the largest number of species of all groups of plants. The seed encloses an embryo and remains protected by a seed-coat. The reserve food material is found either in the cotyledons or in the endosperms, which generally surrounds the embryo. There are two main groups of phanerogams: gymnosperms and angiosperms.

The *gymnosperms* are naked seeded plants, *i.e.*, those in which the seeds are not enclosed in the fruit. They may be regarded as lower 'flowering' plants in which the flowers are unisexual (either male of female), simple in construction and primitive in nature.

The *angiosperms* are closed seeded plants, *i.e.* those in which the seeds are enclosed in the fruit. They may be regarded as higher 'flowering' plants in which the flowers are more complicated in construction and more advanced. Angiosperms are the highest forms of plants. There are two big groups of angiosperms: dicotyledons and monocotyledons.

The *dicotyledons* make the bigger group of angiosperms in which the embryo of the seed bears *two* cotyledons, whereas the *monocotyledons* constitute the smaller group of angiosperms in which the embryo of the seed bears only *one* cotyledon.

Fruit, seed and embryo of angiosperms. The seed is a complex structure and develops from the ovule only after the processes of *pollination* and *fertilization*. Fertilization acts as a powerful stimulus with the result that a series of changes follow in the ovary, etc. The fertilized egg gives rise to the *embryo*, the ovule to the *seed*, and the ovary as a whole to the *fruit*. The embryo lies dormant in the seed, and the seed lies embedded in the fruit. The seed and fruit give proper protection to the embryo and store up food for it. When the seed germinates the embryo grows into a *seedling* which gradually converts into a mature plant.

ALTERNATION OF GENERATIONS
(Life cycle of Angiosperms)

In the life-cycle of angiosperms there is alternation of nutritionally independent and more complex sporophyte with the inconspicuous, reduced and parasitic gametophytes. The sporophyte, which may be *herb, shurb* or a *tree* is differentiated into roots, stem and leaves each with a vascular tissue with the highest degree of perfection. In addition it bears floral leaves or sporophylls organised into structure called the *flower*. The sporophylls are of two kinds: *microsporphylls* and

Embryology of Angiosperms

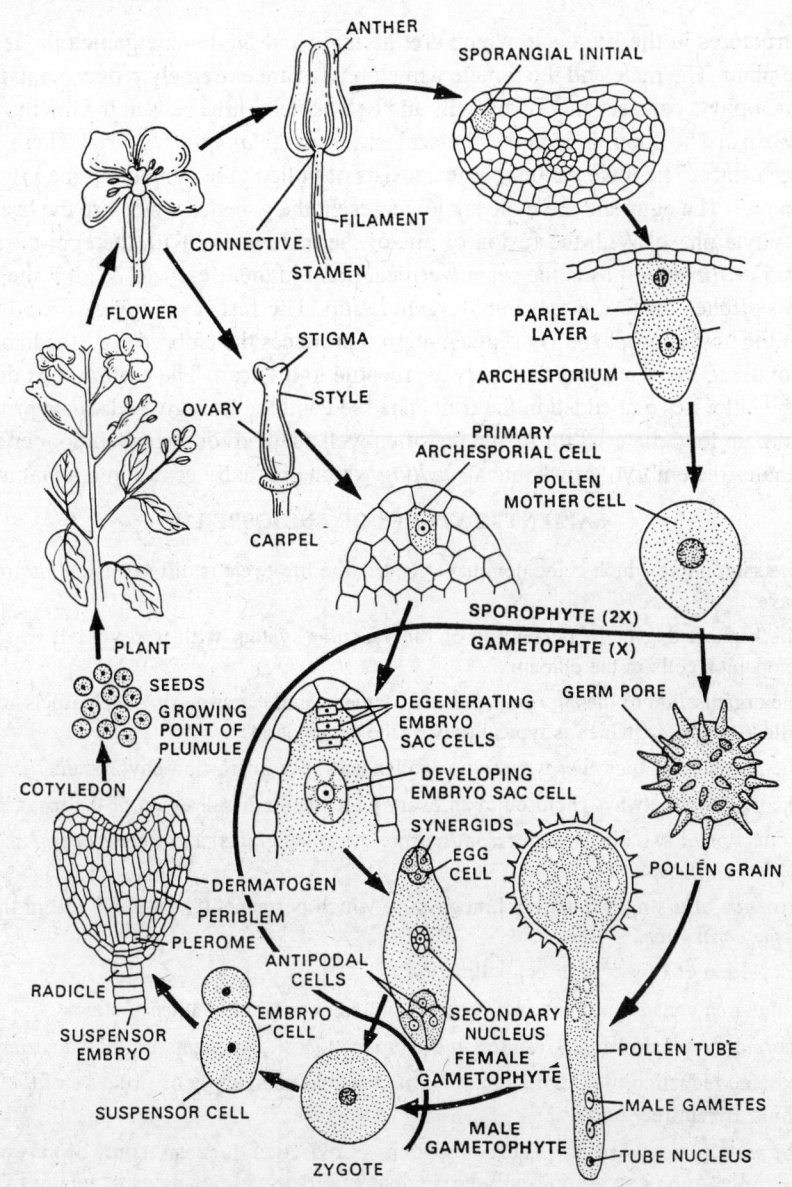

Fig. 21.1. Life-history of a typical angiosperm.

megasporphylls. Each microsporophyll bears four *microsporngia*. The latter contains *microspores* which are differentiated by meiosis from the diploid *microspore mother cells*. The megasporophyll is a flower part, which is called the *ovary*, contains one or more *ovules*. The ovule consists of one or two integuments enclosing the *nucellus* or *megasporagium*. Towards the micropylar end of the megasporangium is differentiated, by meiosis, a single functional *megaspore* which is haploid. The *microspores* and *megaspores* are the first structures of the gametophyte generation. The *sporophyte generation* in the angiosperms, thus consists of the sporophyte plant, flowers, microsporophylls and megasporophylls, microsporangia, megasporangia, microspore mother cells and the megaspore mother cells.

With meiosis the sporophyte phase switches on to the gametophyte phase. The first structures of the gametophyte phase are the *microspores* and the *megaspores*. On germination they produce the alternate structures in the life-cycle which are the male and the female gametophytes and not the sporophyte plant. The male and the female gametophytes are extremely reduced and parasitic. The female gametophyte consists only of six cells and two nuclei. Three cells at the micropylar end form the *egg appartus*. The other three at the chalazal end from the *antipodal group*. There are two *polar nuclei* in the centre. The male gametophyte consists of pollen tube containing the *tube nucleus* and two *male nuclei*. The eggs and the male nuclei represent the gametes which are the last structures of the gametophyte phase. With the fusion of one of the male nuclei with the egg, the gametophyte phase ends. *Fertilization* thus is the second critical point in the life-cycle. With it the gametophyte generation switches on to the sporophyte generation. The fertilized egg or *oospore* is the first structure of the next sporophyte. By segmentation it produces the embryo (*i.e.*, the baby plant in the seed); the ovule to the seed, and the ovary as a whole to the fruit. The embryo lies dormant in the seed, and the latter lies embedded in the fruit. The seed and the fruit give adequate protection to the embryo, store up food material for it, and are often well adapted for dispersal. Sooner or later as the seed germinates the embryo grows into a *seedling* which gradually grows into a mature plant.

SALIENT FEATURES OF ANGIOSPERMS

1. The sporophyte which is the dominant plant in the life-cycle is differentiated into roots, stem and leaves.
2. The highest degree of perfection of the vascular system with true vessels in the xylem and companion cells in the phloem.
3. The organisation of the *microsporophylls* (stamens) and *megasporophylls* (carpels) into a structure called the *flower*, which is typical only of the angiosperms.
4. The presence of four *microsporangia* (pollen sacs) per microsporophyll (stamen).
5. The *ovules* are always enclosed in an ovary which is the basal region of the megasporophyll.
6. Production of two kinds of spores, *microspores* (pollen grains) and *megaspores*. Angiosperms thus are *heterosporous*.
7. Presence of a single functional megaspore which permanently retained within the *nucellus* or *megasporangium*.
8. Adaptation of flower to insect pollination.
9. Pollination consists in the transference of pollen grains from anther to stigma.
10. Spore dimorphism having resulted in the production of gametophytes, male and female.
11. Extreme reduction in size, duration of existence and complexity of structure of the gametophytes which are entirely parasitic.
12. The *male gametophyte* has reached the limits of reduction. It consists only of the pollen grain and the pollen tube containing the tube nucleus and two male gametes or nuclei. The male cells (gametes) are non-ciliated.
13. The *female gametophyte* lacks any extensive development of vegetative tissue. It consists of three egg apparatus cells, three antipodal cells and two polar nuclei in the centre of the embryo sac.
14. The non-motile male cells or nuclei are carried bodily to the neighbourhood of egg apparatus by the pollen tube.
15. The seed or seeds remain enclosed in the ripened ovary called the *fruit*.
16. The phenomenon of *double fertilization* or *triple fusion* is the characteristic of the angiosperms.
17. The endosperm develops after fertilization. It is triploid.
18. The angiosperms are completely adapted to life on land.

THE FLOWER AND ITS PARTS

The flower is a highly specialized reproductive shoot. Each typical flower consists of four distinct types of members arranged in four separate but closely set whorls, one above the other, on the top of a long or short stalk. The lower two whorls are called *accessory* whorls, and the upper two *essential* or *reproductive* whorls because only these two are directly concerned in reproduction. The essential whorls consist of two kinds of sporophylls—microsporophylls or *stamens* and megasporophylls or *carpels*. Both kinds of sporophylls may be present in a flower (hermaphrodite flower), or only one (unisexual flower) may be seen in some types.

PARTS OF A FLOWER

The flower is commonly borne on a short or long axis. This axis consists of two regions, *viz*, the *pedicel* and the *thalamus*. The pedicel is the stalk of the flower. It may be short or long or even absent. The thalamus is the swollen end of the axis, to which the floral leaves are attached. These floral leaves are sepals, petals, stamens and carpels, and the respective whorls consisting them—calyx, corolla, androecium and gynoecium.

Calyx. This is the first or the lowermost whorl of the flower or the first accessory whorl of it and, is composed of a number of green leafy *sepals*. The primary function of the calyx is to enclose the flower in its bud and protect it from sun and rain.

Corolla. This makes the second whorl of the flower or the second accessory whorl of it, and consists of a number of usually brightly coloured *petals*. The main function of the petals is to attract insects for *pollination*. In the bud stage of the flower the corolla encloses the essential organs-stamens and carpels, and protects them from external heat and rain.

Androecium. (*andros* = male). This is the third or male whorl of the flower. It consists of *stamens* or *microsporphylls* which are regarded as the male organs of the flower. Each stamen consists of three parts—*filament, anther* and *connective*. The anther filament is the selender stalk of the stamen, and the anther is the expanded head borne by the filament at its tip. Each anther consists usually of two lobes connected together by a sort of mid-rib, the *connective*. The anther bears four chambers or *pollen sacs* each filled with *pollen grains* or *microspores*. Pollen grains are produced in large quantities in the pollen-sacs.

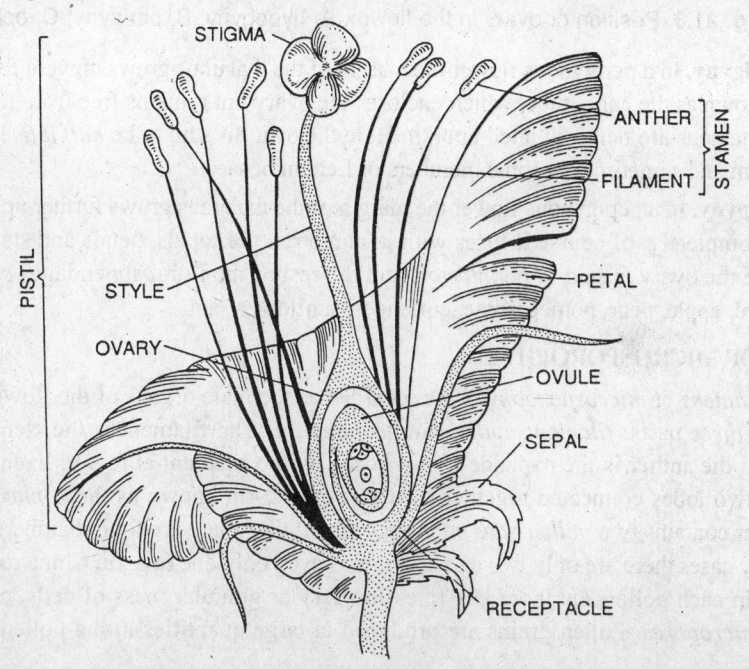

Fig. 21.2. Flower. Sectional view (L.S.) of typical flower.

The small rounded head of the gynoecium is known as the stigma, and the slender stalk supporting the stigma is called the style. The swollen basal part of the gynoecium which forms one or more chambers is known as *ovary*. The ovary contains one or more little, roundish or oval, egg, like bodies which are the rudiments of seeds and are known as *ovules*. Each ovule encloses a large oval cell known as the embryo sac. On maturation, the ovary gives rise to the fruit and the ovules to the seeds.

Position of Floral Leaves on the Thalamus

With respect to their ovaries there is considerable variation in the relative positions of the different whorls of the flowers. These relationships may be of *three* kinds: *hypogyny, perigyny* and *epigyny*.

1. Hypogyny. In a hypogynous flower the thalamus is conical, dome-shaped or flat, and the ovary is being situated at the top on the thalamus, whereas, the stamens, petals and sepals are separately and successively inserted below the ovary. Here, the ovary is said to be *superior* and rest of the floral appendages *inferior, e.g.,* in mustard *Solanum, Magnolia, Hibiscus,* etc.

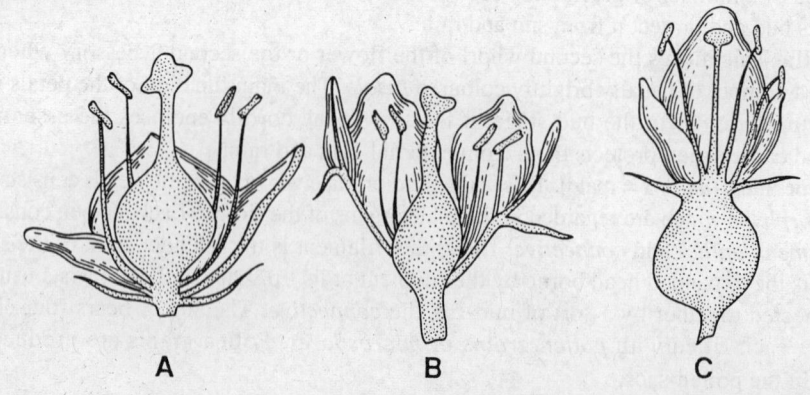

Fig. 21.3. Position of ovary in the flower. A, hypogyny. B, perigyny; C, epigyny.

2. Perigyny. In a perigynous flower the margin of the thalamus grows upward forming a cupular structure known as the calyx tube, which encloses the ovary but remains free from it, and the sepals, petals and stamens are being situated upon it. Here the ovary is said to be *half inferior, e.g.,* in rose, *Prunus*, plum and sometimes in some members of Leguminosae.

3. Epigyny. In an epigynous flower the margin of the thalamus grows further upward, to enclose the ovary completely, of course it fuses with it, and bears the sepals, petals and stamens above the ovary. Here the ovary is said to be *inferior*, and the rest of the floral appendages *superior, e.g.,* in guava, gourd, apple, pear, pomegranate, coriander, sunflower, etc.

STAMEN OR MICROSPOROPHYLL

The *stamens* or *microsporophylls* are regarded as the male organs of the flower. Each stamen consists of *three* parts: *filament, anther* and *connective*. The filament is the slender stalk of the stamen, and the anther is the expanded head borne by the filament at its tip. Each anther consists usually of two lobes connected together by a sort of mid-rib known as the *connective*. Each lobe of the anther contains two *pollen-sacs microsporangia*; thus there are four chambers in each anther. But in many cases there are only two and sometimes even only one (*e.g.,* in China rose, lady's finger etc.). Within each pollen-sac there is a fine, powdery or granular mass of cells, called the *pollen grains* or *micropores*. Pollen grains are produced in large quantities in the pollen-sacs.

Embryology of Angiosperms 403

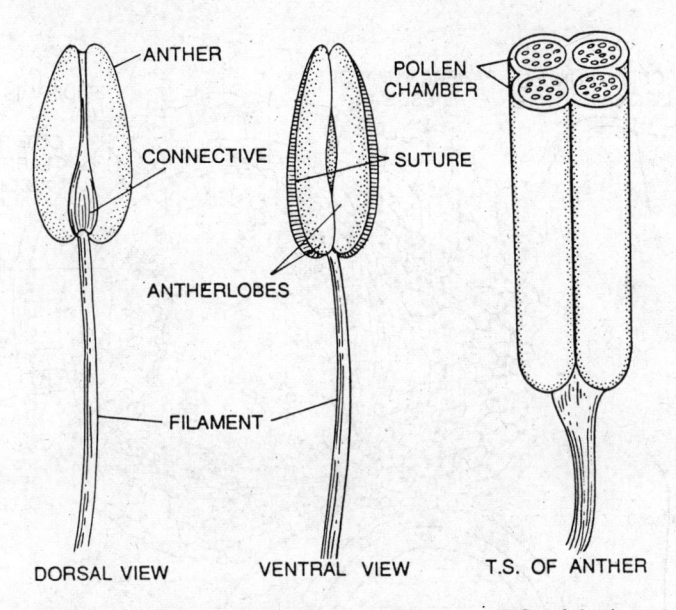

Fig. 21.4. Anther-dorsal view; ventral view; T.S. of Anther.

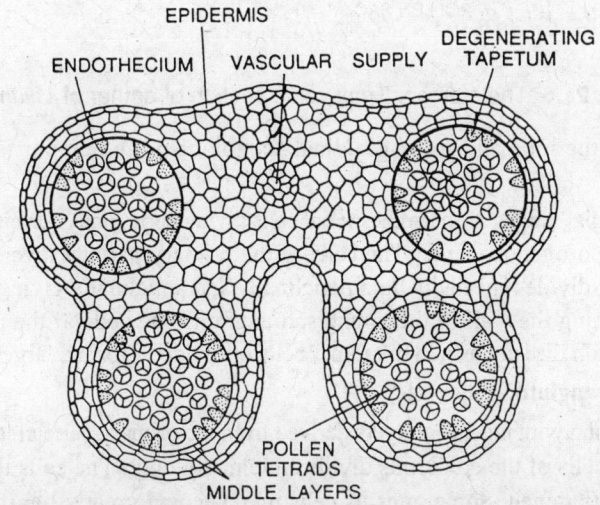

Fig. 21.5. Anther. T.S. of an anther showing tetrads of pollen grains.

ANTHER OR MICROSPORANGIUM (POLLEN SAC)

Each stamen arises as a small papillate outgrowth of meristematic tissue from the growing tip of the floral primordium. It grows actively and soon gets differentiated into an apical broader portion, the anther, and the lower slender part, the filament.

The cross-section of a very young anther consists of a homogeneous mass of meristematic cells surrounded by an epidermal layer. Further growth of the anther makes it four-lobed.

Development of microsporangium (pollen-sac)

In each lobe a few cells in the hypodermal region become differentiated by their large size, radial growth, dense cytoplasm and conspicuous nuclei. They make the *archesporium*. There is much variation in the number of cells of archesporium. Generally, the archesporium consists of a two to there cell wide plate running along the entire length of the lobe. The extent of archesporium varies both lengthwise and breadthwise. In *Boerhaavia* and *Dionaea,* there is only a single archesporial cell

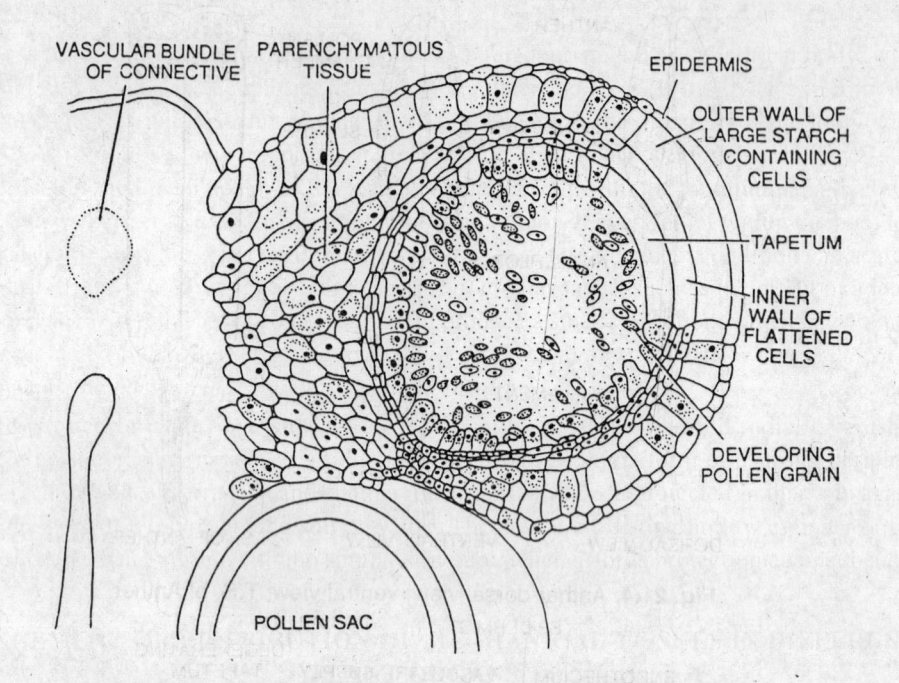

Fig. 21.6. The stamen. Transverse section of anther of *Lilium*.

in a cross section of the lobe and in longitudinal section also the number is only two and one respectively.

The microsporangial initials or the archesporial cells divide periclinally forming a primary parietal layer and a primary sporogenous layer. The cells of the primary parietal layer, lying immediately beneath the epidermis, divide repeatedly both periclinally and anticlinally giving rise to three to five concentric layers forming the wall of the young sporangium. The cells of the primary sprogenous layer may either function directly as pollen mother cells or divide to form a large number of cells.

The wall of microsporangium (pollen-sac)

The epidermis along with the 3 to 5 layers derived from the primary parietal layer form the wall of the sporangium. The cells of the epidermis divide anticlinally only. The cells thus formed become greatly stretched and flattened. Sometimes its cells may become greatly lignified or cutinized. In plants of dry habitats on the other hand, the epidermal cells become so much stretched that they lose contact with each other and at a maturity only their withering remains may be seen.

The layer next to the epidermis is the *endothecium* or the *fibrous layer*. The walls of the cells of endothecium become radialy elongated and from their inner tangential walls fibrous bands develop upwards and terminate near the outer walls in each cell. As a rule, by the development of fibrous bands of thickening the endothecium becomes hygroscopic and is, therefore, mainly responsible for the dehiscene of the mature anther.

The cells of the endothecium are thin walled along the line of dehiscene of each anther lobe. The opening through which the pollen grains are discharged from the pollen sac is called *stomium*. On the maturity of the anther, a strain is exerted on the stomium due to the loss of water by the cells of endothecium, with the result the stomium ruptures and the anther dehisces. However, in *Solanum* and certain grasses, the dehiscence of the anthers takes place through the pores at the tip of the anthers.

Embryology of Angiosperms

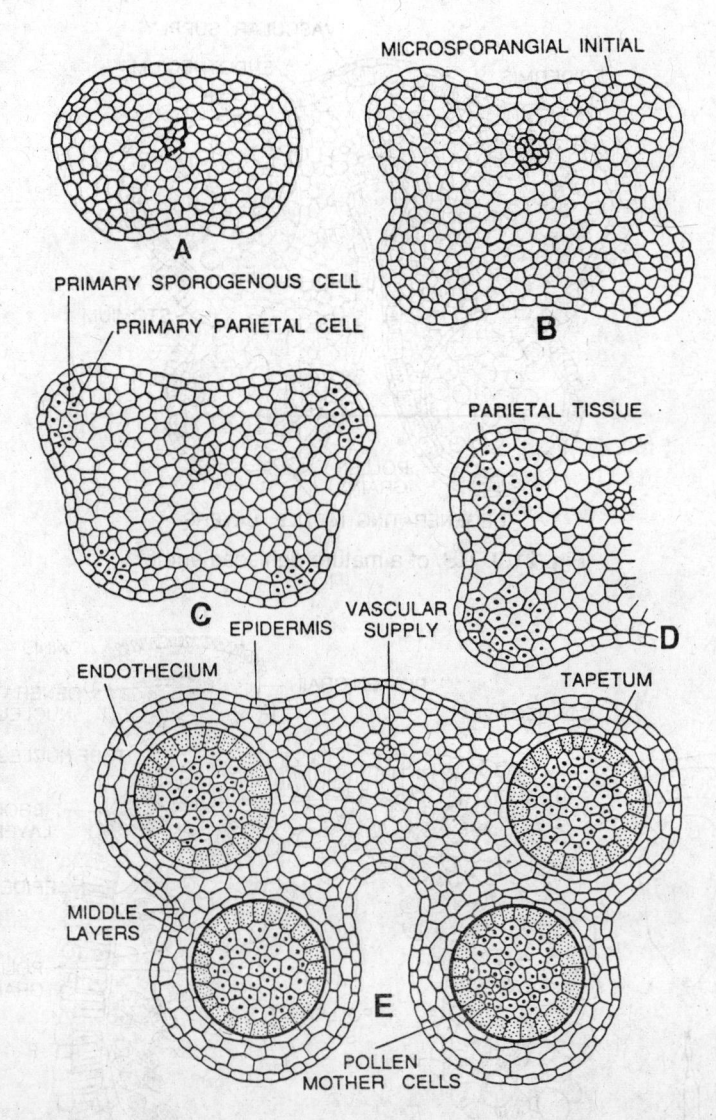

Fig. 21.7. Various stages of the development of microsporangium (pollen sac) upto the formation of pollen mother cells.

The innermost layer of the wall layers develops into a single layered *tapetum*. The cells of tapetal layer have dense cytoplasm and conspicuous nuclei. The tapetal nuclei may divide once or more and sometimes these divisions are accompanied by nuclear fusions resulting in large polyploid nuclei which may divide again. The tapetal layer is of great physiological significance since all the food material entering into the sporogenous tissue diffuses through this layer. Usually the tapetum is single layered but in some cases it is 2-3 layered. Ultimately the cells of the tapetal layer disorganise. Thus, tapetum makes a nutritive layer for the developing microspores.

SPOROGENOUS TISSUE

In the meantime the primary sporogenous layer cells give rise to the *microspore mother cells* or the *pollen mother cells*. The sporogenous cells in normal way divide several times mitotically before functioning as pollen mother cells. In certain cases the primary sporogenous cells may either show

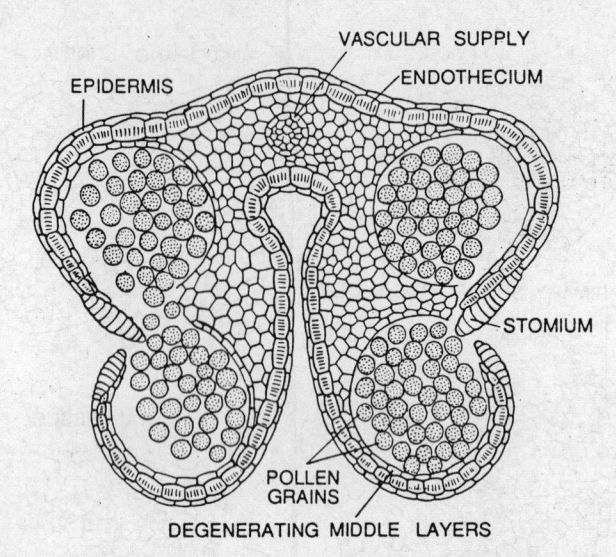

Fig. 21.8. T.S. of a mature dehisced anther.

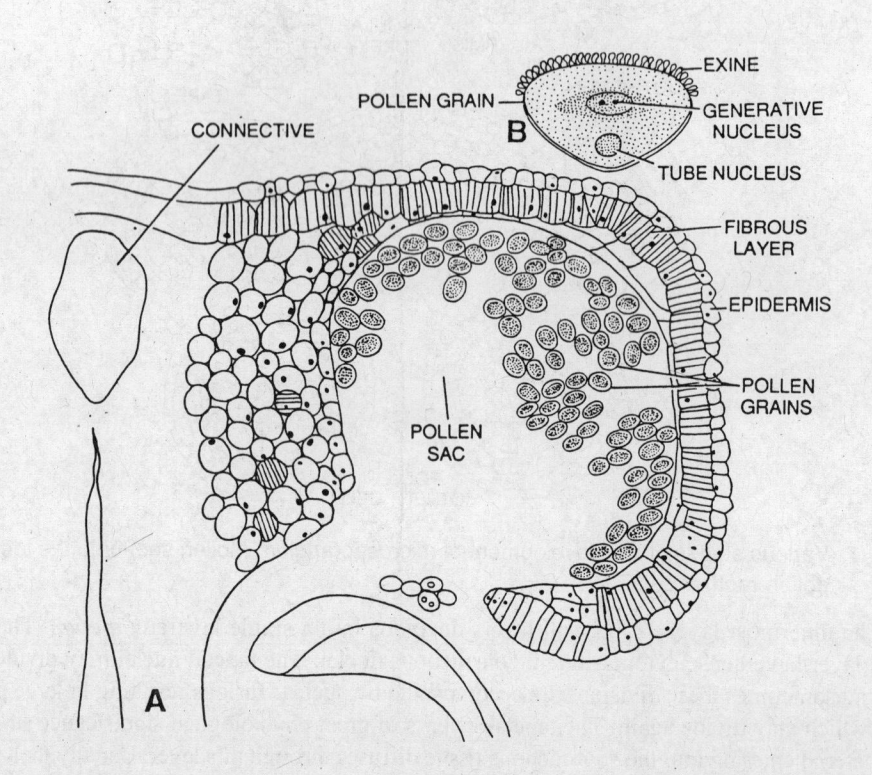

Fig. 21.9. The stamen. A, transverse section of dehiseed anther of *Lilium* showing pollen grains in pollen sac; B, enlarged grain of the same.

only few divisions or no divisions at all. Thus giving rise to a very small group of sporogenous cells or the primary sporogenous cells themselves function directly as the microspore mother cells. In the beginning of their formation the microspore mother cells or the pollen mother cells remain closely

Embryology of Angiosperms 407

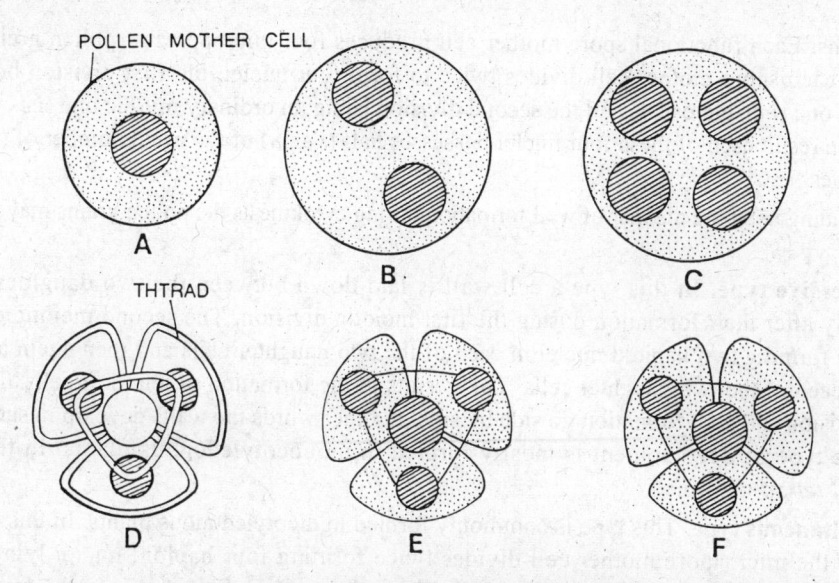

Fig. 21.10. A—F, stages showing development of pollen grains from pollen mother cell.

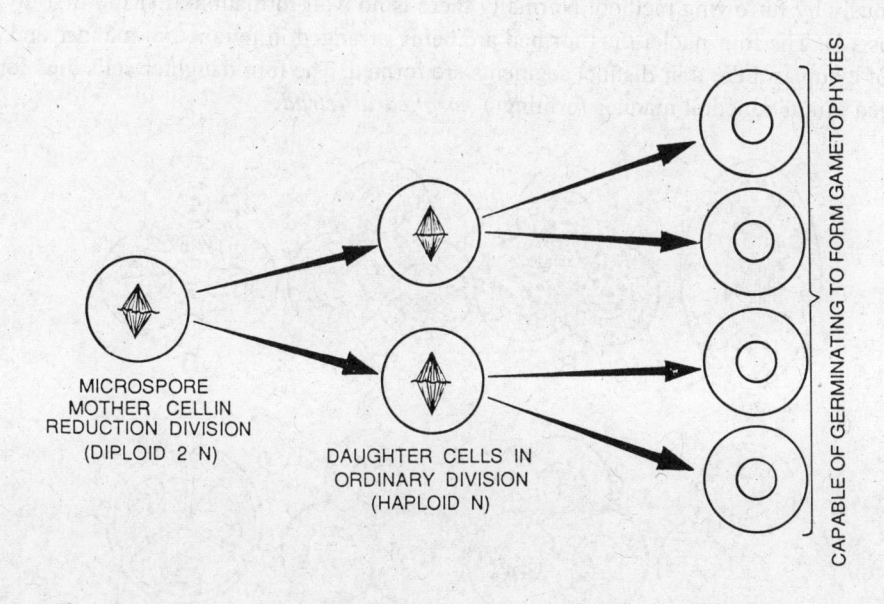

Fig. 21.11. Formation of microspores from microspore mother cells.

packed but as the anther enlarges in size, the pollen sac also increases in size, the microspore mother cells also enlarge in size, become spherical in shape and get loosely arranged. In many cases, some of the sporogenous cells are non-functional and serve as the food material for the functional microspore mother cells.

MICROSPOROGENESIS

The microspore mother cells (pollen mother cells) which are at first polygonal and closely packed gradually become rounded and loosely arranged in the rapidly enlarging microsporangium (pollen sac). Although all the mother cells in an anther are capable of giving rise to pollen grains but some of them may degenerate and serve as food material for the remaining cells which give rise to pollen grains. Each functional spore mother cell produces four micropores or pollen grains. The nucleus of each spore mother cell divides twice to form four nuclei, the first division being the reductional one (*i.e.*, meiosis I) and the second division being an ordinary mitotic one (*i.e.*, meiosis II). With the result each of these four nuclei possesses half (*i.e.*, n) of the usual number of (*i.e.*, 2n) chromosomes.

Depending upon the manner of wall formation during cytokinesis the pollen grains may develop as follows:

Successive type. In this type a cell wall is laid down between the two daughter nuclei immediately after their formation during the first meiotic division. The second meiotic division takes place forming two haploid nuclei in each of the two daughter cells and then again a wall is formed in each of the two daughter cells. This results in the formation of tetrad of cells. Thus the nuclear division and wall formation go side by side. In other words the walls develop in successive steps. This type of development is mostly formed in monocotyledonous plants. In this case *isobilateral tetrad* is formed.

Simultaneous type. This type is commonly formed in dicotyledonous plants. In this case the nucleus of the microspore mother cell divides twice forming four haploid nuclei lying in the common mass of cytoplasm of the mother cell. The walls are laid down between the four nuclei simultaneously by furrowing method. Normally there is no wall formation after the first division (*i.e.*, meiosis I). The four nuclei thus formed are being arranged in tetrahedral manner and after cleavage of cytoplasm the four distinct segments are formed. The four daughter cells thus formed are arranged in a tetrahedral manner forming a *tetrahedral tetrad*.

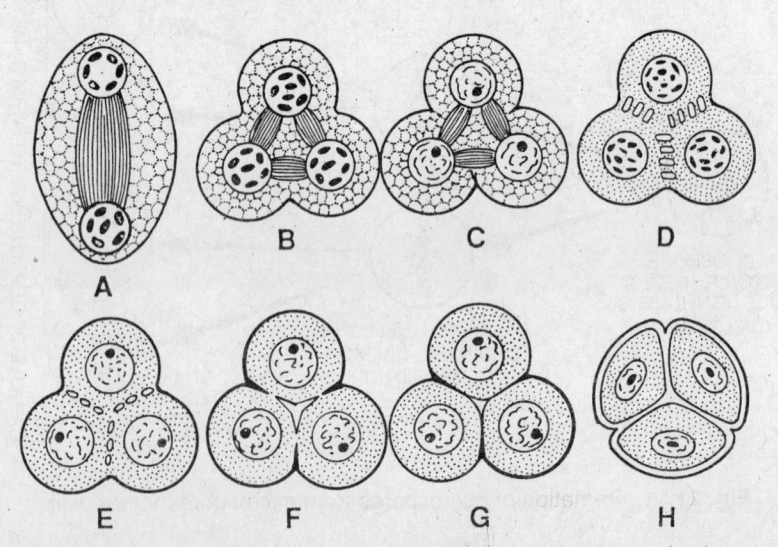

Fig. 21.12. Development of microspores in *Melilotus alba* showing simultaneous type of cytokinesis.

Other types. Usually the young pollen grains are arranged in an isobilateral or tetrahedral manner as already mentioned. In still other cases the spore mother cell may divide transversely giving rise to a *linear tetrad*. *T-shaped tetrads* have been reported from *Butomopsis* and *decussate* tetrads from *Magnolia*.

In *Typha, Juncus, Drosera* and few other cases, the four cells formed in a group do not separate, but remain more or less coherent. In *Calotropis* and orchids the pollen cells of each pollen sac instead of reporting into loose pollen grains are united into a mass known as the *pollinium*.

Embryology of Angiosperms

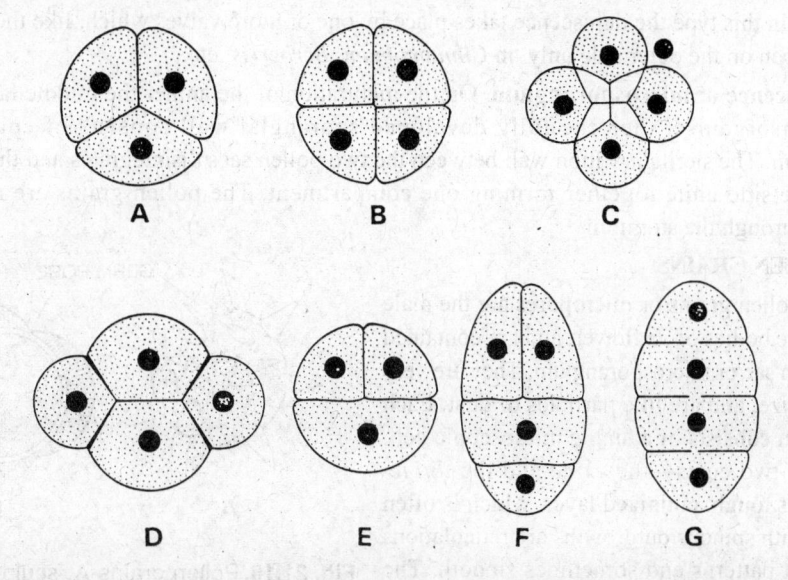

Fig. 21.13. Arrangement of angiospermic micropores in a tetrad. A, tetrahedral; B, isobilateral; C, cross; D, rhomboidal; E, decussate; F, T-shaped; G, linear.

Dehiscence of the anther. When the pollen grains mature they exert some pressure from within on the wall of the anther, with the result the anther bursts and sets free the pollen grains. The dehiscence of anther takes place in four different ways: (A) *Longitudinal*—In this type the anther lobes burst in a longitudinal direction, *e.g.*, in *Datura*, *Annona squamosa*, *Helianthus annuus*, *Hibiscus*, *Gossypium*, etc. (B) *Transverse*—In this type the anther lobes dehisce breadthwise, as in *Ocimum sanctum* (C) *Porous*—In this type, the dehiscence is effected by one or more apical pores, as in species of *Solanum*, (D)

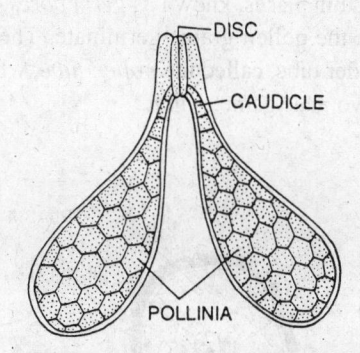

Fig. 21.14. Pollinia of *Calotropis*.

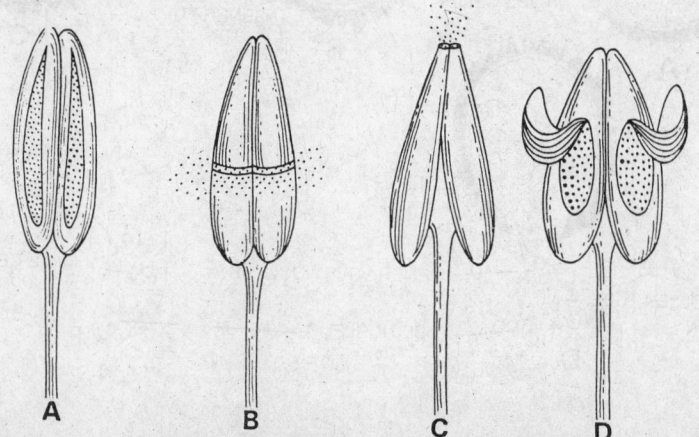

Fig. 21.15. Different types of dehiscence of anther, A, longitudinal slits; B, transverse slits; C, terminal pores; D, hinged valves.

Valvular—In this type the dehiscence takes place by one or more valves which, like the shutter of a window, open on the outer side only, in *Cinnamomum, Berberris,* etc.

Dehiscence of microsporangium. On the maturation of the anther the middle layers and the tapetum disorganise, then the fully developed sporangial wall consists of epidermis and endothecium. The sterile partition wall between the two pollen sacs disintegrates and the two pollen sacs of one side unite together forming one compartment. The pollen grains are released out generally through the stomium.

THE POLLEN GRAINS

The pollen grains or micropores are the male reproductive bodies of a flower, and are contained in the pollen sac or microsporangium. They are very minute in size, and are like particles of dust. Each pollen grain consists of a single microscopic cell, possessing two coats—the *exine* and the *intine*. The exine is tough, cutinized layer, which is often provided with spinous outgrowths are reticulations of different patterns and sometimes smooth. The intine, is a thin, delicate, cellulose layer laying internal to the exine. The exine possesses one or more thin places, known as *germ pores*. The pollen tubes make their way through these germ pores, when the pollen grains germinate. The pollen grains germinate on the stigma, and each forms a selender tube, called the *pollen tube*, which elongates through the tissue of the gynoecium carrying the two male gametes in it.

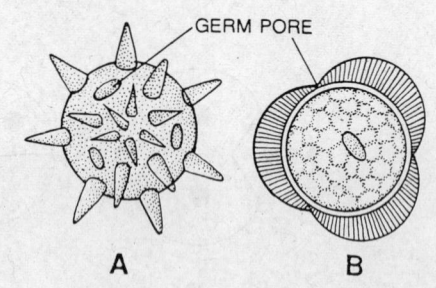

Fig. 21.16. Pollen grains-A, sculptured pollen grain; B, sectional view.

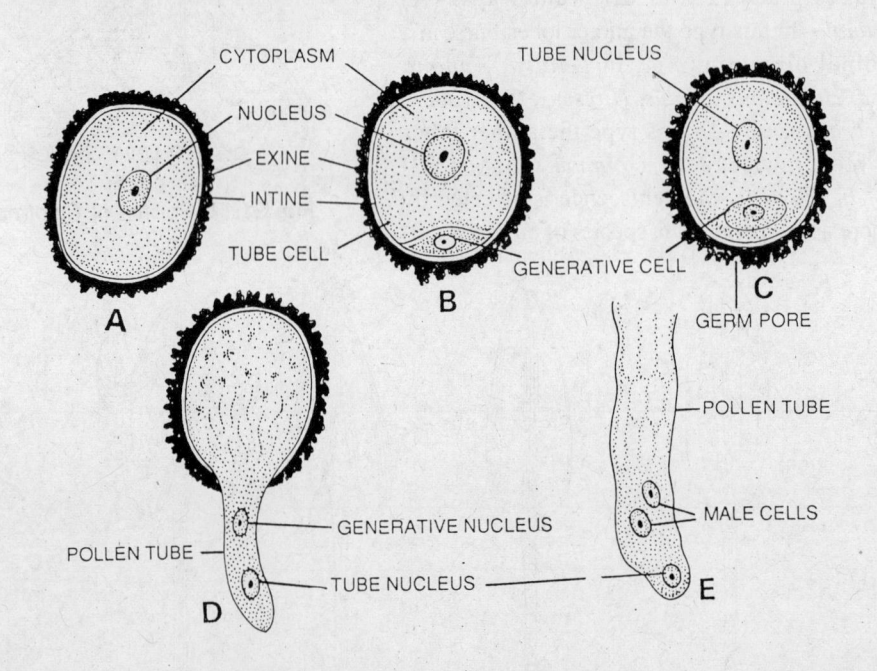

Fig. 21.17. Pollen grain and its germination. A, pollen grain: B—E different stages in the development of male gametophyte.

Embryology of Angiosperms

CARPEL OR MEGASPOROPHYLL

Gynoecium or *pistil* is the female reproductive whorl of the flower. It is composed of one or more *carpels*. The carpels are modified leaves which bear the ovules, and are also called the *megasporophylls*. When the pistil consists of only one carpel (*e.g.*, in pea flower), it is known as *simple*, and when it consists of two or more carpels, the pistil is said to be *compound*. The compound pistil may be *apocarpous* (free carpels) or *syncarpous* (united carpels). Each carpel consists of *three* parts—*stigma, style* and *ovary*. The *stigma* is the terminal end of the style upon which the pollen grains fall, and is generally knob-like and sticky. The *style* is the slender projection of the ovary and bears stigma at its terminal end. The surface of the style may be smooth or covered with hairs. In many cases these hairs collect the pollen grains. The swollen basal part of the pistil which is single or many chambered, is termed *ovary*. The ovary contains one or more roundish or oval, egg-like bodies, known as the *ovules*. Each ovule contains a large oval cell known as the *embryo-sac*. The ovary-gives rise to the fruit and ovules to the seeds.

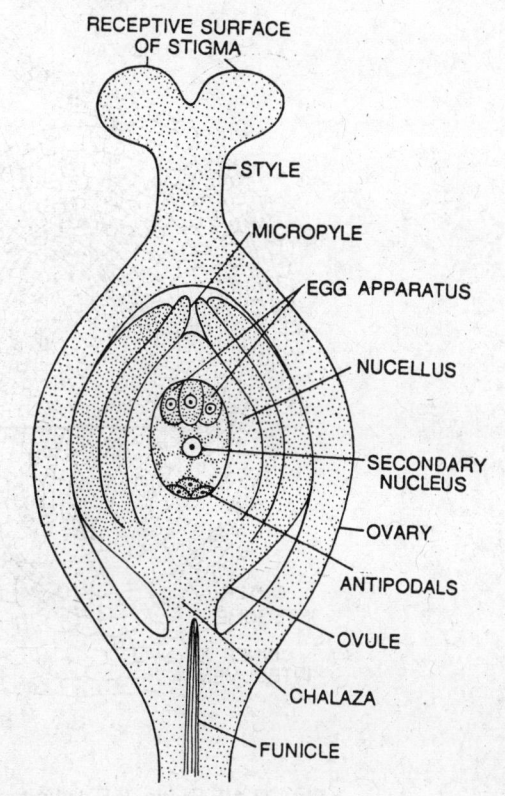

Fig. 21.18. Pistil in longitudinal section showing different parts of mature ovule attached to a basal placenta.

THE MEGASPORANGIUM OR OVULE

An ovule or megasporangium develops from the base or the inner surface of the ovary. It is a small generally oval structure and consists chiefly of a central body of tissue, the *nucellus* and one or two *integuments*. Each ovule is attached on the placenta by a small stalk called the *funiculus*. The place of attachment of the stalk with main body of the ovule is called the *hilum*. In an inverted ovule, the funicle fuses with the main body of the ovule, forming a sort of ridge, known as the *raphe*. The upper end of the raphe which is the junction of the integuments and the nucellus is called the *chalaza*. The nucellus makes the main body of the ovule, which is made up of parenchyma tissue. Nucellus is the megasporangium proper, and it is surrounded by two coats the *integuments*. In Compositae and few other families of Gamopetalae there is only one integnment. In *Santalum* and *Dendrophthoe* there are no integuments. A small opening is left at the apex of the integuments; this is called the *micropyle*. When there are two integuments then the inner integument is formed first and followed by the formation of the outer integument. A large oval cell lying embedded in the nucellus towards the micropylar end is the *embryo sac*. This makes the most important part of the mature ovule. It is the embryo sac, which bears the embryo later on.

Development of ovule. The ovule primordium appears as a small protuberance on the surface of the placenta. It grows rapidly and develops into a prominent conical structure with rounded tip. This

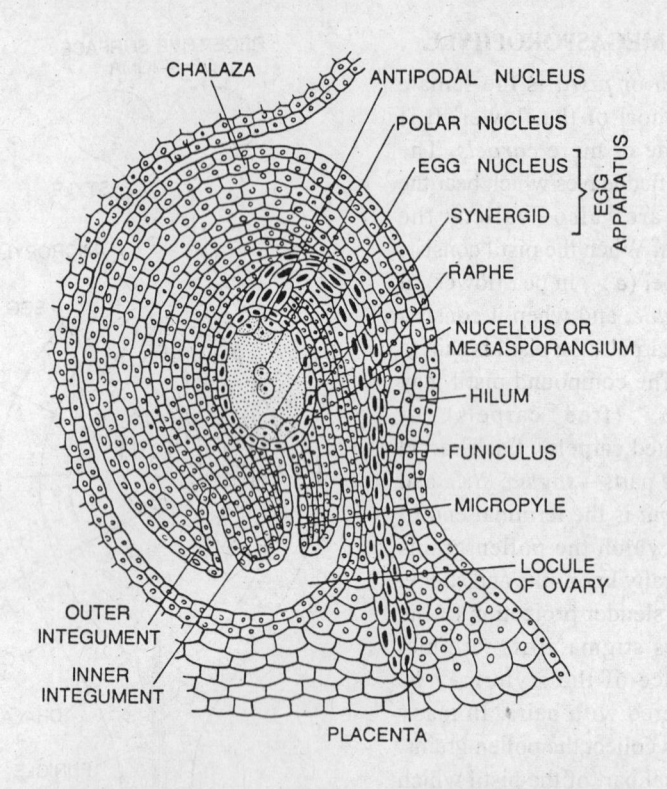

Fig. 21.19. Ovule, L.S., of a typical ovule of angiosperm.

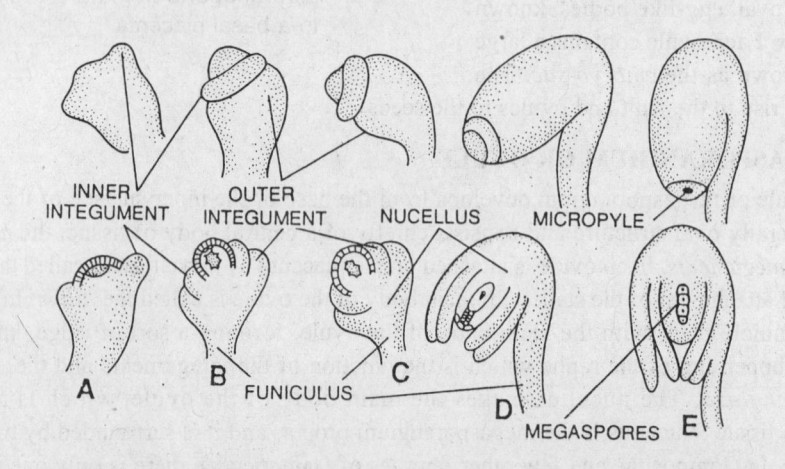

Fig. 21.20. Various stages of the development of anatropous ovule (A-E).

rounded tip is the fore runner of the nucellus. It gradually develops into a projecting mass of tissue by the growth and division of its cells. As the development continues either one or two layers of the tissue develop from the base of the nucellus forming the integuments. The inner integument develops first and thereafter the outer one develops. The growth of the integuments is much faster than the nucellus and thus they completely enclose the nucellus except for a narrow opening, the *micropyle*.

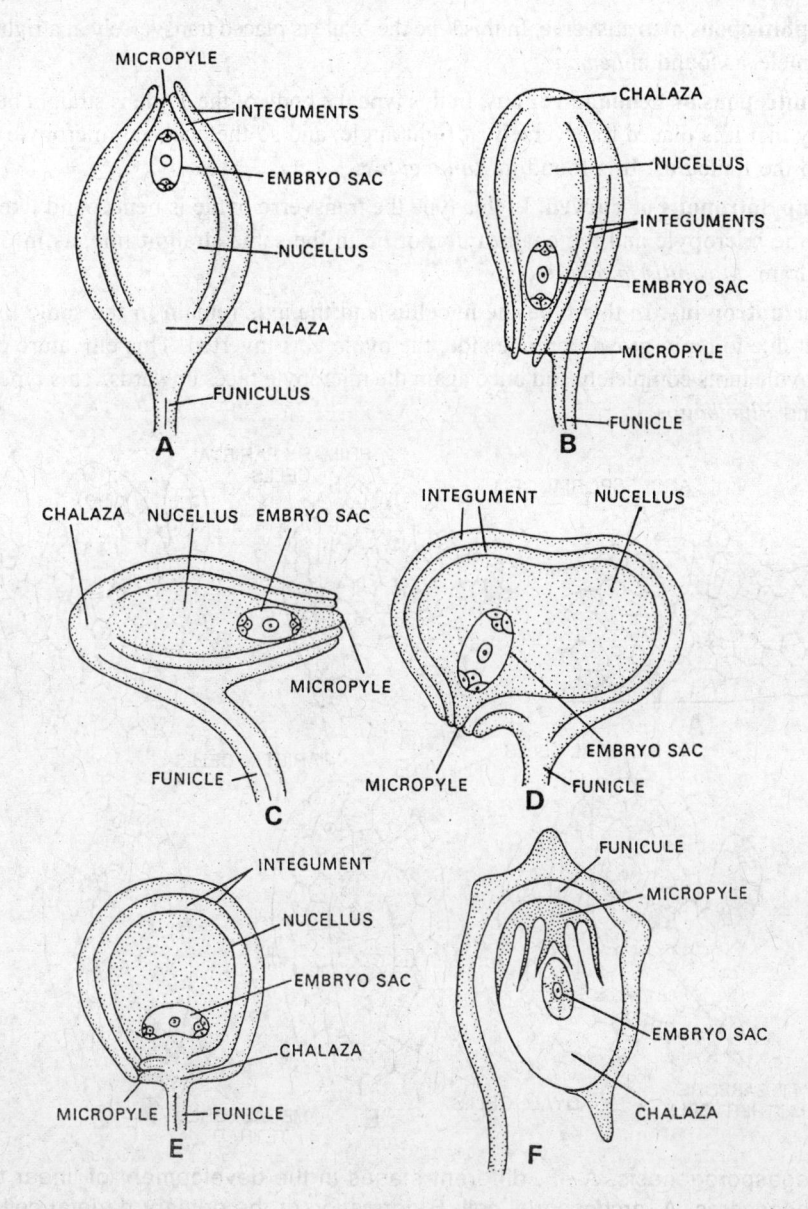

Fig. 21.21. Different forms of the ovule in longitudinal section. A, orthotropous; B, anatropous; C, hemianatropous; D, campylotropous; E, amphitropous; F, circinotropous.

FORMS OF OVULES

On the basis of the relative position of the micropyle, chalaza and funiculus the mature ovules can be categorizsed as follows:

(A) **Orthotropous or straight.** In this type the ovule is erect or straight so that the funicle, chalaza and micropyle lie in one and the same vertical line, as in members of Polygonaceae (*e.g., Polygonum, Rumex*, etc.) and Piperaceae (*e.g., Piper nigrum, Piper betle*).

(B) **Anatropous or inverted.** In this type the ovule bends along the funicle so that the micropyle lies close to the hilum. The chalaza lies at the other end. This is the commonest type of ovule found both in dicots and monocots.

(C) **Amphitropous or transverse.** In this type the ovule is placed transversely at a right angle to its stalk or funicle, as found in *Lemna*.

(D) **Hemitropous or hemianatropous.** In this type the body of the ovule is straight but twisted in such a way that it is placed transversely at right angle, and so the chalazal micropyle line is at right angle to the funiculus. It is found in *Ranunculus*.

(F) **Campylotropous or curved.** In this type the transverse ovule is bent round like a horse-shoe so that the micropyle and the chalaza do not lie in the same straight line, as in *Capparis*, Cruciferae, Gram, *Mirabilis jalapa*, etc.

(G) **Circinotropous.** In this type the nucellus and the axis remain in the same line in the beginning but due to rapid growth on one side, the ovule gets inverted. This curvature continues and thus the ovule turns completely and once again the micropyle faces upwards. This type is found in *Opuntia* and *Plumbago*..

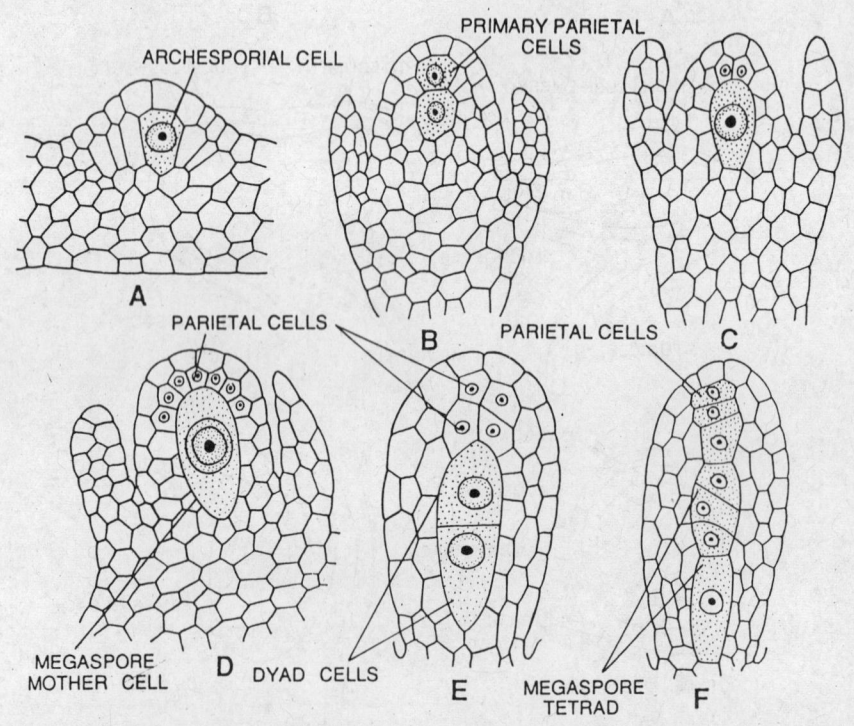

Fig. 21.22. Megasporogenesis. A—F, different stages in the development of linear tetrad of megaspores. A, archesporial cell; B, formation of the primary parietal cell and the megaspore mother cell; C—D, formation of parietal cells; E—D, formation of linear tetrad of megaspores from megaspore mother cell.

MEGASPOROGENESIS

Archesporium and megaspore formation. The archesporium is hypodermal in origin. At some early stage in the development of the ovule, usually at the time of the initiation of the integumentary primordia, a single hypodermal cell, known as *primary archesporial* cell, becomes differentiated at the apex of the nucellus beneath the epidermis. It can be distinguished from other neighbouring cells owing to its large size, conspicuous deeply staining nucleus, and dense cytoplasm. Usually this primary archesporial cell divides periclinally forming an outer *primary parietal cell* and an inner *primary sporogenous cell*. The primary parietal cell may divide further several times both by anticlinal

Embryology of Angiosperms

and periclinal divisions forming a variable amount of parietal tissue, or sometimes it remains undivided. The primary sporogenous cell usually does not divide further and functions directly as the *megaspore mother cell*. In still other cases (*e.g.*, in the families of Sympetalae) the primary archesporial initial directly behaves as megaspore mother cell.

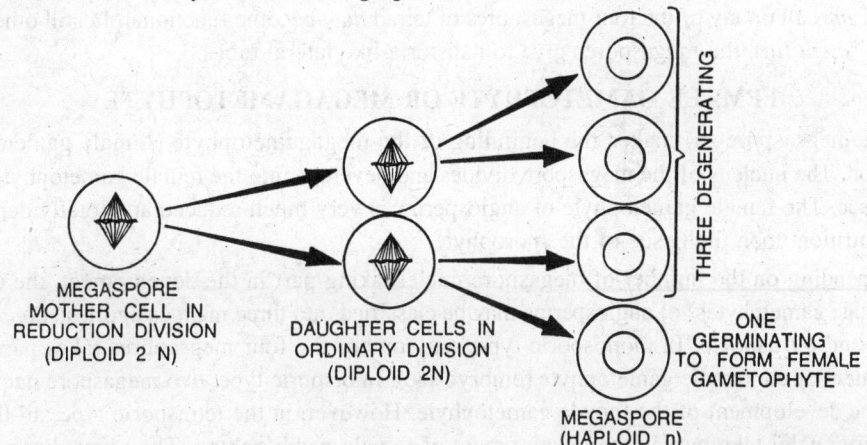

Fig. 21.23. Formation of Megaspores from a megaspore mother cell. only one megaspore is functional.

Usually the megaspore mother cell divides meiotically forming a tetrad of four megaspores. This usual process of meiotic division is termed *megasporogenesis*. Here the first division (*i.e.*, meiosis I) is always transverse and gives rise to two cells. The second division is also transverse (*i.e.*, meiosis II), and thus in total four cells are being formed. The four megaspores thus formed lie in an axial row within the nucellus forming a *linear tetrad*. Sometimes the tetrad may be T-shaped, inverted T-shaped or decussate. Tetrahedral tetrads or isobilateral tetrads of megaspores are rarely as abnormal cases in angiosperms.

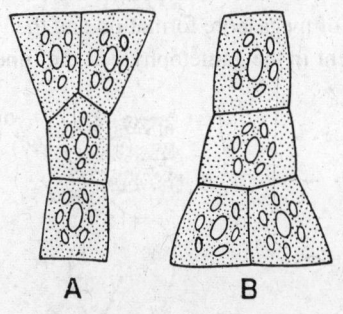

Fig. 21.24. Megaspores. A, arranged in T-shaped tetrad; B, arranged in T-shaped tetrad.

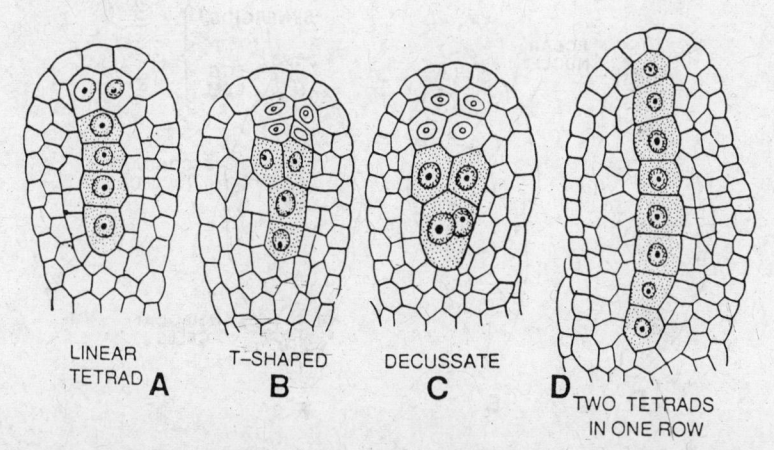

Fig. 21.25. Various types of megaspore tetrads. A, linear tetrad; B, T-shaped; C, decussate tetrad; D, two tetrads arranged in one row.

Of the four megaspores, so formed, each with half (n) the usual number (2n) of chromosomes, the three upper ones degenarate and appear as dark caps, while the lowest one functions and gives rise to the embryo sac. The developing megaspore encroaches upon and absorbs the outer three degenarating megaspores of tetrad and the neighbouring cells of the nucellus. In certain plants (*e.g.*, in *Casuarina*) all or any of the four megaspores of tetrad may become functional. In still other cases (*e.g.*, in *Potentilla*), the megaspores give to haustoria like lateral tubes.

FEMALE GAMETOPHYTE OR MEGAGAMETOPHYTE

The megsaspore (n) makes the beginning of the megagametophyte (female gametophyte) generation. The nucleus of the megaspore divides and develops into the female gametophyte or the embryo sac. The female gametophyte of angiosperms is very much reduced and totally dependent for its nutrition upon the tissue of the sporophyte.

Depending on the number of megaspore nuclei taking part in the development, the embryo sacs (female gametphytes) of angiosperms may be classified into three main categories; *monosporic*, *bisporic* and *tetrasporic*. In monosporic type, only one of the four megsapores takes part in the development of the female gametophyte (embryo sac). In bisporic type, two megaspore nuclei take part in the development of the female gametophyte. However, in the tetrasporic type, all the four megaspore nuclei take part in the development of female gametophyte. They have been further subdivided into *ten* types on the basis of the number of nuclear divisions taking place between the time of megaspore formation and the type of differentiation of the egg, and the total number of nuclei present in the gametophyte at the time when such organization takes place.

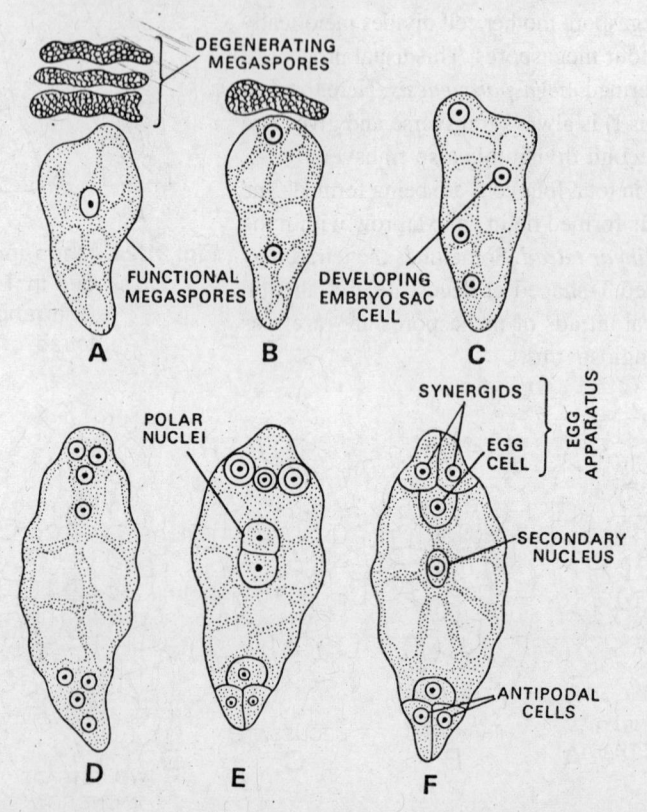

Fig. 21.26. Female gametophyte. A—F, development of the embryo sac (female gametophyte) of normal type (*Polygonum* type.)

Embryology of Angiosperms

Monosporic type. The female gametophytes or the embryo sacs of this type, may be 8 nucleate and 4 nucleate.

Polygonum or normal type. In the development of 8 nucleate embryo sac the nucleus of the functional megaspore divides to form two nuclei: the primary micropylar and the primary chalazal nuclei. These nuclei again divide so that the number is increased to four. Each of these nuclei divides again so that altogether eight nuclei are formed in the embryo-sac, four at each end. The female gametophyte (embryo sac) increases in size. Now one nucleus from each end or pole passes inwards, and the two polar nuclei fuse together somewhere in the middle of the embryo sac, forming the *secondary nucleus* (2n). The remaining three nuclei at the micropylar end, each surrounded by a very thin wall, form the *egg appartus*. The other three nuclei at the opposite or chalazal end, lying in a group, often surrounded by very thin walls; from the *antipodal cells*.

This type of embryo sac is the most common or generally known as the *normal type*. It has also been called *Polygonum type*, because for the first time in 1879, this type was reported in *Polygonum divaricatum* by Strasburger.

Oenothera type. Another variation of monosporic type of embryo sac is known as the *Oenothera type*, and has been reported only in the family Onagraceae. In this type the megaspore nucleus divides twice and thus produces four nuclei at the micropylar end. These nuclei give rise to a normal egg appartus and a single polar nucleus. The second polar nucleus and the antipodal nuclei are absent.

Bisporic type. The bisporic embryo sacs are typically 8-nucleate. They are also known as *Allium type* embryo sacs. Such embryo sacs arise from one of the two dyad cells formed after Meiosis I. Since there is no wall formation at the end of Meiosis II and both the megaspore nuclei formed in the functional dyad cell take part in the development of the embryo sac, only two further divisions of these nuclei give rise to 8-nucleate stage.

Endymion type. This type of embryo sac was reported for the first time in *Endymion hispanicus* (Battaglia, 1958). In this type, after the formation of the dyad, the lower cell may distintegrate or its nucleus may divide once or twice producing 4-nuclei. The 8-nucleate embryo sac is formed by the nuclear divisions of the upper cell of the dyad.

Tetrasporic type. In this type of embryo sac there are several variations. In several cases 16 nuclei are formed with the result of two divisions after megasporogensis. These are further subdivided into the following types:

Peperomia type. In this sub type each of the 4 megaspore nuclei divides twice, forming 16 nuclei which are uniformly distributed at the periphery of the embryo sac. Two nuclei at the micropylar end form an egg and a synergid; 8 are used to form the secondary nucleus, and the remaining 6 nuclei are cut off at the periphery of the embryo sac.

Penaea type. In this sub-type 16 nuclei lie in four distinct quarters which are arranged crosswise, once at each end of the embryo sac and two at the sides. Now three nuclei of each quarter become cut off as cells, while the fourth one remains free and moves towards the centre. Thus there are four *triads* and four polar nuclei. Here, the egg cell of the micropylar triad alone is functional. Such embryo sacs have been found in many members of Malpighiaceae.

Drusa type. In this type, a sixteen-nucleate embryo sac was recorded (Hakansson, 1923) in *Drusa oppositifolia* (family Umbelliferae). Here when the meiotic divisions are over, three of the megaspore nuclei pass down to the basal end of the embryo sac, and only one remains at the micropylar end. This is followed by two successive divisions, forming four micropylar nuclei and twelve antipodals. The four micropylar nuclei give rise to the egg appartus and upper polar nucleus, and the twelve chalazal nuclei to a lower polar nucleus and eleven antipodal cells. This type has been recorded in *Mallotus, Mainthemum, Crucianella, Rubia, Ulmus* and a few other plants.

Crysanthemum cinerariafoluim type. Martinoli (1939) has described a peculiar mode of development in this plant. In this type the four megaspore nuclei show 1 + 2 + 1 arrangement, *i.e.*, one nucleus lies at each pole and two in the centre. The two nuclei of the centre remain quite close to each other but do not fuse together. Now the megaspore nucleus of micropylar end divides twice forming four nuclei, but there is no regularity in the division of the nucleus of chalazal end. Thus, the embryo sac may have six, nine or ten nuclei.

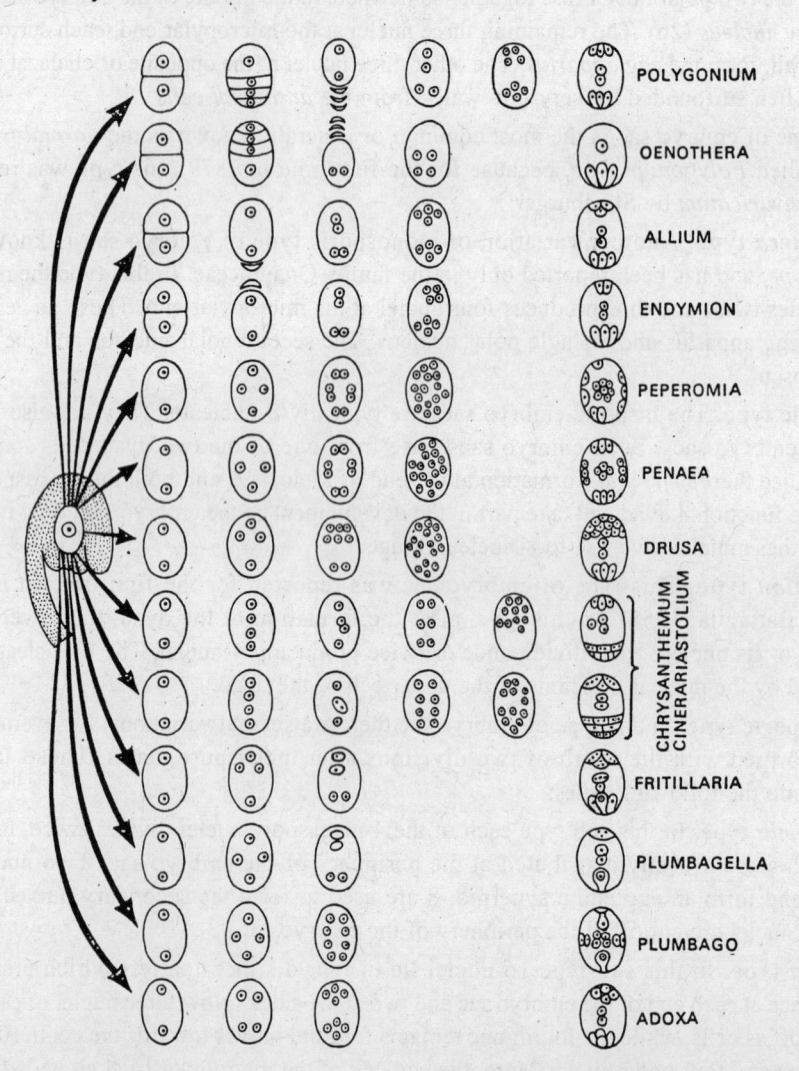

Fig. 21.27. Female gametophyte. Development of different types of embryo sacs in angiosperms.

Sometimes the two central nuclei out of the four megaspore nuclei are fused together forming a single diploid (2n) nucleus; the next division produces six nuclei, one haploid (n) pair at micropylar end; one haploid pair at chalazal end and one diploid pair in the centre. The next division of these six nuclei produces three groups of four nuclei each. At the micropylar end the three nuclei make the egg apparatus, and one migrates in the centre forming upper haploid polar nucleus. The four haploid

Embryology of Angiosperms

antipodal cells are formed at chalazal end. One of the diploid nuclei of the central quarter behaves as the lower polar nucleus and the remaining three organize them as additional antipodal cells. Thus, there are twelve nuclei in the embryo sac. Sometimes less than twelve nuclei (*i.e.*, ten or seven) are developed, because of the failure of certain divisions.

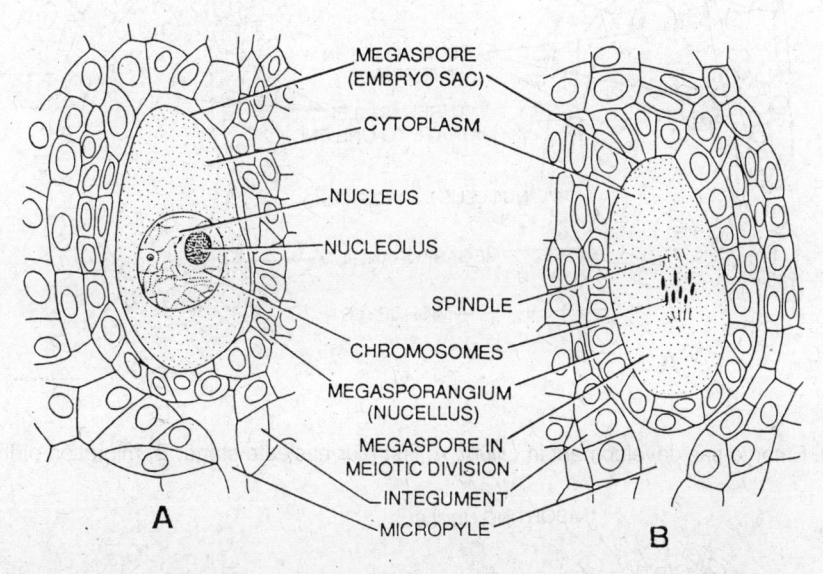

Fig. 21.28. Embryo sac development in *Lilium*, uninucleate stage, B, first division.

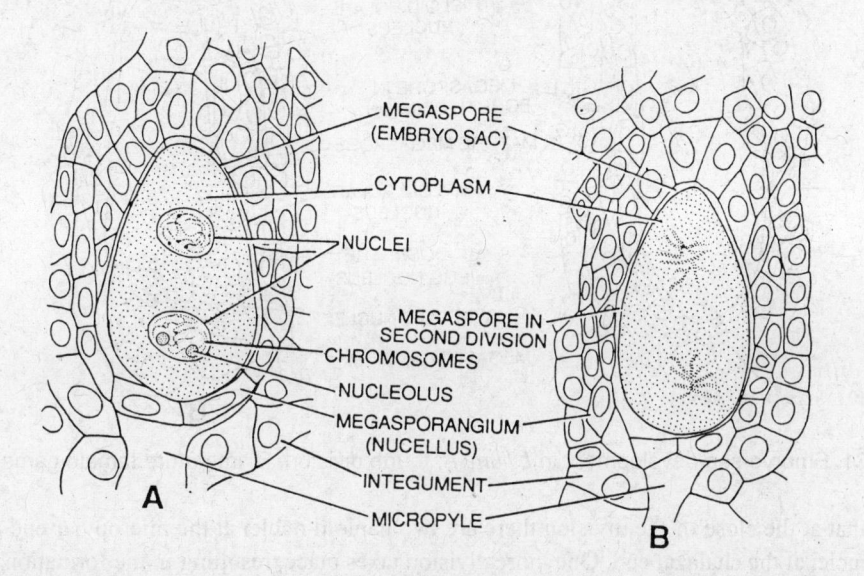

Fig. 21.29. Embryo-sac development in *Lilium*. A, first two-nucleate stage; B, second division.

Fritillaria type. This type of development of embryo sac has been reported in many angiosperms including *Lilium*. In *Fritillaria* and *Lilium* the behaviour of the four megaspore nuclei is peculiar. Here three of the megaspore nuclei go to the chalazal end and the fourth one goes to the micropylar end. The three chalazal nuclei come to lie very close to each other. During the next stage the micropylar nucleus divides normally but the three chalazal nuclei fuse together forming a triploid (3n) nucleus, which then divides mitotically forming two triploid nuclei at the chalazal

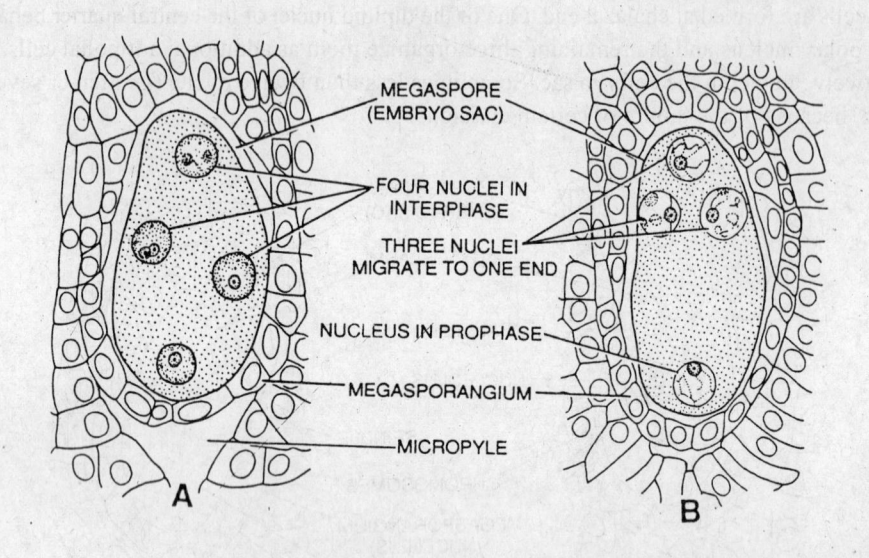

Fig. 21.30. Embryo-sac development in *Lilium*, A, first four-nucleate stage; B, migration of three nuclei.

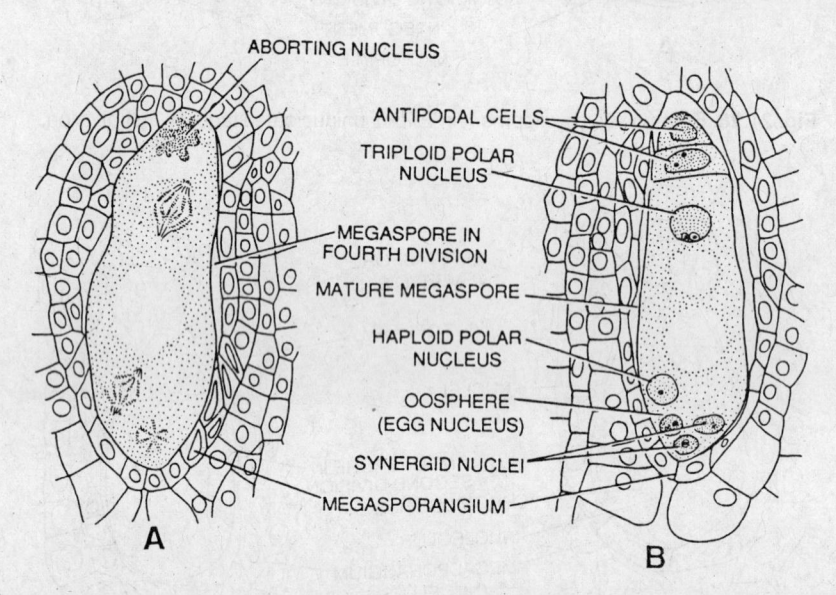

Fig. 21.31. Embryo-sac development in *Lilium*. A, fourth division, B, immature female gametophyte.

end; so that at the close of the division there are two haploid nuclei at the micropylar end and two tripoid nuclei at the chalazal end. One more division takes place, resulting in the formation of eight nuclei, of which the four chalazal nuclei are triploid and the four micropylar nuclei are haploid. Thus, a mature embryo sac consists of haploid cells of egg apparatus, three triploid antipodal cells and a tetraploid (4n) secondary nucleus formed by the fusion of two polar nuclei, one haploid and the other triploid. The endosperm nucleus, when formed after fertilization will be pentaploid (5n) in such type of embryo sac.

***Plumbagella* type.** This type of development of embryo sac has been reported only in *Plumbagella micrantha* (Fagerlind, 1938; Boyes, 1939). In this type the 4 megaspore nuclei take

Embryology of Angiosperms

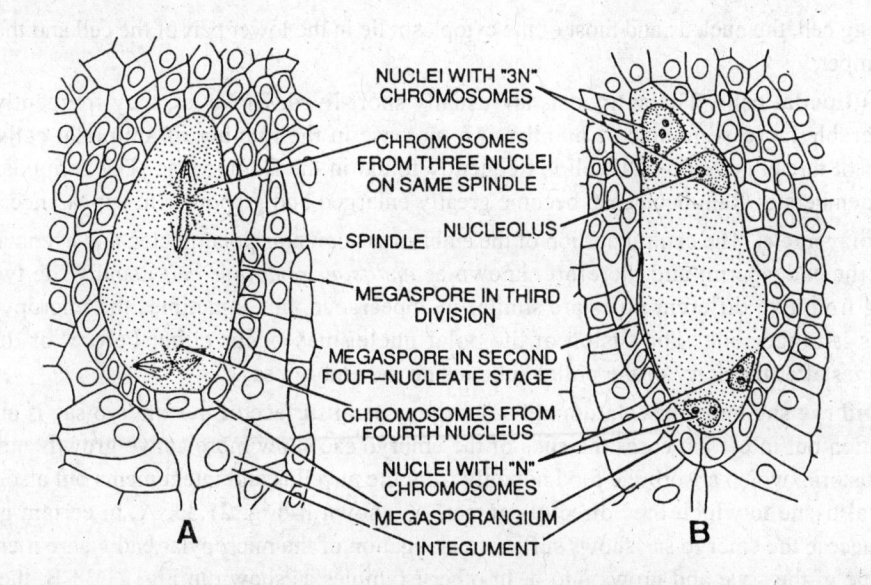

Fig. 21.32. Embryo-sac development in *Lilium*. A, third division; B, second four nucleate stage.

up a 1 + 3 arrangement, *i.e.*, one nucleus at the micropylar end and three nuclei at the chalazal end. The three nuclei of the chalazal end fuse together forming a single triploid nucleus. Now the developing embryo sac is in binucleate stage. Both the nuclei divide once mitotically forming a second four nucleate stage, two micropylar haploid nuclei and two chalazal triploid nuclei. There are no further divisions. The nucleus situated near the micropylar end organizes into the egg; the tripolid nucleus of the chalazal end forms the single antipodal cell, and the remaining two nuclei, one halpoid and the other triploid, fuse to form a tetraploid secondary nucleus in the centre.

Plumbago **type.** This type of development of embryo sac has been reported in *Plumbago capensis* (Haupt, 1934). Here the four megaspore nuclei arrange crosswise and divide once to form eight nuclei in four pairs. One nucleus of the micropylar pair is now cut off to form the lenticular egg cell, and one nucleus from each of the four pairs approach each other in the centre and fuse to form a tetraploid secondary nucleus. The remaining three nuclei degenerate.

Adoxa **type.** This type of development of embryo sac was described for the first time by Jonsson (1879) in *Adoxa moschatellina*. In this type the four megaspore nuclei divide to form eight nuclei which form a normal type of 8-nucleate embryo sac, *i.e.*, a normal egg appartus, three antipodal cells and two polar nuclei. This type of embryo sac is regular feature in *Adoxa* and *Sambucus*.

ORGANIZATION OF MATURE EMBRYO SAC

In majority of angiosperms, the eventual organization of embryo sac shows a uniform pattern, whereas the origin of the mature embryo sac may differ. The *Polygonum, Allium, Fritillaria* and *Adoxa* types have similar appearance at the time of fertilization (*i.e*, three-celled egg apparatus, three antipodal cells, and two polar nuclei). However, in few genera such as *Peperomia, Plumbago, Plumbagella*, etc. the basic plan of the embryo sac is different.

The egg apparatus. Typically the egg apparatus consists of an egg and two synergids. Usually the synergids are ephemeral structures which degenerate and disappear soon after fertilization or sometimes before it. As a rule, each of the synergids is notched and possesses a prominent hook. The nucleus lies in or just below the hook and the lower part of the cell contains a large vacuole.

In the egg cell, the nucleus and most of the cytoplasm lie in the lower part of the cell and the vacuole in the upper.

Antipodal cells. The antipodals are usually short-lived. However, they frequently show a considerable increase in size or number. An increase in the number of antipodal cells and the number of nuclei per antipodal cell is frequently found in the Compositae. The antipodal cells of some members of Ranunculaceae become greatly enlarged and glandular in appearance.

Polar nuclei. The central portion of the embryo sac contains polar nuclei, and eventually gives rise to the endosperm, and therefore, known as *endosperm mother cell* usually the two nuclei coming from two different poles are similar in appearance, but sometimes the micropylar polar nucleus is bigger one. The fusion of the polar nuclei may occur either before, or during, or sometimes after the entry of the pollen tube inside the embryo sac.

Embryo sac haustoria. In normal cases the general surface of the embryo sac is absorptive in function but in certain cases the ends of the embryo sac show more active growth and convert into haustoria, which absorb the food not only from the nucellus and integuments but also from the placental tissue to which they directly contact. As shown in Fig. 21.33, A, in certain genera of Loranthaceae the embryo sac shows sufficient elongation of the micropylar end where it enters into the tissue of the style and grows into it. In other examples as shown in Fig. 21.33 B, the embryo sac grows downward. Here the chalazal end of the embryo sac acts as haustorium and digests its way through the nucellus.

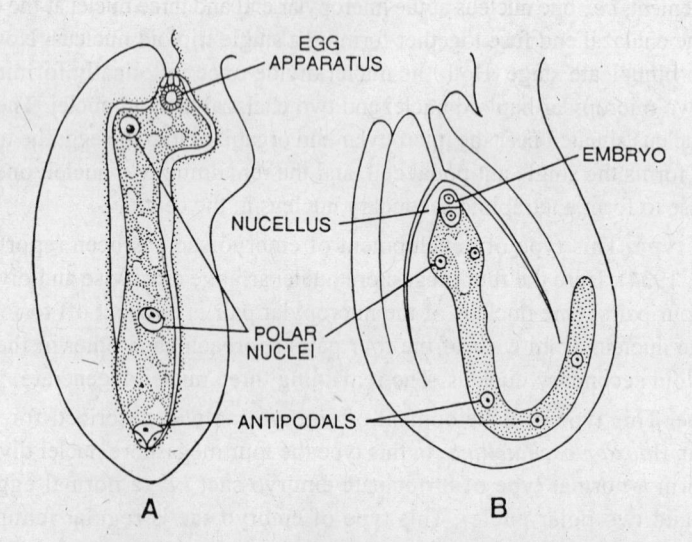

Fig. 21.33. Embryo sac haustoria. A, micropylar end of the embryo sac protrudes out of the ovule, breaking down completely the nucellus above it; B, the chalazal end of the embryo sac enlarges in size and digests its way through the nucellus and integument.

MALE GAMETOPHYTE OR MICROGAMETOPHYTE

The development of the male gametophyte (microgametophyte) is remarkably uniform in angiosperms. The microspore (pollen grain), which makes the first cell of the male gametophyte, undergoes only two divisions. With the result of first division two cells are formed: a large vegetative cell and a small generative cell. The second division is concerned with generative cell only. This division may take place either in the pollen grain or in the pollen tube, and gives rise to two male gametes.

Embryology of Angiosperms

Microspore and its germination. The microspore or the pollen grain (n) represents the beginning of the male gametophyte. The structure of microspore or pollen grain has already been given under the head entitled *Stamen* or *Microsporophyll*. The newly formed microspore possesses a very dense cytoplasm with a central nucleus, but as the cell increases in size, a vacuole develops and the nucleus shifts from centre to a place adjacent to the wall.

The germination of microspore starts while it is still within the microporangium or pollen sac.

Formation of vegetative and generative cells. The first division of the microspore gives rise to the *vegetative* and *generative* cells. The first formed walled and peripheral cell is the *generative cell*, while the larger, naked, central cell, which fills the remainder of the spore-wall cavity, is the *vegetative* or *tube cell*. The nuclei of generative and vegetative cells differ in size, structure and in staining qualities. The nucleus of vegetative cell possesses a prominent nucleolus, while the nucleus of generative cell contains a small nucleolus. The cytoplasm of the generative cells is hyaline and is almost without RNA, whereas that of vegetative cell is rich in RNA. The DNA contents of both the nuclei are same in the beginning but later on they increase in the generative nucleus. There is considerable variation in the form of the generative cell. Usually it is elliptical, or lenticular or spindle-shaped. The starch and fat are the most conspicuous substances found in the vegetative cell. Certain proteinaceous bodies have also been reported in microspores.

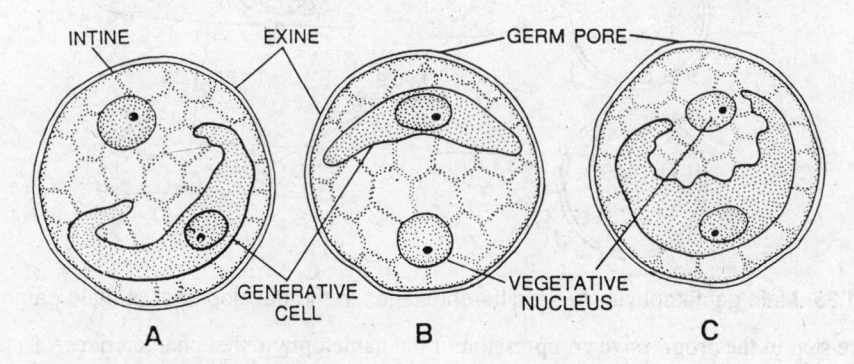

Fig. 21.34. Male gametophyte. Pollen grains showing vegetative nucleus and generative cells. In A and B, the generative cells are much elongated, whereas in C, the generative cell is of amoeboid **nature**.

Eventually the generative cell loses contant with the microspore wall, and is being shifted into the vegetative cell, where it may lie in any part of it. It then becomes oval or spindle-shaped. Thus the microspore becomes two-celled. Generally the microspores (pollen grains) are being shed from the microsporangium (pollen sac) in two-celled stage for pollination. It is only in rare cases that the pollen grain (microspore) becomes three-celled before its liberation from the pollen sac. It seems somewhat difficult to make clear cut demarcation between two-celled and three-celled pollen grains, because the generative cell may be in prophase or metaphase stage at the time the pollen grains are liberated from the pollen sac, and the process of division is continued in the pollen tube. In some cases (*e.g., Viola, Dionaea, Nicotiana,* etc.) both two and three celled pollen grains have been reported in the same plant. However, the division of the generative cell, whether it takes place in the pollen grain or in the pollen tube, occurs in a regular way.

It seems quite clear that in most of angiosperms the pollen grain has two nuclei at the time of shedding from the anther. One of these is the generative nucleus, which later on divides and thus two sperm nuclei are formed. In several plants, the generative nucleus divides before the pollen is shed and thus the pollen grains are trinucleate. The early division of the generative nucleus represents

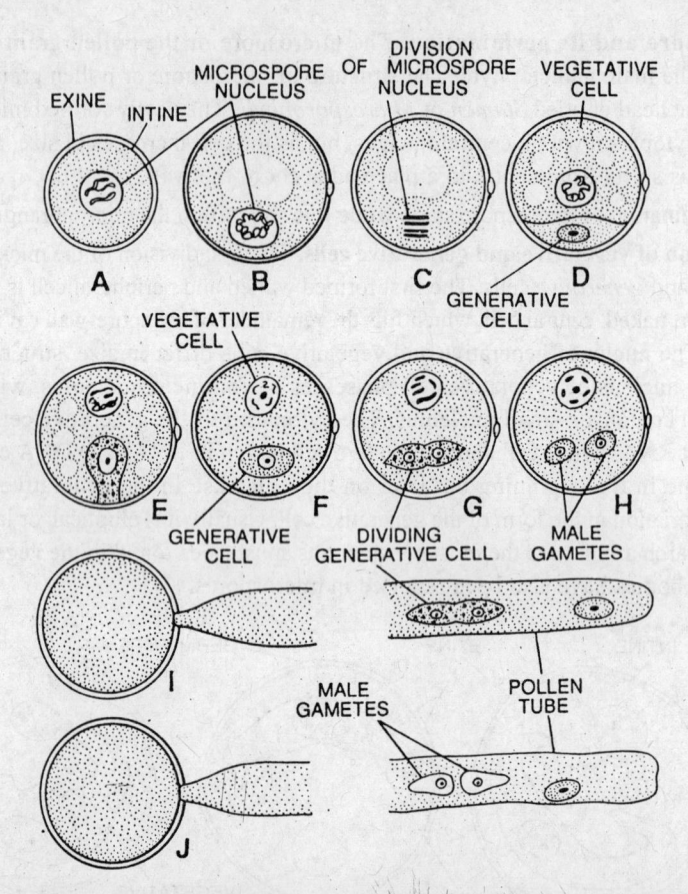

Fig. 21.35. Male gametophyte. A—J, different stages in the development of male gametophyte.

one more step in the progressive compression of the gametophyte that characterizes vascular plants in general. In considering binucleate and trinucleate pollen types, it should be noted that both types produce mature male gametpohytes with three nuclei; they may be differentiated at the time of the division of generative nucleus.

Male cells or male nuclei. It was believed formely that whenever the generative cell divides in the pollen grain, the sperm cells are formed, but if the division takes place in the streaming cytoplasm of the pollen tube only *nuclei* are formed. However, Schnarf (1941) reported, that in all cases the male gametes are definite cells and the cytoplasmic sheath persists throughout their course in the pollen tube. In *Lilium* and other plants, the cytoplasmic sheath around the male nuclei has been followed up to the time of their discharge in the embryo sac.

Vegetative nucleus or tube nucleus. According to earlier workers the vegetative nucleus or tube nucleus played an important role in directing the growth of the pollen tube. It is not always found in the distal end of the pollen tube but frequently lies considerably behind the male gametes. However, in *Ulmus, Senecio, Crepis* and *Secale* it degenerates even before the germination of the pollen grain, and does not enter the tube at all, and the tube continues to function normally. Poddubnaja—Arnoldi (1936) regards the vegetative nucleus as a vestigeal structure without any important function in the growth of the pollen tube.

Pollen tube. On the stigma the pollen grain is caught in a sugary solution or even in water. The pollen grain swells up and the exine ruptures at the germ pore. The intine and contents come out in the form of a *pollen tube*. The pollen tube grows through the style and reaches the ovule in the ovary. The entrance of the pollen tube is either through micropyle, chalaza or side ways. It carries the tube nucleus and the two male gametes as its tip, as it enters the ovule. After discharging its contents in the embryo sac, the pollen tube contains some enzymes such as: smylase, invertase, phosphotase, pectinase, lipase, etc. The distal part of the pollen tube contains some amount of cytoplasm.

FERTILIZATION

In angiosperms, the pollen grains are being transferred from the anther to the stigma, and is termed *pollination*. Several agencies are indulged in bringing about this transfer of pollen from the anthers to the stigma. The pollination may be of two types — *self pollination* and *cross pollination*. The transfer of the pollen from the anther of a flower to the stigma of the same flower is known as self pollination, whereas cross pollination is the transference of the pollen from one flower to another flower. In the condition in which the pollen are discharged from the anther, they show considerable resistance to environmental changes. Sometimes they remain viable for several weeks.

The fusion of two dissimilar sexual reproductive units or gametes is termed fertilization. In angiosperms, the fertilization is being completed as follows:

Germination of pollen grain. After being deposited on the stigma the pollen grain absorbs liquid from the moist surface of a the stigma, expands in size, and the initine protrudes out through a germ pore. The small tubular structure also known as pollen tube continues to elongate, and makes its way down the tissues of the stigma and style. Only the distal part of the pollen tube possesses living cytoplasm.

In most of cases a single pollen tube comes out from each pollen grain. In other cases (*e.g.*, in Malvaceae, Cucurbitaceae and Campanulaceae) several pollen tubes may emerge from a single pollen grain. But usually only one of them is functional. In certain cases (*e.g.*, in Amentiferae) the pollen tube may be branched.

The stigma plays an important role in the germination of pollen grain. The stigma secretes a fluid containing lipids, gums, sugar and resins. The chief function of the stigmatic secretion is to protect the pollen as well as the stigma from desiccation.

Path of pollen tube. After the emergence of the pollen tube from the pollen grain, it makes its way between the stigmatic papillae into the tissues of the style. The style varies in length. In *Zea mays* the style known as silk attains even a length of 50 cm.

Histologically the styles have been classified into three main types—*open, half-closed* and *closed*. In open type the transmitting tissue of the style is dissolved by the action of pectinase enzyme (found in many monocots, Papaveraceae,

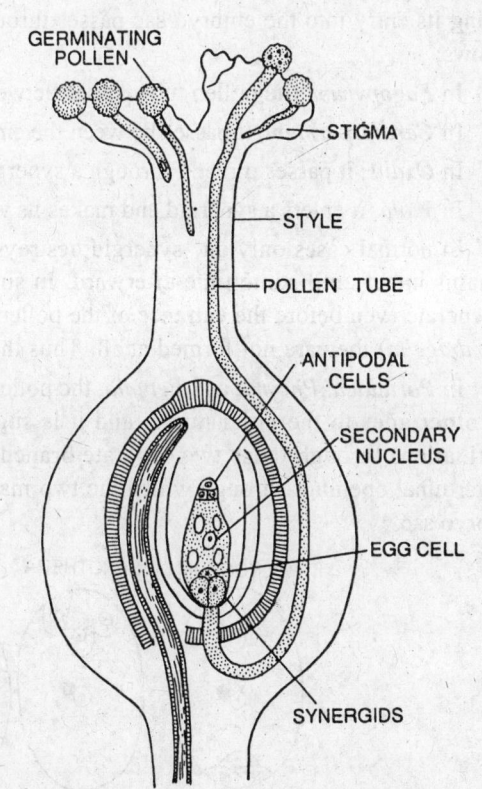

Fig. 21.36. Fertilization. Pistil in longitudinal section showing porogamy.

Aristolochiaceae). In half-closed type the stylar canal remains surrounded by a rudimentary transmitting tissue of two or three layers (found in several members of Cactaceae). In the closed type (found in *Datura* and *Gossypium*), there is no open channel but a solid core of tissue, through which the pollen tube grows downward in order to reach the ovary.

After arriving the wall of the ovary, the pollen tube may enter the ovule either through the micropyle or by some other route. The entrance of the pollen tube through the micropyle is usual condition and is known as *porogamy*. In some cases the pollen tube enters the ovule through the chalaza. This condition is known as *chalazogamy* (*e.g.*, in *Casuarina, Juglans regia, Betula*, etc.)

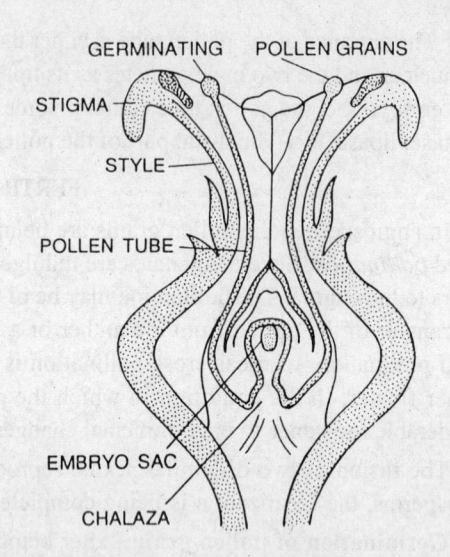

Fig. 21.37. Pistil in longitudinal section showing chalazogamy.

Entry of pollen tube into embryo sac. After entering the wall of the embryo sac the pollen tube during its entry into the embryo sac passes through the nucellar cells, and the synergids. It is as follows;

In *Fagopyrum*, the pollen tube passes between the egg and one synergid.

In *Cardiospermum*, it passes between the embryo sac wall and a synergid.

In *Oxalis*, it passes directly through a synergid.

In *Viola*, it enters a synergid and makes its way through the base of the synergid.

In normal cases only one synergid destroys by the impact of the pollen tube and the other remains intact until some time afterward. In some cases (*e.g., Tacca, Nelumbo*) the synergids degenerate even before the entrance of the pollen tube. In certain cases (*e.g., Plumbago, Vogelia, Plumbagella*) they are not formed at all. Thus they are not essential for fertilization.

In *Portulaca, Phryma* and *Petunia*, the pollen tube bifurcated; one branch goes to the egg and the other goes to the polar nuclei and it is suggested that the two male gametes reach their destinations through these two separate branches. However, in *Coffea arabica*, there are two subterminal openings through which the two male gametes are discharged into the cavity of the embryo sac.

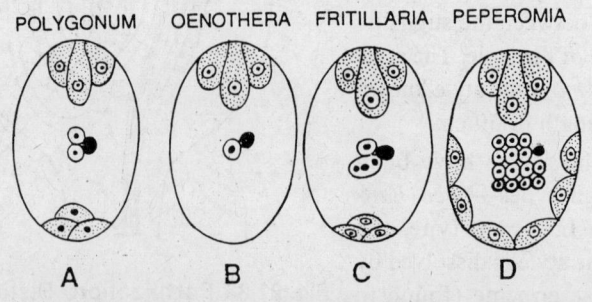

Fig. 21.38. Fusion of the one of the male gametes and the polar nuclei in the embryo sacs in different types of plants.

Embryology of Angiosperms

Gametic fusion. After the discharge of the contents of the pollen tube into the embryo sac, one male gamete fuses with the egg (syngamy) and the other with the two polar nuclei (triple fusion).

The male gametes may be spherical (*e.g.*, in *Erigeron*), ellipsoidal (*e.g.*, in *Levisticum*), rodshaped (*e.g.*, in *Urtica*), or vermiform (*e.g.*, in *Lilium* and *Fritillaria*), in structure. In *Orobanche cumana* the sperm nucleus which fuses the egg is spherical and is larger than the other which is oval in the form.

In normal case one male gamete unites with the egg to form the zygote and the second travels a little farther and unites with the secondary nucleus. This process is known as *double fertilization* (Navaschin, 1898; *Guignard*, 1899). This is also termed *triple fusion*. The process of double fertilization is characteristic of angiosperms. Actual course of fusion of the gametic nuclei, has been studied in *Crepis capillaris* by Gerassimova (1933). In this case, at the time of the approach to the egg nucleus, the male nucleus becomes like a continuous thread rolled into a ball and it soon begins to unwind and spread out with its entire surface adjacent to the nuclear membrane of the egg. Thereafter, it gradually immerses itself within the egg nucleus and thus the phenomenon of syngamy completes.

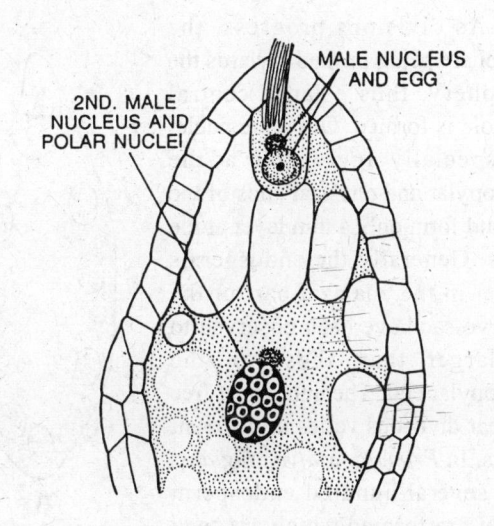

Fig. 21.39. Fertilization. Fusion of the male gametes with egg and secondary nucleus (double fertilization).

Similarly the second sperm fuses with the secondary nucleus thus completing the triple fusion. Gerassimova (1933) concluded that triple fusion is accomplished more quickly than syngamy. With the result of syngamy a zygote (2n) is formed, whereas the triple fusion results in the formation of endosperm nucleus (3n).

THE ENDOSPERM

The endosperm makes the main source of food for the embryo. In gymnosperms the endosperm is haploid (n) and forms a continuation of the female gametophyte. On the other hand, in angiosperms it is formed mostly as the result of a fusion of the two polar nuclei and one of the male gametes. Since all the three nuclei taking part in the fusion are haploid, the endosperm becomes triploid (3n). In normal cases the endosperm is triploid but haploid, tetraploid and polyploid endosperms are also known. Generally the endosperm nucleus divides after the division of the oospore, but in several cases the endosperm is formed to a great extent even before the first division of the oospore.

TYPES OF ENDOSPERM FORMATION

There are three general types of endosperm formation: (*a*) nuclear type, (*b*) cellular type and (*c*) helobial type.

Nuclear type. In this type, the first division and usually several of the following divisions are unaccompanied by wall formation. The nuclei may either remain free or in later stages they may become separated by walls.

As divisions progress, the nuclei are being pushed towards the periphery, thus a large central vacuole is formed. Often the nuclei are specially aggregated at the micropylar and chalazal ends of the sac and form only a thin layer at the sides. Generally the endosperms nuclei in the chalazal part of the embryo sac have been observed to be larger than those in the micropylar end. The number of free nuclear divisions varies in different plants. In *Primula, Malva, Juglans,* etc., several hundred endosperm nuclei are formed which are seen lining the wall of the embryo sac. In *Trapaeolum, Melastoma,* etc., there is no wall formation. In *Asclepias, Calotropis, Rafflesia,* etc., the wall formation occurs at a very early stage when only 8 or 16 nuclei are formed, and in *Coffea* at the 4-nucleate stage. The wall formation generally progresses from the periphery of the embryo sac towards the centre or from its apex towards the base.

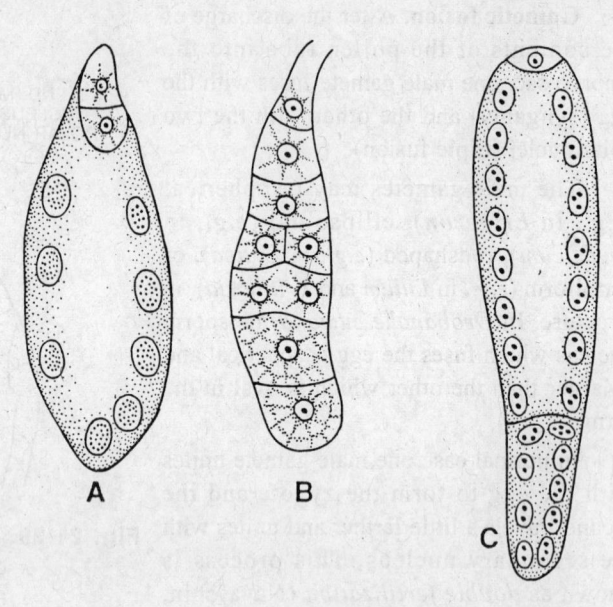

Fig. 21.40. Development of endosperm. A, nuclear type; B, cellular type and C, helobial type.

Cellular type. In this type, the first and most of the following divisions are accompanied by wall formation, and thus the sac is divided into several chambers, some of which may contain more than one nucleus. The first wall is usually transverse but sometimes vertical or oblique, and in some other cases the plane of division is not constant.

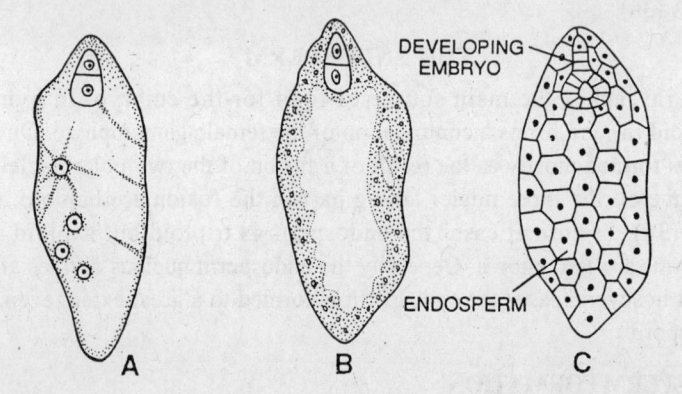

Fig. 21.41. Embryosac showing formation of nuclear type of endosperm. Developing embryo is also seen.

Helobial type. This type is frequently found in the members of the order Helobiales. This type is intermediate between the *nuclear* and the *cellular* types. In this type the first division is followed by a transverse wall resulting in a micropylar and chalazal chamber. Further divisions are generally free nuclear and may take place in both chambers, but the main body of the endosperm is formed by the micropylar chamber only.

Embryology of Angiosperms

HISTOLOGY OF THE ENDOSPERM

The cells of the endosperm are generally isodimetric and store food materials in them. The walls of the cells are thin and devoid of pits. In the grasses and some other plants, the peripheral layer of the endosperm acts like a cambium and produces on its inside a series of thin-walled cells which become filled up with starch. On the maturation of the seed, the outermost layer ceases to divide and its cells become filled with aleurone grains and the walls become somewhat thickened. According to Haberlandt (1914) the chief function of this aleurone layer is the secretion of diastase and other enzymes so that the food materials stored in the endosperm may be converted into the soluble form and be available to the developing embryo.

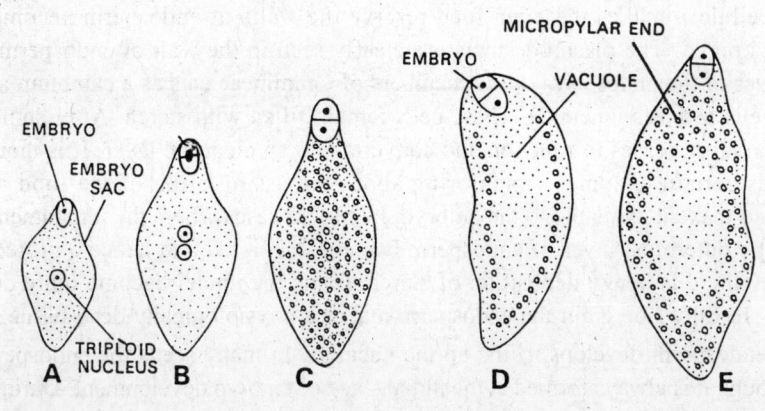

Fig. 21.42. Endosperm. Some more stages of the development of nuclear type endosperm from a triploid nucleus.

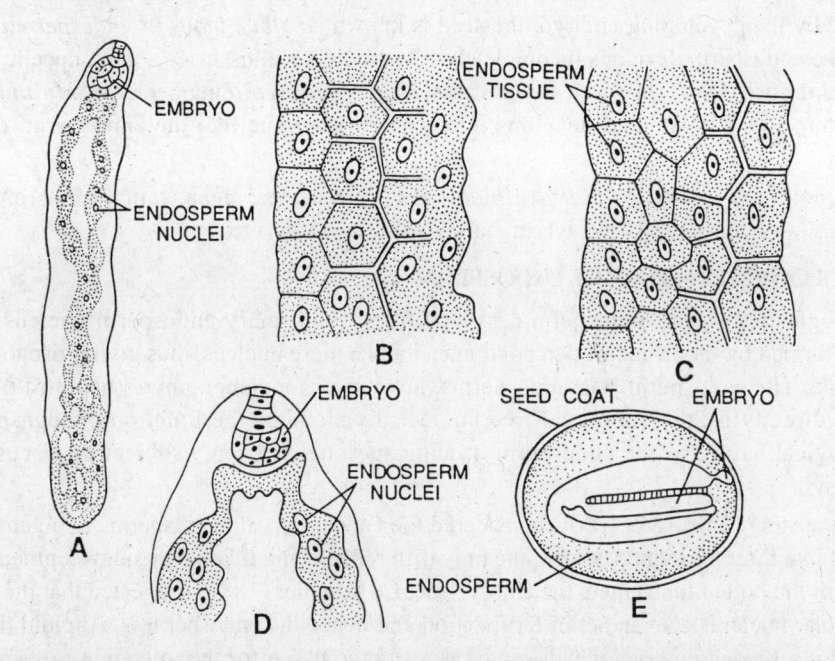

Fig. 21.43. Endosperm. A—D, details of the stages in the development of the nuclear type of endosperm; E, fully formed endosperm which has completely replaced the nucellus; embryo is embedded in it.

Ruminate endosperm. In the families such as Annonaceae, Myristicaceae, Aristolochiaceae, Rubiaceae, Palmae and Eupomatiaceae, *the endosperm is irregularly ridged and furrowed* and is termed *ruminate endosperm*. Periasamy (1962) has defined the ruminate endosperm as one which exhibits some degree of irregularity and unevenness in its contour which inturn is conditioned by the irregular inner surface of the seed coat. The rumination of the endosperm develops as a result of invaginations of the outer tissues, which penetrate deeply and appear as dark wavy bands in the mature seeds, or sometimes it develops as a result of an unequal elongation of the cells of any one layer or the only layer of the seed coat as in *Passiflora*.

Structure and fate of endosperm. Since the endosperm is meant for the storage of food, and therefore, its cells are usually thin-walled, large, isodiametric and without pits. But in certain cases where hemicellulose makes the main food reserve the walls of endosperm become very much thickened and pitted. The plasmodesmata are clearly seen in the wall of endospermic cells. The outermost layer of the endosperm of the members of Gramineae acts as a cambium and produced thin walled cells towards innerside. These cells remain filled with starch. After sometime this so called cambial layer ceases to function and converts into an aleurone layer. It is thought that this layer secretes certain enzymes which bring about the assimilation of the food stored in the endosperm and make it available to the embryo. In certain cases where the integuments are absent the cells of the outermost layer of endosperm become suberized and hence a protective layer is formed. Since there is heavy deposition of starch grains the nuclei become quite deformed and disorganized. In fully ripe grain the endosperm makes a physiologically dead tissue.

As the endosperm develops it fills up the nucellus. In many seeds the endosperm is seen at maturity although it is always formed at the initial stage of embryo development. During the process of the development of the embryo, the food stored up in the endosperm is continuously drawn up by the developing embryo and thus completely exhausted. Such seeds are known as *exalbuminous* or *non-endospermic*. In other cases, where the endosperm grows vigrously and is not completely exhausted by the developing embryo, the seed is known as *albuminous* or *endospermic*. In most cases as the endosperm develops it completely fills up the nucellus space and no nucellus is found in the seed in such cases. However, in few cases, as in *Nymphaea*, Zingiberaceae, *Mirabilis jalapa*, etc., the nucellus persists and develops into a nutritive tissue like the endosperm, called the *perisperm*.

The common examples are: *exalbuminous*—gram, pea, bean, tamarind, orchids, etc.; *albuminous*—castor, poppy, rice, wheat, barley palms, grasses, etc.

MORPHOLOGICAL NATURE OF ENDOSPERM

In angiosperms, the endosperm develops from the primary endosperm nucleus which is normally formed by the fusion of two polar nuclei and a more nucleus, thus giving rise to a triploid (3 n) tissue. The endosperm in gymonsperms, however, is a gametophyte (haploid) tissue as it develops directly by the continued free nuclear divisions of the functional megaspore. The morphological nature of the endosperm of angiosperms has been a subject of discussion and controversy.

Hofmeister (1858, 1859, 1861) considered the endosperm of angiosperms as a gametophytic tissue just like those of gymnosperms, the only difference being that here its development remains checked till the pollen tube enters the embryo sac. Le Monnier (1887) suggested that the fusion of the two polar nuclei is also an act of fertilization and hence the endosperm is a diploid tissue, and he treated it a *second embryo*, which serves as a storage tissue for the true embryo. Nawaschin's (1898) discovery of double fertilization and triple fusion changed the views regarding the morphological mature of the endospherms. Sargant (1900) supported Nawaschin's view of triple fusion and double fertilization. Strasburger (1900) however, opposed this view and suggested that triple fusion was not true fertilization but a vegetative fertilization which was of the nature of growth stimulus only. He considered the endosperm as belated gameophytic tissue. According to him the female gametophyte in angiosperms is much reduced and has no reserve food. However, this was supported by Coulter and Chamberlain (1912).

Several workers like Nemec (1910), Brink and Cooper (1947) suggested that the fusion of the second male gamete with the polar nuclei serve two functions—1. it stimulates the development of endosperm and 2. it helps in the formation of a tissue which is physiologically more suitable for the nourishment of the embryo.

This can be concluded that the endosperm in angiosperms is neither haploid (n) nor diploid (2 n) but it is an undifferentiated tissue which shows different degrees of polyploidy from 2 n to 15 n. For example 2 n in *Oenothera* and *Butomopsis*; 3n in *Polygonum* and most of angiosperms; 4 n in *Ditepalanthus*; 5n in *Fritillaria, Plumbagella, Plumbago* and *Penaea*; 9 n in *Peperomia, Acalypha*; and 15 n in *Peperomia hispidula*.

XENIA

This term was coined by Focke (1881) for representing the *direct effect of pollen on the characters of the tissues surrounding the embryo*. But now this term is used in strict sense to indicate the appearance of the endosperm only. For example, two different races of *Zea mays*, one with yellow endosperm and other with white endosperm may be considered. If the pollen of yellow endosperm race are placed on the stigma of white endosperm race, the hybrid embryo which develops, exhibits the dominant character of yellow endosperm race in the next season when it grows into a mature plant. But the yellow colour appears in the endosperm of the same ovule. This happens because the pollen of yellow dominant race (YY) when pollinates the white recessive race (yy) the primary endosperm nucleus is formed by the fusion of two polar nuclei with y and y set of chromosomes and a male nucleus with Y set of chromosomes. Thus the endosperm contains three sets of chromosomes (Yyy) and since it possesses a dominant factor for yellow colour therefore it develops an endosperm of yellow colour, although the ovule belongs to the white parent.

METAXENIA

The metaxenia may be defined as *the effect of pollen on the seed coat of pericap lying outside the embryo sac.*' Swingle (1928) found that in *Phoenix dactylifera* that the time of maturity of the fruits and their size can be made to vary according to the type of pollen used in fertilization. According to him the embryo and endosperm secrete hormones, which diffuse out in the pericarp and seed coat and exert a specific influence on them resulting in certain specific variations.

MOSAIC ENDOSPERM

In some angiospems, there is the *lack of uniformity in the tissues of the endosperm*, which is termed mosaic enderm, For example, in *Zea mays*, the patches of two different colours are seen forming a sort of irregular mosaic pattern.

According to Webber (1900) the second male nucleus fails to fuse with the polar nuclei and both divide separately giving rise to nuclei of two distinct characters. These nuclei are intermingled during free nuclear division and thus the mosaic endosperm develops.

According to P. Maheswari (1950), it is possible that sometimes only one of the polar nuclei is fertilized while the other divides independently, resulting in the formation of mosaic endosperm.

Clark and Copeland (1940) have given most reasonable explanation for the development of mosaic endosperms. According to them there is either aberrant behaviour of the chromosomes in mitosis of these are somatic mutations which explain the problem.

EMBRYO AND ITS DEVELOPMENT (EMBRYOGENESIS)

After fertilization, the fertilized egg is called zygote or oospore which develops into an *embryo*. The oospore before it actually enters into the process undergoes a period of a rest which may vary from few hours to few months. Generally the zygote (oospore) divides immediately after the first division of the primary endosperm nucleus but sometimes it divides earlier than the primary endosperm nucleus. Unlike gymonsperms where the early stages of the development show free nuclear divisions the first division of zygote is always followed by a wall-formation resulting in a two-celled proembryo. Practically there are no fundamental differences in the early stages of the development of the embryos of monocots and dicots. But in later stages there is a marked difference between the embryos of dicotyledonous and monocotyledonous plants, hence their embryogenesis has been considered here separately.

DEVELOPMENT OF EMBRYO IN DICOTS

According to Soueges the mode or origin of the four-celled proembryo and the contribution made by each of these cells makes the base for the classification of the embryonal types. However, Schnarf (1929), Johansen (1945) and Maheshwari (1950) have recognized *five* main types of embryos in dicotyledons. They are as follows:

I. The terminal cell of the two-celled proembryo divides by longitudinal wall.

(*i*) *Crucifer type*. Basal cell plays little or no role in the development of the embryo.

(*ii*) *Asterad type*. Basal and terminal cells play an important role in the development of the embryo.

II. The terminal cell of the two celled proembryo divides by a transverse wall.

(*a*) Basal cell plays a little or no role in the development of the embryo.

(*iii*) *Solanad type*. Basal cell usually forms a suspensor of two or more cells.

(*iv*) *Caryophyllad type*. Basal cell does not divide further.

(*v*) *Chenopodiad type*. Both basal and terminal cells take part in the development of the embryo.

Here citing the example of *Capsella bursa pastoris* (Shepherd's purse), the detailed study of *Crucifer type* of the development of the embryo has been given:

Development of dicot embryo in *Capsella bursa pastoris* (Crucifer type). For the first time Hanstein (1870) worked out the details of the development of embryo in *Capsella bursa pastoris*, a member of Cruciferae. Later on, this was also studied and confirmed by Faminitizin (1879) and Soueges (1914, 1919). Several other workers such as Schnarf (1922), Soueges (1938), Johansen (1950), Maheshwari (1950) and Wardlaw (1955) also worked out the same.

The oospore divides transversely forming two cells, a *terminal cell* and a *basal cell*. The cell towards the micropylar end of the embryo sac is the *suspensor cell* (*i.e.*, basal cell) and the other one makes the *embryo cell* (*i.e.*, terminal cell). The terminal cell by subsequent divisions gives rise to the embryo while the basal cell contributes the formation of *suspensor*. The terminal cell divides by a vertical division forming a 4-celled ⊥-shaped embryo. In certain plants the basal cell also forms the hypocotyl (*i.e.*, the root end of the embryo) in addition to suspensor. The terminal cells of the four-celled proembryo divide vertically at right angle to the first vertical wall forming four cells. Now each of the four cells divides transversely forming the *octant stage* (8-celled) of the embryo. The four cells next to the suspensor are termed the *hypobasal* or posterior octants while the remaining four cells make the *epibasal* or anterior octants. The epibasal octants give rise to *plumule* and the *cotyledons,* whereas the hypobasal octants give rise to the hypocotyl with the exception of

Embryology of Angiosperms

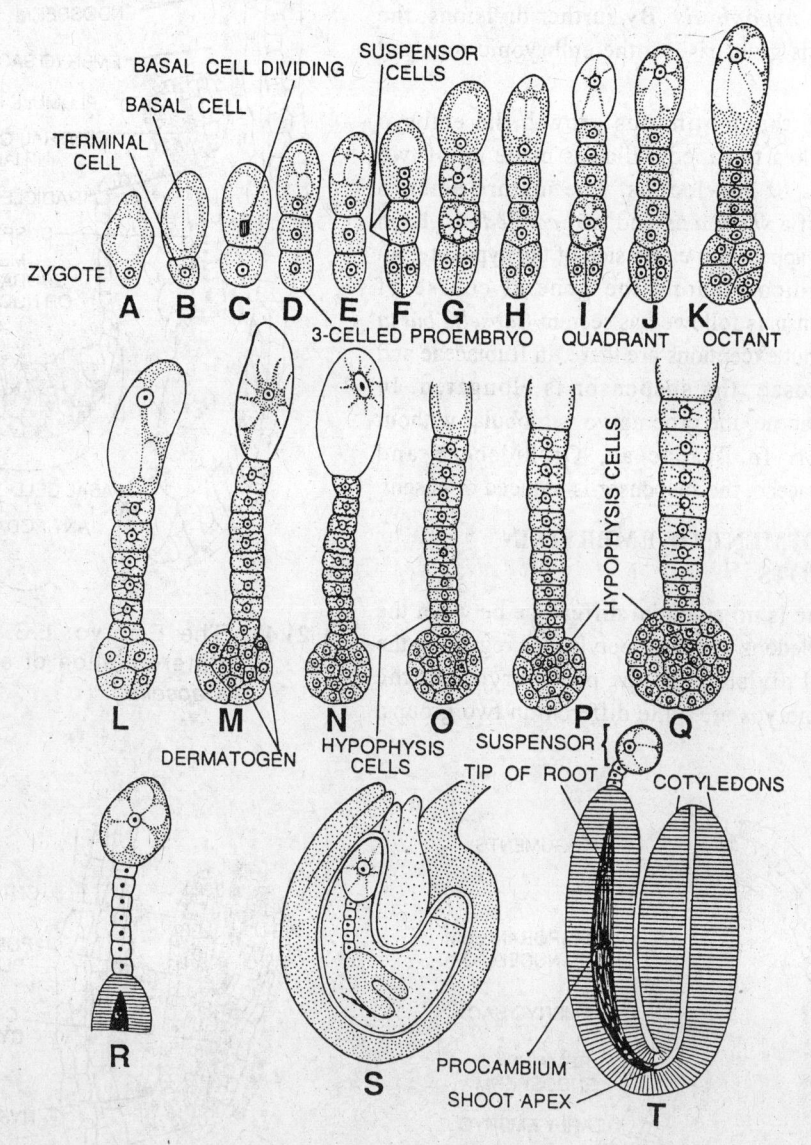

Fig. 21.44. Stages in the development of a typical dicto embryo in *Capsella bursa-pastoris*.

its tip. Now all the eight cells of the octant divide periclinally forming outer and inner cells. The outer cells divided further by anticlinal division forming a peripheral layer of epidermad cells, the *dermatogen*. The inner cells divide by longitunal and transverse divisions forming *periblem* just beneath the dermatogen and *plerome* in the central region. The cells of periblem give rise to the cortex while that of plerome form the stele.

At the time of the development of the octant stage of embryo the two basal cells divide transversely forming a 6-10 celled filament, the suspensor which attains its maximum development by the time the embryo attains globular stage. The suspensor pushes the embryo cells down into the endosperm. The distal cell of the suspensor is much larger than the other cells and acts as a *haustorium*. The lowermost cell of the suspensor is known as *hypophysis*. By further divisions, the hypophysis gives rise to the embryonic root and root cap.

With the continuous growth the embryo becomes heart shaped which is made up of two primordia of cotyledons. The mature embryo consists of a *short axis* and two *cotyledons*. Each cotyledon appears on either side of the hypocotyl. In most of dicotyledons the general course of embryogensis is followed as seen in *Capsella bursa pastoris*, but exceptions are there. In Rubiaceae and Leguminosae, the suspensor is elongated. In Nymphaeaceae, the proembryo is globular without suspensor. In Trapaceae, Orchidaceae and Podostemaceae, the suspensor is reduced or absent.

DEVELOPMENT OF EMBRYO IN MONOCOTS

There is no essential difference between the monocotyledons and the dicotyledons regarding the early cell divisions of the proembryo, but the mature embryos are quite different in two groups.

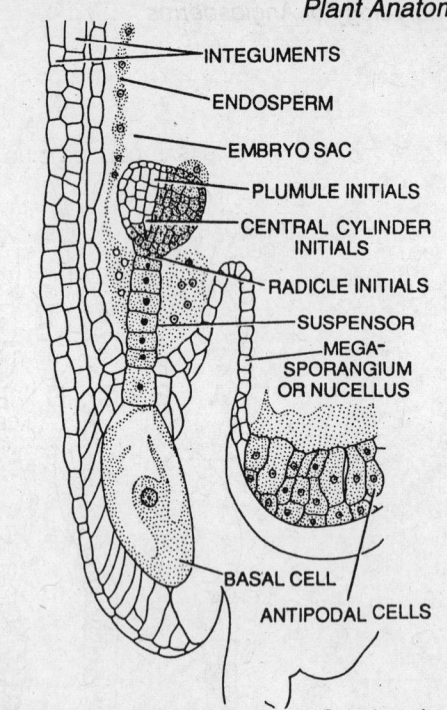

Fig. 21.45. The Embryo. L.S. showing differentiation of embryo in *Capsella*.

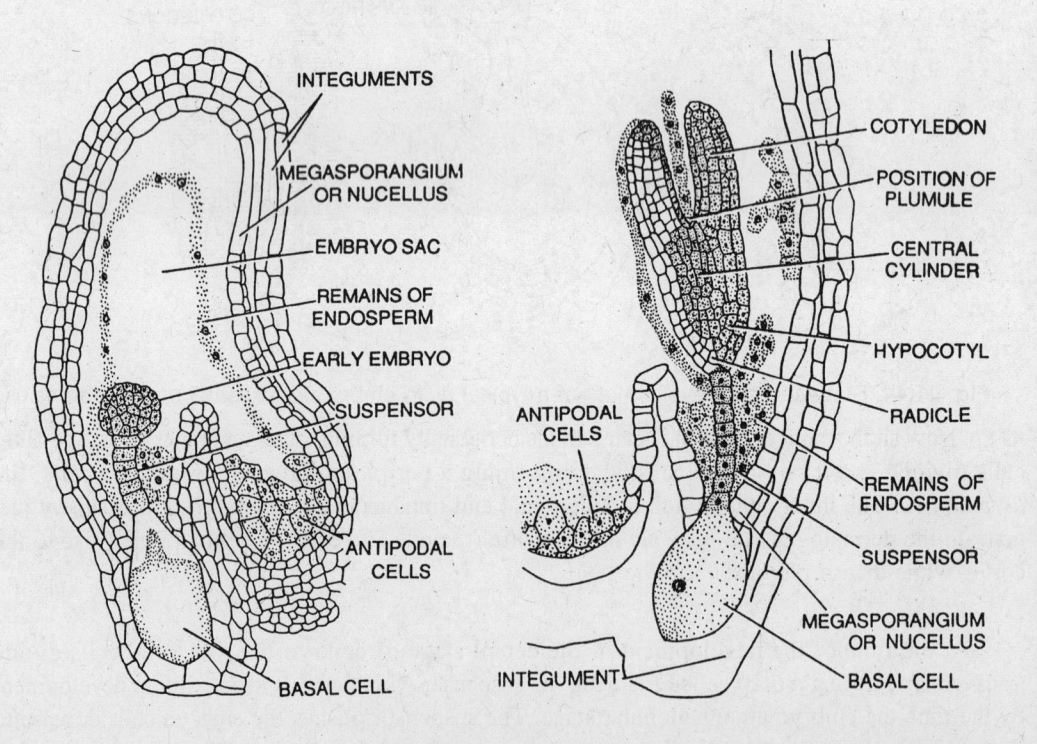

Fig. 21.46. The Embryo. L.S. of early embryo of *Capsella*.

Fig. 21.47. The embryo. L.S. showing development of cotyledons in *Capsella*.

Embryology of Angiosperms

Here the embryogeny of *Sagittaria sagittaeifolia* has been given as one of the examples.

The zygote divides transversely forming the terminal cell and the basal cell. The basal cell, which is the larger and lies towards the micropylar end, does not divide again but becomes transformed directly into a large vesicular cell. The terminal cell divides transversely forming the two cells. Of these, the lower cell divides vertically forming a pair of juxtaposed cells, and the middle cell divides transversely into two cells. In the next stage the two cells once again divide vertically forming quadrants. The cell next to the quadrants also divides vertically and the cell next to the upper vesicular cell divides several times transversely. The quadrants now divide transversely forming the octants, the eight cells being arranged in two tiers of four cells each. With the result of periclinal division, the dermatogen is formed. Later the periblem and plerome are also differentiated. All these regions, formed from the octants develop into a single terminal cotyledon afterwards. The lowermost cell of the three celled suspensor

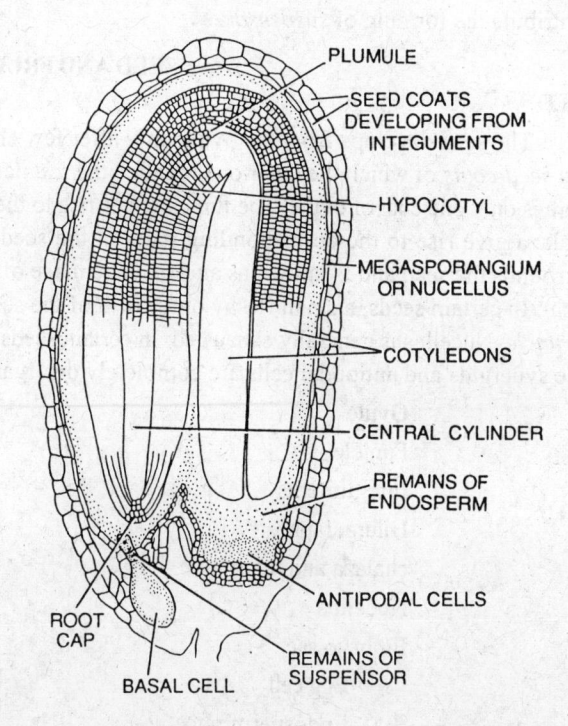

Fig. 21.48. The Embryo L.S. of well-differentiated embryo of *Capsella*.

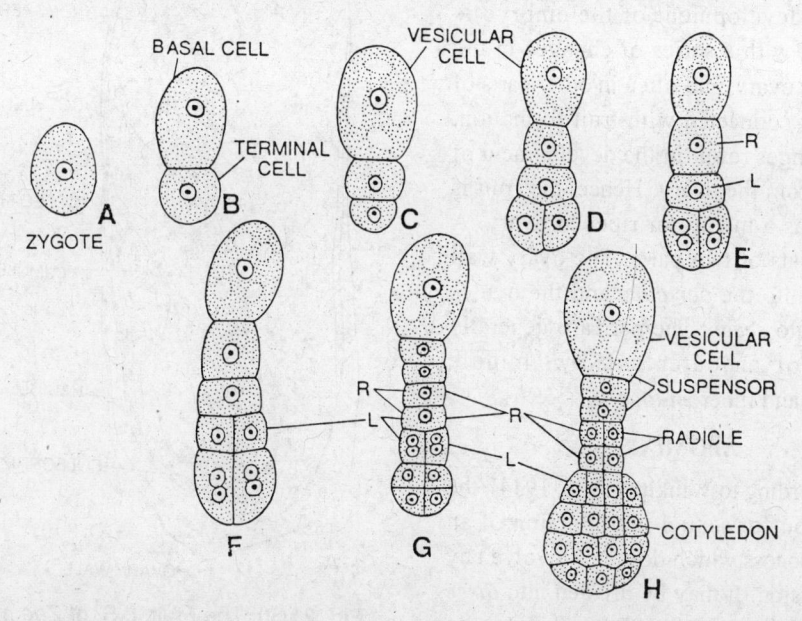

Fig. 21.49. Stages in the development of a typical monocot embryo in *Sagittaria*.

divides vertically to form the *plumule* or stem tip. The cells *r* form the *radicle*. The upper 3—6 cells contribute the forming of *suspensor*.

SEED AND FRUIT

SEED

The seed develops from the ovule only after few changes. The two integuments develop into two *seed-coats* of which the outer one is called the *testa* and the inner one the *tegmen*. In some seeds there is only one coat of ovule. The funicle gives rise to the stalk, and the hilum, micropyle, raphe and chalaza give rise to the corresponding parts in the seed. In some cases an additional investment surrounds the seed and is known as an *aril*. The mace of nutmeg is the aril, and so also the flesh of litchi. In certain seeds, a small fleshy outgrowth of the seed is formed at the micropyle and is called *caruncle*. Nucellus is generally exhausted. In certain seeds, it remains as a thin layer, called *perispem*. The synergids and antipodal cells are completely disorganized.

Ovule	Seed
Funicle	stalk
Integuments	seed coats
Hilum, raphe, chalaza and micropyle	hilum, raphe, chalaza and micropyle.
Nucellus	perisperm
Embryo sac	
Egg cell	embryo
Endosperm nucleus	endosperm
synergids	disorganized
antipodal cells	

disorganized

FRUIT

The development of the embryo is associated with a series of changes in the wall of the ovary, and often in other parts of the flower connected with fruit formation. These changes result in the development of the fruit from the ovary. Hence, the fruit is regarded as a mature or ripened ovary. A fruit consists of two parts. The ovary wall develops into the *pericarp* and the ovules develop into *seeds*. The pericarp generally consists of an outer *epicarp*, middle *mesocarp* and inner *endocarp*.

APOMIXIS

According to Winkler (1908, 1934), the substitution for sexual reproduction of an asexual process which does not involve any nuclear fusion. It may be divided into *three* groups (Maheshwari, 1950), They are as follows:

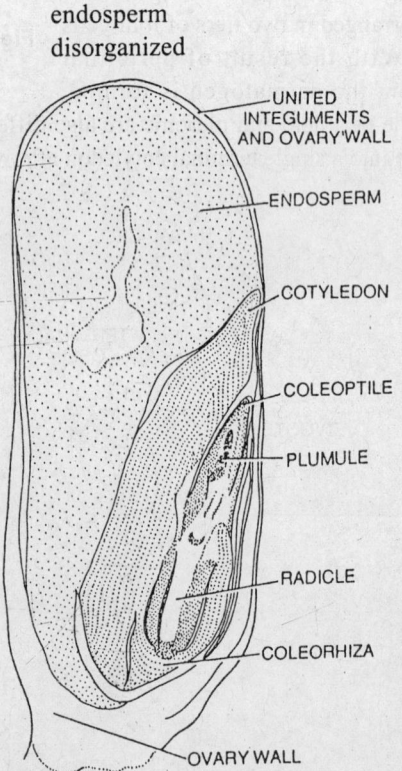

Fig. 21.50. The Fruit. L.S. of *Zea mays* fruit.

Embryology of Angiosperms

1. Nonrecurrent apomixis. In this type the megaspore mother cell undergoes the usual meiotic divisions and a haploid embryo sac is formed. Here the embryo arises either from the egg (*haploid parthenogenesis*) or from other cell of the gametophyte (*haploid apogamy*). The plants produced by the method are, haploid and generally sterile and do not reproduced sexually any more. This type of apomixis has seen in several species such as *Solanum nigram, Lilium* spp., *Bergenia, Erythraea centaurium, Orchis maculata, Nicotiana tabacum*, etc.

2. Recurrent apoximis. In this type, the embryo sac generally arises either from an archesporial cell (*generative apospory*) or from some other part of the nucellus (*somatic apospory*). Here all the nuclei of the embryo sac are diploid, and there is no meiotic division. The embryo arises either from the egg (*diploid parthenogenesis*) or from some other cell of the gametophyte (*diploid apogamy*). Generative apospory has been observed in *Eupatorium glandulosum, Parthenium argentatum*, etc. Somatic apospory has been reported in *Hierarcium excellens, H. flagellare* and *H. aurantiacum*.

3. Adventive embryony. This type of apomixis is also known as *sporophytic budding*. Here, the developed embryo sacs may be haploid or diploid, but the embryos do not arise from the cells of the gametophyte, and they arise only from the cells of nucellus or the integument. There is no alternation of generations, because the diploid tissue of the present sporophyte directly give rise to the new embryo. Adventive embryony has been frequently reported in *Citrus, Euphorbia dulcis; Capparis frondosa, Mangifera indica* and *Hiptage madablota*.

In another type the flowers are replaced by bulbils or other vegetative propagules which generally germinate while still on the plant. However, this is only a kind of vegetative reproduction. The bulbils have been reported in *Globba bulbifera, Allium sativum, Agave, Dioscorea bulbifera, Oxalis*, Pine-apple, etc.

Parthenocarpy. In certain cases of angiosperms the ovary normally develops into a fruit without pollination and fertilization. This type of free development of fruit is known as *parthenocarpy*. Such fruits (parthenocarpic fruits) are always seedless. Sometimes the fruit formation may be induced by artificial pollination by foreign pollen from another species, but without subsequent fertilization. The parthenocarpy may also be induced by the spraying of growth promoting substances such as naphthalene acetic acid NAA. This is called *induced parthenocarpy*. The examples of parthenocarpy are commonly found in banana, papaw, pine apple, guava, grapes, apple, pear, *Thalicturm, Alchemilla*, etc.

POLYEMBRYONY

The occurrence of more than one embryo in the seed is known as *polyembryony*. This phenomenon was initially discovered in the orange by Leeuwenhoek (1719). Ernst (1918) and Schnarf (1929) have classified it into two types—*true* and *false*—depending on whether the embryos arise in the same embryo sac or in different embryo sacs in the ovule. Later on Gustafsson (1946) proposed that the term false polyembryony should be restricted to those cases only where two or more nucelli, each with its own embryo sac, fuse at an early stage. He included all other types under true polyembryony.

Polyembryony is quite common among conifers (gymnosperms), but many species of both dicotyledons and monocotyledons (angiosperms) exhibit this phenomenon. In the following paragraphs the types of true polyembryony have been considered.

Cleavage polyembryony. This is simplest type of polyembryony. Here the increase in the number of embryos is due to the cleavage of the zygote or embryo. This type of polyembryony is quite common in gymnosperms and sporadic in angiosperms. Jeffrey (1895) reported *cleavage polyembryony* in *Erythronium americanum*. Here the zygote divides to form a small group of cells, which continues to increase in volume, and outgrowths are given out at its lower end which function later on as independent embryos. The other examples of cleavage polyembryony are met with in *Nymphaea advena, Empetrum nigrum,Nicotiana rustica, Eulophia epidendraea*, etc.

Origin of embryos from synergids or antipodal cells. The embryos may also be produced from other parts of the embryo sac such as synergids and antipodal cells. In most cases the synergids become egg-like to form the embryos with or without fertilization. The twin embryos have been recorded in *Naias, Tellima, Peltiphyllum* and *Fragaria*, where the fertilization of the synergids by two male gametes is found but endosperm is not formed. Where as in *Crepis, Sagittaria, Poa* and *Aristolochia* the fertilization of the egg and one or two synergids is followed by the formation of endosperm. Production of the embryos from antipodal cells is rare. Shattuk (1905) reported such embryos in *Ulmus americana*. Mauritzon (1933), Fagerlind (1944) have also noted the similar cases in *Allium odorum, Sedum fabaria*, etc.

Origin of embryos from endosperm. Treub (1898) in *Balanophora*, Woodworth (1930) in *Alnus* and others have reported the embryos developed from endosperm. However, Ernst (1913) reported that the embryo develops normally from the egg, embedded in the cellular endosperm and does not develop from the endosperm proper.

Origin of embryos from cell outside embryo sac. The embryos also develop from the cells of the nucellus and integument. Such development of embryos has been reported in *Citrus, Syzygium* and *Mangifera*.

22

Morphogenesis : Tissue And Organ Culture

The smallest viable unit of plant one can at present envisage as reproducing, growing and developing in culture is a single cell. As long ago as 1902, Haberlandt in Germany attempted to grow single higher plant cells in sterile culture, but for various reasons which have been understood, his attempts were unsuccessful.

Following Haberlandt, other workers established methods which would allow the growth of isolated plant organs and tissues in *vitro*, in culture. Excised roots were the first plant organs to be successfully brought into sterile culture, and the work done by White in the 1930's demonstrated that given appropriate nutrients, such roots, would grow and differentiate into normal root tissues. Work by Gautheret (1939) and others established that isolated portions of storage tissue, *e.g.*, carrot roots, could be kept alive and grown in sterile culture, much like micro-organisms. Callus culture derived from such isolated tissues lend themselves to studies of the effects of nutrients, vitamins and hormones upon cell division, differentiation of vascular tissues, and the inception of organized meristematic regions within the predominantly parenchymatous tissue. By definition, a **callus** is a

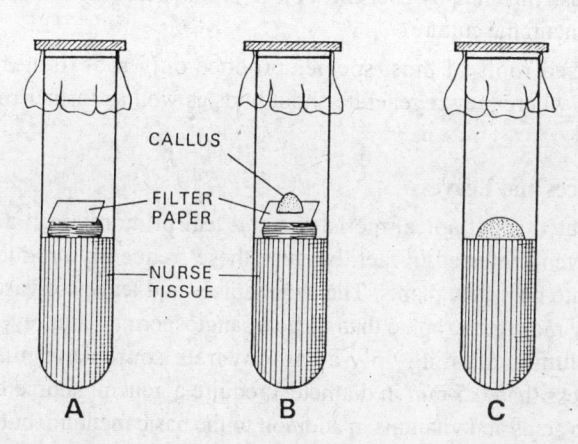

Fig. 22.1. Nurse-tissue technique to raise single-cell clones. A, a single cell from the callus is placed on filter paper lying on the top of a large callus (nurse tissue); B, the cell divides and forms a small tissue; C, this tissue is transferred from the filter paper to the medium directly where it is able to grow further.

mass of proliferating tissue consisting of predominantly parenchymatous cells, but, in which differentiation may occur under suitable conditions.

ORGAN CULTURE

It has been possible to grow several types of plant organs in sterile culture including roots, shoot apices (apical meristem culture), leaves, floral parts and fruits. The nutrient requirements for such organ culture vary considerably from species to species and according to the type of organ in question, but certain general requirements can be recognized.

Intact higher plants are autotrophic, that is, they are able to synthesize all the organic substances required for their own life from water, carbon dioxide, oxygen and mineral nutrients. But since most sterile cultures are unable to carry on photosynthesis, it is clear that they will require at least a carbon source, cultures require the same mineral nutrients as the intact plant, including both macronutrients (*e.g.*, nitrogen, phosphorus, potassium and calcium) and micronutrients or trace elements (*e.g.*, Mg, Fe, Mn, Zn, etc.). In addition to the requirements for a carbon source and mineral nutrients, it is found that most isolated organs have also a requirement for certain special organic substances, such as vitamins, nicotinic acid, etc.

Root Culture

As already indicated, excised roots were the first plant organs of higher plants to be successfully brought into sterile culture. In addition to the usual requirements for a carbon source and mineral nutrients, most isolated roots grown in sterile culture require to be supplied with certain vitamins. This is because, in the intact plant, certain vitamins are synthesized in the leaves and the roots are dependent upon shoots for the supply of these substances which they are unable to synthesize themselves. For example, tomato roots require only sucrose, mineral nutrients and thiamin, and given all these, will grow successfully in culture for many years. But excised roots of some monocotyledonous plants fail to grow even when supplied with a full complement of β – vitamins and other vitamins. In some of these (*e.g.*, rye), an exogenous auxin supplement to the nutrient medium allows growth to proceed. In order to maintain the culture, the excised roots must be regularly subcultured on to a fresh medium, by excising a piece of root bearing a lateral which then proceeds to grow rapidly and maintain the culture.

As a rule, excised roots of most species produce only root tissues in culture. There are exceptions, however, where they regenerate shoot buds as well as further roots, *e.g.*, *Convolvulus*, *Taraxacum* and *Rumex*.

Culture of Shoot Apices and Leaves

Like roots, isolated shoot apical meristems and leaf primordia can also be grown in sterile culture. These are often preferred in fact, because these frequently produce roots and can hence eventually develop into complete plants. The shoot apices and leaves of vascular cryptogams, such as ferns, are relatively more autotrophic than those of angiosperms. Thus, even a small fern apex can be grown on a medium containing only a carbohydrate source and mineral nutrients. Small angiosperm apices (less than 0.5 mm in diameter) require a general source of organic nitrogen and certain specific amino acids and vitamins in addition to the basic medium, but larger apices will grow on a simple medium. The simpler requirements of larger apices may be due to the fact that they carry larger leaf primordia, which apparently can supply some of the requirements of the apex for vitamins and other organic nutrients.

Isolated young leaves of the fern *Osmunda cinnamomea* and of the sunflower (*Helianthus annuus*) and tobacco (*Nicotiana tabacum*) have been successfully grown on a simple medium containing only sucrose and inorganic salts. Such isolated leaf primordia continue to grow and develop into normally differentiated leaves, although the latter are usually very much smaller than normal leaves developed in *vivo, i.e.,* on the plant.

Culture of Reproductive Organs

Culturing the reproductive organs of higher plants can have potentially greater advantages under certain circumstances. Thus, for example, while embryo culture can find applications in plant hybridization, another culture and production of haploids are useful in genetics and plant breeding for rapid production of homozygous diploids.

Whole ovaries can be cultured *in vitro* using suitable media. Alternatively, fertilized ovules can be dissected out of them and cultured.

Ovary culture. For the first time La Rue (1942) cultured the flowers of several angiosperms and reported extensive growth of ovaries. Nitsch (1949) also cultured the excised ovaries of tomato. In India, Ananta Swamy Rau (1956) for the first time cultivated pollinated ovaries *in vitro*. The credit for establishing the technique of ovary culture goes to Nitsch (1951) who experimented with the ovaries of *Phaseolus vulgaris, Cucumis anguria, Nicotiana tabacum* and *Lycopersicon esculentum*. Unpollinated ovaries failed to develop in the nutrient medium unless growth substances were added.

Ovule culture. However, it is easier to maintain ovules in sterile culture when they are left *in situ* within the ovary. The physiological requirements of fertilized ovules do not appear to be species specific, since young ovules of widely different species have grown to mature seeds following transplantation on to the placenta of *Capsicum* fruits.

White (1932) was the pioneer to culture the ovules. White and La Rue cultured ovules of *Erythronium* and *Antirrhinum* on White basic medium containing indole acitic acid (IAA).

Anther culture. For the first time Simakura (1934) of Japan cultured the anthers *in vitro*. Later on this work was taken up by Guha and Maheshwari (1964, 1966). They reported the origin of the embryoids from pollen of *Datura innoxia*. The plantlets of pollen were haploid. They achieved this by culturing the anthers in a nutrient medium contained kinetin, coconut milk or grape juice. Haploids obtained by this technique are of potential significance in basic and applied genetics and plant breeding.

Pollen culture. The pollen population within an anther is genetically heterogeneous. The plants which develop from anther will make heterogeneous population, and the problem arises. This problem can be solved by growing the isolated pollen. Sharp *et al* (1972) have progressed to some extent and they raised haploid tissue clones from pollen of tomato plant by using the nurse culture technique.

Anther and pollen culture. (Pollen embryos, embryoids, haploid pollen plants, anther androgenesis). The first successful pollen cultures were established by Tulecke in the 1950s using mature pollen grains of certain gymnosperms. However, he as well as others failed to culture pollen grains of

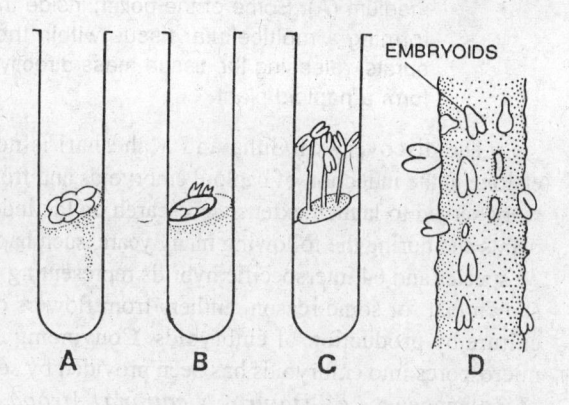

Fig. 22.2. Embryoid formation in tissue cultures of butter cup. A, an unorganized callus; B, six week old culture showing numerous embryoids arising from the callus; C, the embryoids have developed into plantlets, and a fresh crop of embryoids is seen arising from the surface of the hypocotyledonary region of these plants; D, the embryoid-bearing portion enlarged from Fig. C.

angiosperms. In 1964, Guha and Maheshwari of the Botany department of Delhi University, discovered that when excised anthers of *Datura innoxia* were cultured intact in a mineral salt medium in conjunction with coconut milk and other complex organic substances and growth hormones, embryo-like outgrowths appeared from the sides of the anther in about 6–7 weeks. In subsequent studies, they confirmed the haploid nature of the embryoids and their origin from microspores (Guha and Maheshwari, 1966, 1967).

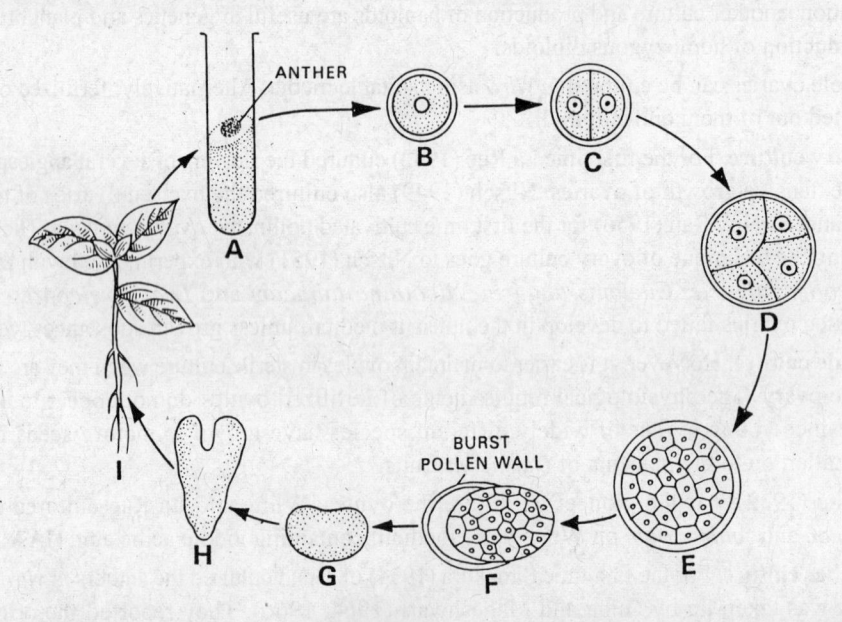

Fig. 22.3. Diagram showing the development of haploid plants from the pollen grains of tobacco. The anther of the right stage is excised from the flower bud and cultured on a suitable medium (A). Some of the pollen inside the anther undergoes repeated division (B – D), forming a multicellular tissue within the parent wall (E). Eventually, the pollen wall bursts, releasing the tissue mass directly develops into an embryoid and germinates to form a haploid plant.

The discovery of Guha and Maheshari is now widely recognized as the principal cornerstone in the induction of haploid embryoids and from them haploid plants in great frequency. Their work served to launch extensive research on the induction of embryoids in anther cultures of other plants and during the following many years, such haploid production has been reported in 17 genera, 23 species and 64 interspecific hybrids representing five angiospermic families. Investigations have shown that for some reason, anthers from flowers of the Solanaceae respond best to excision and culture by production of embryoids. Convincing demonstration of the direct transformation of microspores into embryoids has been provided by several workers in different species of the genera of Solanaceae, *i.e.*, *Datura, Nicotiana, Atropa, Lycium, Petunia, Solanum, Capsicum* and *Hyoscyamus*. It has been observed that embryoids go through the globular, heart-shaped and torpedo stages typical of the ontogeny of normal diploid zygotic embryos before they finally elongate and form shoot and root meristems. Embryoids have also been obtained from cultured anthers of certain cereals like rice (Guha *et al* 1970). All in all therefore, the capacity of microspores in cultured anthers to form haploid embryoids and plants can now scarcely be questioned.

In some other plants, haploids do not arise directly from the microspores, but do so through the intervention of a callus. The miscrospores first develop into multi-cellular bodies which later give rise to an exceedingly dense callus. Subculture of the callus in an embryogenesis – inducing medium containing specific concentration of auxins and a cytokinin led to the initiation of haploid seedlings. There, thus seem to exist the following alternative pathways for haploidy in anther cultures:

1. Microspore → embryogenesis → plantlet, *e.g.*, Solanaceae.
2. Microspore → callus → embryogenesis → plantlet, *e.g.*, *Oryza, Brassica, Hordeum, Coffea, Populus*.
3. Microspore → callus → plantlet, *e.g.*, *Datura metelloides*.

In 1974, in a major extension to haploid production in anther cultures, C. Nitsch was able to induce embryogenesis in free microspores of tobacco grown in a fully defined medium. In this technique, whole anthers are first cultured in a regular medium for 4 days. The microspores are then squeezed out of the anthers and are transferred to the embryogenesis – inducing medium consisting of glutamine, serine, inositol, mineral salts and sucrose. In this case, the anther wall probably transferred some substance to the microspores during the 4-day conditioning period itself. In all cases, plantlets resulting from anther culture are teased out when the anther opens. When adventitious root system is developed they are potted in compost or a mixture of peat and sand and maintained for a few days under a moist chamber before transfer to the nursery or field.

(1) Production of homozygous lines from haploids derived from anther culture.

Nitsch and Nitsch (1969) reared a large number of haploid plants from cultured anthers of four different species of *Nicotiana*. Although the plants flowered profusely, flowers were smaller than those formed on diploid plants, and further, did not set seeds. This was to be expected because of upsetting of meiosis due to the haploid state. In order for haploids to be useful therefore, their chromosome numbers need to be doubled.

Spontaneous doubling of chromosomes in haploids of *Nicotiana* was reported by Sunderland (1970). About 1% of plants gave mixed inflorescenes with haploid and diploid flowers, the latter, probably, a result of endomitosis at some stage of development to give a diploid lineage. If a large number of haploid pollen plants can be produced, even a low frequency of these natural diploids will give a working number of pure lines. Plants doubled spontaneously in this way are preferred, because they are less likely to show nuclear aberrations as compared to those produced by either of the following two methods.

(2) Artificially induced doubling with colchicine.

The simplest way is to immerse the pollen plants in 0.4% colchicine solution for upto 96 hours, as soon as they emerge from the anther, or apply lanolin paste to axillary buds of decapitated plants. Chromosome number can be checked from leaf squashes. Many of the tobacco plants treated in this manner are aberrant.

(3) Regeneration from callus.

Nitsch *et al* produced callus from stem segments or other parts of the pollen plants and then transferred to a regular medium. The high rate of endomitosis in callus tissue does the diploidisation. During homozygote production by this method, the seed needs to be tested vigorously for genetical and chromosome defects.

APPLICATION

The traditional method of producing homozygous pure lines was by controlled inbreeding over several generations. This is a time consuming process even in the case of annual agricultural plants. In perennial tree species, which take several years to complete their juvenile phase before attaining flowering maturity, such repeated selfing over eight or more cycles is a near impossible task, an expensive lengthy process that will take several decades, if not, hundreds of years. Further, inbreeding leads to low seed set and reduced growth and in dioecious species such as poplars, no selfing is possible. It is in this context that haploids and their homozygous derivatives offer an invaluable tool for plant and tree breeding. Because haploid plants can be obtained in quantity by anther culture techniques, and transformed into true-breeding, isogenic diploid lines, by simple procedures, experimental systems are becoming available for rising such lines for breeding programmes in crop improvement and development of new varieties. Screening of haploid progenies from heterozygous F1 individuals, has already been applied in tobacco genetics with respect to leaf colour and resistance to black shank, tobacco mosaic and wildfire diseases. The isolation of nulli-haploid plants from cultured anthers of monosomic tobacco plants and the regeneration of nulli-somics provide a tool for determining linkage relationships and development of genetic maps in this plant. In some cases, induced androgenesis has opened the way to the evolution of disease resistant selections as in *Geranium*.

The selection of haploid mutants from populations of androgenic haploids by irradiation techniques has been pursued by some workers.

A great potential of haploid cell lines is that they provide a foot-hold within higher plants for applying the powerful techniques of molecular biology and microbial genetics in uncovering and selectively recovering defined bio-chemical and temperature sensitive mutants. By using the chemical mutagen Ethyl-Methyl-Sulphonate (EMS), Carlson (1970) isolated six auxotrophs having impaired ability to synthesize hypoxanthine, biotin, p-aminobenzoic acid, arginine, lysine and proline. Plants regenerated from the mutant cells showed different leaf shapes and growth characteristics very much like morphological mutants. The likelihood of emergence of agriculturally significant haploid mutants in tobacco has been enhanced by the discovery that plants regenerated from mutant cell cultures resistant to methionine sulfoximine, which is a doubtful structural analogue of the toxin produced by the wildfire disease pathogen (*Pseudomonas tabaci*), showed milder disease symptoms when inoculated with the pathogen. At the genetic engineering level, pollen derived haploid plants of *Arabidopsis thaliana* and *Lycopersicon esculentum* have been successfully used for the transfer and subsequent expression of three systems of genes from the bacterium *Escherichia coli*. This study is of strategic importance, since it strengthens the case of the use of genetic engineering techniques for information transfer between plant cells to produce desirable strains. Such experiments could have important implications for morphogenesis because they can reveal to what extent we can control or predict the subsequent development of an individual cell by input of information at specific stages during its development and maturation.

CONCLUDING REMARKS

Since its discovery some forty years ago, considerable progress has been made in the field of pollen embryogenesis. Generally, less than 5% of the total microspore population of an anther gives rise to embryoids. Future research should aim at: (1) bringing a higher percentage of microspores into the embryogenic pathway: (2) we need to understand the molecular mechanisms by which the microspores switch their assigned developmental route and form embryoids instead of pollen tubes and gametes. There, no doubt, is some important relationship between pollen embryogenesis and gene activity, but this remains to be elucidated.

Culture of embryos. The easiest method of culturing an embryo is to allow it to develop *in situ* within the ovary or dissected out ovule. The ovule, if placed on a suitable nutrient medium, is able to support the development of the zygote to maturity. The mature embryo is autotrophic and grows well if provided with the normal conditions necessary for germination such as adequate water supply, oxygen and favourable temperature. In 1941, Van Overbeck introduced a technique for the culture of young embryos. According to him the embryos can be grown from a quite immature stage by supplementing sucrose and mineral salts with coconut milk (liquid endosperm). In India, work is going on in some laboratories (especially in the University of Delhi) for culturing the embryos to obtain new hybrids. The embryo culture technique has been successfully employed by Abraham and Ramchandran in *Colocasia esculenta* where the seeds fail to germinate in nature. They excised the embryo and grew on synthetic media and obtained seedlings which when transplanted in soil gave rise to normal healthy plants.

Plant embryo culture involves the excising of young embryos from the ovules and growing them to maturity in artificial media – a process not unlike the well known Caesarian section, in which an immature animal embryo is removed from the body of the mother and grown in an incubator.

Hannig (1904) was the first worker to try this method. Using certain crucifiers as the objects of his study, he tried a variety of nutrient media containing sugars, mineral salts, plant decoctions, certain amino acids and gelatin. Mature plants were reared from embryos that were 1.2 mm in length at the time of excision, but presumably the radicle, plumule and cotyledons had already been formed at this stage. Following Hannig, Stringe (1907) grew embryo of several cereals, but instead of placing them in culture media he transferred to the endosperms of other genera of the family. Still later, Dietrich (1924) found that Knop's solution with 2.5 to 5 per cent sugar and 1.5 per cent of agar enabled prompt and regular growth of embryos removed from immature seeds of several species. He further observed that cultured embryos tended to skip the stages of development which had not been completed at the time of excision and grew directly into seedlings. His efforts to cultivate embryos less than one-third of their mature size were, however, unsuccessful. This was Laibach (1925, 1929) who first showed the possibilities of using embryo culture to economic advantage. While making some interspecific crosses in the genus *Linum*, he found that the cross *L. perenne* X *L. austriacum*, yield fruits of approximately normal size, but the seeds were greatly shrunken and only about half as heavy as the normal ones. By dissecting out the embryos and placing them on damp blotting paper, he was able to induce their germination and the resulting plants flowered and fruited abundantly.

Laibach's brilliant exposition led to more intensive studies on the *in vitro* culture of embryos and during recent years a number of papers have appeared on the subject many of which are from Indian workers. In several crosses which were formerly unsuccessful, the hybrid embryos have been successfully reared to maturity. In crosses involving the early ripening varieties of *Prunus* (cherry, plum and peach genus) as female parents, the embryos abort and the seeds are not viable. Tukey (1944) attempted artificial culture of the embryos and succeeded in obtaining mature plants from them. The stony endocarp was split using a scalpal, the integuments, nucellus and endosperm were cut through, the embryos were removed and dropped under aseptic conditions into bottle containing nutrient agar. The seeding, arising in the bottles, were first transplanted to sterile sand, watered with a nutrient solution, then to soil and finally grown in the field. In time they developed into vigorous fruiting trees. A similar technique was also used for embryos of *Iris*, a garden plant.

Similar artificial culture of hybrid embryos have been recorded in various plants. To mention a few examples, Jorgensen (1928) used this method to obtain hybrids between *Solanum nigrum* and *S. luteum*; Beasley (1940) between *Gossypium hirsutum* and *G. herbaceum*; S. Kirm (1942) between some species of *Prunus* and *Lilium*; Smith (1944) between *Lycopersicon esculentum* and *L. peruvianum* and Sanders (1948) between several species of *Datura*. When the embryo is of a rather large size, it can often be dissected out with a needle, while the seed is held between the fingers. With small embryos, a dissecting microscope is necessary. The excised embryos are transferred to previously prepared culture bottles containing a nutrient medium. Since conditions favourable for the growth of the embryos are also favourable for the growth of various bacteria and fungi, even a slight carelessness may cause the cultures to become contaminated, resulting in death of the embryos. Suitable precautions must therefore, be taken both at the time of dissection of the embryos and during their transfer to the culture medium. Various chemicals are available for sterilising the seeds.

Contamination in the culture room from air borne spores is reduced by using lamellar flow chambers, air filters, or simply spraying the tables and walls with a one per cent solution of carbolic acid. The dissecting instruments are dipped in 70 per cent alcohol and flamed. The embryos being delicate are not treated with any solution but are dropped immediately after dissection, into the sterile culture medium. The composition of the medium is naturally a most important factor. Older embryos are largely autotrophic and usually present little difficulty. The problem is usually with the younger embryos which are largely heterotrophic. As many hybrid embryos frequently abort at a very early stage, before the differentiation of the cotyledons, it is necessary to develop proper methods of culturing them before degeneration commences.

Since embryos are nourished inside the seed by the endosperm, Van Overbeek *et al* (1942) thought of using coconut milk (which is the liquid endosperm), as one of the ingredients of the culture medium. By adding this, they succeeded in growing embryos of *Datura stramonium* which were still in the heart - shaped stage and measured only 0.15 mm in length (the mature embryo is approximately 0.6 mm long).

The technique of artificial culture of embryos has proved its importance in several ways:

(1) It is the method par excellence for understanding the nutritive requirements of the developing embryo.
(2) It gives us an insight into the factors that influence embryonic differentiation.
(3) It offers a means of achieving a much wider range of hybrid combinations than has been possible upto this time. These combinations are of potential economic value in agriculture, horticulture and forestry.

TISSUE CULTURE

In contrast to the techniques of organ culture where whole organs are used, tissue culture involves the aseptic culture of an isolated homogeneous mass of cells. It thus represents a simpler system, which is especially useful for studies on differentiation and for research in physiology and biochemistry. Fortunately, tissues from many sources can be maintained in culture for an indefinite period.

When small pieces of root phloem parenchyma of carrot (*Daucus carota*) or pith parenchyma of tobacco (*Nicotiana tabacum*) stem, or even chlorophyll containing cells from cells of *Arachis hypogea*, and *Crepis capillaris* are placed on a suitable medium, they cannot only be kept alive but can be induced to grow. In general, mature parenchymatous or mesophyll cells, which if left undisturbed in the plant body, would undergo no further cell division, can be made to divide mitotically, giving rise to an undifferentiated callus. An extreme example of retention of the capacity for cell division in mature plant cells was provided by cultivation of a callus from medullary ray tissue, excised from a region adjacent to the pith in 50 year old time (*Tilia*) tree. These cells had matured a full half-century earlier, yet they still retained the potential for active cell division and expressed it under suitable conditions in culture.

Any living, nucleated, plant tissue is potentially capable of proliferating into an undifferentiated callus when excised and placed on a suitable culture medium. There is, however, considerable diversity in the nutritional requirements of tissues from different species, or even from different locations within the same species.

It is comparatively easy to culture tissues consisting of non-green paranchyma, such as phloem or pith parenchyma. Success with establishing green photosynthesizing, callus growths from chloroplast containing leaf cells was achieved much later.

Morphogenesis : Tissue and Organ Culture

In addition to the usual macro and micro-inorganic nutrients, and an organic carbon source, isolated tissues are frequently found to require:

(1) An organic source of reduced nitrogen, which may be supplied as amino acids or in some species, as L-glutamine.
(2) Vitamins including thiamin, nicotinic acid and pyridoxine; and
(3) The sugar alcohol, myo-inositol.

In addition, an auxin, such as 2–4 D and sometimes a cytokinin, are required.

The fact that it is necessary to supply hormones to callus cultures, whereas they are not normally required by organ cultures, may indicate that the organized meristems of organ cultures may be centres of hormone bio-synthesis, whereas the parenchymatous tissues from which callus cultures are derived, do not have the capacity for hormone synthesis.

Vitamins are produced in the leaves of plants and supplied to other organs. Understandably, photosynthetic callus growth from palisade mesophyll cells of *Arachis hypogea* does not need an external supply of any vitamins.

Repeated subculturing of some tissue cultures, such as those of carrots, grapes, tobacco and other plants leads to a spontaneous and irreversible change in that they acquire the capacity to synthesize excess quantities of auxin. Thus, a tissue when first brought into culture requires an exogenous supply of auxin, later on, after subculture, becomes autotrophic for auxin. Such long established callus cultures are then said to be habituated or energized and closely resemble tumorous as opposed to normal plant tissues. For example, plants infected with the crown-gall bacterium (*Agrobacterium tumefaciens*) exhibit tumorous (callus-like) growths at the points of infection. By appropriate heat-treatment, the bacterial can be killed and removal of portion of one of these treated tumours into sterile culture leads to the production of a massive undifferentiated callus, which is completely self-sufficient in auxin. Thus, infected or habituated cells have undergone a permanent change in that they are able to synthesize substances which they were unable to produce before. This capacity is transferred from one cell generation to the next.

FREE CELL CULTURE (Suspension culture).

Suspension culture of free cells have also been obtained. The principal characteristic of free cells in culture are:

(1) Numerous and large vacuoles, even in cells capable of division.
(2) Prominent cytoplasmic strands which show active streaming movements.
(3) A large nucleus with nucleolus.

Suspension cultures are obtained in the following ways:

(*i*) Placing pieces of a friable tissue such as callus in a moving liquid medium; (*ii*) breaking up sterile seedlings or embryos in a homogenizer and subculturing the suspension after sedimentation. Various kinds of mechanical shakers and spinners are used to agitate the liquid medium to secure separation of cells. Where single cell clones are desired the suspensions are plated out on agar plates in petri dishes.

Free cells of higher plants when cultured do not have the morphology of cells in the tissue from which they are derived. They look very much alike no matter from what species they originated. They evidently have a different pattern of metabolism. Typical storage products are absent from them. This is because they are "leaky" that is, lose substances easily to the medium. This is particularly because of their large surface area exposed to the medium, in contrast to callus cells which are surrounded by other cells. Consequently, the nutritional requirements of free plant cells may be more complex than those for callus cultures of the same species, since it is necessary to supply the substances which tend to be lost by the cells into the medium. It is for this reason that it has not been possible to grow free cell cultures on a defined medium. It is necessary to add coconut milk, which must therefore, include certain as yet unknown, special nutrient factors. What is striking is the very close similarity between nutritional requirements for free cell culture and for embryogenesis in isolated young embryos. In the intact plant, these requirements are obviously met by surrounding tissues – particularly, in the case of embryos, by the endosperm.

In a suspension culture of isolated cells some cells divide and give rise to clusters of smaller, more dense cells. Others increase in size and divide with the formation of internal cross walls to produce either a filament of cells, or in some cases, a new free cell by a process analogous to "budding" of yeast cells in culture.

Protoplast Culture (Culture of isolated protoplasts).

An even more advanced technique of free cell culture is that involving the culture of isolated protoplasts. Essentially, the method consists of, first, breaking down the cell wall mechanically or chemically using enzymes and thus freeing the protoplasts which are then cultured like whole cells using appropriate culture media. The cells are, in other words, rendered naked and then cultured.

The isolation methods devised by Takebe *et al* (1968, 1969), consists of the following steps:
1. Stripping of the lower epidermis of the leaves mechanically such as by peeling, to expose the mesophyll.
2. Vacuum infiltration of the mesophyll in a hypertonic solution containing a crude pectinase, mannitol and potassium dextran sulphate.
3. Agitating gently to separate the cells. The first cells which are released are discarded and the medium is replaced by fresh solution.
4. Cells are harvested at intervals, concentrated and washed by low speed centrifugation.
5. Incubation of the isolated cells in a solution containing a crude cellulase from *Trichoderma viride*, mannitol and calcium ion. The protoplasts which result are harvested by gentle centrifugation and are washed several times to remove the cellulase.

Variations of this technique include use of sorbitol in place of mannitol, and cutting of leaf tissue instead of removal of the epidermis. Also "one step" procedures have been deviced by which protoplasts are prepared directly from the leaves without first harvesting cells. Here, leaf tissue is incubated in a mixture containing both pectinase and cellulase which dissolve the cell walls and release the protoplasts.

It is possible to isolate 10^6 to 10^7 protoplasts from 1 gm. of leaf tissue.

Isolated protoplasts in culture is a system that has been used for study of many aspects of pure and applied botany and plant virology.

One of the problems of genetic transformation has been the cell wall which serves as an effective barrier to the introduction artificially into the cell of exogenous DNA. The difficulty is removed with the isolation of the protoplasts from the impeding cell wall and thereby introduction of the DNA into the cell is rendered easy.

Organelle isolation is easier from protoplasts than from whole cells or plants, especially in case of sensitive compounds such as mRNA and tRNA. Transfer of nuclei (somatic fusion), plastids, mitochondria, ribosomes, tonoplasts, etc., is also easy. So also the transfer of micro-organisms (such as blue-green algae for *nif* gene research, fungi, bacteria and viruses in pathological research and in genetic engineering). Isolated protoplasts are essential for parasexual somatic hybridization. Using them, distantly related, unrelated and even plant and animal cells have been fused to give cybrids and hybrids as described in greater detail under somatic hybrids.

Like cells, protoplasts can be induced to produce fertile "test tube" plants. This has been successfully achieved so far with five different genera: *Nicotiana, Daucus, Petunia, Asparagus* and *Brassica*.

REGENERATION STUDIES WITH STERILE CULTURE

Whole plants can be regenerated from isolated pieces of shoot, root or leaf and even from relatively unorganised tissues such as callus, if the proper nutrients and growth regulators are provided in the culture.

In callus culture, cell division occurs randomly in all directions and gives rise to an unorganized mass of tissue; thus, as indicated earlier as there is no clearly defined axes of polarity in a callus.

Skoog discovered that by varying the proportions of auxin and cytokinin in the medium, he could induce initiation of organised meristems in tobacco pith derived callus cultures. A higher concentration of cytokinin relative to auxin caused cells to differentiate into shoot apical meristems, whereas with higher auxin concentration root primordia were initiated. Since this first discovery, control of root or bud-formation in callus cultures by variations in auxin – cytokinin balance has now been demonstrated for tissues of several origins. The interacting effects of the hormones in this phenomonen can be modified by other factors, such as sugar and phosphate levels, sources of nitrogen and other constituents of the medium such as purines. Light can also influence it. For example, initiation of lateral roots in pea stem segments is inhibited by red light. Thus a phytochrome based mechanism may be involved in the initiation of root apical meristems.

The differentiation of normal plant embryos in tissue cultures was first observed by Reinert in 1959. By manipulating the auxin-cytokinin balance in the nutrient media, he obtained perfectly normal embryos in callus cultures of carrot root phloem parenchyma. These adventive embryos, transferred to a suitable medium, developed into whole carrot plants. Since this first report, a number of other workers have obtained adventive embryos in callus cultures from a number of plant sources. Greatest success, however, has been achieved with the wild strain of *Daucus carota*. Steward demonstrated that scrapings of tissue from immature ovular embryos of this plant will form a friable callus in culture, which when transferred to an agitated liquid medium containing coconut milk breaks up to form a suspension of cells. Plating out the cell suspension on to agar containing the same nutrient medium resulted in the formation of literally thousands of carrot embryos each of which could develop into a whole plant. It was noticed that in some cases the embryos formed on the agar medium were derived from what were originally single cells – the first unequivocal demonstration of totipotency in plant cells. Embryos formed in this way pass through the typical stages of embryogenesis seen in normal ovular embryos and eventually grow into adult carrot plants.

Callus derived from the root, hypocotyl, stem and petiole of wild carrot have all been shown to be capable of giving rise to adventive embryos. But those derived originally from embryo tissues show a greater tendency to form adventive embryos than cultures from the above mentioned older tissues. This suggests that previously matured tissues even when "dedifferentiated" by being brought to meristematic state in culture, revert less readily to the embryonic condition.

It is not that coconut milk alone contains some embryo inducing factors, because, Halperin in 1964 succeeded in inducing adventive embryo formation in carrot root callus in a complex, fully defined medium, without coconut milk.

Regeneration of Vascular Tissue

Wounding or grafting causes regeneration of vascular tissue in whole plants. It has been shown that auxin can induce differentiation of vascular tissue in callus mass. Probably wounding or the grafted bud naturally produces auxin which causes vascular strands to differentiate thus helping to establish the graft.

Regeneration in Shoot and Root Cuttings

The stimulatory effect of buds and leaves upon rooting of cuttings is probably due to the production of auxin by these organs. Auxins always travel in a basipetal fashion, but at present we know little of the translocation patterns of cytokinins in plants, although we know that they are synthesized in the roots and translocated up the shoot system. In addition to auxin, root formation requires a supply of sugar as well as other nutrients.

Auxins stimulate lateral root formation, but in higher concentrations, inhibit it. Thus a root immersed in a solution of auxin becomes stunted and possesses rows of newly emerged but suppressed lateral roots.

The phenomenon of regeneration provides strong evidence that the process of differentiation in many types of plant cells does not involve any loss in their genetic potentialities, so that they remain "totipotent".

Cells growing *en masse* in a callus do not form embryos, but will do so if they are separated in a free cell culture. So a cell when it becomes part of a tissue system becomes restricted. Probably neighbouring cells influence and control it in some way through the plasmodesmata.

Generally speaking, in tissue cultures there is first dedifferentiation a regressive change, involving a progressive return to a meristematic or ground stage. Eventually, however, there is the formation of differentiated structures, which supersede the regressive changes and a new course of development is initiated, which is marked by increasing cellular differentiation.

APPLIED ASPECTS OF ORGAN, TISSUE, CELL AND PROTOPLAST CULTURE

Sterile or aseptic culture technique of organs, tissue, cell and isolated protoplasts finds several applications. It can profitably be utilised for the improvement of agricultural crops, in horticulture and forestry as well. This subject will be considered at some length in the following pages.

METABOLIC AND BIOSYNTHESIS STUDIES

Whilst admitting that metabolic patterns observed in excised material grown in culture are not bound to be identical with those in intact plants, many workers have seen in tissue cultures an attractive system in which to investigate metabolic inter-relationships and pathways of biosynthesis. It offers many of the technical advantages or work with micro-organism to the plant investigator. Auxin metabolism of tumour tissue in plant 'Cancers', photomorphogenesis, aminoacid metabolism, synthesis of secondary metabolites like steroids, *e.g.*, diosgenin in *Dioscorea*, Cholesterol in *Nicotiana*, Cardiac glycosides in *Digitalis*, indole alkaloids in *Catharanthus* are all facilitated in tissue cultures. For example, studies of tea callus culture show that caffeine arises from the breakdown of nucleic acids rather than from purine pools.

Application of tissue culture to the controlled production of such compounds as alkaloids, steroids, vitamins, antibodies and enzymes on a commercial scale is a possibility, although this ambition has not so far been realized. For such production systems, cell suspension cultures seem ideal. It may be recalled here that penicillin and other antibiotics and some genetically engineered products such as humulin are in fact obtained through such *in vitro* systems, using specific micro-organisms. So, similar factory production of other drugs, etc., from higher plants is within the realm of future possibilities. Animal tissue culture techniques are now regularly being used in the commercial production of certain vaccines, such as those against poliomyelitis, measles and adenovirus. In sharp contrast, the contribution of plant tissue culture techniques to the study of plant-virus physiology has been, so far, very limited. However, tissue culture has played an important role in the production of virus-free plant, by meristem tip culture. Most commercial crop plants, particularly those which are propagated vegetatively, contain systematic viruses which affect performance or depress yield.

Tissue culture *in vitro* enables investigation of certain aspects of cellular physiology regeneration and morphogenesis which would be otherwise inaccessible for study *in vivo*. In tree physiology, it enables early prediction of tree growth from callus characters and hence early selection of fast growing genotypes without waiting for their phenotype to grow to maturity which will take decades.

MICROPROPAGATION

Tissue culture serves as an alternative to vegetative propagation by conventional methods, such as by rooting of cuttings. This is particularly valuable in the case of difficult-to-root species and those species like the palms which can neither be rooted by cuttings nor grafted because they lack a cambium. Recently, genetically uniform oil palms (*Elaeis guineensis*) have been produced by tissue culture (Jones, 1974; Ahu *et al*, 1981). Attempts are being made in the same direction with the coconut palm also. Vegetative propagation through tissue culture is called micropropagation, because it uses very small quantities of starter material, say, a single cutting or even a few grams of tissue to regenerate hundreds of plants. However, seedlings from tissue cultures are relatively more expensive than those produced by other means of vegetative propagation such as cuttings. Therefore, it is applied only to high value species like oil palm where, moreover, other alternative vegetative propagation methods do not exist. For low-value species such as *Eucalyptus* rooted coppice shoot cuttings are still used commercially. Plants raised from tissue cultures require a longer hardening period before they become fit for transplanting, especially in harsh forest environments. Genetic instability of callus cultures is another drawback. Thus chromosome abnormalities are common in tissue cultures of Gymnosperms which constitute important components of temperate and arboreal forests.

VIRUS-FREE PLANTS

Shoot meristem culture for obtaining virus-free healthy plants has been utilized in the horticulture of some ornamental plants such as *Geraniums*, *Tulips*, *Orchids* and for controlling mosaic in *Cassava* plantations. For some as yet little understood reason even in infected plants, the virus stays clear of the apical meristem of the shoots. So, if a susceptible variety has other desirable characters, we can simply clone apical meristem through aseptic culture *in vitro*, each time planting has to be done a new.

GERMPLASM EXCHANGE

Tissue culture has potential applications in germplasm exchange and plant quarantine. The international transfer of plant propagation material for research, breeding, collection, and conservation involves the concurrent risk of large scale spread of plant pests and pathogens. Aseptic tissue culture can serve as an additional quarantine safeguard in this context, which effectively eliminates such risks. The exchanged culture materials can be multiplied by micropropagation in the respective recipient countries. Meristem tip culture of potato and cassava germplasm have been exchanged in this manner using standard mail and freight services.

IN VITRO CONSERVATION OF GERMPLASM

In recent years the necessity for applying *in vitro* techniques to plant genetic conservation of wild germplasm and endangered species has become apparent. Storage of *in vitro* culture of higher plants was first reported in 1973, when, Nag and Street described the regeneration of plants from a frozen and thawed embryogenic culture of *Daucus carota*. In subsequent years the list of species that have been successfully cryopreserved has grown to more than 50. All types of culture are included from protoplast to cell, callus, organ and embryo. Liquid nitrogen is used to create the required low temperature. Special liquid nitrogen storage refrigerators are now available for organised storage of up to 3600 ampoules of 1 to 9 ml capacity. These are compact, simple to maintain, require no power supply and need only periodic topping up with liquid nitrogen. The technique of cryopreservation is well established in animal breeding where frozen sperm of choice bulls is stored in sperm banks, exchanged and used for artificial insemination to produce new breeds. This method is now so well standardized that it can be used even in remote locations.

One of the chief advantages of cryopreservation of tissues rather than seeds or other propagation material is its compactness. Because of this, very large number of accessions are possible with the minimum space requirements. It is analogous to the preservation of bulky books and archives safely as microfilms or computer discs.

TISSUE CULTURES AS SOURCES OF SOMACLONAL VARIATION

When tissues and cells are removed from their stabilized internal environments of the intact organism and plunged into the alien environment of the culture vessel, they show nuclear irregularities – chromosome and gene mutations. The former are relatively the easier to detect by cytological techniques. There is in the culture medium, a much less severe sieving process to eliminate the unwanted variants than there is *in vivo*. While such genetic instability may be undesirable for certain purposes such as mass cloning of uniform genotypes, *via* tissue culture, such somaclonal variation offers the only practical way to generate extensive and possibly useful phenotypic variability in asexually propagated species like potato and sugarcane.

The old, but particularly valuable potato cultivar Russet Burbank is seed sterile and hence been excluded from potato breeding programmes. Further this cultivar is highly susceptible to late blight (*Phytophthora infestans*). Among plants regenerated from cultured protoplasts of the cultivar, Shepard *et al* found that about 2 per cent of somaclonal derivatives displayed enhanced resistance and some were resistant to both common races of the pathogen. Somaclones were also screened for resistance to early blight (*Alternaria solani*). Among a population of 500 somaclones, five were more resistant than the parent and subsequently four proved to possess field resistance. Shepard concluded that cell culture-generated variation was sufficient to provide enough variability to facilitate the selective improvement of Russet Burbank. Similar level of somaclonal variation has been observed in many potato cultivars and is being actively explored as a viable potato breeding option in both the USA and Europe. Several independent studies have shown that in sugarcane too disease resistant plants can be recovered from callus cultures. These have been summarised by Larkin and Scowcroft (1981) and include resistance to Fiji virus disease and downy mildew (*Sclerospora sacchari*) and culmi-colous smut (*Ustilago scitaminea*). Callus regenerated plants were assayed for their reaction to the host-specific pathotoxin. This assay is rapid, repeatable and discriminates between resistant and susceptible cultivars. Toxin tolerance remained stable through a second culture cycle.

In tobacco, somaclonal variation in cell culture, has been utilised to isolate herbicide resistant plants.

APPLICATION IN GENETIC ENGINEERING

Tissue and protoplast cultures have been used in genetic engineering for the transfer of DNA and extra-chromosomal bodies-plasmids, mitochondria, chloroplasts, nif (nitrogen fixing) genes from the nitrogen fixing bacterium *Klebsiella pneumoniae* to a strain of the colon bacterium *Escherichia coli*. Isolated protoplasts have great advantage in all the aforementioned uses. For transformation purposes cultured apical meristems are also usable because these can easily by regenerated into whole plants and also because intact DNA taken up by plants appears to be rapidly transported to meristematic regions, where growth and differentiation are centred.

The production of haploid plants by anther cultures and their use in developing in a single jump homozygous isogenic lines have already been described at length. Somatic hybridization and its use in overcoming crossing barriers is described in the following paragraphs.

Somatic Hybrids (Somatic or parasexual hybridization).

Hybridization in plants involves sexual crossing, following natural or controlled pollination It involves fusion of specialised cells, the gametes which is soon followed by the fusion of gametic nuclei resulting in the formation of the sexual hybrid. Advanced biotechnological techniques are now available by which somatic cells of different plants can be induced to fuse like gametes to produce hybrids. The technique is called somatic hybridization and the resulting hybrids are known as somatic hybrids. The field is sometimes called somatic cell genetics. To distinguish sexual from somatic hybrids in the former case X is used to denote a sexual cross and + to denote a parasexual union of somatic cells. Thus, if A and B are the parents, a sexual cross between the two is written as A × B and a parasexual one as A + B.

Somatic hybridization necessitates the use of naked cells. Hence the first step in this process is the isolation of protoplasts from the desired plant tissue. As we have already seen, this is achieved by enzymatic degradation of the cellulose wall from plasmolysed cells using specific commercial enzymes. The most commonly used in Onozuka cellulase enzyme complex, a patented product of All Japan Biochem. Co., Japan.

This enzyme preparation is very efficient in giving a high yield of protoplasts within 1–2 hrs with different cell cultures and leaf parenchyma cells.

For bringing about fusion of protoplasts, polyethylene glycol (PEG) is used as an inducing agent. PEG solution severe dehydration and shrinkage of protoplasts. The latter adhere closely and during the subsequent gradual dilution of the PEG, fusion takes place. When PEG is added to a suspension of protoplasts, within a minute the protoplasts aggregate and fusion follows within a few more minutes.

Fusion of protoplasts results first in a heterokaryon – a fused protoplast with two nuclei. These may remain as such without fusion and give rise to a chimaeral aggregate. But in heterokaryons where extensive mixing together of the cytoplasms has occurred and when the nuclei are close together, their fusion is expected. Synchrony of division of the nuclei would greatly increase the changes of common spindle formation, consequent nuclear fusion and formation of a homokaryon and somatic hybrid.

Successful somatic hybridization strategy will depend on a knowledge of the behaviour in culture of heterokaryons and an ability to select and identify, somatic hybrids arising in these cultures.

CULTURE TECHNIQUE

Different cells in a plant do different functions. For example, in a leaf some cells do photosynthesis, while others are responsible for the function of transport. This shows that the fate of a cell mainly depends on its location. The differentiation of zygote into different types of cells is one of the most challenging problems in developmental biology. Experimental embryology has established intimate contacts between embryology and other disciplines of botany, especially genetics and physiology.

Applied embryology has shown a way by which one can study components of plant body separately and understand the relationship between the parts and the whole. As already mentioned the technique of plant tissue culture enables us to study the cells, tissue or organs by isolating them from the plant body and growing aseptically, in suitable containers, on an artificial nutrient medium, under controlled environmental conditions. Thus (*i*) *nutrient medium,* (*ii*) *aseptic conditions* and (*iii*) *aeration of the tissue* are important aspects of the technique of *in vitro* culture.

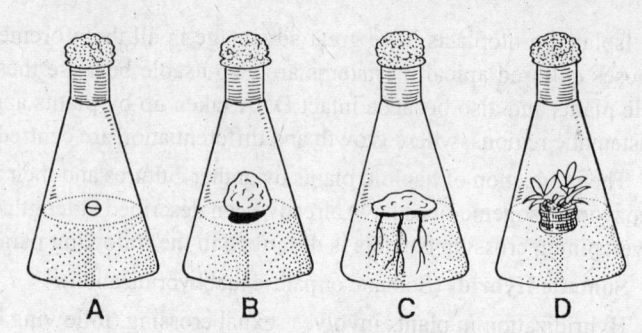

Fig. 22.4. An experiment to demonstrate the effect of growth regulators (auxins and cytokinin) on growth and differentiation in plant tissue culture. A – On the nutrient medium lacking a growth regulator the tobacco pith tissue showed very poor growth; B – In the presence of a cytokinin a vigorously growing callus was formed; C – When transferred to a medium containing 3 mg/1 of indoleacetic acid and 0.2 mg/1 of kinetin this callus differentiated only roots; D – If a portion of the same callus was planted on a medium with higher concentration of kinetin (1 mg/1) and lower concentration of indoleacetic acid (0.03 mg/1), it differentiated only shoots.

Nutrient medium. Every tissue and organ has its special requirements for optimal growth and these need to be worked out when starting work with a new system. However, most of the media contain inorganic salts of major and minor elements, vitamins, and sucrose. A medium with these ingredients will be referred to as basal medium. Sometimes, growth regulators, such as auxins, gibberellins and cytokinins, may also be added to the basal medium. These all constituents are dissolved in distilled water. If necessary, the medium is solidified with about 0.8% agar. The pH of the medium is adjusted around 5.8 (slightly acidic). Now equal quantities of the medium are dispersed in culture vials, which are usually glass tubes or flasks. The culture vials, containing medium, are plugged with non-absorbent cotton wrapped in cheese cloth. Such a closure allows the exchange of gases but does not permit the entry of microorganisms, into the culture vials.

Aseptic conditions. The sugar content of the nutrient media may support a luxuriant growth of many micro-organisms, like bacteria and fungi. It is, therefore, extremely important to maintain a completely aseptic environment inside the culture vials. Micro-organisms can contaminate the medium in at least three ways:

(*a*) The micro-organisms present in the medium right from the beginning may be destroyed by sterilizing the properly plugged culture vials. It can be done by maintaining the temperature at 120°C for about 15 minutes.

(*b*) The micro-organisms may also be carried along with tissue that is being cultured. To prevent this, the plant material from which the tissue is to be excised is surface-sterilized. The material may be surface sterilized with saturated chlorine water and then thoroughly washed with sterilized distilled water and to remove all traces of chlorine. If the material is fairly hard, as are some fruits and seeds, it may be surface-sterilized by rinsing in alcohol.

(*c*) Finally, precautions must also be taken to prevent the entry of micro-organisms while the plug of a culture vial is removed to transfer the tissue to the nutrient medium (inoculation). For this, all operations from surface-sterilization of the tissue up to inoculation are done in an aseptic environment.

Morphogenesis : Tissue and Organ Culture

Aeration. Proper aeration of the cultured tissue is also an important aspect of culture technique. If the tissue is grown on the surface of a semi-solid medium it acquires enough aeration without special device for liquid medium, special device "filter paper bridge" is used. Here, two legs remain dipping in the medium, and the horizontal part carrying the tissue is raised above the level of the medium.

The idea of technique for maintaining fragments of plant tissue alive after their removal from the main body occurred to Gottlieb Haberlandt (1902), a famous German botanist, but he could not get much success. Thereafter Robbins and Kotte worked independently and got a little success. In case of ovule culture, White (1932) isolated ovules and cultured them under aseptic conditions in *Antirrhinum majus*. However, the technique of ovule culture got perfection by Maheshwari (1958).

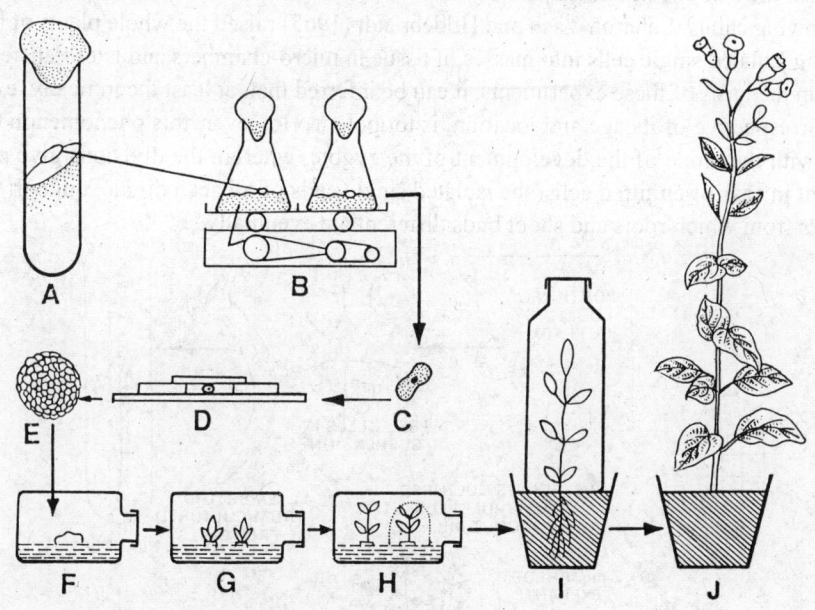

Fig. 22.5. Development of tobacco plants from single cells. A callus is raised from a small piece of tissue excised from the pith (*A*). By transferring it to a liquid medium and shaking the culture flasks (*B*), the callus is dissociated into a single cell (*C*), is mechanically removed from the flask, and is placed on a slide in a drop of culture medium (*D*). A microchamber is formed around the culture drop using the three-cover slips (*D*). A small tissue (*E*) derived from the cell through repeated divisions is then transferred to a semi-solid medium where it grows into a large callus (*E*) and eventually differentiated plants (*F, G*). When transferred to soil (*H*), these plants grow to maturity, flower and set seeds (*I*).

A culture made properly for the occasion (every tissue and organ has its special requirement for optimal growth) provides suitable medium for the organs with intact meristems to grow in organized fashion. A shoot segment with an intact apex may develop into a full plant with adventitious roots. Similarly, pollinated ovaries have also been grown into mature fruits in test-tubes. But in the case the fully differentiated tissues excised from the mesophyll, pith, endosperm or secondary phloem usually show an altogether different pattern of growth. Here, the cells divide in an irregular fashion forming an unorganized and undifferentiated mass of tissue called **callus**. In 1939, Gautheret, Nobecourt, and White succeeded in raising callus cultures from isolated plant parts over an indefinite period of time.

The differentiation of callus into shoots, roots or embryo-like structure (embryoids) depends on the nature of the plant tissue cultured. In case of undifferentiated callus, the treatment with hormones can cause their induction. Skoog and Miller have discovered the role of auxins and cytokinins in tissue culture for controlling the shoot and root formation.

Cellular totipotency. This is the capacity of mature cells showing that when freed from the plant body, they had the ability to reorganize themselves into the new individuals. Steward and his co-workers, showed this phenomenon in the carrot cultures. Here the small pieces of mature carrot root, were grown in a liquid medium supplemented with coconut milk, in special containers. These cultures were shaken generally which freed all the cell clusters into the medium. Some of them developed into rooting clumps. When these were transferred to the tubes containing a semi-solid medium, they gave rise to whole plant that flowered and set seeds. These were Muir and co-workers (1958) who demonstrated the test of cellular totipotency in a real way by the formation of a plant body from an actively growing callus. Later on Vasil and Hildebrandt (1965) raised the whole plants of tobacco by first growing isolated single cells into masses of tissue in micro-chambers and later transferring them to flasks. On the basis of these experiments, it can be inferred that, at least theoretically, every living plant cell, irrespective of its age and location, is totipotent. However, this phenomenon can not be compared with the mode of the development of the zygote, wherein the divisions give predictable manner. But in case of cultured cells, the isolated single cells of tobacco divide, quite irregularly to form a mass from which roots and shoot buds differentiate eventually.

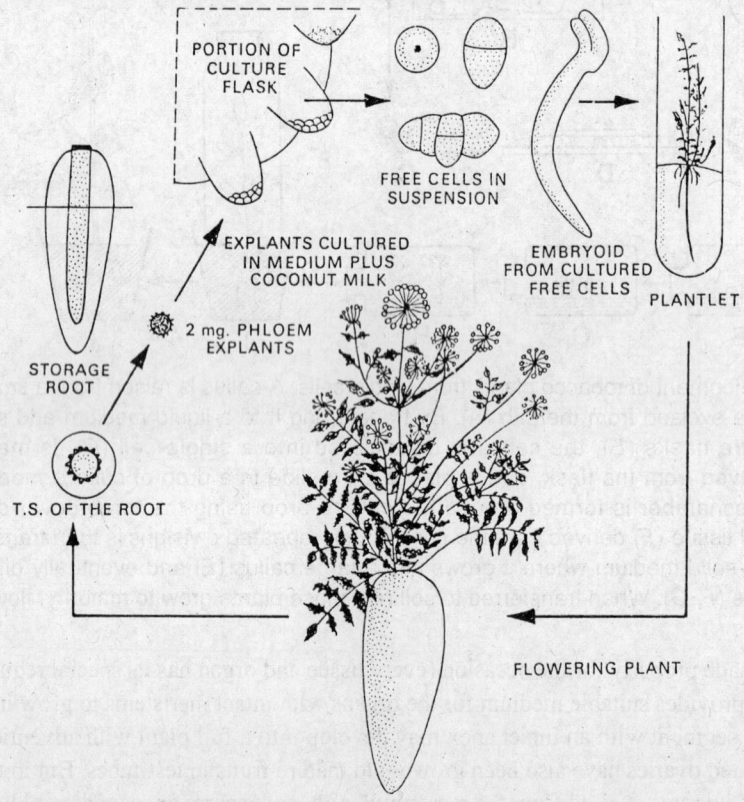

Fig. 22.6. An experiment to demonstrate the totipotency of plant cells. Slices of the carrot root were cut and small pieces of tissue were taken from the phloem region. These were inoculated into a liquid medium in special flasks, the tissue grew actively and single cells and small cell aggregates dissociated into the medium. Some of the cell clumps developed roots, and, when transferred to a semisolid medium, these rooted nodules formed shoots. These plants could be transferred to soil.

In 1964, Guha and Maheshwari, while culturing mature anthers of *Datura innoxia* with an aim of understanding the physiology of meiosis, accidentally noticed that on a basal nutrient medium containing kinetin, coconut milk, or grape juice numerous embryo – like structures appeared from the inside of anthers, which eventually, developed into plantlets. In 1966, these workers confirmed the origin of the embryoids from pollen grains. As expected, the plantlets of pollen – origin were haploid.

APPLICATIONS OF TISSUE-CULTURE

(*a*) The technique provides a way for rapid multiplication of desirable and rare plants.

(*b*) As the experiments reveal, we may obtain healthy stocks from virus infected plants through shoot-tip culture.

(*c*) The development of haploids through the techniques of anther culture is having its potential significance in basic and applied genetics, and plant breeding. During the past so many years the technique has been successfully extended to about 20 species including some economic plants, *e.g.*, *Atropa, Brassica, Hordeum, Lycopersicon, Nicotiana, Oryza* and *Triticum*.

(*d*) The embryo culture has been useful in overcoming seed dormancy. It is also utilized for producing viable plants from crosses which normally fail due to the death of immature embryos, *e.g.*, jute, rice.

(*e*) The embryo tissue culture is also applied for the propagation of rare plants, *e.g.*, "makapuna". Here, some coconuts develop soft, solid, fatty tissue in place of the liquid endosperm (Mohan Ram, 1976). These are rare and very expensive, served only at special banquets in the Phlippines. Under normal conditions the coconut seeds fail to germinate. Using the technique of *in vitro* culture of excised embryos, De Guzman (1969) succeeded in plantlets from mukapuno nuts.

(*f*) Another important use of embryo culture is in obtaining some rare hybrids. It is possible to raise complete hybrid plants through embryo culture. This method has been profitably used for many interspecific crosses.

Abnormal growths. In dicotyledonous plant, an injury made by a sharp blade causes the surrounding cells to divide and form a tissue mass. In case of small wounds, the wounded cells die and dry up, while outer walls or cells of the underlying uninjured layers become impregnated with protective substances. In larger wounds, the outermost uninjured layer of living parenchymatous tissue forms a meristem (phellogen) which produces one or more layers of cork – the wound cork; the cork then protects the wounded – healing is brought about by certain chemical substances hypothetically called the "wound hormone."

The wound healing tissue functions differently when associated with certain micro-organisms. Here, the cell divisions become irregular and a tissue with the ability of unlimited growth is formed which is referred to as **tumours.** Bacteria, viruses, insects or fungi, and hybridization are the causitive factors for the formation of these tumours. *Agrobacterium tumefaciens* causes the well-grown tumours. Here, the bacterium releases tumour-inducing principle, (TIP) which alters the normal tissue to undergo irregular divisions. Secondary tumours appear at a distance from the primary tumour and lack the bacterium but grow as actively as the primary tumour. The tissue in tumours possesses a high concentration of auxins along with other growth substances.

In plants, like *Nicotiana, Kalanchoe*, a high degree of regeneration ability is found. Here, moderately virulent strains of the bacterium (such as Walnut T37 strain) induce tumours which have abnormal shoot like structure. These are partially organized tumours and called *teratoma*.

Sometimes the species of *Nicotiana, Lilium* and *Lycopersicon* give up tumours spontaneously without a causative organism. This generally happens because of the genetic basis of the tumours.

Viruses, mycoplasma, bacteria, fungi and parasitic angiosperms cause "witches broom" which is another kind of abnormal growth. Mango provides the most common example of it when the entire inflorescences become modified into dense clusters of small, highly branched, fleshy shoots with short internodes and reduced leaves. It causes a great loss in north India.

THE PLASTOCHRON

Plastochron and plastochron index. Leaf primordia are formed in regular sequence below the shoot apex, and it is here that many structural characters of the plant seem to be determined. The time period between the initiation of two primordia (or two pairs, if the leaves are opposite) is termed a **plastochron**. These periodic changes in the meristem can be seen best in opposite leaved forms. As the two primordia begin to appear, the apical dome between them becomes relatively flat. When they have developed further, but before another pair appears, the dome bulges upward again and reaches its maximal surface area. In the lower vascular plants (ferns etc.,) the primordium arises from one or more of the surface cells of the meristem but in the higher ones, it develops as a swelling on the side of the apex at the base of the dome, generally, as a result of periclinal divisions in one or more layers below the surface one. The interval between corresponding stages of successive leaves is called **plastochron index**. This index is a better measure of the stage of development of a growing shoot than its chronological age. In many plants, the shape and structure of the meristem change somewhat with the season also.

The ratio of the radial distances from the centre of the meristem to two consecutive primordia is known as the **plastochron ratio**.

23

Comparative Account

TABLE: 23.1

COMPARISON OF SIMPLE PITS AND BORDERED PITS

Simple pits	Bordered pits
1. Simple pit pairs occur in parenchyma cells, in medullary rays, in phloem fibres, companion cells, and in tracheids of several flowering plants.	1. The bordered pits are abundantly found in the vessels of many angiosperms, and in the tracheids of many conifers.
2. In the simple pits, the pit cavity remains of the same diameter and the pit or closing membrane also remains simple and uniform in its structure. There is no overarching pit border.	2. The bordered pits are more complex and variable in their structure than simple pits. The overarching secondary wall which encloses a part of the pit cavity, is called the **pit border**, which opens outside by a small rounded mouth known as **pit aperture**.
3. The simple pit may be circular, oval, polygonal, elongated or some-what irregular in its facial view.	3. The pit aperture may be circular, lenticular, linear or oval in its facial view.
4. The simple pits occurring in the thin walls are shallow whereas in thick wall the pit cavity may have the form of a canal passing from the lumen of the cell towards the closing or common pit membrane.	4. In the case of thick secondary walls, the border divides the cavity into two parts; the space between the closing membrane and the pit aperture is called the **pit chamber**, and the canal leading from pit chamber to the lumen of the cell is termed **pit canal**.
5. Generally the thickening, called **torus** is not present.	5. The thickening called **torus** is present on the pit membrane.

TABLE: 23.2

COMPARISON OF LATEX CELLS AND LATEX VESSELS

Latex cells	Latex vessels
1. The latex cells are also known as **nonarticulate latex ducts**.	1. The latex vessels are also known as **articulate latex ducts**.

2. These ducts are independent units which extend as branched structures for long distances in the plant body.	2. These ducts or vessels are the result of anastamosing of many cells together.
3. They originate as minute structures, elongate quickly and ramify in all directions of the plant body by repeated branching, but they do not fuse together, thus no netted structures are formed.	3. They originate in the meristems from rows of cells by the absorption of the separating walls early in the ontogeny of the cells; they grow more or less as parallel ducts which by means of branching and frequent anastamoses form a complex net work.

TABLE : 23.3

COMPARISON OF SAP WOOD AND HEART WOOD

Sap wood	Heart wood
1. The outer region of the old trees is **sap wood** or **alburnum**.	1. The central region of the old trees is called **heart wood** or **duramen**.
2. The sap wood consists of recently formed xylem elements.	2. It is filled up with tannins, resins, gums and other substances.
3. It is not hard and durable.	3. It is hard and durable.
4. This is of light colour and contains some living cells also in the association of vessels and fibres.	4. It looks black or dark brown due to the presence of various substances in it.
5. Generally the vessels are not plugged with tyloses.	5. Usually the vessels remain plugged with tyloses.
6. This part of the stem performs the physiological activities, such as conduction of water and nutrients, storage of food, etc.	6. The function of heart wood is not of conduction, it gives only mechanical support to the stem.

TABLE : 23.4

DIFFERENCES BETWEEN SIEVE CELLS AND SIEVE TUBES

Sieve cells	Sieve tubes
1. The sieve cells are found in pteridophytes and gymnosperms.	1. The sieve tubes are found in angiosperms.
2. The sieve cells have no companion cells.	2. The sieve tubes always have companion cells.
3. In sieve cells the sieve areas do not form sieve plates.	3. In sieve tubes the sieve areas are confined to sieve plates.
4. In sieve cells the sieve areas are not well differentiated.	4. In sieve tubes the sieve areas are well differentiated.
5. They are elongated cells and are quite long with tapering end walls.	5. They consist of vertical cells placed one above the other forming long tubes connected at the end walls by sieve pores.

TABLE : 23.5
DIFFERENCES BETWEEN TRACHEIDS AND VESSELS

Tracheids	Vessels
1. The tracheid is a fundamental cell type in xylem.	1. In the phylogenetic development of the tracheid the diameter of the cell increases, and the wall becomes perforated, and thus tracheid converts into vessel.
2. A tracheid consists of a single cell which is elongated and possesses tapering end walls.	2. A vessel consists of a row of cells placed one above the other in a tubular form with total or partial disappearance of the cross walls.
3. The tracheids are elongated dead cells, with walls that are thick in some places and thin in others. They serve both as water conducting and as strengthening cells. The walls of tracheids are impregnated with lignin.	3. The vessels are composed of rows of tracheary cells. The diameter of vessels is usually much greater than that of tracheids. They form long tubes and make the principal water conducting elements of the dicotyledonous stem.
4. The tracheids are short and are usually upto one mm in length and rarely upto 12 cm. In some conifers they attain greater lengths.	4. The vessels are longer and may be as long as 10 cms. In rare cases they attain the length upto 2–6 metres, (*e.g.*, in *Quercus, Eucalyptus,* etc.).

TABLE : 23.6
DIFFERENCES BETWEEN PROTOXYLEM AND METAXYLEM

Protoxylem	Metaxylem
1. It is the first formed xylem, and develops in those parts of plant body which have not completed their growth and differentiation as yet.	1. The metaxylem develops later, when the protoxylem has already started developing.
2. The protoxylem matures before the plant organs have completed their elongation.	2. The metaxylem matures after the completion of the elongation and growth of the plant organs.
3. The elements of protoxylem are smaller in diameter.	3. The elements of metaxylem are broader in diameter than the elements of protoxylem.
4. The elements of protoxylem are subjected to compression and stretching and is sometimes completely destroyed.	4. The elements of metaxylem are not subjected to any compression and stretching force. They are not destroyed.
5. Tyloses are not found in protoxylem vessels.	5. Tyloses have been reported from metaxylem vessels.
6. The protoxylem elements are less prominent.	6. The metaxylem vessels are very much pronounced.
7. They have annular and spiral vessels.	7. They possess generally scalariform, reticulate and pitted vessels.

TABLE : 23.7

DIFFERENCES BETWEEN PROTOPHLOEM AND METAPHLOEM

Protophloem	Metaphloem
1. This is the first formed phloem, and found in an active and functional state in the young and actively growing portions of the plant.	1. The metaphloem differentiates later when the growth has already taken place, and generally found in the mature parts of the plant.
2. In the protophloem the sieve elements are usually thin and inconspicuous.	2. In metaphloem the sieve elements are wide, long and conspicuous.
3. The sieve areas are inconspicuous, and absent in the gymnosperms.	3. In the metaphloem, the sieve areas are distinct and conspicuous.
4. The protoplasts of protophloem are conspicuously vacuolate and without nuclei.	4. Here the protoplasts possess one or more nuclei, which usually disappear in mature sieve elements.
5. The companion cells may be absent or present.	5. The companion cells are present in all cases.
6. The elements of protophloem are short lived. They are obliterated and become non-functional as the plant grows towards maturity.	6. The metaphloem elements remain functional for a longer period. In certain plants which have no secondary growth, the metaphloem is the only conducting tissue for the food material.

TABLE : 23.8

DIFFERENCES BETWEEN PRIMARY AND SECONDARY PHLOEM

Primary phloem	Secondary phloem
1. The primary phloem is derived from the procambium of the apical meristem.	1. This is derived from the vascular cambium, which is a lateral meristem.
2. The protophloem and metaphloem elements are clearly demarcated.	2. The secondary phloem is not differentiated into proto-and metaphloem.
3. The sieve tubes are long and narrow.	3. The sieve tubes are short and wide.
4. The phloem parenchyma is less developed and scanty.	4. The phloem parenchyma is well developed and abundant.
5. Sclereids are generally not found in primary phloem.	5. In many plants the sclereids are found in secondary phloem.
6. The phloem fibres, when present are restricted to the outermost part of the tissue.	6. The phloem fibres are generally found among the phloem parenchyma cells.

TABLE: 23.9
DIFFERENCES BETWEEN PRIMARY AND SECONDARY XYLEM

Primary xylem	Secondary xylem
1. The primary xylem is derived from the procambium of the apical meristem.	1. The secondary xylem is derived from the vascular cambium which is a lateral meristem.
2. The primary xylem is differentiated into proto-and metaxylem.	2. The secondary xylem is not differentiated into proto-and metaxylem.
3. The primary xylem may be endarch, mesarch or exarch.	3. There is no such distinction in the secondary xylem.
4. The tracheids and vessels are narrow and long.	4. The vessels and tracheids are short and wide.
5. The medullary rays are derived from the apical meristem.	5. The secondary meduallay rays are derived from ray initials of the cambium.
6. The vessels of primary xylem do not contain tyloses.	6. Here the vessels contain tyloses.
7. There are no annual rings.	7. The annual rings are well demarcated.
8. The primary xylem is not differentiated into sap wood and heart wood.	8. There is clear cut demarcation of sap wood and heart wood in woody trees.
9. The xylem fibres are few in number or absent.	9. The xylem fibres are abundant.

TABLE: 23.10
COMPARATIVE ANATOMY OF *HELIANTHUS* AND *CUCURBITA*

Helianthus	Cucurbita
1. The outline of the stem in transverse section is almost circular.	1. The outline of the stem in transverse section is provided with ridges and furrows.
2. The multi-layered collenchymatous hypodermis is evenly developed. The cells contain chloroplasts.	2. The collenchyma lies just beneath the epidermis and consists of many layers of the cells in the ridges, while in furrows it is few layered or sometimes absent.
3. There is no chlorenchyma in the cortex.	3. There is distinct chlorenchyma just beneath the collenchyma. This region consists of two to many layers of cells.
4. The oil cavities are found in the cortical region.	4. There are no oil cavities in the cortex.
5. There is single layered endodermis.	5. The endodermis is not very much conspicuous.
6. The pericycle consists of thin walled and sclerenchymatous patches.	6. Just beneath the endodermis there is a multilayered zone of sclerenchymatous pericycle.

7. The vascular bundles are arranged in a single ring.	7. The vascular bundles are arranged in two rings. Each ring consists of five vascular bundles. The vascular bundles of outer ring correspond to the ridges and of the inner to the furrows.
8. The vascular bundles are conjoint, collateral, endarch and open.	8. The vascular bundles are conjoint, bi-collateral, endarch and open.
9. The pith is parenchymatous.	9. The pith is hollow. The vascular bundles lie in ground tissue around the pith cavity.

24

Microscopy and Micrometry

The cells of plants, bacteria and most of animals are microscopic in size and are measured by a unit of measurements, the **micron** or μ. One μ is equal to 1/1,000 mm (*i.e.*, 1000 μ = 1 mm). The various organelles and other components of the cells are still smaller, and are measured by **millimicrons** (m μ) or **angstroms** (Å). One Å is equal to 1/10,000 mm or 1000 Å = 0.1 μ, or 10,000 Å = 1 μ. Most of the cells range from 0.1 μ to 1 mm in size.

The cells and other organelles, that are quite minute in size and cannot be observed by human naked eye. Such components are microscopic and can be seen only with the aid microscope. The human eye possesses limited distinguishing or resolving power and therefore, the naked eye cannot distinguish the objects smaller than 0.1 mm (100 μ). Moreover, the living cells are transparent in ordinary light and cannot be distinguished among various cellular components. To overcome such practical difficulties in the observation of living or dead cells, the cell biologists of the past investigated various methods of staining and magnifying the cells and their components. The object of magnification of cells and their components was achieved by the **lenses** of various types. These lenses could magnify the minute objects upto a particular limit, and thereafter so many lenses were combined together to form an instrument known as the **microscope** (Gr. *micros* = small; *skopein* = to see, to look).

RESOLVING POWER OF MICROSCOPE

The microscopes are the important tools in the plant anatomy and cells because they possess the characteristics of magnification and resolving power. Important in magnification is the ability of the microscope to resolve or distinguish, between two objects lying close together. The microscopes cannot resolve objects less than one half the wavelength of light. The average wavelength of white light is about 0.55 μ, and therefore, since the white light is ordinarily used in the laboratory microscope, light microscope using white light cannot resolve object less than about 2,500 Å or 0.25 μ. From this it becomes evident, that the resolving power of a microscope can be increased by using light of shorter wave lengths and achromatic lenses. For example, when violet light which has the wave length (λ 4,000 Å is used as source of illumination, then the limit of resolution of light microscope remains 1,700 Å 0.17λ). Similarly a microscope using infra-red radiation of wave-length (λ) = 800 Å would have a limit of resolution of 0.4 μ. Because the glass lenses are not transparent any longer, therefore, another refractive media such as quartz lenses and reflecting optical instruments are used in improved microscopes.

THE LIGHT MICROSCOPE

The most important scientific tool for a student of biology is the **light microscope.** This was invented by Anton Van Leeuwenhoek (1632–1723), a Dutch scientist. He ground pieces of glass into lenses and made microscopes with a magnification of 150–300 times. The light microscope contains lenses and depends on light for its magnification. With his microscopes, Leeuwenhoek was the first scientist to have recorded observations of microorganisms. The microscope enables biologists to magnify and observe the very tiny parts of plant and animals. An American Robert B. Tolles (1824–1833) made many microscopes with improved objectives and he invented oil immersion objectives.

The student microscope or the compound microscope of twentieth century is the microscope of much improved and modified type. This microscope contains an **objective lens**. The objective lens is a convex lens close to the object to be observed. Generally the microscopes of this type have two objective lenses, a low power objective and a high power objective. The low and high power objective lenses are attached to a piece which

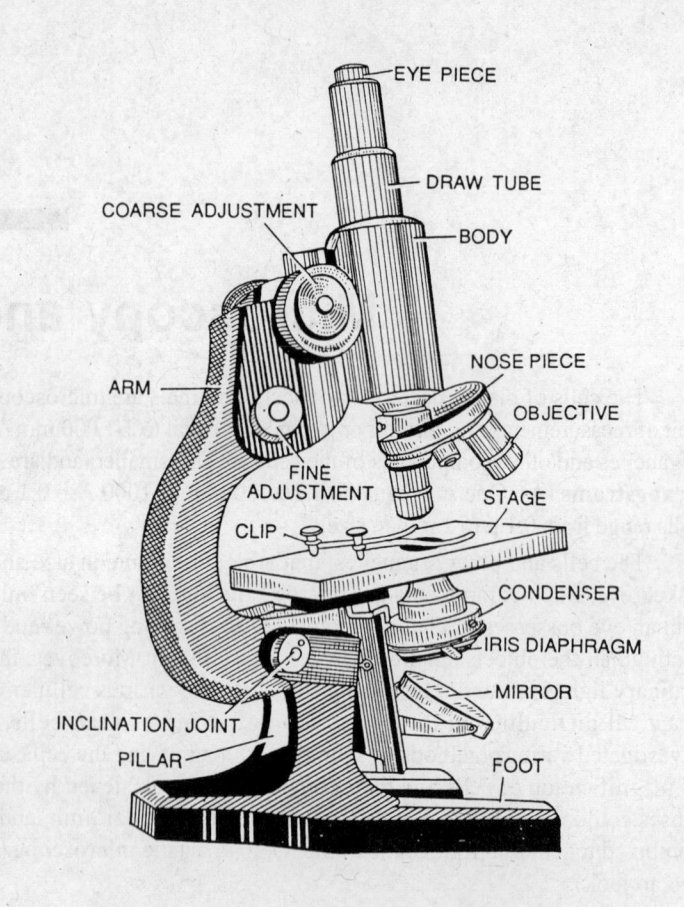

Fig. 24.1. Microscopy. The compound microscope.

turns, called the nosepiece, so that either objective can be rotated into position for viewing. The low power objective lens differs in magnification from the high power objective lens. The magnification of a low power objective lens in a typical student light microscope is about 10 X, ten times. A high power objective lens in a student microscope has a magnification about 45 X forty-five times.

When one views an object with a microscope he/she looks through the eye piece. The lens in the eye piece, is a convex lens. The eyepiece in a student microscope usually have a magnification of 10 X, ten times.

The total magnification of the microscope is determined by multiplying the magnification of the eyepiece lens by the magnification of the objective lens. For example, if one views an object with the low power objective lens, the magnification would be 100 X 10 × 10 = 100) and with the high power objective lens, would be 450 X (10 × 45 = 450).

PHASE-CONTRAST MICROSCOPE

This type of microcsope has the resolving power similar to ordinary compound light microscope and it enables one to see structures in the living cell. It is based on the principle of phase-contrast. The phase-contrast microscope exaggerates the small difference between phases or refractive indices of different cellular components and enables one to distinguish adjacent structures. There is no need of staining.

Microscopy and Micrometry

ELECTRON MICROSCOPE

Magnification by the electron microscope is more than 100,000 X. In electron microscopes the beams of electrons are used rather than light. An **electron** is a particle of negative electricity. Instead of glass lenses the magnets are used to focus the electron beam in electron microscope. Electron rays are not visible, and therefore, the magnified image of an object is viewed on a fluorescent screen similar to the screen in a television set. An electron microscope can also make a picture on a photographic film. The picture on the film has the advantage of being a permanent record that can be stored and studied. Tremendous magnification by the electron microscope has extended man's power of observation. Scientists and biologists can investigate many things with the electron microscope which cannot be seen with the light microscope. (See fig. 24.3)

MICROMETRY

Measuring With the Microscope

The micrometry is the subject in which we have some measurement of the dimensions of an object being observed use under the microscope. The method employs some special types of measuring devices which are so oriented that these can well be attached to or put into the microscope and observed. The object to be measured is calibrated against these scales.

Once an object is being observed under a microscope by the 10 X objective the 10 X eye piece; this way, the image that is perceived is 100 times of the object. We get the magnified view no doubt, and also that it is perfect coordination of the dimensions but to find out the absolute size of the object will need precision and which is achieved through the application of micrometers. Usually the micrometers are of two types—the ocular micrometer disc and the stage micrometer.

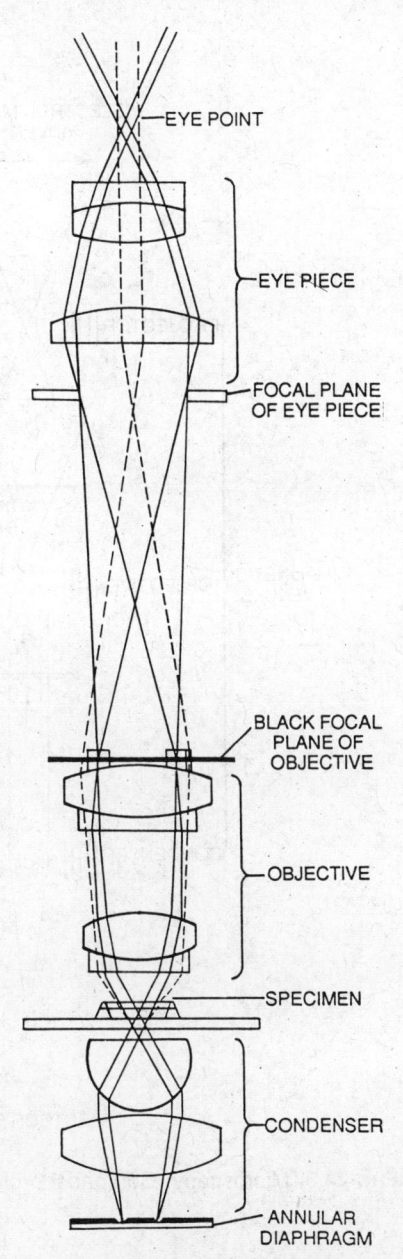

Fig. 24.2. Microscopy. Light path in phase contrast microscope.

The ocular micrometer disc. Before inserting the disc, clean it gently with distilled water and wipe it with a soft lifeless tissue. Examine the disc with a hand magnifier by light reflected from each surface to detect adhering particles and remove them with a clean camel's hair brush. Select the ocular (preferably not stronger than 10 X) and unscrew the eye lens mount. Position the disc over the eyepiece diaphragm with the engraved side down (figures should appear erect), and replace the eye lens. With the micrometer disc in place, insert the ocular into the illuminated microscope and bring the

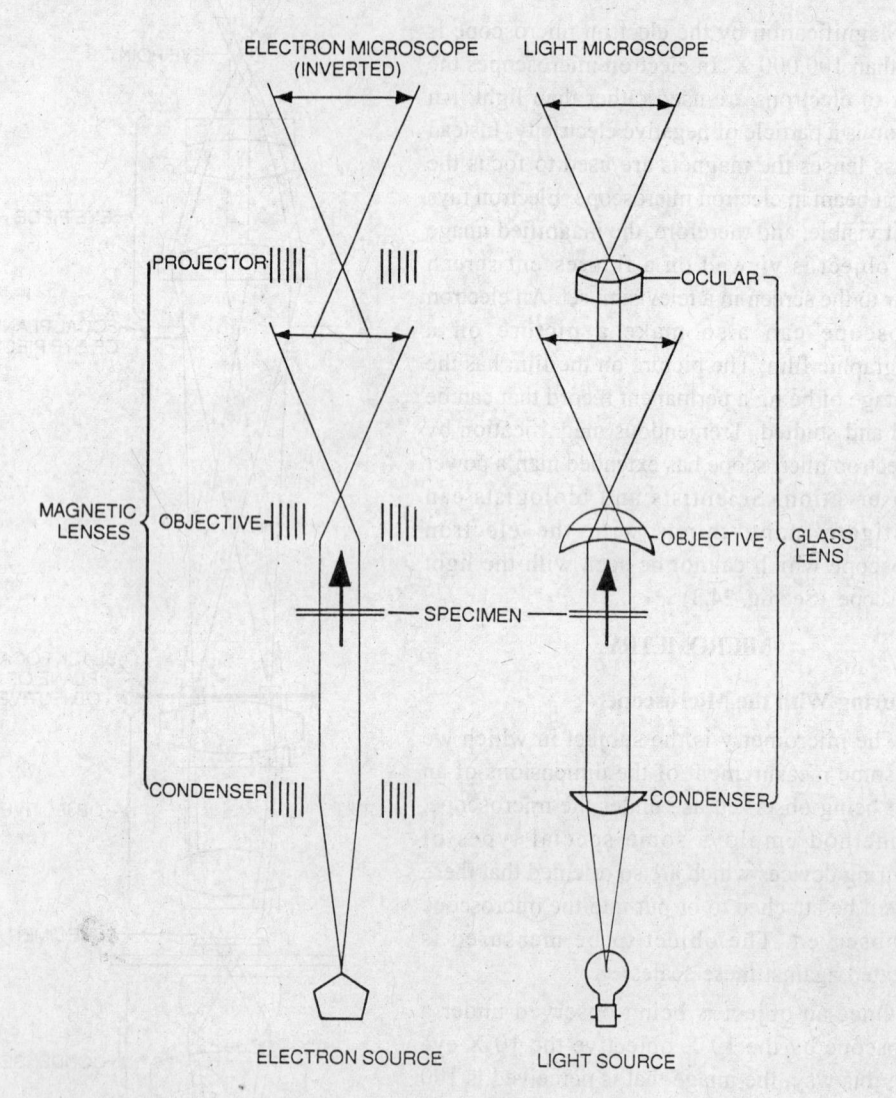

Fig. 24.3. Microscopy. Comparison of optical pathways in light and electron microscopes.

lines into sharp focus unscrewing the upper eyelens for a sufficient distance. There are usually 50 or 100 divisions in the ocular meter which are engraved on the glass.

The stage micrometer. The scale, mounted like a microscope specimen on a glass slide which is generally thicker than a specimen slide. The micrometer has a mount of very finely gradulated scale. The scale measures only one millimeter and has a least count of .01 mm, *i.e.*, 1 millimetre region is divided into 100 divisions. As 1 mm has 1000 μ one division of stage micrometer is equal to 10 μ.

Method. The stage micrometer is kept in low power under microscope and is observed through the eye piece hang ocular micrometer disc. Suppose we have 10 X objective and 5 X eye piece in the microscope with a tube of 170 mm. length. At this magnification the number of ocular divisions coinciding the stage micrometer are observed and then calculated for microns per ocular division, *e.g.*, ocular divisions coincide 8 stage micrometer divisions.

Microscopy and Micrometry

OCULAR MICROMETER IN MICROSCOPE EYEPIECE DIVISIONS ARBITRARY

A

STAGE MICROMETER DISTANCE ABSOLUTE EACH DIVISION EQUALS 10. MICROMETERS

B

THE OCULAR MICROMETER SUPERIMPOSED ON THE STAGE MICROMETER

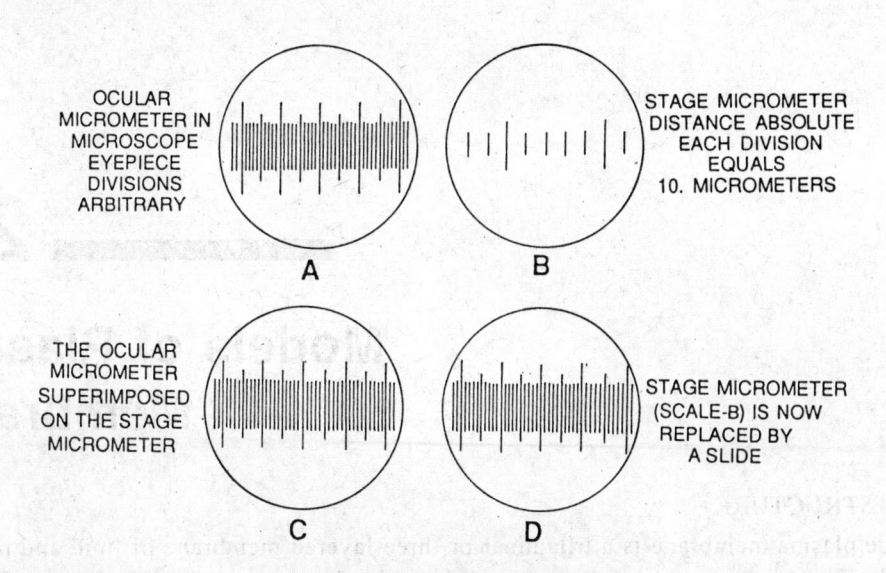

C

STAGE MICROMETER (SCALE-B) IS NOW REPLACED BY A SLIDE

D

Fig. 24.4. Micrometry. Standardization of the ocular micrometer, and its use in measuring microcells.

i.e., 6 ocular divisions = 8 stage micrometer divisions

or 6 ,, ,, = .08 mm.

(since 1 division of stage micrometer is equal to .01 mm.)

$$1 \text{ ,, division} = \frac{0.08}{6} \text{ mm.} = \frac{0.08 \times 1000}{6} = 13.3 \, \mu$$

Thus the microscope is calibrated for different combinations of eye pieces and objective lenses and is kept for record.

Measurement of an object. When the microscope is calibrated, then the object to be measured is kept on the stage of the microscope and is observed through eye piece of ocular. The object is measured in the particular magnification by ocular divisions and then is being changed into micron by multiplying ocular divisions with calibrated value of one ocular division in that particular magnification, *e.g.*, the length of a bacterial cell is equal to five divisions of ocular that means the length in microns will be 5 ocular divisions × 13.3 = 66.5 μ..

This way the object can be measured in any of the magnification.

25
Models of Plasma Membrane

ULTRASTRUCTURE

The plasma membrane is a trilaminar or three-layered membrane of lipid and protein molecules. The trilaminar nature of plasma membrane has been proposed by Danielli and Davson (1935) by certain indirect evidences. Harvey and Danielli (1938) produced a hypothetical model

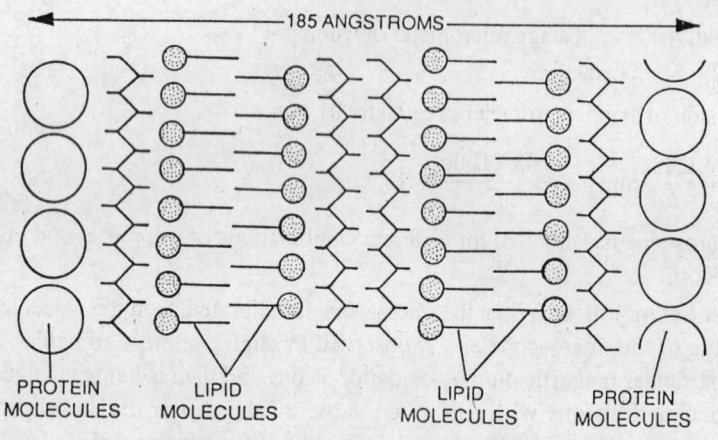

Fig. 25.1. The plasma membrane in molecular terms showing protein and lipid molecules.

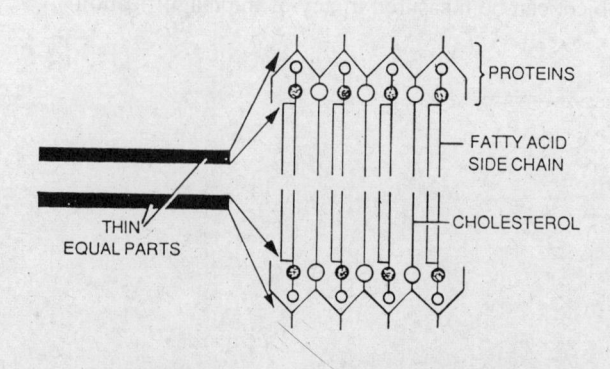

Fig. 25.2. Symmetrical pattern of plasma membrane.

Models of Plasma Membrane

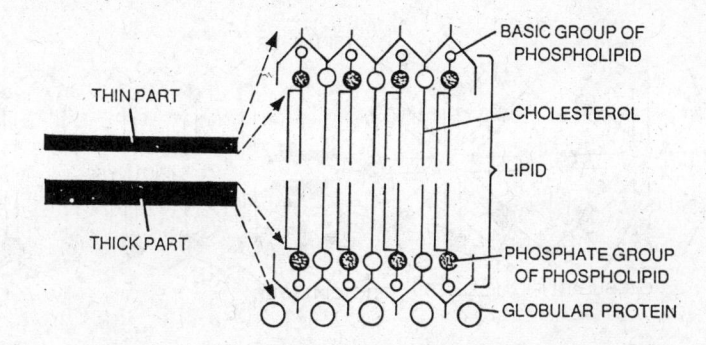

Fig. 25.3. Asymmetrical pattern of plasma membrane.

of the plasma membrane which has shown a bimolecular lipid structure in between two outer and inner layers of protein molecules. The studies by electron microscope have revealed the protein-lipid-protein arrangement in the plasma membrane. Robertson (1959) found wide occurrence of such trilaminar composition in most of the membranes of cellular organelles, and with the result he propounded the concept of the **unit membrane.** A unit membrane possesses the protein-lipid-protein arrangement. The unit membrane has been found in plasma membrane, endoplasmic reticulum, Golgi complex, lysosomes and plastids. It has been found, that both the outer and inner membrances of the mitochondria and the nucleus are the unit membranes.

The plasma membrane of most cells varies from 100 to 215 Å in thickness. The plasma membrane is not a continuous layer but contains certain minute pores of 8 to 15 Å diameter. The opening and closing of these pores depends on the metabolic phase of the cell.

MOLECULAR STRUCTURE

The plasma membrane consists of two layers of protein molecules and two layers of lipid molecules. The lipid molecules occur in chains. The two molecular chains of lipids remain parallel to each other and form a bimolecular or double-layered structure. Both lipid layers remain linked with each other by the inner ends of lipid molecules which are non-polar and hydrophobic (Gr. *hydra* = water; *phobe* = hate) in nature. Both the layers of lipids are held together due to Vanderwaal's forces at these non-polar ends.

The lipid layers remain enclosed by outer and inner layers of proteins. The lipid molecules are linked with the molecules of protein layers by their outer, polar and hydrophilic (Gr. *hydra* = water; *phil* = loving) ends. In the hydrogen bonds, ionic linkages, or electrostatic forces bind the molecules of lipids and proteins together. The carbyhydrate molecules are found in the association of protein molecules and provide stability to lipo-protein complex. The protein layers provide elasticity and mechanical resistance to the plasma membrane.

There is a dispute regarding the precise pattern of molecular sub-units (*i.e.*, lipids,

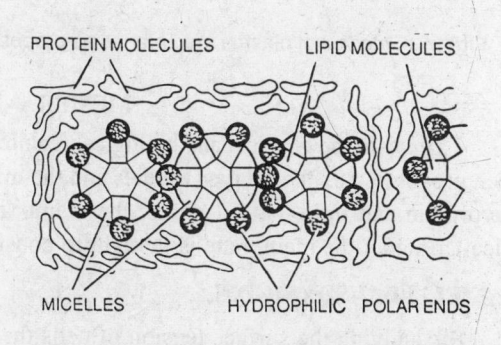

Fig. 25.4. Plasma membrane, based on Micellar theory (diagrammatic).

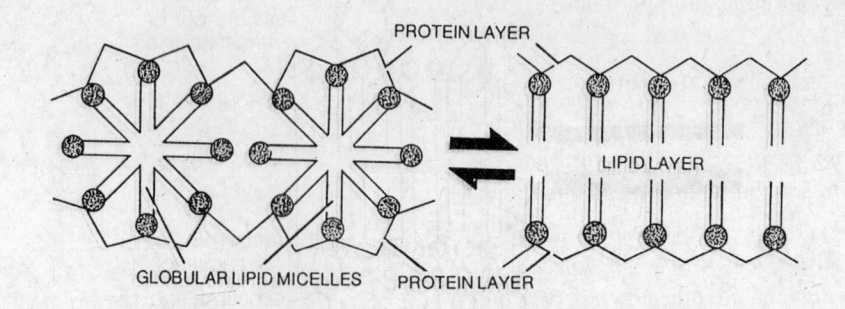

Fig. 25.5. Plasma membrane. Transformation between micellar and lamellar states (globular bilayer transformation).

proteins and carbohydrates) in the plasma membrane. Following three kinds of theories and models, *viz.*, **lamellar theory, micellar theory** and **fluid mosaic theory**, have been suggested for the possible molecular structure of plasma membrane.

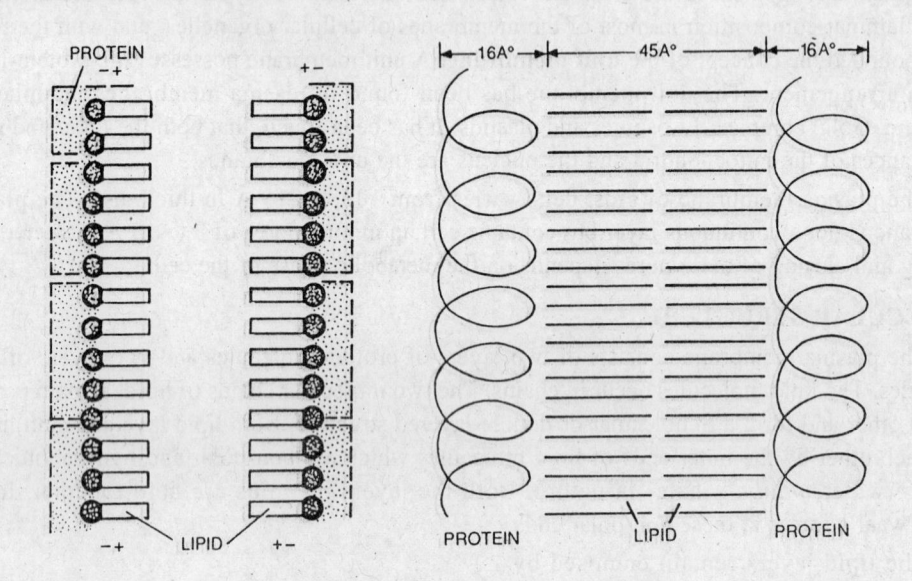

Fig. 25.6. Models of plasma membranes as proposed by (A) Davson and Danielli and (B) Robertson.

MEMBRANE MODELS

To explain the physical and biological features of plasma membranes two main categories have been proposed, *i.e.*, the **bilayer models** and the **micellar** or **sub-unit models**. In the bilayer models the protein and lipid that constitute the membrane are believed to occur in layers, while in the micellar model the membrane is thought to consist of a number of similar units.

DANIELLI-DAVSON MODEL

By studying the surface tension of cells the existence of protein was indicated (Harvey and Cole, 1931; Danielli and Harvey, 1935). This led Davson and Danielli (1935) to propose a lipoprotein model of the cell membrane. According to this model the bimolecular liquid layer consists of two layers of molecules with their polar regions on the outer side. Globular proteins are thought to be associated with the polar groups of the lipid.

Models of Plasma Membrane

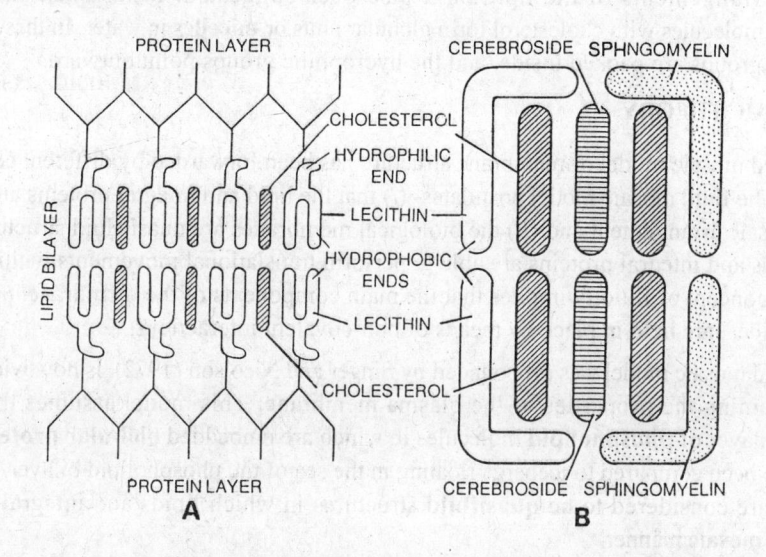

Fig. 25.7. A phospholipid-chlosterol complex of cell membrane.

The lipids in the membrane consist mainly of phospholipids, with their non-polar groups near each other and their polar groups directed outwards. The lipid layer in many cases consists of a phospholipid **lecithin**, alternating with a steroid molecule **cholesterol**. The lecithin molecule consists of two lipid chains of glycerol and a polar head containing **phosphate** and **choline**. According to Finean (1953) the head is bent around like a walking stick, and forms a bond with **cholesterol**.

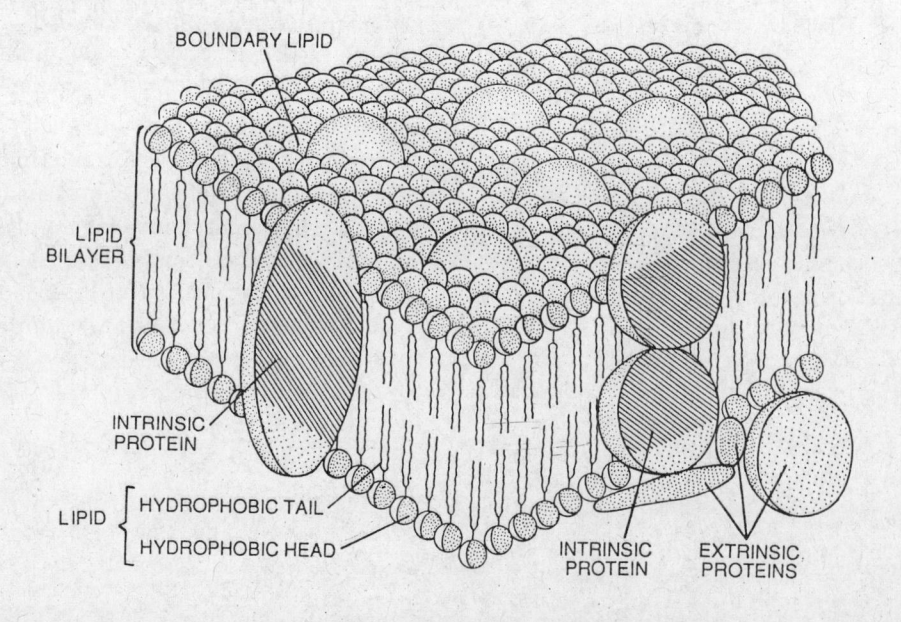

Fig. 25.8. The fluid mosaic model of plasma membrane

Other arrangements of the lipid have also been suggested. It has been suggested that phospholipid molecules with cholesterol form globular units or **micelles** in water. In these micelles the hydrophobic groups are packed inside, and the hydrophilic groups point outwards.

FLUID MOSAIC THEORY

The fluid mosaic model of membrane structure has been forwarded by different cell biologists, after, 1970. The fluid mosaic model postulates- (*i*) that the lipid and integral proteins are disposed in a kind of mosaic arrangement and (*ii*) the biological membranes are quasi-fluid structures in which both the lipids and integral proteins are able to perform translational movements within the overall bilayer. The concept of fluidity implies that the main components of the membrane, *i.e.*, lipids and oligosaccharides, are held in place by means of non-covalent interactions.

The fluid mosaic model was propounded by Singer and Nicolson (1972), is now widely accepted as best explaining the properties of the plasma membrane. This model assumes that there is a continuous bilayer of **phospholipid** molecules in which are embedded **globular proteins**. Here the proteins have been compared to icebergs floating in the sea of the phospholipid bilayer. Thus plasma membranes are considered to be **quasifluid** structures in which lipids and integral proteins are arranged in a **mosaic** manner.

26

Diversity and Morphology of Flowering Plants

Angiosperms or flowering plants form the largest group of plant kingdom, including about 300 families, 8,000 genera and 300,000 species. They are considered to be highest evolved plants on the surface of the earth. Geologically, they are young. From Cretaceous age, the angiosperms eclipsed all other vegetation and now they are dominant. They are found almost every where in each possible type of habitat and climate. They occur in deep lakes, deserts, in beds of seas and even on high peaks of mountains. The species of *Opuntia* (a desert plant-xerophyte) can survive without water in acute desert conditions, whereas on the other hand the species of *Hydrilla* (an aquatic plant) are extremely sensitive to drought conditions. Some species are found on rocks, some in water falls and some are marine. The species of *Rhizophora*, popularly known as mangrove vegetation, are found near the water of the sea. The epiphytes (*Vanda*- an orchid), parasites (*Cuscuta*), saprophytes (*Monotropa*) and even insectivorous plants (*Utricularia, Drosera*, etc.) are also not uncommon.

They may be annual, biennial or perennial herbs, shrubs, trees, climbers, twiners and lianes. On one hand angiosperms may be as minute in size as a pin head, *e.g., Wolffia microscopica*, on the other extremity like eucleptiles (*Eucalyptus*) may reach upto 300 feet in height. For example, plants like pea and gram live only for a few weeks, whereas the famous **Bodhi** tree (*Ficus religiosa*) at Gaya is about 2,500 years old.

The study of various external features of the plants is known as **plant morphology**. The morphology (*morpho*, form; *logos*, discourse or study) deals with the study of forms and features of different plant organs, such as roots, stems, leaves, flowers, seeds and fruits.

DIVERSITY AND MORPHOLOGY IN THE PLANT BODY

The plant body of an angiosperm consists of a number of organs, *i.e.,* roots, stem, leaf and flower. The flower consists of sepals, petals, stamens, carpels and sometimes also sterile members. Each organ, is made up of a number of tissues. Each tissue consists of many cells of one kind.

A vascular plant begins its existence as a morphologically simple unicellular zygote (2n). The zygote develops into the embryo and thereafter into the mature sporophyte. The development of the complex multicellular body of the seed plant is a result of evolutionary specialization of long duration. The specialization has given rise to the establishment of morphological and physiological differences between the various parts of the plant body and also caused the development of the concept of **plant organs.**

As regards the morphologic nature of the flower it is thought that the flower is homologous with a shoot and the floral parts with leaves.

Fundamental parts of the plant body. The axis, consists of two parts — that portion which is normally aerial is known as the **stem**, and that portion which is subterranean is called the **root**. There are three types of appendages arising from the axis. **1. Leaves**—The strands of vascular tissue pass through the leaves. The leaves are characteristic of the stem and do not occur on the root. The leaves are found to be arranged on the stem in a definite manner, and bear an intimate structural relation to the skeleton of the axis. The leaf is looked upon as the lateral expansion of the stem, continuous with it. All fundamental parts of the stem are concerned with the formation of the leaf. **2. Emergences**—In the appendages of the second rank only the outermost layers of stem, the cortex and the epidermis, are usually present which are known as emergences. The prickles of the rose make a good example of it. **3. Hairs.**—The appendages of the third rank are hairs. These are projections of the outermost layer of the cells. The emergences and hairs occur on both axis and leaves, usually without definite arrangement.

DIVERSITY AND MORPHOLOGY IN THE ROOT

The seed plants possess a radicle or simply a root meristem at the root end or root pole of the embryo from which the first root of the plant develops upon germination of the seed. The roots are generally divided into two categories: (*i*) **primary**, normal roots, which originate from the embryo and usually persist throughout life, and (*ii*) **adventitious roots**, which arise secondarily from stem, leaf or other tissues and which may be either permanent or temporary.

Characteristic Features of the Roots

1. The root is the descending portion of the axis of the plant, and grows away from light. Usually, the root is not green in colour. However, in certain cases when the root is exposed to light for a prolonged period it turns green in colour, *e.g.,* in *Tinospora*, some orchids and water chestnut.

2. Root growth is generally directed towards gravity, (*i.e.*, positively geotropic) except for the breathing roots (pneumatophores) of mangrove vegetation.

3. The root does not commonly bear buds. However, in certain cases the roots are seen to bear vegetative buds for vegetative propagation, *e.g.*, in *Aegle*, *Trichosanthes*, sweet potato, etc. Such plants are sometimes propagate by root cuttings.

4. The root ends in and is protected by a cap-like structure, known as the root cap. A distinct multiple root cap is seen in the aerial root of screwpine.

5. The root bears unicellular hairs. Root hairs occur in a cluster in the tender part of the root a little behind the apex. Root hairs possess very thin walls made of cellulose. Root hairs absorb water and minerals from the soil.

6. Lateral roots always develop from an inner layer, and are called endogenous. They are produced endogenously from pericycle.

7. Roots are much variable in their shape and structure. It is related either to their function or environmental conditions.

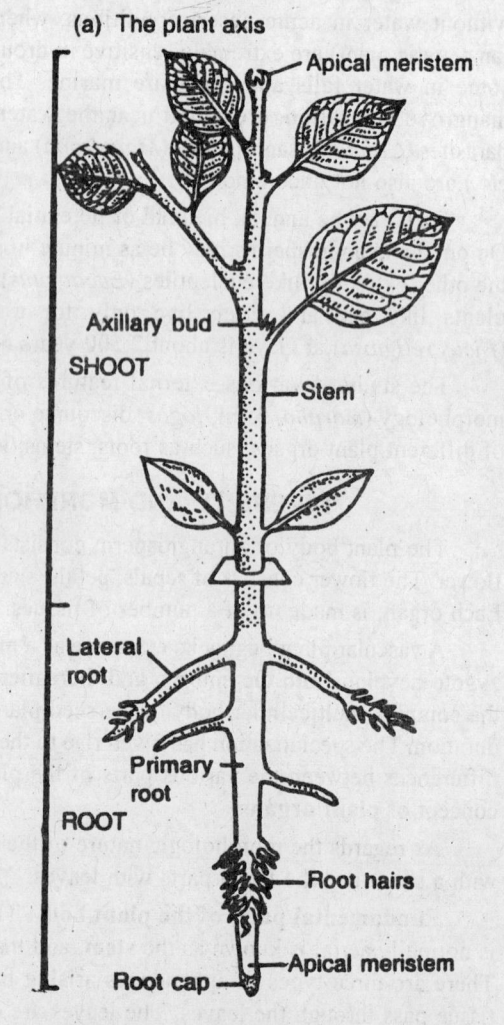

Fig. 26.1. The plant body showing fundamental parts.

Diversity and Morphology of Flowering Plants

ROOT SYSTEMS

The whole extent of the roots of a plant is called the root system. The development of the root system differs fundamentally in vascular plants and may be classified in two categories: **1. the tap root system** and **2. the fibrous root system.** The tap root system is normally found in dicotyledons and gymnosperms whereas the fibrous root system is commonly met with in monocotyledons. The direct prolongation of the radicle forms the **primary root.** If it persists and continues to grow it is called the **tap root.**

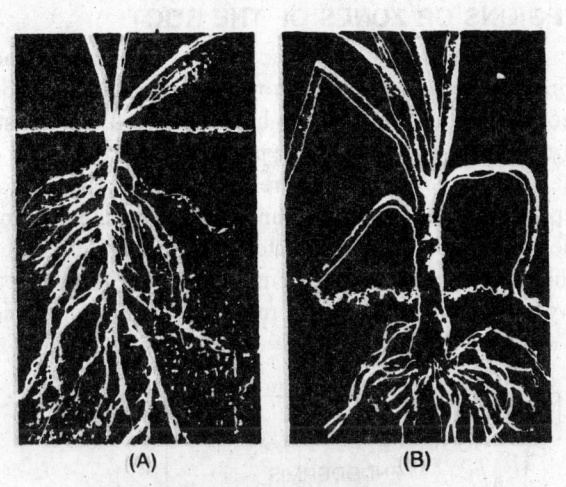

Fig. 26.2. Root systems. A, tap root system; B, fibrous root system of maize.

1. Tap root system. The tap root produces lateral branches which are known as the **secondary roots**, and these in turn produce the **tertiary roots**, and so on. All these lateral roots are produced in **acropetal** succession, *i.e.*, the older and longer roots are away from the tip, and the younger and shorter ones are towards it. The tap root normally grows vertically downwards to a shorter or longer depth while the secondary or tertiary roots grow obliquely downwards, or in many cases horizontally outwards. The tap roots absorb water and mineral salts from the soil and give proper anchorage to the plant. The tap root system may be regarded as characteristic of dicotyledons.

2. Fibrous root system. In monocotyledons the radicle also gives rise to the primary root, but this does not develop any further and soon perishes and is replaced by many thin roots developed from the base of the stem. These are known as the **fibrous roots**. Such roots also develop from nodes of stems, as in sugarcane, bamboo and other grasses. Thus fibrous root system may be regarded as characteristic of monocotyledons.

Table 26.1. Differences Between Tap Root and Fibrous Root Systems

Tap root system	Fibrous root system
1. The main tap root develops directly from the radicle.	1. The radicle does not dominate and its growth stops early during germination.
2. They reach very deep in soil.	2. They are comparatively shallow and spread in a large area than depth.
3. The primary root continues to grow and remains distinct.	3. The primary root stops growing and is not distinct.
4. The primary root gives off the lateral roots which grow horizontally and downwards.	4. Numerous lateral roots arise from the stem base and give rise to a mass of branching adventitious roots.
5. Mostly the examples are found in dicots, *e.g.*, carrot, turnip, radish, etc.	5. The examples are common in monocots, *e.g.*, grasses.

REGIONS OR ZONES OF THE ROOT

The tip of each root is covered by a protective root cap, a thimble-shaped covering of cells which fits over the rapidly growing meristematic region (the calyptrogen). The outer part of the root cap is rough and uneven because its cells are constantly being worn away as the root pushes through the soil. The growing point consists of actively dividing meristematic cells from which all the other tissues of the root are formed. The growing point also gives rise to new root cap cells to replace the ones worn away. Immediately behind the growing point is the **zone of elongation**, here the cells remain undifferentiated but grow rapidly in length by taking in large amounts of water. The growing point is about 1 mm. in length and the zone of elongation is 3 to 5 mm. long; these two are the only parts of the root that account for the continued elongation of the root.

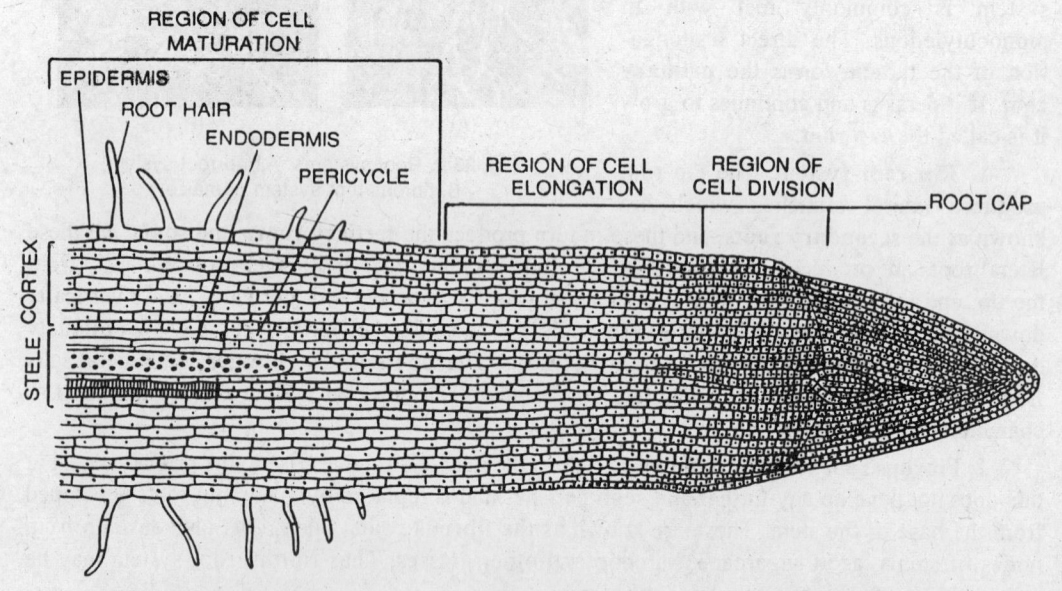

Fig. 26.3. Longitudinal section of root tip showing various regions of growth and root cap.

Above the zone of elongation is the **zone of maturation** characterized externally by a downy covering of whitish root hairs. In this zone the cells differentiate into the permanent tissues of the roots. Each root hair is a slender, elongated, lateral projection from a single epidermal cell, through which most of water and minerals are absorbed. Root hairs are delicate and short-lived; new hairs are constantly formed just behind the zone of elongation and old hairs farther back wither and die as the root elongates. Only a short segment of the root, perhaps one to six cm. long has root hairs.

KINDS OF ROOTS

On the basis of place of origin there are two types of roots; 1. Tap roots and 2. Adventitious roots.

1. Tap roots. The root that develops directly from the radicle is known as **primary root**. In most of the cases primary root persists and becomes stronger to form tap root. The tap root normally grows vertically downwards in a shorter or longer depths. Tap root usually produces lateral branches known as **secondary roots**. The secondary roots may further branch to give rise to **tertiary roots**. The primary root may be sparingly or profusely branched according to the need of the plant.

2. Adventitious roots. The roots that develop from any part of the plant other than the radicle are known as **adventitious roots**. They may develop from the base of the stem replacing the

Diversity and Morphology of Flowering Plants

primary root or in addition to it, or from any node and internode of the stem or the branch or even from the leaf. On the basis of the nature of their development the adventitious roots may be categorized as follows:

(*i*) **Fibrous roots.** These roots may be given off in clusters from the base of the stem, *e.g.*, in onion, or from the nodes of creeping branches of grasses, or from the lower nodes of the stem, *e.g.*, in maize, sugarcane, etc. Fibrous roots of monocotyledons are all adventitious roots.

(*ii*) **Foliar roots.** The roots that develop from the leaf known as foliar roots, *e.g.*, in *Bryophyllum, Begonia, Pogostemon,* etc.

(*iii*) **True adventitious roots.** Such roots are given off by many plants from their nodes and sometimes from the internodes as they creep on the ground, *e.g.*, in Indian pennywort. The adventitious roots are also produced from branch cuttings when they are put into the soil, *e.g.*, in rose, sugarcane, tapioca, etc.

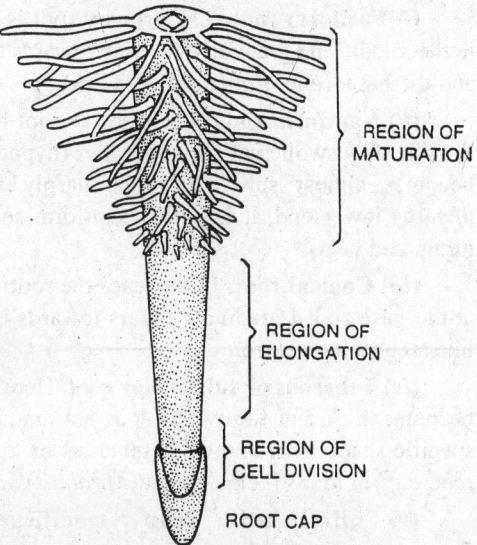

Fig. 26.4. Regions of the root.

Table 26.2. Differences Between Tap Root and Adventitious Root.

Tap root	Adventitious root
1. The tap root develops directly from the radicle.	1. The adventitious root develops from any other part of the plant, other than the radicle.
2. The primary root continuously grows.	2. The primary root stops to grow.
3. They always develops below the soil in the ground.	3. They may be underground or aerial.
4. The single main root gives rise to the fine lateral roots.	4. They form a cluster of roots which may arise from the same point.

MODIFICATIONS OF ROOTS

Many roots of peculiar form and function are known among angiosperms. All these are derived from normal type by the specialized development of one or other of its characteristic functions or structures, in a manner suitable to special circumstances. When the 'modification' of the roots are being considered, it is implied that the simple type of absorbing root in the soil was the earliest type to be evolved and that specialized forms have been derived from it.

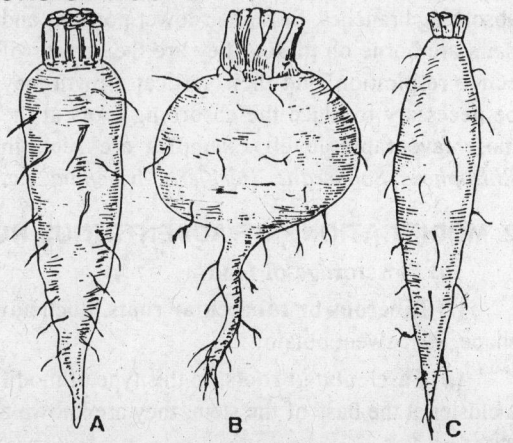

Fig. 26.5. Storage tap roots. A, conical; B, napiform; C, fusiform.

1. MODIFICATIONS OF TAP ROOT

For storage of food

(*i*) **Fusiform root.** The primary root is swollen in the middle and gradually tapering towards the apex and the base, (*e.g., Raphanus* = radish).

(*ii*) **Napiform root.** The primary root becomes considerably swollen at the upper part (hypocotyl). It becomes almost spherical, and sharply tapering towards lower end, it is called napiform root, *e.g.,* turnip and beet.

(*iii*) **Conical root.** In this case the root is broad at the base and it gradually tapers towards the apex like a cone, *e.g.,* carrot.

(*iv*) **Tuberous or tubercular root.** Here the root becomes thick and fleshy but does not maintain any specific shape. It is known as tuberous or tubercular root, *e.g.,* in four 'o' clock plant (*Mirabilis*).

(*v*) **Branched root modification-Pneumatophores.** Many plants growing in marshy places especially in the tidal swamp forests produce special roots which grow up into the air from beneath the slime are known **pneumatophores** or beathing roots. They arise from long horizontal roots, and as they are negatively geotropic they grow vertically upwards, projecting several inches into the air. They often occur in large numbers around the trunk of the tree. There is a corky layer over each root, which covers even the apex, and the portion in the air bears numerous lenticels. Root hairs are not produced either by the pneuma-tophores themselves or by the horizontal roots from which they arise, but the pneumatophores form short absorbing branches, from their lower portions and hairs are borne on these. They are the centres of active respiration, and their vertical growth may be necessary to place the absorbing roots at the most favourable level. Examples are seen in *Rhizophora, Sonneratia, Heritiera, Avicennia,* etc.

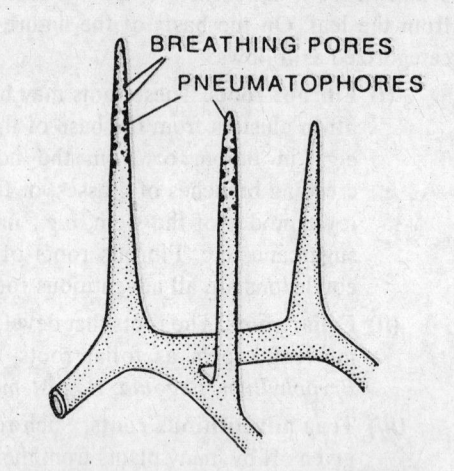

Fig. 26.6. Roots. Pneumatophores arising vertically upwards from an underground root.

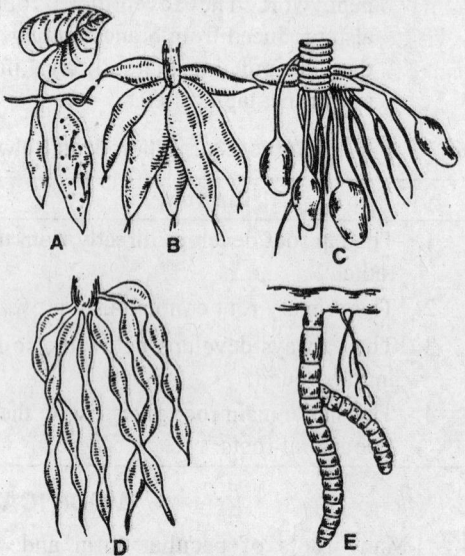

Fig. 26.7. Adventitious modified roots. A, Tuberous roots of sweet potato; B, Fasciculated tuberous roots of *Dahlia*; C, modulated roots of mango ginger; D, Moniliform roots of *Momordica*; E, Annulated roots.

2. MODIFICATIONS OF ADVENTITIOUS ROOTS

(*a*) **For storage of food**

(*i*) **Tuberous or tubercular roots.** Such adventitious roots are swollen without any definite shape, *e.g.,* sweet potato.

(*ii*) **Fasciculated roots.** In this type of modification of roots several tubercular roots occur in a cluster at the base of the stem, they are known as fasciculated roots, *e.g.,* in *Asparagus, Dahlia,* etc.

Diversity and Morphology of Flowering Plants

(*iii*) **Nodulose roots.** In such modification the slender root becomes suddenly swollen near the apex, and is known as nodulose root, *e.g.*, in mango ginger.

(*iv*) **Moniliform or beaded roots.** In such modification of root there are some swellings at frequent intervals, and the roots are termed moniliform or beaded, *e.g.*, in *Momordica*.

(*v*) **Annulated roots.** When the root possesses a series of ring like swellings on its body, it is known as annulated root, *e.g.*, in ipecacuanha.

(*b*) **For mechanical support**

(*vi*) **Prop roots.** Such roots are those which develop from the main branches of a tree. The best example of this is *Ficus benghalensis*, the Banyan tree. The prop roots grow straight downwards to the soil, which they enter and form underground branch roots. As the roots grow old they increase in thickness due to secondary growth. As they become thickened they give support to the branches which are thus able to continue their horizontal growth, producing more prop roots at intervals.

(*vii*) **Stilt roots.** Adventitious roots sometimes form supporting stilts. The maize (*Zea mays*) plant has a tall slender stem. It develops a cluster of roots from the first one or two nodes above ground level. They grow obliquely downwards into the soil and give added support to the stem. Stilt roots are also well known in *Pandanus*. The stilt roots are less commonly found in Mangroves such as *Rhizophora*, *Sonneratia* and *Avicennia*. Besides the functions of absorption such roots provide additional support to the stem.

(*viii*) **Climbing roots.** Many tropical climbers like *Piper betle*, *P. longum*, *P. nigrum*, *Pothos*, etc., produce roots from their nodes and often from the internodes, by means of which such plants attach themselves to their

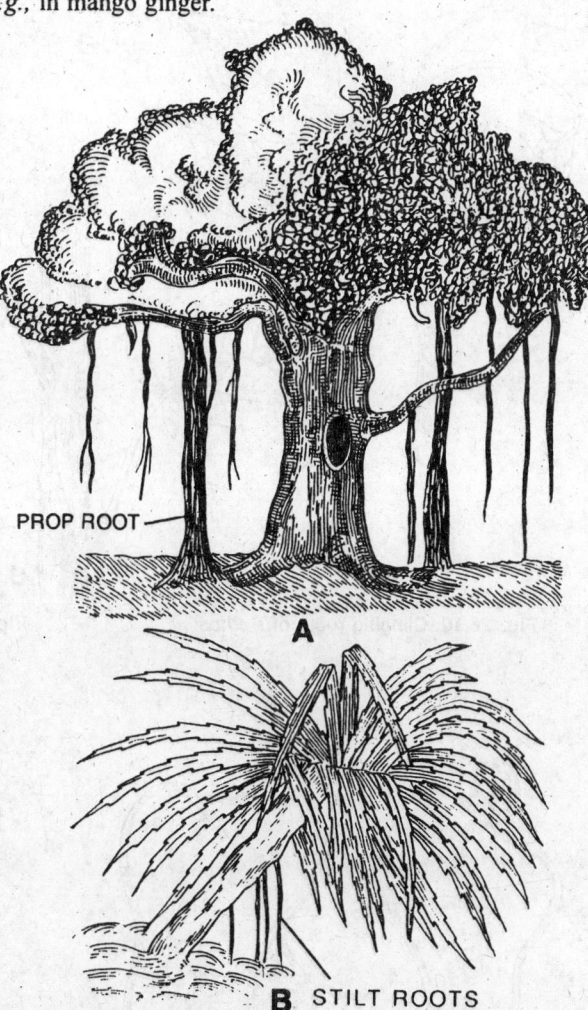

Fig. 26.8. Adventitious modified roots. A, prop root of banyan tree; B, stilt roots of *Pandanus*.

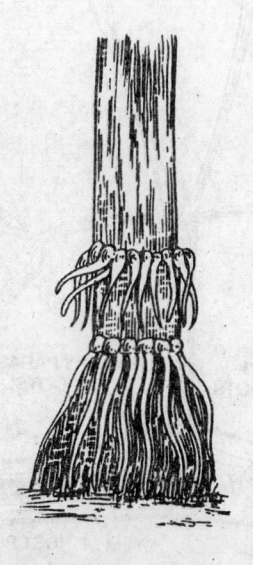

Fig. 26.9. Stilt roots of *Zea mays*.

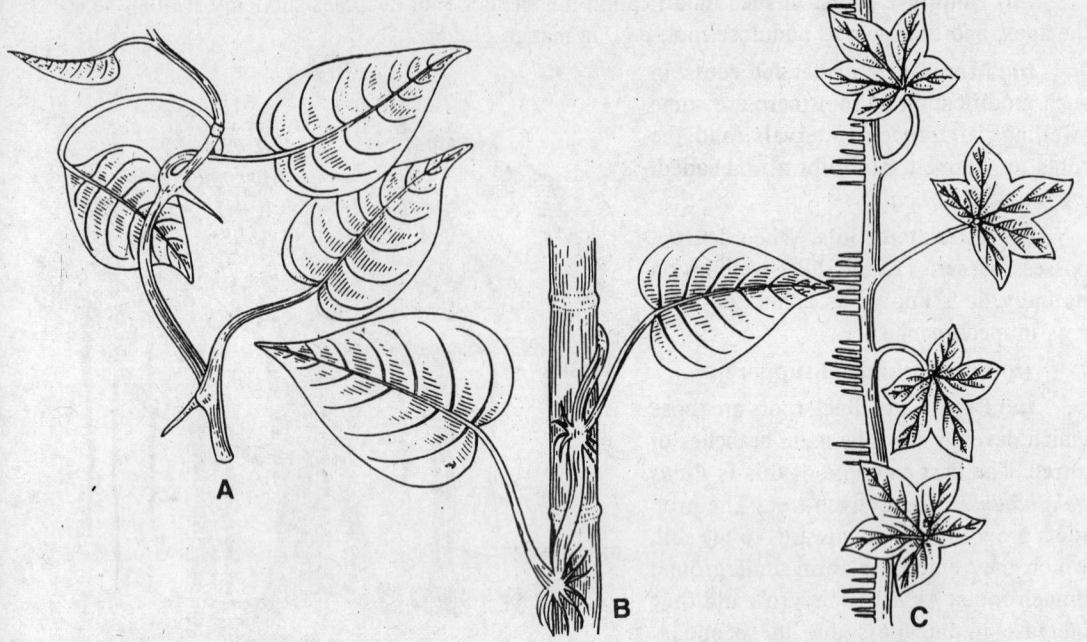

Fig. 26.10. Clinging roots of *Pothos*.

Fig. 26.11. Clinging roots of *Hedera helix*.

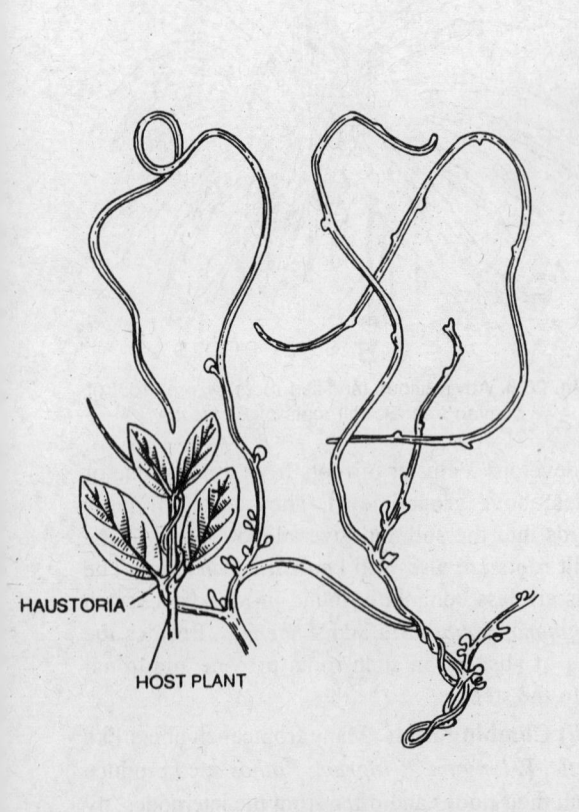

Fig. 26.12. Parasitic roots of *Cuscuta*.

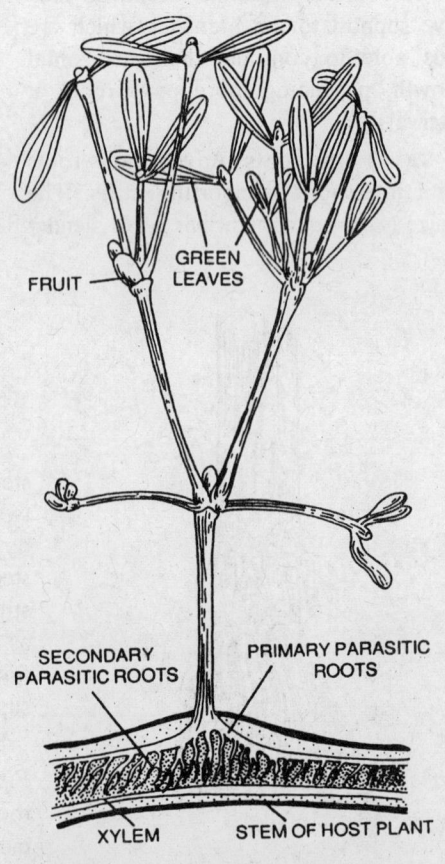

Fig. 26.13. Parasitic roots of *Viscum*.

Diversity and Morphology of Flowering Plants

support and climb it. These roots cling closely to the bark of the supporting tree. To ensure a foothold such roots secrete a kind of sticky juice which quickly dries up in the air as seen in *Hedera helix* and *Ficus pumila*. In *Hedera* adventitious roots are formed in great numbers on the side of the stem next to the support. They attach themselves to the support by the formation of mucilage from the surface cells of the apex. Often the roots form at their apex a disc like structure or a sort of claw for firm foothold. Such roots are also called **clinging roots**.

(c) **For vital functions**

(*ix*) **Sucking roots or haustoria.** Parasites among the flowering plants make use of modified roots as a means of penetrating the tissues of the host plant. In *Orobanche*, the primary root of the seedling performs this function. It bears no root cap and is the only root formed by the parasite, which serves only for fresh attachments and may be modified roots. *Cuscuta*, has only a temporary root, in the seedling state. The stem of *Cuscuta* twines around the stem of the host and obtains its nourishment by means of numerous penetrating suckers which spring from the surface of its stem nearest to the host. A strand of xylem and phloem now differentiates in the sucker and makes connection between the vascular tissues of the host and corresponding tissues in the *Cuscuta*. Such sucking roots are known as **haustoria**.

(*x*) **Respiratory roots.** In some aquatic plants such as *Jussiaea* the floating branches develop a kind of adventitious roots which are soft, light, spongy and colourless. They usually develop above the level of water and serve to store up air. Such respiratory roots facilitate respiration.

(*xi*) **Epiphytic roots.** In many orchids, which grow on branches of trees, the epiphytic roots are found. Such roots do not absorb any food from the supporting plant. The epiphytic roots are aerial and hang freely in the air. Each hanging root is surrounded by a spongy tissue, known as **velamen**. With the help of velamen the hanging root absorbs moisture from the atmosphere, *e.g.*, *Vanda* (a common orchid).

(*xii*) **Assimilatory roots.** In some plants, the aerial adventitious roots form chlorophyll in them and become assimilatory in function. They synthesize food by photosynthesis. The common examples are: *Tinospora, Trapa*, etc.

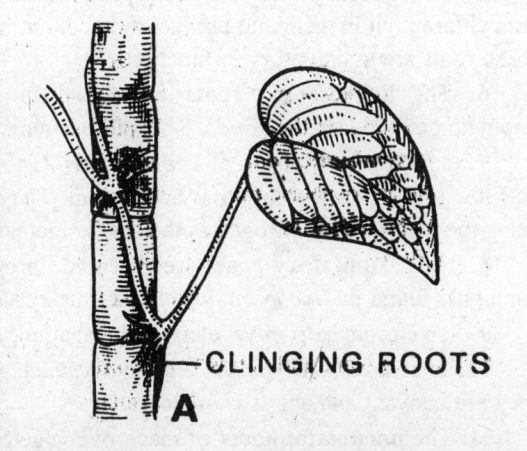

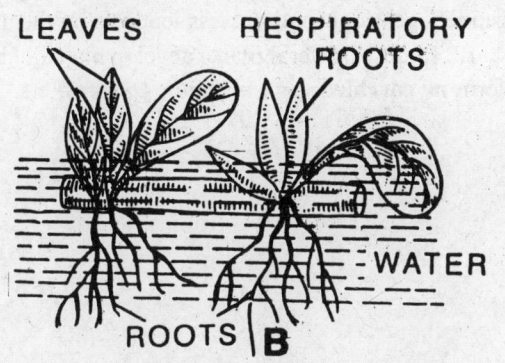

Fig. 26.14. Respiratory roots of *Jussiaea*.

Functions of the Root

The root performs many functions. They may be mechanical, *e.g.*, **fixation**, and physiological, *e.g.*, **absorption, conduction** and **storage**. The details are as follows:

1. Fixation. This is mechanical function. Here the main root goes deep into the soil and the lateral roots spread out in different directions, with the result the root system as a whole firmly anchors the plant. In monocots this is done by the fibrous roots.

2. Absorption. The absorption of water and minerals from the soil is done with the help of root hairs which develop in a cluster at a little distance behind the root apex. These root hairs adhere to the soil particles and absorb water and soluble mineral salts from them.

3. Conduction. The root is also concerned in the conduction of water and mineral salts sending them upwards into the stem and the leaves.

4. Storage. There is a certain amount of food stored up in the root, usually in its mature region. As the root grows further the stored food is utilized.

5. Assimilation. On branches of *Tinospora*, long, slender, hanging roots are developed, which form chlorophyll in them and turn green in colour. These roots are assimilatory in function.

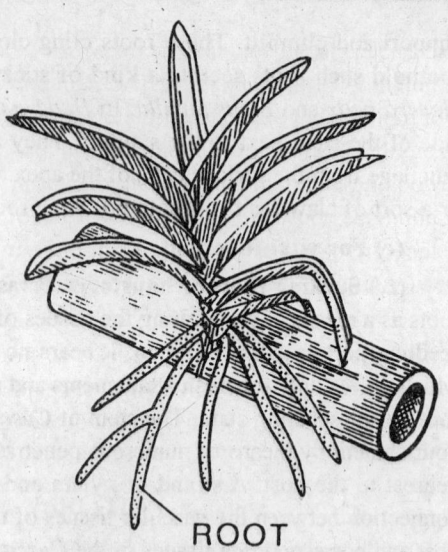

Fig. 26.15. Epiphytic roots of *Vanda*.

6. The **hygroscopic roots** are found in epiphytic orchids, *e.g.,* in *Vanda*. They absorb moisture directly from the atmosphere.

7. In most of the genera of Papilionaceae (Fabaceae) (*e.g.,* in gram, pea, many other pulses, etc.), the **symbiotic root nodules** are present. They contain nitrogen fixing bacteria, and help in fixing the atmospheric nitrogen in the soil to increase its fertility.

8. The **assimilatory roots** are green and prepare food by photosynthesis. Such roots are commonly found in *Tinospora*, water chestnut, certain orchids, etc.

9. Some roots help in vegetative propagation, *e.g.,* in sweet potato.

10. In some aquatic plants, *e.g.,* in *Jussiaea*, roots are spongy and store much air in them. They are buoyant and act as **root floats.**

11. The **pneumatophores** of mangrove vegetation (*e.g.,* in *Avecinnia*, *Sonneratia*, etc.) are negatively geotropic and possess lenticels on them for exchange of gases.

12. Roots of several plants develop mutually beneficial partnership with some fungal hyphae to form **mycorrhiza.** *e.g.,* in *Pinus*, *Quercus*, etc.

Table 26.3. Classification of Roots

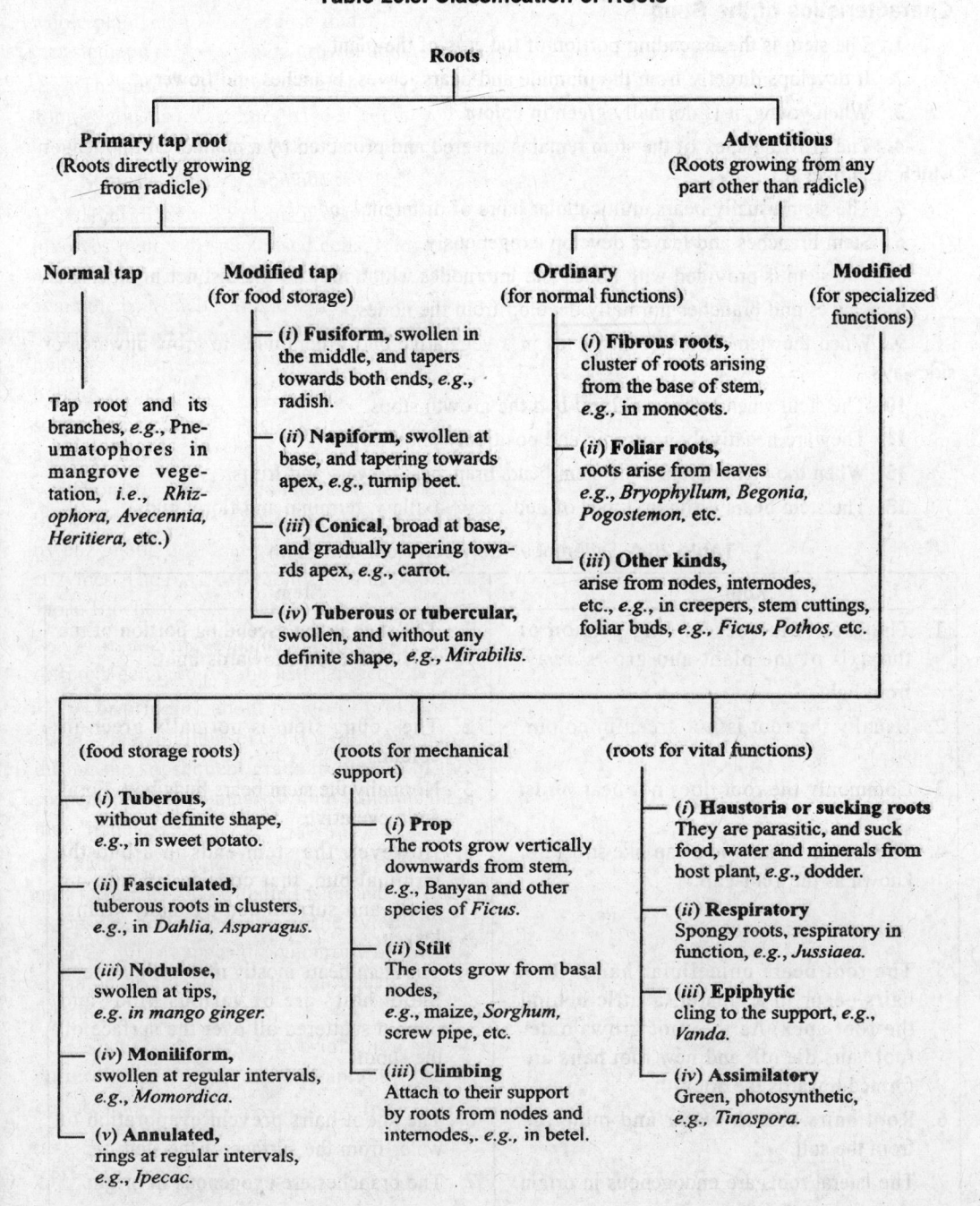

Roots

- **Primary tap root** (Roots directly growing from radicle)
  - **Normal tap**: Tap root and its branches, e.g., Pneumatophores in mangrove vegetation, i.e., *Rhizophora, Avecennia, Heritiera*, etc.)
  - **Modified tap** (for food storage)
    - (*i*) **Fusiform**, swollen in the middle, and tapers towards both ends, e.g., radish.
    - (*ii*) **Napiform**, swollen at base, and tapering towards apex, e.g., turnip beet.
    - (*iii*) **Conical**, broad at base, and gradually tapering towards apex, e.g., carrot.
    - (*iv*) **Tuberous** or **tubercular**, swollen, and without any definite shape, e.g., *Mirabilis*.

- **Adventitious** (Roots growing from any part other than radicle)
  - **Ordinary** (for normal functions)
    - (*i*) **Fibrous roots**, cluster of roots arising from the base of stem, e.g., in monocots.
    - (*ii*) **Foliar roots**, roots arise from leaves e.g., *Bryophyllum, Begonia, Pogostemon*, etc.
    - (*iii*) **Other kinds**, arise from nodes, internodes, etc., e.g., in creepers, stem cuttings, foliar buds, e.g., *Ficus, Pothos*, etc.
  - **Modified** (for specialized functions)

Modified branches:

(food storage roots)
- (*i*) **Tuberous**, without definite shape, e.g., in sweet potato.
- (*ii*) **Fasciculated**, tuberous roots in clusters, e.g., in *Dahlia, Asparagus*.
- (*iii*) **Nodulose**, swollen at tips, e.g. in mango ginger.
- (*iv*) **Moniliform**, swollen at regular intervals, e.g., *Momordica*.
- (*v*) **Annulated**, rings at regular intervals, e.g., *Ipecac*.

(roots for mechanical support)
- (*i*) **Prop** The roots grow vertically downwards from stem, e.g., Banyan and other species of *Ficus*.
- (*ii*) **Stilt** The roots grow from basal nodes, e.g., maize, *Sorghum*, screw pipe, etc.
- (*iii*) **Climbing** Attach to their support by roots from nodes and internodes,. e.g., in betel.

(roots for vital functions)
- (*i*) **Haustoria or sucking roots** They are parasitic, and suck food, water and minerals from host plant, e.g., dodder.
- (*ii*) **Respiratory** Spongy roots, respiratory in function, e.g., *Jussiaea*.
- (*iii*) **Epiphytic** cling to the support, e.g., *Vanda*.
- (*iv*) **Assimilatory** Green, photosynthetic, e.g., *Tinospora*.

DIVERSITY AND MORPHOLOGY IN THE STEM

Characteristics of the Stem

1. The stem is the ascending portion of the axis of the plant.
2. It develops directly from the plumule and bears leaves, branches and flowers.
3. When young, it is normally green in colour.
4. The growing apex of the stem remains covered and protected by a number of tiny leaves which arch over it.
5. The stem usually bears multicellular hairs of different kinds.
6. Stem branches and leaves develop exogenously.
7. The stem is provided with nodes and internodes which may not be distinct in all cases.
8. Leaves and branches normally develop from the nodes.
9. When the stem or the branch ends in a vegetative bud it continues to grow upwards or sideways.
10. The stem when ends in a floral bud the growth stops.
11. They are negatively geotropic and positively phototropic.
12. When the plant matures the stem bears branches, flowers and fruits.
13. The stem bears different kinds of buds, *e.g.*, axillary, terminal and floral buds.

Table 26.4. Differences Between Root and Stem

Root	Stem
1. The root is the descending portion of the axis of the plant and grows away from light.	1. The stem is the ascending portion of the axis and it turns towards light.
2. Usually the root is not green in colour.	2. The young stem is normally green in colour.
3. Commonly the root does not bear **buds.**	3. Normally the stem bears buds both floral and vegetative.
4. The root terminates in a cap like structure, known as the **root-cap.**	4. However, the stem ends in a bud-the terminal bud, that consists of growing apex, and surrounded by many young leaves.
5. The root bears unicellular hairs. Root hairs occur in a cluster, a little behind the root apex. As the root grows older root hairs die off, and new root hairs are formed towards the apex.	5. The stem bears mostly multicellular hairs. Shoot hairs are of various kinds and remain scattered all over the surface of the shoot.
6. Root hairs absorb water and minerals from the soil.	6. The shoot hairs prevent evaporation of water from the surface of the stem.
7. The lateral roots are endogenous in origin.	7. The branches are exogenous in origin.
8. The nodes and internodes are absent.	8. The nodes and internodes are present.

Nodes and Internodes. The place on the stem or branch where one or more leaves arise is known as the **node**, and the space between two successive nodes is called the **internode**. In some cases they are very conspicuous, *e.g.*, in bamboos and grasses, while in others they are not clearly marked.

Diversity and Morphology of Flowering Plants

DIVERSITY IN HABIT OF THE PLANT

The plants can be classified into different categories according to their height or according to their life-cycle.

1. According to the height

They may be (*i*) **herbs,** (*ii*) **shrubs** and (*iii*) **trees.**

(*i*) **Herbs.** These are small plants with soft stems. They may be **annuals**, *e.g.*, mustard, pea, rice, etc., **biennials**, *e.g.*, beet, carrot, turnip, etc., **perennials**, *e.g.*, canna, ginger, banana, etc.

The **annuals** are those herbs that attain their full growth in one season, living for few months or at the most for one year only. The **biennials** are those herbs that live for two years and they produce flowers and seeds in the second year after which they die off. The **perennials** are those herbs that persist for a number of years. The aerial parts of such plants die every year at the end of the season but new shoots develop again from the underground stem.

(*ii*) **Shrubs.** They are medium sized plants with hard and woody stems. They branch profusely from near the ground, and thus the plants become bushy in habit without a clear trunk. The examples are *Hibiscus rosa-sinensis*, night jasmine (*Nyctanthes arbortristis*), garden croton, etc.

(*iii*) **Trees.** They are tall plants with a clear trunk and possess hard and woody stem and branches, *e.g.*, mango, sissoo, nim, teak, jack, etc.

The shrubs and trees are perennials

2. According to the life-cycle

They may be (*i*) **annuals,** (*ii*) **biennials** and (*iii*) **perennials.**

(*i*) **Annuals.** Complete their whole life, from seed to fruit in one year or less. In some cases even in a few weeks such as in *Senecio vulgaris*, so that several generations may be passed through in one summer while nothing but the seeds remain through the winter. Biennials and annuals are therefore, typically **monocarpic**, fruiting but once.

(*ii*) **Biennials.** Last only for two years. In the first season they produce at soil level a very contracted stem bearing a rosette of leaves. During the second season the stem elongates and bears the flowers and fruit, after which the whole plant dies, *e.g.*, carrot, radish, turnip, etc.

(*iii*) **Perennials.** May be woody, either tree like with one main trunk, or else shrub-like with a cluster of stems. They may also be herbaceous lying down to the ground level each winter and persisting only by under ground organs. The duration of perennials is very variable. Some herbs live only five to six years. Large trees on the other hand may take twenty five to thirty years to attain flowering. The common factor in all perennials is that they are **polycarpic**, *i.e.*, they flower and fruit again and again.

Certain monocotyledonous perennials, on the other hand, are naturally monocarpic. For example, bamboos flower every twenty to thirty years and then die. *Agave americana* flowers once but only after hundred years producing a 50 feet inflorescence and thereafter dies.

DIVERSITY IN BRANCHING OF THE STEM

The mode of arrangement of the branches on the stem is known as **branching**. The branches develop exogenously from the lateral vegetative buds. The lateral branching is of two main types, *i.e.*, **racemose** and **cymose**.

1. Racemose type. In this type of branching the growth of the main stem is indefinite, *i.e.*, it continues to grow indefinitely by its terminal bud and gives off branches in acropetal succession.

Here the lower branches are older and longer than the upper branches, *e.g., Polyalthia, Casuarina, Eucalyptus*, etc.

2. Cymose type. In such type of branching the main axis or the stem does not grow indefinitely due to the limited growth of the terminal or apical bud. Here the growth is definite, and the main stem produces one or more lateral branches which grow more vigorously than the terminal one. The process may be repeated again and again.

The cymose branching may be subdivided as follows: Uniparous cyme and biparous cyme.

A. Uniparous cyme. In such type, only one lateral branch is produced at a time and the branching is known as uniparous or monochasial. It is further subdivided into two types, *i.e.,* helicoid cyme and scorpioid cyme.

(*a*) **Helicoid cyme.** Here the successive branches develop on one side only, *e.g.,* in *Saraca indica*.

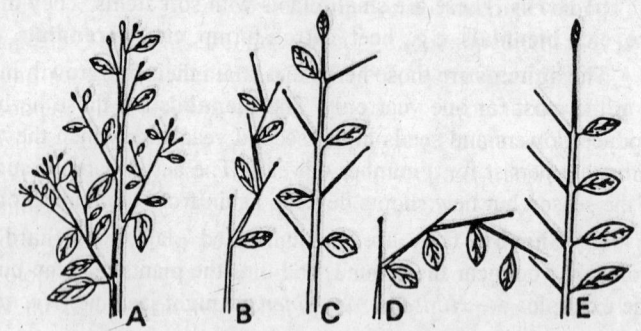

Fig. 26.16. Branching of stem. A, racemose branching; B—C, scorpioid-uniparous cymose branching; D—E, helicoid-uniparous cymose branching.

(*b*) **Scorpioid cyme.** Here the successive lateral branches develop on alternate sides, forming a zig-zag, *e.g.,* in *Vitis vinifera*.

(*c*) **Biparous cyme.** In such type of cymose branching, two lateral axes develop at a time, and it is called biparous or dichasial cyme, *e.g.,* in *Mirabilis, Viscum, Stellaria*, etc.

DIVERSITY IN MODIFICATIONS OF STEMS

In most of the plants the stem is aerial, vertically upward and bears leaves, flowers and fruits. However, some stems are modified into various shapes to carry on special functions, such as perennation, vegetative propagation, and synthesis and storage of food, etc. The various modifications of the stems may be categorized as follows: **1.** Underground modifications of stems, **2.** Sub-aerial modification of stems and **3.** Aerial modifications of stems.

1. Underground Modifications of stems

Some stems develop underground in the soil for the purpose of perennation, vegetative propagation and storage of food. These modifications of stems can be differentiated from the roots as having (*a*) nodes and internodes, (*b*) scale leaves and (*c*) buds of axillary and terminal types. Various types of underground stems are as follows:

(*i*) **Rhizomes.** They are plagiotropic underground stems of the most varied kind. It is a prostrate, thickened stem, creeping horizontally under the

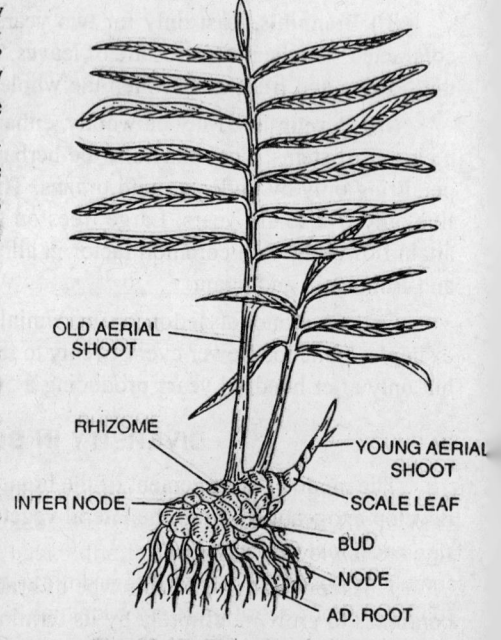

Fig. 26.17. Rhizome of ginger.

Diversity and Morphology of Flowering Plants

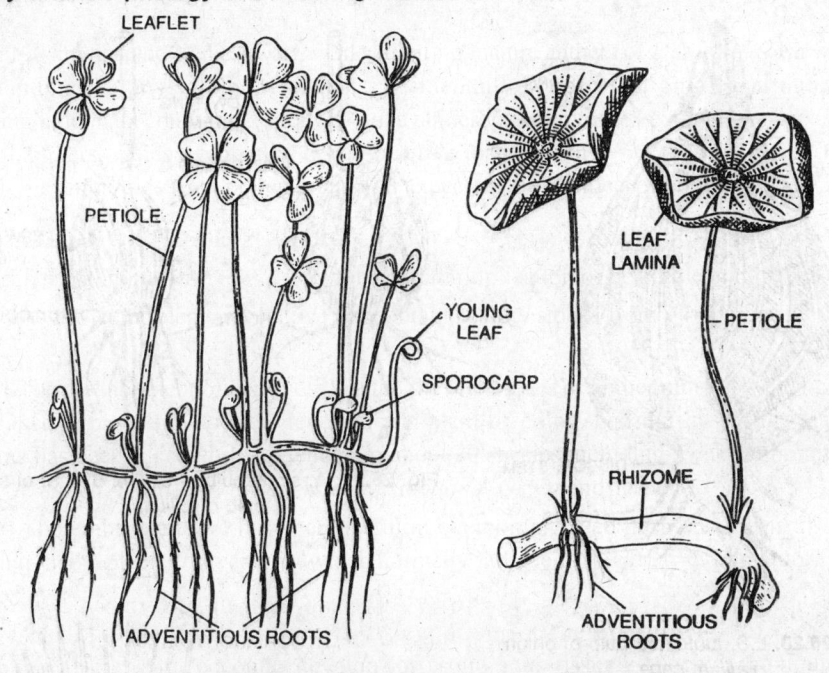

Fig. 26.18. Rhizome of Marsilea and *Nelumbo*.

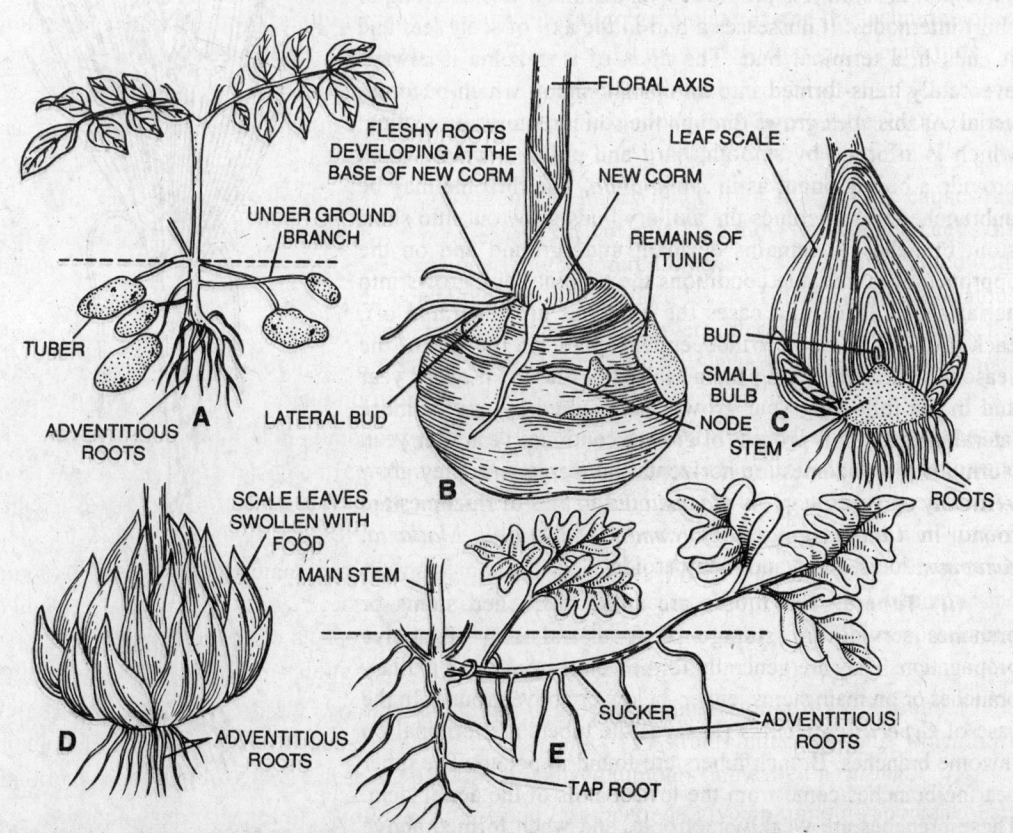

Fig. 26.19. Modified underground stems. A, tuber of potato; B, corm of *Colocasia*; C, tunicated bulb of onion; D, scaly bulb of *Oxalis*; E, sucker of *Chrysanthemum*.

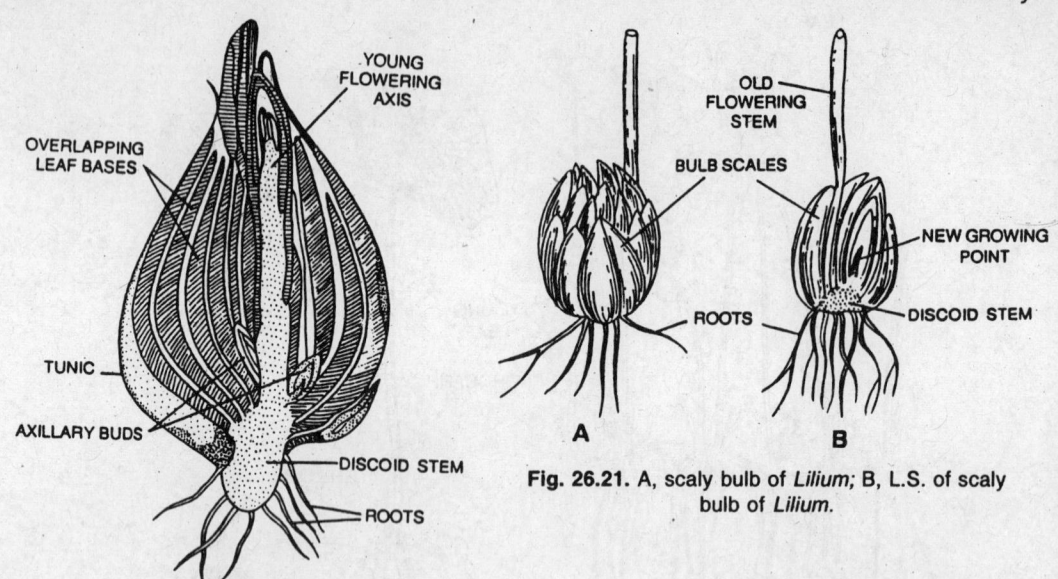

Fig. 26.20. L.S. tunicated bulb of onion (*Allium cepa.*)

Fig. 26.21. A, scaly bulb of *Lilium;* B, L.S. of scaly bulb of *Lilium*.

surface of the soil. It is provided with distinct nodes and long or short internodes. It possesses a bud in the axil of scaly leaf and it ends in a terminal bud. The apex of a rhizome is always eventually trans-formed into an upright shoot, which becomes aerial. As this apex grows through the soil it requires protection, which is afforded by smooth, hard and pointed scales which provide a boring point, as in *Ammophila*. The rhizome may be unbranched or sometimes the axillary buds grow out into short stout branches. It remains dormant underground and on the approach of favourable conditions the terminal bud grows into aerial shoot. In certain cases the branches are separated off, each growing into a new independent plant. In the end of the season or after flowering the aerial parts die down every year and in the following year growth is resumed by one or more lateral buds and this process of growth continues year after year. Normally the rhizomes are horizontal but sometimes they grow vertically upwards, *e.g.,* in *Alocasia*. Examples of rhizomes are found in *Canna*, ginger, *Curcuma*, water lily, *Elattaria*, *Amomum*, lotus, ferns and many aroids.

(*ii*) **Tubers.** The tubers are solid, thickened stems or branches serving for storage of food and also vegetative propagation. They are generally formed on rhizomes, on axillary branches or on main stems, either below or above ground. In the case of *Cyperus esculentus* **(kaseru)** the tubers are terminal on rhizome branches. Branch tubers are found in potato. The tuber bearing branches come from the lowest axils of the aerial stem. These branches are weakly geotropic, and when formed above ground they bury themselves and swell at their apices to form the

Fig. 26.22. A laminate or tunicated bulb of *Crocus*.

massive tubers. These tubers bear temporary scale leaves with buds in their axils called 'eyes'. The tuber remains dormant for sometime. On its germination, orthotropic shoots are developed from the axiliary buds, which become new aerial stems. Even a single detached bud may grow into a new plant. The tubers of *Helianthus tuberosus* are developed like those of potato, but comparatively they may borne on short branches. They contain inulin instead of starch.

(*iii*) **Bulbs.** The bulb is also a short specialized underground shoot or stem. In this case, the stem remains, comparatively small, and the food material is stored in the large, fleshy scales which invest and overlap the stem. However, the stem is no more than a thick disc or very flat cone, but it has an apical bud on the upper side, and adventitious roots are formed in an annual plant from the marginal portion of the underside. The bulb develops in the resting period or two to four months after flowering of the plant. The fleshy scales are set very close together and their bases of insertion partially or completely surround the stem. In scaly bulbs (*e.g., Lilium*), the fleshy scales composing the main bulk overlap at their margins. In **tunicated bulbs** (*e.g., Allium*, tulip, etc.) the outer leaves are large and completely ensheath the inner portion of the bulb.

The bulbs are organs of perennation and of vegetative propagation. While still underground, the organs of the aerial shoot, including the parts of the flower, are already formed in their buds.

Table 26.5. Differences Between Rhizome and Tuber

Rhizome	Tuber
1. This is a prostrate thick stem creeping horizontally under surface of the soil.	1. This is the swollen end of a special underground branch that arises from the axil of a lower leaf, and grows horizontally.
2. Provided with distinct nodes and internodes.	2. A number of eyes are present on its surface which grow in new plants.
3. The slender adventitious roots are given off from its lower side.	3. Here the adventitious roots are absent, and develop only when tuber is planted from the base of aerial shoots.
4. Example: ginger	4. Example: potato

Table 26.6. Differences Between Corm and Bulb

Corm	Bulb
1. It consists of a stout, solid, fleshy underground stem growing vertically.	1. It consists of a short convex disc, from the upper surface of which fleshy scale leaves arise.
2. Food is stored in the stem.	2. Food is stored in the fleshy leaves.
3. Buds arise from the external surface of the fleshy corm.	3. Buds are internal and remain covered by fleshy leaves.
4. It is differentiated into circular nodes and internodes.	4. Being discoid, it is not differentiated into nodes and internodes.
5. Adventitious roots develop from the base and sometimes also from the sides.	5. Adventitious roots develop downward from the bulbous disc.
6. Examples: *Colocasia, Crocus*.	6. Examples: onion, garlic.

Table 26.7. Differences Between Rhizome and Corm

Rhizome	Corm
1. This is a prostrate, thickened underground stem.	1. This is stout, solid, fleshy underground stem growing vertically.
2. It is usually branched and rarely unbranched.	2. It is usually unbranched.
3. It is elongated in shape.	3. It is spherical in shape.
4. It contains less amount of food material.	4. It contains a heavy deposit of food material and often grows to a large size.
5. They are generally perennial and live for several years.	5. They can live only for few (1-3) years.
6. Growth is continuous so that succession of rhizome is not produced.	6. New corm is produced every year either above or on the sides of old corm.
7. Examples: ginger, water lily, etc.	7. Examples: *Amorphophallus, Colocasia, Crocus*, etc.

(*iv*) **Corms.** The corm is a condensed form of rhizome and consists of a stout, solid, fleshy, underground stem growing in the vertical direction. The swollen stem bears a number of loose, more or less sheathing scale leaves. The size of the stem is due to the large amount of food material stored in it. On top it bears an apical bud, from which summer shoots with leaves and flowers are produced, and it bears the annual adventitious roots, usually from the lower end. The examples are seen in *Crocus* and *Gladiolus*. In these cases the corm is considerably flattened and its length remains shorter than its diameter. The axillary buds are occasionally found at the nodes on the corm, from which side shoots and new corms are given out. After flowering, the leaves of *Crocus* remain active for some time, and the material assimilated by them is stored at the base of the flowering stem, producing a new corm, on top of the old one, from an axillary bud. Other examples are seen in *Colocasia, Amorphophallus, Colchicum*, etc.

2. Sub-aerial Modifications of stems

In some plants for the purpose of vegetative propagation some of the lower dormant buds of the stem grow out into lateral branches which according to their origin, nature and mode of propagation are known by different names, such as runner, stolon, offset and sucker. They may be subaerial or partly subterranean.

(*i*) **Runner.** This type of slender, prostrate branch possesses distinct nodes and long internodes. It creeps on the ground and the roots are produced from the nodes. The runner develops

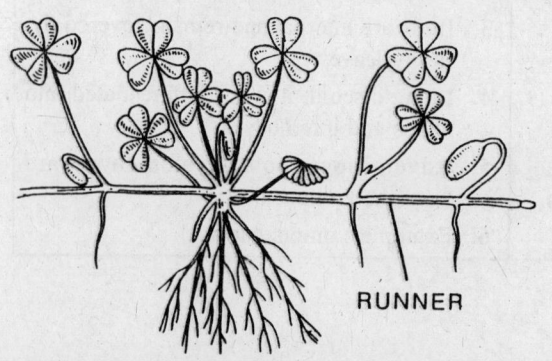

Fig. 26.23. Sub-aerial modification of stem. Runner of *Oxalis*.

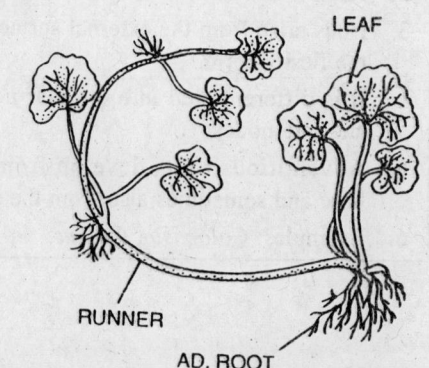

Fig. 26.24. Runner of *Centella asiatica*.

Diversity and Morphology of Flowering Plants

as an axillary bud and creeps some distance away from the mother plant. The roots are given from the nodes and thus new plants develop, *e.g.*, Oxalis, Cynodon, Fragaria, Centella, etc.

(*ii*) **Stolon.** It is a weak, slender, lateral branch that arises from the base of the shoot buried in the soil. It grows horizontally outwards, and where it touches the ground, the adventitious roots are given out. The branches grow out in different directions, and its terminal bud emerges out of the soil and develops into a new plant. The branch is subterranean. Examples are seen in *Alocasia*, *Colocasia*, jasmine, etc.

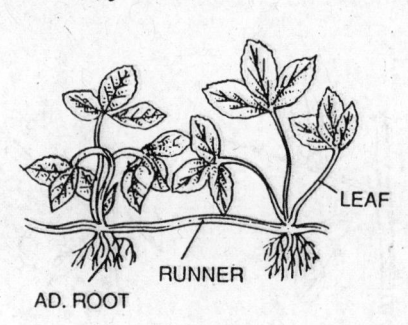

Fig. 26.25. Runner of *Fragaria vesca*.

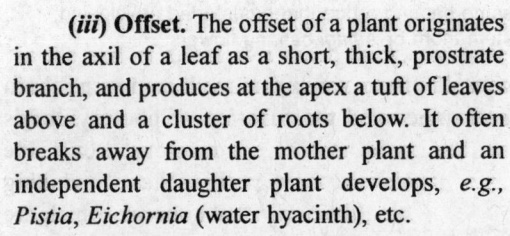

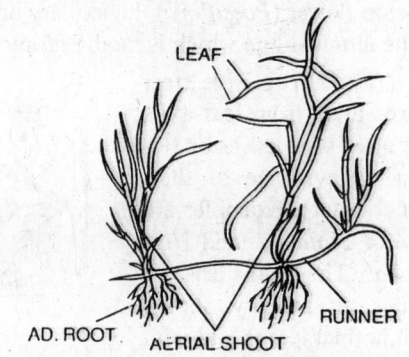

Fig. 26.26. Runner of doob grass.

(*iii*) **Offset.** The offset of a plant originates in the axil of a leaf as a short, thick, prostrate branch, and produces at the apex a tuft of leaves above and a cluster of roots below. It often breaks away from the mother plant and an independent daughter plant develops, *e.g.*, *Pistia*, *Eichornia* (water hyacinth), etc.

(*iv*) **Sucker.** Like the stolon the sucker also develops as a lateral branch from the underground part of the stem. It grows obliquely upwards and develops into a new plant. It is always much shorter than a stolon. The sucker develops roots at the base. This is device for vegetative propagation, as seen in *Chrysanthemum*, mint, etc.

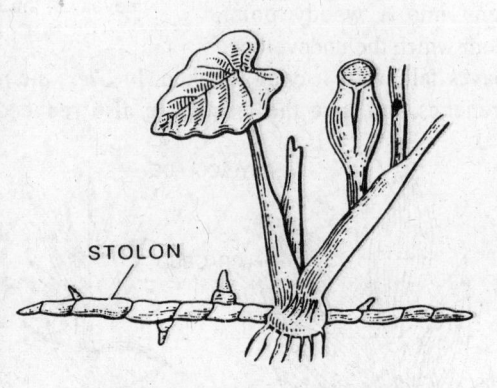

Fig. 26.27. Stolon of *Colocasia*.

3. Aerial Modifications of stems

Vegetative and floral buds found on the stem develop into branches and flowers respectively. However very often this aerial stem under- goes extreme degree of modification in certain plants for definite purposes. Such metamorphosed organs may be stem

Fig. 26.28. Sub-aerial modification of stem. Offset of *Pistia*.

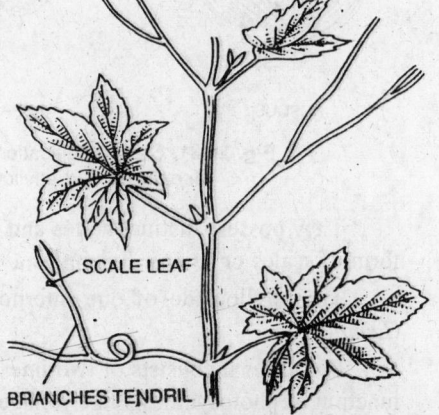

Fig. 26.29. Leaf opposed tendrils of vine— *Vitis vinifera*.

tendril for climbing, thorn for protection, phylloclade for food manufacture, and bulbil for vegetative propagation.

(*i*) **Stem-tendril.** These are highly specialized climbing organs. They are slender often branched and may bear small scale leaves. The climbers by means of these tendrils attach themselves to neighbouring objects and climb them. The stem tendril is the modification of a stem is evident from the fact that it arises in the axil of a leaf or at the apex of a branch. For example in passion flower (*Passiflora*) the axillary bud is modified into the tendril, whereas in vine (*Vitis*) it is the terminal bud which is modified into tendril.

(*ii*) **Thorns.** True stem thorns arise from leaf-axils and often bear leaves or flowers as evidence of their branch nature. Examples are seen in *Crataegus* and *Hippophae*. The thorns arise in spring as axillary shoots with normal leaves and with an apical bud. The apex soon stops growth and hardens into a woody point, from which the undeveloped leaves fall away, leaving it naked.

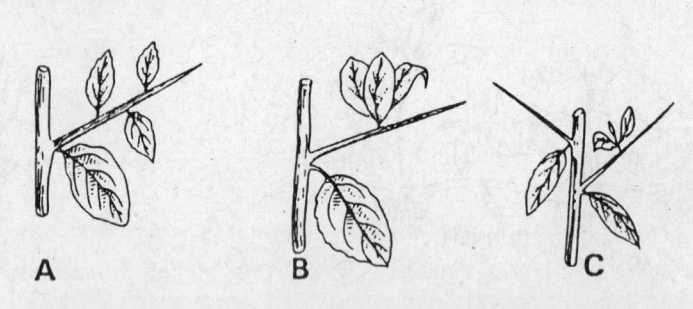

Fig. 26.30. Stem thorns. A, stem thorn of *Pyrus coronaria* bearing leaves; B, the same bearing a bud that has started developing; C, stem thorn of *Alhagi* bearing flowers.

In *Ulex*, the main shoots are thickly set with compound thorn branches, but here the leaves are also reduced to spines and the function of photosynthesis develops on the branches including the thorns, which are all green.

(*iii*) **Phylloclade.** This is a green, flattened rounded stem, usually found in the plants of dry and arid habitats. This stem structure has taken on the general appearance and functions of a leaf. Usually the phylloclades represent lateral branches.

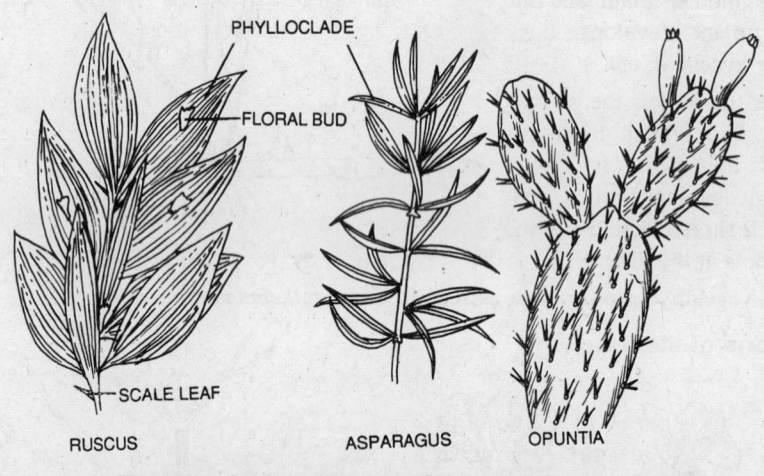

Fig. 26.31. Stem modifications, Cladodes of *Ruscus*, *Asparagus* and phylloclade of *Opuntia*.

They possess distinct nodes and internodes and sometimes they bear modified leaves in the form of scales or spines. Examples are found in *Opuntia*, *Cocoloba*, *Euphorbia*, etc.

The phylloclade of one internode is known as **cladode**, *e.g.*, in *Asparagus*, duckweed (*Lemna*).

In *Ruscus* it consists of two internodes. Here the cladodes are flat, leaf like and perform the function of photosynthesis like foliage leaves.

(*iv*) **Bulbils.** They may be described as axillary buds, which become large and fleshy owing to the storage of food material in their leaves. it is a special multicellular body essentially meant

Diversity and Morphology of Flowering Plants 495

for the reproduction of the plant. The bulbils differ from ordinary buds in the fact that they separate from the parent plant, fall to the ground and produce new plants, thus serving for reproduction. The examples are found in *Dioscorea, Globba, Agave,* onion, etc.

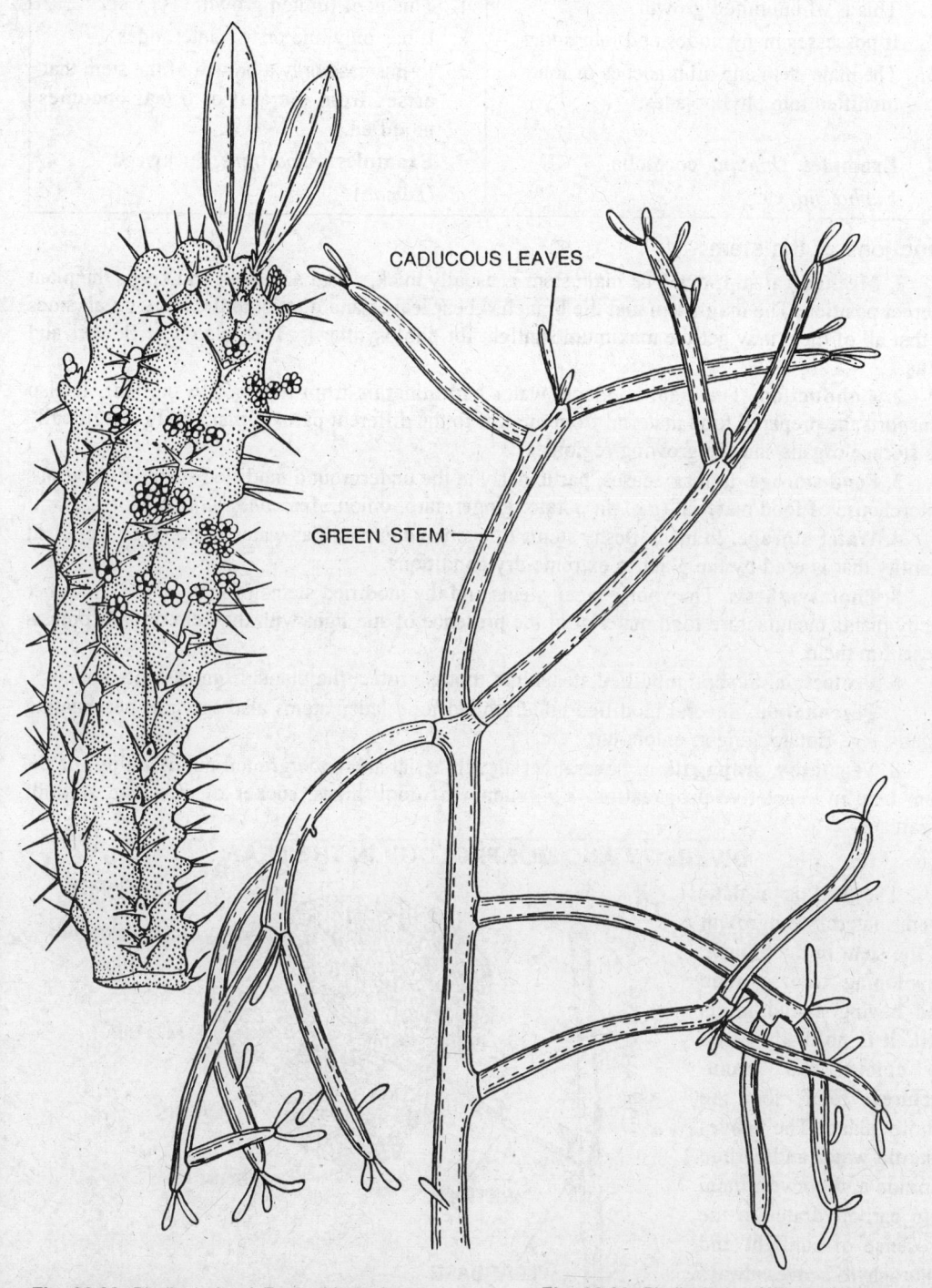

Fig. 26.32. Phylloclade of *Euphorbia royleana*. **Fig. 26.33.** Phylloclade of *Euphorbia tirucalli*.

Table 26.8. Differences Between Phylloclade and Cladode

Phylloclade	Cladode
1. This is of unlimited growth.	1. This is of limited growth.
2. It possesses many nodes and internodes.	2. It has only one or two internodes.
3. The main stem and all branches of it are modified into phylloclades.	3. In this case only a branch of the stem that arises from the axil of a leaf becomes modified.
4. **Examples.** *Opuntia*, cocoloba, *Euphorbia*, etc.	4. **Examples:** *Asparagus*, duckweed (*Lemna*).

Functions of the stem

1. Mechanical support. The main stem is usually thick, aerial and stout and keeps the plant in erect position. The main stem and the branches bear leaves and distribute them out on all sides so that all of them may get the maximum sunlight for photosynthesis. They also bear flowers and fruits.

2. Conduction. The stem conducts water and minerals from the root to the leaf. It also transports the prepared food material from the leaf to the different parts of plant body, specifically the storage organs and the growing regions.

3. Food storage. In many cases, particularly in the underground modified stems, it serves as a storehouse of food material (*e.g.,* in potato, ginger, taro, onion, *Amorphophallus*, etc.

4. Water storage. In many fleshy stems of cacti and euphorbias water is stored in sufficient quantity that is used by the plant in extreme dry conditions.

5. Photosynthesis. The young green stems and the modified stems of many cacti and other fleshy plants manufacture food material in the presence of sun light with the help of chloroplasts present in them.

6. Protection. Several modified stems into thorns protect the plants from grazing animals.

7. Perennation. Several modified underground food laden stems also serve as perennating organs, *e.g.,* potato, ginger, onion, taro, etc.

8. Vegetative propagation. Several aerial, sub-aerial and underground modifications of the stem help in vegetative propagation, *e.g.,* runner of doob grass, sucker of mint and setts of sugarcane.

DIVERSITY AND MORPHOLOGY IN THE LEAF

The leaf is a flattened, lateral outgrowth of the stem in the branch, developing from a node and having a bud in its axil. It is normally green in colour and manufactures food for the whole plant. The leaves take up water and carbon dioxide and convert them into carbohydrates in the presence of sunlight and chlorophyll. Leaves always follow an **acropetal** development and are exogenous in origin.

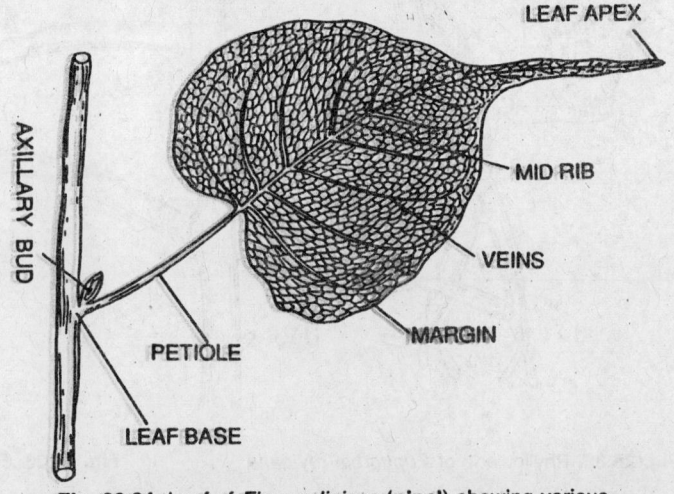

Fig. 26.34. Leaf of *Ficus religiosa* (**pipal**) showing various parts of the leaf.

Diversity and Morphology of Flowering Plants

Parts of a leaf

A typical leaf of *Ficus religiosa* (pipal) has a broad thin, flat structure called the **lamina**. The thin stalk below the lamina is the **petiole**. The lamina possesses a network of **veins**. The veins have both xylem and phloem elements which are continuous with similar tissues of the stem through those of the petiole. A strong vein, known as the **midrib**, runs centrally through the leaf-blade from its base to the apex; this produces thinner lateral veins which in their turn give rise to still thinner veins or veinlets. The lamina is the most important part of the leaf since this is the seat of food manufacture for the whole plant.

Types of a Leaf

Radical. Proceeding from or near the root, *e.g.*, onion, radish, etc.

Cauline. Pertaining to the stem, *e.g.*, palms.

Cauline and ramal. Pertaining to the main stem as well as its branches, *e.g.*, mango.

Phyllotaxy of Leaves

Alternate. A single leaf arising at each node, *e.g.*, *Hibiscus rosa-sinensis*.

Opposite. On different sides of the axis with the bases at the same level.

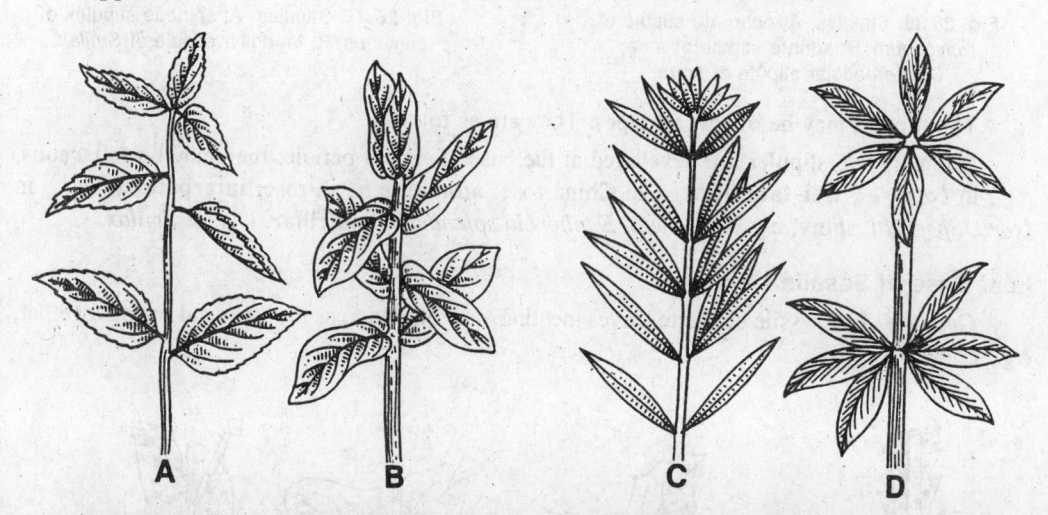

Fig. 26.35. Phyllotaxy of leaves. A, alternate; B, opposite decussate; C, whorled leaves of *Nerium*; D, whorled leaves of *Alstonia*.

Opposite decussate. In pairs at right angles to one another, *e.g.*, *Calotropis*.

Opposite superposed. A pair of leaves that stands directly over the lower pair in the same plane, *e.g.*, guava.

Whorled. More than two leaves arranged in a circle round an axis, *e.g.*, *Spergula*, *Alstonia*.

Petiole

Petiolate. The leaf blade is situated on the petiole, *e.g.*, *Hibiscus*, *Ficus*, etc.

Sessile. Without a petiole or stalk, *e.g.*, *Ixora*

Sub-sessile. Having a short stalk, *e.g.*, *Polygonum*.

Stipules

Stipulate. The leaf with stipules, *e.g.*, rose, *Ixora*,

Exstipulate. The leaves having no stipules, *e.g.*, *Ipomoea*.

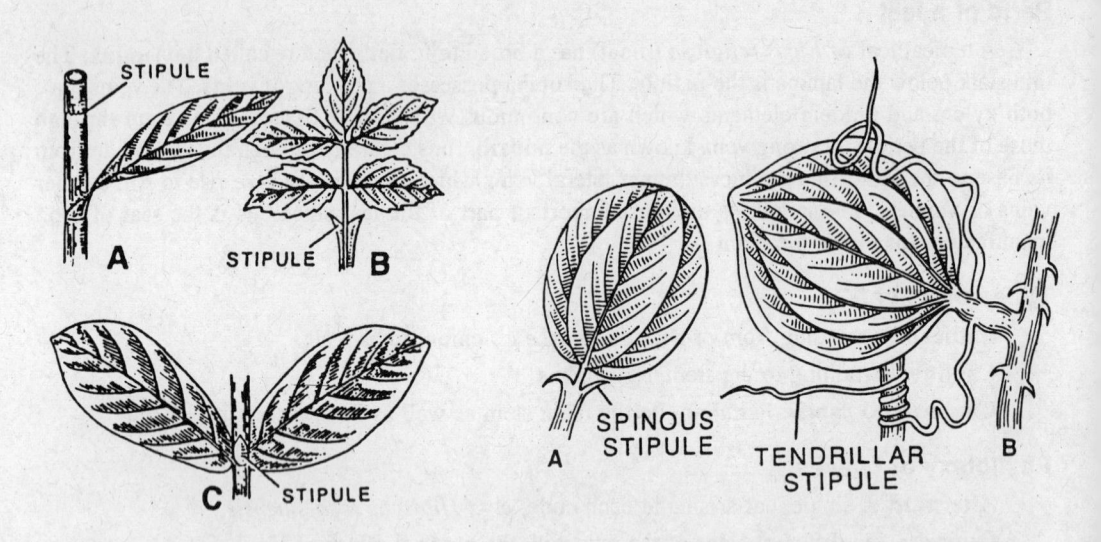

Fig. 26.36. Stipules. A, ochreate stipule of *Polygonum;* B, adnate stipule of rose; C, interpetiolar stipule of *Ixora*.

Fig. 26.37. Stipules. A, spinous stipules of *Zizyphus;* B, tendrillar stipule of *Smilax*.

The stipules may be of several types. They are as follows:

Normally two stipules are developed at the base of a leaf petiole; they may be **foliaceous**, *e.g.,* in *Lathyrus*; **free lateral**, *e.g.,* in China rose; **adnate**, *e.g.,* in rose; **interpetiolar**, *e.g.,* in *Ixora, Spergula*; **spiny**, *e.g.,* in *Acacia, Euphorbia splendens*; **tendrillar**, *e.g.,* in *Smilax*.

Leaf Base of Sessile Leaves

Connate. Two sessile opposite leaves meeting each other across the stem and fusing together, *e.g., Lomicera flava.*

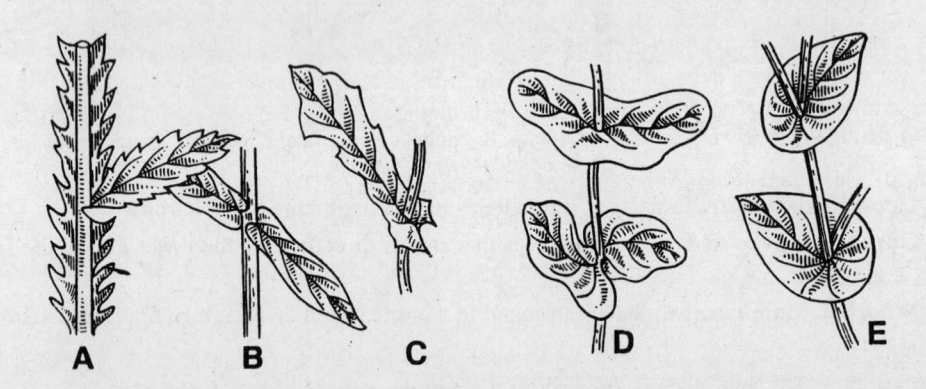

Fig. 26.38. Leaf base of sessile leaves. A, decurrent leaf of *Laggera;* B, auriculate leaf of *Calotropis;* C, amplexicaul leaf of *Emilia;* D, connate leaf of *Lomicera;* E, perfoliate leaves.

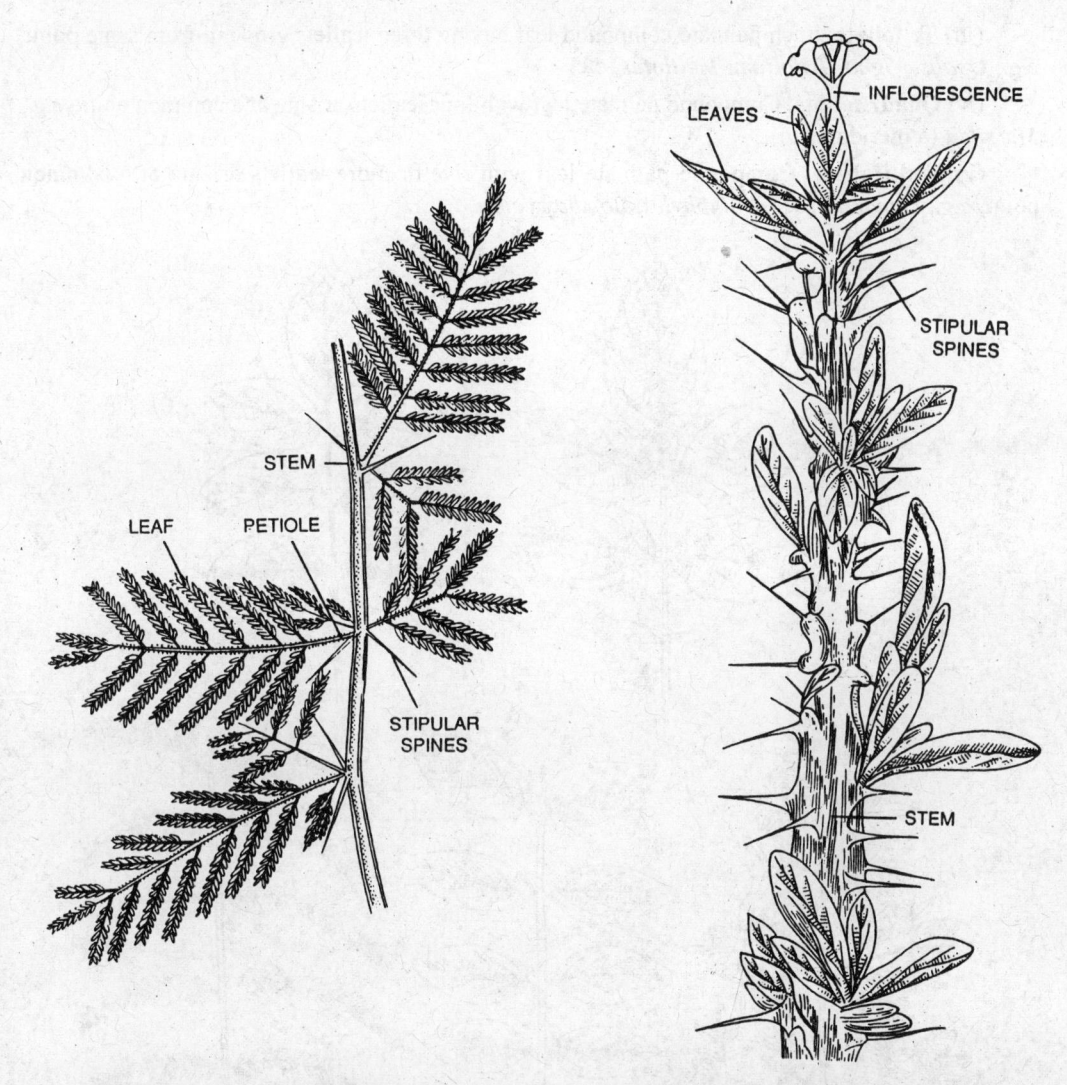

Fig. 26.39. *Acacia nilotica.* The stipules are modified into spines.

Fig. 26.40. *Euphorbia splendens.* The stipules are modified into spines.

Amplexicaul. Clasping or surrounding the stem, as base of leaf, *e.g., Sonchus.*

Auriculate. Leaf with expanded bases surrounding stem, *e.g., Calotropis.*

Decurrent. Having leaf base prolonged down stem as a winged expansion or rib, *e.g., Laggera pterodonta.*

Perfoliate. A leaf with basal lobes so united as to appear as if stem ran through it, *e.g., Aloe perfoliata.*

Simple and Compound Leaves

Simple. A leaf which may be entire or incised to any depth, but not down to the midrib or petiole.

Compound. A leaf made up of two or more leaflets, *e.g.,* pea, and several other members of Leguminosae. The compound leaves may be of several types. They are as follows:

A. Palmately compound Leaf

(*i*) **Unifoliate.** Having one leaflet only, *e.g., Citrus.*

(*ii*) **Bifoliate.** Palmate compound leaf with two leaflets, *e.g., Prinsepia, Balanites.*

(*iii*) **Trifoliate.** Such palmate compound leaf having three leaflets growing from same point, e.g., *Oxalis*, *Vigna*, *Trifolium*, *Melilotus*, etc.

(*iv*) **Quadrifoliate.** Compound palmate leaf with four leaflets arising at a common point, e.g., *Marsilea* (a pteridophyte).

(*v*) **Multifoliate.** Compound palmate leaf with five or more leaflets arising at a common point, e.g., *Gynandropsis pentaphylla*, *Bombax ceiba*.

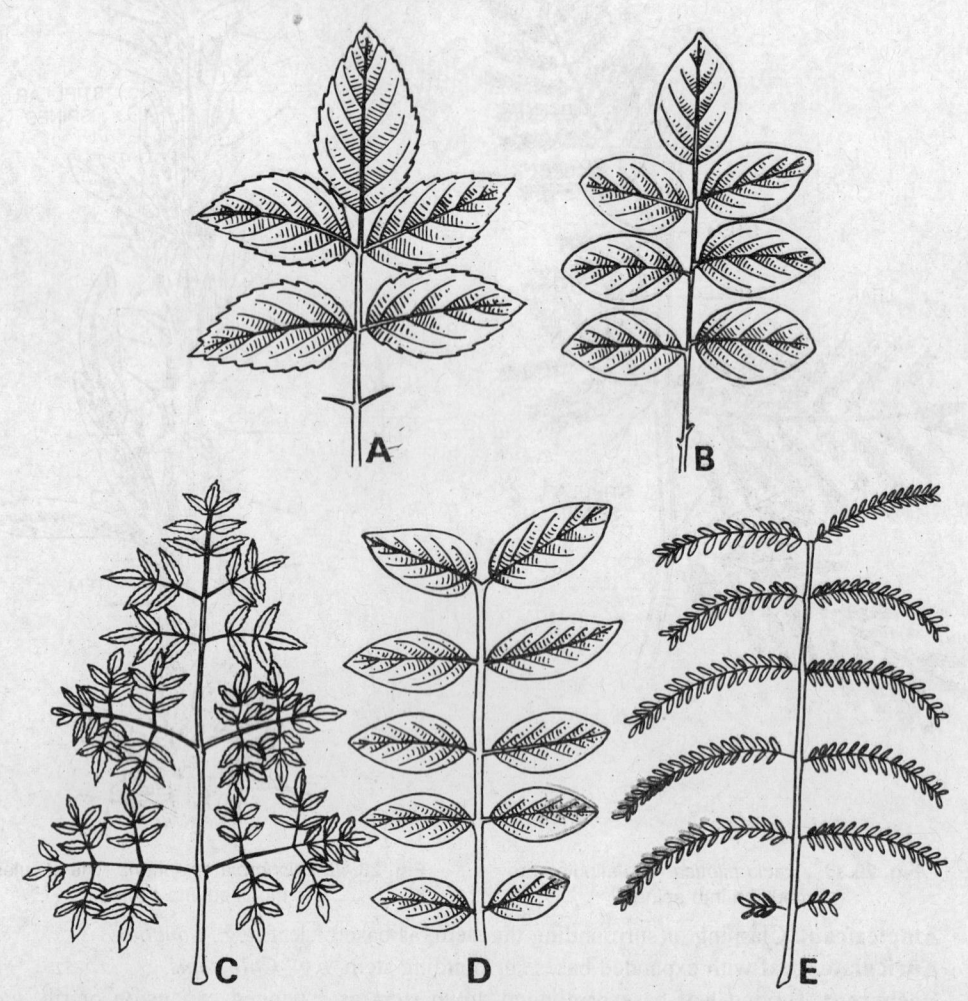

Fig. 26.41. Compound leaves. A, imparipinnate compound leaf of rose; B, imparipinnate compound leaf with alternate leaflets in *Murraya*; D, paripinnate compound leaf of *Cassia fistula*; E, bipinnate compound leaf of *Acacia*; C, tripinnate compound leaf of *Melia azadarach*.

B. Pinnately Compound Leaf

(*a*) **Pinnate.** A compound leaf having leaflets on each side on an axis or midrib.

(*b*) **Unipinnate.** Having leaflets on each side of an axis, e.g., *Cassia*.

(*c*) **Bipinnate.** The central axis produces secondary axis which bears the leaflets, e.g., *Acacia*.

(*d*) **Tripinnate.** The secondary axes produce the tertiary axis which bear the leaflets, e.g., *Moringa*.

Diversity and Morphology of Flowering Plants

(e) **Decompound.** More than thrice pinnate, *e.g.,* old leaves of coriander.
(f) **Paripinnate.** Pinnately compound without a terminal leaflet, *e.g., Cassia.*
(g) **Imparipinnate.** Pinnately compound leaf with an odd terminal leaflet, *e.g.,* pea.

Table 26.9. Differences between compound leaf and leafy branch

Compound leaf	Leafy branch
1. A compound leaf never bears a terminal bud.	1. A branch bears a terminal bud.
2. A compound leaf, like a simple leaf, always bears an axillary bud in the axil of leaf, but itself does not arise in the axil of another leaf.	2. However, a branch does not bear an axillary bud, but itself occupies the axillary position of a leaf-simple or compound.
3. The leaflets of a compound leaf have no axillary buds.	3. The leaves (simple) born on a branch have a bud in their axil.
4. The rachis of a compound leaf has no nodes and internodes.	4. Whereas a branch is always provided with nodes and internodes.

Margin of Lamina

Entire. With continuous margin, *e.g., Psidium,* mango, madar.
Dentate. With large saw like teeth on the margin, *e.g., Nymphaea,* water melon.

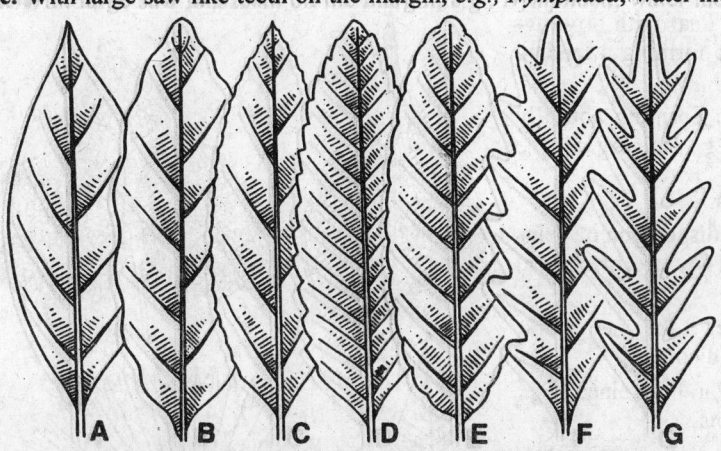

Fig. 26.42. Leaf margins. A, entire; B, undulate; C, serrate; D, dentate; E, crenate; F, lobed; G, parted.

Serrate. With serrate edges themselves toothed, *e.g.,* China rose, nim.
Undulate. The margin is wavy, *e.g., Polyalthia.*
Convolute. Rolled together.
Crenate. With obtusely toothed margin, *e.g., Bryophyllum, Centella.*
Lacerate. Having margin or apex deeply cut into irregular lobes, *e.g.,* many members of Ranunculaceae.
Laciniate. Irregularly incised, fringed.
Laciniolate. Minutely incised or fringed.
Ciliate. Bearing fine hairs on the margin, *e.g., Cleome viscosa.*
Crispate. Curled or extremely undulate margin.
Spinous. Bearing many spines, *e.g., Argemone.*
Pectinate. Comb like margin.
Lobed. Leaf margin divided into many lobes, *e.g., Ranunculus.*

Leaf Apex

Acute. Ending in a sharp point forming an acute angle, *e.g.,* mango.

Acuminate. Drawn out into long point; tapering; pointed, *e.g., Ficus religiosa.*

Obtuse. With blunt or rounded end, *e.g.,* Banyan.

Emarginate. Having a notch at apex, *e.g., Bauhinia.*

Truncate. Terminating abruptly, as if tapering end were cut off, *e.g., Caryota urens,*

Mucronate. Abruptly terminated by a sharp spine, *e.g.,* apex of leaflet of *Cassia obtusifolia.*

Cuspidate. Terminating in a point.

Aristate. Provided with awns or with a well developed bristle.

Retuse. Obtuse with a broad shallow notch in middle, *e.g., Oxalis.*

Cirrhose. Leaf with prolongation or mid-rib forming a tendril, *e.g., Gloriosa.*

Apiculate. Forming abruptly to a small tip, *e.g., Dalbergia.*

Leaf Venation

System or disposition of veins in the leaves. They are of several types.

Reticulate. Like net work, *e.g.,* in most of dicots.

Parallel. Parallel veined, *e.g.,* most of monocots.

(a) Unicostate reticulate. Having only one principal vein, *e.g.,* mango, banyan, etc.

(b) Multicostate reticulate. Having many principal veins, *e.g.,* castor, cucumber, etc.

Shape of Leaf Lamina

Linear. Long and narrow leaf, *e.g.,* many grasses.

Lanceolate. Lance-shaped leaf, *e.g.,* bamboo, *Nerium,* etc.

Round or orbicular. Leaf with a circular leaf blade, *e.g.,* lotus, garden nasturtium, etc.

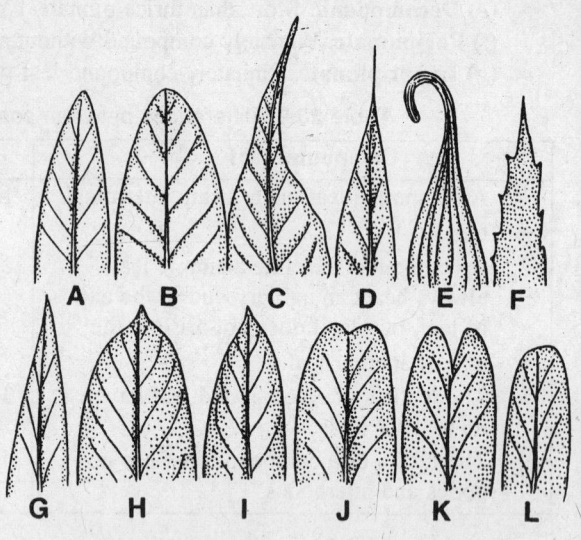

Fig. 26.43. Leaf apices. A, acute; B, obtuse; C, acuminate; D, aristate; E, cirrhose; F, spinous or cuspidate; G, caudate; H, mucronate; I, same as H; J, retuse; K, emarginate; L, truncate.

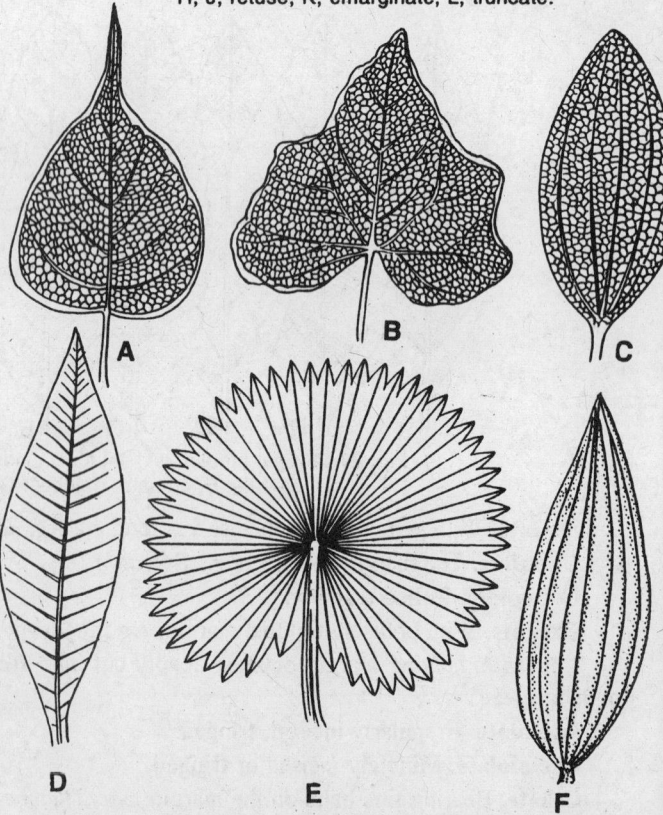

Fig. 26.44. Leaf venation. A, reticulate unicostate venation; B, reticulate multicostate—divergent venation; C, reticulate multicostate—convergent venation; D, parallel unicostate; E, parallel; F, multicostate—convergent.

Elliptical. An ellipse-shaped leaf, *e.g.,* guava, jack, etc.

Ovate. Leaf with an egg-shaped leaf lamina, *i.e.,* slightly broader at the base than at the apex, *e.g.,* banyan, China rose, etc.

Spathulate. Spatula-shaped leaf, *i.e.,* broad and round at the top and narrower towards the base, *e.g., Calendula* and *Drosera.*

Oblique. Leaf with two unequal halves, *e.g., Begonia.*

Oblong. Leaf with wide and long leaf lamina. Here the two margins run more or less straight up, *e.g.,* banana.

Reniform. Kidney-shaped leaf, *e.g.,* Indian pennywort.

Cordate. Leaf with heart shaped leaf lamina, *e.g.,* betel.

Obcordate. Inversely heart-shaped leaf blade, *e.g.,* wood-sorrel.

Sagittate. Leaf with an arrow shaped leaf blade, *e.g.,* arrow-head and some aroids.

Hastate. Sagittate leaf with its two lobes directed outside, *e.g.,* water bindweed and *Typhonium.*

Lyrate. Lyre-shaped leaf lamina, *i.e.,* with a large terminal lobe and some smaller lateral lobes, *e.g.,* radish, mustard, etc.

Acicular. Long, narrow and cylindrical leaf, *i.e.,* needle-shaped, *e.g.,* pine (a gymnosperm).

Cuneate. Wedge shaped leaf, *e.g.,* water lettuce.

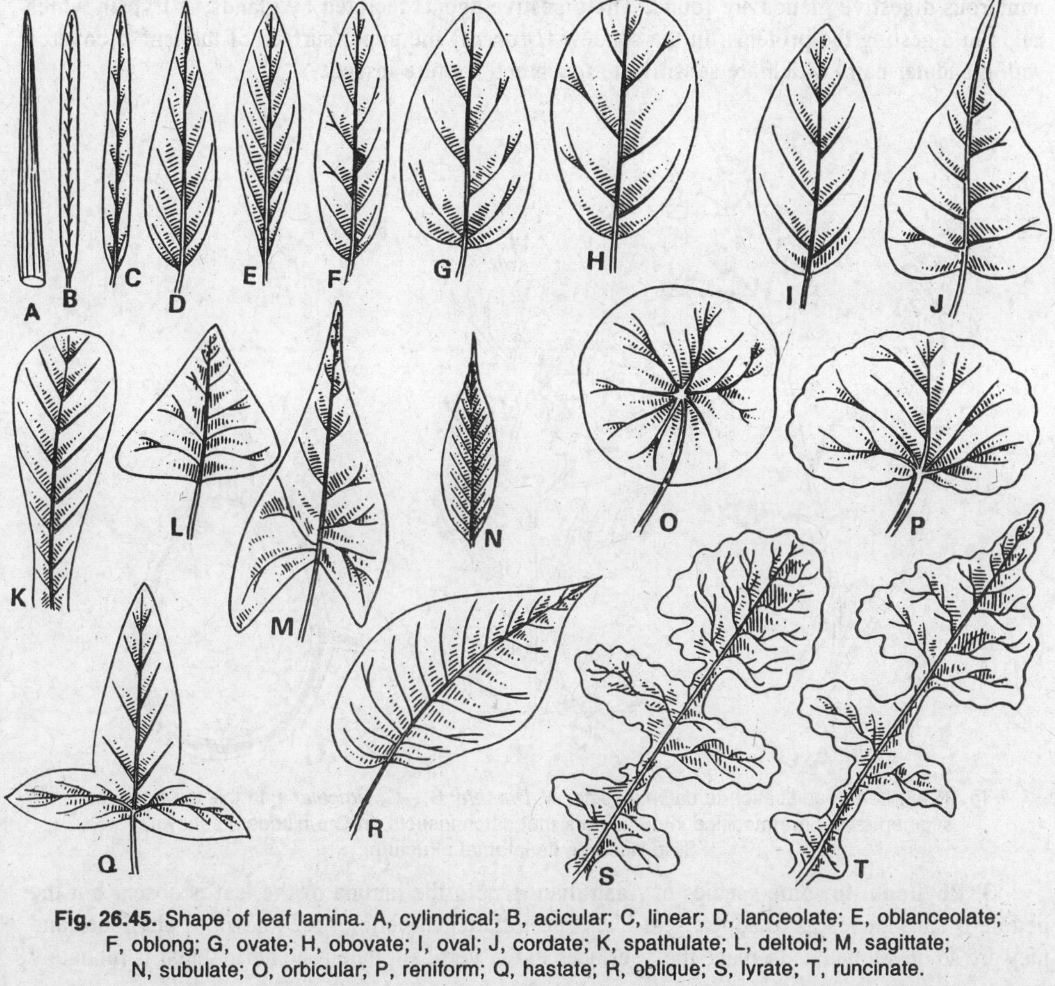

Fig. 26.45. Shape of leaf lamina. A, cylindrical; B, acicular; C, linear; D, lanceolate; E, oblanceolate; F, oblong; G, ovate; H, obovate; I, oval; J, cordate; K, spathulate; L, deltoid; M, sagittate; N, subulate; O, orbicular; P, reniform; Q, hastate; R, oblique; S, lyrate; T, runcinate.

DIVERSITY IN MODIFICATIONS OF LEAVES

The leaves of several plants are metamorphosed into certain special structures which carry on special functions. They are as follows:

1. Bladder. In bladderwort (*Utricularia*) the leaves are very much segmented and they simulate roots excepting that they are green in colour. Some of these segments become modified into **bladders.** Each bladder is about 3 mm in diameter and is provided with a trapdoor entrance. The trap door acts as a short of valve which can be pushed open inwards from outside, but never from inside to outside. This trap-door entrance allows aquatic animalcules to pass in, but never to come out. The inner surface of the bladder is dotted all over with numerous digestive glands. These glands secrete the digestive agent and absorb the digestive products.

2. Pitcher. In the pitcher plant (*Nepenthes*) the leaf becomes modified into a **pitcher.** The morphology of the leaf of pitcher plant is that the pitcher itself is the modification of leaf blade, the tendrillar stalk supporting the pitcher is the modification of the petiole, and the laminated structure that of the leaf base. The inner surface of the pitcher corresponds to the upper surface of the leaf and the lid of the pitcher arises as an outgrowth of leaf apex. The function of the pitcher is to capture and digest insect. When young the mouth of the pitcher remains closed by its lid which later on opens and stands erect. The inner side of the pitcher remains covered with numerous, smooth and sharp hairs, all pointing downwards. Lower down the inner surface numerous digestive glands are found. The digestive agent, secreted by glands, is trypsin which helps in digesting the proteins. In the sundew (*Drosera*) the upper surface of the leaf is covered with glandular hairs which are sensitive to touch and capture insects.

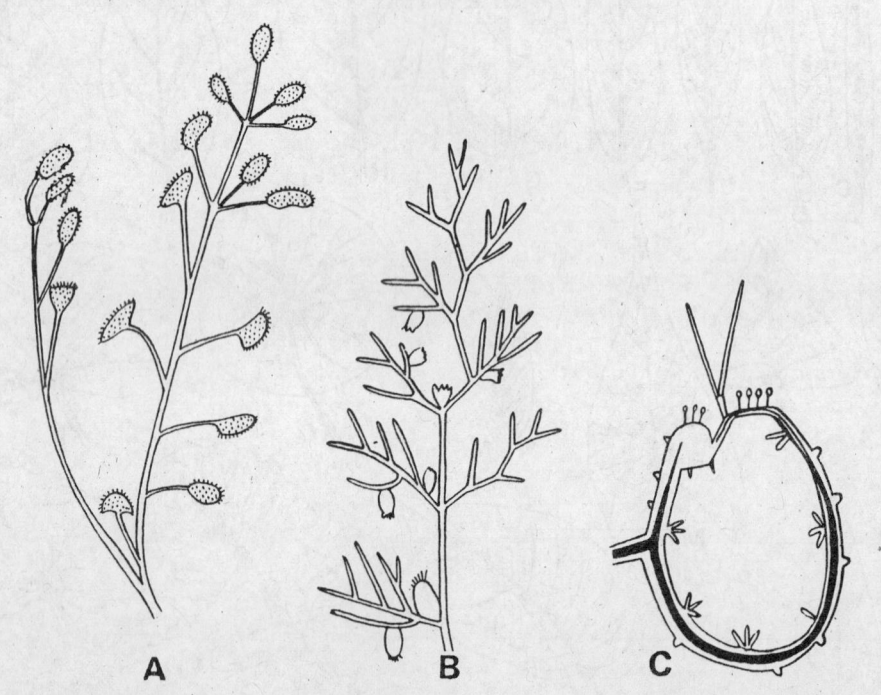

Fig. 26.46. Leaf modification to catch insects. A, *Drosera*; B,—C, *Utricularia*. In this case the leaf segments become modified into bladders that catch insects. In C a bladder is shown in L.S. to illustrate its internal structure.

3. Phyllode. In some species of Australian *Acacia* the lamina of the leaf is absent but the petiole is so flattened as to appear leaf-like. These flattened petioles are known as phyllodes and they are so developed as to place their surfaces in the vertical plane. The normal leaf is pinnately

Diversity and Morphology of Flowering Plants

compound and only develops in the seedling stage. The phyllode then carries all the functions of the leaf.

4. Leaf-tendrils. In certain plants the leaves become modified into slender, wire-like-coiled structures known as **tendrils**. The leaf may be partially or wholly modified into tendrils.

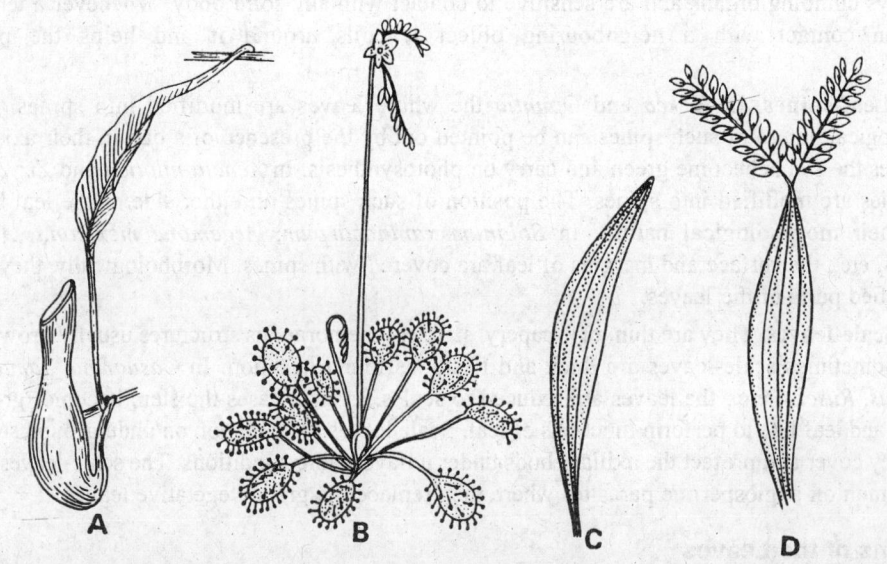

Fig. 26.47. Modification of leaves to conserve matter and to catch insects. A, *Nepenthes khasiana* or the pitcher plant. It catches insects in the pitcher; B, *Drosera* sp. leaves modified to catch insects; C—D, Australian *Acacia*. The petiole becomes flattened, green and leaf like. It is called phyllode.

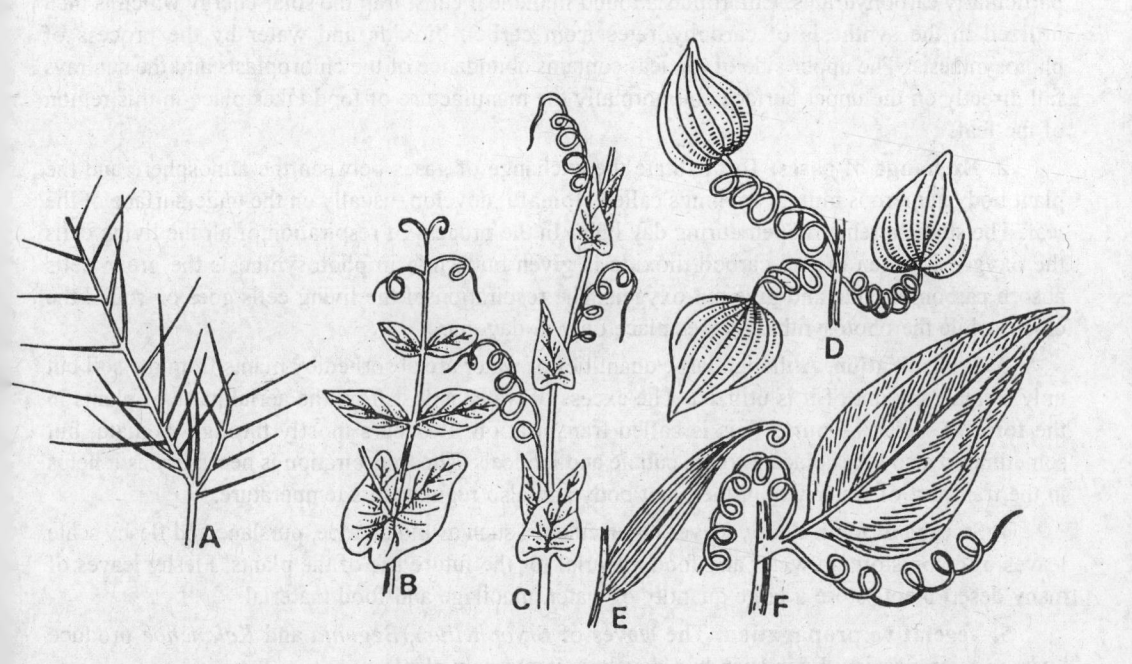

Fig. 26.48. Leaf tendrils and spines. A, leaf spines of *Hakea* ; B, leaflet tendril of *Pisum sativum;* C, leaf tendril or *Lathyrus aphaca*. Note the leaf like stipules; D, petiolar tendrils of *Clematis;* E, tendrillar leaf tip of *Gloriosa superba* ; F, stipular tendrils of *Smilax*.

For example, in pea only the upper leaflets are modified into tendrils. In *Naravelia* and *Bignonia* the terminal leaflet converts into a tendril. In *Gloriosa* the leaf apex becomes modified into a tendril. In *Nepenthes* the petiole acts as a tendrillar structure. In *Lathyrus aphaca* the whole leaf is being converted into a single tendril while the two foliaceous stipules act like the leaves. Tendrils are always climbing organs and are sensitive to contact with any solid body. Whenever a tendril comes in contact with a neighbouring object it coils around it and helps the plant to climb.

5. Leaf-spines. In *Hakea* and *Opuntia* the whole leaves are modified into spines. The morphological nature of such spines can be pointed out by the presence of a bud in their axis. In such cases the stems become green and carry on photosynthesis. In *Acacia nilotica* and *Zizyphus* the stipules are modified into spines. The position of such spines on either side of the leaf base shows their morphological nature. In *Solanum xanthocarpum, Argemone mexicana, Aloe, Acanthus*, etc., the surface and margins of leaf are covered with spines. Morphologically, they are the modified parts of the leaves.

6. Scale-leaves. They are thin, dry, papery, stalkless membranous structures usually brown in colour. Sometimes scale-leaves are thick and fleshy as found in onion. In *Casuarina, Tamarix, Asparagus, Ruscus*, etc., the leaves are reduced to scales. In such cases the stem becomes green, flattened and leaf like to perform functions of leaf. Scale leaves are common on underground stems where they cover and protect the axillary buds under unfavourable conditions. The scale-leaves are also common on angiospermic parasites where they replace the green vegetative leaves.

Functions of the Leaves

The functions of the leaves are as follows:

1. Manufacture of carbohydrates. The main function of the leaf is to manufacture food particularly carbohydrates. Chloroplasts found in the leaf cells, trap the solar energy which is then utilized in the synthesis of carbohydrates from carbon dioxide and water by the process of photosynthesis. The upper side of the leaf contains abundance of the chloroplasts and the sun rays fall directly on the upper surface and normally the manufacture of food takes place in this region of the leaf.

2. Exchange of gases. To facilitate the exchange of gases between the atmosphere and the plant body numerous minute openings called **stomata**, develop, usually on the undersurface of the leaf. The stomata remain open during day light. In the process of respiration of all the living cells the oxygen is taken in and carbon dioxide is given out while in photosynthesis the green cells absorb carbon dioxide and give out oxygen. The respiration of the living cells goes on round the clock, while the photosynthesis takes place only in daytime.

3. Transpiration. Although large quantities of water are absorbed by plants from the soil but only a small amount of it is utilized. The excess of water is lost from the aerial parts of plants in the form of water vapours. This is called transpiration. It occurs mostly through stomata, but sometimes it also takes place through cuticle and lenticels. The transpiration is necessary as it helps in the transport of water within the plant body and also regulates its temperature.

4. Storage of food. Fleshy leaves of succulents, such as Indian aloe, purslane and fleshy scale leaves of onion store up water and food material for the future use of the plants. Fleshy leaves of many desert plants store a large quantity of water, mucilage and food material.

5. Vegetative propagation. The leaves of *Bryophyllum, Begonia* and *Kalanchoe* produce buds on their margins. Each such bud develops into a new plant.

6. Protection. The leaves also give necessary protection to the axillary bud. The leaves modified into thorns and spines (*e.g.*, in *Berberis, Aegle*), give protection to the plants from animals.

DIVERSITY AND MORPHOLOGY IN THE INFLORESCENCE

In many cases the vegetative axis bears a solitary flower either at its apex (*i.e.*, terminal flower) or in the axil of a leaf (*i.e.*, axillary flower). In so many other cases the floral region remains quite distinct from the vegetative region. Thus, the floral region consisting of a collection of flowers is known as the **inflorescence.** The inflorescence may be terminal or axillary, and may be branched in several ways. This way, depending on the type of branching several kinds of inflorescence have been developed, and, they may be classified into two distinct groups— **(1) Racemose** (*i.e.*, indefinite) and **(2) Cymose** (*i.e.*, definite)

1. Racemose Inflorescence

In this type of inflorescence the main axis does not end in a flower, but it grows continuously and develops flowers on its lateral sides in acropetal succession (*i.e.*, the lower or outer flowers are older than the upper or inner ones). The various forms of racemose inflorescence may be described under three heads. They are as follows:

(*i*) **With the main axis elongated,** *i.e.*, (*i*) **raceme**; (*ii*) **spike**; (*iii*) **spikelets**; (*iv*) **catkin** and (*v*) **spadix.**

(*ii*) **With the main axis shortened,** *i.e.*, (*i*) **corymb** and (*ii*) **umbel.**

(*iii*) **With the main axis flattened,** *i.e.*, **capitulum or head.**

(*i*) **Main Axis Elongated**

(*a*) **Raceme.** In such cases the main axis remains elongated and it bears laterally a number of stalked flowers. The lower or older flowers possess longer stalks than the upper or younger ones, *e.g.*, radish (*Raphanus sativus*), mustard (*Brassica campestris*), etc. When the main axis of raceme is branched and the lateral branches bear the flowers, the inflorescence is known as **compound raceme** or **panicle**, *e.g.*, neem (*Azadirachta indica*), **gul-mohar** (*Delonix regia*), etc.

The main axis of the inflorescence together with the latest axes, if present, is termed as the **peduncle.** The stalk of the individual flower of the inflorescence is called the **pedicel.**

(*b*) **Spike.** In this type of racemose inflorescence the main axis remains elongated and the lower flowers are older, *i.e.*, opening earlier than the upper ones, as found in raceme, but here the flowers are sessile, *i.e.*, without pedicel or stalk, *e.g.*, amaranth (*Amaranthus* spp.), **latjira** (*Achyranthes aspera*), etc.

Table 26.10. Differences Between Racemose and Cymose Inflorescences

Racemose	Cymose
1. The main axis is of unlimited growth.	1. The main axis is of limited growth.
2. It does not terminate into a flower.	2. It terminates into a flower.
3. May be branched or unbranched.	3. It is usually branched.
4. The flowers are arranged in acropetal succession. The lower or outermost flowers are older and open earlier than the upper or inner ones.	4. The flowers develop in basipetal succession. The terminal flower is older and opens earlier than the lateral ones.
5. The opening of flowers is centripetal.	5. The opening of flowers is centrifugal.

(*c*) **Spikelets.** Each spikelet may bear one to several flowers (florets) attached to a central stalk known as **rachilla.** Spikeletes are arranged in a spike inflorescence which is composed of several to many spikelets which are combined in various manners on a main axis called the **rachis.** Some are in compound spikes (*i.e.*, in wheat—*Triticum aestivum*), others are in racemes (*e.g.*, in *Festuca*), while some are in panicles (*e.g.*, in *Avena*). The usual structure of spikelet is as follows–

—There is a pair of sterile **glumes** at the base of spikelet, the lower, outer glume called the **first**, and the upper, inner one called the **second**. Just above the glumes, there is series of florets, partly enclosed by them. Each floret has at its base a **lemma** and **palea**. The lemma is the lower, outer bract of the floret. Usually the lemma also known as **inferior palea** bears a long **awn** as an extension of the mid-rib at the apex or back. The floral parts borne in the axil of lemma. The palea (also known as **superior palea**) often with two longitudinal ridges (keels or nerves), stands between the lemma and the rachilla. Flowers and glumes are arranged on the spikelet in two opposite rows. Spikeletes are characteristic of Poaceae (Gramineae) or Grass family, *e.g.*, grasses, wheat, barley, oats, sorghum, sugarcane, bamboo, etc.

(*d*) **Catkin.** This is a modified spike with a long and drooping axis bearing unisexual flowers, *e.g.*, mulberry (*Morus alba*), birch (*Betula* spp.), oak (*Quercus* spp.), etc.

(*e*) **Spadix.** This is also a modification of spike inflorescence having a fleshy axis, which remains enclosed by one or more large, often brightly coloured bracts, the **spathes**, *e.g.*, in members of Araceae, Musaceae and Palmaceae. This inflorescence is found only in monocotyledonous plants.

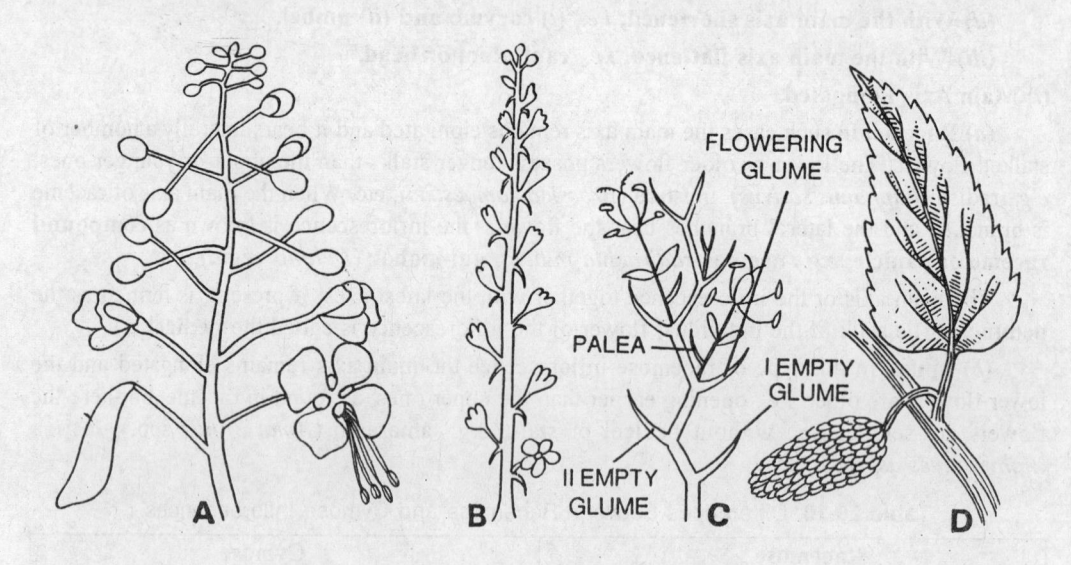

Fig. 26.49. Inflorescence-racemose. A, raceme of **gul-mohar** ; B, spike ; C, spikelet of a grass; D, female catkin of mulberry.

(*ii*) **Main axis shortened**

(*a*) **Corymb.** In this inflorescence the main axis remains comparatively short and the lower flowers possess much longer stalks or pedicels than the upper ones so that all the flowers are brought more or less to the same level, *e.g.*, in candytuft (*Iberis amara*).

(*b*) **Umbel.** In this inflorescence the primary axis remains comparatively short, and it bears at its tip a group of flowers which possess pedicels or stalks of more or less equal lengths so that the flowers are seen to spread out from a common point. In this inflorescence a whorl of bracts forming an involucre is always present, and each individual flower develops from the axil of a bract. Generally the umbel is branched and is known as **umbel of umbels (compound umbel)**, and the branches bear flowers, *e.g.*, in coriander (*Coriandrum sativum*), fennel, carrot, etc. Sometimes, the umbel is unbranched and known as **simple umbel**, *e.g.*, **Brahmi** (*Centella asiatica*). This inflorescence (umbel) is characteristic of Apiaceae (Umbelliferae) family.

Diversity and Morphology of Flowering Plants

(iii) Main axis flattened

Capitulum or head. In this type of inflorescence the main axis or receptacle becomes suppressed, and almost flat, and the flowers (also known as florets) are sessile (without stalk) so that they become crowded together on the flat surface of the receptacle. The florets are arranged in a centripetal manner on the receptacle, *i.e.,* the outer flowers are older and open earlier than the inner ones.

The individual flowers (florets) are bracteate. In addition the whole inflorescence remains surrounded by a series of bracts arranged in two or three whorls. The flowers (florets) are usually of two kinds—*(i)* **ray florets** (marginal strap-shaped flowers) and *(ii)* **disc florets** (central tubular flowers). The capitulum (head) may also consist of only one kind of florets, *e.g.,* only tubular florets in *Ageratum* or only ray or strap-shaped florets in *Sonchus*. A capitulum or head is characteristic of Asteraceae (Compositae) family, *e.g.* sunflower (*Helianthus annuus*), marigold (*Tagetes indica*), safflower (*Carthamus tinctorius*), *Zinnia, Cosmos, Tridax, Vernonia,* etc. Besides, it is also found in *Acacia* and sensitive plant (*Mimosa pudica*) of Mimosaceae family.

The capitulum inflorescence has been considered to be the most perfect. The reasons are as follows:

The individual flowers are quite small, and massed together in heads, and therefore, they add to greater conspicuousness to attract the insects and flies for pollination.

At the same time there is a considerable saving of material in the construction of the corolla and other floral parts.

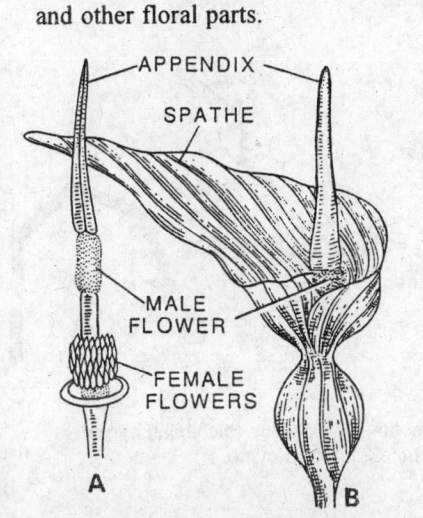

Fig. 26.50. Inflorescence-spadix–A, spadix of an aroid without spathe; B, same with spathe.

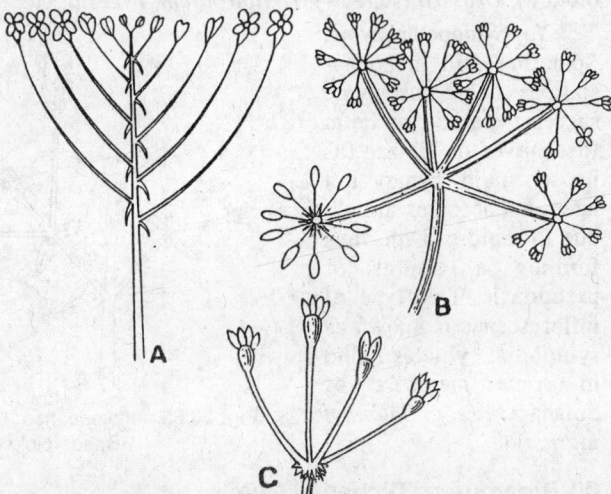

Fig. 26.51. Inflorescence, A, corymb; B, compound umbel (umbel of umbels); C, simple umbel.

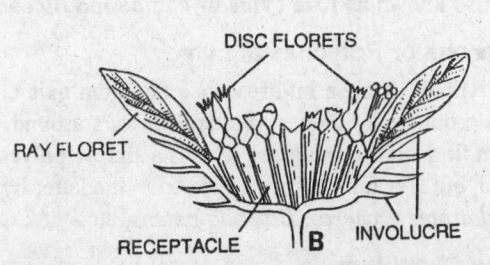

Fig. 26.52. Inflorescence. Head (capitulum)—A, a head; B, a head in L.S.

A single insect may pollinate flowers in a short time without flying from one flower to another.

2. Cymose Inflorescence

In this type of inflorescence the growth of the main axis is ceased by the development of a flower at its apex, and the lateral axis which develops the terminal flower also culminates in a flower and its growth is also ceased. The flowers may be pedicellate (stalked) or sessile (without stalk). Here the flowers develop in basipetal succession, *i.e.,* the terminal flower is the oldest and the lateral ones younger. This type of opening of flowers is known as **centrifugal**. The cymose inflorescence may be of four main types (*i*) Uniparous or monochasial cyme; (*ii*) Biparous or dichasial cyme; (*iii*) Multiparous or polychasial cyme and (*iv*) Cymose capitulum.

(i) Uniparous or Monochasial cyme

Here the main axis ends in a flower and it produces only one lateral branch at a time ending in a flower. The lateral and succeeding branches again produce only one branch at a time like the primary one. There are three forms of uniparous cyme — (*a*) **helicoid**, (*b*) **scorpioid**, and (*c*) **sympodial**

(*a*) **Helicoid cyme.** When the lateral axes develop successively on the same side, forming a sort of helix, the cymose inflorescence is known as **helicoid** or **one-sided cyme**, *e.g.,* in *Begonia, Juncus, Hemerocallis* and some members of Solanaceae.

(*b*) **Scorpioid cyme.** When the lateral branches develop on alternate sides, forming a zigzag, the cymose inflorescence is known as **scorpioid** or **alternate-sided cyme**, *e.g.,* in *Gossypium* (cotton), *Drosera* (sundew), *Heliotropium, Freesia,* etc.

(*c*) **Symopodial cyme.** Sometimes, in monochasial or uniparous cyme successive axes may be at first curved or zig-zag (as in scorpioid cyme) but later on it becomes straight due to rapid growth, thus forming a central or pseudoaxis. This type of inflorescence is known as **sympodial cyme** as found in some members of Solanaceae (*e.g., Solanum nigrum*).

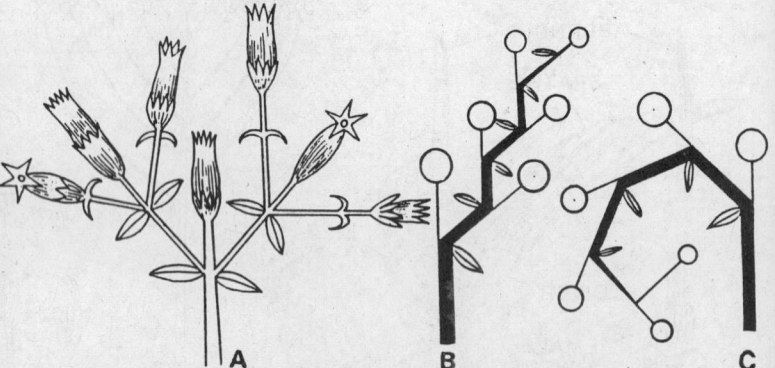

Fig. 26.53. Inflorescence. Cymose—A, biparous (dichasial) cyme; B, scorpioid cyme; C, helicoid cyme.

(ii) Biparous or Dichasial Cyme

In this type of inflorescence the peduncle bears a terminal flower and stops growing. At the same time the peduncle produces two lateral younger flowers or two lateral branches each of which terminates in a flower. There are three flowers; the oldest one is in the centre. The lateral and succeeding branches in their turn behave in the same manner, *e.g.,* jasmine, teak, *Ixora, Saponaria,* etc. This is also known as **true cyme** or **compound dichasium**.

(iii) Multiparous or Polychasial cyme

In this type of cymose inflorescence the main axis culminates in a flower, and at the same time it again produces a number of lateral flowers around. The oldest flower is in the centre and ends the main floral axis (peduncle). This is a **simple polychasium.** The whole inflorescence looks like an umbel, but is readily distinguished from the latter by the opening of the middle flower first, *e.g.,* **Ak** (*Calotropis procera*), *Hamelia patens,* etc.

(iv) Cymose Capitulum

This type of inflorescence is found in *Acacia, Mimosa* and *Albizzia*. In such cases the **peduncle** is reduced or condensed to a circular disc. It bears sessile or sub-sessile flowers on it.

Diversity and Morphology of Flowering Plants

The oldest flowers develop in the centre and youngest towards the periphery of the disc, such arrangement is known as **centrifugal**. The flowers make a globose head, which is also called **glomerule**.

Compound Inflorescence

In this type of inflorescence the main axis (peduncle) branches repeatedly once or twice in racemose or cymose manner. In the former case it becomes a compound raceme and in the latter case it becomes a compound cymose inflorescence. The main types of compound inflorescence are as follows.

1. Compound raceme or panicle. In this case the raceme is branched, and the branches bear flowers in a racemose manner, *e.g., Delonix regia, Azadirachta indica, Clematis buchaniana, Cassia fistula*, etc.

2. Compound umbel. Also known as **umbel of umbels.** Here the peduncle (main axis) is short and bears many branches which arise in an umbellate cluster. Each such branch bears a group of flowers in an umbellate manner. Usually a whorl of leafy bracts is found at the base of branches and also at the bases of flowers arranged in umbellate way. The former whorl of bracts is called **involucre** and the latter **involucel**. Typical examples of compound umbel are—*Daucus carota* (carrot), *Foeniculum vulgare* (fennel), *Coriandrum sativum* (coriander), etc.

3. Compound corymb. Also known as **corymb of corymbs.** Here the main axis (peduncle) branches in a corymbose manner and each branch bears flowers arranged in corymbs. Typical example-cauliflower.

4. Compound spike. Also known as **spike of spikelets.** The typical examples are found in Poaceae (Gramineae) family such as-wheat, barley, sorghum, oats, etc. This type has already been described under sub-head spikelets.

5. Compound spadix. Also known as **spadix of spadices.** Here the main axis (peduncle) remains branched in a racemose manner and each branch bears sessile and unisexual flowers. The whole branched structure remains covered by a single spathe. The examples are common in Palmaceae (Palmae) family.

6. Compound head. Also known as **head of heads** or **capitulum of capitula.** In this case many small heads form a large head. The typical example is globe thistle (*Echinops*). In this plant the heads are small and one-flowered and are arranged together forming a big compound head.

Special Types of Inflorescence

1. Cyathium. This type of inflorescence is found in genus *Euphorbia* of family Euphorbiaceae; also found in genus *Pedilanthus* of the family. In this inflorescence there is a cup-shaped involucre, often provided with nectar secreting glands. The involucre encloses a single female flower, represented by a pistil, in the centre, situated on a long stalk. This female flower remains surrounded by a number of male flowers arranged centrifugally. Each male flower is reduced to a solitary stalked stamen. It is evident that each stamen is a single male flower from the facts that it is articulated to a stalk and that it possesses a scaly bract at the base. The examples can be seen in poinsettia (*Euphorbia*), *Pedilanthus*, etc.

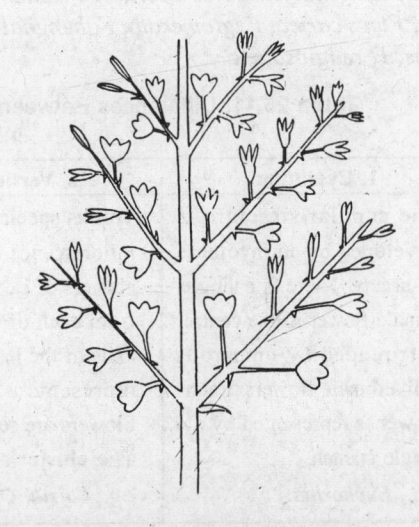

Fig. 26.54. Inflorescence. Panicle.

2. Verticillaster. This type of inflorescence is a condensed form of dichasial (biparous) cyme with a cluster of sessile or sub-sessile flowers in the axil of a leaf, forming a false whorl of

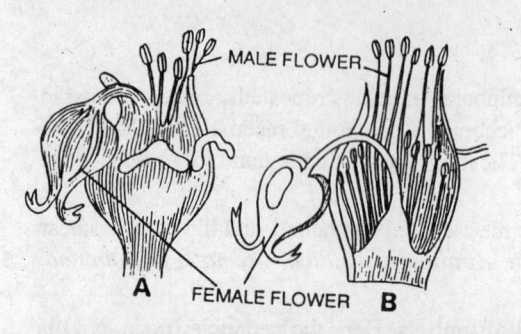

Fig. 26.55. Inflorescence-special. Cyathium—A, cyathium of *Euphorbia*; B, L.S. of same.

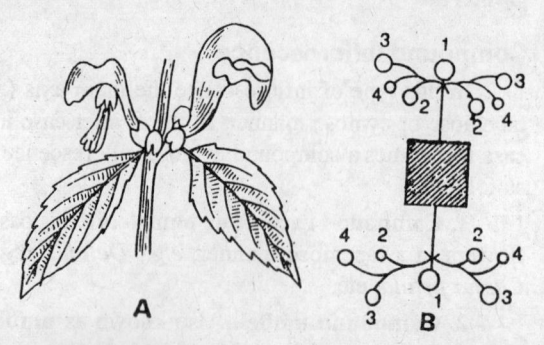

Fig. 26.56. Inflorescence—special verticillaster—A verticillaster of Labiatae; B, diagram of same.

flowers at the node. The first of main floral axis gives rise to two lateral branches and these branches and the succeeding branches bear only one branch each on alternate sides. The type of inflorescence is characteristic of Lamiaceae (Labiatae) family. Typical examples, are—*Ocimum, Coleus, Mentha, Leucas,* etc.

3. Hypanthodium. In this type of inflorescence the receptacle forms a hollow cavity with an apical opening guarded by scales. Here the flowers are borne on the inner wall of the cavity. The flowers are unisexual, the female flowers develop at the base of the cavity and the male flowers towards the apical pore. The examples, are found in genus *Ficus* of Moraceae family, e.g., *Ficus carica, F. glomerata, F. benghalensis, F. religiosa,* etc.

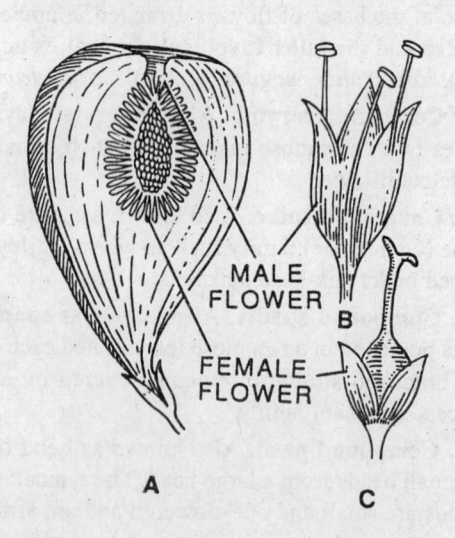

Fig. 26.57. Inflorescence. A, hypanthodium of *Ficus*; B, male flower; C, female flower.

Table 26.11. Differences Between Cyathium, Verticillaster and Hypanthodium Inflorescence.

1. Cyathium	2. Verticillaster	3. Hypanthodium
The cupular structure is developed by an involucre of bracts. There is a single female flower in the centre, surrounded by numerous stalked male flowers. Each flower is represented by a single stamen, e.g., *Euphorbia*.	This is special type of cymose inflorescence. It consists of two clusters of flowers that develop from each of the two opposite axils of the leaves. Each cluster represents a dichasial cyme. Flowers are sessile and appear as a cluster around the node, e.g., *Salvia, Ocimum,* etc.	The main axis develops in a cupular receptacle, with an apical opening at the apex. Both male and female flowers are present on the inner wall of cavity enclosed in a cupular structure, e.g., *Ficus*.

Diversity and Morphology of Flowering Plants

Table 26.12. Main Types of Inflorescences

Racemose

- **Raceme.** An elongated axis bearing stalked flowers (*e.g., Mustard*).
- **Spike.** Same as raceme, but flowers have no stalks, (*e.g., Achyranhes*).
- **Catkin.** Spike with unisexual flowers, (*e.g., Morus*).
- **Spadix.** Spike with fleshy axis enclosed by one or more large bracts, (*e.g., Colocasia, Musa,* etc.).
- **Corymb.** The axis is short, and the lower flowers have longer stalks than the upper ones. Thus all flowers come to the same level, (*e.g.,* Candytuft.)
- **Umbel.** The axis is short and bears a cluster of flowers with stalks of equal length arsing from a common point, (*e.g.,* Coriander)
- **Head or Capitulum.** The main axis is a flattened, more or less convex structure on which sessile flowers (florets) are arranged in a centripetal order. The inflorescence is surrounded by prominent bracts, (*e.g.,* Sunflower, marigold, etc.)

Cymose

- **Monochasial Cyme** The main axis terminates in a flower and one lateral branch axis develops from its base which also ends in a flower, (*e.g., Begonia, cotton,* etc.)
- **Dichasial Cyme.** Two lateral branches develop on either side of the terminal flower of the main axis; the lateral branches also end in a flower and may again branch similarly (*e.g.,* jasmine, *Dianthus*, etc.)
- **Polychasial Cyme.** More than two lateral branches arise from the base of the terminal flower, (*e.g., Calotropis*).

Special Types

- **Hypanthodium.** The main axis forms a cup-shaped receptacle with a small opening at the top. Flowers are enclosed within the cup in cymose groups. (*e.g. Ficus*).
- **Cyathium.** Here the involucre forms a cup. Single female flower without perianth arises in the middle surrounded by a large number of male flowers represented by stalked stamens. *e.g., Euphorbia*.
- **Verticillastor.** In this type of inflorescence, typical of plants with opposite leaves, a cyme arises in each leaf axil. The first axis ends in a flower. Two branches arise below it bearing branches in an alternating manner. Flowers are sessile and appear as a cluster around the node, *e.g., Ocimum*.

DIVERSITY AND MORPHOLOGY IN THE FLOWER

The flower, in its basic design, is a shoot in which the internodes (the regions of stem between nodes) do not elongate. As a result, the floral parts that occur successively along this shoot, develop as a closely packed cluster. The enlarged portion of the stalk on which the flower develops is known as the **receptacle**, and the stalk of flower is the **pedicel**.

Flower develops from a bud known as **flower bud.** This bud usually develops in the axil of a leaf called the **bract.** Sometimes the floral bud arises from the apical bud when the bract is absent. The flower buds are homologous with the vegetative buds. It consists of a stem tip, the receptacle, resembling in ontogeny and fundamental structure a vegetative tip. It consists of nodes and internodes, and bears appendages. The nodes are crowded and brought together by the suppression of internodes. The apical growth is determinate in its growth and typically only four nodes are formed bearing floral leaves at the nodes, the calyx, corolla, androecium and gynoecium, from the base towards the apex respectively. The calyx and corolla are more leaf like, and are known as the **non-essential organs** as they are only indirectly concerned in the process of reproduction and the other two, the androecium and gynoecium are more unlike leaves, and are known as the **essential organs**, as they are directly required in the process of reproduction.

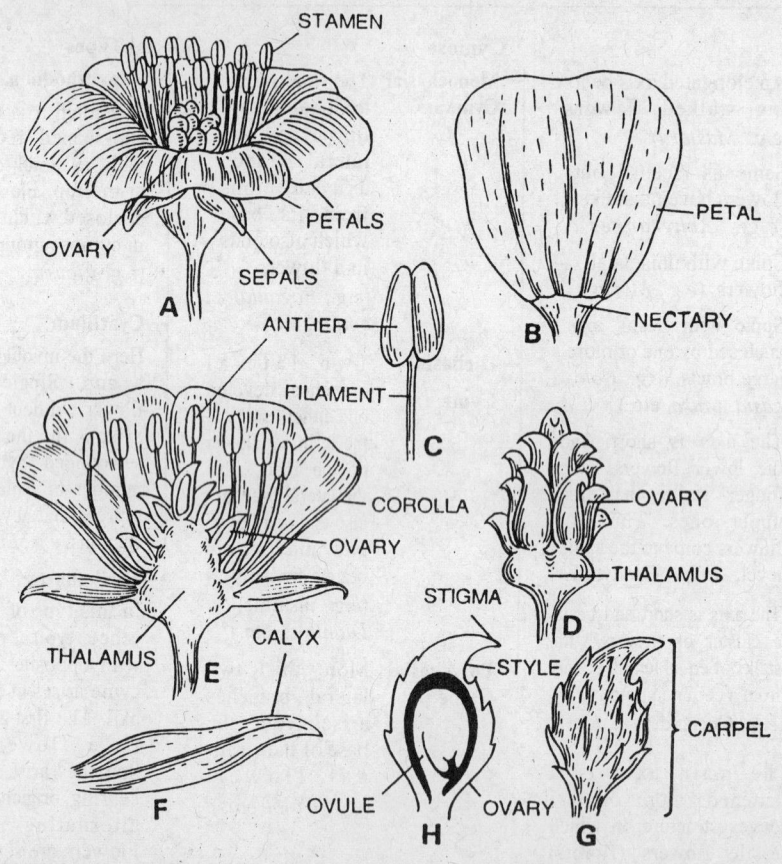

Fig. 26.58. Various parts of a flower of *Ranunculus*. A complete flower; B, a petal; C, stamen; D, thalamus with apocarpous gynoecium; E, a sepal; F, V.S. of flower, G, a carpel; H, V.S. of carpel.

The various groups of floral parts develop one after another from the receptacle. The first to form are **sepals**, which are usually green in colour and leaf-like in structure. Collectively the sepals are known as **calyx**. The calyx encloses the flower bud. Next to form are the **petals**, which are also usually shaped somewhat like leaves but are often brightly coloured. Collectively the petals make up the **corolla**. The special function of the corolla is to advertise the presence of the flower to potential pollinators. In some flowers, such as the lily and tulip, the sepals and petals cannot be distinguished separately, in such cases, they are known as **tepals**. This situation is commonly found in monocotyledons, in which the flower parts are often borne in three's or multiples of three.

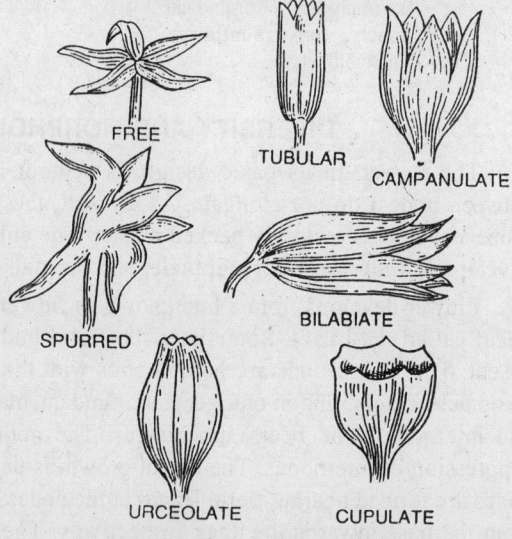

Fig. 26.59. Calyx. Forms of calyx.

Diversity and Morphology of Flowering Plants 515

Collectively the tepals are known as the **perianth**. The perianth consists of the sterile, or non-reproductive, parts of the flower.

The next primordia to form on the floral apex are the **stamens**. Collectively the stamens form the **androecium**. The number of stamens varies from one to many, depending on the species. The stamens are regarded as the male organs of the flower. Each stamen consists of three parts—**filament, anther** and **connective**. The anther bears four chambers or **pollen sacs**, each filled up with **pollen grains**. The pollen when ripe, are released from within the anther, usually through narrow slits. The haploid male nuclei, which function like sperm but do not swim, differentiate within the pollen tube, which grows from the pollen into the ovary.

The upper and innermost parts of the flower, and usually the last floral parts to mature, are the **carpels**. The carpels of a flower make up the **gynoecium** or the **pistil**. This is the female whorl. The carpels of most angiosperms consist of a hollow expanded base, or ovary, which later forms the fruit; **stigma**, which is a flat, often hairy and sticky surface specialized to receive the pollen; and connecting stigma and ovary, a slender stalk, the **style**, down which the pollen tube travels. A flower may have one (*e.g., Pisum* spp), or several carpels (*e.g., Ranunculus* spp.) and they may occur singly or fused. The ovary bears some minute egg-like bodies known as the **ovules**. Each ovule encloses a large oval cell known as the **embryo sac**. The haploid egg nucleus is formed within the embryo sac.

On the maturation of the flower, the stigma becomes receptive to pollen. A pollen grain, after being deposited on the stigma, produces a long

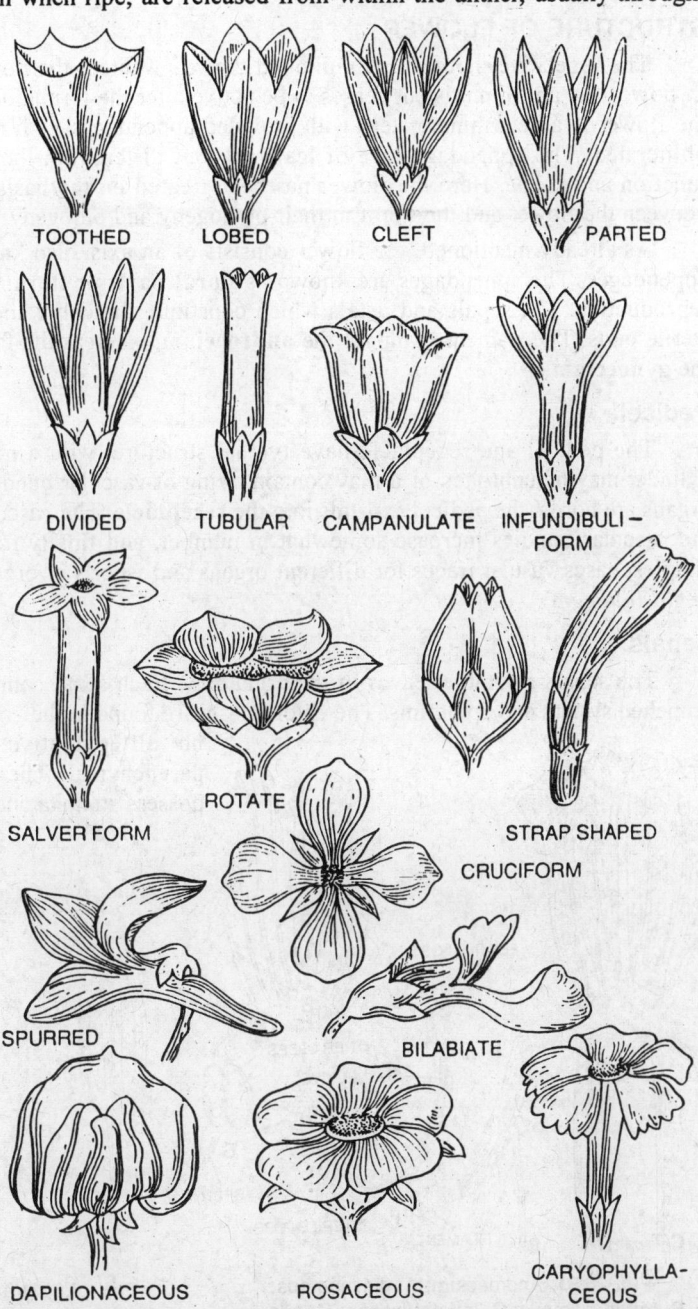

Fig. 26.60. Forms of corolla.

tube which grows down through the stigma and the style and enters one of the ovules in the ovary. Two sperm nuclei are found at the end of this tube. One of these enters to egg of an ovule and fuses with the nucleus of the egg to form the zygote. This fusion of male and female nuclei is called **fertilization**, and the flower is said to be **fertilized** when this has taken place. After fertilization the ovule develops into a seed, while the whole ovary becomes a **fruit.** The mature fruit contains one to many seeds, each with an **embryo** inside it. From these embryos grow the young plants of the next generation, and the life cycle of angiosperm begins again.

STRUCTURE OF FLOWER

The vegetative shoot shows unlimited growth, whereas the flower shows the limited growth. In flower, the apical meristem ceases to be active after the formation of floral parts. Anatomically the flower is a determinate stem with crowded appendages with internodes much shortened or obliterated. The appendages are of leaf rank but differ from those of the vegetative stem, in function and shape. Here the flower has been treated on the basis of the concept of homology between the flower and the shoot in their phylogeny and ontogeny.

As already mentioned, the flower consists of an axis, also known as receptacle and lateral appendages. The appendages are known as **floral parts** or **floral organs.** They are sterile and reproductive. The sepals and petals which constitute the calyx and corolla respectively are the sterile parts. The stamens compose the **androecium,** whereas the free or united carpels compose the **gynoecium.**

Pedicel

The pedicel and receptacle have typical structure, with a normal vascular cylinder. The cylinder may be unbroken or it may contain a ring of vascular bundles. In the region where floral organs are borne, the pedicel expands into the **receptacle.** The vascular cylinder also expands and the vascular bundles increase somewhat in number, and finally traces begin to diverge. In the simplest cases vasular traces for different organs and whorls of organs arise quite independently, (*e.g., Aquilegia*).

Sepals

The sepals resemble leaves in their anatomy. Each sepal consists of ground parenchyma a branched system and epidermis. The chloroplasts are found in the green sepals but usually there is no differentiation into palisade and spongy parenchyma. The epidermis of sepals may possess stomata and trichomes.

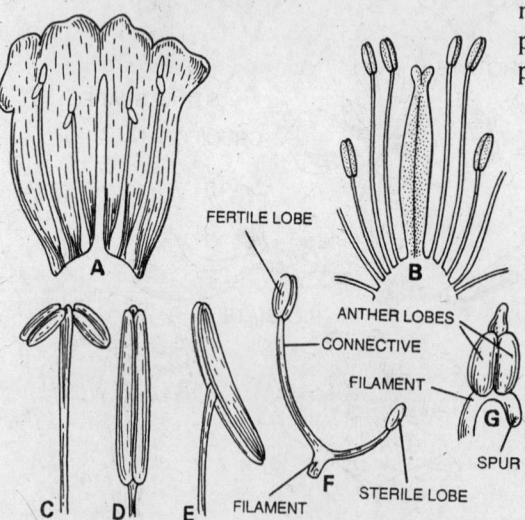

Fig. 26.61. Androecium; A, didynamous; B, tetradynamous; C, innate fixation; D, adnate fixation; E, varsatile; F, stamen of *Salvia* with an elongated connective; G, spurred stamens of *Viola*.

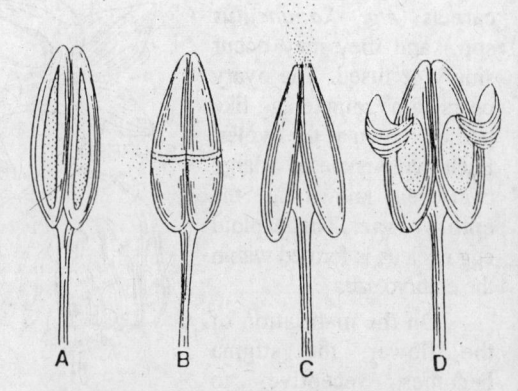

Fig. 26.62. Various types of dehiscence of anther. A, longitudinal slit; B, transverse slit; C, apical pore (porous); D, valvular.

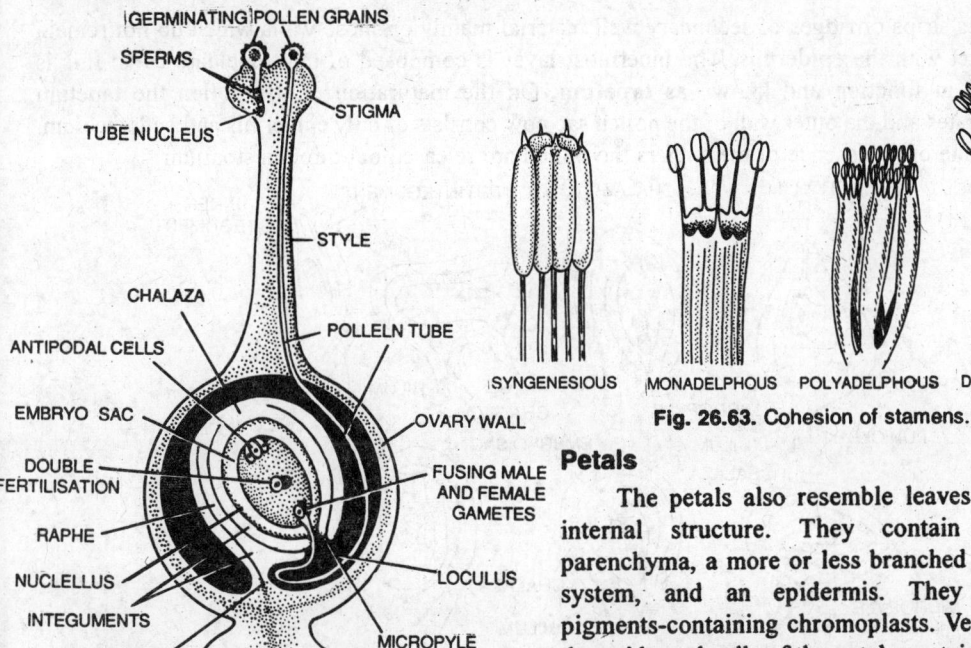

Diversity and Morphology of Flowering Plants

Fig. 26.64. A monocarpellary pistil cut vertically to show its external and internal parts. It also shows pollen grains on stigma.

Fig. 26.63. Cohesion of stamens.

Petals

The petals also resemble leaves in their internal structure. They contain ground parenchyma, a more or less branched vascular system, and an epidermis. They contain pigments-containing chromoplasts. Very often, the epidermal cells of the petals contain volatile oils which emit the characteristic fragrance of the flowers.

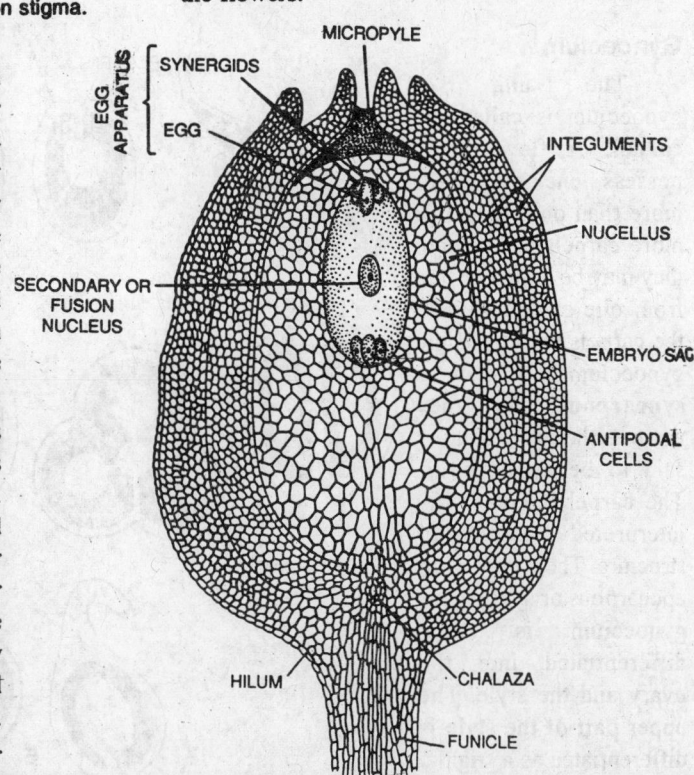

Stamen

Commonly the stamen consists of a two-lobed four loculed anther. The anther is found to be situated on a slender filament which bears vascular bundle. In certain primitive dicotyledonous families the stamens are leaf-like and possess three veins, whereas in advanced types they are single veined.

The structure of filament is quite simple. The epidermis is cutinized and bears trichomes. The stomata may also be found in the epidermis of both anther and filament. The vascular bundle is found throughout the filament and culminates blindly in the connective tissue situated in between the two anther lobes.

The outermost wall layer of the anther is the epidermis. Just beneath the epidermis there is endothecium which usually

Fig. 26.65. V.S. ovule showing its structure just before fertilization.

possesses strips or ridges of secondary wall material mainly on those walls which do not remain in contact with the epidermis. The innermost layer is composed of multinucleate cells; this is nutritive in function and known as **tapetum**. On the maturation of the pollen the tapetum disintegrates and the outer wall of the pollen sac now consists of only epidermis and endothecium. At the time of dehiscence of the anthers the pollen are released out through stomium.

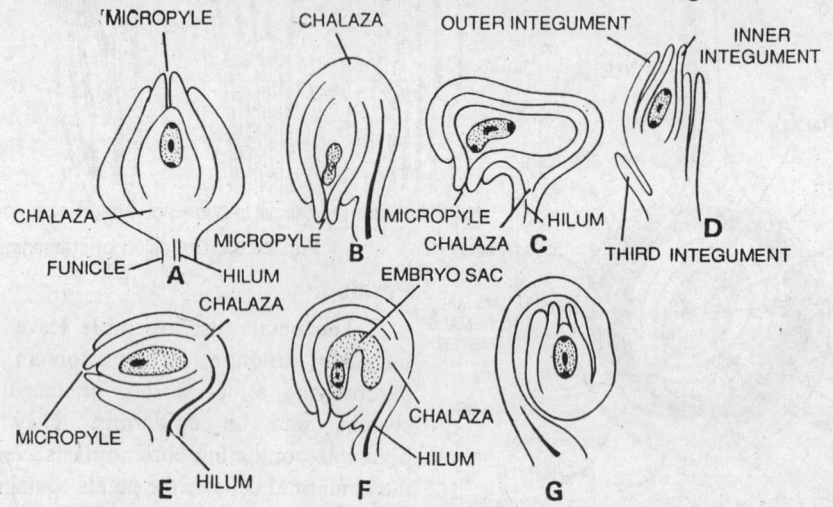

Fig. 26.66. Types of ovules. A, orthotropous; B, anatropous; C, campylotropous; D, ovule with three integuments; E, hemianatropous; F, amphitropous; G, circinotropous.

Gynoecium

The unit of gynoecium is called the **carpel**. A flower may possess one carpel or more than one. If two or more carpels are present they may be united or free from one another. When the carpels are united the gynoecium is known as **syncarpous**; when they are free the gynoecium is said to be **apocarpous**. The carpel is commonly interpreted as foliar structure. The carpel of an apocarpous or syncarpous gynoecium is being differentiated into the ovary and the style. The upper part of the style is differentiated as a stigma. The stigma is sessile.

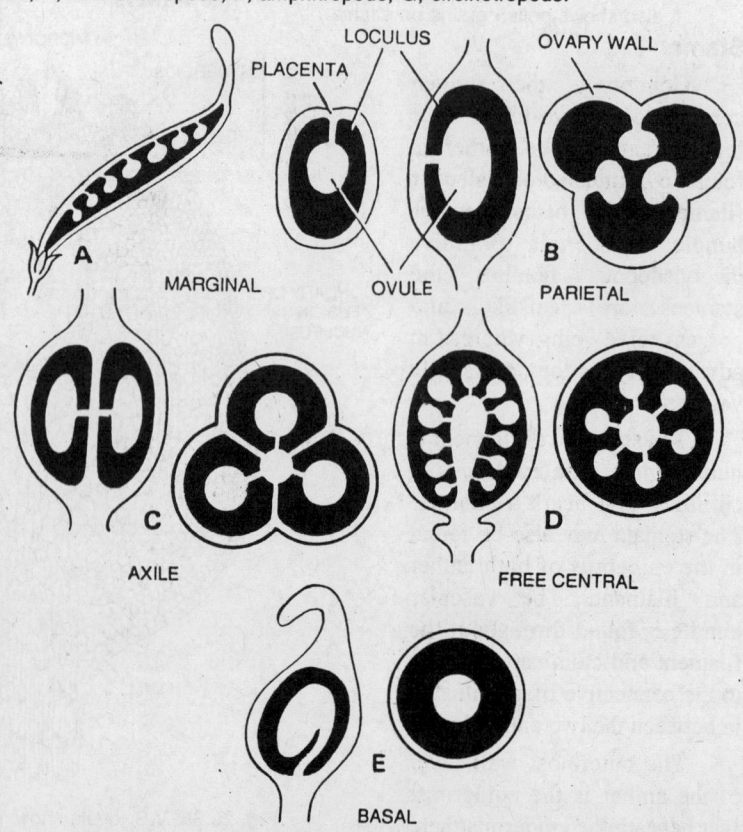

Fig. 26.67. Various types of placentation in angiosperms (A—E).

Diversity and Morphology of Flowering Plants

Placentation

The ovary consists of the ovary wall, the locule or locules, and in a multilocular ovary, the partitions. The ovules are found to be situated on the inner or adaxial (ventral) side of the ovary wall. The ovule-bearing region forms the **placenta.** In a carpel the placenta occurs close to the margin. Since there are two margins, the placenta is double in nature. The two halves may be united or separate. The number of double placentae in compound ovaries is equal to number of carpels. When the carpels are folded, the ovary is multilocular and the placentae occur in the centre of the ovary where the margins of the carpels meet. This is **axile placentation.** When the partitions of the ovary disappear, it becomes **free-central placentation.** When the carpels are joined margin to margin and the placentae are found to be situated on the ovary walls, the placentation is **parietal.**

Flower is a Modified Shoot

The undermentioned facts prove that the thalamus is a modified branch, whereas sepals, petals, stamens and carpels are modified vegetative leaves, and the flower as a whole is a modified vegetative bud.

1. The thalamus represents the axis of the floral whorls with internodes between them normally remaining undeveloped or suppressed. However, in some flowers the thalamus becomes elongated showing distinct nodes and internodes, *e.g.,* in *Gynandropsis, Capparis,* etc.

2. Sometimes the thalamus becomes elongated upwards and bears ordinary foliage leaves, and behaves as a branch, *e.g.,* in rose.

3. A floral bud like a vegetative bud is either terminal or axillary in position.

4. The arrangement of sepals, petals, stamens and pistil, etc., on the thalamus is the same as that of the leaves on the stem or the branch, *i.e.,* whorled, alternate, spiral or opposite.

5. The arrangement of sepals and petals with respect to each other (*i.e.,* aestivation) is similar to that of the foliage leaves (*i.e.,* prefoliation).

6. The foliar nature of sepals and petals is evident from their similarity to leaves as regards structure, form and venation, *e.g.,* in *Mussaenda* one of the sepals becomes modified into a distinct cream coloured leaf. In green rose the petals are leaf-like in structure and green in colour. The water lily shows a gradual transition from sepals to petals and from petals to stamens. In cultivated roses many stamens gradually change into petals. In *Canna* the stamens and the style become petaloid.

7. In *Agave americana* some of the floral buds become modified into vegetative buds, called bulbils for vegetative propagation. Such bulbils thus show a reversion to ancestral forms, from which they have been derived.

DIVERSITY AND MORPHOLOGY IN THE FRUIT

Development of the fruit. With the result of fertilization the embryo develops and a series of other changes takes place in the ovule and as a result the **seed** is formed. The development of the embryo remains associated with a series of changes in the ovary wall and other floral parts. These changes result in the development of the **fruit** from the ovary. In other words the mature or ripened ovary is known as a fruit. Whenever the fertilization does not take place, the ovary simply withers and falls off. A fruit mainly consists of two parts-(*i*) the **pericarp** developed from the ovary wall and (*ii*) the **seeds** developed from the ovules. In certain cases (*e.g.,* oranges, grapes, bananas, etc.) the ovary may develop into the fruit without fertilization. Such fruit is seedless and known as **parthenocarpic** fruit. In most of cases the pericarp consists of three parts (*i*) **epicarp** - the outer part, which makes the skin of the fruit; (*ii*) **mesocarp** the middle part, which makes the pulpy part of the fruit, as in mango, peach, etc., and (*iii*) **endocarp**, the inner part which is very often thin and membranous, as in orange, or it may be hard and stony as in mango, plums, etc. In some cases the pericarp is not differentiated into three parts.

In the cases where only the ovary of the flower develops into the fruit, it is called the **true fruit**, but in other cases where the other floral parts such as the thalamus, receptacle or calyx also take part in the development of the fruit; the fruit is said to be **false fruit, spurious fruit** or **pseudocarp.** For example, in apple and pear the thalamus grows round the ovary and becomes fleshy. In *Dillenia*, the calyx is persistent and fleshy and makes the main part of the fruit. In rose the enlarged thalamus bears small true fruits on its inner concave surface. In cashewnut the peduncle and thalamus develop into a false fruit or pseudocarp, whereas the true fruit is a nut, and develops from the ovary seated on the swollen peduncle. The aggregate fruit of custard apple having number of small true fruits fused together, makes a spurious fruit. Lastly, the fruits which develop from an inflorescence as those of mulberry, pine apple, jack fruit, fig, banyan, peepal, etc., also known as spurious fruits.

In the development of a fruit both cell division and cell expansion are involved. The growth of the ovary after fertilization may be rapid. For example, the pumpkin ovary shows a 20 fold increase in about two weeks time.

All the flowers borne on a plant do not always mature into fruit. In mango for example, fruit-set is extremely low as compared to the total number of flowers produced. Shedding of flowers may occur before or after anthesis; even young fruits may drop. Dropping of blossoms and young fruits is of advantage to the cultivator because it results in the production of fruits of large size. In certain orchard trees, such as apples, pears and plums, flowers thinning by hormonal spray is a regular practice.

Fruit ripening. As soon as the growth of the ovary wall due to cell division and cell enlargement stops, the fruit is said to be mature. This is followed by the final phase of fruit development called fruit ripening. During this period the conversion of starch into sugar, the reduction in the concentration of acids, the production of various esters, and the breakdown of chlorophyll lead to changes in colour, texture, taste, and flavour of the fruit. A mature fruit of mango is hard and green and its edible portion white and sour. On ripening, the mesocarp becomes yellow-orange, juicy and sweet. A characteristic feature of ripening of some fruit (like banana) is a sudden increase in respiration which is known as the **climacteric.** After this stage, the fruit starts decaying. Although from the point of view of the fruit, ripening leads to death, for man and other animals it is a stage which is most useful.

Parthenocarpy. Some plants are able to form fruits without fertilization. Such fruits are called parthenocarpic and the phenomenon is known as **parthenocarpy.** Parthenocarpic fruits are either seedless or contain empty or nonviable seeds. In these fruits, the "seed-factor" for fruit growth is provided by the tissue of the ovary wall itself. Seedless varieties of grapes and oranges have been reported to have up to seven times as much auxin in the ovaries of unpollinated flowers as seeded varieties. This may explain the normal development of parthenocarpic fruits even in the absence of healthy seeds. Most commonly cultivated varieties of banana are parthenocarpic. Parthenocarpic oranges and watermelons are also very common. Even in those plants which normally bear seeded fruits, parthenocarpy could be induced by the application of low concentrations of axuin or gibberellin.

Parthenocarpy is of great commercial value. Parthenocarpic fruits are ideal for consumption as such, or in preparation of jams and fruit juices on a commercial scale. You can imagine the annoyance caused by seed while eating a watermelon or guava.

Biological Significance of Fruit Formation

Fruits (of all categories) were used by the ancient man as his main food and even today they form an important part of man's diet. However, we might ask the question: "why does a plant invest so much of its food in the production of fruit?" Certainly not for human consumption. The fruit serves the plant in various ways. It protects the immature seeds against hostile climatic conditions and animals. The seeds remain enclosed in the fruit until they are ready to germinate or, at least withstand the possible unfavourable environmental conditions. The function of seed protection by the fruit wall is also achieved through its colour. When young, most fruits are green, and remain hidden in the green foliage. As the seeds mature, the fruit acquires bright colours and attract the seed dispersing agents. Immature fruits also offer chemical defence against the animals as they often contain such unpalatable and repelling substances as astringents, tannins, bitter alkaloids and sour acids which disappear upon ripening.

Another important role played by the fruit tissues is in the dispersal of seeds to distant localities which is of great biological significance.

Classification of Fruits (Types of Fruits)

The fruits are classified into three main groups, *viz.,* **(1) simple (2) aggregate** and **(3) multiple or composite.**

1. Simple Fruits

When the ovary of a flower with or without other accessory floral parts converts into a single fruit, the fruit is said to be a **simple fruit.** It may be **dry** or **fleshy.** The dry fruits may further be classified into (*i*) **dehiscent,** (*ii*) **indehiscent** or (*iii*) **schizocarpic** fruits.

(*i*) **Dry dehiscent fruits**

(*a*) **Legume or pod.** This type of simple dry fruit is monocarpellary, developing from a superior, one chambered ovary. It dehisces by both the sutures. Typical examples are found in Leguminosae family, *e.g.,* pea, bean, pulses, gram and many others.

(*b*) **Follicle.** This type of simple dry fruit is monocarpellary, developing from a superior, one-chambered ovary like the legume fruit, but it dehisces by one suture only as in *Calotropis* (**Ak**), *Asclepias, Vinca rosea* (**Sadabahar**), *Michelia*, etc.

(*c*) **Siliqua.** This is a long, narrow, many seeded fruit which develops from a superior bicarpellary ovary with two parietal placentae. It dehisces from below upwards by both the sutures. Here the ovary is one-chambered at first, but later on it becomes two-chambered because of the development of a flase septum, the **replum**, which extends from one placenta to another. This type of fruit is commonly found in Brassicaceae (Cruciferae), *e.g.,* mustard, radish, *Eruca*, etc.

(*d*) **Silicula.** When a siliqua fruit becomes much shorter and flattened and as broad as it is long with a few seeds, it is called **silicula**, *e.g., Capsella* and candytuft (*Iberis amara*).

(*e*) **Capsule.** This is a many-seeded, uni-or multilocular fruit which develops from a superior bi-or polycarpellary ovary. It dehisces in many ways. The dehiscent fruits which develop from a syncarpous ovary are commonly called capsule fruits. A capsule may dehisce by pores, as in poppy; or transversely in cock's comb; or loculicidally, as in cotton and **Bhindi** septicidally, as in linseed and septifragally as in *Datura*. In *Datura* the dehiscence lines are irregular and expose the central column bearing seeds.

(*ii*) **Dry indehiscent or achenial fruits.** These fruits are dry and one-seeded. Here the pericarp does not split or break open to allow the seed to escape. They develop from monocarpellary to polycarpellary syncarpous pistil with one chamber and one ovule. The main kinds of such fruits are as follows:

(*a*) **Caryopsis.** This is a very small, dry, one-chambered and one-seeded fruit which develops from a superior monocarpellary ovary. Here the pericarp remains fused with the seed coat. The well known examples are found in Poaceae (Gramineae) family, *e.g.,* maize, wheat, barley, millets, etc.

(b) **Achene.** An achene is a small, dry indehiscent fruit. It develops from a superior monocarpellary, one-chambered and single ovuled ovary. The pericarp is thin and leathery and encloses a single seed. The pericarp remains free from the testa or seed coat. The achenes are commonly developed from an apocarpous pistil, and therefore a single flower produces as many achenes as many free carpels are there, *e.g., Clematis, Naravelia, Ranunculus*, etc.

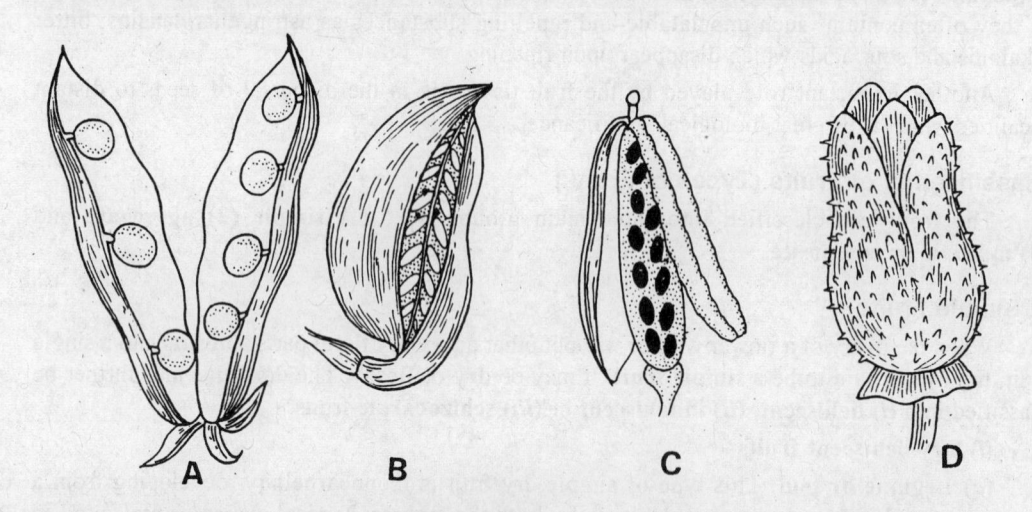

Fig. 26.68. Fruit. A, legume or pod of pea; B, follicle of *Calotropis;* C, siliqua of mustard; D, capsule of *Datura*.

(c) **Cypsela.** This is a dry, one-chambered and one-seeded fruit. It develops from an inferior, bicarpellary ovary. The pericarp and the seed coat remain free from each other. The examples are commonly found in Asteraceae (Compositae family, *e.g.,* sunflower, *Cosmos, Tagetes, Ageratum, Sonchus*, etc.

(d) **Nut.** This is a dry, one-chambered and one-seeded fruit with hard and woody epicarp. It develops from a superior bi-or polycarpellary ovary, *e.g.,* chestnut, oak, walnut, etc.

In *Litchi* the nut develops from a tricarpellary, syncarpous, superior and trilocular ovary with more than one ovules. On the maturity of the fruit only one ovule develops into a mature seed. The juicy aril which is an outgrowth of the testa from micropylar end surrounds the whole seed. The fleshy and juicy aril is edible. The pericarp is hard and leathery and bears numerous echinulate tubercles on it. It becomes brownish red at maturity.

In *Juglans regia* (English walnut) the fruit is a nut which develops from bicarpellary, syncarpous, inferior and one-chambered ovary with one basal ovule. Here the fruit remains surrounded by an **exocarp** of fused involucre and perianth surrounding the ovary. It makes the husk of the fruit and is removed by burning it before marketing the fruits. The thalamus and pericarp are hard and woody. The seed is large with convoluted cotyledons, which make the edible part of the nut. The stony pericarp is produced into ingrowths.

In cashewnut (*Anacardium occidentale*) the fruit (nut) develops from a monocarpellary superior, single-chambered, single-ovuled ovary. The thalamus becomes fleshy and edible. The kernel of true nut is also edible.

In *Chenopodium* the nut remains surrounded by persistent perianth and is known as **utricle**.

The fruit (nut) develops from a bicarpellary, syncarpous superior and unilocular ovary with single basal ovule.

Diversity and Morphology of Flowering Plants

In oak (*Quercus* spp.) the nut develops from a tri-carpellary syncarpous and inferior ovary. Here only one ovule develops into a seed. The pericarp is smooth, brown and hard. It remains enveloped at its base by a cupular structure, the **cupule**. The cupule is formed by the fusion of hard and persistent bracts.

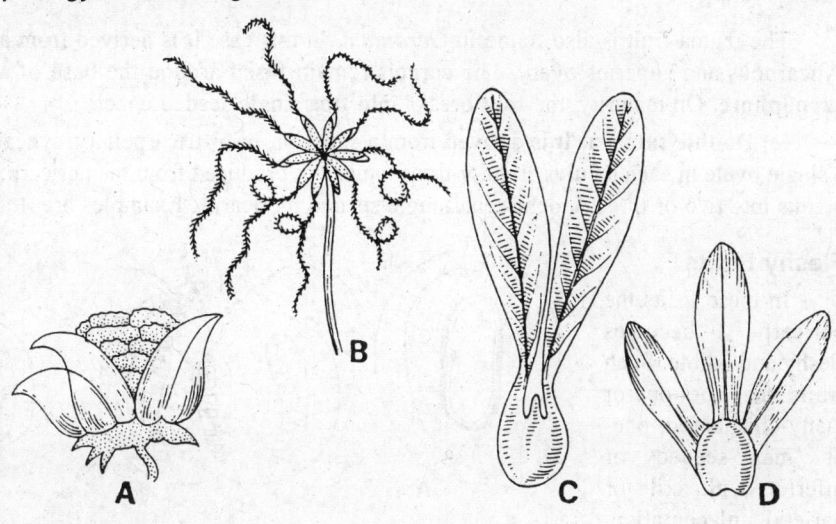

Fig. 26.69. Fruit. A, capsule of cotton; B, achenes of *Naravelia* with persistent feathery styles; C, samara of *Dipterocarpus;* D, samara of *Hiptage*.

In chestnut, the cupule develops small spine-like outgrowths. Here the cupule surrounds three nuts.

(*e*) **Samara.** This fruit is dry, indehiscent, one-or two seeded and develops from a superior, bi-or tricarpellary ovary, with flattened wing-like outgrowths, *e.g., Hiptage, Dipterocarpus, Acer*, etc. Here the wings always develop from the pericarp, and the fruit breaks into its component parts, each enclosing a seed.

The fruit of *Shorea* is also winged, but here the wings are dry and persistent sepals. Such winged fruits are called **samaroid** fruits.

(*iii*) **Splitting or schizocarpic fruits.** They are dry, many seeded, indehiscent fruits. The fruits generally break into many or few one-seeded portions, the **mericarps**. The mericarps are indehiscent. In certain schizocarpic fruits, *e.g.,* castor the one seeded parts of the fruit are dehiscent, and are called **cocci**. They are of following types.

(*a*) **Lomentum.** It is characteristic of Mimosaceae family, *e.g., Acacia, Mimosa*, etc., the fruit is derived from a monocarpellary superior, unilocular ovary with marginal placentation. Here the legume is constricted or partitioned, into a number of one-seeded **mericarps**.

(*b*) **Cremocarps.** This type of fruit is dry indehiscent and two-chambered. It develops from an inferior, bicarpellary ovary. On maturity the fruit splits apart into two indehiscent, one seeded portions, the **mericarps**. The mericarps remain attached to the prolonged end of the axis, known as **carpophore**. This type of fruit is characteristic of Apiaceae (Umbelliferae) family, *e.g.,* coriander, fennel, carrot, cumin, etc.

(*c*) **Carcerulus.** This type of fruit is derived from a syncarpous and superior ovary with many chambers and axile placentation. Each loculus may possess one or two ovules. The mature fruit splits into many single seeded and indehiscent mericarps, *e.g., Althaea rosea* (hollyhock), *Abutilon indicum* (**Kanghi**), *Salvia, Ocimum*, etc.

(*d*) **Regma.** This type of fruit is characteristic of castor plant (*Ricinus communis*). It develops from tricarpellary, syncarpous, superior, trilocular ovary with axile placentation. The fruit wall bears spine like tubercles. The locules are separated from each other on maturity by an explosive mechanism. These dehiscent locules are known as **cocci**, and remain attached to a central column, the **carpophore**.

The regma fruit is also found in *Geranium*. In this case it is derived from a pentacarpellary, syncarpous and superior ovary. The carpels remain fused around the base of a central axis the **carpophore**. On maturity, the fruit breaks into free single seeded cocci.

(*e*) **Double samara.** It is derived from a superior, bi-or-tricarpellary, syncarpous ovary with a single ovule in each loculus. Two or three wings are produced from the pericarp. The mature fruit breaks into two or three, indehiscent, single seeded mericarps. Examples are *Acer, Dodonaea*.

Fleshy Fruits

In these fruits the pericarp becomes fleshy and edible. Such fruits may be one-or many chambered, one- or many-seeded or inferior, with axile or parietal placentation. Commonly they are indehiscent fruits, and their seeds may be liberated only on the decay of the pulp. The pericarp of fleshy fruits is generally distinguished into three parts, (*a*) **epicarp,** (*b*) **mesocarp** and (*c*) **endocarp**.

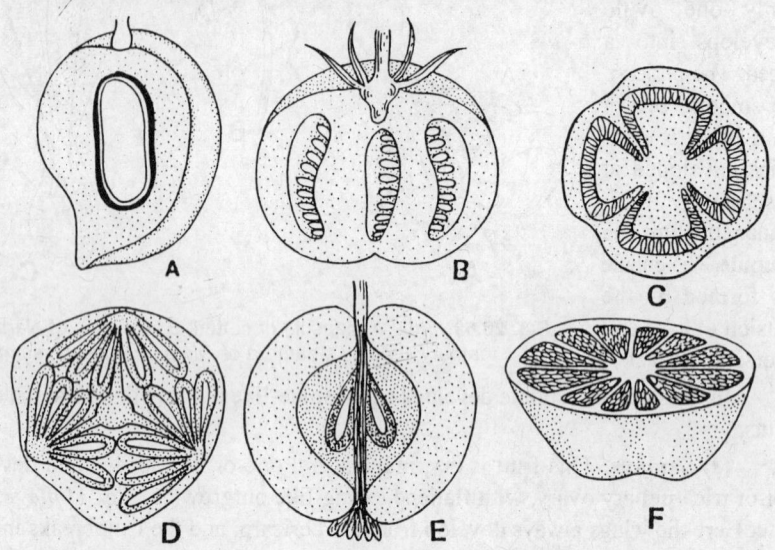

Fig. 26.70. Fruits. A, drupe of mango; B, berry of tomato, C; same in T.S.; D, pepo of cucumber in T.S.; E, pome of apple; F, hesperidium of orange.

Epicarp is the outermost layer of the pericarp. It is the protective layer and may be thin or smooth or thin and papery. In *Aegle marmelos* it becomes hard and stony. Just beneath the epicarp, the **mesocarp** is found. In some (*e.g.,* mango) it is fleshy, juicy and edible, whereas in others it is fibrous (*e.g.,* coconut). In orange and banana it is thread like. The endocarp is the inner most layer of the pericarp. It is hard and stony in mango; fleshy and edible in banana and tomato. In orange the endocarp is thin and papery and bears juicy hair on its inner surface. Here the juicy hairs are edible.

1. Drupe. This is a fleshy, one-or more chambered and one-or more seeded fruit. It develops from a monocarpellary or syncarpous pistil. The pericarp of the fruit remains differentiated into thin epicarp, fleshy mesocarp and stony endocarp. Because of the presence of stony endocarp the fruit is also known as **stone fruit**, *e.g.,* mango, peach, plum, coconut, almond, etc.

2. Berry or **Bacca**. This is a superior or inferior, indehiscent, many seeded and fleshy fruit. It develops from a single carpel or from a syncarpous pistil with axile or parietal placentation, *e.g.,* tomato, grapes, brinjal, guava, papaw, etc. In this type of fruit the seeds at first remain attached to the placentae, but later on they are detached from the placentae and lie free in the pulp. The date-palm and *Artabotrys* are the examples of one-seeded berry. The pericarp of berry is also differentiated into epicarp, mesocarp and endocarp but like drupe the endocarp is not stony.

3. Pepo. This type of fruit is characteristic of Cucurbitaceae family, *e.g.,* gourd, cucumber, melon, watermelon, squash, etc., the fruit is fleshy or pulpy and many seeded. It develops from an inferior, syncarpous pistil with parietal placentation. Here the seeds remain embedded in the pulp and attached to the placentae.

4. Pome. This type of fruit is found in Rosaceae family, *e.g.,* apple, pear, etc., the fruit develops from pentacarpellary, syncarpous and inferior ovary with many seeds. The fruits remain surrounded by fleshy thalamus. The fleshy thalamus makes the edible part, whereas the actual fruit lies within it. The outer portion of the pericarp (*i.e.,* epicarp and mesocarp) is fleshy and fused with the thick edible portion (thalamus). The inner layer of pericarp (*i.e.,* endocarp) is cartilaginous and forms the core surrounding the seeds. Each part of the core represents one of the five carpels. The remains of the perianth may be seen at the top of the fruit.

5. Hesperidium. This type of berry is derived from a polycarpellary, syncarpous superior, multilocular ovary with axile placentation and many ovules. Here the epicarp is thick and leathery and contains oil cavities. Next to epicarp there is white thread-like fibrous layer of mesocarp. The endocarp projects inwards forming distinct chambers. The inner walls of the endocarp give rise to numerous juicy outgrowths which makes the edible part of the fruit, *e.g.,* orange, lemon, etc.

2. Aggregate Fruits

Such fruit develops from a single flower having an apocarpous pistil. In such case the carpels are free, and each of them develops into a simple fruitlet. A collection or a group of simple fruitlets makes an aggregate fruit. As many fruitlets are developed in a group as carpels are there in a flower. An aggregate of simple fruitlets on a single flower is termed as etaerio-such as an etaerio of achenes, an etaerio of follicles, an etaerio of drupes, an etaerio of berries, etc.

(*i*) **An etaerio of achenes.** This type of fruit develops from polycarpellary apocarpous gynoecium. For example, in *Clematis* and *Naravelia* the achenes are provided with feathery and persistent styles, in *Rosa* the achenes remain enclosed in a hollow, receptacular thalamus, in *Fragaria* the achenes are found on the fleshy thalamus and in *Nelumbo* the fruitlets occur on a flat, top-shaped spongy thalamus.

(*ii*) **An etaerio of follicles.** In this case fruit develops from a bicarpellary apocarpous (*e.g.,* in *Calotropis*) or tricarpellary (*e.g.,* in *Aconitum*) or polycarpellary apocarpous (*e.g.,* in *Michelia*) gynoecium. In *Asclepias* and *Calotropis* each etaerio consists of a pair of follicles. In *Aconitum* an aggregate of three follicles is produced on the top of the thalamus, whereas in *Michelia* numerous follicles are developed on an elongated thalamus.

(*iii*) **An etaerio of drupes.** The characteristic example of this type is found in *Rubus* (raspberry). It develops from superior, polycarpellary and apocarpous gynoecium. The carpels are borne on a conical thalamus. Here a number of small drupelets develop from separate carpels of a flower and they are grouped together on a fleshy thalamus.

(*iv*) **An etaerio of berries.** In this type of fruit each carpel of an apocarpous pistil develops into a berry. During development the margins of the carpels may be fused (*e.g.,* in *Annona squamosa*) or they may remain separate from each other (*e.g.,* in *Artabotrys odoratissimus*). In *Annona squamosa* the mesocarp of berries is edible.

3. Composite or Multiple Fruits

A composite or multiple fruit develops from the entire inflorescence. Here the flowers as well as the peduncles on which they are borne take part in the development of the fruit. Such a fruit is also known as **infructescence.** There are two main types of such fruits.

(*i*) **Sorosis.** This type of multiple fruit develops from a spike or spadix. Here the flowers fuse together by their perianth lobes and simultaneously the axis bearing them becomes fleshy and juicy, and as a result the whole inflorescence forms a compact mass, *e.g.,* pine apple, jack fruit, etc. In *Morus* (mulberry), the perianth lobes become thick and juicy and are edible. Here the ovules do not mature into seeds and therefore, the carpels develop into small nutlets that are seedless.

(*ii*) **Syconus.** This type of composite fruit develops from a hollow, pear-shaped, fleshy receptacle which encloses numerous minute male and female flowers. The receptacle develops further and converts into the so-called edible fleshy fruit. It really encloses a number of true fruits or achenes which develop from the female flowers of the hypanthodium inflorescence, *e.g.,* fig, banyan and many species of *Ficus*.

Table 26.13. Common Edible Fruits

English name	Botanical name and family	Type of fruit	Edible part
Apple	*Malus sylvestris* Rosaceae	pome	thalamus
Banana	*Musa paradisiaca* Musaceae	berry	mesocarp and endocarp
Cashew-nut	*Anacardium occidentale* Anacardiaceae	nut	peduncle and cotyledons
Coconut	*Cocos nucifera* Palmaceae	fibrous drupe	endosperm
Cucumber	*Cucumis sativus* Cucurbitaceae	pepo	mesocarp, endocarp and placentae
Custard apple	*Annona squamosa* Annonaceae	etaerio of berries	fleshy pericarp of individual berries
Date palm	*Phoenix sylvestris* Palmaceae	one-seeded berry	pericarp
Fig	*Ficus carica* Moraceae	syconus	fleshy receptacle
Jack fruit	*Artocarpus heterophyllus* Moraceae	sorosis	bracts, perianth and seeds
Grape	*Vitis vinifera* Vitaceae	berry	pericarp and placentae
Guava	*Psidium guava* Myrtaceae	berry	thalamus and pericarp
Indian plum	*Zizyphus jujuba* Rhamnaceae	drupe	mesocarp and epicarp
Litchi	*Litchi chinensis* Sapindaceae	one-seeded nut	fleshy aril
Wheat and other cereals	*Triticum aestivum* Poaceae	caryopsis	starchy endosperm
Mango	*Mangifera indica* Anacardiaceae	drupe	mesocarp
Melon	*Cucumis melo* Cucurbitaceae	pepo	mesocarp
Orange	*Citrus aurantium* Rutaceae	hesperidium	juicy placental hair
Papaw	*Carica papaya* Caricaceae	berry	mesocarp
Pea	*Pisum sativum* Papilionaceae	legume	cotyledons
Pear	*Pyrus communis* Rosaceae	pome	fleshy thalamus
Pine-apple	*Ananas comosus* Bromeliaceae	sorosis	outer portion of receptacle bracts and perianth
Pomegranate	*Punica granatum* Punicaceae	special	juicy testa (balusta)
Strawberry	*Fragaria vesca* Rosaceae	etaerio of achenes or nuts	succulent thalamus
Tomato	*Lycopersicon esculentum* Solanaceae	berry	pericarp and placentae
Wood apple	*Aegle marmelos* Rutaceae	berry	mesocarp, endocarp and placentae

Fig. 26.71. Some edible fruits. A, Pomegranate (*Punica granatum*) ; B, Strawberry (*Fragaria vesca*) ; C, Apple (*Malus sylvestris*); D, Peach (*Prunus persica*) ; E, Sapodilla (*Achras zapota*) ; F, Mulberry (*Morus alba*) ; G, Guava (*Psidium guava*) ; H, Pineapple (*Ananas comosus*). ; I, Custard apple (*Annona squamosa*); J, Litchi (*Litchi chinensis*).

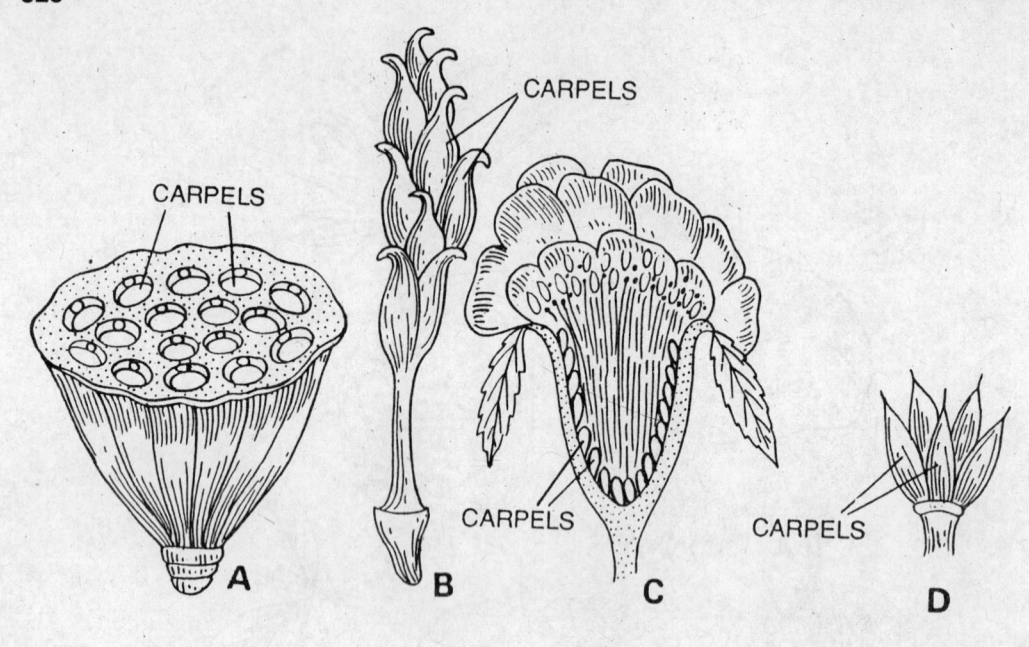

Fig. 26.72. Apocarpous pistil. A, lotus; B, *Michelia;* C, Rose; D, *Sedum*.

Functions of Fruits

The fruits are meant for the protection of the seeds. They are the main instruments of seed dispersal. The fruit wall is variously modified in many fruits for dispersal by wind, animals and water. The fleshy fruits by their decay enrich the soil with inorganic acids and thus provide a good start to the young seedings which develop from the seed. Many fruits are edible and make the main food of human beings.

DIVERSITY AND MORPHOLOGY IN SEED

A true seed is defined as a fertilized mature ovule that possesses embryonic plant, stored material, and a protective coat or coats. Seed is the reproductive structure characteristic of all phanerogams. The structure of seeds may be studied in such common types of pea, gram, bean, almond or sunflower. They are all built on the same plan although there may be differences in the shape or size of the seed the relative proportion of various parts.

There are hundreds of variations in the seed size, shape, colour and surface. The seeds range in size from tiny dust particles, as found in some orchids, to large double-coconuts. The seed surface may be smooth, wrinkled, striate, ribbed, furrowed, reticulate, tuberculate, alveolate, hairy, pulpy or having patterns like finger prints.

In the seed, life activities are temporarily suspended in order to enable the plant to successfully pass through unfavourable and injurious climatic conditions. On the approach of favourable conditions, the seed resumes active life and grows into full plant. In the form of seeds, a plant can be carried to long distances without special precautions.

Structure of Seed

The various parts of a seed may be easily studied after it has been soaked in water for a day or so varying according to the nature of the seeds. A mature seed contains an embryonic plant (with a radicle and plumule), and is provided with reserve food materials and protective seed coats. A

Diversity and Morphology of Flowering Plants

mature pod of pea (*Pisum sativum*) has a number of seeds arranged in two rows. The seeds are attached to the fruit wall by a small stalk, the **funiculus**. At maturity, on one side of the seed coat a narrow, elongated scar representing the point of attachment of seed to its stalk is distinctly seen, this is the **hilum**. Close to the hilum situated at one end of it there is a minute pore, **micropyle**. During seed germination, water is absorbed mainly through this pore, and the radicle comes out through it. Continuous with the hilum there is sort of ridge in the seed coat, the **raphe**. The seed is covered by two distinct **seed coats**; the outer whitish one is the **testa**, while the other inner thin, hyaline and membranous covering is the **tegmen**. The seed coats give necessary protection to the embryo which lies within. The whitish fleshy body, as seen after removing the seed-coats is the embryo. It consists of two fleshy cotyledons and a short **axis** to which the cotyledons remain attached. The position of the axis lying outside the cotyledons, bent inward and directed towards the micropyle is the **radicle** and the other portion of the axis lying in between the two cotyledons is the **plumule**. The plumule is crowned by some minute young leaves. The radicle gives rise to the root, the plumule to the shoot and the cotyledons store up food material. Since the reserve food material is stored in the massive cotyledons and the seed lacks a special nutritive tissue, the **endosperm**. The seeds which lack endosperm at maturity are called **non-endospermous** or **exalbuminous**. On the other hand in several other plants such as castor bean (*Ricinus communis*), coconut (*Cocos nucifera*) and cereals, food is stored in the endosperm. Such seeds where endosperm persists and nourishes the seedling during the initial stages are called **endospermous** or **albuminous**.

On the basis of the number of cotyledons in the embryo the angiosperms have been divided into two large groups;

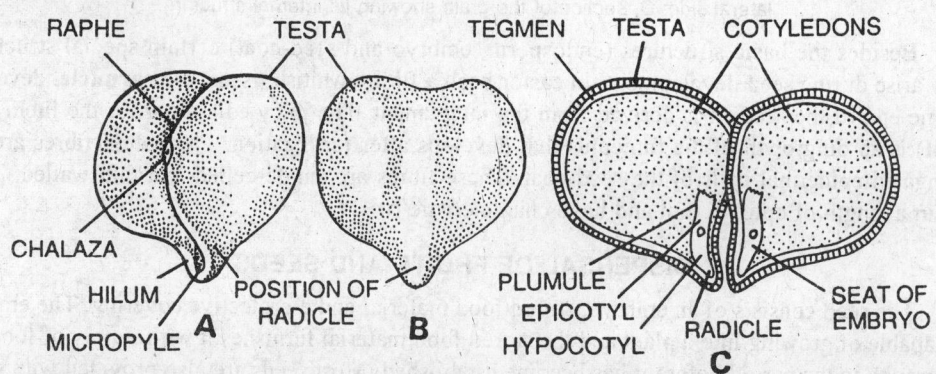

Fig. 26.73. Structure of gram seed. A, complete seed showing various parts; B, seed as seen after removing the testa; C, L.S. of seed.

1. **Dicotyledons**, having embryos with two cotyledons, and
2. **Monocotyledons**, with only one cotyledon.

A maize grain is a single-seeded fruit in which the seed coat and the fruit wall are unseparable. On one side of the grain a small, opaque, whitish, deltoid area is seen to be distinctly marked out from the region. The embryo lies embedded in this area. There is only a thin layer surrounding the whole grain. This layer is made up of the seed-coat and the wall of the fruit fused together. The grain remains divided into two unequal portions by a definite layer known as the **epithelium**. The bigger portion is the **endosperm**, and the smaller portion, the **embryo**. The endosperm, is the food storage tissue. The embryo consists of one shield shaped cotyledon, known as the **scutellum** and **axis**. The upper portion of the axis, with minute leaves arching over it, is the **plumule**, and the lower portion provided with the root cap the **radicle**. The plumule is surrounded by a leaf-sheath or coleoptile and the radicle is surrounded by a root sheath or **coleorhiza**. These are the protective sheaths of the plumule and the radicle respectively.

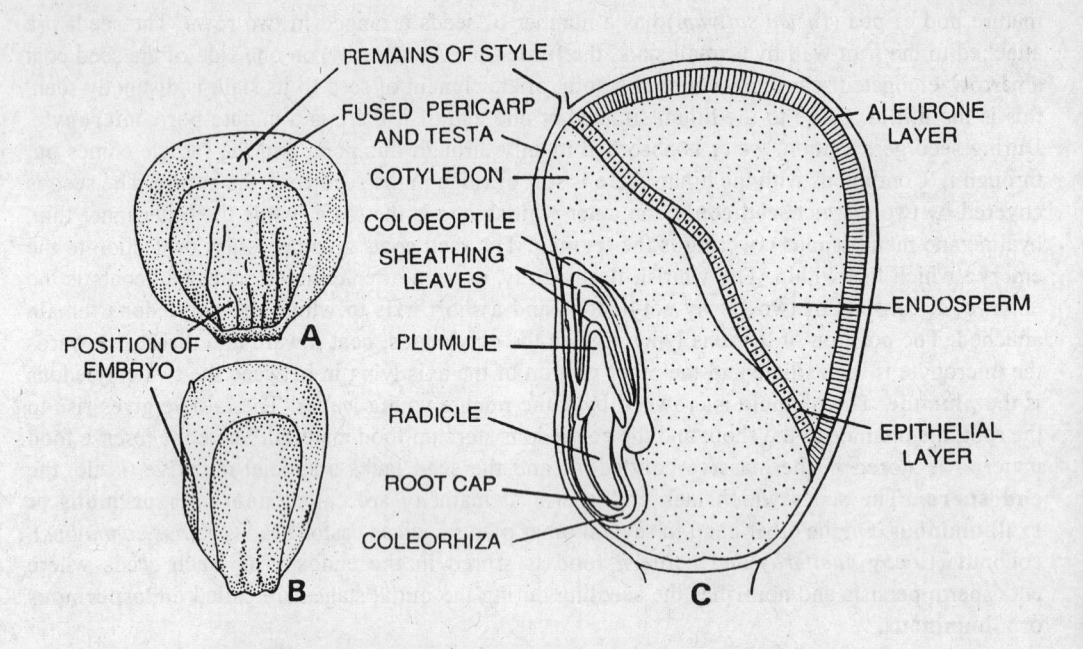

Fig. 26.74. Structure of maize grain. A, grain illustrating of external characters; B, grain as seen from lateral side; C, section of the grain showing its internal structure.

Besides the basic structures (endosperm, embryo and seed-coat) certain special structures may arise during seed development. In castor bean a fleshy whitish tissue, the **caruncle**, develops at one end of the seed. It is derived from the integument. The juicy edible part of the litchi fruit (**aril**) is an outgrowth of the funiculus that develops after fertilization. The cotton fibres are the elongated epidermal cells of the seed-coat. These fibres are single-celled and thin walled. They attain a length of upto 45 mm and have characteristic twists.

DISPERSAL OF FRUITS AND SEEDS

The seed consists of an embryo, stored food material and a protective covering. The embryo is capable of growing into a plant and the stored food material furnishes it with supply of food for its growth in the period before it has become established. Most seeds are also provided with some means of dispersal. Without this the great majority of seeds would fall under the parent plant and not be carried to a location favourable to germination and growth.

Dispersal by Wind

Either the whole fruit or the individual seeds may be suited to dispersal by wind. Seeds that are thus disseminated are characteristically light. The means of adaptation to wind dispersal may be grouped under the headings of **minute seeds, flattened fruits** or **seeds, wing like outgrowths, feathery appendages** and the so-called **censor mechanisms.**

Minute seeds. The seeds of orchids are very small and besides, have a light, inflated outer covering. These dust-like seeds can be blown by the wind for great distance.

Wings. In many cases, seeds and in others, whole fruits are flattened or have wing-like outgrowths or they may be both flattened and winged. This type of structure results in the scattering of the seed by the wind, (*e.g.*, the fruits of *Terminalia*, ash, maple, *Macrozanonia, Pterocymbidium, Dipterocarpus*, linden, etc.)

Diversity and Morphology of Flowering Plants

Feathery appendages. Seeds or fruits may have feathery appendages, (*e.g., Asclepias, Vernonia*, etc.) which greatly increase their buoyancy, so that they are frequently carried by the wind to considerable altitudes. These feathery appendages are very characteristic of the seeds of *Asclepias* and of the achenes or many Compositae. Commercial cotton consists of trichomes which grow from the epidermal cells of the seed of the cotton plant. These trichomes form a flossy mass round the seed.

Censor mechanisms. The capsules of many plants open in such a way that the seeds can escape only when the capsules are violently shaken as by a strong wind, (*e.g., Aristolochia*, Poppy, etc.). This has a tendency to scatter the seeds. The seeds may in addition have a flat shape or winged outgrowths (as in *Aristolochia*); and as they are likely to escape when there is a strong wind, they may be blown for considerable distances.

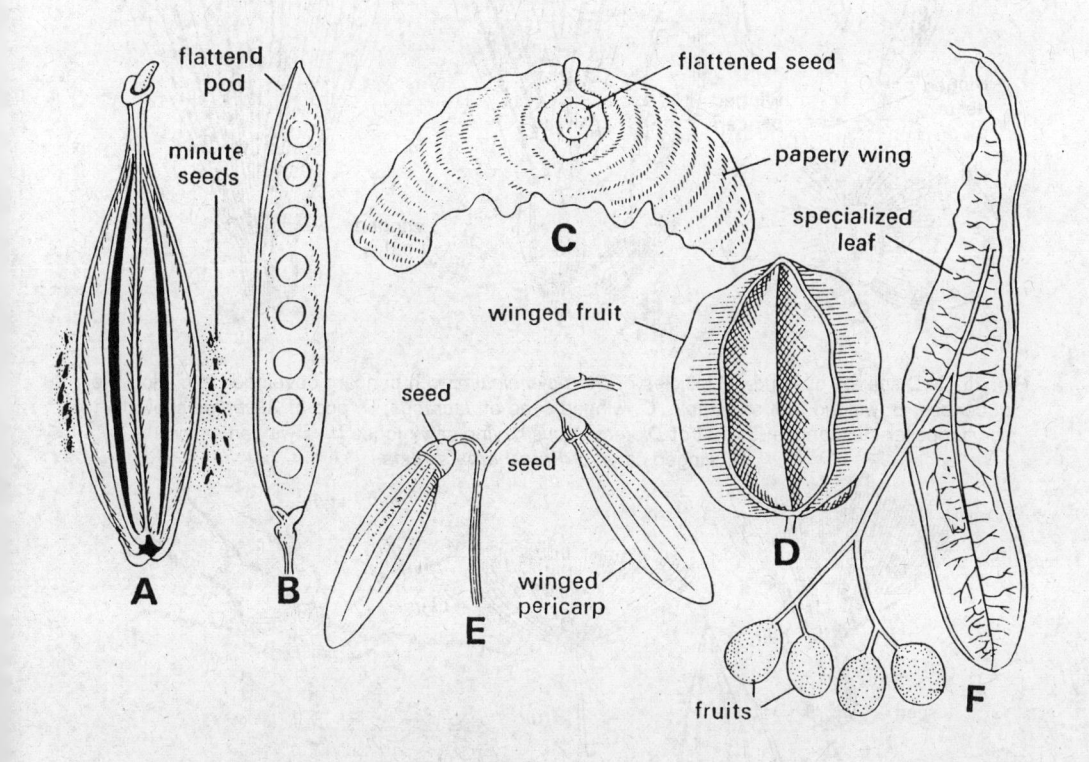

Fig. 26.75. Dispersal by wind. A, capsule of an orchid (*Cymbidium* spp.) releasing minute seeds; B, flattened pod of *Albizzia* spp.; C, winged seed of *Macrozanonia;* D, winged fruit of *Terminalia arjuna;* E. fruits of *Tilia americana* attached to a flattened bract that help in wind dispersal; F, the fruit with seeds attached to a specialized leaf for dispersal.

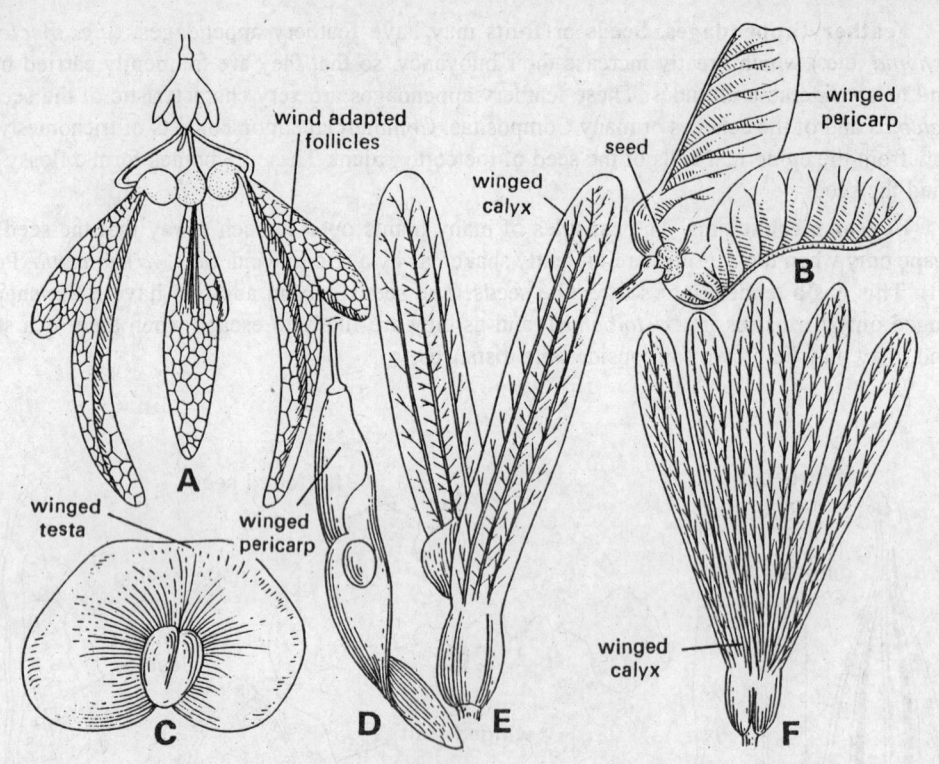

Fig. 26.76. Dispersal by wind. A, follicles of *Pterocymbidium,* with pericarp developed into wing like structures; B, winged fruit of mapple; C, winged seed of *Jacranda;* D, pod of *Alianthus* showing propellar mechanism; E, fruit of *Dipterocarpus* having calyx lobes (*i.e.,* winged calyx); F, winged calyx of *Parashorea stellata.*

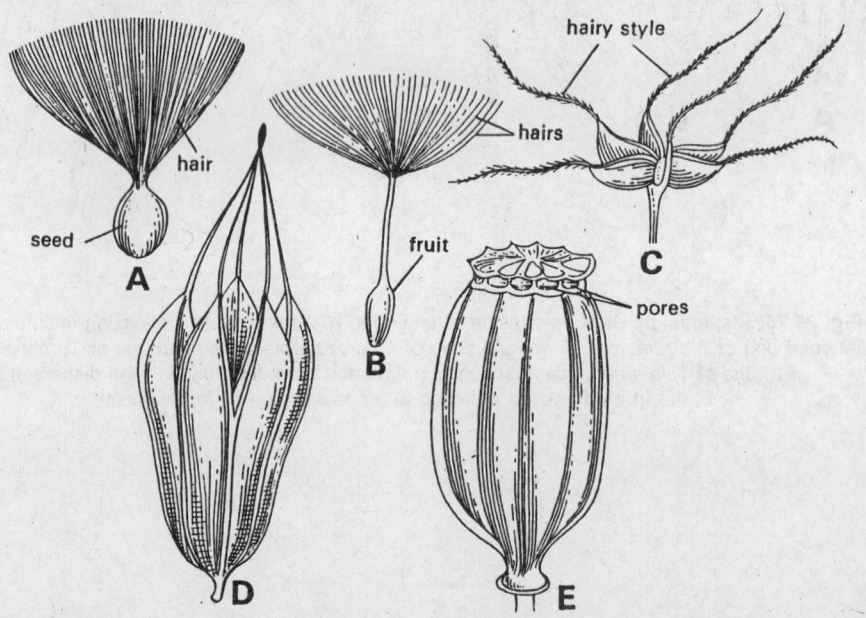

Fig. 26.77. Dispersal by wind. A, plumed seeds of *Calotropis procera;* B, fruit of *Taraxacum* with hairy pappus; C, an eaterio of achenes of *Nigella;* D, capsule of *Aristolochia,* the seeds are being dispersed when the capsule is shaken by wind; E, capsule of poppy, the minute seeds are released through pores in capsule and are dispersed by wind.

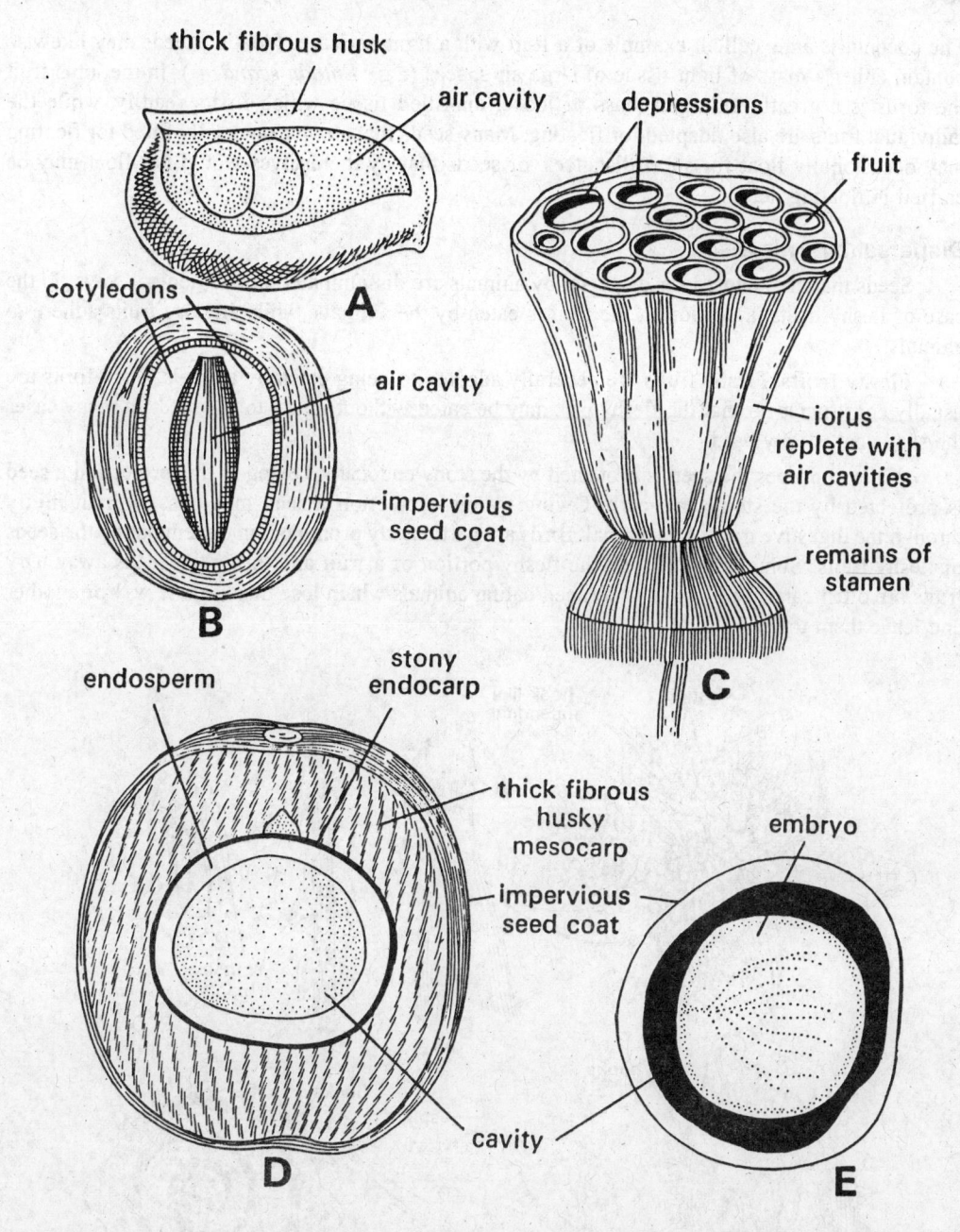

Fig. 26.78. Dispersal by water. A, fruit of *Heritiera littoralis* covered with a fibrous husk and having an air cavity around the seeds; B, seed of *Entada scandens* having an air cavity between the cotyledons; C, fruit of *Nelumbo* having single-seeded achenes, the thalamus has lot of air cavities, which floats in water; D, fruit of coconut with fibrous mesocarp and can float in water; E, seed of *Mucuna gigantea* having an air cavity around the embryo and thus the seed floats in water.

Dispersal by Water

Adaptations for dispersal by water occur in many seashore and aquatic plants. Either the whole fruit or the seed may be adapted for floating. The pericarp of a fruit may be composed of light tissue, (*e.g.,* in *Terminalia catappa*) or the fruit may be inflated, (*e.g.,* in *Heritiera littoralis*).

The coconut is an excellent example of a fruit with a light pericarp. Floating seeds may likewise contain either a mass of light tissue of large air spaces (*e.g., Entada scandens*). In the lotus fruit the torus is a greatly enlarged mass of loose, air-filled tissue which floats readily, while the individual fruits are also adapted for floating. Many seeds that are not especially fitted for floating may occasionally float for short distances, or seeds that by themselves would not float may be carried in floating debris.

Dispersal by Animals

Seeds that are adapted for dispersal by animals are disseminated in two general ways: in the case of fleshy fruits, a portion of the fruit is eaten by the animals, while any dry fruits adhere to animals.

Fleshy fruits. Fleshy fruits are generally adapted to being eaten by animals. Such fruits are usually constructed so that the fleshy part may be eaten without injury to the seed. In many cases the seed coat is very hard.

While in drupes the seed is protected by the stony endocarp. Owing to this protection a seed is protected by the stony endocarp. Owing to this protection a seed may pass without injury through the digestive tract of an animal. Birds are particularly prominent in disseminating the seeds of fleshy fruits. Sometimes they eat the fleshy portion of a fruit and throw the seeds away. Dry fruits are often carried off for food by seed-eating animals which lose them in one way or another and leave them to grow.

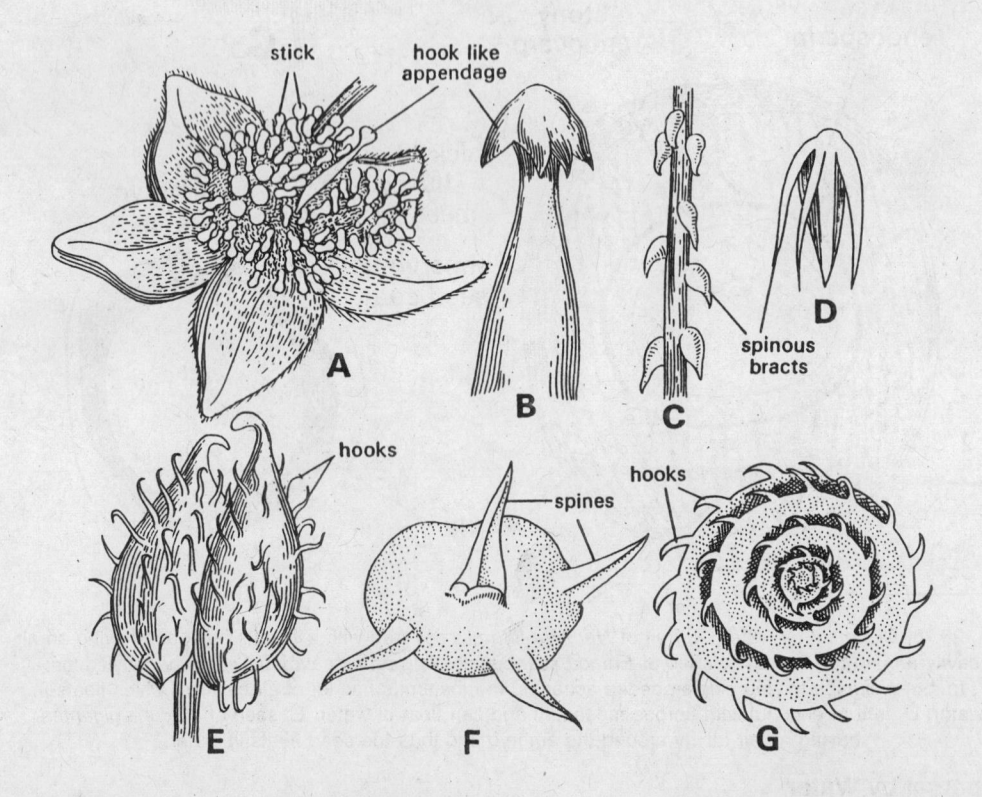

Fig. 26.79. Dispersal by animals. A, fruit of Chinese forget-me-not bearing sticky glands; B, hook-like appendage of A; C, fruits of *Achyranthes aspera* bearing spinous bracts; D, spinous bract of C; E, fruit of *Xanthium* bearing hooks on the outer covering of fruit, made up of thalamus; F, fruit of *Tribulus terrestris* bearing spinous outgrowths; G, spirally coiled pod of *Medicago* bearing hook-like out-growths.

Diversity and Morphology of Flowering Plants 535

Adhesive fruits. Many dry fruits have hooklike appendages (*e.g., Cosmos, Bidens, Mimosa pudica, Triumfetta*, etc.). which are particularly fitted for grasping the hair of animals. Animals to which the fruits adhere carry them about and thus distribute the seeds. In a similar way fruit may adhere to clothing and thus be disseminated by man. Some seeds and fruits have a sticky covering which will adhere to the hair of an animal. Fruits of Chinese forget-me-not adhere by sticky hooklike appendages. Feathery appendages are usually capable of adhering to fur as well as of flying on the wind.

Minute seeds. Many plants have minute seeds which are disseminated by being caught in mud that adhere to the feet or other parts of birds or other animals.

Dispersal by Explosive Mechanism

Many fruits are so constructed that they explode when ripe and scatter the seeds. This method of dispersal is frequently conspicuous in members of the bean family, where the explosive forces are due to stresses arising from the drying of the valves of the pod. The balsam has somewhat fleshy capsules which are very turgid. When these are disturbed by contact the segments of the pericarps roll up with considerable force and in such a way that they scatter the seeds. An unusual explosive mechanism is found in the squirting cucumber (*Ecaballium elaterium*).

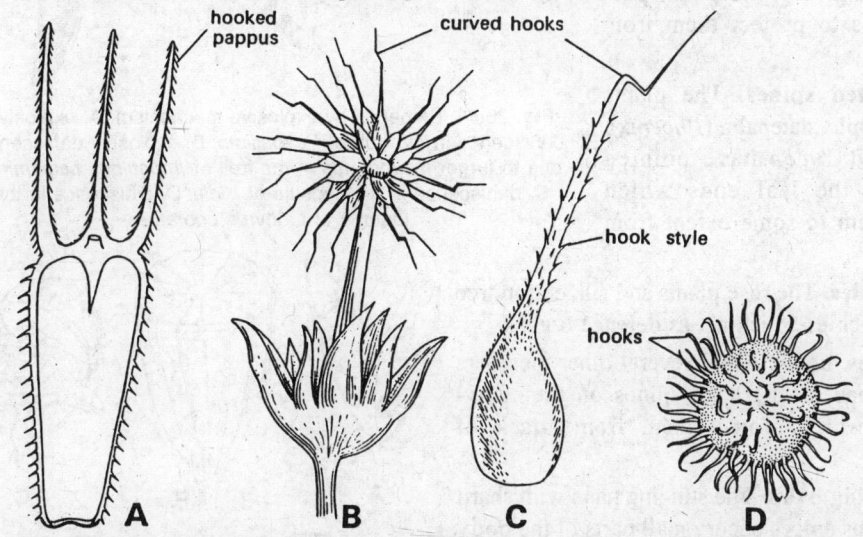

Fig. 26.80. Dispersal by animals. A, cypsella of *Bidens* bearing pappus; B, fruit of a grass bearing curved styles covered with hooks; C, hooked style of B; D, hooked fruit of *Triumfetta*.

Dehiscence of Fruits

The dehiscent fruits rupture to liberate their seeds in several ways. They are as follows:

Transverse. The fruits of *Portulaca* and *Celosia* liberate their seeds through transverse rupture.

Porous. In poppy (*Papaver*) and *Luffa* the seeds are liberated through pores in fruits.

Sutural. In pea (*Pisum*) and bean the seeds are liberated through sutural valves.

Loculicidal. The fruits of cotton (*Gossypium*) and lady's finger (*Abelmoschus*) the seeds are liberated through locules.

Septicidal. In linseed (*Linum*) and mustard (*Brassica*) the seeds are liberated through septa or partition wall.

Septifragal. In *Datura* and other such plants, the fruits rupture loculicidally throwing valves away from fruit and leaving seeds attached to the central axis.

Defense Mechanisms in Plants

Many plants develop certain devices or special organs which give protection to them from their enemies. Some devices are mentioned here.

Thorns. Many plants, such as lemon (*Citrus*), pomegranate (*Punica*), bael (*Aegle marmelos*), *Duranta*, and ber (*Zizyphus*) possess thorns to protect them from their enemies.

Pointed spines. The plants like pineapple, datepalm (*Phoenix*), *Agave* and *Yucca* have pointed spines at the leaf ends which protect them to some extent from enemies.

Prickles. The rose plants and silk cotton tree possess prickles which act as defence organs.

Spines. *Opuntia* and several other members of Cactaceae (cacti) have spines on their phylloclades which protect them from attack of animals.

Stinging hairs. The stinging hairs with sharp and siliceous apices occur on all parts of the body in nettles (*Laportea*), *Urtica dioica* and several other members of Urticaceae which protect the plants from their expected enemies.

Glandular hairs. Glandular hair with sticky substances are found on plant body of *Jatropha*, *Boerhaavia* and tobacco (*Nicotiana*) which protect them from outer agencies.

Stiff hair. The dense coating of such stiff hair is repulsive to attacking animals, they are found on cud weed (*Gnaphalium*) and many cucurbits.

Besides, above mentioned special organs certain other defense devices like poisonous and irritating substances are found in many plants. They are as follows.

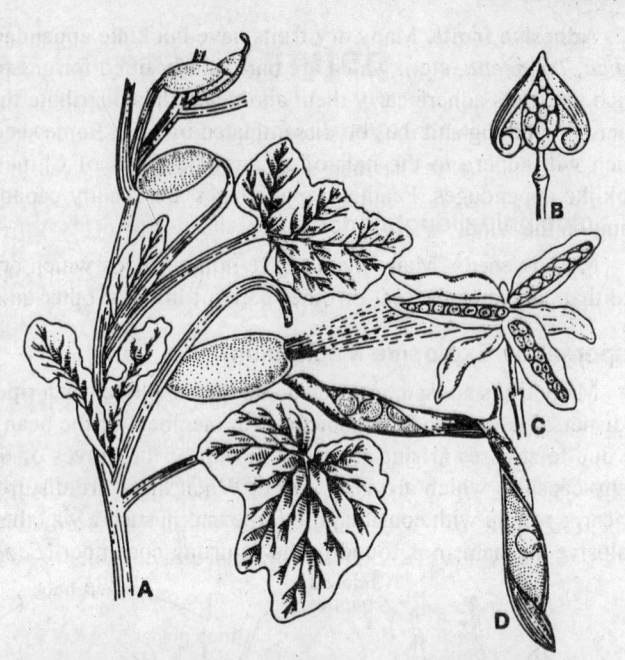

Fig. 26.81. Dispersal by explosive mechanism. A, explosive dehiscence in the fruit of *Ecaballium*; B, explosive dehiscence due to turgor movements in the fruit of *Impatilens balsamina*; C, dehiscence in the capsule of *Viola*; D, dihiscence in the legume of *Lathyrus odoratus*.

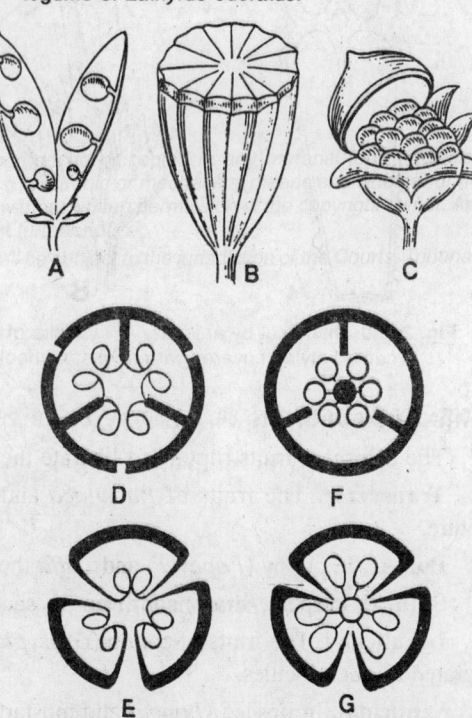

Fig. 26.82. Dehiscence of fruits, A, sutural; B, porous; C, transverse; D, loculicidal; E, septicidal; F and G, septifragal.

Latex. Several species of plants, such as *Nerium*, *Ficus* and *Euphorbia* have latex in their different parts. The latex protect these plants to some extent from his enemies.

Alkaloids. Several plants, such as Poppy (*Papaver*), *Datura* and tobacco (*Nicotiana*) possess alkaloids in them which are poisonous and give protection to the plants.

Raphides. Calcium oxalate crystals are also known as raphides which are irritating and poisonous. They are found in *Colocasia* and several other members of Araceae (aroids). They help in defense of these plants.

Smell and bitter taste. Several plants of Lamiaceae, such as *Ocimum* and mint have typical smell and bitter taste which protect the plants from attack of insects. Neem and Karela have also bitter taste.

Tannin. Several plants such as *Acacia nitotica*, *Acacia catechu*, etc., possess tannins in them which act as substances of protection.

Resin and essential oils. In many angiosperms resins and essential oils are present. In *Pistacea* and other members of Anacardiaceae the gums and resins are found which protect them from several injurious agencies.

Geophilous habit. Zinger (*Zingiber officinale*), turmeric (*Curcuma longa*), colocasia and onion are main representatives of this habit. This underground habit protects them from several injurious agencies.

Myrmecophily. Some plants, such as guava (*Psidium* spp.), mango (*Mangifera indica*) have a habit of harbouring ants which save the plants from the damage by other organisms.

Mimicry. This is habit of initiating the general appearance, colour and shape of other plants, generally disliked by attackers. For example, the aroids (*Cladium*) and *Sansevieria* resemble spotted snakes and thus by this habit keep away from plant-eating animals. The species of *Arisaema* (aroid) possess hood like inflorescence which resembles cobra in appearance and thus protect them from invading animals.

MODULAR GROWTH

Variation in organisational form that arises from size dependent relationship among parts is a fundamental aspect of development and evolutionary change. **Allometric analysis** is one of the earliest morphometric tools that help in revealing ontogenetic or evolutionary changes in shape as a consequence of changes in organ or body size (Gayon, 2000).

In plant population the individuals change widely in size may be due to asymmetric competition for light (Weimer, *et al*, 2001) or meagre distribution of other resources. The variability is dependent on the way plants are constructed. Plants grow through the repeated addition of similar morphological sub-units called **metamers** and **modules**. This mode of development contributes to size variation by creating the potential for both **indeterminate** and **exponential** growth.

Modularity presents special challenge and opportunities for allometric analysis. In biological system there are different levels of organisation, *e.g.*,

genetic families → individuals → population → communities

Modular organisms have additional levels of organisation due to their clear within individual substructure. The modular growth provides serial adjustment to the phenotype through the addition of new parts during development.

Module and **metamer** may be defined as follows :

Module is a product of an apical meristem; while **metamers** are serially homologous repeated units along an axis and are generally sub-units of modules. A vegetative metamer consists of a leaf, the segment of stem subtending it and its axillary meristem.

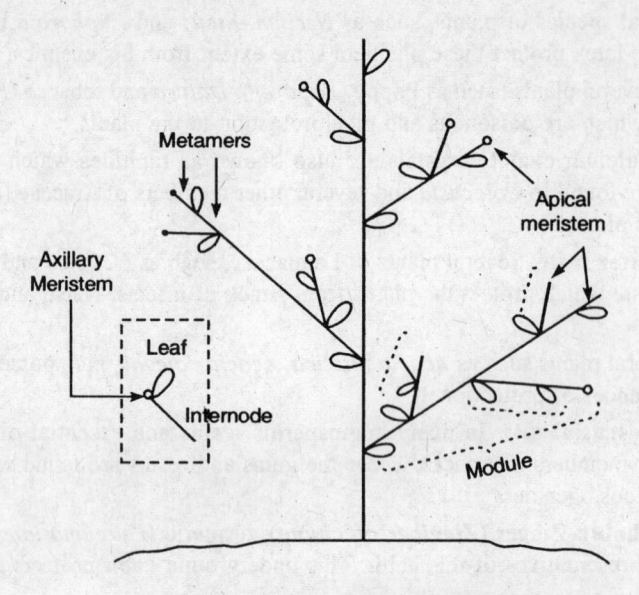

Fig. 26.83. Modular construction of plant

Types of Modularity

On developmental genetics modules refer to largely autonomous developmentally and functionally integrated units (Carrol, 2001). According to comparative morphology the modular growth occurs through the retention of the same few types of structures, *e.g.*, **branches** and **flowers** in plants. In contrast to developmental genetic modules, the modular subunits comprising an organism are similar to each other, *i.e.*, serially homologous and loosely analogous to individuals in population.

Modularity in Plants

In plants, module is the product of apical meristem, *e.g.*, branches, cones or flowers. Vegetative modules produce new meristem that may give rise to additional vegetative or reproductive modules, creating a modular hierarchy.

Usually modules themselves are composed of repeated units or **metamers**. For example, a typical vegetative metamer consists of a node and an internode, a leaf and an axillary meristem (Fig. 26.83).

Modular Development

In seed plants, *e.g.*, gymnosperms or angiosperms, the plant body is essentially **modularly** organised. The smallest elementary unit of construction in the shoot system is called a **metamer**. Metamers make branches and branches construct branching system. Each shoot apex is a potential point of further development. Development is iterative or repetitive process. The finite bit of programme is repeated to add new **metamers**. The number of iterations can be infinite in some species. In annual plants iterations certainly terminate, but their number is not strictly defined and may depend on environment (Schlichting and Smith, 2002). The number of leaves of a tree can vary with age and habitat conditions. It is due to capability of formation of organs continuously. Organs, metamers and branches are disposable without death of organism.

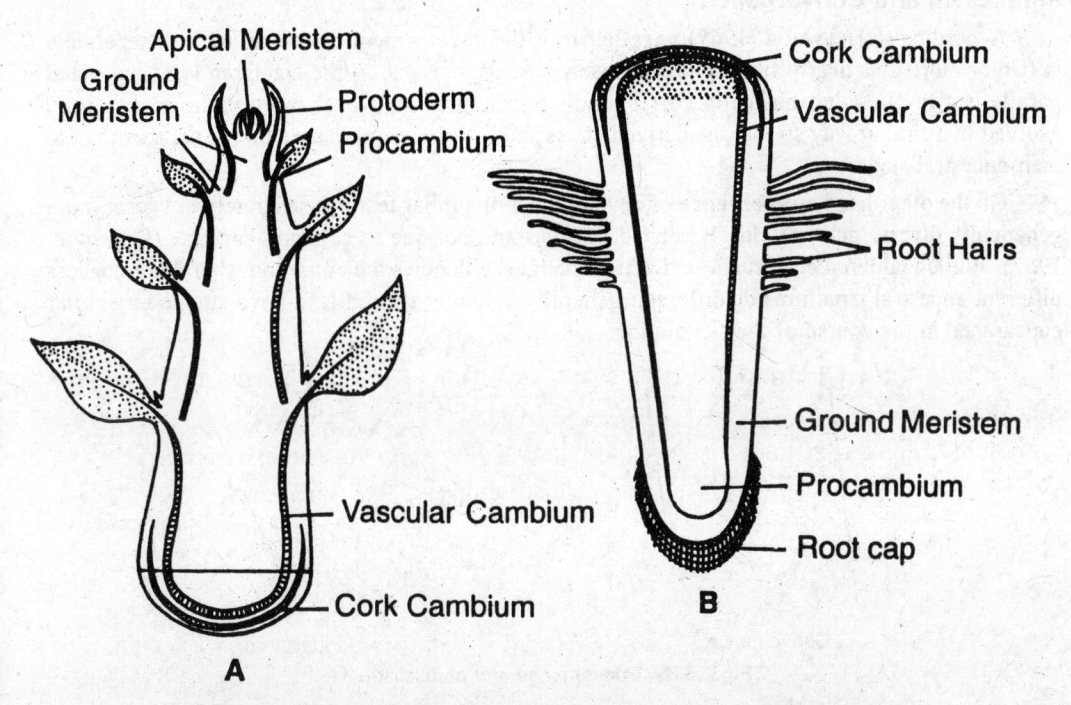

Fig. 26.84. Modular development. Shoot system and root system modules.

Shoot System Modules

For the first time, the term **metamer** was used in animals. **Metamers** unite to form **module**. Modules divide due to outgrowths of axillary buds or adventitious buds. When the axillary bud grows continually without undergoing rest period is called **sylleptic sub-module** while the axillary bud that undergoes some dormancy period is called **proleptic sub-module**.

The activity of module is directly related with that of apical meristem. After getting a stimulus, the axillary bud activates and module branches. New modules may arise some time playing an important role in **regeneration**, *e.g.*, in *Eucalyptus*, after fire new modules grow out rapidly from hidden buds produced by modification of the base of stem.

Any module or sub-module terminates in a flower or inflorescence depending upon the portion or orientation of metamers in shoot apex.

Root System Modules

Primary and lateral roots are modules of root system. Lateral roots are well developed in herbs. Various types of roots, such as respiratory roots, storage roots, etc., are types of modules.

Convergence is a phenomenon in nature found in all organisms. It is observed tendency of living forms that are quite unrelated phylogenetically, to respond to similar contingencies of life by developing similar structures. In the process of evolution this similarity way arise along two evolutionary lines, *i.e.*, **parallelism** and **convergence** (Fig. 26.85.).

Influence of environmental factors may also create similarities by structural modification, *e.g.*, in desert habitat most of the plants show cactus like habit and known as **cacti**. But only on flowering they can be separated in euphorbias, asclepias and cacti. Members of distantly related taxa attained similar structural features which are attained independently by various groups of plants, is called **convergent evolution**.

Parallelism and Convergence

According to Heywood (1967) **parallelism** is the development of similar features separately in two or more genetically similar, fairly closely related lineages. While Dahlgren *et al* suggested parallelism implies that basically corresponding structures in two or more plant groups have evolved in a similar way so that the derived states show more resemblances to one another than to their ancestral stages.

On the other hand **convergence** of development of similar features separately in two or more genetically diverse and not closely related lineages and not due to common ancestry (Heywood, 1967). While Dahlgren *et al* suggested that convergence denotes the condition where more or less different ancestral structures or different, generally distantly related taxa have attained a similar appearance in the course of their evolution.

Convergence   Parallelism

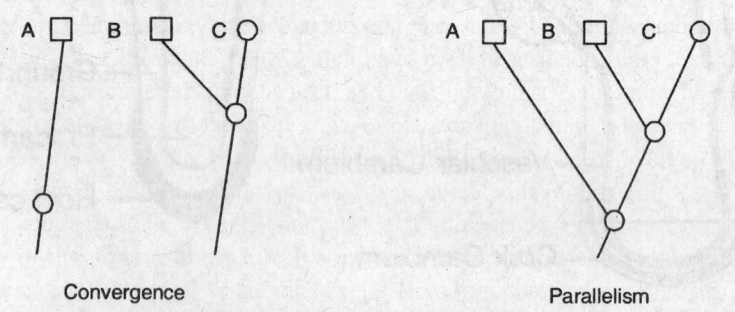

Fig. 26.85. Convergence and parallelism.

Thus, when two or more closely related groups tend to evolve by acquiring similar characters is called **parallelism** and when distantly related groups acquire similar characters is called **convergence**.

Thorne (1973) suggested that germination, nutrition, pollination, fertilisation, dispersal, etc., are the channels causing many diverse unrelated groups to proceed along the same evolutionary channel leading to **convergence**. According to Dahlgren *et al.* (1985), in homologous structures convergence results in superficial similarity of the whole plant, *e.g.*, succulents; and in non homologous structures it leads to superficially similar organs in different taxa, *e.g.*, presence of spines and tendrils in different groups of plants.

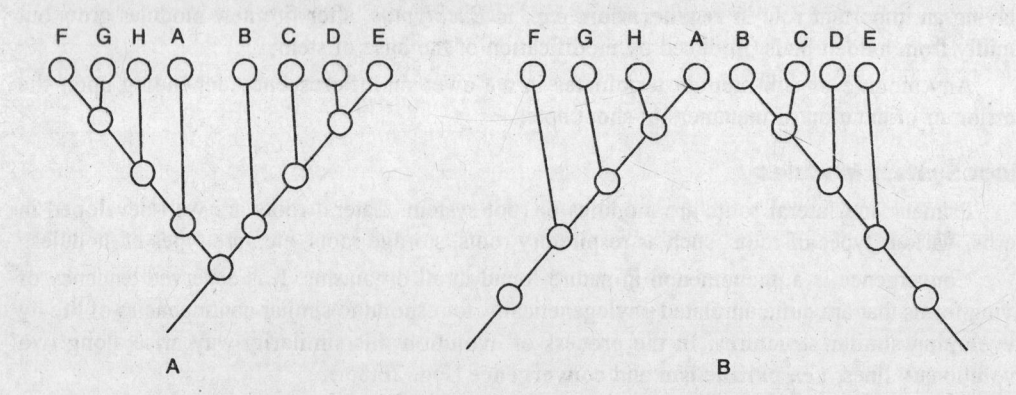

A   B

Fig. 26.86. Patristic relationship.

Convergence is generally brought about by adaptation to similar climate or habitats mostly affecting the vegetative parts, or even to similar methods of pollination or dispersal affecting the flower, fruit or seed.

Diversity and Morphology of Flowering Plants

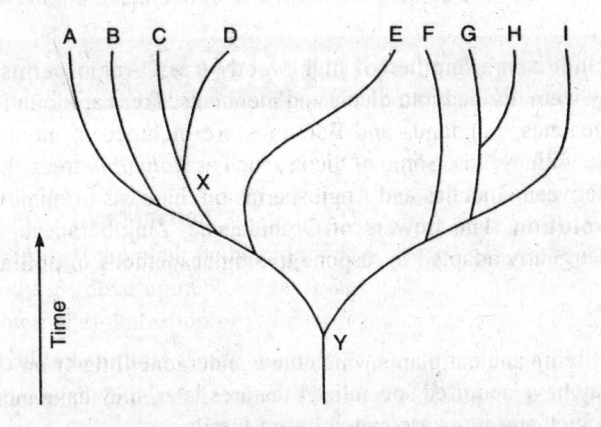

Fig. 26.86. (a) Cladistic relationship.

For example, *Arenaria* and *Minuartia* form natural groups of species which were earlier placed in genus *Arenaria*. However, *Arenaria leptocladus* and *Manuartia hybrida* show much similarities. If similarity is **patristic**, *i.e.*, from common ancestry; the two species are most primitive members and show a single monophyletic group. But on the other hand, if they are most advanced members, they show distinct line of ancestry and similarity between them will be superficial and purely due to convergence. Thus, similarity between two plants due to a known or informed common ancestry is called **patristic** similarity in terms of phyletic lines, the plants are directly related. The pathways by which a given similarity is attained or closeness of relationship in terms of phyletic lines is called **cladistic** similarity.

Parasitic habit in different families, such as Lauraceae (*Cassytha*), Convolvulaceae (*Cuscuta*), Scrophulariaceae (*Striga*), Orobanchaceae (*Orobanche*) and Balanophoraceae (*Balanophora*); Carnivorous accessory nutrition in Droseraceae (*Drosera*), Utriculariaceae (*Utricularia*), Sarraceniaceae (*Sarracenia*), Nepenthaceae (*Nepenthes*) and Lentibulariaceae are some good examples of convergence.

Convergence of Evolution of Tree Habit

Trees are perennials. The groups of trees looking alike are few, *e.g.*, palms (monocots), conifers (gymnosperms), etc. Generally, the convergence in tree habit can be seen in their height. The primitive pteridophytes, gymnosperms and angiosperms, etc., were trees. Progymnosperms were present in upper Devonian and lower Carboniferous period of Palaeozoic era. Many palaeobotamists believe in origin of gymnosperms from pteridosperms.

The link between fern like *Archaeopteris* and gymnospermous *Callixylon* was discovered in 1960. *Archaeopteris* was a 18 m high tree with crown of leaves. The members of Lyginopteridaceae, *e.g.*, *Lyginopteris oldhamia* were large fern like trees with well branched weak stems. *Medullosa*, a member of Medullaceae (fossil gymnosperm) was large fern like but small tree. The fossil gymnosperms had fern like leaves.

However, members of Cycadales are medium sized trees with fern like leaves. *Macrozamia hopei* is tallest cycad 60 feet in height. Mexican cycad *Dioon spinulosum* is 50 feet in height while 1000 year old *D. edule* is only 6 feet high. *Ginkgo biloba*, a member of Ginkgoales is called living fossil. This is oldest living seed plant and there is great similarity between fossil gymnosperm *Ginkgoites* and *Ginkgo biloba* of today.

The members of Coniferales are of great heights, *e.g.*, *Cedrus, Cupressus, Abies, Picea, Pinus, Taxodium, Taxus, Podocarpus*, etc.

Gnetales include *Welwitschia*, *Gnetum* and *Ephedra* which make link between gymnosperms and angiosperms.

Angiosperms include some families of high woody trees. Angiosperms first appeared in Cretaceous period; they were divided into dicots and monocots. Trees are found in both dicots and monocots. Palms, Dracaenas, Pandanus and Bamboos are included in monocots. Palms show similarity in appearance with cycads. Some of dicots, such as *Polyalhia* trees show similarity with conifers. Similarities between Gnetales and Angiosperms on the basis of number of cotyledons is due to **convergent evolution**. The flowers of Orchidaceae, Zingiberaceae, Polygalaceae and Papilionaceae are convergently adapted in respone to similar methods of pollination.

Living Fossils

There are several living ancient plants which have undergone little or no change through the ages. Although they may have acquired specialized features later, they have more or less retained their original features. Such survivors are called **living fossils.**

These include the maiden-hair tree, *Ginkgo biloba* which has continued to live for over one hundred million years. *Cycas revoluta*, a cycad, and conifers such as Downredwood (*Metasequoia glyptostroboides*), California Big Tree (*Sequoia sempervirens*) are also considered living fossils.

Perhaps consistency of living habitat and isolation of these forms (living fossils) from their competitors can be two factors that may account for their survival.

It is to be noted that a tree starts its life as a sapling growing from a seed as in most angiosperms. Forests are big assemblage of trees and other plants. In tropical rain forests, typically, the forest canopy is composed of three stories characterized by different types of trees. The trees of the top, or dominant, story form a nearly closed canopy which is frequently 60 metres or more

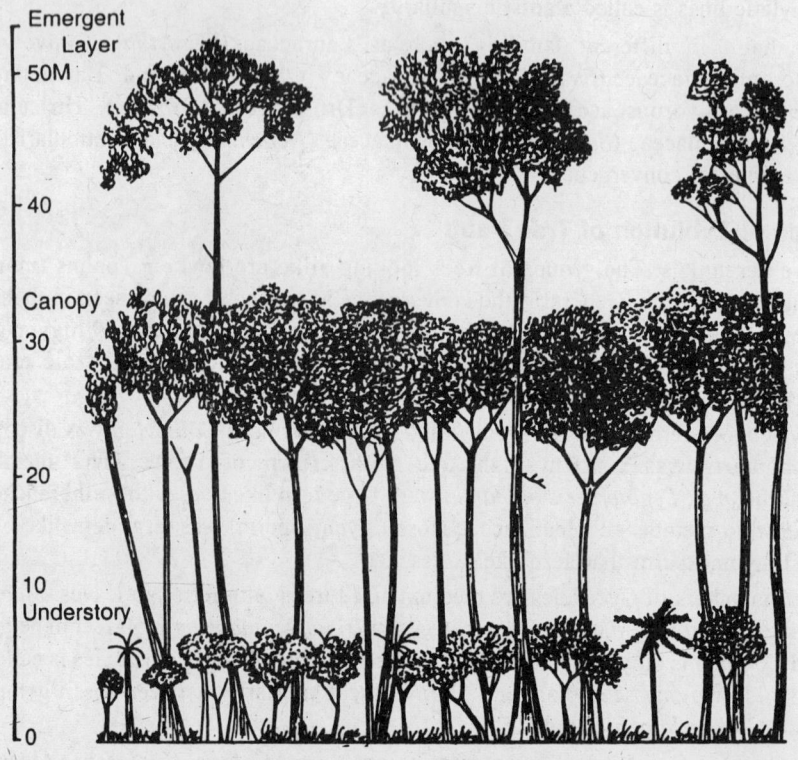

Fig. 26.87. Different strata of canopy

Diversity and Morphology of Flowering Plants

in height. The crowns of the second story are beneath those of the dominant story and like those of the dominant story, frequently form a nearly closed canopy. The trees of the third or lowest story are usually small and slender and have small open crowns. The presence of these three stories of different trees is not usually evident on casual observation for the composition of all the stories is very complex and few of the trees present any striking peculiarities. Moreover, smaller trees of a higher story always occur in a lower story as well as between the different stories, while the different species of a story have different heights. Erect palms are frequently numerous in the lower stories.

Beneath the tree stories there is a ground covering. In dry areas, it may consist largely of woody plants, while in moist situation herbs and ferns are abundant. Some individual tall trees are called **emergents** which protrude above the general level of the **canopy** in tropical rain forests.

There are certain forests which are dominated by single species of trees such as oak (*Quercus*), beech (*Fagus*), pine (*Pinus*) or spruce (*Picea*), etc.

Structural diversity of trees can also be noticed. For example Banyan tree (*Ficus benghalensis* Linn., Family-Moraceae) is an enormous tree, 70 to 100 feet high, sending down roots from the branches, which enter the ground and form trunks, thus extending the growth of the tree indefinitely. Banyan is a common tree planted near temples and shrines, and an open ground near villages. The roots of trees reaching ground act as pillars to support the branches. The trees have the biggest crowns of any in the world and with their aerial prop roots, a single tree looks like a small grove. The Banyan tree of botanical garden, Kolkata, has about 3000 trunks of which more than 2000 are 2 to 3 feet in diameter. About ten thousand people can get shelter under this tree at a time.

Fig. 26.88. The Banyan tree. Botanical garden Kolkata.

There are several other trees, *e.g.*, Pipal tree (*Ficus religiosa*), Shisham (*Dalbergia latifolia*), Indian coral tree (*Erythrina indica*), Mahua tree (*Madhuca indica*), Jamun tree (*Syzygium cumini*), Tamarind tree (*Tamarindus indica*), Silk cotton tree (*Bombax ceiba*), etc., which are commonly found in north Indian plains and peninsular India.

In strict sense, all trees which are not cone bearing are flowering trees (angiosperms) while cone bearing trees are gymnosperms, *e.g.*, *Pinus, Cedrus, Cupressus, Picea, Abies, Araucaria, Cryptomeria, Ginkgo* and Cycads.

There are several species of palms (Monocots: family palmaceae), such as oil palm (*Elaeis guineensis*), date palm (*Phoenix dactylifera*), coconut palm (*Cocos nucifera*), palmyra palm (*Borassus flabellifer*), betelnut palm (*Areca catechu*), Toddy palm (*Caryota urens*), etc. All palms have unbranched stems and crown of leaves on their tops.

The Baobab tree (*Adansonia digitata* Linn., Family–Bombacaceae) is known in India for many centuries, particularly along the western coast of this country. This is a strange looking tree. The great size attained by the trunk makes it strange looking. A soft-wooded tree, its trunk is about thickest in the world, suddenly tapering into thick branches.

In those parts of tropical Africa where the tree is indigenous, the trunk reaches 10 metres or more in diameter and this makes the trunk about 30 m in girth. This is an erect tree, seldom going over 20 m high.

Fig. 26.89. The Baobab tree (*Adansonia digitata*)

Oldest habit of seed plants : Tree habit is oldest habit of seed plants. They are perennial in habit living for many years. Due to secondary growth in dicots and gymnosperms they increase in thickness or girth.

Diversity and Morphology of Flowering Plants

California Coastal – redwood (*Sequoia sempervirens*) is considered a living fossil, the tallest tree of world, 112.6 m high. California Big Tree (*Sequoiadendron giganteum*) is also considered a living fossil, popularly known as 'general sherman' of 83.82 m height and 24 m girth or circumference.

The tree with greatest girth is sweet chestnut (*Castanea sativa*) of Mount Etna, Sicily. The tree is about 4000 years old. The girth or circumference is about 60 m or more.

Taxodium distichum var. *mucronatum* is also a tree with large girth of about 40 m, situated in St. Maria, Oaxaca.

Abies nobilis (noble fir) is another big tree about 84.73 m high and situated in Gifford Natural Forest, Washington.

Among angiosperms *Eucalyptus regnans* is tallest tree. The tallest tree, 150 m high is situated in Victoria, Australia.

Trees: Longest-lived organisms. Trees are longest-lived organisms (plants). Some of long lived trees are mentioned here : *Shorea laprosula*, a tree of Dipterocarpaceae is more than 250 years old. Average age of *Parashorēa* spp., is considered about 300 years. Most trees of tropical rain forests are 100 years old or so. Generally gymnosperms, such as *Sequoiadendron giganteum* are long-lived having average age of 3000 years.

Fig. 26.90. *Dracaena draco* of Sumatra.

Betula spp., (Birch tree) about 100–200 yrs; *Acer rubrum* (red maple) 100 years or more; *Fraxinus* spp., (Ash tree) about 150 years ; *Quercus* spp., (Oak tree), 800 to 1000 years; *Ficus benghalensis* (Banyan tree) 500 years or more ; *Sequoia sempervirens* (California Coastal redwood) more than 2000 years ; *Sequoiadendron giganteum* (California Big Tree) more than 3000 years ; *Pinus aristata* (Bristle cone fern) more than 4500 years; *Taxus baccata* (Yew) more than 5000 years ; Dragon trees (*Dracaena draco*) of Sumatra, more than 10,000 years of age.

Taxus baccata (Yew tree) of Scotoland is supposed to be of 5000 years age with a girth of about 17 metres. English Yew (*Taxus baccata*) lives upto 1000 years. English Oak (*Quercus ruber*) is also a long-lived tree and lives upto 400–500 years. Gymnosperms, such as coast and mountain redwoods of Western United States, some conifers and dragon trees of Sumatra are well establised long-lived trees.

In some trees, during process of ageing in natural life cycle the branches and leaves fall off. For example in Dougluss fir or Xmas tree (*Pseudotsuga mengies*) The leaves and branches fall off as shown in fig. 33.5. A, 8 year old tree; B, 30-40 years old; C, 50-60 years; D, 100 year old tree and E, 200 year mature tree, which may attain a height of 325 feet; F – G, after about 400 years where branches and leaves fall off and H, shows toppled down tree.

Fig. 26.91. Ageing of Dougllus fir tree (*Pseudotsuga menzies*).

Diversity and Morphology of Flowering Plants

Different Kinds of Trees

On the basis of their leaves, trees may be divided into three important categories. They are as follows :

1. Conifers. They possess needle and scale like leaves, *e.g., Pinus, Cedrus, Cupressus,* etc.

2. Broad-leaved trees. They have flattened simple or compound leaves. Most of Dicotyledons, such as *Bombax, Ficus, Butea,* etc.

3. Monocotyledonous trees. They are palms with crown-leaves, feathery in appearance, *e.g., Borassus, Roystenia, Phoenix,* etc.

Today broad leaved trees are dominant in tropical rain forests. Palms, generally occur in coastal regions. They are supposed to be ancient plants.

Conifers make the gymnospermous trees of today. However, they were dominant 300 million years ago but today they are found only in colder regions. Cycadales and Ginkgoales were dominant in Mesozoic era. Today *Cycas revoluta* and *Ginkgo biloba* are typical examples of living fossils.

Tree ferns are now rare and most of them are found in fossilised forms.

Tree cacti are common plants of deserts.

Glossary

Abaxial, That surface of any structure which is remote or turned away from the axis.

Absciss. Layer of meristematic cells just outside cork layer, to whom fall of leaves, floral parts, fruits and certain branches is due.

Abscission. The separation of parts.

Abscission layer. Layer at base of leaf stalk in woody dicotyledons and gymnosperms, in which the parenchyma cells become separated from one another through dissolution of middle lamella before leaf-fall.

Accessory bud. An additional axillary bud; a bud formed on a leaf.

Accessory cells. Auxiliary cells.

Acicular. Like a needle in shape; sharp pointed.

Acrogenous. Increasing in growth at summit or apex.

Acropetal. Development of organs in succession towards apex, the oldest at base, youngest at tip, *e.g.*, leaves on a shoot.

Acuminate. Drawn out into long point; tapering; pointed.

Adaxial. Turned towards the axis.

Adenine. A compound occurring in many cells; $C_5H_5N_5$.

Adventitious. Tissues and organs arising in abnormal positions; secondary.

Aerenchyma. Cortex of submerged roots of certain swamp plants; aerating cortical tissue in floating portions of some aquatic plants.

Aerial. Roots growing above ground, from stems.

Aerophyte. A plant growing attached to an aerial portion of another plant; epiphyte.

Agranular. Without granules.

Air-cells. Air spaces in plant tissue.

Alburnum. Sap-wood or splint-wood, soft white substance between inner bark and true wood; outer young wood of dicotyledon.

Aleurone. Protein grains found in general protoplasm, and used as reserved food material; aleurone.

Aleuroplast. Colourless plastid, storing protein.

Alkaloid. Basic nitrogenous organic substance with poisonous or medicinal properties produced in certain plant species as caffeine, morphine, nicotine, etc.

Ameiosis. Occurrence of only one division in meiosis instead of two.

Ameiotic. Parthenogenesis in which meiosis is suppressed.

Glossary

Amino acids. Compounds containing amino (NH_2) and carboxyl (COOH) groups, constituents of proteins, synthesized in autotrophic organisms.

Amitosis. Direct cell-division and cleavage of nucleus without thread-like formation of nuclear material.

Amphicribral. Amphiphloic.

Amphicribral bundle. Concentric bundles are amphicribral, when phloem surrounds xylem, *e.g.*, in some ferns.

Amphiphloic. With phloem both peripheral and central to xylem.

Amphivasal. With primary xylem surrounding or on two sides of centric phloem, vascular bundle; amphixylic, perixylic. Opp. amphicribral, amphiphloic, periphloic.

Amphivasal bundle. Concentric bundles are amphivasal, when xylem surrounds phloem, *e.g.*, in *Dracaena*.

Amyloplast. A leucoplast of colourless starch forming granule in plants; amyloplastid.

Amylum. Vegetable starch.

Anaphase. A stage in mitosis during divergence of daughter chromosomes; the stages of mitosis upto division of chromatin into chromosomes; kataphase.

Anatomy. The science which treats of the structure of plants and of animals, as determined by dissection.

Angienchyma. Vascular tissue.

Angiosperm. Having seeds in a closed case, the ovary.

Angular. Collenchyma with cell-walls thickened in the angles of the cells.

Annual ring. One of the rings, seen in transverse sections of dicotyledons, indicating the secondary growth during a year.

Annular. Ring-like, certain vessels in xylem, owing to ring-like thickenings in their interior.

Anomaly. Any departure from type characteristics.

Anthocyanin. One of the blue, reddish or violet pigments of flowers, leaves, fruits and stems.

Anticlinal. Line of division of cells at right angles to surface of apex of a growing point.

Apical. At tip; cell at tip of growing point; meristem.

Apical meristem. Growing point (zone of cell division) at tip of root and stem in vascular plants, having its origin in a single cell (initial), *e.g.*, Pteridophyta, or in a group of cells (initials), *e.g.*, Spermatophyta.

Apposition. The formation of successive layers in growth of a cell wall.

Aquatic. Living in water. An aquatic plant.

Arachnoid. Consisting of fine entangled hairs.

Artefact. An appearance, or apparent structure, due to preparation and not natural.

Articulate. Jointed; articulated.

Astrosclereid. A multiradiate sclereid or stone cell.

Autoplast. Chloroplast.

Axial. Axis or stem.

Axillary. Growing in axil, as buds.

Axis. The main stem or central cylinder.

Bark. The tissues external to the vascular cambium, collectively; phloem, cortex and periderm; outer dead tissues and cork.

Bast. The inner fibrous bark of certain trees.

Bicollateral. Having the two sides similar, *e.g.*, a vascular bundle with phloem on both sides of xylem, as in Cucurbitaceae and Solanaceae.

Bifacial. Leaves with distinct upper and lower surfaces; dorsiventral.

Bilateral. Having two sides symmetrical about an axis.

Blind pit. A cell wall pit which is not backed by a complementary pit.

Biometry. Application of mathematics to the study of living things.

Bordered pit. A form of pit developed on walls of tracheids and wood-vessels, with overarching border of secondary cell-wall.

Botany. The branch of biology dealing with plants; phytology.

Brachysclereids. A stone-cell.

Branch gaps. Gaps in the vascular cylinder of a main stem, subtending branch traces.

Branch traces. The vascular bundles connecting those of a main stem to those of a branch.

Brownian movements. (R. Brown, Scottish botanist). The passive vibratory movements of fine particles when suspended in a fluid.

Bud. A rudimentary shoot, or flower.

Bulbous. Like a bulb.

Bulliform. Thin-walled cells which cause rolling, folding or opening of leaves by turgor changes.

Bundle-sheath. A layer of large parenchymatous cells surrounding vascular tissue of leaf-vein.

Callose. An occasional carbohydrate or periodic component of plant cell walls; as on sieve-plates.

Callus. Tissue that forms over cut or damaged plant surface; deposit of callose on sieve-plates.

Cambial. Pertaining cambium.

Cambiform. Similar to cambium cells.

Cambium. The tissue from which secondary growth arises in stem and roots.

Canada balsam. Gum commonly used, dissolved in xylene, for making permanent microscopical preparations. The object is placed in a thin layer of balsam solution between cover-slip and slide. The balsam dries hard, and because its refractive index is like that of proteins and other constituents of biological objects, it makes them very transparent.

Capitate. Swollen at tip.

Carbohydrates. Compounds of carbon, hydrogen and oxygen, aldehydes or ketones constituting sugars, or condensation products thereof.

Carotene. A yellow pigment synthesized by plants.

Carotenoids. Pigments occurring in plants, including carotenes xanthophylls and other fat-soluble pigments.

Casparian band. (R. Caspary. German botanist). A cork or wood-like strip encircling radial walls of endodermis cells; Casparian strip.

Cauline. Vascular bundles not passing into leaves.

Cell. A small cavity or hollow; a unit mass of protoplasm, usually containing a nucleus or nuclear material; originally, the cell-wall.

Cell Division. Division of cell, both cytoplasm and nucleus, into two.

Cell lineage. Developmental history in terms of descent by cell division of later cells from earlier cells. The cell lineage of an organ traces the succession of cells, from the zygote onwards, which culminates in the group of cells constituting that organ.

Glossary

Cell membrane. A bimolecular layer of lipoids and proteins enveloping the protoplasm of a cell; plasma membrane; ectoplast and tonoplast of a plant cell.

Cell plate. Equatorial thickening of spindle fibres from which partition wall arises during division of plant cells.

Cell sap. Fluid in vacuoles of plant cell.

Cell theory. Theory, initiated by Schleiden and Schwann, 1838-39; that all animals and plants are made up of cells and their products and that growth and reproduction are fundamentally due to division of cells.

Cellular. Consisting of cells.

Cellulose. A carbohydrate forming main part of plant cell-walls. $(C_6H_{10}O_5)x$.

Cell-wall. Investing portion of cell.

Cement. A uniting substance, as between cells or animals.

Central body. Centrosome.

Central cylinder. Stele.

Centrifugal. Turning or turned away from centre of plants or axis.

Centriole. The central particle of the centrosome; the centrosome itself.

Centripetal. Turning or turned towards centre or axis.

Centromere. The part of the chromosome located at the point lying on the equator of the spindle at metaphase, and dividing at anaphase controlling chromosome activity.

Centrosome. A cell-organ, the centre of dynamic activity in mitosis, consisting of centriole and attraction-sphere.

Chitin. Nitrogen-containing polysaccharide with long fibrous molecules, forming material of considerable mechanical strength and resistance to chemicals. Present in cuticle.

Chlorenchyma. Tissues collectively, or stem tissue, or mesophyll, containing chlorophyll.

Chlorophyll. The green colouring matter found in plants and in some animals.

Chloroplast. A minute granule or plastid containing chlorophylls a and b found in plant cells exposed to light.

Chloroplast pigment. Chlorophylls, carotene, and xanthophyll.

Chondriosomes. Mitochondria, chondriomites, chondrioconts, chondriospheres, chondrioplast.

Chromatid. A component of tetrad in meiosis; a half chromosome between early prophase and metaphase in mitosis, or between diplotene and second metaphase in meiosis.

Chromatin. A substance in the nucleus which contains nucleic acid, proteids, and stains with basic dyes.

Chromatophore. A coloured plastid of plants and animals; a colourless body in cytoplasm and developing into a leucoplast, chloroplast or chromoplast.

Chromomere. One of the chromatin granules of which a chromosome is formed and which corresponds to a gene.

Chromoplast. A coloured plastid or pigment body; coloured plastid other than a chloroplast; chromoplastid.

Chromosome. One of deeply straining bodies, the number of which is constant for the cells of a species, into which the chromatin resolves itself during karyokinesis, and meiosis.

Cluster-crystals. Globular aggregates of calcium oxalate crystals in plant cells; sphaeraphides.

Collateral. Side by side; bundles with xylem and phloem in the same radius.

Collenchyma. Parenchymatous peripheral supporting tissue with cells more or less elongated and thickened, either at the angles or on walls adjoining intercellular spaces, or tangentially.

Colloid. A gelatinous substance which does not readily diffuse through an animal or vegetable membrane.

Columnar. Cells longer than broad.

Companion cell. A narrow cell, retaining its nucleus, derived, from a cell giving rise also to a sieve-tube element, in phloem of angiosperms.

Complementary. Non-suberized cells loosely arranged in cork tissue and forming air passages.

Concentric. Having a common centre.

Concentric bundle. The vascular bundle, with one tissue surrounding the other, *i.e.*, amphicribral and amphivasal.

Conjunctive. Tissue, mesocycle and pericycle in a stele.

Cork. A tissue derived usually from outer layer of cortex in woody plants.

Cork-cambium. Phellogen.

Corpus. Body; core of apical meristem within the tunica.

Cortex. The extrastelar fundamental tissue of the sporophyte.

Cortical. Pertaining the cortex.

Cristae. Folds of the inner membrane of a mitochondrion.

Crystalloids. A protein crystal found in certain plant cells.

Crystal sand. A deposit of minute crystals of calcium oxalate, as in Solanaceae.

Cuticle. An outer skin or pellicle.

Cuticularisation. Cutinisation in external layers of epidermal cells.

Cutin. A mixture of substances associated with cellulose, found in external layers of thickened epidermal cells of plants.

Cutinisation. The deposition of cutin in cell-wall, thereby forming cuticle.

Cyanin. The blue pigment.

Cystolith. A mass of calcium carbonate occasionally of silica, formed on in growths of epidermal cell walls in some plants.

Cytokinesis. In a dividing cell, division of cytoplasm as distinct from division of nucleus.

Cytology. Study of cells.

Cytoplasm. Substance of cell-body exclusive of nucleus.

Cytosine. A cleavage product of nucleic acids, $C_4H_6N_3O$.

Deoxyribose nucleic acid, DNA. Stable nucleic acid component of kinetoplasts, chromosomes, bacterial cells and phages, which consists structurally of two spirals linked transversely and constitutes a pattern or template for replication.

Dermatogen. The young or embryonic epidermis in plants.

Dextrose. Grape sugar or glucose, the end product of starch digestion, $C_6H_{12}O_6$.

Diakinesis. The later prophase stage of meiosis, between diplotene and prometophase; movement of chromosomes between metaphase and telophase.

Diarch. With two xylem and two phloem bundles; root in which protoxylem bundles meet and form a plate of tissue cross cylinder with phloem bundle on each side.

Diastase. An enzyme which acts principally in converting starch into sugar.

Glossary

Dicotyledon. A plant with two seed-leaves.

Dictyosome. Unit of Golgi apparatus several of which occur as discrete bodies in cells of plants.

Dictyostele. Amphiphloic siphonostele that is broken up by crowded leaf gaps into a net work of distinct vascular strands (meristeles), each surrounded by a endodermis. Present in stems of certain ferns.

Diploid. Having the 2n number of chromosomes.

Diplotene. Stage in meiosis at which bivalent chromosomes split longitudinally.

Discoid. Flat and circular; disc-shaped.

DNA. Deoxyribonucleic acid, a compound consisting of a large number of nucleotides attached together in single file to form a long strand.

Dorsiventral. With upper and lower surfaces distinct; bifacial.

Duct. Any tube which conveys fluid or other substance.

Duplication. A translocated chromosome fragment attached to one of normal set.

Dynamic. Producing or manifesting activity.

Ecology. Study of the relations of plants, particularly of plant communities, to their surroundings.

Ectoplast. The protoplasmic film or plasma-membrane just within the true wall of a cell.

Elaioplast. A plastid in a plant cell, which forms or helps to form oil globules.

Embryo. A young organism in early stage of development.

Emergence. An outgrowth from sub-epidermal tissue.

Endarch. With central protoxylem, or with several surrounding a central pith.

Endodermis. Innermost layer of cortex in plants; layer surrounding pericycle.

Endoplasmic reticulum (ER). Complex mesh work of tubular channels, often extended into slit like cavities (*cisternae*) together with more or less flattened vesicles, all bounded by unit membranes, occurring in the cytoplasm of many eucaryotic cells; usually only visible by electron microscopy.

Enzyme. A catalyst produced by living organism and acting on one or more specific substrates.

Epiblema. The outermost layer of root tissue; piliferous layer.

Epicotyl. That portion of seedling stem above cotyledons.

Epidermis. The outermost protective layer of stems, roots and leaves.

Epiphyte. Plant which lives on surface of other plants.

Epithelium. Layer of cells living schizogenously formed secretory canals and cavities, *e.g.*, in resin canals of pine.

Ergastic. Life-less cell inclusions, as fat, starch.

Ergastoplasm. Endoplasmic reticulum.

Essential oils. Volatile oils, composed of various constituents and contained in plant organs, with characteristic odour.

Eumeristem. Meristem composed of isodiametric thin-walled cells.

Eumitosis. Typical mitosis.

Exarch. With protoxylem strands outside metaxylem, or in touch with pericycle.

Excentric. One sided.

Exodermis. A specialized layer below the piliferous layer.

Extracellular. Occurring outside the cell.

Extracortical. Not within the cortex.

Extranuclear. Situated outside the nucleus.

Extrastelar. Occurring outside the stele.

Extraxylary. On the outside of the xylem.

Fibre. Elongated plant cell for mechanical strength.

Fibre tracheids. Fibres of a nature intermediate between that of libriform fibres and of tracheids.

Fibrous. Composed of fibres.

Fibrovascular. Bundle of vascular tissue surrounded by non-vascular fibrous tissues.

Filament. A thread-like structure.

Flower. Specialized reproductive shoot, consisting of an axis (*receptacle*), on which are inserted four different sorts of organs.

Fructose. Fruit sugar.

Gland. Superficial discharging secretion externally, *e.g.*, glandular hair, nectary, hydathode; or embedded in tissue, occurring as isolated cells containing the secretion, or as layer of cells surrounding intercellular space (secretary cavity) into which secretion is discharged, *e.g.*, resin canal of pine.

Gland cell. An isolated secreting cell.

Glandular. With glands.

Glandular tissue. Tissue of single or massed cells, parenchymatous and filled with granular protoplasm, adapted for secretion of aromatic substances in plants.

Globoid. A spherical body in aleurone grains, a double phosphate of calcium and magnesium.

Globose. Spherical.

Globule. Any minute spherical structure.

Glucose. The grape sugar, dextrose.

Glycogen. A carbohydrate storage product of plants.

Golgi apparatus or complex. (C. Golgi, Italian, histologist). Cell-constituents, located or diffused, often consisting of separate elements, the Golgi bodies, batonettes, dictyosomes or pseudochromosomes, containing lipoprotein, and concerned with cellular synthesis and secretion.

Grana. Minute particles consisting of a pile of thin double platelets, probably containing chlorophyll, in chloroplasts.

Granum. Singular of grana.

Ground tissue. Conjunctive parenchyma.

Growing point. A part of plant body at which cell-division is localised, generally terminal and composed of meristematic cells.

Guanine. A purine base found in some plants; $C_6H_5ON_5$.

Guard cells. Two cells surrounding stomata of aerial epidermis of plant tissue.

Gum. An exudation of certain plants and trees; vegetable mucilage.

Hair. Any epidermal filamentous outgrowth consisting of one or more cells, varied in shape.

Halophyte. A shoreplant; plant capable of thriving on salt impregnated soils.

Haploid. Having a single set of unpaired chromosomes in each nucleus.

Haustorium. Specialized organ of parasitic plant, *e.g.*, *Cuscuta*, which penetrates into and withdraws food material from tissues of host plants.

Heart-wood. The darker, harder, central wood of trees; duramen.

Helix. A spiral.

Hemicellulose. One of several polysaccharides, chemically unrelated to cellulose, occurring as cell-wall constituents in cotyledons, endosperms, and woody tissues, and serving as reserve food.

Herb. Plant with no persistent parts above ground, as distinct from shrubs and trees.

Herbaceous. Being a herb.

Hereditary. Transmissible from parent to offspring, as characteristics.

Hexarch. Having six radiating vascular strands; *e.g.*, roots.

Hilum. Nucleus of starch grain.

Hinge-cells. Large epidermal cells which, by changes in turgor, control rolling and unrolling of a leaf.

Histogens. Tissue producing zones or layers, plerome, periblem, dermatogen, and calyptrogen.

Histology. The science which treats of the detailed structure of animal or plant tissues; microscopic morphology.

Histone. A basic protein constituent of cell nuclei.

Homologous. Resembling in structure and origin; chromosomes with the same sequence of genes.

Hormones. Substances normally produced in cells and necessary for the proper functioning of other distant cells to which they are conveyed and of the body as a whole.

Hydathode. An epidermal structure specialized for secretion, or for exudation of water, water stoma.

Hydrophyte. An aquatic plant.

Hypocotyl. Part of seedling stem below cotyledons.

Hypodermis. The cellular layer lying beneath the epidermis.

Idioblast. Plant cell containing oil, gum, calcium carbonate or other product and which differs from the surrounding parenchyma.

Intercalary. (Of a meristem) situated between regions of permanent tissue, *e.g.*, at base of nodes and leaves in many monocotyledons.

Intercellular. Among or between cells.

Interfascicular. Situated between the vascular bundles.

Internode. The part between two successive nodes or joints, as of plant stem.

Interphase. Resting stage between first and second mitotic divisions; interkinesis.

Interxylary. Between xylem strands; interxylary phloem.

Intracellular. Within a cell.

Intrafasicular. Within a vascular bundle.

Intrastelar. Within the stele of a stem or root; ground tissue, bundles.

Intraxylary. Within wood or xylem.

Inulin. A carbohydrate occurring in rhizomes and roots of many plants; dahlia starch; $(C_6H_{10}O_5)x$.

Invertase. A plant enzyme which converts cane sugar into dextrose and laevulose.

Irritability. *e.g.*, protoplasmic movements. A universal property of living things.

Isobilateral. A form of bilateral symmetry where a structure is divisible in two planes at right angles.

Isodiametric. Having equal diameters; cells or other structures.

Isolateral. Having equal sides; leaves with palisade tissue on both sides.

Karyokinesis. Indirect cell-division; mitosis.

Karyolymph. Nuclear sap.

Kataphase. The stages of mitosis from formation of chromosomes to division of cell; anaphase.

Lacunate. Collenchyma with cell-walls thickened where bordering intercellular spaces.

Lamella. Any thin plate.

Latex. A milky, or clear, sometimes coloured, juice or emulsion of diverse composition found in some plants, as in spurges, rubber trees.

Laticifer. Any latex containing cell, series of cells or duct.

Laticiferous. Conveying latex; cells, tissue, vessels.

Leaf-gap. Gap in vascular cylinder of stem, a parenchymatous region associated with leaf-traces.

Leaf trace. Scar making position where a leaf was formerly attached to the stem.

Leaf scar. Vascular bundles extending from stem bundles of leaf base.

Lenticel. Ventilating pore in angiosperms stems or roots.

Leptotene. Stage in early prophase of first division of meiosis.

Leucoplastids. Colourless plastids from which amylo-, chloro- and chromoplastids arise.

Leucoplasts. Colourless granules of plant cytoplasm.

Lianes. Climbing plants found in tropical forests, with long woody rope like stems of anomalous anatomical structure.

Libriform. Resembling bast.

Lignification. Wood formation; thickening of plant cell-walls by deposition of lignin.

Lignin. A complex substance which, associated with cellulose, causes the thickening of plant cell-walls, and so forms wood.

Lipoid. Resembling a fatty substance.

Lumen. Central cavity of a plant cell.

Lysosomes. Particle in cytoplasm, smaller than mitochondria, consisting of a membrane enclosing several enzymes.

Macerate. To wear away or to isolate parts of a tissue or organ.

Macrosclereids. Relatively large columnar sclereids, as in coat of certain seeds.

Matrix. Ground substance of connective tissue.

Medullary. In region of medulla.

Medullary rays. A number of strands of connective tissue extending between pith and pericycle.

Meiosis. Process of reduction division of germ-cell chromosomes from diploid to haploid number at maturation.

Meristele. See *dictyostele*.

Meristem. Tissue formed of cells all capable of diversification, as found at growing points; meristematic tissue.

Meristematic. Consisting of meristem; tissue cells of growing point.

Mesarch. Xylem having metaxylem developing in all directions from the protoxylem, characteristic of ferns.

Mesophyll. The internal parenchyma of a leaf.

Glossary

Mesophyte. A plant thriving in temperate climate with normal amount of moisture.

Messenger RNA. RNA molecule that conveys from the DNA the information that is to be translated into the structure of a particular polypeptide molecule.

Metabolic. Chemical change, constructive and destructive, occurring in living organism.

Metaphase. The stage in mitosis or meiosis in which chromosomes are split up in equatorial plate.

Metaphloem. The phloem of secondary xylem.

Metaxylem. Secondary xylem with many thick walled cells.

Micron. One thousandth part of a millimetre symbol: μ

Microscope. The ordinary microscope of the laboratory is a compound microscope with two sets of lenses (objective and eye-piece) which magnify the object in two steps.

Microtome. Machine for cutting extremely thin sections of tissue (usually 3 to 20 μ).

Microtomy. The cutting of thin sections of objects, as of tissues, or cells, in preparing specimens for microscopic or ultramicroscopic examination.

Middle lamella. The layer derived from the cell plate, and covered on both sides by cellulose in formation of the wall of a plant cell.

Mid-rib. The large central vein of a leaf, constitution of the petiole.

Mitochondria. Granular, rod shaped, or filamentous self-replicating organellae in cytoplasm, consisting of an outer and inner membrane containing phosphates and numerous enzymes, varying in different tissues and functioning in cell respiration and nutrition; chondriosomes.

Mitochondrion. Singular of mitochondria.

Mitosis. Indirect or karyokinetic nuclear division, with chromosome formation, spindle formation, with or without centrosome activity.

Mitotic. Produced by mitosis.

Monocotyledonous. Having one cotyledon.

Morphogenesis. The development of shape; origin and development of organs or parts of organisms.

Morphology. The science of form and structure of plants and animals, as distinct from consideration of functions.

Mucilage. A substance of varying composition, hard when dry, swelling and slimy when moist, produced in cell-walls of certain plants.

Mucilaginous. Composed in mucilage.

Mycorrhiza. Association of fungal mycelium with roots of a higher plant.

Nectar. Sweet substance secreted by special glands, nectaries, in flowers and certain leaves.

Nectary. A group of modified subepidermal cells of no definite position in a flower, less commonly in leaves, secreting nectar; a nectar gland.

Nodal. Pertaining a node or nodes.

Node. The knob or joint of a stem at which leaves arise.

Nuclear membrane. Delicate membrane bounding a nucleus, formed for surrounding cytoplasm.

Nuclear sap. Karyolymph.

Nuclei. Plural of nucleus.

Nucleic. Acids containing phosphorus, found in nuclei of cells.

Nucleolus. A dense rounded mass in a cell-nucleus, consisting of protein and ribonucleic acid granules, and functioning in RNA and protein synthesis controlled by a special region or nucleolar organiser in the chromosome; plasmosome or a karyosome.

Nucleoplasm. Reticular nuclear substance; karyoplasm.

Nucleoprotein. A compound of protein and nucleic acid, a constituent of cell nuclei.

Nucleotide. Compound formed from sugar (with 5 carbon atoms) phosphoric acid, and a nitrogen-containing base (purine or pyrimidine).

Nucleus. Complex spheroidal mass essential to life of most cells.

Obtuse. With blunt or rounded end.

Oil gland. A gland which secretes oil.

Oil immersion objective. Objective of light microscope, space between which and the cover slip over the object examined is filled with drop of oil of same refractive index as glass; system used for highest magnification with the light microscope.

Ontogeny. The whole course of development during an individual's life-history.

Organ. Multicellular part of a plant which forms a structural and functional unit, *e.g.*, leaf.

Organelle. A persistent structure with specialized function forming part of a cell, *e.g.*, a mitochondrion.

Osteosclereid. A sclereid with both ends knobbed.

Oxalates. Salts of oxalic acid, occurring as metabolic by-products in various plant tissues.

Pachytene. Stage in prophase of first division of meiosis, follow zygotene.

Palisade. Tissue, the layer or layers of photosynthetic cells beneath the epidermis of many foliage leaves.

Papain. A proteolytic enzyme in fruit juice of the tree *Carica papaya*.

Parasite. An organism living with or within another to its own advantage in food or shelter.

Parenchyma. Plant tissue, generally soft and of thin-walled relatively undifferentiated cells, which may vary in structure and function as pith, or mesophyll, etc.

Parenchymatous. Pertaining or found in parenchyma; a kind of cell.

Passage cells. Thin-walled endodermal or exodermal cells of root, which permit passage of solutions.

Pectic. Substances in cell walls and cell sap of plants, including pectic acid and its salts, pectin and pectose; enzymes; pectosinase, pectase and pectinase, which hydrolyse pectic substances.

Pectose. A carbohydrate constituent of plant cell-walls, converted into pectin and cellulose by action of pectosinase.

Pedicel. Stalk of individual flowers of an inflorescence.

Pentarch. With five alternating xylem and phloem groups.

Periblem. Layers of ground or fundamental tissue between dermatogen and plerome of growing points.

Periclinal. System of cells parallel to surface of apex of a growing point.

Pericycle. The external layer of stele, the layer between endodermis and conducting tissues.

Periderm. The outer layer of bark; phellogen, phellem and phelloderm collectively; epiphloem.

Permanent tissue. Tissue consisting of cells which have completed their period of growth and subsequently change little until they lose their protoplasm and die.

Glossary

Permeability. Of membrane, extent to which molecules of a given kind can pass through it.
Phellem. Cork; cork and non-suberized layers forming external zone of periderm; phellem.
Phelloderm. The secondary paranchymatous suberous cortex of trees, formed on inner side of cork cambium.
Phellogen. The cork-cambium of tree stems arising as a secondary meristerm and giving rise to cork and phelloderm.
Phloem. Bast tissue; the soft bast to vascular bundles, consisting of sieve-tube tissue.
Phloem parenchyma. Thin-walled parenchyma associated with sieve tubes of phloem.
Phloem sheath. Pericycle, together with inner layer of a bundle sheath later consists of two layers.
Phloic. Pertaining to phloem.
Photosynthesis. In green plants, synthesis of organic compounds from water and carbon dioxide using energy absorbed by chlorophyll from sunlight.
Phylloclade. A green flattened or rounded stem functioning a leaf as in *Cactus;* flattened axillary bud, as in *Ruscus;* phyllocladium, cladode, cladophyll.
Phyllode. Winged petiole with flattened surfaces placed laterally to stem, functioning as leaf.
Phylogeny. Evolutionary history.
Physiology. Study of the processes which go on in living organism.
Pigment. Colouring matter in plants.
Pigment cell. A chromatophore.
Piliferous. Bearing or producing hair; outermost layer of root or epiblema which gives rise to root hairs.
Pit. A depression formed in course of cell-wall thickening of plant tissue.
Pit chamber. The cavity of a bordered pit below the overarching border.
Pit-fields. Areas of depressions in primary cell-walls.
Pith. Central part of an organ of plant; central core of usually parenchymatous tissue in those stems in which the vascular tissue is in the form of a cylinder.
Pit membrane. Middle lamella of plant cell-wall forming floor of pits of adjacent cells.
Plasmalemma. Plasma membrane (external plasma membrane in plants).
Plasma membrane. The membrane forming the surface of cytoplasm and consisting of a bimolecular phospholipid layer between an inner and outer layer of protein molecules.
Plasmodesma. Cytoplasmic threads penetrating cell wall and forming intercellular bridge; plural plasmodesmata.
Plastid. A cell-body other than nucleus or centrosome.
Plastidome. In a cell, the plastids as a whole, cytoplasmic inclusions which give rise to plastids.
Plastochondria. Mitochondria..
Plerome. The core or central part of an apical meristem.
Polyarch. Having many protoxylem bundles.
Pore. A minute opening or passage, as of sieve plates, stomata, etc.
Primary. First; principal; original.
Primary meristem. Ground meristem, procambium and protoderm.
Procambium. The tissue from which vascular bundles are developed.
Promeristem. Meristem of growing point, and primary meristem.
Prometaphase. Stage between prophase and metaphase in mitosis and meiosis.

Prophase. The preparatory changes, the first stage in mitosis, or in meiosis.

Proplastid. An immature plastid, as in meristematic cells.

Protein. A nitrogenous compound of cell protoplasm; a complex substance characteristic of living matter and consisting of aggregates of amino-acids, and generally containing sulphur.

Protoderm. The outer cell layer of apical meristem; primordial epidermis of plants; superficial dermatogen.

Protophloem. The first phloem elements of a vascular bundle.

Protoplasm. Living cell substances; cytoplasm and karyoplasm.

Protoplasmic. Pertaining or consisting of protoplasm.

Protoplast. Protoplasm of a plant cell.

Protoxylem. Primary xylem lying next pith of stems.

Pteridophyta. Division of plant kingdom comprising ferns, horse-tails, club-mosses, etc.

Purine. Bases—adenine and guanine.

Pyrimidine. Bases—thymine and cytosine.

Quantasomes. Regularly arranged sub-units observed by electron microscopy in thylakoid lamellae. Estimated each contains about 300 chlorophyll molecules which are believed to function as a photosynthetic units in absorption of light quanta.

Quiescent centre. Inactive or passive region of cells.

Raphides. Needle shaped crystals of calcium oxalate occurring in bundles in certain plant cells.

Radial. Pertaining to radius.

Ray. A parenchymatous band penetrating from cortex towards centre of stem.

Replication. Duplication of a molecule or aggregate by copying from a pre-existing molecule or structure of the same kind of mitochondria, chlorplast, etc.

Reticulate. Like network; thickening of cell-wall.

Rhizome. Underground stem, bearing buds in axils of reduced scale-like leaves.

Rhytidome. The outer bark.

Ribonucleic acid. RNA, a nucleic acid containing adenine, guanine, cytosine and uracil, in nucleolus, mitochondria, and ribosomes, and taking part in cytoplasmic protein synthesis

Ribosomes. Spherical granules or microsomal particles containing ribo-nucleic acid, on nuclear membrane and membranes of endoplasmic reticulum, and taking part in protein synthesis.

RNA. Ribonucleic acid, a molecule consisting of a large number of nucleotides attached together to form a long strand one nucleotide thick. Each nucleotide contains the sugar ribose, and one of the four different bases found in DNA except that uracil replaces thymine.

Root. Descending portion of a plant, fixing it in soil, and absorbing moisture and nutrients.

Root-cap. A protective cap of tissue at apex of root.

Root-hairs. Unicellular epidermal outgrowths from roots, of protective and absorbent function.

Rosette. A cluster of crystals, as in certain plant cells.

Saprophyte. The plant that obtains organic matter in solution from dead and decaying tissues of plants or animals, *e.g., Monotropa.*

Saprophytic. A plant which lives on dead and decaying organic matter.

Sap-wood. The more superficial, softer wood of trees; alburnum.

Scalariform. Ladder-shaped; vessels or tissues having bars like a ladder.

Scalariform thickening. Internal thickening of wall of a xylem vessel or tracheid, in the form of more or less transverse bars, suggestive of ladder rungs.

Glossary

Scape. Leafless flowering stem arising from ground level, *e.g., Canna*.
Schizogenous. Cavity originating by separation of cells, *e.g., Citrus*.
Sclereid. Any cell with a thick lignified wall; a sclerenchymatous cell; a stone cell.
Sclerenchyma. Plant tissue of thickened end of the hard cells or vessels.
Scleroid. Hard.
Sclerotic. Containing lignin.
Secondary. Arising not from growing point, but from other tissue.
Secondary cortex. Phelloderm.
Secondary growth. Development of secondary meristem or cambium producing new tissue on both sides, as in woody ditocyledons.
Secondary meristem. Phellogen, Region of a active cell division.
Secondary wood. Wood formed from cambium.
Sepal. One of the parts forming calyx of dicotyledonous flowers.
Septum. Partition or wall.
Shoot. Stem of a vascular plant derived from the plumule.
Sieve area. Perforated area of cell wall of sieve elements, with groups of pores surrounded by callose.
Sieve cell. A phloem cell having perforated areas of cell-wall; a cell of sieve tubes.
Sieve elements. The conducting parts of phloem; sieve cells and sieve tube cells.
Sieve plate. Part of the wall of a sieve cell, containing simple or compound sieve areas; the perforated and thickened end of a sieve tube cell.
Sieve tubes. Phloem vessels, long slender structures consisting of elongated cells placed ends to end, forming lines of conduction.
Somatic cells (Soma). The cells of an organism, other than the germ cells.
Spermatophyta. Seed plants. Division of plant kingdom providing dominant flora of present day, including most trees, shrubs, herbs, grasses, etc.
Spindle. A structure formed of a chromatin fibres during mitosis.
Spiral. Thickening of cell wall.
Spireme. Thread-like appearance of nuclear chromatin during prophase of mitosis.
Spongy. Parenchyma of mesophyll.
Sporophyte. Phase of life cycle of plants which had diploid nuclei, and during which spores are produced.
Stamen. Organ of flower which produces pollen grains.
Starch. The common carbohydrate formed by plants and stored in seeds. $(C_6H_{10}O_5)x$.
Starch sheath. Endodermis with starch grains.
Stele. A bulky strand or cylinder of vascular tissue contained in stem and root of plants, developed from plerome.
Stellate. Star shaped hair.
Stem. Main axis of a plant.
Steroids. Complex hydrocarbons, chemically similar, occurring in plants and animals.
Sting. Stinging hair and cell.
Stoma. A small orifice; minute openings with guard cells, in epidermis of plants, especially on under surface of leaves, or, the stomatic pores only.
Stomata. Plural of stoma.

Stone cells. Sclerotic cells are rounded sclerenchymatous elements, as found in pear; brachysclereids.
Suberisation. Modification of cell-walls due to suberin formation.
Suberin. The waxy substance developed in a thickened cell-wall, characteristic of cork tissues.
Subsidiary cells. Additional modified epidermal cells lying outside guard cells.
Substomatal. Hypostomatic.
Sucrose. Cane sugar $C_{12}H_{22}O_{11}$.
Sulcate. Furrowed, grooved.
Superficial. On or near the surface.
Tabular. Flattened, as certain cells.
Tannins. Group of astringent substances of wide occurrence in plants, dissolved in cell sap; particularly common in the bark of trees, unripe fruits, leaves and galls. Complex organic compounds containing phenols, hydroxy-acids, or glucosides.
Tap root. Root system with a prominent main root, directing vertically downwards and bearing smaller lateral roots.
Taxonomy. Study of the classification of organisms according to their resemblances and differences.
Telophase. Terminal stage of mitosis or meiosis during which nuclei revert to resting stage.
Terminal. Situated at the end, as terminal bud at end of twig.
Tetrarch. With four protoxylem bundles.
Thylakoid. In photosynthetic organisms, vesicle, wall of which bears photosynthetic pigments. Thylakoids vary in form arrangement in different groups of organisms.
Tissue. The fundamental structure of which animal and plant organs are composed; an organization of like cells.
Tissue culture. A technique for maintaining fragments of animal or plant tissue or separated cells alive after their removal from the organism.
Tonoplast. A vascular membrane; a plastid with distinct vacuole walls, a special form of vacuole-producing plastid.
Torus. Thickened centre of a bordered pit membrane.
Trachea. Spiral or annular vascular tissue of plants; wood vessel.
Tracheid. One of the cells with spiral thickening or bordered pits, conducting water and solutes, and forming wood tissue.
Tracheophyta. Division including all vascular plants (Pteridophyta and Spermatophyta).
Transection. Cross section; transverse section.
Transfer RNA. A relatively small molecule of RNA, whose function is to place the amino acids that will be linked into a polypeptide molecule in the specific sequence specified by a molecule of *Messenger RNA.*
Transfusion Tissue. Tissue of empty cells with pitted and occasionally, internally thickened walls, and protein containing, parenchyma cells, accompanying vascular tissue in leaves of most gymnosperms, lying on either side of the vascular bundles of the single vein.
Triarch. Having three xylem bundles uniting to form the woody tissue of root.
Trichoblast. A cell, of plant epidermis which develops into a root hair.
Trichome. An outgrowth of plant epidermis, either hairs or scales; a hair tuft.
Tunica. Apical meristematic cell giving rise to protoderm.
Tunica corpus concept. An interpretation of the shoot apex which recognizes two tissue zones in the promeristem, *tunica* consisting of one or more peripheral layers, in which planes of cell

Glossary

division are predominantly anticlinal, enclosing *corpus*, or central tissue of irregularly arranged cells in which the planes of cell division vary.

Tylosis. Development of irregular cells in a cell cavity; a cellular intrusion into vessel through pit of parenchyma cells.

Ultrastructure. Structure, at the molecular or electron microscopical level.

Unciform. Shaped like a hook or barbed.

Uncinate. Unciform; hook like.

Unit membrane. The common form, as seen in the electron microscope, of the membrane of the cells: *i.e.*, of the plasma membrane and of the membranes organelles such as mitochondria, nucleus and endoplasmic reticulum.

Vacuole. One of spaces in cell protoplasm containing air, sap, or partially digested food.

Vascular. Containing, or concerning vessels.

Vascular bundle. A group of special cells consisting of two parts, xylem or wood and phloem or bast portion; many have in addition a thin strip of cambium separating the two parts.

Vascular cylinder. Stele.

Vascular plant. Plant possessing vascular system, member of Tracheophyta. (Pteridophyta and Spermatophyta).

Vascular system. Plant tissue consisting mainly of xylem and phloem which forms a continuous system throughout all parts of higher plants. It functions in conduction of water, mineral salts, and synthesized food materials and for mechanical support.

Vascular tissue. Specially modified plant-cells, usually consisting of either tracheae or sieve cells for circulation of sap.

Veins. Strands vascular tissue of leaf.

Velamen. A specialized moisture absorbing tissue.

Ventral. Situated on lower surface.

Vesicle. Small globular or bladder-like air space in tissues.

Vessel. Any tube or canal with properly defined walls.

Water stomata. Pores on surfaces of leaves for excretion of water; hydathodes.

Wax. A substance soluble in fat solvents.

Wood. The hard substance of a tree stem, xylem of vascular bundles.

Wood vessel. An element of tracheal tissue, a long tubular structure formed by cell-fusion.

Xanthophylls. Yellow or brown carotenoid pigments found in plastids.

Xerophyte. A plant growing in desert or alkaline or physiologically dry soil.

Xylary. Pertaining xylem.

Xylem. Woody tissue; lignified portion of vascular bundle.

Xylem-parenchyma. Short lignified cells surrounding vascular cells or produced with other xylem cells toward the end of the growing season.

Xylem-ray. Ray or plate of xylem between two medullary rays.

Zygotene. Stage in prophase of first division of meiosis following leptotene, in which pairing (synapsis) of homologous chromosomes occurs with formation of bivalents.

Further Reading

Carlquist, S. Comparative Plant Anatomy. *Hold, Rinehart and Wintson, New York*. 1961.

Clowes, F.A.L. Apical Meristems. *Blackwell Oxford*. 1961.

Clowes, F.A.L. and Juniper, B.E. Plant Cells. *Blackwell Oxford*. 1968.

Cutter, E.G. Plant Anatomy : Experiment and Interpretation. Part I. Cells and Tissues. *Edward Arnold, London*. 1969.

Cutter, E.G. Plant Anatomy : Experiment and Interpretation. Part II Organs. *Edward Arnold, London*. 1971.

Eames, A.J. Morphology of the Angiosperms. *Mc-Graw-Hill, New York*. 1961.

Eames, A.J. and MacDaniels, L.H. An Introduction to Plant Anatomy. 2nd edition. *McGraw Hill, New York and London*. 1947.

Esau, K. Anatomy of Seed Plants. *Wiley, New York*. 1960.

Esau, K. Plant Anatomy. 2nd edition. *Wiley, New York*. 1965.

Esau, K. Vascular Differentiation in Plants. *Holt, Rinehart and Wintson, New York*. 1965.

Fahn, A. Plant Anatomy (Translated from the Hebrew by Sybil Broido-Altman). *Pergamon Press, Oxford*. 1967.

Foster, A.S. Practical Plant Anatomy. 2nd edition. *Van Nostrand, New York*. 1949.

Haberlandt. G. Physiological Plant Anatomy. English translation by M. Drummond, *London*, 1914.

Metcalfe, C.R. Anatomy of the Monocotyledons. I. Gramineae. *Clarendon Press, Oxford*. 1960.

Metcalfe, C.R. and Chalk, L. Anatomy of the Dicotyledons. Vols. I and II. *Clarendon Press, Oxford*. 1950.

Popham, R.A. Laboratory Manual for Plant Anatomy. *C.V. Mosley Company, Saint Louis*. 1966.

Tomlinson, P.B. Anatomy of the Monocotyledons. II. Palmae. General Editor, C.R. Metcalfe. *Clarendon Press, Oxford*. 1961.

Tomlinson, P.B. Anatomy of the Monocotyledons. Commelinales — Zingiberales. General Editor, C.R. Metcalfe. *Clarendon Press, Oxford*. 1969.

Wardlaw, C.W. Organization and Evolution in Plants. *Longmans, London*. 1965.

Wardlaw, C.W. Morphogenesis in Plants. *Methuen, London*, 1968.

References

1. Adamson, R.S. on the comparative anatomy of leaves of certain species of *Veronica. J. Linn. Soc. (Bot.)*, **40**, 247-74; 1912.
2. ——Anomalous secondary thickening in Compositae *Ann. Bot. Lond.* **48**, 505-14; 1934
3. ——Anomalous secondary thickening in *Osteaspermum. Trans roy. Soc. S. Afr.* **24**, 303-12; 1937.
4. Addicot, F.T. The anatomy of leaf abscission and experimental defoliation in guayule. *Amer. Bot.* **32**, 250-6; 1945.
5. ——Physiology of abscission. *Handle. Pfi Physiol.*, **15, 2,** 1904-1126, 1965.
6. Adkinson, J. Some features of the anatomy of the Vitaceae. *Ann. Bot. Lond.* **27**, 133-9; 1913.
7. Allsopp, A., and Misra, P. The Constitution of the cambium the new wood and the mature sap wood of the common ash, the common elm and the stock pine. *Biochem. J.* **34,** 1078-84; 1940.
8. ——Shoot morphogenesis. *A. Rev. Pl. Physiol.* **15,** 225-254; 1964.
9. Arber, A. Water plants. *Cambridge*, pp. 436; 1920.
    ——Goethe's botany. *Chron. Bot.* **10,** 63-126; 1946.
10. ——Root and shoot in the angiosperms. A study of morphological categories. *New Phytol.* **29, 5.** 297-315; 1930.
11. ——The interpretation of leaf and root in the angiosperms. *Biol. Rev. Camb. Phil. Soc,* **16,** pp. 24; 1941.
12. Artschwager, E.F. on the anatomy of *Chenopodium album L. Amer. J. Bot.* **7,** 252-60; 1920.
13. ——Anatomy of the vegetative organs of the sugar beet. *J. Agric. Res.* **33,** 143-76; 1926.
14. Ashplant, H. Investigations into *Hevea* anatomy. *Bull. Rubb. Gr.* **Ass. 10.** 484-90; 796-800; 1928.
15. Avery, G.S., Jun. Structure and germination of tobacco seed and the developmental anatomy of the seeding plant *Amer. J. Bot.* **20,** 309-27; 1933 (a).
16. ——Jun. Structure and development of the tobacco leaf. *Amer. J. Bot.* **20,** 365-92; 1933 (b).
17. Bailey, I.W. Notes on the wood structure of Betulaceae and Fagaceae. *For. Quart.* **8,** 175-85; 1910.
18. ——The relation of the leaf-trace to the formation of the compound rays in the lower dicotyldons. *Ann. Bot. Lond.* **25,** 225-41; 1911.
19. ——Investigations on the phylogeny of the angiosperms. *Bot Gaz.* **58,** 36-60; 1914.
20. ——The problem of differentiating and classifying tracheids, fibres-tracheids, and libriform wood fibres. *Trop. Woods.* **45,** 18-23; 1936.
21. ——The development of vessels in angiosperms and its significance in morphological research. *Amer. J. Bot.* **31,** 421- 8; 1944.

——The cambium and its derivative tissues.

22. II. Size variations of cambium initials in gymnosperms and angiosperms. *Amer. J. Bot.* **7,** 355-67; 1920.
23. III. The increase in girth of the cambium. *Amer. J. Bot.* **10,** 499-506; 1923.

Bailey, I.W., and Howard, R.A. The comparative morphology of the Icacinaceae.

24. I. Anatomy of the node and internode. *J. Arnold Arbor.* **22,** 125-32; 1941.
25. II. Vessels. *J. Arnold Arbor.* **22,** 171-87; 1941.

26. III. Imperforate tracheary elements and xylem parenchyma. *J. Arnold Arbor* **22,** 432-42; 1941.
27. IV. Rays of the secondary xylem. *J. Arnold Arbor.* **22,** 556-66; 1941.
28. Bailey, I.W., and Richard. H.P. The significance of certain variations in the anatomical structure of wood. *For Quart.* **14,** 662-70; 1916.
29. Bailey. I.W., and Thompson, W.P. Additional notes upon the angiosperms, *Tetracentron, Trochodendron,* and *Drimys*, in which vessels are absent from the wood. *Ann. Bot, Lond.* **32,** 28: 503-12; 1918.
30. Bailey, I.W. and Tupper, W.W. Size variations in tracheary cells: I. a comparison between the secondary xylems of vascular cryptogams, gymnosperms and angiosperms. *Proc. Amer. Acad. Arts Sci.* **54,** 149-204; 1918.
31. Balatinecz, J.J. and Farrar, J.L. Pattern of renewed cambial activity in relation to exogenous auxin in detached woody shoots. *Can. J. Bot.* **44,** 1108-1110; 1966.
32. Ball, E. Cell division in living shoot apices. *Phytomorphology.* **10,** 377-396; 1960.
33. Balwant Singh. The origin and distribution of inter-and intraxylary phloem in *Leptadenia*. *Proc. Indian. Acad. Sic.* 18. Sect. B, 14-19; 1943.
34. ———A contribution to the anatomy of *Salvadora persica* L., with special reference to the origin of the included phloem *J. Indian bot. Soc.* **23,** 71-8; 1944.
35. Bancroft, H. The arboresent habit in the angiosperms. A review. *New phytol.* **29,** 153-69 and 227-75; 1930.
36. Barghoorn, E.S., Jun. Origin and development of the uniseriate ray in the Coniferae. *Bull. Torrey Bot.* Cl. **67,** 303-28; 1940.

———The ontogenetic development and phylogenetic specialization of rays in the xylem of dicotyledons:
37. I. The primitive ray structure. *Amer. J.Bot.* **27,** 918-28; 1940.
38. II. Modification of the multiseriate and uniseriate rays. *Amer. J. Bot.* **28,** 273-82; 1941.
39. Barker, W.G. Growth and development of the banana plant. Gross leaf emergence *Ann. Bot.* H.S. **33,** 523-535; 1969.
40. Barkley, G. Differentiation of vascular bundles of *Trichoxanthes anguina. Bot, Gaz.* **83,** 173-84; 1927.
41. Barratt, K. The origin of the endodermis in the stem of *Hippuris. Ann. Bot. Lond,* **30,** 91-9; 1916.
42. Bentley, N.J., and Wolf, F.A. Glandular leaf hairs of oriental tobacco. *Bull Torrey bot.* Dl. **72,** 345-60; 1945.
43. Bews, J.W. Preliminary note on a peculiarity in the pith of a species of cucurbit. *Trans Bot. Soc. Edinb* **23,** 246-8; 1905.
44. Bienfait, J.L., and Pfeiffer, J.Ph. A scheme for identification of woods with the aid of a hand lens. *J. For.* **22,** 724-61; 1924.
45. Blaser, H.W. Anatomy of *Cryptostegia grandiflora,* with special reference to the latex system. *Amer. J. Bot.* **32,** 135-41; 1945.
46. Bliss, M.C. The vessel in seed plants. *Bot. Gaz.* **71,** 314-26; 1921.
47. Boke, N.H. Histogenesis and morphology of the phyllode in certain species of *Acacia. Amer. J. Bot.* **27,** 73-90; 1940.
48. ———Development of the adult shoot apex and floral initiation in *Vinca rosea. Amer. J. Bot.* **34,** 433-9; 1947.
49. Bond, G. The stem-endodermis in the genus *Piper. Trans. roy. Soc. Edinb.* **56,** 695-724; 1931.
50. Bonnett. H.T., Jun. The root endodermis; fine structure and function. *J. Cell. Biol* **40,** 144-159; 1969.

References

51. Boodle, L.A. Lignification of phloem in *Helinathus*. *Ann. Bot. Lond.* **20**, 319-21; 1906.
52. Bowes, B.G. The Ultrastructure of the shoot apex and young shoot of *Glechyma hederacea* L. *Cellub,* **65**, 351-356; 1965.
53. ——The structure and development of the vegetative shoot apex in *Glechyma hederacea* L. *Ann Bot.* N.S., **27**, 357-364; 1963.
54. Bremerkamp, C.E.B. Controversial questions in taxonomy, I. *Chronica bot.* **7**, 255; 1942.
55. Brierley, W.G. Cambial activity in the red raspberry cane in the second season. *Proc. Amer. Soc. hort. Sci* 278-80; 1929.
56. Brown, A.B. Cambial activity, root habit and sucker shoot development in two species of poplar. *New Phytol.* **34**, 163-79; 1935.
57. Brown, F.B.H. Scalariform pitting a primitive feature in angiospermous secondary wood. *Science*, **48**, 16-18; 1918.
58. Brown. H.P. An elementary manual on Indian wood technology. *Calcutta*, pp. 121; 1925.
59. ——Commercial timbers of United States. *New York*, pp. 554; 1940.
60. Brown H.P. and Panshin, A.J. Identification of the commericial timbers of the United States. *New York*, pp. 223; 1934.
61. Brown, W.V. The morphology of the grass embryo. *Phytomorphology*, **10**, 215-223; 1960.
62. Bunning, E. Morphogenesis in plants. *Surv. Biol. Prog,* **2**, 105-140; 1952a.
63. Burgess, C.E. An abnormal stem of *Lonicera priclymenum*.
64. Cameron, D. An investigation of the latex system in *Euphorbia marginata* with particular attention to the distribution of latex in the embryo, *Trans. bot, Soc. Edin.* **32**, 187-94; 1936.
65. Campbell, D.H. The phylogeny of the angiosperms. *Bull Torrey bot.* Cl. **55**, 79-89; 1928.
66. Chakravarty, H.L. Physiological anatomy of the leaves of Cucurbitaceae. *Philipp J. Sci.* **63**, 4. 409-31; 1937.
67. Chalk. L. The phylogenetic value of certain anatomical features of dicotyledonous woods. *Ann. Bot. Lond.* N.S.I, 1937.
68. ——A note on the meaning of the terms early wood and late wood. *Proc Leeds Phil. Lit Soc.* **3**, 325-6; 1937.
69. ——On the taxonomic value of anatomical structure of the vegetative organs of the dicotyledons. 2. The taxonomic value of wood anatomy. *Proc. Linn. Soc. Lond.* **155**, 3. 214-18; 1944.
70. Chalk, L., and Chattaway, M.M. Identification of wood with included phloem. *Trop. Woods,* **50**, 1-31; 1937.
71. Chattaway, M.M. the wood anatomy of the Sterculiaceae. *Philas. Trans. Roy Soc.* **228**, 313-66; 1937.
72. Cheadle, V.I. The occurrence of the vessels in the monocotyledons. *Amer J. Bot*, **26**, 95; 1939.
73. ——Secondary growth by means of a thickening ring in certain monocotyledons. *Bot Gaz.*, **98**, 535-555; 1937.
74. ——The role of anatomy in phylogenetic studies of the Monocotyledoneae. *Chronica Bot.* **7**, 253; 1942.
75. Chibber, H.M. The morphology and histology of Piper betle Linn. (the betel vine). *J. linn. Soc. (Bot).* **41**, 357-83; 1912.
76. Chowdhury, K.A. The identification of important Indian sleeper woods. *(Indian) For. Bull.* No. **77**, pp, 18; 1932.
77. ——The identification of the commercial timbers of the Punjab. *(Indian) For. Bull.* **84**, pp, 70; 1934.

——The formation of growth rings in Indian trees:
78. *I Ind. For. Rec. Util,* **2,** I. 1-39; 1939.
79. *II Ind. For. Rec. Util,* **2,** 2. 40-57; 1940.
80. *III Ind. For. Rec., Util.* **2,** 3. 59-75; 1940.
81. ——How to distinguish between the sapwood and heartwood of sal (*Shorea robusta*). (*Indian For. Bull.*) No, 115. pp. 3; 1942.
82. ——Initial parenchyma cells in dicotyledonous woods. *Nature. Lond.*, No. **4070,** 609; 1947.
83. Clarke, S.H. Fine structure of the plant cell wall. *Nature, Lond.* **142,** 3603. 899-904; 1938.
84. Clowes, F.A.L. Root apical meristems of *Fagus sylvatica, New Phytol.* **49,** 48-268; 1950.
85. ——Chimeras and meristems. *Heredity,* 11, 141-148; 1957.
86. ——Development of quiescent centres in root meristems *New Phytol.* **57,** 85-88; 1958a.
87. ——Apical meristems of roots. *Biol. Rev.* 34, 501-529; 1959a.
88. ——Apical meristems. *Blackwell, Oxford.* 1961a.
89. ——The DNA content of the cells of quiescent centre and root cap of *Zea mays, New Phytol.,* **67,** 631-639; 1968.
90. Clowes, F.A.L., and Juniper, B.E. The fine structure of the quiescent centre and neighbouring tissues in root meristems. *J. exp. Bot.* **15,** 622-630; 1964.
91. Cockrell, R.A. A comparatative study of the wood structure of several South American species of *Strychnos. Amer. J. Bot.* **28,** 32-41; 1941.
92. ——An anatomical study of eighty Sumatran woods. Univ. *Microfilm Publ.* 384; 1942.
93. Cockerham, G. Some observations on cambial activity and seasonal starch content in sycamore (*Acer pseudoplatanus* L.) *Proc. Leeds. phil. lit. Soc.* **2,** 64-80; 1930.
94. Cochen, W.E. The identification of wood by chemical means. *Tech. Pap. For., Prod. Res. Aust.,* No. **15,** pp. 23; 1935.
95. Collins. M.I. on the structure of resin secreting glands in some Australian plants. *Proc. Linn. Soc.* N.S.W. **45,** 329-36; 1920.
96. Conard, H.S. The water lilies. *Washnigton,* pp. 279; 1905.
97. Cooke, G.B. Cork and Cork products. *Econ. Bot.,* **2,** 393-402; 1948.
98. Cooper, D.C. The development of the peltate hairs of *Shepherdia canadensis. Amer. J. Bot.* **19,** 423-8; 1932.
99. Corson, G.E. Jun. Cell division studies of the shoot apex of *Datura stramonium* during transition to flowering. *Amer. J. Bot.,* **56,** 1127-1134; 1969.
100. Crafts, A.S. Phloem anatomy, exudation and transport of organic nutrients in cucurbits. *Plant Physiol.* **7,** 183-225; 1932.
101. Croizat, L. *Trochodendron, Tetracentron* and their meaning in phylogeny. *Bull. Torrey bot.* Cl. **70,** 60-76; 1947.
102. Crooks, D.M. Histological and regenerative studies on the flax seedling. *Bot Gaz.* **95,** 2. 209-39; 1933.
103. Cutter, E.G. Recent experimental studies of the shoot apex and shoot morphogenesis. *Bot. Rev.* **32,** 7-113; 1965.

D'Almeida, J.F.R. A contribution to the study of the biology and physiological anatomy of Indian marsh and aquatic plants.
104. Part I. Malvaceae. *J. Bombay nat. Hist. Soc.* **42,** 298-304; 1941.
105. Part II. Elatinaceae. *J. Bombay nat. Hist. Soc.* **43,** 92-6; 1942.
106. Dastur, R.H. The origin and course of vascular bundles in *Achyranthes aspera* L. *Ann. Bot. Lond.,* **39,** 539-45; 1925.

References

107. Davis, E.L. Medullary bundles in the genus *Dahlia* and their possible origin. *Amer. J. Bot.* **48**, 108-113; 1961.
108. Davy, A.J. Note on the structure of epicotyl in *Juglans nigra*. *New Phytol.* **34**, 201-10; 1945.
109. Denne, M.P. Leaf development in *Trifolium repens*. *Bot. Gaz.* **127**, 202-210; 1966c.
110. Desct. H.E. Anatomical variation in the wood of some dicotyledonous trees. *New Phytol.* **32**, 73-118; 1932.
111. ——'Latex canals' of the Apocynaceae. *Malayan Forester*, **3**, 219; 1934.
112. Dobbins, D.R. Studies on the anomalous cambial activity in *Doxantha unguis-cati* (Bignoniaceae). I. Development of the vascular pattern. *Can. J. Bot.* **47**, 2101-2106; 1969.
113. Dormer, K.J. Some examples of correlation between stipules and lateral leaf traces, *New Phytol* **43**, 151-3; 1944.
114. ——An investigation of the taxonomic value of shoot structure in angiosperms with special reference to Leguminosae. *Ann. Bot. Lond.*, N.S. **9**, 141-53; 1945.
115. ——Anatomy of the primary vascular system in dicotyledonous plants. *Nature, Lond.* **158**, 737-9; 1946.
116. Esau, K. Ontogeny of phloem in the sugar beet (*Beta vulgaris* L.). *Amer. J. Bot.* **21**, 632-44; 1934.
117. ——Developmental anatomy of the fleshy storage organ of *Daucus carota*. *Hilgardia*, **13**, 5, 175-209; 1940.
    ——Vascular differentiation in the vegetative shoot of *Linum*:
118. I. The procambium. *Amer. J. Bot.* **29**, 738-47; 1942.
119. II. The first phloem and xylem. *Amer. J. Bot.* **30**, 248-55; 1943.
120. ——Origin and development of primary vascular tissues in seed plants. *Bot. Rev.* **9**, 125-206; 1943.
121. Esau, K., and Cheadle, V.I. Secondary growth in *Bougainvillaea*. *Ann. Bot.* N.S. **33**, 807-819; 1969.
122. Fisher, J.B. Development of the intercalary meristem of *Cyperus alternifolius*. *Americ. J. Bot.* **57**, 691-703; 1970a.
123. ——Control of the internodal intercalary meristem of *Cyperus alternifolius*. *Americ. J. Bot.* **57**, 1017-1026; 1970b.
124. Foster, A.S. Structure and growth of the shoot apex in *Ginkgo biloba*. *Bull. Torrey bot. club.* **65**, 531-556; 1938.
125. Fritsch, F.E. The use of anatomical characters for systematic purposes, *New Phytol.* **2**, 177-84; 1903.
126. Frost, F.H. Bordered pits in parencyma. *Bull. Torrey bot. Cl.* **56**, 259-63; 1930.
127. ——Specialzation in secondary xylem dicotyledons. I Origin of vessels. *Bot. Gaz.* **89**, 67-94; 1930.
128. Gerry, E. Tyloses: their occurrence and practical significance in some American woods. *J. Agric. Res.* I, 445-69; 1914.
129. Ghosh, E. On the microstructure of the stem of Bengal Cucurbitaceae with reference to its value in taxonomy. *J. Indian bot. Soc.* **11**, 259-70; 1932.
130. Gifford, E.M., Jr. The shoot apex in angiosperms. *Bot. Rev.*, **20**, 477-529; 1954.
131. ——Developmental studies of vegetative and floral meristems. *Brookhaven Symp. Biol.*, **16**, 126-137; 1964.
132. Gifford, E.M., Jr., and Stewart, K.D. Ultrastructure of the shoot apex of *Chenopodium album* and certain other seed plants *J. Cell Biol.*, **33**, 131-142; 1967.

133. Gifford, E.M. Jr. and Tepper, H.B. Ontogenetic and Histochemical changes in the vegetative shoot tip of *Chenopodium album. Americ. J. Bot.*, **49,** 902-911; 1962 b.
134. Griffioen, K.A. study of the dark coloured duramen of ebony. *Rec. Trav. bopt. neerland.* **31,** 780-809; 1934.
135. Groom, P. The evolution of the annual ring and meduallary rays of *Quercus. Ann. Bot. Lond.* **25,** 985-1003; 1911.
136. ——The medullary rays of Fagaceae. *Ann. Bot. Lond.* **26,** 1124-5; 1912.
137. Gupta. K.M. On the wood anatomy and theoretical significance of homogylous angiosperms. *J. Indian bot. Soc.* **13,** 71-101; 1934.
138. Gwynne-Vaughan, D.T.. On some points in the morphology and anatomy of the Nymphaeceae. *Trans. Linn. Soc. lond. Bot.* **5,** 287-99; 1987.
139. Haberlandt, G. Physiological plant anatomy. English translation by M. Drummond, *London*, pp. 777; 1914.
140. Hamilton, A.G. The xerophilous characters of *Hakea dactylodies. Cav. Proc. Linn. Soc. N.S.W.* **39,** 152-6; 1914.
141. Handa, T. Anomalous secondary growth in *Bauhinia japonica* Maxim. *Jap. J. Bot.* **9,** 37-53; 1937.
142. ——Anomalous secondary growth in the axis of *Bauhinia championi* Benth. *Jap. J. Bot.* **9,** 3, 303-11; 1937.
143. Harris, W.M. and Spurr, A.R. Chromoplasts of tomato fruits. I Ultrastructure of low pigment and high-beta mutants. Carotene analyses. *Americ. J. Bot.,* **56,** 369-379; 1969.
144. Hasselberg, G.B.E. The vascular system in the stem-axis of the genus *Fagraea. Svensk. bot Tidskr.* **25,** 220-37; 1931.
145. Havis, L. Anatomy of the hypocotyl and roots of *Daucus carota. J. Agric. Res.* **58,** 557-64; 1939.
146. Hayden, A. The ecologic foliar anatomy of some plants a prairie province in Central Iowa. *Amer. J. Bot.* **6,** 69-86; 1919.
147. Hayward, H.E. The Structure of Economic Plants. *New York,* pp. 674; 1938.
148. Hector, J.M. Introduction to the botany of field crops. Vol. II Non-cereals. *Johannesburg. S. Africa*, 1936.
149. Heimsch, C., Jun., and Wetmore, R.H. The significance of wood anatomy in the taxonomy of the Jugalandaceae. *Amer. J. Bot.* **26,** 8, 651-60; 1939.
150. Hess. R.W. Occurrence of raphides in wood. *Troop Woods,* **46,** 22-31; 1936.
151. Hill, T.G. Stelar theories, *Sci. Progress,* No. **2,** 1-19; 1906.
152. Hodge, A.J. Mclean, J.D., and Mercer, F.V. Ultrastructure of the lamellae and grana in the chloroplasts of *Zea mays L.J. biophys. biochem. Cytol.* **1,** 605-614; 1955.
153. Holden, R. Some features in the anatomy of Sapindales. *Bot. Gaz.* **53,** 50-7; 1912.
154. Inamdar, J.A. Epidermal structure and ontogeny of stomata in some Verbenaceae. *Ann. Bot. N.S.,* **33,** 55-56; 1968a.
155. ——Ontogeny of stomata in some Oleaceae. *Proc. Indian Acad. Sci.,* **67,** 157-164; 1968b.
156. International Association of wood Anatomists. Glossary of terms used in describing woods. *Trop. Woods,* **36,** 1-13; 1933.
157. Jane, F.W. Aspects of the study of wood anatomy. *Sci. Progr. Twent. Cent.* **111,** 439-55; 1934.
158. ——The Structure of Wood. *A. and C. Black London*; 1962.
159. Jeffrey, E.C. The anatomy of woody plants. *Chicago*, pp. 478; 1922.
160. Johnson, M.A., and Truscott, F.H. On the anatomy of *Serjania*. I. Path of the bundles. *Americ. J. Bot.,* **43,** 509-518; 1956.

161. I. Anatomy of *Alternanthera sessilis* R. Br. *J. Indian bot. Soc.* **10,** 221-31; 1931.
162. II. Primary vascular system of *Achyranthes aspera* L., *Cyathula prostrata* Blume and *Pupalia lappacca* Juss. *J. Indian bot. Soc.* **10,** 265-92; 1931.
163. Joshi, A.C. Variations in the medullary bundles of *Achyranthes aspera* L. and the original home of the species. *New Phytol.* **33,** 53-7; 1934.
164. ——Some salient points in the evolution of the secondary cylinder of Amaranthaceae and Chenopodiaceae. *Amer. J. bot.* **24,** 3-9; 1937.
165. Kanehira, R. Anatomical notes on Indian woods. *Bull. Dept. For. Formosa,* No. **4,** pp. 40; 1924.
166. Kaplan, D.R. Floral morphology, organogenesis and interpretation of the inferior ovary in *Downingia bacigalupii. Americ. J. Bot.*, **54,** 1274-1290; 1967.
167. Kaufman, P.B., Cassell, S.J. and Adams, P.A. On nature of intercalary growth and cellular differentiation in internodes of *Avena sativa. Bot. Gaz.,* **126,** 1-3; 1965.
168. Koehlar, A. The identification of furniture woods. *Misc. Circ. U.S. Dept. Agric.;* No. **66,** pp. 76; 1926.
169. Kribs, D.A. Salient lines of structural specialization in the wood rays of dicotyledons. *Bot. Gaz.* **96,** 547-57; 1935.
170. ——Salient lines of Structural specialization in the wood parenchyma of dicotyledons. *Bull. Torrey bot. Cl.* **64,** 177-86; 1937.
171. Kumazawa, M. The medullary bundle system in the Ranunculaceae and allied plants. (English summary). *Bot. Mag. Tokyo,* **46,** 327-32; 1932.
172. Kundu, B.C. The anatomy of two Indian fibre plants, *Cannabis* and *Corchorus*, with special reference to fibre distribution and development. *J. Indian bot. Soc.* **21,** 93-128; 1942.
173. Kundu, B.C., and Preston, R.D. The fine structure of phloem fibres. I untreated and swollen hemp. *Proc. Roy. Soc.* B, **128,** 214-31; 1940.
174. Leach, W. An anatomical and physiological study of the petiole in certain species of *Populus. New Phytol.* **33,** 225-39; 1934.
175. Luthura, J.C., and Sharma, M.M.L. Some studies on the conductivity and histology of grafted mango shoots. *J. Indian bot. Soc.* **25,** 221-9; 1946.
176. Lyndon, R.F. Planes of cell division and growth in the shoot apex of *Pisum. Ann. Bot.*, N.S., **34,** 19-28; 1970b.
177. Maheswari, P. Contribution to the morphology of *Boerhaavia diffusa. J. Indian bot. Soc.* **9,** 42-61; 1930.
178. Majumdar, G.P. Hetero-archic roots in *Enhydra fluctuans* Lour. *J. Indian bot. Soc.* **11,** 225-7; 1932.
179. ——Anomalous structure of the stem of *Nyctanthes arbortristis* L.J. Indian. bot. Soc. **20,** 119-22; 1941.
180. ——Some aspects of anatomy in modern research. *Pres. Address, Indian Sci. Congress, Nagpur,* pp. 15; 1945.
181. Marco, H.F. Systematic anatomy of the woods of the Rhizophoraceae. *Trop. Woods* **44,** 1-20; 1935.
182. Matthews, J.R. Floral morphology and its bearing on the classification of angiosperms. *Proc. bot. Soc. Edinb.* **33,** 69-82; 1941.
183. McCormick, F.A. Notes of the anatomy of the young tuber of *Ipomoea batatas* Lam. *Bot. Gaz.* **61,** 388-98; 1916.
184. Metcalfe, C.R. The 'aerenchyma' of *Sesbania* and *Neptunia. Kew Bull.* No. **3,** 151-4; 1931.

185. ——On the taxonomic value of the anatomical structure of the vegetative organs of dicotyledons. 1. An introduction with special reference to the anatomy of leaf and stem. *Proc. Linn. Soc. Land.* **155,** 210-14; 1943.
186. ——The role of microscope in botanical identification. *J. Quekett. Micr.* Cl., Ser. **4, 2,** 68-75; 1945.
187. ——Comparative anatomy as a modern discipline. *Adv. Bot. Res.* **1,** 101-147, 1963.
188. Miksche, J.P., and Greenwood, M. Quiescent centre of the primary root of *Glycine max. New Phytol.,* **65,** 1-4, 1966.
189. Morre, D.J. cell wall dissolution and enzyme secretion during leaf abscission. *Pl. Physiol., Lancaster,* **43,** 1545-1559; 1968.
190. Moss, E.H. Interxylary cork in *Artemisia* with a reference to its taxonomic significance. *Amer, J. Bot.* **27,** 762-8; 1940
191. Mullan, D.P. Observation on the biology and physiological anatomy of some Indian halophytes, *J. Indian bot. Soc.* **11,** 103-18 and 185-202; 1932 and **12,** 165-82; 1933 and **12,** 235-53; 1933.
192. Myers, L. Tyloses in *Menispermum. Bot. Gaz.* **78,** 453-6; 1924.
193. Newman. I.V. Pattern in meristems of Vascular plants. I. cell partition in living apices and in the cambial zone in relation to the concepts of initial cells and apical cells. *Phytomorphology,* **6,** 1-19; 1916.
194. ——Pattern in the meristems of vascular plants. III. Pursuing the patterns in the apical meristem where no cell is a permanent cell. *J. Linn. Soc. (Bot.),* **59,** 185-214; 1965.
195. Pant, D.D. and Kidwai, P.F. Structure and ontogeny of stomata in some Caryophyllaceae. *J. Linn. Soc. (Bot),* **60,** 309-314.; 1968.
196. Philipson, W.R. The ontogeny of the shoot apex in dicotyledons, *Biol, Rev.* **24,** 21-50; 1959.
197. Philipson, W.R. and Balfour, E.E. Vascular patterns in dicotyledons. *Bot. Rev.,* **29:** 382-404; 1963.
198. Popham, R.A. Principal types of vegetative shoot apex organization in vascular plants. *Ohio. J. Sci.,* **51,** 249-270; 1951.
199. Priestly, J.H. The meristematic tissues of the plant. *Biol. Rev.* **3,** 1-20; 1928.
200. Priestley, J.H. and Scott. L. The vascular anatomy of *Helianthus annuus L .Proc leeds Phil. lit. Sco.* **3,** 159-73; 1936.
201. Priestley J.H., Scott, L. and Malins, M.E. Vessel development in angiosperms. *Proc. Leeds. Phil lit. Soc.* **3,** 42-54; 1935.
202. Record, S.J. Cystholiths in wood. *Trop. Woods,* **3,** 10-12; 1925.
203. ——Spiral tracheids and fiber-tracheid. *Trop Woods,* **3,** 12-16; 1925.
204. ——Role of wood anatomy in taxonomy. *Trop Woods.* **37,** 1-9; 1934.
205. ——Importance of the study of wood anatomy. *Rodriguesia,* **11,** 319-22; 1937.
206. Record, S.J., and Chattaway, M.M. List of anatomical features used in classifying dicotyledonous woods. *Trop. Woods.,* **57,** 11-16; 1939.
207. Rosso, S.W. The ultrastructure of chromoplast development in red tomatoes. *J. Ultrastruct. Res.,* **25,** 307-322; 1968.
208. Sabnis, T.S. The physiological anatomy of the plants of the Indian desert. *J. Indian bot Soc* **I** 42 and 65-83; 1919 **I,** 2. 97-113; 1919. I, -7; 1919. **1,** 33-43; 1920 I, 65-84; 1920; **1,** 183-205; 1920. I, 277-95; 1920, **2,** 1-19; 1931. **2,** 93-115; 1921, **2,** 157-73; 1921.
209. Saxena, N.P. Studies in the family Saxifragaceae I. A contribution to the morphology and embryology of *Saxifraga diversifolia* wall. *Proc. Ind. Acad. Sci.,* **60,** 38-51; 1964b.

References

210. ——Studies in the family Saxifragaceae. IX. Anatomy of the flower of some members of Saxifragaceae. *Jour. Ind. Bot. Soc.,* **52,** 251-266; 1973.
211. ——Morphological studies in the family Saxifragaceae. PhD. Thesis, Agra Univ., 1966.
212. Singh, V. Morphological and anatomical studies in Helobiae. VI. Proc. natu. Acad. Sci. India. Sect. B, **36,** 329-344; 1966.
213. Sayeedud-Din, M, and Suxena, M.R. On the anatomy of some of the Asclepiadaceae. *Proc. nat. Acad. Sci. Wash.* **10,** 129-32; 1940.
214. Scott, F.M. Cystoliths and plasmodesmata in *Beloperone, Ficus,* and *Boehmeria, Bot, Gaz.* **107,** 3, 372-8; 1946. Sinott, E.W., *et al.* Investigations on the phylogeny of the angiosperms:
215. I. The anatomy of the node as an aid in the classification of the angiosperms. *Amer. J. Bot.* 1, 303-22; 1914.
216. III. Nodal anatomy and the morphology of stipules. *Amer. J. Bot.* **1,** 441-53; 1914.
217. Small, J. The identification value of hairs. *Pharm. J. Ser.* **4,** 36, 587-91; 1913.
218. Solereder, A. Systematic anatomy of the dicotyledons English edition, translated by L.A. Boodle and F.E. Fritsch. *Oxford* 2 vols., pp. 1183; 1908.
219. Tepfer, S.S. Floral anatomy ontogeny in *Aquilegia formosa* var. *truncata* and *Ranunculus repens. Univ. Calif. Publs. Bot.* **25,** 513-648; 1953.
220. Tippo, O. The role of wood anatomy in phylogeny. *Amer. Midl. Not.* **36,** 362-72; 1946.
221. Tomlinson, P.B. and Zimmermann, M.H. vascular bundles in palm stems - their bibliographic evolution. *Proc. Americ. Phill. Soc.,* **110,** 174-181; 1966a.
222. ——Anatomy of the palm *Rhapis excelsa.* III. Juvenile phase *J. Arnond Arbor.,* **47,** 301-312; 1966 b.
223. ——The 'wood' of monocotyledons. *Bull. Ass. Wood. Anat.* **2,** 4-24; 1967.
224. Vascular anatomy of monocotyledons with secondary growth - an introduction. *J. Arnond Arbor.,* **50,** 159-179; 1969.
225. Trotter, H. The common commercial timbers of India and their uses. *Calcutta,* pp. 234; 1941.
226. Turrell, F.M. Citrus leaf Stomata; structure composition, and pore size in relation to penetration of liquids. *Bot. Gaz.* **108,** 476-83; 1947.
    Turril, W.B. Taxonomy and Phylogeny.
227. Pt. I. *Bot. Rev.* **8,** 247-70; 1942.
228. Pt. II. *Bot. Rev.* **8,** 473-532; 1942.
229. Vestal, P.A. Wood anatomy as an aid to classification and phylogeny. *Chronica bot.* **6,** 53-4; 1940.
230. Wagner, K.A. Notes on the anomalous stem structure of a species of *Bauhinia. Amer. Midl. Nat.* **36,** 251-6; 1946.
231. Wardlaw, C.W. Comparative observation on the shoot apices of vascular plants. *New phytol.,* **52,** 195-205; 1953.
232. ——The organization of the shoot apex. *Handb Pfl Physiol* **15,** 1, 966-1076; 1965c.
233. Webster, B.D. Anatomical aspects of abscission. *Pl. Physiol. Lancaster,* **43,** 1512-1544; 1968.
234. ——Abscission, In Mc Graw—Hill year book of science and Technology, 1969, 85-88. *McGraw Hill, New York;* 1969.
235. Wieland, C.R. Wood anatomy and angiosperm origin. *Trop. Woods,* **39,** 1-11; 1934.
236. Wilson, C.L. Medullary bundle in relation to primary vascular system in Chenopodiaceae and Amaranthaceae. *Bot. Gaz.* **78,** 175-99; 1924.
237. Zimmermann, M.H., and Tomlinson, P.B. The vascular system in the axis of *Dracaena fragrans* (Agavaceae). I. Distribution and development of primary strands. *J. Arnold Arbor.,* **50,** 370-383; 1969.

Question Bank

LONG ANSWER QUESTIONS

1. Compare and constrast the anomalous secondary growth in dicot stems and monocot stems.
2. Starting from a cambial cell demonstrate from labelled sketches how a vessel is formed.
3. Develop with illustration how secondary growth takes place in the stem of sunflower until it reaches *two* years old.
4. Describe any *three* theories with reference to organization of the apical meristems.
5. Describe any *two* types of nodal vasculature met with in Angiosperms.
6. What is cambium? What is its function?
   Explain the abnormal function and position of the cambium in the following:
   1. *Boerhaavia* stem,   2. *Dracaena* stem,   3. *Beta vulgaris* root.
7. Giving diagrams, describe the anatomy of a dorsiventral leaf. How does it differ from that of an isobilateal leaf?
8. Describe the cells of the shoot apex and give an account of the ways in which they are modified, in the formation of parenchyma, collenchyma, epidermis, fibre, cambium, xylem and phloem.
9. Discuss the distinction between protoxylem, metaxylem and secondary xylem, illustrate your answer by reference to known examples of the various organs of angiosperms.
10. Where are vascular rays located in the trunk of a tree? What would one of these rays look like if you could remove it intact from the surrounding tissues?
11. Write on the recent advances in our knowledge about the structure of phloem.
12. Discuss the role of anatomy in phylogeny.
13. Discuss the modern trends in plant anatomy.
14. With the help of a labelled sketch describe the anatomy of a grass leaf.
15. Describe the process of secondary thickening in a dicotyledonous root: give a brief account of an anomalous type of secondary thickenings in any one dicot root.
16. Describe important events seen during secondary growth in thickness in dicotyledonous stem, in stelar, and extrastelar region.
17. Describe the different theories relating to the growth and development in the stem apex and root apex.
18. Enumerate any three types of mechanical tissues and describe their distinguishing features.
19. Describe the secondary growth in *Boerhaavia* stem and comment on abnormal features.

SHORT ANSWER QUESTIONS

1. Give the structure of fibro-vascular bundle of any monocot. Label and mention the function of each.
2. What is Ring bark and Scaly bark and give one example for each.
3. Explain the terms, Exarch, Endarch and Mesarch with reference to xylem and mention one example for each.
4. What is the function of companion cell in phloem?

Question Bank

5. Show how the leaf traces are different from leaf gaps.
6. How do you differentiate an early metaxylem vessel from late metaxylem vessel?
7. What is "quiescent centre"?
8. What kind of epidermis do you find in Gramineae?
9. At 12 noon, on a hot day, the grass leaves show a kind of involution and folding. Which cells are involved in this kind of folding?
10. Phloem helps in the conduction of food materials. Why not xylem do the same function?
11. State how the fall of the leaves takes place? Disprove the statement that the endodermis and pericycle are morphologically similar.
12. How will you differentiate the histogen theory from Tunica corpus Theory?
13. Draw the C.S. of a dicot leaf and explain the structure.
14. Describe any two types of stomata and indicate the function.
15. What controls the initial differentiation of the cell as a tracheary element?
16. If xylem which consists of tubular extensions of xylem vessels can conduct water upwards, why not the phloem which has sieve tubes do the same function?
17. Write the significance of double fertilization and triple-fusion in Angiosperms.
18. Differentiate vascular cambium from procambium.
19. Explain with reasons why heart wood is dark in colour and is not used for ascent of sap?
20. Write short note on Annual rings.
21. Draw labelled diagram of vessel in L.S.
22. Draw labelled diagram of a stone cell.
23. Draw labelled diagram of sclerenchyma cells in T.S.
24. Draw labelled diagram of a patch of wood fibres in T.S.
25. Draw detailed diagrams of:
    (a) Bordered pit (surface and sectional view)   (b) Mature sieve tube (L.S.)
26. Write critical note on the Anatomy of monocot leaf.
27. By means of brief notes, illustrations and examples distinguish between: diacytic and paracytic stomata.
28. By means of brief notes, illustrations and examples, distinguish between: Sieve tube and xylem vessel.
29. Distinguish between: vascular cambium and cork cambium.
30. Distinguish between: Sieve tube and Sieve cell.
31. Illustrate the special features of interest in the anatomy of hydrophytic stem.
32. What are the principal changes that occur in the sap wood as it develops into heartwood? How would you recognise the former from the latter?
33. List four gross morphological features that are common to monocots and dicots, and four not common to them.
34. Epidermis covers the entire surface of the plant body. Does this mean that it has the same organization all through? Explain.
35. How do phloem island arise in the xylem?
37. Describe the anatomy of the following:
    (a) Stem of *Dracaena*,   (b) Leaf of *Nerium*   (c) Petiole of *Nymphaea*
    (d) Stem of *Amaranthus*   (e) Stem of *Aristolochia*.

38. What are growth rings? How are they formed?
39. How does secondary thickening occur in monocotyledons?
40. Write a critical note on the Anatomy of the stem of *Leptadenia*.
41. Trace the origin of a lateral root in dicotyledonous plant.
42. Xylem and phloem are derivatives of the cambium. Can these two vascular tissues occur independent of one another? Explain.
43. Write critical note on the anatomy of the root of *Beta*.
44. Name two vesselless angiosperms.
45. Write short note on the anatomy of *Peperomia* stem.
46. What are tyloses and how are they formed?
47. Write what is most unique about Abscission layer.
48. Write briefly on the following: cork.
49. Write briefly on the following: Tunica corpus concept.
50. Write what is most unique about the following: Bicollateral vascular bundle.
51. Distinguish between: Tracheid and Vessel.
52. Write short note on the following: Laticiferous tissue.
53. Write note on the following: Sclerenchyma.
54. Why a stem which has undergone secondary growth shows more xylem than phloem?
55. Why will the bark of a twig separate quite easily from its wood in the rainy season but not in winter?
56. Draw a section of a leaf cut parallel to the surface and passing through the palisade cells. Draw another passing through the spongy tissue.
57. Write notes on tentacles of *Drosera*.
58. Large air spaces or lacunae sometimes arise in the root cortex, particularly in aquatic plants. What is their function or how would you distinguish in transverse section a schizogenous lacuna from one which lysigenous?
59. What function do vascular rays serve in the stem?
60. The lumber cut from heart wood is usually more resistant to decay than lumber cut from sap wood. Explain.
61. By means of labelled illustrations comment on the differences between collenchyma and sclerenchyma as seen in longitudinal and transverse sections.
62. Distinguish between: Heartwood and sap wood.
63. What are the changes taking place in the stelar regions during the growth in thickness in Dicot stem?
64. Distinguish between: Hard wood and soft wood.
65. What are the basic concepts of the "Cell theory"?
66. Distinguish between:
    (a) growth rings and annual rings. (b) phloem fibres and xylem fibres.
67. Write short note on: Companion cells.
68. Write critical notes on: Summer and spring wood.
69. Write short notes on: Cystolith.
70. Where is phloem located in a dicotyledonous stem? Describe its structure and functions.
71. Where are the phloem fibres located? What is their value to the plant? What use do we make of them?

Question Bank

72. Write critical notes on: Cambium.
73. Write notes on: Sclereid.
74. Write notes on: Lenticel.
75. Write notes on: Callus.
76. Suppose you are asked to conduct a practical class in plant anatomy which local plants and their parts would you select to demonstrate: Scattered arrangement of vascular bundles, exarch xylem, sunken stomata, closely packed palisade, sclerenchyma, bicollateral bundles.
77. Which tissues contribute to the mechanical strength of the plants? Where do they occur?

SIMPLE QUESTIONS

1. Where can you look for the primary xylem in a mature wood?
2. Name a part of a plant where concentric starch grains are found.
3. Which is the Enucleate element of the phloem?
4. Which the living element found in xylem?
5. When is the xylem called endarch?
6. Name the types of stomata found in angiosperms.
7. Name three types of sclereids.
8. Name any one plant having hydathodes.
9. Name any one plant having stem with medullary vascular bundles.
10. Write the one-word technical term for the following:
    Outgrowth of ray of axial parenchymatous cell into a vessel occluding it partially or completely.
11. What are hydathodes? Give two examples.

MULTIPLE CHOICE QUESTIONS

1. The outermost layer of stele in a dicot stem is
   (a) Cortex   (b) Endodermis   (c) Pericycle   (d) Hypodermis
2. If you cut the old trunk of a tree transversely you will observe the outer region of secondary wood is lighter in colour. This region of the wood is known as
   (a) Heartwood   (b) Sapwood   (c) Spring wood   (d) Autumn wood
3. Commercial cork is nothing but a dead tissue with thickened walls by the deposition of
   (a) Cuticle   (b) Cellulose   (c) Lignin   (d) Suberin
4. Bicollateral vascular bundles are common in
   (a) *Aristolochia*   (b) *Helianthus*   (c) *Pongamia*   (d) *Cucurbita*
5. Conjoint collateral, closed vascular bundles are present in
   (a) *Aristolochia* stem   (b) Sunflower stem   (c) *Telia* root   (d) Grass stem
6. In the case of dicot root cambium is derived from
   (a) Hypodermis   (b) Epidermis   (c) Pericycle   (d) Cortex
7. A dicot root differs from monocot root in one of the following
   (a) Presence of piliferous layer   (b) Presence of exodermis
   (c) Presence of ill developed pith   (d) Separate radial vascular bundles
8. Numerous vascular bundles are arranged in a scattered manner in
   (a) monocot stem   (b) monocot root   (c) monocot leaf   (d) monocot seed

9. The cells with excessive length are common in the tissue of:
    (a) parenchyma   (b) sclerenchyma   (c) collenchyma   (d) epidermis
10. Phylloclades are characteristic of xerophytes in which normally a photosynthetic stem is seen, a good example from monocots is
    (a) *Opunita*   (b) *Casuarina*   (c) *Muehlenbeckia*   (d) *Ruscus*
11. Pith is well developed in
    (a) Monocot root   (b) Dicot root   (c) Both the above   (d) None of the above
12. The formation of distinct annual rings during secondary growth, mainly depends upon
    (a) marked seasonal variations   (b) more or less uniform climate
    (c) formation of unequal quantity of xylem and phloem   (d) none of the above
13. In which meristem do you see cell division occurring in all planes
    (a) File meristem   (b) Plate meristem   (c) Mass meristem   (d) Ground meristem
14. A waterproofing substance secreted by the cork and endodermal cells is
    (a) Cutin   (b) Tannin   (c) Suberin   (d) Lignin
15. What are the parts of periderm
    (a) Periblem, Phellogen, Phelloderm   (b) Periblem, Phellogen, Plerome
    (c) Phellem, Phellogen, Phelloderm   (d) Cortex, Dermatogen, Phelloderm
16. In which family members have silica crystals deposition
    (a) Cyperaceae   (b) Palmae   (c) Gramineae   (d) Scitaminae
17. When the cell wall deposition is layer upon layer, the process is said to be
    (a) intussusception   (b) apposition   (c) centripetal growth
18. The histogen theory was proposed by
    (a) Haberlandt   (b) Schmidt   (c) Bailey   (d) Hanstein
19. The lateral roots take their origin from
    (a) endodermis   (b) pericycle   (c) cortex
20. Which is the first wall layer formed by dividing plant cells and occurring between the subsequently formed cell walls of daughter cell
    (a) Primary wall   (b) Middle lamella   (c) Secondary wall   (d) Cellulose layer
21. The bark is made up of
    (a) Cork cambium   (b) Epidermis   (c) Cork   (d) Phelloderm
22. Which are the special openings seen on a tree trunk through which respiration is made possible
    (a) Water stomata   (b) Stomata   (c) Lenticels   (d) Cracks of the bark
23. A tumour like tissue of thin-walled cells developing over the wounds
    (a) Tylosis   (b) Gall   (c) Callose   (d) Callus
24. The chief function of velamen tissue is to
    (a) synthesize food   (b) store food materials
    (c) conduct food materials   (d) absorb moisture
25. Periderm is composed of
    (a) Cork-cambium and bark   (b) Cork-cambium and secondary cortex
    (c) Cork-cambium, secondary cortex and cork   (d) Cork, bark and secondary cortex

Question Bank

26. Nature of vascular bundle in dicot stem generally is
    (a) Radial   (b) Concentric   (c) Collateral   (d) Eccentric
27. Epidermal tissue acts generally as
    (a) a protective tissue   (b) absorbing tissue
    (c) photosynthetic tissue   (d) conducting tissue
28. Submerged leaves possess
    (a) stomata on both surfaces   (b) stomata on the upper surface
    (c) stomata on the lower surface   (d) No stomata on any surface
29. Vascular bundles are scattered in
    (a) Monocot root   (b) Dicot root   (c) Monocot stem   (d) Dicot stem
30. Which of the following can be seen in monocot root
    (a) A large pith   (b) No pith   (c) Medullary ray   (d) Endarch xylem
31. Parenchyma cells are characterized by:
    (a) presence of thickenings at the corner   (b) presence of uniform thickening
    (c) presence of intercellular spaces   (d) presence of lignified walls
32. Sclerenchyma help in
    (a) photosynthesis   (b) conduction   (c) support   (d) absorption
33. Growth rings are formed due to activity of
    (a) Extrastelar cambium   (b) Intrastelar cambium
    (c) Primary cambium   (d) Intercalary cambium
34. Root hair arises from
    (a) Pericycle   (b) Endodermis   (c) Cortex   (d) Epiblema
35. Multiple epidermis is found in
    (a) Banyan leaf   (b) *Nerium* leaf   (c) Maize leaf   (d) *Hibiscus leaf*
36. In guava, the bark is peeled off every now and then. Will a layer of bark be formed. How?
    (a) The bark tissues are peeled off because they are not needed any more.
    (b) Every season the cork produces the bark, which can be regenerated.
    (c) Every season or regularly the bark is regenerated from the periderm even if peeling occurs.
37. A History student observes the bark of tamarind tree and wonders how they appear dead tissue. He says that it is dead tissue. What is your answer?
    (a) The bark cells are dead because they are exposed all the time.
    (b) The bark cells are living because they are produced from a living tissue called phellogen
    (c) The bark is protective and dead tissue produced from the phellogen
38. A symbiotic association of Fungus with the subterranean part of a plant is said to be
    (a) Mycorrhiza   (b) Endophyte   (c) Exophyte   (d) Parasite
39. A stele with solid xylem core surrounded by phloem is said to be
    (a) Siphonostele   (b) Dictyostele   (c) Protostele   (d) Polystele
40. Cortex is the region found between
    (a) epidermis and stele   (b) hypodermis and endodermis
    (c) endodermis and pith   (d) endodermis and vascular bundles
41. A solenostele when cut up by crowded leaf gaps, the stele is said to be
    (a) Meristele   (b) Dictyostele   (c) Actinostele   (d) Protostele

42. Vascular part of a dictyostele between two leaf gaps in a T.S. is called
    (a) Haplostele   (b) Actinostele   (c) Plectostele   (d) Meristele
43. Trapa is a
    (a) Xerophyte   (b) Hydrophyte   (c) Halophyte   (d) Mesophyte
44. The arrangement of xylem in a stem is said to be:
    (a) Exarch   (b) Collateral   (c) Bicollateral   (d) Endarch
45. What are the parts of a periderm
    (a) Periblem, phellogen, phelloderm   (b) Periblem, phellogen, plerome
    (c) Phellem, phellogen, phelloderm   (d) Cork, cortex, phelloderm
46. Pear fruit is full of
    (a) Parenchyma cells   (b) Stone cells   (c) Fibre cells   (d) Aeranchyma cells
47. Tunica Corpus theory was proposed by
    (a) Hofmeister   (b) Nageli   (c) Strasburger   (d) Schmidt
48. In angiosperms the leaf gap is found in
    (a) Internode   (b) Node   (c) Shoot apex   (d) Petiole
49. The primary pit field is seen as
    (a) Depression on the secondary wall   (b) Pore in the primary wall
    (c) Pore in the nuclear membrane
50. A xylem vessel is considered to be primitive if it has
    (a) simple perforation   (b) reticulate perforation
    (c) foraminate perforation   (d) scalariform perforation
51. Knots are formed in wood because
    (a) branches get buried in the main stem   (b) branches fall off and leave scars
    (c) of insect bites   (d) of injuries caused by animals
52. Hydathodes are:
    (a) Pores in the bark
    (b) Pores through which transpiration takes place
    (c) Special openings in the leaf through which liquid water is forced out
    (d) Stomata in water plants
53. A stamen is usually a .... structure
    (a) 1-traced   (b) 2-traced   (c) 3-traced   (d) multi-traced
54. In the land plants guard cells differ from other epidermal cells in the possession of—
    (a) mitochondria   (b) chloroplasts   (c) nucleus   (d) ER
55. That the cells in the quiescent zone in root tips divide slowly has been demonstrated by the use of
    (a) Feulgen stain   (b) rate of respiration   (c) tritiated thymidine   (d) microscopic study
56. The presence of Casparian strips is a characteristic feature of
    (a) pericycle   (b) periblem   (c) endodermis   (d) endosperm
57. The product of second fertilization (fusion) will lead to the formation of
    (a) Endosperm   (b) Embryo   (c) Perisperm   (d) Cotyledons
58. The name perisperm is given to the
    (a) peripheral endosperm   (b) remnant of the nucellus
    (c) disintegrated antipodal cells   (d) disintegrated synergids

Question Bank

59. Double fertilization is characteristic of
    (a) pteridophytes (b) bryophytes (c) gymnosperms (d) angiosperms
60. The youngest layer of secondary xylem in a woody stem is located just
    (a) outside the cambium (b) outside the pith
    (c) inside the cambium (d) inside the epidermis
61. The lateral branch in a root arises out of the
    (a) epidermis (b) cortex (c) pericycle (d) phloem
62. A collateral vascular bundle is one
    (a) which has either phloem strand or a xylem strand
    (b) in which both xylem and phloem are present with the xylem towards the centre
    (c) in which both xylem and phloem are present with the xylem towards the periphery
    (d) in which both xylem and phloem are present, with the xylem on both the sides
63. Choose the correct phrase for completing the statement, a linear tetrad of four cells lying in an axial row develops during the development of the
    (a) ovary (b) ovule (c) embryo-sac (d) pollen-sac
64. Germ pores are found on the
    (a) ovules (b) seeds (c) pollen grains (d) embryo-sac
65. The chief factor in successful fertilization is:
    (a) availability of water (b) long style
    (c) long pollen tube (d) size of the pollen grain
66. Among the following kinds of ovules which kind of ovule is found in the family Cruciferae.
    (a) orthotropous (b) anatropous (c) amphitropous (d) campylotropous
    (e) cincinnus
67. The following question consists of four terms one of which does not belong to the other three. You are to select that term which does not belong:
    (a) parthenogenesis (b) apogamy
    (c) parthenocarpy (d) sporophytic budding
68. The double fertilization in angiosperms was first discovered by:
    (a) Darwin (b) Nawachin (c) Mendel (d) Linnaeus
69. The nuclear division during gametogenesis is:
    (a) mitotic (b) meiotic (c) amitotic (d) free nuclear
70. In Angiosperms the free-nuclear division takes place during:
    (a) gamete formation (b) endosperm formation
    (c) flower formation (d) embryo-sac formation
71. Which of the following constitutes the best definition of a fruit? A fruit is:
    (a) a product of the flower (b) a product of ovary
    (c) post-fertilization product of the pistil (d) a body that contains seed
72. An embryo may sometimes develop from a cell of an embryo-sac other than egg. This is called:
    (a) apospory (b) parthenogenesis (c) parthenocarpy (d) apogamy
73. The formation of distinct annual rings during secondary growth of stem mainly depends upon:

(a) contrasting seasonal variations
(b) more or less uniform climate, throughout the year
(c) formation of unequal quantities of xylem and phloem
(d) formation of cork cambium

74. With increasing secondary growth, which will increase in diameter:
    (a) sap-wood (b) heart wood
75. Secondary growth in anomalous in:
    (a) *Dracaena* (b) *Annona* (c) Sunflower (d) *Pinus*
76. Plerome gives rise to:
    (a) epidermis (b) hypodermis (c) endodermis (d) stele
77. The branch of botany which deals with study of form and features of different plant organs is known as:
    (a) Histology (b) Morphology (c) Anatomy (d) Embryology
78. Sieve tubes are found in
    (a) Tracheid (b) Wood fibres (c) Phloem (d) Xylem
79. The morphology of the perisperm is the remaining
    (a) Nucellus (b) Endosperm (c) Female gametophyte (d) Integument
80. The morphology of aril is
    (a) Integumentary outgrowth
    (b) Innermost pericarp
    (c) Third integument
    (d) Outgrowth of nucellus
81. The mechanical tissue construction of the leaves is done on the basis of withstanding
    (a) longitudinal compression
    (b) radial or crushing pressure
    (c) longitudinal tension
    (d) shearing stresses
82. The important constituent of cork cell wall is
    (a) Lignin (b) Suberin (c) Cutin (d) Pectin
83. Collenchyma in which cells are arranged compactly and with vigorously thickened tangential walls so that they appear like bands is called
    (a) Lacunate or tubular collenchyma
    (b) Plate or lamellar collenchyma
    (c) Angular collenchyma
    (d) Collenchyma
84. Rod-shaped elongated sclereids found in the seed coats of pulses are known as:
    (a) Brachysclereids
    (b) Astrosclereids
    (c) Macrosclereids
    (d) Trichosclereids
85. Water secreting glands in plants are
    (a) Nectaries
    (b) Digestive glands
    (c) Hydathodes
    (d) Epithelium cells
86. When simple pits break into two or more cavities they are referred to as:
    (a) Bordered pits
    (b) Half bordered pits
    (c) Ramiform pits
    (d) Vestured pits
87. A stele with a central core of xylem surrounded by phloem is called
    (a) Protostele (b) Actinostele (c) Siphonostele (d) Dictyostele
88. The lateral roots take their origin from
    (a) Endodermis (b) Pericycle (c) Cortex (d) Epidermis

Index

A

Abnormal growths, 262
Abscission of leaves, 313
Absence of vessels in xylem, 282
Abutilon, 97
Acer, 135
Achyranthes, 273. 274
Adiantum, 166
Adventive embryony, 437
Aeration, 455
Aesculus hippocastanum, 118
Agave, 331
Agropyron, 131
Ailanthus, 48
Alkaloids, 59
Allium cepa, 193
　　　　sativum, 365
Aloe, 292, 306, 330
Alternation of generations, 398
Amaranthus, 272
Amino-compounds, 51
Ammophila arenaria, 335
Anatomy and Taxonomy, 375-378
Anatomy of Banyan root, 198
　　of cotyledon, 357
　　of *Cucurbita* stem, 19
　　of dicot leaf, 295, 296
　　of dicot roots, 181
　　of dicot stem, 213-219
　　of embryo and young seedling, 357-360
　　of epiphytic roots, 196
　　of floral parts, 349-356
　　of gymnosperm leaf, 307
　　of hypocotyl, 359
　　of leaf and petiole, 295-315
　　of mesocotyl, 358
　　of monocot leaf, 304
　　of monocot roots, 185
　　of monocot stem, 226-234

　　of orchid root, 196
　　of petiole, 309
　　of phylloclade, 234, 235, 236
　　of phyllode, 312
　　of seedling, 360
　　of seedling of monocot, 360
　　of seedling root, 360
　　of sheath cotyledon, 359
　　of storage roots, 199, 200
　　of *Tinospora* root, 199, 200
Andromeda, 36
Anemarshena, 211
Angelica, 59
Angiosperms, 121, 383, 398
Anigozanthos, 138
Anomalous secondary growth, 262
Annual rings, 242
Anther, 403
　　dehiscence of, 409
Apical cell theory, 123, 126
Apomixis, 436
Aquilegia, 351
Archesporium, 403
Argyereia, 287
Arisaema, 107
Aristolochia, 221, 240, 263, 265, 266, 365
Arnica, 149
Artemisia, 148
Asclepias, 94, 290
Aseptic conditions, 454
Asparagus, 232, 236, 395
Atropa, 148
　　belladona, 56
Aubrietia, 146
Australian *Acacia,* 312
Avena, 146
　　sativa, 191, 231
Axillary buds, 130, 131
　　initiation, 130, 131

B

Banana, 95
Banksia serrata, 310, 333
Bark, 252
Bast, 158
Bauhinia, 263
    langsdorffiana, 263
    rubiginosa, 263
Beet root, 202, 203
Beta vulgaris, 202, 203, 358, 396
Betula, 105
Bignonia, 263, 264
Boehmeria, 365
    nivea, 366
Boerhaavia diffusa, 270, 271, 280
Bougainvillaea, 267, 268
Branch gaps, 164, 168
    Traces, 164, 168
Brassica, 165
Bryonia, 225
Bryophyllum, 72

C

Calendula, 34
Calotropis, 332, 409
Cambium, 170-174, 237
    abnormal behaviour of, 264
    cell division in, 172
    cellular structure of, 172
    duration of, 171
    fascicular, 171
    function of, 172
    growth about wounds, 173
    in budding, 174
    in grafting, 174
    in monocots, 174
    interfascicular, 171
    origin of, 170
    structure of, 172
Canna, 95, 233
Cannabis, 148
Capparis decidua, 339
Capsella, 433, 434, 435
Capsicum, 139, 289
Carica papaya, 56
Carpel, 411
Carya glabra, 56
    ovata, 107
Cassia, 149
Castanea, 314
Casuarina equisetifolia, 336

Celery petiole, 362
Cell-division, 79–88
    meiosis, 79–83
    mitosis, 83–87
Cell-structure and components, 11-59
    historical account, 11
    membrane, 16
    structure, 9, 18, 19
    Theory, 11
    tissues, 6
    types, 6
Cell-wall, 60–78
    chemical nature of, 71
    formation of, 75
    growth of, 77
    intercellular spaces, 72
    microscopic and submicroscopic, 73
    structure of, 60–78
    Thickening of, 65
    wall layers, 60
Cellular totipotency, 456
Cellulose, 49
Ceratophyllum, 322, 329
Chenopodium, 80, 275, 276
    album, 80
Chlorophyll, 37, 38
Chloroplast, 35, 36, 37–39
Citrus, 135
    sinensis, 135
Cocoloba, 234
Cocos, 98
Collenchyma, 361
Colocasia 55
Colour in cell, 52
Combretum, 277
Commelina, 16, 188
Commercial cork, 251
Companion cells, 111
Comparative account, 459–464
Complex tissues, 101
Controlling centre, 21
Convalvulus floridus 309
Copaifera, 58
Coreopsis, 146
Cornus, 135
Cortex, 149
Cotton fibre, 64
Crystals, 54, 55
    calcium carbonate, 54
    calcium oxalate, 56
    cluster, 57
    mineral, 54

Index

prismatic, 57
rosette, 56, 57
sand, 57
Cucumis, 146
Cucurbita, 97, 109, 160, 211, 223, 224, 362
Culture technique, 453
Cuscuta, 286, 340, 341
Cuticle, 135, 136
Cydonia, 64
Cynodon dactylon, 227
Cyperus, 138
Cystolith, 54

D

Dahlia, 49
Darbya, 355
Dasylirion serratifolium, 135, 334
Datura, 128, 149
Daucus, 201
Dendrobium, 197
Dendrochronology, 244
Derivatives, 117
Dermal tissue system, 135-148
Development of secondary tissue, 204
Dicot root, 181-185
Dicot stem-three dimensional, 212, 214, 215
Differences between dicot and monocot roots stems, 261
Digitalis, 148
Disopyros, 65
Dirca palustris, 370
DNA, 25-30
  chemical analysis of, 26
  chemical strucrture of, 26
  classification of, 28
  functions of, 30
  molecular structure of, 26
  replication, 27, 29
  structure, 26
Dracaena, 98, 135, 174, 292
Drimys, 381
Drosera, 115
Druses, 56, 57
Dryoptersis filix-mas, 148

E

Ecological anatomy, 316-348
Eicohhornia crassipes, 310
Elatine alsinastrum, 319
Embryo, 432
Embryo development, 432
  in dicots, 432
  in monocots, 434
Embryogenesis, 432
Embryoid formation, 441
Embryology of angiosperms, 398-438
Embryo sac, 421
  development of, 416
  haustoria, 422
  normal type, 417
  Polygonum type, 417
  organization of, 421
Empetrum nigrum, 336
Endarch, 158
Endodermis, 149-152
Endoplasmic reticulum, 40
Endosperm, 427
Endothecium
Entada, 277
Enzymes, 52
Ephedra, 105
Epidermal tissue system
Epidermis, 135
Epigaea, 146
Epiphytes, 338
Equisetum, 117
Ergastic substances, 46-59
Erica cinerea, 334
  tetralix, 334
Erythroxylon, 300
Essential oils, 58
Eucalyptus, 298
Euphorbia pulcherrima, 52
  tirucalli, 338
Exarch, 157
Exodermis, 177

F

Fats, 51
Fatty oils, 51
Female gametophyte, 416
  development of, 416
Fertilization, 425
Fibres, 98, 102, 365
  bast, 99
  classification of, 366
  cortical, 99
  development of, 366
  extraxylary, 98
  function, 366
  pericyclic, 99
  perivascular, 99
  phloem, 99
  tracheids, 101, 102

wood, 107
xylem, 101
Ficus benghalensis, 198
    elastica, 54, 137
Flax, 153
Flower, 401
Foeniculum, 114
    vulgare, 114
Formation of adventitions roots, 195
    cambium, 204
    lateral roots, 192
Food-products, 46
Fragaria, 53
Fraxinus, 246
Fruit, 388, 436
Fruit wall, 388
    dry, 388
    fleshy, 391
Funaria, 211
Further reading, 492

G

Gametic fusion, 425
Ginkgo biloba, 119
Girder, 371
Glands, 57, 58, 115
    digestive, 116
    gums, 58, 115
    secreting resins, 115
    water secreting, 116
Glandular tissue, 114
Glossary, 475–491
Glyoxysomes, 42
Gnetum, 53, 365
Golgi complex, 43, 44
Gram root, 183
Grass root, 191
Growth rings, 242
Gums, 58
Gymnosperms, 120, 256, 386

H

Hairs, 143
    glandular, 145
    scale or peltate, 145
    stinging, 145
    various types, 146
Hakea 332, 333
    dactyloides, 311
Halophytes, 341
Hamamelis, 149
Hedychium coronarium, 189

Helianthus, 219
    annuus, 219
Heliotropium, 146
Hemicellulose, 49
Heteroarchy in roots, 203, 204
Hevea brasiliensis, 113
Hibiscus cannabinus, 366
Hippuris vulgaris, 323, 324
Histogenic layer concept,
Histogen theory, 126
Historical sketch, 1
Homalocladium platycladum, 235
Hordeum vulgare root, 193
Humulus, 362
Hyascyamus, 148
Hydthodes, 116
Hydrilla, 192, 285
Hydrophytes, 316
Hypercum, 130, 131, 329

I

Idioblasts, 57
Incompressibility, 372
Inextensibility, 371
Inflexibility, 371
Initials, 117
Interxylary phloem, 276
Intraxylary phloem, 286
Introduction, 1
Inulin, 49
Ipomoea batatas, 200, 201, 202
Iris, 139, 160, 187
Isobilateral lily leaf, 304
Ixora, 301

J

Juglans nigra, 55

K

Korper-kapper theory, 125

L

Lactuca, 97, 390, 391, 392
    sativa, 4
Lateral bud initiation, 131
Lathyrus, 211
Latex, 57
    cells, 113
    vessels, 114
Laticiferous tissue, 113
Leaf buttress, 130
    closing gaps, 169

gaps, 163
initiation, 130
three dimensional, 300
traces, 163
Lenticel, 254, 255
Lepidium, 394
Leptadenia, 277, 280, 281
Lilium, 353, 404, 406, 419, 420, 421
Lily leaf, 304
Limnanthemum, 325
Lingustrum, 148
Linum, 7, 155, 170
usitatissimum, 7, 153, 156
Liriodendron, 70, 105, 109, 112, 242
Lobelia, 150
Lomasomes, 42
Lycopersicon, 139, 394
Lysosomes, 42

M

Magnolia, 70, 394
Maize, 230
grain, 389
Male gametophyte, 422
development of, 422
Malus, 52, 53, 55, 105, 112
pumila, 52, 53, 55, 103, 109, 112, 135, 354
Mangifera indica, 256, 309
Mantle core concept, 122
Marantia, 47
Mechanical tissue, 361–374
support in leaf, 299
Medicago sativa, 210
Megagametophyte, 416
Megasporangium, 411
Megaspores, 414, 415
Megasporogenesis, 414
Megasporophyll, 411
Melilotus, 408
Mentha, 362
Meristems, 89, 117
and growth of plant body, 89
and permanent tissues, 93
apical, 92, 117, 118, 127, 128
classification of, 90
Mesarch, 158
Mesophytes, 323
Metaxenia, 431
Michelia, 378
Microgametophyte, 422
Micrometry, 467
Microsporangium, 403

dehiscence of, 406
Microspore mother cells, 405–408
Microscopy, 465, 466
Microspores, 423
germination, 423
Microsporogenesis, 407
Microsporophyll, 402
Microtubules, 43
Mirabilis, 269
nyctaginea, 309
Mitochondria, 31–34
development, 33
functions, 34
Models of plasma membrane, 470–472
Molinia, 340
Monocotyledonous root, 185–192
Morphogenesis, 439–458
Morus alba, 253
Mosaic endosperm, 431
Muehlembeckia platyclada 235
Musa, 47, 48, 136, 150
textilis, 48
Mycorrhiza, 192

N

Nacre wall, 110
Nectar, 52
Nectaries, 52, 116
septal, 116
Neottia, 196
Nerium, 296
Nicotiana, 66, 108, 110, 144, 164
tabacum, 302
Nitrogenous products, 50
Nodal anatomy, 163–167
in wheat, 166
of *Triticum* stem, 168
Non-protoplasmic components, 45
Nucleoproteins, 14
Nucleotides, 26
Nucleus, 21, 22, 23, 24
structure of, 21–31
Nuphar luteum, 328
Nutrient medium, 454
Nyctanthes arbortristis, 288
Nymphaea, 203, 330
stellata, 311

O

Oil glands, 114
Onion, 127
Onopordum, 146

Organic acids, 59
Orientation of vascular tissue, 301
Origin of branches, 130
    of lateral root, 192, 194, 195
    of leaves, 130
    of reproductive shoot apex, 132
    of stem, 210
Oryza sativa, 190
Osmophors, 52
Ovule, 411
    development of, 411
    forms of, 413
Oxalis, 139

P

Palm, 294
Parasites, 340
Parthenocarpy, 437
Passage cells, 177
Pastinaca, 362
Pear leaf, 299
Pelargonium, 365
Peperomia, 284
Pericycle, 153, 154
Periderm, 8, 209, 250, 254
Peristrophe, 222, 223
Peroxisomes, 42
Phaseolus, 47, 48
    radiatus, 184, 185
    vulgaris, 19
Phelloderm, 252
Phellogen, 250
Phloem, 108, 158, 370
    fibres, 111, 370
    parenchyma, 112
Phoenix, 65
Phormium tenax, 366
Phryma, 146
Picea, 166
Pine leaf, 307
Pinus, 53, 58, 145, 150, 260
    merkusii, 145
    nigra, 331
    strobus, 101, 119
Pistia, 55
Pistil, 411
Pisum, 48
Pith, 154, 155
Pits, 67
Plant anatomy, 1, 1–520
    historical sketch, 1
    in India, 383 - 387
    introduction, 1–10

Plant body, 2, 3
    development of, 2, 3
    fundamental parts of, 3
Plant tissue and organ culture, 440–458
    applications of, 457
Plantago, 383
Plasma membrane, 16
Plasmodesmata, 65, 66
Plastochron, 458
Platanus, 146
Plastids, 34–39
    amyloplast, 48
    chloroplasts, 35–59
    chromoplasts, 35
    leucoplasts, 34
Pollen grains, 410
    germination of, 425
    mother cells, 407
    sac, 403
    tube, 424–426
Pollinia, 409
Polyembryony, 437
Polygonum, 416
Polypodium, 94, 150
Pontederia, 327
Position of lateral roots, 194
Potamogeton, 320, 321
    epihydrus, 318
    pectinatus, 319
Presence of cortical bundles, 286
    of exclusive phloem bundles, 284
    of exclusive xylem bundles, 284
Primary vascular system, 156, 164
    tissue, 156
    xylem, 217
Procambium, 156
Promeristem, 125
Protoplasm, 13
    carbohydrates, 47
    composition, 13
    fats, 13
    nucleoproteins, 14
    other chemical substances, 14
    physical nature of, 14
    properties of, 15
    proteins, 14, 50
Prunus, 218, 239, 250, 252, 255, 393
Pteridium, 105, 109, 117
Pteriodophytes, 119, 387
Pteris, 122, 158
Punica, 289
Pyrus, 98, 252

Index

Q

Quercus alba, 101, 103, 107
Question bank, 503–512
Quiescent centre, 125

R

Radish root, 201
Ranunculus, 4, 171, 179, 180
Raphanus sativus, 201
Raphides, 55, 56
References, 493–502
Resins, 58
    ducts, 58, 258
Retting, 367
Rhizophora, 342
    mucronata, 342, 343, 344
Rhytidome, 253
Ribosomes, 41
Ricinus communis, 216
RNA, 30, 31
    classification of, 31
    comparison, 30
    molecular structure of, 31
Robinia, 109, 112
Root, 175–209
    apex, 122, 123, 124, 127
    cap, 175
    development, 138
    function of, 177
    hairs, 177
    primary and secondary structure, 175–209
    stem-transition, 211
Rubus, 146, 392
Rumex, 165
Ruscus, 237

S

Saccharum officinarum, 228, 229
Sagittaria, 435
Salix, 70, 112, 164, 165, 167
    nigra, 55
Salvadora, 277, 282, 283
Sambucus, 362
Samolus floribundus, 355
Saprophytes, 339
Sapwood and heart wood, 246
Sarjania ichthyoctona, 263
Scattered vas. bs. in dicots, 282
Sclerenchyma, 364
Sclereids, 99, 367
    astrosclereids, 100
    brachysclereids, 99
    classification of, 367
    macrosclereids, 99
    ontogeny of, 368
    osteosclereids, 100
Secondary growth in dicot roots, 204–209
    in roots, 206–209
    in dicots stems, 236–259
    in monocots, 259, 291
    phloem, 247, 258
    phloem in conifers, 256
    xylem in conifers, 256
Secretory products, 52
Secretory tissue, 113–116
Securidaca lanceolata, 263
Sedum, 139
Seed, 393, 436
    coat, 393
Seedling, 210, 360
Selaginella, 164
Sequoia, 72, 145
    sempervirens, 72
Shoot apex, 118, 119, 126
    organization, 121
Sieve cells, 109
    elements, 108
    plates, 109, 110
    tube elements, 109, 110
Simple tissues, 94
    collenchyma, 96
    parenchyma, 94
    sclerenchyma, 97
Smilacina racemosa, 56
Smilax, 135, 150, 154, 179, 188
Solanum tuberosum, 97, 289
Sonneratia, 345
    apetala, 345, 346, 347
Spartina, 337
Spiraea, 165
Sphaerosomes, 42
Sporogenous tissue, 405
Stamen, 402
Starch, 47
    grains, 47, 48
    sheath, 149
Stelar system, 160, 161, 162, 163
Stem-Anomalous structure, 262–294
    development of, 91
    of *Pinus*, 260
    primary and secondary structure, 210–261
Stomata, 136
    development of, 143
    water, 116

Stomium, 404, 406
Strophanthus, 149
Structural development theories, 125
Strychnas, 278, 279
    nuxvomica, 149
Subject index, 513–520
Succisa, 133
Sugars, 49
Swietenia, 103

T

Tamus communis, 292
Tannin, 53
Tapetum, 405
Terminalia, 378
Termination of veins, 298
Tetracentron, 381
Thickening of palms, 293
Thinouia scandens, 263
Thuja occidentalis, 257
Thymus, 148
Tilia, 365
    americana, 55
Tinospora, 152, 199
    cordifolia, 200
Tissue, 6, 89–116
    andorgan culture, 439–458
    meristematic, 89–93
    permanent, 93–113
Tissue system, 134–169
    dermal, 135–148
    fundamental, 148–155
    ground, 155–163
Tobacco, 176
Tomato flower, 352
Tracheids, 101, 102, 368
    fibre, 102, 369
    gelatinous, 103, 369
    septate fibre, 103
    of pine, 258, 259
Trapa bispinosa, 303, 319, 326
Trichomes and taxonomy, 146
    development, 146
    functions of, 147
    of various kinds, 148
Triticum, 168, 365
    aestivum, 234, 305
Trochodendron, 56
Tunica-corpus, 120
    organization, 120
    theory, 127

Turnip root, 201
Tyloses, 244, 245
Tylosoids, 246
Typha, 317
    latifolia, 318

U

Urtica dioica, 148

V

Vaccinum, 135
Vacuoles, 45, 46
Vallisneria, 16
Variations in stem structure, 218
Vascular bundles, 157, 159
    aranged in a ring in monocots, 291
    bicollateral, 159, 160
    collateral, 159. 160
    concentric, 159, 160
    conjoint, 159, 160
    radial, 159, 160
Vascular cryptogams, 119
Vascular plant, 6
    internal organization of, 6, 7
Vascular system, 181
    tissue, 148
    tissue system, 148
Vegetative shoot apex, 118
Vein of leaf, 302, 303
Velaman, 177, 178
Verbascum, 149
Vessel, 103, 104, 105, 380
    ontogeny, 106
Viburnum lentago, 56
Vicia faba, 141
Victoria regia, 328
Vinca, 128
Viola tricolor, 394
Vitis, 53, 105, 139
    vinifera, 53, 247, 248
Volatile oils, 58

W

Waste products, 53
    nitrogenous, 59
    non-nitrogenous, 53
Wheat, 359
Wood, 157
    anatomy and phylogeny, 379–382
    fibres, 369
    libriform fibres, 369

X

Xanthium, 222
    stromarium, 383
Xenia, 431
Xerophytes, 323
Xylem, 101, 368
    centrifugal, 157
    centripetal, 157
    primary, 106
    proto, 106, 107
    secondary, 239
    vessel, 103, 104, 105, 106

Y

Yucca, 174, 366

Z

Zanichellia, 320, 329
Zea, 50, 94, 107
    mays, 50, 123, 124, 152, 186, 304, 305, 436

NOTES

NOTES

NOTES

NOTES

NOTES

NOTES